ES

ANNUAL REVIEW OF
CELL BIOLOGY

ANNUAL REVIEW OF CELL BIOLOGY

VOLUME 2, 1986

GEORGE E. PALADE, *Editor*
Yale University School of Medicine

BRUCE M. ALBERTS, *Associate Editor*
University of California, San Francisco

JAMES A. SPUDICH, *Associate Editor*
Stanford University

ANNUAL REVIEWS INC.   4139 EL CAMINO WAY   P.O. BOX 10139   PALO ALTO, CALIFORNIA 94303-0897

ANNUAL REVIEWS INC.
Palo Alto, California, USA

International Standard Serial Number : 0743-4634
International Standard Book Number : 0-8243-3102-8

Annual Review and publication titles are registered trademarks of Annual Reviews Inc.

Annual Reviews Inc. and the Editors of its publications assume no responsibility for the statements expressed by the contributors to this Review.

TYPESETTING BY AUP TYPESETTERS (GLASGOW) LTD., SCOTLAND
PRINTED AND BOUND IN THE UNITED STATES OF AMERICA

PREFACE

The first volume of the *Annual Review of Cell Biology* was received with considerable interest by the scientific community within and outside the boundaries that can be ascribed to cell biology. The Editorial Committee is gratified by this encouraging response and assumes that it indicates the timeliness and usefulness of this new Annual Reviews series.

The second volume of the *Annual Review of Cell Biology* follows the example set by its predecessor. It considers that cell biology and molecular biology, molecular genetics included, are merging into a continuum of knowledge and research activity dependent on a common body of basic unifying concepts and a common set of powerful technologies. The ultimate goal of this merging field is to arrive at a full understanding of basic cell processes in terms of the molecular interactions that proceed within the structural framework of the cell and its organs. Full understanding should eventually include the regulatory mechanisms of these processes and their integration into the overall activities of the cell.

The chapters selected for Volume 2 of the *Annual Review of Cell Biology* cover some of the rapidly evolving areas of this broad and remarkably active field of research. There is no dearth of new knowledge and excitement at the growing tips or advancing fringes of cellular and molecular biology. But there is keen competition for researchers' time, and because of this factor it is not easy to achieve a proper balance of contents that will satisfy most, if not all, cell biologists, irrespective of their primary interests. The multiple facets of cellular and molecular biology are more fully represented in the second volume than in the first. This volume covers topics related to chromatin, ribonucleoprotein particles, cellular membranes, signal transduction, protein traffic control, cell motility, and cell–extracellular matrix interactions. It reviews protein traffic control in both prokaryotic and eukaryotic cells, and it covers alternative signal transduction mechanisms. In addition, it includes a number of chapters on developmental biology and cell differentiation in the immune system.

The Editorial Committee is aware of the importance of a proper balance in covering this vast and rapidly advancing field and will continue its sustained efforts towards this goal, using both its own resources and advisory input from the American Society of Cell Biology.

GEORGE E. PALADE
EDITOR

ANNUAL REVIEWS INC. is a nonprofit scientific publisher established to promote the advancement of the sciences. Beginning in 1932 with the *Annual Review of Biochemistry*, the Company has pursued as its principal function the publication of high quality, reasonably priced *Annual Review* volumes. The volumes are organized by Editors and Editorial Committees who invite qualified authors to contribute critical articles reviewing significant developments within each major discipline. The Editor-in-Chief invites those interested in serving as future Editorial Committee members to communicate directly with him. Annual Reviews Inc. is administered by a Board of Directors, whose members serve without compensation.

ANNUAL REVIEWS OF
Anthropology
Astronomy and Astrophysics
Biochemistry
Biophysics and Biophysical Chemistry
Cell Biology
Computer Science
Earth and Planetary Sciences
Ecology and Systematics
Energy
Entomology
Fluid Mechanics
Genetics
Immunology

Materials Science
Medicine
Microbiology
Neuroscience
Nuclear and Particle Science
Nutrition
Pharmacology and Toxicology
Physical Chemistry
Physiology
Phytopathology
Plant Physiology
Psychology
Public Health
Sociology

SPECIAL PUBLICATIONS

Annual Reviews Reprints:
  Cell Membranes, 1975–1977
  Immunology, 1977–1979

Excitement and Fascination
  of Science, Vols. 1 and 2

Intelligence and Affectivity,
  by Jean Piaget

Telescopes for the 1980s

A detachable order form/envelope is bound into the back of this volume.

Annual Review of Cell Biology
Volume 2, 1986

CONTENTS

viii CONTENTS (*continued*)

SOME RELATED ARTICLES IN OTHER *ANNUAL REVIEWS*

From the *Annual Review of Biochemistry*, Volume 55 (1986):

Lectins as Molecules and as Tools, H. Lis and N. Sharon

Lysosomal Enzymes and Their Receptors, K. von Figura and A. Hasilik

Genetics of Mitochondrial Biogenesis, A. Tzagoloff and A. M. Myers

Bacterial Periplasmic Transport Systems, G. Ferro-Luzzi Ames

Proteoglycan Core Protein Families, J. R. Hassell, J. H. Kimura, and V. C. Hascall

Acidification of the Endocytic and Exocytic Pathways, I. Mellman, R. Fuchs, and A. Helenius

Eukaryotic DNA Replication, J. L. Campbell

The Transport of Proteins into Chloroplasts, G. W. Schmidt and M. L. Mishkind

Actin and Actin-Binding Proteins, T. D. Pollard and J. A. Cooper

Splicing of Messenger RNA Precursors, R. A. Padgett, P. J. Grabowski, M. M. Konarska, S. Seiler, and P. A. Sharp

From the *Annual Review of Biophysics and Biophysical Chemistry*, Volume 15 (1986):

Relationships Between Chemical and Mechanical Events During Muscular Contraction, M. G. Hibberd and D. R. Trentham

Active Transport in Escherichia coli, H. R. Kaback

Identifying Nonpolar Transbilayer Helices in Amino Acid Sequences of Membrane Proteins, D. M. Engelman, T. A. Steitz, and A. Goldman

The Role of the Nuclear Matrix in the Organization and Function of DNA, W. G. Nelson, K. J. Pienta, E. R. Barrack, and D. S. Coffey

From the *Annual Review of Genetics*, Volume 20 (1986):

Homeotic Genes and the Homeobox, W. J. Gehring and Y. Hiromi

Pre-mRNA Splicing, M. R. Green

Control of Multicellular Development, D. Kaiser

From the *Annual Review of Immunology*, Volume 4 (1986):

Lymphadenopathy-Associated-Virus Infection and Acquired Immunodeficiency Syndrome, J. C. Gluckman, D. Klatzmann, and L. Montagnier

Regulation and B-Cell Differentiation, T. Hamaoka and S. Ono

The Regulation and Expression of c-myc *in Normal and Malignant Cells*, K. Kelly and U. Siebenlist

The Membrane Attack Complex of Complement, H. J. Müller-Eberhard

The Molecular Genetics of the T-Cell Antigen Receptor and T-Cell Antigen Recognition, M. Kronenberg, G. Siu, L. E. Hood, and N. Shastri

From the *Annual Review of Microbiology*, Volume 40 (1986):

Organization of the Genes for Nitrogen Fixation in Photosynthetic Bacteria and Cyanobacteria, R. Haselkorn

Invasion of Erythrocytes by Malaria Parasites, T. J. Hadley, F. W. Klotz, and L. H. Miller

Microbial Ecology and Evolution: A Ribosomal Approach, G. J. Olsen, D. J. Lane, S. J. Giovannoni, N. R. Pace, and D. A. Stahl

From the *Annual Review of Neuroscience*, Volume 10 (1987):

Calcium Action in Synaptic Transmitter Release, G. J. Augustine, M. P. Charlton, and S. J. Smith

Molecular Biology of Visual Pigments, J. Nathans

Developmental Regulation of the Nicotinic Acetylcholine Receptor, S. M. Schuetze and L. Role

From the *Annual Review of Physiology*, Volume 48 (1986):

Regulation of Transepithelial H^+ Transport by Exocytosis and Endocytosis, G. J. Schwartz and Q. Al-Awqati

The Role of Osmotic Forces in Exocytosis from Adrenal Chromaffin Cells, R. W. Holz

Mimicry and Mechanism in Phospholipid Models of Membrane Fusion, R. P. Rand and V. A. Parsegian

ATP-Driven H^+ Pumping into Intracellular Organelles, G. Rudnick

Neuronal Receptors, S. H. Snyder

From the *Annual Review of Plant Physiology*, Volume 37 (1986):

Photoregulation of the Composition, Function, and Structure of Thylakoid Membranes, J. M. Anderson

Ann. Rev. Cell Biol. 1986. 2: 1–26

ACTIVATION OF SEA URCHIN GAMETES

James S. Trimmer and Victor D. Vacquier

Marine Biology Research Division A-002, Scripps Institution
of Oceanography, University of California at San Diego, La Jolla,
California 92093

CONTENTS

INTRODUCTION

Sea urchins are plentiful, easy to obtain, and provide a rich source of gametes for the study of fertilization. The sexes are separate, and fertilization can be readily manipulated as it occurs external to the adult body. Sea urchins are normally fertilized in seawater, a well-defined medium whose composition can be duplicated or altered. Gametes can be obtained in large numbers by intracoelomic injection of adults with 0.5 M KCl, which yields as many as 10^{12} spermatozoa from a single male or 10^7 eggs from a single female. Except for the presence of a small number of phagocytic coelomocytes, suspensions of sea urchin gametes are nearly homogeneous; neither spermatozoa nor eggs are shed with accessory cells. The most widely used sea urchin for cell biological research is *Strongylocentrotus purpuratus*, found on the west coast of North America. This species exhibits seasonality in gamete production, with maximum production in the winter months.

1

0743–4634/86/1115–0001$02.00

Eggs of *S. purpuratus* are 80 μm in diameter, with a 10–30-nm-thick fibrous extracellular matrix, known as the vitelline coat, bonded to the plasma membrane (Kidd 1978; Chandler & Heuser 1980, 1981). Surrounding the vitelline coat is a second extracellular matrix, the egg jelly layer, which can be up to 40 μm thick (Figure 1). Solubilized jelly coat induces profound changes in sperm morphology and physiology. Unlike almost all other animal eggs, sea urchin eggs have completed meiosis when they are released from the ovary. They have a haploid interphase nucleus of decondensed chromatin surrounded by a complete nuclear envelope. The egg cytoplasm is maintained at pH 6.8, which represses macromolecular synthesis and keeps the eggs metabolically dormant (Epel 1978; Shen 1983).

Spermatozoa consist of a conical head of 1 μm in diameter and 4 μm in length, a single donut-shaped mitochondrion, and a single flagellum 0.1 μm in diameter and 50 μm long (Figure 2). The head contains the acrosomal granule, a G-actin pool in the nuclear fossa, the haploid nucleus containing densely packed chromatin, and a pair of centrioles. There is very little cytoplasm, and the machinery for protein synthesis (RNA, ribosomes, and endoplasmic reticulum) has been lost during spermatogenesis. The terminally differentiated sperm cell is specialized for two functions: flagellar motility and the acrosomal reaction. The acrosomal reaction (AR) (Dan 1954, 1967) is the exocytosis of the acrosomal granule and the

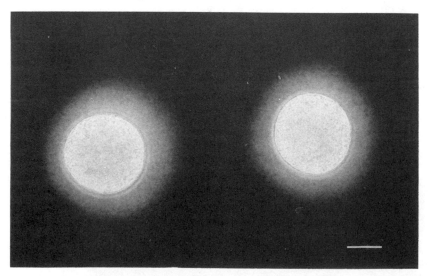

Figure 1   Phase micrograph of *S. purpuratus* eggs suspended in india ink to visualize the egg jelly layer. (Bar = 50 μm.)

A   B

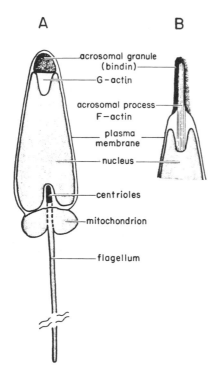

acrosomal granule
(bindin)
G-actin

acrosomal process
F-actin

plasma
membrane

nucleus

centrioles

mitochondrion

flagellum

Figure 2 Diagram of *S. purpuratus* spermatozoa (A) before and (B) after under-
going the acrosomal reaction. (From Vacquier 1986. Reproduced from *Trends in
Biochemical Science* with permission of
Elsevier Press.)

formation of the acrosomal process by the polymerization of G-actin in
the nuclear fossa into F-actin. The acrosomal reaction is required for
fertilization as it exposes the membrane that will fuse with the egg.

This review describes the activation events of both spermatozoon and
egg during fertilization. Since a three volume work on fertilization was
recently published (Metz & Monroy 1985), we concentrate this review on
the discoveries reported since the completion of these volumes.

SPERMATOZOA

Activation of Motility and Respiration

Sea urchin spermatozoa are stored in the testis for months at a time. The
inactive state of these cells is maintained by a low intracellular pH (pH_i).
Evidence indicates that high CO_2 tension in semen in the testes maintains
the pH_i at about 7.2 (Johnson et al 1983), which keeps motility and
respiration repressed (Schackmann et al 1981; Christen et al 1982; Lee et
al 1983). Upon release of spermatozoa into seawater, a large Na^+-depen-
dent acid extrusion occurs (Nishioka & Cross 1978; Christen et al 1982);
much of the acid is in the form of H^+ (Johnson et al 1983). This elevates

pH$_i$ by 0.4–0.5 pH units to a pH$_i$ of 7.5–7.6 (Christen et al 1982, 1983a; Lee et al 1983). This increase in pH$_i$ results in the initiation of motility. Activation of motility cannot occur in Na$^+$-free media, but the addition of either Na$^+$ or NH$_4$Cl to sperm in Na$^+$-free media initiates motility (Schackmann et al 1981; Christen et al 1982, 1983c; Johnson et al 1983; Lee et al 1983; Bibring et al 1984). Lee has studied Na$^+$ and H$^+$ transport in preparations of isolated flagella surrounded by intact plasma membranes (Lee 1984a). He has described an amiloride-resistant Na$^+$-H$^+$ exchanger in which both directions of the exchange are sensitive to membrane potential (Lee 1984b). This exchanger activity is isolated with flagellar plasma membrane vesicles, and the voltage sensitivity of this exchanger is modulated by Mg^{2+} (Lee 1985). Evidence has been presented (Clapper et al 1985) that Zn^{2+} may also alter the activity of this Na$^+$-H$^+$ exchanger. Thus, it appears that regulation of pH$_i$ by the sperm membrane may be under the control of several ions in seawater. Na,K-ATPase activity appears to be involved in the maintenance of pH$_i$. Sperm exhibit an additional alkalinization when K$^+$ is added in the presence of Na$^+$ (Gatti & Christen 1985). Alkalinization is blocked by the Na,K-ATPase inhibitor ouabain at levels inhibiting ^{86}Rb uptake into sperm or by cellular ATP depletion. These observations indicate the presence of an active Na,K-ATPase in sperm (Gatti & Christen 1985). The Na,K-ATPase is thought to maintain the low intracellular Na$^+$ necessary for Na$^+$-H$^+$ exchange. We found that monoclonal antibodies to the rat hepatocyte Na,K-ATPase (Schenk & Leffert 1983) inhibit the activity of this ATPase in isolated *S. purpuratus* sperm plasma membranes (J. S. Trimmer, unpublished observations). The activity of the Na,K-ATPase is enriched eighteenfold in these membranes, which suggests that sperm membranes may be a rich starting material for future purification of this enzyme (J. S. Trimmer, unpublished observations).

The absolute correlation between pH$_i$ and motility results from the sensitivity of the axonemal dynein ATPase to pH. This enzyme complex hydrolyzes ATP to provide energy for motility and is inactive below a pH$_i$ of 7.3; activation increases linearly from pH$_i$ 7.4 to 8.0. At a pH$_i$ of 7.2 or lower, mitochondrial respiration is inactive because of high ATP and low ADP concentrations. When dynein ATPase is activated, ATP levels drop, ADP levels increase, and respiration increases about fiftyfold (Christen et al 1983b). ATP production is primarily the result of fatty acid oxidation (Mita & Yasumasu 1983). Energy transport from the mitochondrion to the flagellum occurs via a phosphorylcreatine shuttle. Distinct isozymes of creatine kinase in the head and tail are involved in this shuttle (Tombes & Shapiro 1985). Treatment of sperm with the creatine kinase inhibitor 1-fluoro,2,4-dinitrobenzene paralyzes distal portions of the sperm flagellum, which supports the idea that the phosphorylcreatine shuttle is necessary for energy transport away from the mitochondrion (Tombes & Shapiro 1985; reviewed in Shapiro et al 1985; Shapiro & Tombes 1985). Upon release from the testis into seawater, the activation of motility and respiration in sea urchin spermatozoa is regulated by the elevation of pH_i, which activates dynein ATPase. Consequently, by a decrease in ATP and an increase in ADP levels, mitochondrial respiration is activated. Sperm motility is dependent on continuous activity of the Na^+-H^+ exchanger acting in concert with the Na,K-ATPase to maintain pH_i at permissive levels for dynein ATPase activity. Without this, H^+ produced by ATP hydrolysis rapidly acidifies the cell, resulting in cessation of motility.

Interaction of Spermatozoa with the Egg Jelly Layer

Spermatozoa must pass through the egg jelly layer before contacting the egg surface (Figure 1). Egg jelly induces the sperm acrosomal reaction (Dan 1954, 1967), which consists of the exocytosis of the acrosomal granule and the extension of the acrosomal process by polymerization of actin (G-to-F transformation) (reviewed in Shapiro et al 1981, 1985; Lopo 1983). The acrosomal reaction requires the net influx of Ca^{2+} and Na^+ and the net efflux of H^+ and K^+ (Dan 1954, 1967; Schackmann et al 1978; Tilney et al 1978; Schackmann & Shapiro 1981; Christen et al 1983b). Egg jelly also induces changes in sperm respiration and motility (Ohtake 1976), changes in sperm cyclic nucleotide levels (Watkins et al 1978; Garbers & Kopf 1980; Garbers et al 1983), and increases in cAMP-dependent protein kinase activity (Garbers et al 1980). Recent work has focused on the factors in egg jelly responsible for these profound effects on sperm and the signal transduction mechanism that turns these signals into metabolic changes.

ETHANOL-SOLUBLE COMPONENTS OF EGG JELLY   The bulk of the egg jelly layer can be dissolved from the cells by treatment with pH 5 seawater, although even with extensive washing jelly molecules remain bound to the egg vitelline layer (Vacquier et al 1979). Solubilized egg jelly can be divided by ethanol precipitation into two fractions: a precipitate containing a large, acidic glycoconjugate and a sialoglycoprotein, and a soluble fraction of peptides. The ethanol-soluble peptides activate sperm respiration and motility at pH 6.6 (Kopf et al 1979) and increase fatty acid oxidation and cyclic nucleotide metabolism (Hansbrough et al 1980). These activations by egg jelly peptides are all Ca^{2+}-independent. Several peptides have been isolated from the ethanol-soluble fraction of sea urchin egg jelly; the most extensively studied is speract from S. purpuratus (Hansbrough & Garbers 1981a) and Hemicentrotus pulcherrimus (Suzuki et al 1980). The sequence of this peptide is Gly-Phe-Asp-Leu-Asn-Gly-Gly-Gly-Val-Gly (Suzuki et al 1981; Garbers et al 1982). Purified speract, the synthesized peptide, and various analogs have been studied for their effects on sperm respiration (Garbers et al 1982; Nomura & Isaka 1985). A M_r 77,000 (77 kDa) receptor for speract has been identified by the use of speract analogs (Dangott & Garbers 1984). Purified or synthesized speract has half-maximal effects on sperm respiration at 30 pM in seawater at pH 6.6 (Hansbrough & Garbers 1981a; Garbers et al 1982). The stimulation of respiration by speract is Na^+-dependent, can be mimicked by the Na^+-H^+ ionophore monensin (Hansbrough & Garbers 1981b), and is mediated by H^+ efflux (Repaske & Garbers 1983). The changes in cGMP levels induced by speract, however, are Na^+-independent, and 8-bromo-cGMP can produce speract-like changes in respiration without Na^+-H^+ exchange (Hansbrough & Garbers 1981b).

The ethanol-soluble fraction of egg jelly from the sea urchin *Arbacia punctulata* induces the dephosphorylation of a major sperm flagellar membrane protein, resulting in a shift in its mobility on SDS-PAGE from 160 to 150 kDa (Ward & Vacquier 1983). This protein has been purified and identified as guanylate cyclase (Ward et al 1985b). The egg jelly–induced dephosphorylation results in a large decrease in guanylate cyclase activity (Ramarao & Garbers 1985; Ward et al 1985b), and the activity is proportional to the amount of enzyme in the phosphorylated (160 kDa) form (Ward et al 1985b). The dephosphorylation depends on the pH of the surrounding seawater, shows an absolute requirement for Na^+ (Ward 1985), and can be induced by monensin in the absence of egg jelly (Ward et al 1986b). The enzyme can also be dephosphorylated in vivo in the absence of egg jelly by suspending sperm cells in pH 9 seawater. The dephosphorylated form of the cyclase can be rephosphorylated in vivo in these cells by placing them in low pH (7.2) seawater, with a resultant increase in specific activity. This is the first direct demonstration that phosphorylation is responsible for changes in the activity of membrane-bound guanylate cyclase (Ward et al 1986a). The component of the ethanol-soluble fraction of *A. punctulata* jelly that induces the dephosphorylation of guanylate cyclase is the peptide resact (Cys-Val-Thr-Gly-Ala-Pro-Gly-Cys-Val-Gly-Gly-Gly-Arg-Leu-NH_2) (Suzuki et al 1984; Ward 1985). Resact, although structurally unrelated to speract, causes increased respiration and motility in *A. punctulata* sperm at pH 6.6 and causes a transitory increase in cGMP concentration of about tenfold (Suzuki et al 1984). The effects of speract and resact are species specific; speract has no effect on *A. punctulata* sperm and resact has no effect on *S. purpuratus* sperm (Suzuki et al 1984; Ramarao & Garbers 1985).

When spermatozoa are swimming in natural seawater (pH 8.0), respiration and motility are both maximally activated. In pH 6.6 seawater both parameters are repressed. The peptides speract and resact stimulate respiration and motility in sperm of their respective species at pH 6.6 up to the maximal values seen in normal seawater (pH 8.0) (Suzuki et al 1980, 1984; Hansbrough & Garbers 1981a; Nomura et al 1983). It has been suggested that egg jelly is acidic (Ohtake 1976) and that the presence of these peptides in jelly stimulates respiration and motility as the sperm swims through the acidic environment of the jelly layer. However, Holland & Cross (1983) showed with pH microelectrodes that the pH of the egg jelly layer is 8.0. The true biological role of these egg jelly peptides in fertilization thus remained unresolved.

Suzuki & Garbers (1984) reported that the fucose sulfate glycoconjugate (FSG) of egg jelly when free of peptides represses sperm respiration and motility at pH 7.7; they found that the addition of resact to *A. punctulata* sperm, or speract to *Lytechinus pictus* sperm, restores respiration to maximal values. However, these experiments, utilizing cross-species mixtures of sperm and FSG, do not settle the question of the role of egg jelly peptides in fertilization.

Recently, a possible biological role for resact has been demonstrated. When this peptide is microinjected into a suspension of *A. punctulata* sperm, the cells show a chemotactic response to the peptide; they change their flagellar wave form and accumulate as a swarm to the region of resact (Figure 3). This chemoattraction is species specific and Ca^{2+} dependent (Ward et al 1985a). This is the first molecule of known structure shown to

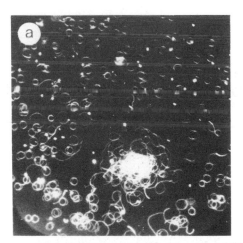 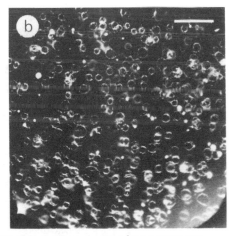

Figure 3   Effects of (a) 100-nM resact and (b) 10-μM speract on *A. punctulata* spermatozoa, 50 sec after peptide was added just below center of field. (Bar = 300 μm.) (From Ward et al 1985a. Reproduced from the *Journal of Cell Biology* with permission of The Rockefeller University Press.)

be a chemoattractant for animal sperm. It is interesting to note that the concentration dependencies of the resact induction of chemotaxis and the resact-induced dephosphorylation of guanylate cyclase are identical. However, direct evidence linking these two events is not yet available (Ward et al 1985a). It is intriguing to note that in the rod outer segment of the vertebrate retina cGMP directly mediates cation channel activity (reviewed in Lewin 1985). Could resact-induced changes in cGMP levels have analogous effects on sperm ion channel activity? These studies await the biophysical characterization of the ion channels of the sperm membrane.

ETHANOL-INSOLUBLE COMPONENTS OF EGG JELLY   The ethanol-precipitable fraction of sea urchin egg jelly also has major effects on sperm physiology that are under the control of a different set of mechanisms than those of the peptides. The ethanol-insoluble macromolecular fraction of egg jelly is excluded from Sepharose CL-4B in 0.5 M NaCl, which indicates its molecular mass is greater than 10^6. The excluded fraction contains fucose, sialic acid, protein, and sulfate. When this fraction is subjected to DEAE chromatography, two peaks are obtained, a component rich in sialic acid and protein (20%) and a fucose- and sulfate-rich fraction (80%) (SeGall & Lennarz 1979). This fucose sulfate glycoconjugate (FSG) is the inducer of the acrosomal reaction (AR) (SeGall & Lennarz 1979, 1981; Kopf & Garbers 1980). The induction occurs within seconds, requires Ca^{2+} and Na^+, and is inhibited by pH lower than 7.2 and by high K^+ (Dan 1954, 1967; Lee et al 1983; Schackmann et al 1978, 1981; Schackmann & Shapiro 1981; Christen et al 1983c; SeGall & Lennarz 1981). The egg jelly-induced AR can be inhibited by the Ca^{2+} channel antagonists D600, verapamil (Schackmann et al 1978), and nitrendipine (Kazazoglou et al 1985), which indicates the presence of slow Ca^{2+} channels analogous to those in mammalian cells (Reuter 1983). Earlier studies on the FSG-induced Ca^{2+} influx in sperm employed a $^{45}Ca^{2+}$ filter assay (Schackmann et al 1978; Kopf & Garbers 1980), which showed that only a small portion of the total $^{45}Ca^{2+}$ uptake occurred within the time frame of the morphological changes of the AR (Schackmann et al 1978). In addition, the late (> 1 min) uptake of $^{45}Ca^{2+}$ was inhibited by agents such as cyanide, FCCP, and oligomycin, which block mitochondrial $^{45}Ca^{2+}$ uptake (Schackmann & Shapiro 1981). It was concluded that the Ca^{2+} influx necessary for induction of the AR could not be determined by this method and that the bulk of the egg jelly–induced $^{45}Ca^{2+}$ uptake (which lasts as long as 10 min after the induction of the AR) was mitochondrial (Schackmann & Shapiro 1981). Recently, the availability of highly fluorescent calcium chelating probes, such as Quin 2 (Tsien et al 1982), Indo 1, and Fura 2 (Grynkiewicz et al 1985), have permitted the real-time measurement of changes in intracellular free Ca^{2+} in sea urchin sperm. These dyes have been used to study changes in cytoplasmic free Ca^{2+} induced by speract and FSG (R. W. Schackmann & P. B. Chock, in preparation). Both of these egg jelly components induce elevated levels of cytoplasmic free Ca^{2+}, although the speract-induced changes are much smaller and more transient. The dye fluorescence is associated with the head and flagellar regions of the sperm cell; little or no fluorescence is associated with the mitochondrion. Using these fluorescent Ca^{2+} indicators we have found that certain monoclonal antibodies (MAbs) to a 210-kDa glycoprotein of the *S. purpuratus* sperm plasma membrane induce an increase in intracellular free Ca^{2+} similar in magnitude and timing to that induced by egg jelly (J. S. Trimmer et al, in preparation). These same MAbs inhibit the egg jelly–induced AR and greater than 90% of the egg jelly–induced $^{45}Ca^{2+}$ uptake into the mitochondrion (Trimmer et al 1985). Although these MAbs induce increases in intracellular free Ca^{2+}, the morphological changes of the AR are not detectable, at least by light microscopy (J. S. Trimmer et al, in preparation). Egg jelly also induces a release of H^+ from the sperm, leading to an additional increase in pH_i of about 0.16 pH units (Schackmann et al 1978, 1981; Lee et al 1983; Collins & Epel 1977; Schackmann & Shapiro 1981). This pH_i change may also be coupled to Na^+ influx, as evidenced by the 1:1 stoichiometry of H^+ efflux and $^{22}Na^+$ uptake (Schackmann et al 1978). Tilney et al (1978) showed that inducing the AR with the Ca^{2+}-H^+ ionophore A23187 in normal seawater (10 mM Ca^{2+}) results in acrosomal granule exocytosis and extension of the acrosomal process, whereas A23187 induction in calcium-free seawater resulted in extension of the process without granule exocytosis. In starfish sperm, treatment with NH_4Cl to raise pH_i also induces extension of the acrosomal process without exocytosis (Schroeder & Christen 1982), which indicates separate roles of Ca^{2+} influx (exocytosis) and intracellular alkalinization (actin polymerization resulting in extension of the acrosomal process) in the morphological changes of the AR. We found that inducing Ca^{2+} uptake with MAbs to the 210-kDa protein or increasing pH_i by addition of either NH_4Cl, monensin, or speract will not induce the AR, but the anti-210-kDa MAbs plus any one of these agents will induce the AR (J. S. Trimmer et al, in preparation).

Sea urchin sperm also exhibit K^+-supported membrane potentials of -90 to -150 mV (Collins & Epel 1977; Schackmann et al 1981). Egg jelly induces rapid depolarization of the sperm membrane, to $+30$ mV; this depolarization is obligatorily linked to the AR (Schackmann et al 1984). It has been hypothesized that this depolarization may lead to opening of voltage-gated Ca^{2+} channels (Shapiro et al 1985). Resolution of this awaits the reconstitution of Ca^{2+} channels.

The ion movements induced by egg jelly during the AR are complex, and the interrelationships among different ionic fluxes are as yet unknown. Use of fluorescent probes to measure intracellular free Ca^{2+} and pH_i and use of probes of the cell surface, such as monoclonal antibodies, should lead to a better understanding of the ionic events involved in this complex cellular activation.

In addition to the ionic movements and morphological changes it induces in the AR, FSG has profound Ca^{2+}-dependent effects on cyclic nucleotide levels in sperm. Treating *S. purpuratus* sperm with egg jelly results in a sixfold increase in cAMP concentrations (Garbers & Hardman 1975, 1976). Addition of the phosphodiesterase inhibitors theophylline or 3-isobutyl-1-methyl xanthine (IBMX) before exposure to egg jelly or purified FSG elevates cAMP concentration 100–400-fold (Garbers & Hardman 1975, 1976). Jelly treatment results in a 50-fold activation of adenylate cyclase (Watkins et al 1978). This activation is Ca^{2+} dependent (Tubb et al 1978). However, A23187, which like egg jelly induces the AR, does not elevate cAMP levels (Garbers & Kopf 1978). Purified FSG is the inducer of elevation of cAMP levels (Kopf & Garbers 1980; Garbers et al 1983) and the tenfold activation of cAMP dependent protein kinase (Garbers et al 1980). The Ca^{2+} dependence of egg jelly–induced increases in cAMP levels and cAMP-dependent protein kinase activity was further demonstrated by the sensitivity of these changes to the Ca^{2+} channel blockers D600 and verapamil (Garbers et al 1980; Kopf & Garbers 1980). Sea urchin sperm cAMP and cGMP phosphodiesterases (Wells & Garbers 1976) and protein phosphatase (Swarup & Garbers 1982) activities were characterized, and the phosphatase activity was shown to undergo a Ca^{2+}-dependent, threefold increase upon treatment of sperm with FSG (Swarup & Garbers 1982). Isolated sperm heads retain the ability to elevate cAMP levels in response to nigericin or seawater of pH 9.0; these two treatments are known to induce the acrosomal reaction in intact sperm (Schackmann et al 1981). These findings indicate that the isolated heads contain the components for elevation of cAMP concentrations (Garbers 1981). However, in these experiments neither egg jelly nor purified FSG was tested (Garbers 1981). Garbers et al (1983) showed that both the fucose-esterified sulfate and the protein components of FSG were necessary to elevate cAMP levels.

FSG also causes phosphorylation of sperm histone H1. Sea urchin sperm contain a unique set of sperm-specific histones. Sperm histone H1 is replaced by egg-derived cleavage stage H1 within 10 min after insemination. It has been postulated that phosphorylation of sperm-specific H1 may play a role in its rapid loss from the male pronuclear chromatin after the sperm nucleus enters the egg cytoplasm (Green & Poccia 1985). In *S. purpuratus*, phosphorylation of sperm H1 on serine is induced by FSG

(Porter & Vacquier 1986). Phosphodiesterase inhibitors such as IBMX and SQ 20009, which elevate sea urchin sperm cAMP levels (Kopf et al 1983), induce the phosphorylation of H1 in the absence of FSG, which indicates that cAMP-dependent protein kinase activation may be responsible for this phosphorylation (Porter & Vacquier 1986). The effects of egg jelly on sperm are numerous and complex, and the interrelationships between changes in ionic permeability, cyclic nucleotide levels, enzymatic activity, and protein phosphorylation state remain unknown. The sea urchin sperm is a unique cell for studying the role of these events in cellular activation.

The Sperm Membrane

The events described above are induced by the interaction of components of the extracellular egg jelly layers with the sperm plasma membrane. There is much interest in discovering which sperm membrane proteins are involved in these signal transduction events and the mechanism by which signal transduction occurs. An obvious starting point for such studies is the isolation of the sperm plasma membrane.

Various methods have been developed to isolate the plasma membranes of sea urchin spermatozoa. Cross (1982) modified the original Gray & Drummond (1976) method of shearing isolated flagella in a hypotonic medium to obtain a membrane fraction enriched tenfold in the specific activity of lactoperoxidase-catalyzed surface-iodinated proteins. Major iodinated proteins were present at 200, 130, 80, and 60 kDa. Podell et al (1984) used high pH to disrupt sperm and obtain a plasma membrane fraction enriched seventeenfold in particulate adenylate cyclase activity. These sperm membrane vesicles exhibited species-specific binding to sea urchin eggs, which was inhibited by soluble egg jelly. This membrane preparation exhibited an electrophoretic profile similar to that of the membranes obtained by Cross (1982). Ward et al (1985b) used nitrogen cavitation and differential centrifugation to obtain a membrane fraction from *Arbacia punctulata* sperm enriched sevenfold in particulate guanylate cyclase activity. We have used this technique with *S. purpuratus* sperm to obtain fractions enriched seventeenfold for Na,K-ATPase activity. Kazazoglou et al (1984) showed that binding of the Ca^{2+} channel antagonists nitrendipine and verapamil is retained in hypotonically isolated membranes that are 9.5-fold enriched in Na,K-ATPase activity. Although these isolation techniques yield membranes enriched in plasma membrane marker enzymes, the isolated membranes do not retain significant $^{45}Ca^{2+}$ sequestering activity compared to in vivo levels. Darszon et al (1984) used the method of Cross (1982) to prepare sperm membranes for reconstitution into liposomes and found that egg jelly increased the permeability of these vesicles to $^{45}Ca^{2+}$ only slightly ($+$jelly/$-$jelly $= 1.51 \pm 0.75$). We found small increases (two- to threefold) in $^{45}Ca^{2+}$ permeability were induced by egg jelly in liposomes reconstituted with the membrane preparation of Ward et al (1985b). Similar levels of Ca^{2+} influx have also been seen in the hypotonically prepared sperm membrane vesicles of Kazazoglou et al (1984) (B. M. Shapiro, personal communication). Further refinement of techniques for reconstituting the components of the Ca^{2+} flux induced by egg jelly during the AR will hopefully yield levels of Ca^{2+} flux nearer those seen in vivo.

Cell surface molecules mediating ion transport can be defined by the use of biochemical or immunological probes that react with specific cell surface molecules. Lopo & Vacquier (1980) raised polyclonal rabbit antisera to *S. purpuratus* membrane proteins excised from SDS polyacrylamide gels. A sperm agglutinating antiserum made to an externally disposed 80-kDa protein inhibited the egg jelly–induced AR, while a sperm agglutinating antiserum to a 60-kDa protein had no inhibitory effects. Podell & Vacquier (1984a) made antisera to the 80-kDa protein of *S. purpuratus* in the same way, and although it showed monospecific characteristics by Ouchterlony assay, it was found to react with the 80-kDa protein and another protein of 210 kDa. When the 210-kDa protein was excised from SDS-poly-acrylamide gels and used as the immunogen, similar antisera that cross-reacted with both the 210-kDa and 80-kDa proteins were obtained (Podell & Vacquier 1984a). Saling et al (1982) and Eckberg & Metz (1982) raised a polyclonal antiserum to *A. punctulata* sperm and found that it too inhibited the jelly-induced AR, but no biochemical analysis of antibody specificity was performed. Wheat germ agglutinin (WGA) agglutinates *S. purpuratus* spermatozoa and binds over the entire sperm plasma membrane (Podell & Vacquier 1984b). It also inhibits the egg jelly–induced AR (Podell & Vacquier 1984b). Western blot analysis detects WGA reaction only with the 210-kDa glycoprotein, further implicating this protein in the induction of the AR (Podell & Vacquier 1984b).

We recently applied monoclonal antibody (MAb) techniques to the study of the sea urchin sperm cell membrane. We used *S. purpuratus* sperm and isolated sperm membranes to immunize mice for use in hybridoma production. Our early efforts to isolate MAbs that react with different membrane proteins were hampered by the tremendous antigenicity of the 210-kDa protein in mice. All the early fusion attempts yielded MAbs that reacted only with this protein. In fact, of the 80 MAbs we have obtained that react with sperm cell surface proteins, 65 are directed to the 210-kDa protein. Repeated attempts to immunize mice with other membrane proteins (i.e. the 60-kDa and 80-kDa bands excised from SDS-poly-acrylamide gels) did not elicit an immune response, and no positive hybridomas were isolated. By repeated immunizations with fractions of sperm membranes, we have finally generated hybridomas that produce MAbs to other sperm surface proteins.

The MAbs to the 210-kDa protein from our early hybridoma attempts proved to be very useful probes of the sperm surface. These MAbs exhibit large differences in their ability to inhibit the egg jelly–induced AR (Figure 4). The strongest inhibitor, MAb J10/14, inhibits the AR as either IgG or Fab fragments, while a noninhibitory MAb, J4/4, has no effect on this event at equivalent levels of cell surface binding (Trimmer et al 1985). All 65 MAbs to the 210-kDa protein exhibit restricted regions of binding to the cell surface (Figure 5), as opposed to the polyclonal antisera that cross-react with the 80- and 210-kDa proteins (Podell & Vacquier 1984a) and WGA (Podell & Vacquier 1984b), which bind to the entire sperm surface. Future work using MAbs may elucidate the pathways of transduction of the signals that lead to the biochemical and morphological changes that occur in sperm during fertilization.

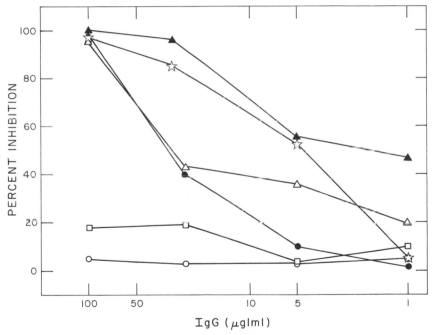

Figure 4 Inhibition of the egg jelly–induced acrosomal reaction in *S. purpuratus* spermatozoa by monoclonal antibodies reacting with the 210-kDa protein. Assay performed as in Trimmer et al 1985. (*Filled triangles* = J10/14 IgG_2a; *open triangles* = J16/5 IgG_1; *filled circles* = J17/18 IgG_1; *open circles* = J4/4 IgG_2a; *squares* = J16/18 IgG_2b; *stars* = J18/2 IgG_3.)

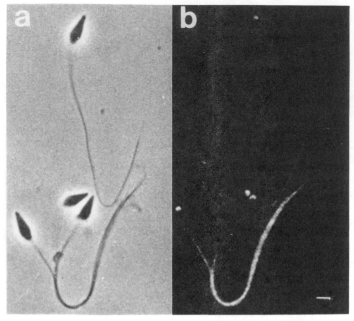

Figure 5 Immunofluorescent localization of the 210-kDa protein of *S. purpuratus* spermatozoa using MAb J10/14: (*a*) phase contrast, (*b*) fluorescence. (Bar = 1 μm.) (From Trimmer et al 1985. Reproduced from *Cell* with permission of MIT Press.)

Bindin

One of the major consequences of the egg jelly–induced AR is the exocytosis of the sperm acrosomal granule. This exocytotic granule contains the protein bindin, a 30.5-kDa protein (Vacquier & Moy 1977; reviewed in Vacquier 1983) that binds the sperm to the vitelline layer of the egg in a species-specific manner (Glabe & Vacquier 1977; Glabe & Lennarz 1979). Bindin released from the acrosomal granule covers the surface of the acrosomal process and is present in the area of attachment of the process to the vitelline layer (Moy & Vacquier 1979). Bindin agglutination of unfertilized eggs can be inhibited by glycoconjugates released from the egg surface by pronase digestion (Glabe & Lennarz 1979, 1981; Kinsey & Lennarz 1981; Rossignol et al 1981, 1984). This agglutination can also be blocked by complex polysaccharides such as fucoidin and xylan. Bindin also is a hemagglutinin whose activity can also be inhibited by fucoidin or glycoconjugates released from the egg surface by pronase digestion, which shows that bindin is a lectin (Glabe et al 1982). Bindin can be purified by DEAE chromatography (Vacquier 1983) and provides a useful system for studying intercellular adhesion. It has been shown to associate specifically with gel-phase phospholipid vesicles (Glabe 1985a), and it will induce the fusion of mixed-phase phospholipid vesicles (Glabe 1985b). These observations suggest that bindin may play a dual role in fertilization: It may mediate the attachment of sperm to egg and the fusion of the sperm and egg plasma membranes. The activation of sperm respiration and motility upon release from the gonad into seawater and the egg jelly–induced AR are activation phenomena of general interest in cell biological research.

ACTIVATION OF EGGS

Mature (postmeiotic) sea urchin eggs are stored in the ovary for months before spawning. In the ovary they are metabolically dormant cells; they do not synthesize nucleic acids, and they synthesize protein at only a very low rate (reviewed in Brandhorst 1985). Fusion with a spermatozoon activates the egg, initiating a stereotypic sequence of events termed the program of metabolic activation of development (Epel 1978). The sequential activations of the *S. purpuratus* egg have been divided into three periods: the "early" activations (0–1 min after addition of sperm to eggs); cytoplasmic alkalinization (1–5 min postinsemination); and the "late" activations (5–30 min postinsemination), all of which have been reviewed in detail by Epel (1978) (see also Shen 1983; Whitaker & Steinhardt 1985). This section reviews only the most recent data on the "early" metabolic activations of sea urchin eggs. The "early" activations we consider here are: the change in egg membrane potential, the changes in polyphosphoinositides which underlie the release of intracellular Ca^{2+} stores, and the propagation of increased intracellular Ca^{2+} as a wave across the egg. We also describe recent advances in the use of cortical granules in the study of the mechanism of exocytosis.

Electrical Changes of the Egg Plasma Membrane

Unfertilized eggs maintain a resting potential of about -70 mV. Within about 3 sec after insemination the egg membrane is very rapidly depolarized to a potential as high as $+20$ mV (Schatten & Hülser 1983). This change has been termed the fertilization potential and is generated by both Na^+ and Ca^{2+} entry (Chambers & de Armendi 1979). In what are now classical experiments, L. A. Jaffe (1976) showed that this positive fertilization potential inhibits the fusion of supernumerary sperm cells with the egg, i.e. it is the long-sought-after "fast block to polyspermy." Such fast, reversible, temporary electrical blocks against polyspermic fertilization have been found in both vertebrate and invertebrate eggs. Extensive reviews on this topic can be found in other articles (Nuccitelli & Grey

1984; Jaffe & Gould 1985; Jaffe & Cross 1986). If the fertilization potential is completely suppressed, and the egg is held at -70 mV with a voltage clamp electrode, sperm are blocked from entry into the cytoplasm. The cortical granule reaction occurs and literally blows the activating sperm off of the egg surface (Lynn & Chambers 1984). Although metabolically activated, such eggs do not cleave because they lack the pair of centrioles provided by the spermatozoon, which seed the formation of the mitotic apparatus (Schatten 1982). The mechanism by which a negative membrane potential inhibits the egg cortex cytoskeletal system involved in sperm entry (Tilney & Jaffe 1980; Schatten 1982) remains completely unknown.

Increase in Inositol Trisphosphate and Ca^{2+} Release from Intracellular Stores

The dramatic visual evidence that a sea urchin egg has been fertilized is the massive exocytosis of cortical granules from the egg cortex, termed the cortical reaction. The secretory proteins released in this exocytosis not only serve to release the vitelline layer from the plasma membrane, but also transform the vitelline layer into a mechanically tough and relatively impermeable fertilization envelope (reviewed in Runnström 1966; Epel & Vacquier 1978; Kay & Shapiro 1985; Schuel 1985). In *S. purpuratus* at 17°C, cortical granule exocytosis begins at the point of sperm fusion 20–30 sec after insemination and spreads as a circular wave around the egg, finishing at the antipode in 10–20 sec. Fusion of the 18,000 cortical granules with the egg membrane greatly increases the area of the egg membrane (Schroeder 1979). This extra membrane is subsequently endocytosed by massive formation of coated pits in the first 5 min following the cortical reaction (Fisher & Rebhun 1983). Direct evidence that the exocytosis of the cortical granules is triggered by elevation of free Ca^{2+} (in micromolar concentrations) has been recently reviewed (Jaffe 1985; Whitaker & Steinhardt 1985).

A large body of evidence from many cell types shows that cellular activations mediated by increases in cytosolic free Ca^{2+} correlate with the hydrolysis of polyphosphatidylinositol 4,5-bisphosphate (PIP_2) by phospholipase C to yield inositol 1,4,5-trisphosphate (IP_3) and diacylglycerol. Arachidonic acid is preferentially hydrolyzed from PIP_2. Direct evidence from a variety of cells shows that IP_3 is the compound that triggers the release of Ca^{2+} from Ca^{2+}-sequestering endoplasmic reticulum (reviewed in Majerus et al 1985; Marx 1985; Michell 1986). Regarding fertilization, the current working hypothesis is that the binding or fusion of sperm with the egg activates a membrane-associated phospholipase C whose activity increases the cytosolic concentration of IP_3, which in turn releases Ca^{2+} from a ramifying tubular reticulum. The increased Ca^{2+} triggers the exocytosis of egg cortical granules, resulting in elevation of the fertilization envelope and creation of a permanent block to polyspermy.

Analysis of the sea urchin egg plasma membrane shows that phosphatidylinositol comprises 25% of the total phospholipids, which is an unusually high proportion compared to other cells (Kinsey et al 1980). Fatty acid compositions of egg plasma membranes and cortical granule membranes are high in arachidonic acid (Decker & Kinsey 1983). A Ca^{2+}-stimulated arachidonic acid peroxidase activity has been found in egg homogenates (Perry & Epel 1985a). The activity increases at fertilization coincident with the rise in cytosolic free Ca^{2+} (Perry & Epel 1985b).

The intracellular concentration of IP_3 increases 40% in 15 sec and 52% in 60 sec postinsemination of sea urchin eggs (Turner et al 1984). Phosphatidylinositol (PI) content decreases 50% by 30 sec postinsemination, but recovers to prefertilization values by 5 min postinsemination (Kamel et al 1985). The labeling of IP_3 with radiophosphate also occurs within the first 60 sec postinsemination (Kamel et al 1985). Cortical granule exocytosis can be induced by the microinjection of IP_3 into sea urchin eggs (Whitaker & Irvine 1984), with exocytosis beginning at the point of microinjection and proceeding in a wave-like manner identical to that triggered by sperm. Electrical discharge has also been used to permeate eggs, and IP_3 introduced by this method will also induce cortical granule exocytosis (Clapper & Lee 1985). Ca^{2+}-induced hydrolysis of PIP_2 occurs in isolated sea urchin egg plasma membrane–cortical granule complexes to yield IP_3 (Whitaker & Aitchison 1985).

Guanyl nucleotide–binding proteins (G-proteins) have also been implicated in regulating IP_3 production in sea urchin eggs by Turner et al (1986). In their study, injection of 28-nM IP_3 induced cortical granule exocytosis. Injection of EGTA before injection of IP_3 inhibited exocytosis, further implicating IP_3 as the mediator of intracellular Ca^{2+} release. Injection of eggs with 3-mM GDP-β-S [guanosine-5'-0(2-thiophosphate)], a metabolically stable analog of GDP that binds G-proteins and blocks their biological effects, prevented sperm from triggering exocytosis. However, injection of GDP-β-S followed by injection of IP_3 resulted in exocytosis. These data suggest that sperm binding or fusion with the egg surface activates a G-protein in the egg which in turn regulates intracellular IP_3 levels. Increased IP_3 then releases Ca^{2+} from intracellular stores triggering cortical granule exocytosis (Turner et al 1986).

The Ca^{2+} Wave and Cortical Granule Exocytosis

As previously mentioned, the exocytosis of cortical granules occurs as a propagating circular wave across the egg cytoplasm from the point of sperm fusion to its antipode (reviewed in Jaffe 1985). Eisen et al (1984) studied the real-time relations among the occurrence of five events of "early" activation in single eggs of *A. punctulata*. These events are: the fertilization potential; surface contraction; increase in cytosolic free Ca^{2+} (as monitored with the Ca^{2+} photoprotein aequorin); increase in NAD(P) reduction; and initiation of elevation of the fertilization envelope. (Their limit of detection of free Ca^{2+} in the unfertilized egg cytoplasm was about 0.1 μM or lower in eggs containing 13 μM aequorin.) In this study, after addition of sperm, time 0 was chosen as the point at which the fertilization potential occurred, an event coincident with the surface contraction of the egg. Increased Ca^{2+} around the point of sperm entry occurred at 23 sec; increased free Ca^{2+} propagated across the egg as a wave in 6–9 sec. Increased reduction of NAD(P) occurs at 51 sec, coincident with the beginning of fertilization envelope elevation. Although the Ca^{2+} wave propagates across the egg in 6–9 sec, peak uniform aequorin luminescence does not occur until 17 sec after initiation of the free Ca^{2+} increase, or 40 sec after the fertilization potential. Peak free Ca^{2+} is maintained for a variable period (15–60 sec) and then decays uniformly to preactivation levels. The propagated wave of cortical granule exocytosis begins between the time of peak uniform free Ca^{2+} (40 sec) and initiation of fertilization envelope elevation (51 sec). Because the propagated wave of cortical granule exocytosis begins at the maximum uniform cytosolic Ca^{2+} concentration (about 1 μM), Ca^{2+} is a necessary but not sufficient component of the mechanism of propagation of the wave of exocytosis (Eisen et al 1984). Factors responsible for propagation of the exocytotic wave await discovery.

Aequorin-injected eggs, centrifuged to displace all organelles except the endoplasmic reticulum (ER), show uniform luminescence at fertilization, which implicates the ER as the source of Ca^{2+} released during early activation (Eisen & Reynolds 1985). The surprising finding of this study is that before fertilization the egg mitochondria do not contain Ca^{2+} that is releasable by uncoupling agents, whereas after fertilization they do. These findings strongly implicate the egg ER as the Ca^{2+} source and the mitochondria as the Ca^{2+} sink. The ER can also act as a Ca^{2+} sink when the mitochondria are blocked from sequestering Ca^{2+} (Eisen & Reynolds 1985). Clapper & Lee (1985) have shown IP_3-induced Ca^{2+} release from sea urchin egg endoplasmic reticulum microsomes, which suggests that this may be a promising system for studying the mechanism of IP_3-induced Ca^{2+} release.

Isolated Cortical Granules

The large number of cortical granules in sea urchin eggs and the rapidity and synchrony of their propagated exocytosis at fertilization make them ideal for studies of the mechanism of exocytosis. Cortical granules (~ 1 μm in diameter) are tightly bonded to the cytoplasmic face of the egg plasma membrane. The isolation of these granules bound to fragments of the plasma membrane was first accomplished by electrostatically binding the eggs to protamine sulfate–coated surfaces and then shearing away the cytoplasm with a jet of isosmotic medium containing a Ca^{2+} chelator (Vacquier 1975, 1976). Such preparations have been termed "cortical granule lawns." When freshly prepared, they retain the ability to undergo in vitro exocytosis in response to micromolar Ca^{2+} (Baker et al 1980; Moy et al 1983). This in vitro exocytosis involves fusion of the granule membrane with the plasma membrane (Whitaker & Baker 1983; Crabb & Jackson 1985). There is evidence for the involvement of calmodulin in regulating this exocytosis (Baker & Whitaker 1979; Steinhardt & Alderton 1982; Moy et al 1983). The granules can be isolated from the plasma membrane, and preliminary biochemical characterization has been performed (Kopf et al 1982; Decker & Kinsey 1983).

Although cortical granule lawns are excellent for certain types of experiments, it is impractical to prepare them in large quantities. Large scale isolations were developed that employ homogenization of eggs and differential centrifugation to isolate fragments of the egg cortex that retain Ca^{2+} sensitivity; these fragments consist of the vitelline layer, plasma membrane, and cortical granules (Detering et al 1977; Vacquier & Moy 1980; Haggerty & Jackson 1983; Sasaki & Epel 1983). Spectrophotometric assays have been developed that measure decreases in the turbidity of suspensions of cortical fragments in response to added Ca^{2+} (Haggerty & Jackson 1983; Sasaki & Epel 1983). Cortical fragments placed in 500-mM (isosmotic) chaotropic anions (Cl^-, I^-) lose much of their sensitivity to Ca^{2+}. Restoration of Ca^{2+} sensitivity occurs upon addition of fresh extracts of eggs or cortices. A 100-kDa protein isolates with this restorative activity (Sasaki 1984). Proteolytic treatment of cortices also results in a loss of sensitivity to Ca^{2+} (Jackson et al 1985).

Ultrastructural studies show each cortical granule is linked to the egg plasma membrane and to 4–6 adjacent granules by filaments of 6 nm in diameter (Chandler 1984). A tubular reticulum with a regular geometric pattern is apparent on the inner face of the egg plasma membrane, between the granules and the plasma membrane (Chandler 1984; Sardet 1984). This rough ER is thought to be the site of the sequestering reticulum that releases Ca^{2+} at fertilization (Eisen & Reynolds 1985).

It is generally agreed that the sensitivity of cortical granule exocytosis to Ca^{2+} drastically decreases when the granules are dissociated from the plasma membrane. Ca^{2+} concentrations in the millimolar range are required to trigger fusion of dissociated granules, compared to the micromolar sensitivity of granule–plasma membrane preparations (Kopf et al 1982; Crabb & Jackson 1985). Crabb & Jackson (1985) found that cortical granules dissociated from the plasma membrane in lawn preparations will reattach to the inner face of egg plasma membranes. Reattachment of the granules reconstitutes the components necessary for regaining sensitivity to micromolar Ca^{2+} (Crabb & Jackson 1985). The breakdown of cortical granules in response to Ca^{2+} is exocytosis and not lysis, as demonstrated by the vectoral expulsion of granule proteins to the external surface of the egg plasma membrane (Whitaker & Baker 1983; Crabb & Jackson 1985). This is the first report of the functional reconstitution of exocytosis in vitro, and further analysis may lead to the identification of the proteins involved in conferring Ca^{2+} sensitivity to the cortical granules (Crabb & Jackson 1985).

Sea urchin cortical granules have also been used to investigate the role of osmotic swelling of secretory vesicles prior to their fusion with the plasma membrane. Exocytosis of cortical granules is blocked if the osmolarity of the medium bathing the eggs is increased. Addition of Ca^{2+} to granule lawns causes the granules to swell. A combination of Ca^{2+} and high osmolarity inhibits exocytosis, but when normal osmolarity is restored, exocytosis proceeds even in the absence of Ca^{2+} (Zimmerberg & Whitaker 1985). The kinetics of cortical granule exocytosis in hyperosmotic conditions further support the hypothesis that osmotic swelling is mandatory for fusion of the granule membrane with the egg plasma membrane (Zimmerberg et al 1985).

The recent discoveries described in this review emphasize the importance of the study of the activation of sea urchin gametes at fertilization. Gamete interaction provides an opportunity not only to detail mechanisms of activation basic to all cells, but also to describe the role of the cell surface in mediating these events.

Literature Cited

Baker, P. F., Knight, D. E., Whitaker, M. J. 1980. The relation between ionized calcium and cortical exocytosis in eggs of the sea urchin, *Echinus esculentus. Proc. R. Soc. London Ser. B* 207: 149–61

Baker, P. F., Whitaker, M. J. 1979. Trifluoperazine inhibits exocytosis in sea urchin eggs. *J. Physiol.* 298: 55

Bibring, T., Baxandall, J., Harter, C. C. 1984. Sodium-dependent pH regulation in active sea urchin sperm. *Dev. Biol.* 101: 425–35

Brandhorst, B. P. 1985. Informational content of the echinoderm egg. In *Developmental Biology. A Comprehensive Synthesis*, ed. L. W. Browder, 1: 525–76. New York: Plenum. 632 pp.

Chambers, E. L., de Armendi, J. 1979. Membrane potential, action potential and activation potential of eggs of the sea urchin, *Lytechinus variegatus. Exp. Cell Res.* 122: 203–18

Chandler, D. E. 1984. Exocytosis in vitro: Ultrastructure of the isolated sea urchin egg cortex as seen in platinum replicas. *J. Ultrastruct. Res.* 89: 198–211

Chandler, D. E., Heuser, J. 1980. The vitelline layer of the sea urchin egg and its modification during fertilization. A freeze fracture study using quick-freezing and deep-etching. *J. Cell Biol.* 84 : 618–32

Chandler, D. E., Heuser, J. 1981. Postfertilization growth of microvilli in the sea urchin egg: New views from eggs that have been quick-frozen, freeze-fractured, and deeply etched. *Dev. Biol.* 82 : 393–400

Christen, R., Schackmann, R. W., Dahlquist, F. W., Shapiro, B. M. 1983a. ^{31}P-NMR analysis of sea urchin sperm activation. Reversible formation of high energy phosphate compounds by changes in intracellular pH. *Exp. Cell Res.* 149 : 289–94

Christen, R., Schackmann, R. W., Shapiro, B. M. 1982. Elevation of intracellular pH activates respiration and motility of sperm of the sea urchin, *Strongylocentrotus purpuratus*. *J. Biol. Chem.* 257 : 14881–90

Christen, R., Schackmann, R. W., Shapiro, B. M. 1983b. Metabolism of sea urchin sperm. Interrelationships between intracellular pH, ATPase activity and mitochondrial respiration. *J. Biol. Chem.* 258 : 5392–99

Christen, R., Schackmann, R. W., Shapiro, B. M. 1983c. Interactions between sperm and sea urchin egg jelly. *Dev. Biol.* 98 : 1–14

Clapper, D. L., Davis, J. A., Lamothe, P. J., Patton, C., Epel, D. 1985. Involvement of zinc in the regulation of pH$_i$, motility, and acrosome reactions in sea urchin sperm. *J. Cell Biol.* 100 : 1817–24

Clapper, D. L., Lee, H. C. 1985. Inositol trisphosphate induces calcium release from non-mitochondrial stores in sea urchin egg homogenates. *J. Biol. Chem.* 260 : 13947–54

Collins, F., Epel, D. 1977. The role of calcium ions in the acrosome of sea urchin sperm. Regulation of exocytosis. *Exp. Cell Res.* 106 : 211–22

Crabb, J. H., Jackson, R. C. 1985. In vitro reconstitution of exocytosis from plasma membrane and isolated secretory vesicles. *J. Cell Biol.* 101 : 2263–73

Cross, N. L. 1982. Isolation and electrophoretic characterization of the plasma membrane of sea urchin sperm. *J. Cell Sci.* 59 : 13–25

Dan, J. C. 1954. Studies on the acrosome. III. Effect of calcium deficiency. *Biol. Bull.* 107 : 335–49

Dan, J. C. 1967. The acrosome reaction and lysins. In *Fertilization*, ed. C. B. Metz, A. Monroy, 1 : 237–93. New York: Academic

Dangott, L. J., Garbers, D. L. 1984. Identification and partial characterization of the receptor for speract. *J. Biol. Chem.* 259 : 13712–16

Darszon, A., Gould, M., De La Torre, L., Vargas, I. 1984. Response of isolated sperm plasma membranes from sea urchin to egg jelly. *Eur. J. Biochem.* 144 : 515–22

Decker, S. J., Kinsey, W. H. 1983. Characterization of cortical secretory vesicles from the sea urchin egg. *Dev. Biol.* 96 : 37–45

Detering, N. K., Decker, E. D., Schmell, D., Lennarz, W. J. 1977. Isolation and characterization of plasma membrane associated cortical granules from sea urchin eggs. *J. Cell Biol.* 75 : 899–914

Eckberg, W. R., Metz, C. B. 1982. Isolation of an *Arbacia* sperm fertilization antigen. *J. Exp. Zool.* 221 : 101–5

Eisen, A., Kiehart, D. P., Wieland, S. J., Reynolds, G. T. 1984. Temporal sequence and spatial distribution of early events of fertilization in single sea urchin eggs. *J. Cell Biol.* 99 : 1647–54

Eisen, A., Reynolds, G. T. 1985. Sources and sinks for the calcium released during the fertilization of single sea urchin eggs. *J. Cell Biol.* 100 : 1522–27

Epel, D. 1978. Mechanisms of activation of sperm and egg during fertilization of sea urchin gametes. *Curr. Top. Dev. Biol.* 12 : 185–246

Epel, D., Vacquier, V. D. 1978. Membrane fusion during invertebrate fertilization. *Cell Surf. Rev.* 5 : 1–63

Fisher, G. W., Rebhun, L. I. 1983. Sea urchin egg cortical granule exocytosis is followed by a burst of membrane retrieval via uptake into coated vesicles. *Dev. Biol.* 456–72

Garbers, D. L. 1981. The elevation of cyclic AMP concentrations in flagella-less sea urchin sperm heads. *J. Biol. Chem.* 256 : 620–24

Garbers, D. L., Hardman, J. G. 1975. Factors released from sea urchin eggs affect cyclic nucleotide metabolism in sperm. *Nature* 257 : 677–78

Garbers, D. L., Hardman, J. G. 1976. Effects of egg factors on cyclic nucleotide metabolism in sea urchin sperm. *J. Cyclic Nucleotide Res.* 2 : 59–70

Garbers, D. L., Kopf, G. S. 1978. Effects of factors released from eggs and other agents on cyclic nucleotide concentrations of sea urchin spermatozoa. *J. Reprod. Fertil.* 52 : 135–40

Garbers, D. L., Kopf, G. S. 1980. The regulation of spermatozoa by calcium and cyclic nucleotides. *Adv. Cyclic Nucleotide Res.* 13 : 251–306

Garbers, D. L., Kopf, G. S., Tubb, D. J., Olson, G. 1983. Elevation of sperm adenosine 3'-5'-monophosphate concentrations by a fucose-sulfate-rich complex associated with eggs: 1. Structural characterization. *Biol. Reprod.* 29: 1211–20

Garbers, D. L., Tubb, D. J., Kopf, G. S. 1980. Regulation of sea urchin sperm cyclic AMP-dependent protein kinases by an egg associated factor. *Biol. Reprod.* 22: 526–32

Garbers, D. L., Watkins, H. D., Hansbrough, J. R., Smith, A., Misono, K. S. 1982. The amino acid sequence and chemical synthesis of speract and of speract analogues. *J. Biol. Chem.* 257: 2734–37

Gatti, J. L., Christen, R. 1985. Regulation of internal pH of sea urchin sperm. *J. Biol. Chem.* 260: 7599–7602

Glabe, C. G. 1985a. Interaction of the sperm adhesive protein, bindin, with phospholipid vesicles. I. Specific association of bindin with gel-phase phospholipid vesicle. *J. Cell Biol.* 100: 794–99

Glabe, C. G. 1985b. Interaction of the sperm adhesive protein, bindin, with phospholipid vesicles. II. Bindin induces the fusion of mixed-phase vesicles that contain phosphatidyl-choline and phosphatidyl serine *in vitro*. *J. Cell Biol.* 100: 800–6

Glabe, C. G., Grabel, L. B., Vacquier, V. D., Rosen, S. D. 1982. Carbohydrate specificity of sea urchin sperm bindin: a cell surface lectin mediating sperm-egg adhesion. *J. Cell Biol.* 94: 123–28

Glabe, C. G., Lennarz, W. J. 1979. Species-specific adhesion in sea urchins: A quantitative investigation of bindin-mediated egg agglutination. *J. Cell Biol.* 83: 595–604

Glabe, C. G., Lennarz, W. J. 1981. Isolation of a high molecular weight glycoconjugate derived from the surface of *S. purpuratus* eggs that is implicated in sperm adhesion. *J. Supramol. Struct. Cell. Biochem.* 15: 387–94

Glabe, C. G., Vacquier, V. D. 1977. Species specific agglutination of eggs by bindin isolated from sea urchin sperm. *Nature* 267: 836–37

Gray, J. P., Drummond, G. I. 1976. Guanylate cyclase of sea urchin sperm: Subcellular localization. *Arch. Biochem. Biophys.* 172: 31–38

Green, G. R., Poccia, D. L. 1985. Phosphorylation of sea urchin sperm H1 and H2B histones precedes chromatin decondensation and H1 exchange during pronuclear formation. *Dev. Biol.* 108: 235–45

Grynkiewicz, G., Poenie, M., Tsien, R. X. 1985. A new generation of Ca^{2+} indicators with greatly improved fluorescence properties. *J. Biol. Chem.* 260: 3440–50

Haggerty, J. G., Jackson, R. C. 1983. Release

of granule contents from sea urchin egg cortices: New assay procedures and inhibition by sulfhydryl-modifying reagents. *J. Biol. Chem.* 258: 1819–25

Hansbrough, J. R., Garbers, D. L. 1981a. Speract. Purification and characterization of a peptide associated with eggs that activates spermatozoa. *J. Biol. Chem.* 256: 1447–52

Hansbrough, J. R., Garbers, D. L. 1981b. Sodium-dependent activation of sea urchin spermatozoa by speract and monensin. *J. Biol. Chem.* 256: 2235–41

Hansbrough, J. R., Kopf, G. S., Garbers, D. L. 1980. The stimulation of sperm metabolism by a factor associated with eggs and by 8-bromoguanosine 3′,5′-monophosphate. *Biochim. Biophys. Acta* 630: 82–91

Holland, L. Z., Cross, N. L. 1983. The pH within the jelly coat of sea urchin eggs. *Dev. Biol.* 99: 258–60

Jackson, R. C., Haggerty, J. H., Ward, K. K. 1985. Mild proteolytic digestion restores exocytotic activity to N-ethylmaleimide-inactivated cell surface complex from sea urchin eggs. *J. Cell Biol.* 101: 6–11

Jaffe, L. A. 1976. Fast block to polyspermy in sea urchin eggs is electrically mediated. *Nature* 261: 68–70

Jaffe, L. A., Cross, N. L. 1986. Electrical regulation of sperm-egg fusion. *Ann. Rev. Physiol.* 48: 191–200

Jaffe, L. A., Gould, M. 1985. Polyspermy-preventing mechanisms. In *Biology of Fertilization*, ed. C. B. Metz, A. Monroy, 3: 223–99. Orlando: Academic

Jaffe, L. F. 1985. Sources of calcium in egg activation: A review and hypothesis. *Dev. Biol.* 99: 265–76

Johnson, C. H., Clapper, D. L., Winkler, M. M., Lee, H. C., Epel, D. 1983. A volatile inhibitor immobilizes sea urchin sperm in semen by depressing intracellular pH. *Dev. Biol.* 98: 493–501

Kamel, L. C., Bailey, J., Schoenbaum, L., Kinsey, W. 1985. Phosphatidylinositol metabolism during fertilization in the sea urchin egg. *Lipids* 20: 350–56

Kay, E. S., Shapiro, B. M. 1985. The formation of the fertilization membrane of the sea urchin egg. In *Biology of Fertilization*, ed. C. B. Metz, A. Monroy, 3: 45–81. Orlando: Academic

Kazazoglou, T., Schackmann, R. W., Fosset, M., Shapiro, B. M. 1985. Calcium channel antagonists inhibit the acrosome reaction and bind to plasma membranes of sea urchin sperm. *Proc. Natl. Acad. Sci. USA* 82: 1460–64

Kidd, P. 1978. The jelly and vitelline coats of the sea urchin egg: New ultrastructural features. *J. Ultrastruct. Res.* 64: 204–15

Kinsey, W. H., Decker, G. L., Lennarz, W. J. 1980. Isolation and partial characterization of the plasma membrane of the sea urchin egg. *J. Cell Biol.* 87: 248–54

Kinsey, W. H., Lennarz, W. J. 1981. Isolation of a glycopeptide fraction from the surface of the sea urchin egg that inhibits sperm-egg binding and fertilization. *J. Cell Biol.* 91: 325–31

Kopf, G. S., Garbers, D. L. 1980. Calcium and a fucose-sulfate-rich polymer regulate sperm cyclic nucleotide metabolism and the acrosome reaction. *Biol. Reprod.* 22: 1118–26

Kopf, G. S., Lewis, C. A., Vacquier, V. D. 1983. Methylxanthines stimulate calcium transport and inhibit cyclic nucleotide phosphodiesterases in abalone sperm. *Dev. Biol.* 99: 115–20

Kopf, G. S., Moy, G. W., Vacquier, V. D. 1982. Isolation and characterization of sea urchin egg cortical granules. *J. Cell Biol.* 95: 924–32

Kopf, G. S., Tubb, D. J., Garbers, D. L. 1979. Activation of sperm respiration by a low molecular weight egg factor and by 8-bromoguanosine 3′,5′-monophosphate. *J. Biol. Chem.* 254: 8554–60

Lee, H. C. 1984a. Sodium and proton transport in flagella isolated from sea urchin spermatozoa. *J. Biol. Chem.* 259: 4957–63

Lee, H. C. 1984b. A membrane potential-sensitive Na^+-H^+ exchange system in flagella isolated from sea urchin spermatozoa. *J. Biol. Chem.* 259: 15315–19

Lee, H. C. 1985. The voltage sensitive Na^+/H^+ exchange in sea urchin spermatozoa flagellar membrane vesicles studied with an entrapped pH probe. *J. Biol. Chem.* 260: 10794–99

Lee, H. C., Johnson, C., Epel, D. 1983. Changes in internal pH associated with the initiation of motility and acrosome reaction of sea urchin sperm. *Dev. Biol.* 95: 31–45

Lewin, R. 1985. Unexpected progress in photoreception. *Science* 227: 500–3

Lopo, A. C. 1983. Sperm-egg interactions in invertebrates. In *Mechanism and Control of Animal Fertilization*, ed. J. F. Hartmann, pp. 269–324. New York: Academic. 561 pp.

Lopo, A. C., Vacquier, V. D. 1980. Antibody to a specific sperm surface glycoprotein inhibits the egg jelly-induced acrosome reaction. *Dev. Biol.* 79: 325–33

Lynn, J. W., Chambers, E. L. 1984. Voltage clamp studies of fertilization in sea urchin eggs. I. Effect of clamped membrane potential on sperm entry, activation, and development. *Dev. Biol.* 102: 98–109

Majerus, P. W., Wilson, D. B., Connolly, T. M., Bross, T. E., Neufeld, E. J. 1985. Phosphoinositide turnover provides a link in stimulus-response coupling. *Trends Biochem. Sci.* 10: 168–71

Marx, J. L. 1985. The polyphosphoinositides revisited. *Science* 228: 312–13

Metz, C. B., Monroy, A. 1985. *The Biology of Fertilization*, Vol. 1–3. Orlando: Academic. 1335 pp.

Michell, B. 1986. Profusion and confusion. *Nature* 319: 176–77

Mita, M., Yasumasu, I. 1983. Metabolism of lipid and carbohydrate in sea urchin spermatozoa. *Gamete Res.* 7: 133–44

Moy, G. W., Kopf, G. S., Gache, C., Vacquier, V. D. 1983. Calcium-mediated release of glucanase activity from cortical granules of sea urchin eggs. *Dev. Biol.* 100: 267–74

Moy, G. W., Vacquier, V. D. 1979. Immunoperoxidase localization of bindin during the adhesion of sperm to sea urchin eggs. *Curr. Top. Dev. Biol.* 13: 31–44

Nishioka, D., Cross, N. 1978. The role of external sodium in sea urchin fertilization. In *Cell Reproduction*, ed. E. R. Dirksen, D. Prescott, C. F. Fox, pp. 403–13. New York: Academic

Nomura, K., Isaka, S. 1985. Synthetic study on the structure-activity relationship of sperm activating peptides from the jelly coat of sea urchin eggs. *Biochem. Biophys. Res. Commun.* 126: 974–82

Nomura, K., Suzuki, N., Ohtake, H., Isaka, S. 1983. Structure and action of sperm activating peptides from the egg jelly of a sea urchin, *Anthrocidaris crassipina*. *Biochem. Biophys. Res. Commun.* 117: 147–53

Nuccitelli, R., Grey, R. D. 1984. Controversy over the fast, partial, temporary block to polyspermy in sea urchins: A reevaluation. *Dev. Biol.* 103: 1–17

Ohtake, H. 1976. Respiratory behavior of sea urchin spermatozoa. I. Effect of pH and egg water on the respiratory rate. *J. Exp. Zool.* 198: 303–12

Perry, G., Epel, D. 1985a. Characterization of a Ca^{2+}-stimulated lipid peroxidizing system in the sea urchin egg. *Dev. Biol.* 107: 47–57

Perry, G., Epel, D. 1985b. Fertilization stimulates lipid peroxidation in the sea urchin egg. *Dev. Biol.* 107: 58–65

Podell, S. B., Moy, G. W., Vacquier, V. D. 1984. Isolation and characterization of a plasma membrane fraction from sea urchin sperm exhibiting species specific recognition of the egg surface. *Biochim. Biophys. Acta* 778: 25–37

Podell, S. B., Vacquier, V. D. 1984a. Inhibition of sea urchin sperm acrosome reaction by antibodies directed against two sperm membrane proteins. *Exp. Cell Res.* 155: 467–76

Podell, S. B., Vacquier, V. D. 1984b. Wheat germ agglutinin blocks the acrosome reaction in *Strongylocentrotus purpuratus* sperm by binding a 210,000-mol-wt. membrane protein. *J. Cell Biol.* 99: 1598–1604

Porter, D. C., Vacquier, V. D. 1986. Phosphorylation of sperm histone H1 is induced by the egg jelly layer in the sea urchin *Strongylocentrotus purpuratus*. *Dev. Biol.* 15: In press

Ramarao, C. S., Garbers, D. L. 1985. Receptor-mediated regulation of guanylate cyclase activity in spermatozoa. *J. Biol. Chem.* 260: 8390–96

Repaske, D. R., Garbers, D. L. 1983. A hydrogen ion flux mediates stimulation of respiratory activity by speract in sea urchin spermatozoa. *J. Biol. Chem.* 258: 6025–29

Reuter, H. 1983. Calcium channel modulation by neurotransmitters, enzymes and drugs. *Nature* 301: 569–74

Rossignol, D. P., Earles, B. J., Decker, G. L., Lennarz, W. J. 1984. Characterization of the sperm receptor on the surface of eggs of *Strongylocentrotus purpuratus*. *Dev. Biol.* 104: 308–21

Rossignol, D. P., Roschelle, A. J., Lennarz, W. J. 1981. Sperm-egg binding: Identification of a species-specific sperm receptor from eggs of *Strongylocentrotus purpuratus*. *J. Supramol. Struct. Cell. Biochem.* 15: 347–58

Runnström, J. 1966. The vitelline membrane and cortical particles in sea urchin eggs and their function in maturation and fertilization. *Adv. Morphol.* 5: 221–325

Saling, P. M., Eckberg, W. R., Metz, C. B. 1982. Mechanism of univalent antisperm antibody inhibition of fertilization in the sea urchin *Arbacia punctulata*. *J. Exp. Zool.* 221: 93–99

Sardet, C. 1984. The ultrastructure of the sea urchin egg cortex isolated before and after fertilization. *Dev. Biol.* 105: 196–210

Sasaki, H. 1984. Modulation of calcium sensitivity by a specific cortical protein during sea urchin egg cortical vesicle exocytosis. *Dev. Biol.* 101: 125–35

Sasaki, H., Epel, D. 1983. Cortical vesicle exocytosis in isolated cortices of sea urchin eggs: Description of a turbidometric assay and its utilization in studying effects of different media on discharge. *Dev. Biol.* 98: 327–37

Schackmann, R. W., Christen, R., Shapiro, B. M. 1981. Membrane potential depolarization and increased intracellular pH accompany the acrosome reactions of sea urchin sperm. *Proc. Natl. Acad. Sci. USA* 78: 6066–70

Schackmann, R. W., Christen, R., Shapiro, B. M. 1984. Measurement of plasma membrane and mitochondrial potentials in sea urchin sperm. Changes upon activation and induction of the acrosome reaction. *J. Biol. Chem.* 259: 13914–22

Schackmann, R. W., Eddy, E. M., Shapiro, B. M. 1978. The acrosome reaction of *Strongylocentrotus purpuratus* sperm. Ion requirements and movements. *Dev. Biol.* 65: 483–95

Schackmann, R. W., Shapiro, B. M. 1981. A partial sequence of ionic changes associated with the acrosome reaction of *Strongylocentrotus purpuratus*. *Dev. Biol.* 81: 145–54

Schatten, G. 1982. Motility during fertilization. *Int. Rev. Cytol.* 79: 59–164

Schatten, G., Hülser, D. 1983. Timing the early events during sea urchin fertilization. *Dev. Biol.* 100: 244–48

Schenk, D. B., Leffert, H. L. 1983. Monoclonal antibodies to rat Na^+/K^+-ATPase block enzymatic activity. *Proc. Natl. Acad. Sci. USA* 80: 5281–85

Schroeder, T. E. 1979. Surface area change at fertilization: Resorption of the mosaic membrane. *Dev. Biol.* 70: 306–26

Schroeder, T. E., Christen, R. 1982. Polymerization of actin without acrosomal exocytosis in starfish sperm. Visualization with NBD-phallicidin. *Exp. Cell Res.* 140: 363–71

Schuel, H. 1985. Functions of egg cortical granules. In *Biology of Fertilization*, ed. C. B. Metz, A. Monroy, 3: 1–44. Orlando: Academic

SeGall, G. K., Lennarz, W. J. 1979. Chemical characterization of the component of the jelly coat from sea urchin eggs responsible for induction of the acrosome reaction. *Dev. Biol.* 71: 33–48

SeGall, G. K., Lennarz, W. J. 1981. Jelly coat and induction of the acrosome reaction in echinoid sperm. *Dev. Biol.* 86: 87–93

Shapiro, B. M., Schackmann, R. W., Gabel, C. A. 1981. Molecular approaches to the study of fertilization. *Ann. Rev. Biochem.* 50: 815–43

Shapiro, B. M., Schackmann, R. W., Tombes, R. M., Kazazoglou, T. 1985. Coupled ionic and enzymatic regulation of sperm behavior. *Curr. Top. Cell. Regul.* 26: 97–113

Shapiro, B. M., Tombes, R. M. 1985. A biochemical pathway for a cellular behavior: pHi, phosphorylcreatine shuttles, and sperm motility. *Bioessays* 3: 100–3

Shen, S. S. 1983. Membrane properties and intracellular ion activities of marine invertebrate eggs and their changes during activation. In *Mechanism and Control of Animal Fertilization*, ed. J. F. Hartmann, pp. 213–67. New York: Academic. 561 pp.

Steinhardt, R. A., Alderton, J. M. 1982. Calmodulin confers calcium sensitivity on secretory exocytosis. *Nature* 295: 154–55

Suzuki, N., Garbers, D. L. 1984. Stimulation of sperm respiration rates by speract and resact at alkaline extracellular pH. *Biol. Reprod.* 30: 1167–74

Suzuki, N., Nomura, K., Ohtake, H. 1980. Sperm activating peptides obtained from jelly coat of sea urchin eggs. *Zool. Mag.* 89: 350

Suzuki, N., Nomura, K., Ohtake, H., Isaka, S. 1981. Purification and primary structure of sperm-activating peptides from the jelly coat of sea urchin eggs. *Biochem. Biophys. Res. Commun.* 99: 1238–44

Suzuki, N., Shimomura, H., Radany, E. W., Ramarao, C. S., Ward, G. E., et al. 1984. A peptide associated with eggs causes a mobility shift in a major plasma membrane protein of sea urchin spermatozoa. *J. Biol. Chem.* 259: 14874–79

Swarup, G., Garbers, D. L. 1982. Phosphoprotein phosphatase activity of sea urchin spermatozoa. *Biol. Reprod.* 26: 953–60

Tilney, L. G., Jaffe, L. A. 1980. Actin, microvilli, and the fertilization cone of sea urchin eggs. *J. Cell Biol.* 87: 771–82

Tilney, L. G., Kiehart, D. P., Sardet, C., Tilney, M. 1978. Polymerization of actin, IV. Role of Ca^{2+} and H^+ in the assembly of actin and in membrane fusion in the acrosomal reaction of Echinoderm sperm. *J. Cell Biol.* 77: 536–50

Tombes, R. M., Shapiro, B. M. 1985. Metabolite channeling: A phosphorylcreatine shuttle to mediate high energy phosphate transport between sperm mitochondrion and tail. *Cell* 41: 325–34

Trimmer, J. S., Trowbridge, I. S., Vacquier, V. D. 1985. Monoclonal antibody to a membrane glycoprotein inhibits the acrosome reaction and associated Ca^{2+} and H^+ fluxes of sea urchin sperm. *Cell* 40: 697–703

Tsien, R. Y., Pozzan, T., Rink, T. J. 1982. Calcium homeostasis in intact lymphocytes: Cytoplasmic free calcium monitored with a new, intracellularly trapped fluorescent indicator. *J. Cell Biol.* 94: 325–34

Tubb, D. J., Kopf, G. S., Garbers, D. L. 1978. The elevation of sperm adenosine 3',5'-monophosphate concentrations by factors released from eggs requires calcium. *Biol. Reprod.* 18: 181–85

Turner, P. R., Jaffe, L. A., Fein, A. 1986. Regulation of cortical vesicle exocytosis in sea urchin eggs by inositol 1,4,5-trisphosphate and GTP-binding protein. *J. Cell Biol.* 102: 70–76

Turner, P. R., Sheetz, M. P., Jaffe, L. A. 1984. Fertilization increases the polyphosphoinositide content of sea urchin eggs. *Nature* 310: 414–15

Vacquier, V. D. 1975. The isolation of intact cortical granules from sea urchin eggs: Calcium ions trigger granule discharge. *Dev. Biol.* 43: 62–74

Vacquier, V. D. 1976. Isolated cortical granules: A model system for studying membrane fusion and calcium mediated exocytosis. *J. Supramol. Struct.* 5: 27–35

Vacquier, V. D. 1983. Purification of sea urchin sperm binding by DEAE-cellulose chromatography. *Anal. Biochem.* 129: 497–501

Vacquier, V. D. 1986. Activation of sea urchin spermatozoa during fertilisation. *Trends Biochem. Sci.* 11: 77–81

Vacquier, V. D., Brandriff, B., Glabe, C. G. 1979. The effect of soluble egg jelly on the fertilizibility of acid-dejellied sea urchin eggs. *Dev. Growth Differ.* 21: 47–60

Vacquier, V. D., Moy, G. W. 1977. Isolation of bindin: the protein responsible for adhesion of sperm to sea urchin eggs. *Proc. Natl. Acad. Sci. USA* 74: 2456–60

Vacquier, V. D., Moy, G. W. 1980. The cytolytic isolation of the cortex of the sea urchin egg. *Dev. Biol.* 77: 178–90

Ward, G. E. 1985. Dephosporylation of sperm guanylate cyclase during sea urchin fertilization. PhD thesis. Univ. Calif., San Diego. 158 pp.

Ward, G. E., Brokaw, C. J., Garbers, D. L., Vacquier, V. D. 1985a. Chemotaxis of *Arbacia punctulata* spermatozoa to resact, a peptide from the egg jelly layer. *J. Cell Biol.* 101: 2324–29

Ward, G. E., Garbers, D. L., Vacquier, V. D. 1985b. Effects of extracellular egg factors on sperm guanylate cyclase. *Science* 227: 768–70

Ward, G. E., Moy, G. M., Vacquier, V. D. 1986a. Phosphorylation of membrane-bound guanylate cyclase of sea urchin spermatozoa. *J. Cell Biol.* In press

Ward, G. E., Moy, G. M., Vacquier, V. D. 1986b. Dephosphorylation of sperm guanylate cyclase during sea urchin fertilization. In *Molecular and Cellular Biology of Fertilization*, ed. J. Hedrick. New York: Plenum. In press

Ward, G. E., Vacquier, V. D. 1983. Dephosphorylation of a major sperm membrane protein is induced by egg jelly during sea urchin fertilization. *Proc. Natl. Acad. Sci. USA* 80: 5578–82

Watkins, H. D., Kopf, G. S., Garbers, D.

L. 1978. Activation of sperm adenylate cyclase by factors associated with eggs. *Biol. Reprod.* 19: 890–94

Wells, J. N., Garbers, D. L. 1976. Nucleoside 3′,5′-monophosphate phosphodiesterases in sea urchin sperm. *Biol. Reprod.* 15: 46–53

Whitaker, M., Aitchison, M. 1985. Calcium-dependent polyphosphoinositide hydrolysis is associated with exocytosis *in vitro. FEBS Lett.* 182: 119–24

Whitaker, M. J., Baker, P. F. 1983. Calcium-dependent exocytosis in an *in vitro* secretory granule plasma membrane preparation from sea urchin eggs and the effects of some inhibitors of cytoskeletal function. *Proc. R. Soc. London Ser. B* 218: 397–413

Whitaker, M., Irvine, R. F. 1984. Inositol 1,4,5-trisphosphate microinjection activates sea urchin eggs. *Nature* 312: 636–39

Whitaker, M. J., Steinhardt, R. A. 1985. Ionic signalling in the sea urchin egg at fertilization. In *Biology of Fertilization,* ed. C. B. Metz, A. Monroy, 3: 168–222. Orlando: Academic

Zimmerberg, J., Sardet, C., Epel, D. 1985. Exocytosis of sea urchin egg cortical vesicles *in vitro* is retarded by hyperosmotic sucrose: Kinetics of fusion monitored by quantitative light microscopy. *J. Cell Biol.* 101: 2398–2410

Zimmerberg, J., Whitaker, M. 1985. Irreversible swelling of secretory granules during exocytosis caused by calcium. *Nature* 315: 581–84

Ann. Rev. Cell Biol. 1986. 2 : 27–47

CELL-MATRIX INTERACTIONS AND CELL ADHESION DURING DEVELOPMENT

Peter Ekblom, Dietmar Vestweber, and Rolf Kemler

Friedrich-Miescher-Laboratorium der Max-Planck-Gesellschaft,
Spemannstrasse 37–39, D-7400 Tübingen, Federal Republic of Germany

CONTENTS

INTRODUCTION

The three-dimensional structure of tissues in multicellular organisms is highly complex, and at first sight it may seem very difficult to analyze the processes that lead to the development of each organ. The various organs have unique morphological and functional characteristics, and the individual cells are highly specialized. In spite of this, morphogenesis in all tissues seems to be dependent on a limited number of primary processes. These include cell proliferation, cell-cell adhesion, cell movement, and controlled cell death. Cell adhesion is widely believed to represent one of the most fundamental histogenetic processes for the development of multicellular organisms. In all solid tissues, cells adhere to each other and

27

0743–4634/86/1115–0027$02.00

to the substratum. The adhesive events are thus instrumental both for embryonic development and for the maintenance of tissue architecture.

MOLECULES THAT MEDIATE CELL ADHESION

Several different classes of molecules seem to mediate the adhesiveness of differentiating cells. They include cell surface components, particularly those involved in the formation of cell junctions, and extracellularly located matrix molecules. Cell junctions and their role in morphogenesis have been reviewed by Cowin et al (1985) and by Gilula (1986). In this review we focus on the developmental role of adhesive glycoproteins of the extracellular matrix and of the cell surface. These are not the only morphogenetically important surface and matrix components, but the developmental role of surface glycolipids has been discussed in depth by Hakomori (1981), and a number of recent articles have reviewed the role of proteoglycans and collagens in histogenesis (Hay 1981; Trelstad 1984).

Extracellular Matrix Proteins

A role for extracellular matrix components in histogenesis was suggested long ago by several investigators (Grobstein 1954; Pierce 1966; Hay 1981). It is now widely thought that the matrix provides attachment sites for the cells and that cell-matrix interactions are required for the maintenance of the proper tissue histoarchitecture. All matrices contain glycoproteins, collagens, and proteoglycans as major structural elements. The basement membrane is a special type of matrix that is deposited in a polar fashion around epithelial and endothelial cells. Around fat cells, muscle, and Schwann cells it is seen on all sides of the cell. Experimental studies on tissue repair have shown that the basement membrane is important for the maintenance of tissue structure, and embryological studies suggest a role for the basement membrane in histogenesis (Timpl & Dziadek 1986). The role of basement membranes in embryonic histogenesis was studied in the sixties (Pierce 1966), but at that time very little was known about the molecules of basement membranes. During the last ten years, several membrane components have been described, and accumulating evidence suggests that cells deposit basement membrane proteins at very early stages of embryonic development.

Although there may be close to fifty components in a typical basement membrane (Timpl & Dziadek 1986), it is thought that most of the major structural constituents have been defined. They are large macromolecules such as type IV collagen, proteoglycans, and the glycoproteins laminin and entactin. The collagenous component of basement membranes, collagen IV, forms a chicken-wire-like network (Kühn et al 1981). This network may be instrumental in basement membrane assembly. Proteoglycans may likewise have a structural role, but in addition they are responsible for the filtration properties of the basement membranes. The adhesive properties of the basement membranes are usually ascribed to the glycoproteins. In vivo, these components must interact with each other to form the tightly packed, thin basement membrane. The intermolecular interactions could alter the biological properties of the individual basement membrane components. The attachment of the glycoproteins and the proteoglycans to the collagen IV network seems to occur extracellularly by self-assembly (Charonis et al 1985; Kleinmann et al 1986). The literature on the composition of basement membranes is vast (Kefalides et al 1979; Martinez-Hernandez & Amenta 1983; Timpl & Dziadek 1986); here we limit ourselves to a discussion of the morphogenetic roles of basement membrane glycoproteins, with special emphasis on epithelial cell development. The significance of basement membrane matrices for epithelial cell development has long been known (Bernfield 1978), and their role in neural development is now obtaining increasing attention (Bunge & Bunge 1983; Sanes 1984).

LAMININ   This glycoprotein was originally purified in an intact form from a mouse tumor (Timpl et al 1979); peptide chains similar to the laminin chains were independently discovered by Chung et al (1979). It is now clear that laminin is a major constituent of all basement membranes. During embryogenesis, laminin is markedly enriched during early stages of epithelial histogenesis. This suggested that laminin is an adhesive protein for the developing epithelium (Ekblom et al 1980). In line with this it was soon reported by others that adhesion of epithelial cells is favored by laminin (Terranova et al 1980; Carlsson et al 1981). Many different cell types can bind to laminin, and this cell-matrix interaction alters cell behavior (Palotie et al 1983; Engvall & Ruoslahti 1983; Goodman & Newgreen 1985). It is therefore no longer believed that laminin acts as an attachment protein only for cells producing a basement membrane. This does not in any way contradict the suggestion that laminin is of morphogenetic importance for epithelial sheets. In fact, a number of studies suggest that laminin strongly promotes the growth, morphology, and differentiation of cells, particularly of epithelial cells (Kleinman et al 1985). The detailed molecular mechanisms of these effects are still unknown. Although laminin alone can be an effective stimulator of cell differentiation (Edgar et al 1984), the effect can be enhanced when laminin is complexed to other basement membrane components (Hadley et al 1985).

    The domains in the laminin molecule responsible for the various effects on the cells are not yet known. Laminin has a multidomain structure with a M_r of approximately 1000k. Morphological studies on pure laminin suggest a cross-shaped structure. The structure is a result of a covalent binding between different chains. Initially, two chains were identified (A, M_r 440k; and B, M_r 220k), but there is now also evidence for two separate B chains. The A, B1, and B2 chains could be genetically and structurally different, and they appear to occur in equimolar amounts in the molecule (Timpl & Dziadek 1986). It is not yet known how this large protein can fit into the rather thin basement membrane. Studies on the protein structure and sequencing of cDNAs for laminin (Kurkinen et al 1983; Barlow et al 1984; Yamada et al 1985) should help to clarify this issue. Many biological studies suggest that laminin has a variety of functions; it will be interesting to know whether these diverse functions can be assigned to different, clearly defined domains of the molecule. Some data on this are already available. A heparin-binding domain seems to promote neurite outgrowth (Edgar et al 1984), another domain interacts with type IV collagen, while the central core region of the laminin cross could be involved in cell binding (Timpl & Dziadek 1986; Liotta 1986).

OTHER GLYCOPROTEINS OF BASEMENT MEMBRANES   To date, only a few other basement membrane glycoproteins have been well defined. Entactin is a 150-kDa sulfated glycoprotein initially found in the matrix of endodermal cell lines (Carlin et al 1981; Hogan et al 1982). Nidogen could be isolated from a tumor, and it was thought to have a M_r of 80k (Timpl et al 1983). However, this seems to be a proteolytic fragment; genuine nidogen apparently has a M_r of 150k, a ratio very similar to that of entactin (Dziadek et al 1985). Comparisons of the sequences of nidogen and entactin should clarify whether the proteins are identical. Studies on the biological functions of these smaller basement membrane glycoproteins are not yet available. Laminin and nidogen do not always codistribute. Laminin seems to appear at early stages of histogenesis, whereas nidogen can only be detected much later. However, when both are present in basement membranes, they are apparently tightly complexed (Dziadek et al 1985). Nidogen is thought to stabilize the basement membrane, but it is nevertheless easily degraded by proteases. This feature could be important during embryogenesis since it could allow a rapid disruption of certain parts of the basement membrane during tissue remodeling (Timpl & Dziadek 1986).

FIBRONECTIN   Another adhesive matrix protein, fibronectin, can occur in basement membranes, but it is also found in many other locations (Vaheri & Mosher 1978). It nevertheless seems to have many biological properties similar to those of laminin. Expression of fibronectin during embryogenesis suggests that it plays a role in adhesion and migration of cells (Hynes & Yamada 1982). The structure of fibronectin is now well known, and the different domains that can bind to collagen, heparin, or the cell surface have now been defined (Vartio & Vaheri 1983). Surprisingly the cell-binding region is very short, a tripeptide sequence of Arg-Gly-Asp (Pierschbacher & Ruoslahti 1984; Yamada & Kennedy 1985). Short synthetic peptides containing the tripeptide sequence can mimic the action of fibronectin, at least in some assays. Addition of such peptides can inhibit embryonic morphogenesis, which suggests that proper binding of cells to fibronectin is required for morphogenesis (Thiery et al 1985; Ruoslahti et al 1985).

RECEPTORS FOR MATRIX PROTEINS   The effect of matrix proteins may be mediated by binding to cell surface receptors. The existence of receptors for matrix proteins was predicted some time ago, and during the past few years several candidates have been characterized. Many authors have shown that surface proteins in the 140 kDa range bind to fibronectin (Tarone et al 1982; Pytela et al 1985) and that laminin binds to a protein of about 70 kDa (Lesot et al 1983; Malinoff & Wicha 1983; Terranova et al 1983). The receptor for laminin was obtained by affinity chromatography on laminin. The isolated receptor has high affinity for laminin, and receptors inserted into liposomes still bind laminin. These data suggest that the laminin-binding protein is an integral plasma membrane protein with a true receptor function. It remains to be seen, however, whether the many effects on cell behavior attributed to laminin (Kleinman et al 1985) are all mediated through binding to this protein. These studies and data on the binding of different collagen isotypes to cells (Goldberg 1979; von der Mark et al 1984; Kurkinen et al 1984; Koda et al 1985) suggest that matrix proteins interact with cells through their own specific surface receptors. In addition, there may exist surface proteins with the capacity to bind several matrix components. A surface protein previously detected by an antibody that perturbs cell substrate adhesion (Damsky et al 1985) binds to both laminin and fibronectin. Its M_r, 140k, is similar to that of the receptor for fibronectin (Horwitz et al 1985).

The biological roles of the receptors for the various matrix proteins are at present unknown. Given the importance of the extracellular matrix in development, insights into this issue will likely emerge from embryological studies on the expression of the receptors. Duband et al (1986) have recently shown that the 140-kDa fibronectin receptor has a widespread distribution during embryogenesis. It is not particularly enriched at sites of cell adhesion.

Cell-Cell Adhesion Molecules

Cell-cell contacts are thought to be initiated and maintained by membrane glycoproteins called cell adhesion molecules (CAMs; Edelman 1983). In addition to these glycoproteins there are other adhesive surface molecules that can participate in adhesion during embryogenesis (Grabel et al 1979; Shur 1983; Bird & Kimber 1984), but these will not be discussed further here. Most of the presently known mammalian CAMs have been detected by generating antibodies that can block adhesion in a functional cell aggregation assay. Thus, the functional role of CAMs is mostly defined indirectly via antibodies. Functionally active anti-adhesion antibodies were first successfully used to detect adhesive cell surface determinants in *Dictyostelium* (Gerisch 1977), and with the use of similar approaches, several CAMs have now been characterized in other species. The number of presently known mammalian cell adhesion molecules is rather limited. This could mean that the adhesion processes mediated by the cell surface in most tissues are mediated by expression and modulation of only a few adhesive molecules (Edelman 1983). It should be kept in mind, however, that the CAMs discovered to date have been characterized by similar immunological strategies. The CAMs are apparently highly immunogenic, and a variable number of as yet unknown adhesion molecules may exist. Cell adhesion is a multistep process, and the stabilization of tissue form may require further components in addition to the presently known CAMs. Nevertheless, it can be concluded that the CAMs characterized to date represent major adhesive components for organogenesis.

The known CAMs can be classified into two groups, calcium-dependent and calcium-independent molecules. The calcium-dependent molecules require calcium for adhesion and calcium also protects these CAMs from proteolysis (Takeichi et al 1983). With one exception (Hatta et al 1985), neural CAMs are calcium-independent; calcium-dependent CAMs are found in nonneural tissues.

NEURAL CELL ADHESION MOLECULES   The best described neural cell adhesion molecule is N-CAM. It was originally isolated from chicken retina cells (for a review see Cunningham 1985). N-CAM analogues have been independently found in mice and rats (Hirn et al 1981; Jørgensen et al 1980). N-CAM is expressed by neurons and myoblasts, and during embryogenesis it is transiently expressed on developing epithelia. N-CAM-positive cells are thought to contact each other by direct binding via N-CAM (Hoffman & Edelman 1983). In mouse, three immunologically related N-CAM polypeptides of approximately 120, 140, and 180 kDa have been described (Gennarini et al 1984). Chicken N-CAM is composed of proteins of 170 and 140 kDa that arise by alternative splicing of mRNA from a single N-CAM gene (Murray et al 1986). An adult form of N-CAM contains less sialic acid, and it has been suggested that the sialic acid content of N-CAM influences the adhesiveness of the molecule (Hoffman

& Edelman 1983). Application of anti-N-CAM antibodies blocks the formation of neurite bundles and disrupts histogenesis in cultured chick retina (Buskirk et al 1980; Rutishauser 1984). N-CAM may also be involved in the heterotypic interaction between muscle and nerve cells (Grumet et al 1982; Rutishauser et al 1983).

Although N-CAM may be a major adhesive molecule for neural histogenesis, some other neural CAMs have now been characterized. They are molecularly distinct from N-CAM, and at least two of them, Ng-CAM (Grumet et al 1984) and the L1 antigen (Rathjen & Schachner 1984), appear rather late during development, which suggests that the primary adhesive events are mediated by N-CAM (Edelman 1985). Ng-CAM is present on neurones, and it has been proposed that it mediates cell-cell contact between neurones and between neurones and glia cells (Grumet et al 1984). Although there are some similarities between Ng-CAM and L1, L1 does not appear to mediate cell-cell contact between neurones and astrocytes (Keilhauer et al 1985). Another CAM, the nerve growth factor–inducible large external glycoprotein, was recently proposed to be immunochemically identical to L1 (Bock et al 1985).

R-Cognin, a cell adhesion molecule on chicken retina cells, is one of the few CAMs originally identified without immunological techniques (Hausman & Moscona 1975). The purified glycoprotein enhances the reaggregation of retina cells, while antibodies against this molecule disturb the histotypic development of the retina in organ cultures (Ophir et al 1984) and inhibit aggregation of retinal membrane vesicles (Trocolli & Hausman 1985).

EPITHELIAL CELL ADHESION MOLECULES   Several CAMs have been isolated and characterized from epithelial cells. However, as can be seen from Table 1, many of them are similar or even identical to uvomorulin, a 120-kDa

Table 1   Summary of 120-kDa epithelial CAMs

Name	Original source	Species	Reference
Uvomorulin	preimplantation embryo	mouse	Hyafil et al 1980
Cadherin	teratocarcinoma cell	mouse	Yoshida-Noro et al 1984
Cell-CAM	epithelial cell line	human	Damsky et al 1983
L-CAM	liver cell	chicken	Gallin et al 1983
Arc-1	kidney epithelial cell line (MDCK)	canine	Imhof et al 1983 Behrens et al 1985
rr-1 antigen	kidney epithelial cell line (MDCK)	canine	Gumbiner & Simons 1986

glycoprotein originally described as involved in compaction during mouse preimplantation development (Kemler et al 1977; Hyafil et al 1980). Uvomorulin is present on most epithelia in adult tissues (Peyrieras et al 1983; Vestweber & Kemler 1984a). Takeichi and coworkers have identified an adhesive glycoprotein from mouse teratocarcinoma cells, and antibodies against this glycoprotein decompact early preimplantation embryos. The protein has recently been named cadherin, but it was established that uvomorulin and cadherin are identical (Ogou et al 1983; Yoshida-Noro et al 1984). Human epithelial cells express an adhesive glycoprotein, cell-CAM 120/80 (Damsky et al 1983). This molecule is most likely identical to uvomorulin, although this has not yet been shown by direct comparison. An adhesive protein of chicken embryonal liver cells, L-CAM (Gallin et al 1983; Cunningham et al 1984), is also very similar to uvomorulin; however the respective antibodies do not recognize the molecules across the species barrier. More recently, a protein similar to uvomorulin has been found in canine epithelia. The anti-Arc-1 monoclonal antibody dissociates canine kidney epithelial cells (Imhof et al 1983) and reacts with an uvomorulin-like protein (Behrens et al 1985). The same protein is recognized by the monoclonal antibody rr-1, which was selected using a transepithelial electrical resistance assay of epithelial cell monolayers (Gumbiner & Simons 1986). Thus many investigators have found uvomorulin-like adhesive proteins of about 120 kDa, although their approaches and starting materials for antigen detection were not the same (Table 1).

Only a few other well-characterized CAMs of epithelia are presently known. A glycoprotein, called cell-CAM 105, required for adhesion of hepatocytes has been described (Ocklind & Öbrink 1982), but it is apparently not related to the 120-kDa CAMs. This 105-kDa glycoprotein is expressed on liver cells and on simple epithelia, but it appears later in development than uvomorulin (Ocklind et al 1984). Cell-CAM 105 is structurally, immunologically, and functionally distinct from uvomorulin. Although the two proteins can be found on the same cell, they are localized to different parts of the surface (Vestweber et al 1985b). Following the nomenclature for neural CAMs (Edelman 1985), uvomorulin could be called a primary and cell-CAM 105 a secondary epithelial CAM.

Most of the molecular features described here for uvomorulin are shared by the other 120-kDa epithelial CAMs listed in Table 1. Trypsin digestion of the intact 120-kDa surface glycoprotein releases a calcium-protected, 84-kDa, soluble fragment, which contains the antigenic sites for the functionally active antibodies (Hyafil et al 1981; Vestweber & Kemler 1984a). Antibodies against the 120-kDa uvomorulin protein and antibodies against the 84-kDa fragment (Peyrieras et al 1983; Vestweber & Kemler 1984b) recognize additional peptides of 102 and 92 kDa. Presently, it is unclear how these peptides are related to the 120-kDa uvomorulin, although preliminary data suggest that these proteins may have similar biological functions as uvomorulin. The 120-kDa and the 102-kDa proteins are both located at the cell surface, but only the first is glycosylated, and the peptide maps obtained from Cleveland digests are different (Vestweber & Kemler 1984b; Peyrieras et al 1985). Interestingly, heterogeneity in polypeptide composition is also seen for N-CAM and for cell-CAM 105. The latter is immunochemically related to another cell surface molecule called CDP-1 (Ocklind et al 1984). It remains to be seen whether such heterogeneity is a general feature of CAMs. Furthermore, it is important to find out how the different forms of the CAMs are associated with the cell membrane. Such information would shed light on the molecular basis of the cell adhesion process. The molecular mechanism of cell adhesion mediated by uvomorulin is still not well understood. The search for a functional cell-adhesive site on uvomorulin has recently been facilitated by the use of three independently selected anti-uvomorulin monoclonal antibodies, each of which disturbs cell-cell contacts. The fact that the epitopes of these monoclonal antibodies react with a 26-kDa fragment led to the conclusion that this fragment could contain an adhesive domain of uvomorulin (Vestweber & Kemler 1985).

An important step towards an understanding of the molecular structure of uvomorulin has been made by the isolation of cDNA sequences for the molecule. Coding sequences for L-CAM (Gallin et al 1985) and for mouse uvomorulin (Schuh et al 1986) are now available. The available cDNA for uvomorulin and L-CAM both recognize a single poly(A)$^+$ RNA species of 4.3 kilobases (kb) for uvomorulin and of 4.0 kb for L-CAM.

ROLE OF MATRIX AND CELL ADHESION MOLECULES IN EMBRYOGENESIS

Preimplantation Development

Adhesive cell-cell interactions are already taking place during the onset of embryogenesis. During preimplantation development there is cell proliferation, but there are also a series of morphological changes necessary for further development. The first clear morphological change is compaction at the morula stage, during which several events take place concurrently. The outer wall blastomeres become flattened (Ducibella et al 1977) and polarized (Johnson 1985), and they establish intercellular junctions (Gilula 1986). The compaction at the morula stage can be inhibited by anti-uvomorulin antibodies (Figure 1A,B). Removal of the antibodies leads to recompaction and to formation of blastocysts. The blastocysts develop into normal mice when reimplanted in pseudopregnant mice

(Kemler et al 1977). These studies suggest that uvomorulin is required for compaction.

No similar studies on the effect of laminin or anti-laminin antibodies have so far been reported. The expression of laminin during these stages of development suggests that it could have a role in the compaction process.

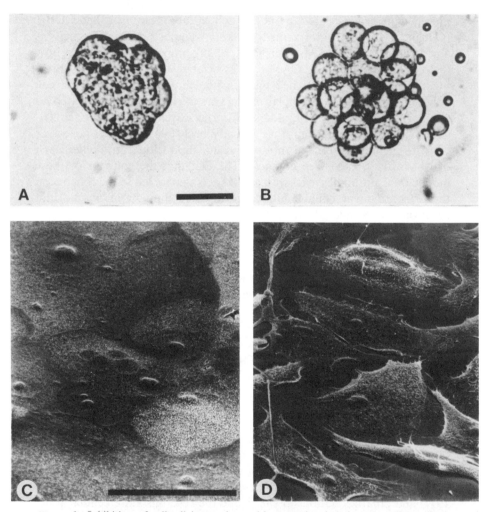

Figure 1 Inhibition of cell-cell interactions with monoclonal anti-uvomorulin antibody DECMA-1. Eight-cell embryos cultured without (A) and with (B) 50 μg/ml DECMA-1. MDCK cells form tight epithelial sheets in the presence of control antibodies (C) but grow as single adhering cells in the presence of 100 μg/ml DECMA-1 (D). Bars represent 50 μm. (From Vestweber & Kemler 1985.)

Initial studies suggested that laminin appears at the sixteen-cell stage (Leivo et al 1980), but it is now known that laminin is present in two-cell embryos (Cooper & McQueen 1983; Dziadek & Timpl 1985). However, at this stage only laminin B chains are synthesized, and they are present only in the cytoplasm. Deposition of laminin (A and B chains) to the extracellular space is first seen in later morulae and in blastocysts, where laminin occupies the intercellular space of the inner morula. Nidogen is first detected at the late morula stage (Dziadek & Timpl 1985). There is no clear basement membrane at this point; laminin is first seen organized within the basement membrane of the primitive endoderm and in Reichert's membrane.

Like laminin, uvomorulin can also be detected very early in development (Hyafil et al 1983; Vestweber & Kemler 1984a; Richa et al 1985). Interestingly, uvomorulin exhibits a different cell surface localization on pre-implantation embryos: It is uniformly distributed on the cell surface of fertilized eggs, and it vanishes progressively from the outer cell membranes during compaction. It is no longer seen on the apical membrane part of trophectodermal cells. Immunofluorescence data obtained with anti-cell-CAM 120/80 antibodies (Damjanov et al 1986) and immunoelectron microscope localization studies of uvomorulin on trophectoderm (Boller et al, manuscript in preparation) show that uvomorulin is restricted to the adjacent lateral membrane parts of trophectoderm cells, while a uniform distribution persists on cells of the inner cell mass.

Although uvomorulin can be detected rather early, it apparently does not express its adhesive properties, since blastomeres are only loosely held together up to the eight-cell stage. The uvomorulin molecules of the early stages may be different from those expressed during compaction, but no data on this have been presented. Another possibility is that clustering of uvomorulin at certain sites on the cells during compaction is important for initiation of adhesion. The assembly of uvomorulin could initiate cell-cell contacts, which might trigger cell polarization and the formation of intercellular junctions at those points. Although this postulate is attractive, there is presently no experimental evidence for it, and it should be kept in mind that other molecules implicated in the compaction process have been reported (Grabel et al 1979; Shur 1983; Silver et al 1983; Bird & Kimber 1984).

A detailed analysis of the molecular processes leading to compaction and the formation of the epithelial sheet of trophectoderm could help to clarify whether a general mechanism for epithelial formation during histogenesis exists. The cell surface localization of uvomorulin is similar on trophectoderm and adult epithelial cells of the small intestine (Boller et al 1985): Uvomorulin is found on the basolateral membranes and is highly concentrated in the intermediate junctions of the zona adherens. Other epithelial sheets may be formed in a fashion similar to that of the trophectoderm. Uvomorulin is distributed randomly all over the cell membrane on proliferating, developing epithelial cells. However, clustering of uvomorulin initiates cell-cell contact and the formation of the zona adherens, which then stabilizes the polarized state. This model is certainly an oversimplification, and the role of the basement membrane in this process should be better defined.

Epithelial Development

Formation of epithelia continues throughout embryogenesis. In most parenchymal tissues, branched epithelia comprise major parts of the functionally active units. These treelike, hollow linings arise from smaller epithelial buds, which undergo folding and growth of certain parts. These events lead to cleft formation and finally to a highly branched morphology. Although slight differences in the growth process are seen in various tissues, the basic strategy for epithelial development is the same. In all cases the surrounding mesenchyme is required for branching (Kratochwil 1983). Because the branching process in all organs occurs in a similar fashion, it has been suggested that organ-specific branching patterns could be generated without an immense amount of genetic information (Bernfield 1978; Wessells 1982). Small variations in the expression of a limited number of molecules could generate the tissue-specific branching characteristics. Considerations of mechanical forces and other theoretical models of epithelial foldings are in agreement with such postulates (Gierer 1977).

In one tissue, the developing kidney, a somewhat different branching strategy is seen. The mesenchyme is still required for branching of the epithelial bud (the ureter bud), but some cells of the mesenchyme are then actually converted to epithelium. The branching ureteric tree recruits these former mesenchymal cells into the epithelium, and therefore major parts of the epithelium in adult kidneys (glomerular epithelium, proximal and distal tubules) are derivatives of mesenchyme (Saxén et al 1968; Ekblom et al 1981b). The conversion of mesenchyme to epithelium provides us with a unique opportunity to study the expression of epithelial adhesion molecules. Morphological studies suggested that an increased adhesiveness could be an important early change during the conversion (Saxén et al 1968), but the molecular basis of the adhesion process was unknown. Recent studies on the expression of CAMs and matrix proteins are now beginning to shed light on this area.

The cells of the nephrogenic mesenchyme are morphologically similar to fibroblasts, and their bias towards epithelial differentiation is not in any way apparent. The matrix composition is also typical for loose mesenchyme, with collagen I and III and fibronectin. When the conversion to epithelium begins, the mesenchyme gradually seems to lose its original matrix. The apparent disappearance of collagen I and III and fibronectin may be related to the condensation of the mesenchyme, which is the first clearly detectable morphological change (Ekblom 1981; Ekblom et al 1981a). This rather passive condensation due to disappearance of the material between the cells is followed by more active adhesion between the cells.

In the loose mesenchyme, adhesive proteins have so far not been detected, but when the condensates begin to form, at least two adhesive proteins, laminin and uvomorulin, can be found by immunofluorescence (Ekblom et al 1980; Vestweber et al 1985a). In addition, the condensates, at least in the chicken mesonephros, contain both N-CAM and L-CAM (Thiery et al 1982, 1984). It is not yet known when nidogen appears during this process. It should be pointed out that biosynthetic studies for laminin and uvomorulin have not yet been done. Thus, we do not know what chains of laminin are present in the condensates. The immunofluorescence studies nevertheless suggest that the conversion of the mesenchyme to epithelium is accompanied by the appearance of epithelial adhesive proteins.

Initially, laminin can be seen in a punctate pattern, and at this stage there is no clearly detectable basement membrane. As development proceeds, it is more clearly localized to the basal side of the cells. (A similar polar expression of uvomorulin is not detected.) When the cells are terminally differentiated, with a clearly distinguishable apical side, uvomorulin (and N-CAM and L-CAM in the chicken) are still rather uniformly distributed on all surfaces.

Many of the antibodies against CAMs can prevent cell adhesion in vitro. It was therefore expected that anti-uvomorulin antibodies would influence kidney tubule formation (Figure 2), but they do not. The failure of anti-uvomorulin antibodies to inhibit kidney tubule development is difficult to explain since these same antibodies inhibit morula compaction (Figure 1A,B), adhesion between embryonic liver cells (Vestweber & Kemler 1984b), and the epithelial sheet formation of adult kidney epithelial cells in culture (Figure 1C,D). The data on embryonic kidney development suggest that epithelial histogenesis in some cases appears to proceed normally even though functionally active anti-adhesion antibodies against one molecule are present. The basement membrane and N-CAM may be more influential in kidney tubule development than uvomorulin, but this has not yet been shown.

As can be seen from kidney development, the formation of a complex epithelium has several similarities with the formation of trophoblast epithelium, one of the first stages of preimplantation development. Notable for the developing kidney epithelium are the changes in the matrix composition and the presence of N-CAM, which is not seen during preimplantation development. Nevertheless, the other similarities support the hypothesis that cells throughout development use basically the same strategies for adhesive interactions during morphogenesis (Wessells 1982; Edelman 1985). Data on the development of epithelia in other organs fit this hypothesis well. The presence of uvomorulin on the cell surface (Vestweber & Kemler 1984a; Thiery et al 1984) and a basement membrane between the epithelium and mesenchyme (Bernfield et al 1984) are common to all epithelia regardless of the germ layer origin of the cells. The variations in epithelial development seen in the different organs may be due to slightly different combinations of the adhesive proteins. The developing feather provides a good example of the relationship between CAM expression and histogenesis. The evidence suggests that the complex pattern of feather morphogenesis could be largely due to variations in the expression of N-CAM and L-CAM (see Edelman, this volume). Changes in the matrix components also occur, but they do not correlate very closely with the known morphogenetic events (Chuong & Edelman 1985). However, in other tissues, the matrix could have a more crucial role. Basement membrane deposition is seen during the onset of epithelial development of most

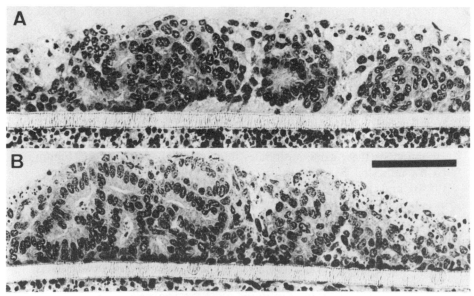

Figure 2 Development of kidney tubules proceeds well both in the absence (A) and presence (B) of 100 μg/ml of rabbit anti-uvomorulin Fab fragments. (From Vestweber et al 1985a.)

organs (Leivo et al 1980), and experimental evidence for a role of its constituents in branching morphogenesis has been presented (Bernfield et al 1984).

In many tissues, development of cell polarity takes place soon after the first cell-cell adhesion. The expression of CAMs and basement membrane proteins during polarization, for instance during kidney morphogenesis, suggests that the basement membrane could be crucial for the establishment of polarity (Ekblom 1981; Vestweber et al 1985a). However, there is no direct evidence for this proposal. In vitro models for the development of cell polarity make it possible to experimentally study this issue. Certain cell lines maintain cell polarity in vitro. The Madin-Darby canine kidney (MDCK) cell line has turned out to be a particularly useful model in this respect.

MDCK Cells and Cell Polarity

The development of polarity in freshly seeded MDCK cells has features in common with the embryonic development of polarized epithelia (Simons & Fuller 1985). The maintenance of the polarity is dependent on a number of factors, but it is noteworthy that the cells do not produce any significant amounts of basement membrane material, and they grow on plastic without matrix. On proper filters they maintain their polarity much better, so the filter in this system may replace the effect of the basement membrane. In any case, the data suggest that at least differentiated epithelial cells under some conditions can be polar in the absence of laminin and other basement membrane constituents.

Uvomorulin, however, seems to be required for polarity. The cells express uvomorulin, and the anti-uvomorulin antibodies can block the formation of the polarized epithelial sheet. This has been shown in three laboratories with independently produced anti-uvomorulin antibodies (Vestweber & Kemler 1985; Behrens et al 1985; Gumbiner & Simons 1986). It is not immediately apparent from the embryological immunolocalization studies why uvomorulin, but not laminin, would be important for the cell polarity (Ekblom 1981; Vestweber et al 1985a). That uvomorulin might also be involved in maintenance of cell polarity was suggested by the finding that the intermediate junctions in some adult epithelia are enriched in this protein (Boller et al 1985). Similar polar expression of uvomorulin can be seen during some stages of trophoblast formation (Damjanov et al 1986) but not during embryological development of the kidney epithelium (Thicry et al 1984; Vestweber et al 1985a).

The development of cell polarity is one of the most fundamental processes in embryogenesis, and the differentiation of all epithelial cells requires that different specialized surface domains emerge. The maintenance of differentiated domains of the cell surface is also instrumental for cell function in adult tissues (Palade 1983). The current data emphasize the role of uvomorulin in maintaining cell polarity. It remains to be seen whether adhesive properties of uvomorulin account for this effect or whether other functional domains are required.

CONCLUDING REMARKS

Histogenesis during embryonic development is often initiated by increased adhesiveness between cells. The available data suggest that basement membrane proteins and CAMs are of major significance for these histogenetic events. Laminin, uvomorulin, and N-CAM are almost invariably expressed when epithelial morphogenesis begins, and there is direct evidence of roles for uvomorulin and N-CAM in many histogenetic events. The different tissues and their epithelia have unique forms, but this could be generated by variations in the amount of a few adhesive molecules. At later stages, secondary CAMs may come into play, and finally tissue-specific products are expressed.

In many developing epithelia, the initial adhesion between the cells is followed by polarization. Our understanding of the development of cell polarity is still fragmentary. The presence of uvomorulin seems to be a prerequisite for initiation of polarization, but it would be premature to conclude that the basement membrane plays no role in this process. Basement membranes are at very early stages deposited in a polar fashion, and in many model systems the matrix has been shown to influence the polarity of epithelial cells (Emerman & Pitelka 1977; Chambard et al 1981; Sugrue & Hay 1981; Ritzki & Ritzki 1983; Grover et al 1983). Therefore basement membrane probably influences epithelial polarity at least during embryogenesis, although its exact role in this process is still imprecisely defined.

Since adhesive molecules influence the earliest stages of embryonic development in most tissues, it is worth asking how the expression of the adhesive molecules is turned on in the first place. In many organs, it has been shown that cell-cell interactions between dissimilar cells during embryogenesis lead to stimulation of the expression of these adhesive proteins. Such inductive interactions influence the expression of CAMs during feather development (Chuong & Edelman 1985) and of CAMs and matrix proteins during the development of kidney tubules (Ekblom & Thesleff 1985; Vestweber et al 1985a), for example. This suggests that the expression of adhesive proteins is regulated by embryonic induction. Use of in vitro models for embryonic induction may therefore lead to an understanding of the mechanisms leading to expression of the adhesive proteins.

Literature Cited

Barlow, D., Green, N. M., Kurkinen, M., Hogan, B. L. M. 1984. Sequencing of laminin B chain cDNA reveals C-terminal regions of coiled-coil alpha helix. *EMBO J.* 3: 2355–62

Behrens, J., Birchmeier, W., Goodman, S. L., Imhof, B. A. 1985. Dissociation of MDCK epithelial cells by the monoclonal antibody anti-arc-1: Mechanistic aspects and identification of the antigen as a component related to uvomorulin. *J. Cell Biol.* 101: 1307–15

Bernfield, M. 1978. Mechanisms of embryonic organ formation. In *Abnormal Fetal Growth: Biological Bases and Consequences*, ed. F. Naftolin, pp. 101–20. Berlin: Springer-Verlag

Bernfield, M., Banerjee, S. D., Koda, J. E., Rapraeger, A. C. 1984. Remodelling of the basement membrane as a mechanism of morphogenetic tissue interaction. See Trelstad 1984, pp. 545–72

Bird, J. M., Kimber, S. J. 1984. Oligosaccharides containing fucose linked α(1–3) and α(1–4) to *N*-acetylglucosamine cause decompaction of mouse morulae. *Dev. Biol.* 104: 449–60

Bock, E., Richter-Landsberg, C., Faissner, A., Schachner, M. 1985. Demonstration of immunochemical identity between the nerve growth-factor inducible large external (NILE) glycoprotein and the cell adhesion molecule L1. *EMBO J.* 4: 2765–68

Boller, K., Vestweber, D., Kemler, R. 1985. Cell-adhesion molecule uvomorulin is localized in the intermediate junctions of adult intestinal epithelial cells. *J. Cell Biol.* 100: 327–32

Bunge, R. P., Bunge, M. B. 1983. Interrelationship between Schwann cell function and extracellular matrix production. *Trends Neurosci.* 6: 499–505

Buskirk, D. R., Thiery, J.-P., Rutishauser, U., Edelman, G. M. 1980. Antibodies to a neural cell adhesion molecule disrupt histogenesis in cultured chick retinae. *Nature* 285: 488–89

Carlin, B. E., Jaffe, R., Bender, B., Chung, A. E. 1981. Entactin, a novel basal-lamina associated sulfated glycoprotein. *J. Biol. Chem.* 256: 5209–14

Carlsson, R., Engvall, E., Freeman, A., Ruoslahti, E. 1981. Laminin and fibronectin in cell adhesion: Enhanced adhesion of cells from regenerating liver to laminin. *Proc. Natl. Acad. Sci. USA* 76: 2403–6

Chambard, M., Gabrion, J., Mauchamp, J. 1981. Influence of collagen gel on the orientation of epithelial cell polarity: Follicle formation from isolated thyroid cells and from preformed monolayers. *J. Cell Biol.* 91: 157–66

Charonis, A. S., Tsilibary, E. C., Yurchenco, P., Furthmayr, H. 1985. Binding of laminin to type IV collagen: A morphological study. *J. Cell Biol.* 100: 1848–53

Chung, A. E., Jaffe, E. R., Freeman, I. L., Vergnes, J. P., Braginski, J. E., Carlin, B. 1979. Properties of a basement membrane–related glycoprotein synthesized in culture by a mouse embryonal carcinoma-derived cell line. *Cell* 16: 277–87

Chuong, C.-M., Edelman, G. M. 1985. Expression of cell-adhesion molecules in embryonic induction. I. Morphogenesis of nestling feathers. *J. Cell Biol.* 101: 1009–26

Cooper, A. R., McQueen, H. A. 1983. Subunits of laminin are differentially synthesized in mouse eggs and early embryos. *Dev. Biol.* 96: 467–71

Cowin, P., Franke, W. W., Grund, C., Kapprell, H. P., Kartenbeck, J. 1985. The desmosome-intermediate filament complex. See Edelman & Thiery 1985, pp. 427–60

Cunningham, B. A. 1985. Structure of cell adhesion molecules. See Edelman & Thiery 1985, pp. 197–218

Cunningham, B. A., Leutzinger, Y., Gallin, W. J., Sorkin, B. C., Edelman, G. M. 1984. Linear organization of the liver cell adhesion molecule L-CAM. *Proc. Natl. Acad. Sci. USA* 81: 5787–91

Damjanov, I., Damjanov, A., Damsky, C. H. 1986. Developmentally regulated expression of the cell-cell adhesion glycoprotein cell-CAM 120/80 in preimplantation mouse embryos and extraembryonic membranes. *Dev. Biol.* In press

Damsky, C. H., Richa, J., Solter, D., Knudsen, K., Buck, C. A. 1983. Identification and purification of a cell surface glycoprotein mediating intercellular adhesion in embryonic and adult tissue. *Cell* 34: 455–66

Damsky, C. K., Knudsen, D., Bradley, C., Buck, C., Horwitz, A. 1985. Distribution of cell substratum attachment (CSAT) antigen on myogenic and fibroblastic cells in culture. *J. Cell Biol.* 100: 1528–39

Duband, J.-L., Rocher, S., Chen, W.-T., Yamada, K. M., Thiery, J.-P. 1986. Cell adhesion and migration in the early vertebrate embryo: Location and possible role of the putative fibronectin receptor complex. *J. Cell Biol.* 102: 160–78

Ducibella, T., Ukena, T., Karnovsky, M., Anderson, E. 1977. Changes in cell surface

44   EKBLOM, VESTWEBER & KEMLER

and cortical cytoplasmic organization during early embryogenesis in the preimplantation embryo. *J. Cell Biol.* 74: 153–67

Dziadek, M., Timpl, R. 1985. Expression of nidogen and laminin in basement membranes during mouse embryogenesis and in teratocarcinoma cells. *Dev. Biol.* 111: 372–82

Dziadek, M., Paulsson, M., Timpl, R. 1985. Identification and interaction repertoire of large forms of the basement membrane protein nidogen. *EMBO J.* 4: 2513–18

Edelman, G. M. 1983. Cell adhesion molecules. *Science* 219: 450–57

Edelman, G. M. 1985. Cell adhesion and the molecular process of morphogenesis. *Ann. Rev. Biochem.* 54: 135–69

Edelman, G. M., Thiery, J.-P. 1985. *The Cell in Contact.* New York: Wiley. 507 pp.

Edgar, D., Timpl, R., Thoenen, H. 1984. The heparin-binding domain of laminin is responsible for its effects on neurite outgrowth and neuronal survival. *EMBO J.* 3: 1463–68

Ekblom, P. 1981. Formation of basement membranes in the embryonic kidney: An immunohistological study. *J. Cell Biol.* 91: 1–10

Ekblom, P., Thesleff, I. 1985. Role of transferrin and extracellular matrix components in kidney differentiation. *Mod. Cell Biol.* 4: 85–127

Ekblom, P., Alitalo, K., Vaheri, A., Timpl, R., Saxén, L. 1980. Induction of a basement membrane glycoprotein in embryonic kidney: Possible role of laminin in morphogenesis. *Proc. Natl. Acad. Sci. USA* 77: 485–89

Ekblom, P., Lehtonen, E., Saxén, L., Timpl, R. 1981a. Shift in collagen type as an early response to induction of the metanephric mesenchyme. *J. Cell Biol.* 89: 276–83

Ekblom, P., Miettinen, A., Virtanen, I., Wahlström, T., Dawnay, A., Saxén, L. 1981b. In vitro segregation of the metanephric nephron. *Dev. Biol.* 84: 88–95

Emerman, J. T., Pitelka, D. 1977. Maintenance and induction of morphological differentiation in dissociated mammary epithelium. *In Vitro* 13: 316–78

Engvall, E., Ruoslahti, E. 1983. Cell adhesive, protein binding, and antigenic properties of laminin. *Coll. Relat. Res.* 3: 359–69

Gallin, W. J., Edelman, G. M., Cunningham, B. A. 1983. Characterization of L-CAM, a major cell adhesion molecule from embryonic liver cells. *Proc. Natl. Acad. Sci. USA* 80: 1038–42

Gallin, W. J., Prediger, E. A., Edelman, G. M., Cunningham, B. A. 1985. Isolation of a cDNA clone for the liver cell adhesion molecule (L-CAM). *Proc. Natl. Acad. Sci. USA* 82: 2809–13

Gennarini, G., Rougon, G., Deagostini-Bazin, H., Hirn, M., Goridis, C. 1984. Studies on the transmembrane disposition of the neural cell adhesion molecule N-CAM. A monoclonal antibody recognizes a cytoplasmic domain and evidence for the presence of phosphoserine residues. *Eur. J. Biochem.* 142: 57–64

Gerisch, G. 1977. Univalent antibody fragments as tools for analysis of cell-cell interactions in *Dictyostelium. Curr. Top. Dev. Biol.* 14: 243–70

Gierer, A. 1977. Physical aspects of tissue envagination and biological form. *Rev. Biophys.* 10: 529–93

Gilula, N. B. Structure and chemistry of gap junctions: Their role in development. *Ann. Rev. Cell Biol.* Manuscript in preparation

Goldberg, B. 1979. Binding of soluble type I collagen to the fibroblast plasma membrane. *Cell* 16: 265–75

Goodman, S. L., Newgreen, D. 1985. Do cells show inverse locomotory response to fibronectin and laminin substrates? *EMBO J.* 4: 2769–71

Grabel, L. B., Rosen, S., Martin, G. M. 1979. Teratocarcinoma stem cells have a cell surface carbohydrate-binding component implicated in cell-cell adhesion. *Cell* 17: 474–84

Grobstein, C. 1954. Tissue interaction in the morphogenesis of mouse embryonic rudiments in vitro. In *Aspects of Synthesis and Order of Growth*, ed. G. Rudnick, pp. 233–56. Princeton Univ. Press

Grover, A., Andrews, G., Adamson, E. D. 1983. Role of laminin in epithelium formation by F9 aggregates. *J. Cell Biol.* 97: 137–44

Grumet, M., Rutishauser, U., Edelman, G. M. 1982. Neural cell adhesion molecule is on embryonic muscle cells and mediates adhesion to nerve cells. *Nature* 295: 693–95

Grumet, M., Hoffman, S., Edelman, G. M. 1984. Two antigenically related neuronal CAMs of different specificities mediate neuron-neuron and neuron-glia adhesion. *Proc. Natl. Acad. Sci. USA* 81: 267–71

Gumbiner, B., Simons, K. 1986. A functional assay for proteins involved in establishing of epithelial occluding barrier: Identification of an uvomorulin-like polypeptide. *J. Cell Biol.* 102: 457–68

Hadley, M. A., Byers, S. W., Suarez-Quian, C. A., Kleinman, H. K., Dym, M. 1985. Extracellular matrix regulates Sertoli cell differentiation, testicular cord formation, and germ cell development in vitro. *J. Cell Biol.* 101: 1511–22

Hakomori, S. 1981. Glycosphingolipids in cellular interaction, differentiation and oncogenesis. *Ann. Rev. Biochem.* 50: 733–64

Hatta, K., Okada, T. S., Takeichi, M. 1985. A monoclonal antibody disrupting calcium-dependent cell-cell adhesion of brain tissues: Possible roles of its target antigen in animal pattern formation. *Proc. Natl. Acad. Sci. USA* 82: 2789–93

Hausman, R. E., Moscona, A. 1975. Purification and characterization of the retina-specific-cell-aggregating factor. *Proc. Natl. Acad. Sci. USA* 72: 916–20

Hay, E. D. 1981. *Extracellular Matrix.* New York: Plenum

Hirn, M., Pierres, M., Deagostini-Bazin, H., Hirsch, M., Goridis, C. 1981. Monoclonal antibody against cell surface glycoprotein of neurons. *Brain Res.* 214: 433–39

Hoffman, S., Edelman, G. M. 1983. Kinetics of homophilic binding by E and A forms of the neural cell adhesion molecule. *Proc. Natl. Acad. Sci. USA* 80: 5762–66

Hogan, B. L. M., Taylor, A., Kurkinen, M., Couchman, J. R. 1982. Synthesis and localization of two sulphated glyco-proteins associated with basement membranes and the extracellular matrix. *J. Cell Biol.* 95: 197–204

Horwitz, A., Duggan, K., Greggs, R., Decker, C., Buck, C. 1985. The cell substrate attachment (CSAT) antigen has properties of a receptor for laminin and fibronectin. *J. Cell Biol.* 101: 2134–44

Hyafil, F., Morello, D., Babinet, C., Jacob, F. 1980. A cell surface glycoprotein involved in the compaction of embryonal carcinoma cells and cleavage stage embryos. *Cell* 21: 927–34

Hyafil, F., Babinet, C., Jacob, F. 1981. Cell-cell interactions in early embryogenesis: A Molecular approach to the role of calcium. *Cell* 26: 447–54

Hyafil, F., Babinet, C., Huet, C., Jacob, F. 1983. Uvomorulin and compaction. See Silver et al 1983, pp. 197–208

Hynes, R., Yamada, K. 1982. Fibronectins: Multifunctional modular glycoproteins. *J. Cell Biol.* 95: 369–77

Imhof, B. A., Vollmers, H. P., Goodman, S. L., Birchmeier, W. 1983. Cell-cell interaction and polarity of epithelial cells: Specific perturbation using a monoclonal antibody. *Cell* 35: 667–75

Johnson, M. 1985. Three types of cell interaction regulate the generation of cell diversity in the mouse blastocyst. See Edelman & Thiery 1985, pp. 27–48

Jørgensen, O. S., Delouvée, A., Thiery, J. P., Edelman, G. M. 1980. The nervous system specific protein D2 is involved in adhesion among neurites from cultured rat ganglia.

FEBS Lett. 11: 39–42

Kefalides, N. A., Alper, R., Clark, C. C. 1979. Biochemistry and metabolism of basement membranes. *Int. Rev. Cytol.* 61: 167–228

Keilhauer, G., Faissner, A., Schachner, M. 1985. Differential inhibition of neurone, neuron-astrocyte, and astrocyte-astrocyte adhesion by L1, L2 and N-CAM antibodies. *Nature* 316: 728–30

Kemler, R., Babinet, C., Eisen, H., Jacob, F. 1977. Surface antigen in early differentiation. *Proc. Natl. Acad. Sci. USA* 74: 4449–52

Kleinman, H. K., Hassell, J. R., Aumailley, M., Terranova, V. P., Martin, G. R., Dubois-Dalq, M. 1985. Biological activities of laminin. *J. Cell Biochem.* 27: 317–25

Kleinman, H. K., McGarvey, M. L., Hassell, J. R., Star, V. L., Cannon, F. B., Laurie, G. W., Martin, G. R. 1986. Basement membrane complexes with biological activity. *Biochemistry* 25: 312–18

Koda, J. E., Rapraeger, A., Bernfield, M. 1985. Heparan sulfate proteoglycans from mouse mammary epithelial cells. Cell surface proteoglycan as a receptor for interstitial collagens. *J. Biol. Chem.* 260: 8157–62

Kratochwil, K. 1983. Embryonic induction. In *Cell Interactions and Development*, ed. K. M. Yamada, pp. 99–122. New York: Wiley

Kurkinen, M., Cooper, A. R., Barlow, D. P., Jenkins, J. R., Hogan, B. L. M. 1983. Gene expression during parietal endoderm differentiation in mouse embryos and teratocarcinoma cells. See Silver et al 1983, pp. 389–401

Kurkinen, M., Taylor, A., Garrels, J. I., Hogan, B. L. M. 1984. Cell surface associated proteins which bind native type IV collagen. *J. Biol. Chem.* 259: 5915–22

Kühn, K., Wiedemann, H., Timpl, R., Risteli, J., Dieringer, H., et al. 1981. Macromolecular structure of basement membrane collagens. Identification of 7S collagen as a crosslinking domain of type IV collagen. *FEBS Lett.* 125: 123–28

Leivo, I., Vaheri, A., Timpl, R., Wartiovaara, J. 1980. Appearance and distribution of collagens and laminin in the early mouse embryo. *Dev. Biol.* 76: 100–14

Lesot, H., Kühl, U., von der Mark, K. 1983. Isolation of a laminin-binding protein from muscle cell membranes. *EMBO J.* 2: 861–65

Liotta, L. A. 1986. Tumor invasion and metastases—role of extracellular matrix. *Cancer Res.* 46: 1–7

Malinoff, H. L., Wicha, M. S. 1983. Isolation

of a cell surface receptor for laminin from murine sarcoma cells. *J. Cell Biol.* 96: 1475–79

Martinez-Hernandez, A., Amenta, P. 1983. The basement membrane in pathology. *Lab. Invest.* 48: 656–77

Murray, B. A., Hemperley, J. J., Prediger, E. A., Edelman, G. M., Cunningham, B. A. 1986. Alternatively spliced mRNA code for different polypeptides chains of the chicken neural cell adhesion molecule (N-CAM). *J. Cell Biol.* 102: 189–93

Ocklind, C., Öbrink, B. 1982. Intercellular adhesion of rat hepatocytes. *J. Biol. Chem.* 257: 6788–95

Ocklind, C., Odin, P., Öbrink, B. 1984. Two different cell adhesion molecules, cell-CAM 105 and calcium-dependent protein, occur on the surface of rat hepatocytes. *Exp. Cell Res.* 151: 29–45

Ogou, S., Yoshida-Noro, C., Takeichi, M. 1983. Calcium-dependent cell-cell adhesion molecules common to hepatocytes and teratocarcinoma stem cells. *J. Cell Biol.* 97: 944–48

Ophir, J., Moscona, A. A., Ben-Shand, Y. 1984. Cell disorganization and malformation in neural retina caused by antibodies to R-cognin: Ultrastructural study. *Cell Diff.* 15: 53–60

Palade, G. E. 1983. Membrane biogenesis. An overview. *Meth. Enzymol.* 96: xxix–lv

Palotie, A., Peltonen, L., Risteli, L., Risteli, J. 1983. Effect of the structural components of basement membranes on the attachment of teratocarcinoma derived endodermal cells. *Exp. Cell Res.* 144: 31–37

Peyrieras, N., Hyafil, F., Louvard, D., Ploegh, H., Jacob, F. 1983. Uvomorulin: A nonintegral membrane protein of the early mouse embryo. *Proc. Natl. Acad. Sci. USA* 80: 6274–77

Peyrieras, N., Louvard, D., Jacob, F. 1985. Characterization of antigens recognized by monoclonal and polyclonal antibodies directed against uvomorulin. *Proc. Natl. Acad. Sci. USA* 82: 8067–71

Pierce, B. M. 1966. The development of basement membranes of the mouse embryo. *Dev. Biol.* 13: 231–49

Pierschbacher, M., Ruoslahti, E. 1984. Variants of cell recognition site of fibronectin that retain attachment-promoting activity. *Proc. Natl. Acad. Sci. USA* 81: 5985–88

Pytela, R., Pierschbacher, M., Ruoslahti, E. 1985. Identification and isolation of a 140 kd cell surface glycoprotein with properties of a fibronectin receptor. *Cell* 40: 191–98

Rathjen, F., Schachner, M. 1984. Immunocytological and biochemical charac-

terization of a new neuronal cell surface component (L1 antigen) which is involved in cell adhesion. *EMBO J.* 3: 1–10

Richa, J., Damsky, C. H., Buck, C. A., Knowles, B. B., Solter, D. 1985. Cell surface glycoproteins mediate compaction, trophoblast attachment, and endoderm formation during early mouse development. *Dev. Biol.* 108: 513–21

Ritzki, T. M., Ritzki, R. M. 1983. Basement membrane polarizes lectin binding sites of *Drosophila* larval fat body cells. *Nature* 303: 340–42

Ruoslahti, E., Hayman, E., Piersbacher, M. B. 1985. Extracellular matrixes and cell adhesion. *Arteriosclerosis* 5: 581–94

Rutishauser, U. 1984. Developmental biology of a neural cell adhesion molecule. *Nature* 310: 549–54

Rutishauser, U., Grumet, M., Edelman, G. M. 1983. Neural cell adhesion molecule mediates initial interactions between spinal cord neurons and muscle cells in culture. *J. Cell Biol.* 97: 145–52

Sanes, J. 1984. Roles of extracellular matrix in neural development. *Ann. Rev. Physiol.* 45: 581–600

Saxén, L., Koskimies, O., Lahti, R., Miettinen, H., Rapola, J., Wartiovaara, J. 1968. Differentiation of kidney mesenchyme in an experimental model system. *Adv. Morphol.* 7: 251–93

Schuh, R., Vestweber, D., Riede, I., Ringwald, M., Rosenberg, U. B., et al. 1986. Molecular cloning of the mouse cell adhesion molecule uvomorulin: cDNA contains a B1 related sequences. *Proc. Natl. Acad. Sci. USA* 83: 1364–68

Shur, B. 1983. Embryonal carcinoma cell adhesion: The role of surface galactosyltransferase and its 90 K lactosaminoglycan substrate. *Dev. Biol.* 99: 360–72

Silver, L. M., Martin, G. R., Stickland, S. 1983. *Teratocarcinoma Stem Cells.* Cold Spring Harbor, Mass: Cold Spring Harbor Lab. 743 pp.

Simons, K., Fuller, S. D. 1985. Cell surface polarity in epithelia. *Ann. Rev. Cell Biol.* 1: 243–88

Sugrue, S., Hay, E. D. 1981. Response of basal epithelial cell surface and cytoskeleton to solubilized extracellular matrix molecules. *J. Cell Biol.* 91: 45–54

Takeichi, M., Yoshida-Noro, C., Ogou, S., Shirayoshi, Y., Okada, T. S. 1983. A cell-cell adhesion molecule involved in embryonic cellular interactions as studied by using teratocarcinoma cells. See Silver et al 1983, pp. 163–71

Tarone, G., Galleto, G., Prat, M., Comoglio, P. 1982. Cell-surface molecules and fibronectin mediated cell adhesion: Effect of proteolytic digestion of membrane proteins. *J. Cell Biol.* 94: 179–86

Terranova, V. P., Rohrbach, D. H., Martin, G. R. 1980. Role of laminin in the attachment of PAM 212 (epithelial) cells to basement membrane collagen. *Cell* 22: 719–26

Terranova, V., Rao, C. N., Kalebic, T. M., Marguiles, T., Liotta, L. A. 1983. Laminin receptor on human breast carcinoma cells. *Proc. Natl. Acad. Sci. USA* 80: 444–48

Thiery, J. P., Duband, J. L., Rutishauser, U., Edelman, G. M. 1982. Cell adhesion molecules in early chicken embryogenesis. *Proc. Natl. Acad. Sci. USA* 79: 6737–41

Thiery, J. P., Delouvée, A., Gallin, W., Cunningham, B. A., Edelman, G. M. 1984. Ontogenetic expression of cell adhesion molecules: L-CAM is found in epithelia derived from three primary germ layers. *Dev. Biol.* 102: 61–78

Thiery, J. P., Duband, J. L., Tucker, G. C. 1985. Cell migration in the embryo. Role of cell adhesion and tissue environment in pattern formation. *Ann. Rev. Cell Biol.* 1: 91–113

Timpl, R., Dziadek, M. 1986. Structure, development and molecular pathology of basement membranes. *Int. Rev. Exp. Pathol.* In press

Timpl, R., Rohde, H., Robey, P. G., Rennard, S. I., Foidart, J. M., Martin, G. R. 1979. Laminin—a glycoprotein from basement membranes. *J. Biol. Chem.* 254: 9933–37

Timpl, R., Dziadek, M., Fujiwara, S., Nowak, H., Wick, G. 1983. Nidogen: A new self-aggregating basement membrane protein. *Eur. J. Biochem.* 137: 455–65

Trelstad, R. 1984. *Role of Extracellular Matrix in Development*. New York: Liss. 642 pp.

Trocolli, N. M., Hausman, R. E. 1985. Vesicle interaction as a model for the retinal cell-cell recognition mediated by R-cognin. *Cell Diff.* 16: 43–49

Vaheri, A., Mosher, D. 1978. High molecular weight, cell surface glycoprotein (fibronectin) lost in malignant transformation. *Biochim. Biophys. Acta* 516: 1–25

Vartio, T., Vaheri, A. 1983. Fibronectin—a chain of domains with diversified functions. *Trends Biochem. Sci.* 8: 42–44

Vestweber, D., Kemler, R. 1984a. Rabbit antiserum against a purified surface glycoprotein decompacts mouse preimplantation embryos and reacts with specific tissues. *Exp. Cell Res.* 152: 169–78

Vestweber, D., Kemler, R. 1984b. Some structural and functional aspects of the cell adhesion molecule uvomorulin. *Cell Differ.* 15: 269–73

Vestweber, D., Kemler, R. 1985. Identification of a putative cell adhesion domain of uvomorulin. *EMBO J.* 4: 3393–98

Vestweber, D., Kemler, R., Ekblom, P. 1985a. Cell-adhesion molecule uvomorulin during kidney development. *Dev. Biol.* 112: 213–21

Vestweber, D., Ocklind, C., Gossler, A., Odin, P., Öbrink, B., Kemler, R. 1985b. Comparison of two cell-adhesion molecules, uvomorulin and cell-CAM 105. *Exp. Cell Res.* 157: 451–61

von der Mark, K., Mollenhauer, J., Kühl, U., Bee, J., Lesot, H. 1984. Anchorins: A new class of membrane proteins involved in cell-matrix interactions. See Trelstad 1984, pp. 67–88

Wessells, N. 1982. A catalogue of processes responsible for metazoan morphogenesis. In *Evolution and Development*, ed. J. T. Bonner, pp. 115–54. Berlin: Springer-Verlag

Yamada, K., Kennedy, D. W. 1985. Amino acid sequence specificities of an adhesive recognition signal. *J. Cell. Biochem.* 28: 99–104

Yamada, Y., Sasaki, M., Kohno, K., Kleinman, H., Kato, S., Martin, G. R. 1985. A novel structure for the protein and gene of the mouse laminin B1 chain. In *Basement Membranes*, ed. S. Shibata, pp. 139–46. Amsterdam: Elsevier

Yoshida-Noro, C., Suzuki, N., Takeichi, M. 1984. Molecular nature of the calcium-dependent cell-cell-adhesion system in mouse teratocarcinoma and embryonic cells studied with a monoclonal antibody. *Dev. Biol.* 101: 19–27

Ann. Rev. Cell Biol. 1986. 2 : 49–80

SPATIAL PROGRAMMING OF GENE EXPRESSION IN EARLY *DROSOPHILA* EMBRYOGENESIS

Matthew P. Scott

Department of Molecular, Cellular and Developmental Biology, University of Colorado, Boulder, Colorado 80309–0347

Patrick H. O'Farrell

Department of Biochemistry and Biophysics, University of California, San Francisco, California 94143

CONTENTS

One of the most fundamental issues of developmental biology is how spatial patterns are generated. Recent developments in the study of gene expression in *Drosophila* have provided clues to the mechanisms whereby genes control pattern. Genes that regulate *Drosophila* embryonic development have been found to be expressed in distinct spatial patterns that precede and make possible the appearance of morphological patterns in the embryo. Therefore, the problem of pattern formation can be stated as

49

0743–4634/86/1115–0049$02.00

two questions: How is position-specific gene expression attained and what are the functions of the position-specific gene products?

The recent advances in *Drosophila* developmental genetics (Mahowald & Hardy 1985; for reviews of earlier work see Lawrence 1976) have depended upon two major lines of research. First, extensive searches of the entire genome for mutations affecting oogenesis and early development have led to the identification of a large number of relevant loci (Nüsslein-Volhard & Wieschaus 1980; Nüsslein-Volhard et al 1984; Jürgens et al 1984; Wieschaus et al 1984a). These loci include a variety of segmentation genes, in which mutations cause deletions of some of the normal body segments. Thus, the wild-type segmentation pattern (Figure 1E) can be changed by mutations in "gap" loci, which delete groups of segments; by "pair-rule" locus mutations, which delete alternate segmental units; or by "segment polarity" mutations, which change the pattern within each segment (Figure 2). In addition, many homeotic genes, in which mutations alter the identity of segments or parts of segments, have been identified. Second, the application of molecular techniques such as gene cloning, in situ hybridization to sectioned tissues, immunofluorescence, and the

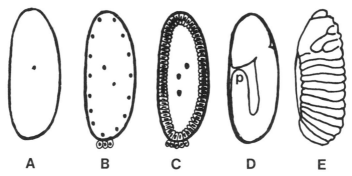

A   B   C   D   E

Figure 1   Stages in early *Drosophila* development. (A) A fertilized egg, showing the asymmetries evident in the shape of an oocyte. (Ventral is to the right, anterior is up.) (B) After a series of syncytial nuclear divisions, most of the nuclei move to the cortex of the egg. A few nuclei at the posterior pole are formed into cells, some of which will become the germ-line precursor cells. (C) After a total of 13 rounds of nuclear division, the approximately 6000 resulting nuclei are formed into cells. Immediately thereafter, gastrulation begins with a series of invaginations. (D) Gastrulation and germ-band elongation produce an embryo with the most-posterior precursor cells moved so that they lie behind the developing head region (P = posterior end of germ band). While the germ band is extended, the segments first appear. (E) After the germ band shortens (by about 10 hr), the lobed head segments, three thoracic segments, eight abdominal segments, and the caudal region are visible. (For a complete description of early development see Foe & Alberts 1983 and Turner & Mahowald 1976, 1977.)

introduction of cloned DNA into the germ line have allowed patterns of gene expression to be analyzed with a precision that was previously unattainable. Most importantly, the gene products being tracked are ideal morphological markers, limited to groups of cells that are otherwise largely indistinguishable from other cells. In addition to their restricted spatial distributions, the molecules encoded by the segmentation and homeotic genes are themselves involved in regulating embryogenesis.

Historically, the dominant idea regarding the origins of biological pattern was that graded variations in the concentrations of morphogens (either prepackaged in the egg or formed by zygotic processes) specify the fates of different cells (Kalthoff 1976; Meinhardt & Gierer 1980; Meinhardt 1982; Slack 1983; Sander 1984; Russell 1985). There is, however, no direct evidence supporting this type of model. It has been supported primarily by theoretical considerations, and other types of models remain equally plausible. As we describe here, recent studies of spatial patterns of embryonic gene expression suggest that combinatorial gene actions and localized cell interactions could provide alternative explanations for how pattern formation is achieved.

GAP   PAIR-RULE   SEGMENT POLARITY

Kr   ftz   en

Figure 2   Classes of segmentation loci: patterns of gene expression and deletion phenotypes. The diagrams show one example from each of the three classes of segmentation genes. *Krüppel (Kr)* is a "gap" locus; null mutant *Kr* embryos lack thoracic and anterior abdominal segments (*gray*). The initial pattern of *Kr* transcripts is a belt near the middle of the embryo (*gray shading*). *Kr* RNA is first detected before cells form, as shown. *fushi tarazu (ftz)* is an example of a pair-rule gene. Mutations in *ftz* cause deletions of alternate segmental boundaries, as indicated (*gray areas*). Shortly before cellularization, *ftz* RNA and protein products are observed in seven stripes (*gray*), the most posterior of which is wider than the others. The *engrailed (en)* gene is required in the posterior of each segment (*gray areas*). The *en* gene is an example of the segment polarity or polarity reversal class of loci. The *en* striped pattern is first seen after cells have formed. There are 14 (later 15) *en* stripes, about twice the number of stripes seen when observing pair-rule gene products. (For further descriptions see text and references therein.)

52   SCOTT & O'FARRELL

AN OUTLINE OF EARLY DEVELOPMENT IN *DROSOPHILA*

Patterns develop very quickly in the early *Drosophila* embryo (Figure 1). After only twelve hours of development at 25°C, the organization of the *Drosophila* embryo is quite complex. In addition to basic features, such as distinct endoderm, mesoderm, ectoderm, and germ cells, the segments have formed a number of segment-specific structures (Figure 3a). Because of its intricate pattern, the presence of the partially formed central and peripheral nervous system dramatically illustrates the detail of the organization (Figure 3b). In this review we concentrate on the events of the first twelve hours of development, during which the basic body plan of the organism arises. Since more is known about anterior-posterior pattern regulation than about dorsal-ventral regulation, our focus is on the events that lead to the anterior-posterior differentiation of the body segments.

Following fertilization, fly development begins with a series of roughly synchronous syncytial mitoses, which occur once every 6–10 min at the cleavage stage (Turner & Mahowald 1976; Foe & Alberts 1983). A cytoskeletal network organizes a zone of cytoplasm around each syncytial nucleus, and together nuclei and cytoplasm go through a programmed series of divisions and movements. During the telophase periods of the eighth and ninth nuclear division cycles, most of the nuclei migrate to the periphery of the egg to form the syncytial blastoderm. About five nuclei arrive at the surface of the posterior pole of the embryo at cycle nine, cause the egg membrane to bulge outward, and become fully enclosed in cell membranes during the next nuclear division. These "pole cells" are the progenitors of the germ cells. Most of the remaining nuclei reach the surface of the egg in cycle ten and undergo four more syncytial divisions at progressively slower rates (9–21 min). Streaming of cytoplasm to and from the anterior and posterior poles of the embryo occurs during the later nuclear divisions (Foe & Alberts 1983). During a 30-min period following the thirteenth division, cell membranes grow in between the nuclei to form individual cells (cellularization stage). Completion of cellularization, at about 2.5 hr of development, produces the cellular blastoderm. Although the resulting shell of about 6000 cells has few visibly distinctive features, it is primed to undergo a rapid development of visible pattern during the gastrulation movements. As is discussed below, at least a rough code of spatial information is already present in the cellular blastoderm.

As soon as the cell membranes are formed, gastrulation begins (Turner & Mahowald 1977). The first movements visibly segregate presumptive mesoderm, endoderm, and ectoderm. The presumptive mesoderm, about 1000 cells along the ventral midline, folds in to produce the ventral furrow and subsequently pinches off to form a ventral tube, which flattens as a layer of mesoderm beneath the ventral ectoderm. Presumptive endoderm invaginates as two pockets at the anterior and posterior extremes of the ventral furrow. Additional early movements of gastrulation substantially

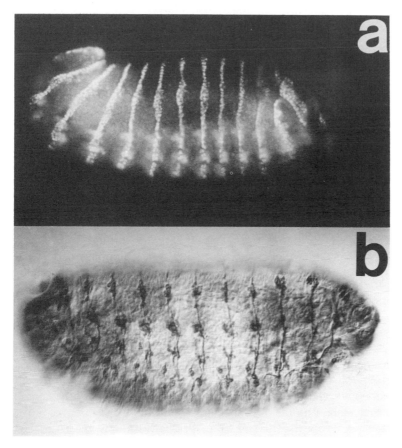

Figure 3 Segmental features of early (about 12 hr) *Drosophila* embryos. (*a*) A whole mount of an embryo after germ-band shortening stained for *en* protein by indirect immunofluorescence. The distribution of this nuclear antigen outlines segmental features. A stripe of staining is seen in the posterior part of each segment. Additional features are seen that do not repeat in the simple segmental array. At this stage, the oral segments (anterior) and the two most posterior segments (9 and 10) have been substantially reshaped by morphogenetic movements (DiNardo et al 1985; picture courtesy of S. DiNardo). (Anterior is to the right and ventral is down in both *a* and *b*.) (*b*) A whole mount embryo stained with an antibody specific for cells of the peripheral nervous system (Jan et al 1985; Ghysen et al 1986). The stained cells are clusters of peripheral sensory cells underlying the epidermis of the 12-hr embryo. The intricacy of the pattern of organization of these cells and its segmental repeat is evident. (Picture courtesy of R. Bodmer.)

influence the shape of the embryo, but their function is unknown. A transverse cephalic furrow begins to form laterally near the anterior extreme of the ventral furrow. This infolding persists for about 3 hr and then unfolds with no clear consequence. Posterior to the cephalic furrow, cells begin to move ventrally around the sides of the embryo. The cells gather along the ventral midline, now called the germ band, and begin to move posteriorly, continuing around the posterior end of the embryo to the dorsal side of the embryo. When the germ band is fully extended, the cells that will form the most posterior larval structures are located just behind the presumptive head region. While the germ band is extended, at about 6 hr of development, the body segments first appear. There are several lobes that represent the early head segments, three thoracic segments, eight obvious abdominal segments, and at least two caudal segments. The epidermal ectoderm, the neural ectoderm, and the mesoderm cells are all arranged in segmental divisions. The germ band reverses its earlier movements to return the posterior segments to the posterior tip of the embryo, a process that is completed by about 10 hr of development. During germ-band extension and retraction, the epidermal ectoderm cells undergo 2–3 cell divisions. These cell cycles are much slower than the cleavage stage nuclear division cycles. In addition, mitoses begin at specific times in several areas of the embryo and then spread to cells in other regions of the embryo. Most cells in the germ band undergo their first postblastoderm mitosis during the latter half of germ-band elongation (Hartenstein & Campos-Ortega 1985). Thus, early and dramatic morphogenetic movements occur without cell division.

Development of the nervous system begins early during gastrulation (Poulson 1950; Hartenstein & Campos-Ortega 1984). Starting at about 4.5 hr, neuronal precursor cells located in two strips along each side of the ventral midline (the "neurogenic" region) enter the interior of the embryo as individual cells, squeezing between and separating from the epidermal cells that remain on the outside. These neuroblast cells are positioned in a pattern that repeats with small but specific variations in each segmental primordium. The neuroblasts follow a stereotyped pattern of divisions to give rise to numerous neurons arranged in an orderly but complex pattern.

Experiments in which cells are killed (Lohs-Schardin et al 1979; Underwood et al 1980), transplanted (Simcox & Sang 1983), or observed through movements of injected peroxidase (Hartenstein et al 1985) have defined an approximate fate map of the cellular blastoderm. At the blastoderm stage the individual cells have learned their relative positions in the embryo. If blastoderm cells are transplanted, they do not form structures characteristic of their new location, but instead form structures that they would have developed in their original positions. On this basis, it has been concluded that the blastoderm cells are spatially determined with respect to segmental identity (Illmensee 1976). It should be emphasized that these experiments have a rather low spatial resolution and that other experiments suggest that determination of fate is a progressive process with some events preceding and some following formation of the cellular blastoderm. Only part of the pattern can be determined at the blastoderm stage, since many new cells are later produced by cell division and incorporated into the pattern (e.g. Hartenstein & Campos-Ortega 1985; Doe & Goodman 1985; Kuwada & Goodman 1985).

Classical genetic analysis of mitotic clones has shown that each segment is composed of cells from two distinct lineages, the anterior and posterior compartments (Garcia-Bellido et al 1976). The distinction between these groups of cells is made at or shortly after the cellular blastoderm stage (Wieschaus & Gehring 1976). The clonal studies anticipated some of the molecular analysis reviewed here, in that they predicted that the blastoderm would be subdivided in a zebra-striped pattern of alternating groups of differently specified cells (anterior and posterior). Two repeat units can be defined in an alternating pattern of anterior and posterior stripes: a segment, which is composed of an anterior and a posterior compartment, and a parasegment, which is defined as the posterior compartment of one segment and the anterior compartment of the next most posterior segment (Martinez-Arias & Lawrence 1985). The subdivision into parasegments and compartments is important because it has been suggested that these groups of cells are the developmental units that function in pattern-refining steps and in events regulating homeotic gene expression (for review see Brower 1985).

MATERNAL CONTRIBUTIONS TO EMBRYONIC PATTERN FORMATION

A number of experimental efforts have been made to address two difficult questions. First, what components needed for pattern formation are packaged into the egg? Second, to what extent do *localized* substances in the oocyte provide information to direct subsequent pattern formation? That some "maternal hardware" is needed for anterior-posterior differentiation has been shown experimentally: Females carrying mutations in genes encoding necessary substances normally provided during oogenesis produce embryos that, regardless of the zygotic genotype, have segmentation defects (e.g. Nüsslein-Volhard 1979; Schüpbach & Wieschaus 1986). However, additional studies are needed to determine whether the *position* of such substances in the oocyte is important to subsequent development. For example, the importance of localized determinants can be revealed by experimental manipulations, such as centrifugations or ligations that disrupt the organization of the egg (e.g. Schubiger & Newman 1982; Kalthoff 1983).

Both genetic and developmental analyses suggest that the egg contributes maternal hardware essential to some of the earliest stages of pattern formation, for example, dorsal-ventral and anterior-posterior axis determination (Nüsslein-Volhard 1979). Thus, screens for maternal effect mutations have led to an estimation of at least 10 maternally provided gene products that are essential for definition of the dorsal-ventral axis and appear to have no other essential function (Anderson & Nüsslein-Volhard 1984). Likewise, at least that many loci are involved in anterior-to-posterior patterning (Lehmann 1985; Schüpbach & Wieschaus 1986; T. Schüpbach, personal communication). However, the rescue of mutant phenotypes by cytoplasmic transfer suggests that some of the maternal gene products required for this patterning are not, or may not need to be, precisely localized in the oocyte (Anderson & Nüsslein-Volhard 1984).

Asymmetries in the egg clearly determine the overall orientation of the anterior-posterior and dorsal-ventral axes. For example, in another family of flies, the midges, the anterior end of the cylindrical egg is specified by UV-sensitive and RNase-sensitive determinants localized near this pole (Kalthoff 1983; Sander 1984). There are also localized determinants near the posterior tip of insect eggs specifying the formation of pole cells, as was first shown in beetles (Hegner 1911). The transplantation of *Drosophila* posterior polar cytoplasm to a new location has been used to demonstrate that cytoplasm at the posterior tip of the oocyte contains information involved in determining cells to become germ-line precursors (Illmensee & Mahowald 1974; Okada et al 1974). The posterior determinants may reside in the polar granules, characteristic structures associated with the posterior cytoplasm.

The results of experiments in which *Drosophila* embryos are divided by ligation at various stages of early development argue against the importance of precisely localized, maternally provided determinants in specifying the fate of cells that form at particular positions along the embryo (Schubiger & Newman 1982; reviewed in Kalthoff 1983; Sander 1984). There are not, for example, localized determinants for each segment: If the anterior or posterior end of an insect embryo is separated by constriction from the rest of the embryo, a cell whose position would have caused it to give rise to a certain segment of the insect instead participates in the formation of a different segment. This plasticity suggests that most of the blastoderm cells are *not* determined simply by responding to prelocalized cortical cytoplasmic materials when the nuclei move into their positions and cellularize during blastoderm formation.

Evidence from experiments with the insect *Smittia* also argues against the importance of prelocalized maternal determinants. Centrifugation and UV irradiation of the *Smittia* egg can result in the formation of normal-looking embryos that are reversed along the anterior-posterior axis relative to the egg case (Kalthoff et al 1982; Kalthoff 1983). Either the normal body pattern of the reversed embryos forms despite any localized determinants that are in the oocyte, or all of these determinants are rearranged in an orderly way by the experimental treatments used, an unlikely proposition. These experiments show that determination is a dynamic and flexible process that can respond to experimental manipulations performed after oogenesis is complete.

THE INITIAL ACTIVATION OF THE ZYGOTIC GENOME

Eggs are unusually large cells, and the synthetic capacity of a diploid nucleus appears inadequate to direct the rapid changes that occur in the initial stages of development. The *Drosophila* egg is about 400 μm long and 160 μm across and contains about 10^9 molecules of mRNA (Hough-Evans et al 1980; Jacobs-Lorena et al 1980). If a single gene were packed with RNA polymerases at the maximum density (McKnight & Miller 1976), the rate of transcriptional elongation would limit transcript production to about 1000 molecules per hour. At this rate, it would take a diploid gene about a year to make a 1% contribution to the pool of mRNA. This type of reasoning suggests that early events must be primarily carried out by products previously packaged into the eggs. Indeed, early embryonic divisions proceed with little detectable transcription, and activity of the genome appears to be unnecessary for many of the early events. At cellularization there is an abrupt transition during which transcriptional activity is dramatically increased and the zygotic nuclei begin to take on their normal, substantial, synthetic responsibilities (Edgar & Schubiger 1986).

Studies of frog embryos have uncovered and described a developmental transition extraordinarily analogous to that of *Drosophila* (Newport & Kirschner 1982). This event, the midblastula transition, may be a fundamental feature of early development in many complex metazoans. In both *Drosophila* and frog, the timing of the transition is determined by the nuclear to cytoplasmic ratio (Edgar et al 1986; Newport & Kirschner 1982), which increases progressively during the rapid nuclear or cell divisions. Perhaps it is the devotion of cellular processes to rapid DNA replication and nuclear division in the early embryo that limits early transcription.

In support of this idea, inhibition of DNA replication during the early cell divisions leads to precocious activation of transcription (Edgar & Schubiger 1986). Some of the genes involved in pattern formation events (see below) show exceptionally early expression that precedes the general activation of transcription. Additionally, electron microscopic analysis of transcription fibers on dispersed chromatin suggests that prior to full transcriptional activation, a special class of short transcription units are very heavily transcribed (McKnight & Miller 1976). It is not at all clear how transcription of these particular sequences is activated during the cleavage stages.

A critical problem in pattern formation is how genes that come to be expressed in spatially specific patterns are initially regulated in the very early embryo. It appears unlikely that localized cortical determinants activate specific patterns of transcription in nuclei upon their arrival at the egg surface. Experiments described above suggest that cortical determinants that are prelocalized in the oocyte do not determine the fates of blastoderm cells, although determinants could become localized in the cortex as the early embryo develops. Another relevant result is that the initial transcription of the *Krüppel* "gap" segmentation locus (Wieschaus et al 1984b) occurs in specific regions along the anterior-posterior axis both in the cells at the surface of the blastoderm and in yolk nuclei located at corresponding positions in the interior of the embryo (Knipple et al 1985). Therefore, nuclei that have never reached the cortex are sensing their position in the embryo. However, another possibility is that radial streaming of the cytoplasm (Foe & Alberts 1983) is responsible for communication between the yolk nuclei and the cortex. A final point is that cellularization does not appear to be a critical event in the establishment of localized transcript distribution. Transcripts encoded by some of the segmentation genes occur in specific positions along the embryo prior to cellularization, as is described below.

A GENETIC HIERARCHY: THE ZYGOTICALLY ACTIVE SEGMENTATION GENES

Recent results suggest a perspective that emphasizes interaction among zygotically active genes as a dynamic process that gradually refines the developing pattern of the early embryonic segments (Raff & Kaufman 1983; Mahowald & Hardy 1985). According to this model, the maternal influences—localized determinants and/or gradients—provide only approximate information about the axes and organization of the embryo. In this view, the burden of precise pattern formation is placed upon the zygotically active gene systems. The maternal genome provides essential functions, but the maternally provided molecules need not be positioned with extreme precision.

Although developmental mutations in *Drosophila* have many origins, one large-scale screen has been an exceptionally fruitful source of mutations affecting early embryonic development (Nüsslein-Volhard & Wieschaus 1980; Nüsslein-Volhard et al 1984; Jürgens et al 1984; Wieschaus et al 1984a). Lethal mutations were screened by examination of embryos in the lethal arrest state. Statistically, the number of lethal mutations examined was sufficient to identify lesions in most of the loci involved in production of embryonic cuticular pattern, assuming the loci are susceptible to ethylmethane sulfonate mutagenesis. In addition, the identified loci accounted for the defects seen when deletions covering much of the genome were examined for their effects on cuticle formation. However, some other relevant loci may have yet to be found. About twenty of the identified loci are involved in the process of segmentation. These are particularly interesting because they appear to represent a group of regulatory genes that interact in a regulatory cascade that specifies the repeating pattern of segments.

The segmentation genes are divided into three phenotypic classes based on the extent of the pattern defects (Figure 2). Mutations in the four "gap" loci delete large parts of the embryonic segment pattern; mutations in the eight "pair-rule" genes cause deletions of patterns spaced at two segment intervals; mutations in the nine "segment polarity" loci affect parts of every body segment (Figure 3a). The gap loci can be viewed as coarse dividers of the embryo, the pair-rule loci as functioning to divide the embryo into segmental units, and the segment polarity loci as involved in forming patterns within each segment. There are indications that a balance between the levels of different segmentation gene products is important. Hyperdosage of the *runt* pair-rule gene causes defects in segments that are unaffected by the absence of *runt* (J. P. Gergen & E. Wieschaus 1986b). In a related study it was found that it is as harmful to express the *fushi tarazu* (*ftz*) gene in places in the embryo where it is normally not expressed as it is to lack *ftz* products where they are normally required (Struhl 1985).

The localized defects seen in segmentation mutants appear to be well-correlated with the spatial patterns of expression of these genes (Figure 2). For example, in situ hybridization to tissue sections shows that as early as division cycle 11, RNA from the *Krüppel* (*Kr*) gap locus is localized in a thick belt around the middle of the embryo (Knipple et al 1985). The position of the pattern deletion in *Kr⁻* embryos roughly corresponds to the position of the early belt of expression (Figure 2), though a larger region is absent from the segmentation pattern later than would be expected from the width of the early belt of expression. Thus, *Kr* may affect cells in which it is not expressed. The pattern of *Kr* expression becomes more complex later in development, spreading into new regions well separated from the first place the gene was expressed. Thus, this gap gene is expressed very early in development, and the transcripts are localized before the embryo is divided into cells.

Mutations in two extensively studied pair-rule loci, *fushi tarazu* (*ftz*) and *hairy* (*h*), cause defects in alternate segments, which are offset from each other. The *ftz* mutations result in the absence of the labial-maxillary, T1-T2, T3-A1, A2-A3, A4-A5, A6-A7, and A8-caudal (T = thoracic, A = abdominal) segment boundaries (Figure 2; Wakimoto & Kaufman 1981), while *h* mutations result in the absence of the segment boundaries that are not affected by *ftz* (Nüsslein-Volhard & Wieschaus 1980; Ish-Horowicz et al 1985). Both *h* and *ftz* transcripts are first detected at about the same time that *Kr* RNA is first detected (Hafen et al 1984a; Ingham et al 1985). Initially, *h* RNA is found uniformly throughout the embryo, while *ftz* RNA is restricted to a broad region between 15 and 65% of the egg length (0% is the posterior tip of the embryo and 65% is approximately the junction of the thoracic and head segments). During the next two nuclear divisions, both of the patterns become more complex, resolving first into several broad bands and then into seven transverse stripes at two-segment intervals (Hafen et al 1984a; Weir & Kornberg 1985; Ingham et al 1985). Thus, the periodic pattern of defects seen in the mutant embryos is preceded by spatially restricted expression in a similarly periodic pattern. The striped patterns are established before cell membranes form, and the *h* stripes are offset from the *ftz* stripes (Ish-Horowicz et al 1985; Ingham et al 1985), as one would predict from the phenotypes. The *h* and *ftz* stripes are not perfectly offset; they overlap by one to two cells. Some cells express both genes, some express neither, and some express just one. The similarity between the offset in the periodic patterns of expression of *h* and *ftz* and the offset in *h* and *ftz* mutant defects argues for a correspondence between the sites of gene function and the localization of RNA and protein products. However, as is discussed below (see section on cells responding to their neighbors), other factors could come into play such that there may not be a precise correspondence between the cells that express a gene product and those affected by the absence of the product. The remarkable patterns of gene expression detected with the molecular probes provide a tantalizing glimpse of pattern formation in progress. The refinement of the pattern from initially uniform expression to stripes is precisely the type of process we would like to understand. It seems likely that regulatory interactions among the gap gene products and the various pair-rule gene products play a role in this process.

Genes of the third class of zygotic segmentation genes, the segment polarity genes, act in every segment. Again, the expression of these genes may be spatially offset; some genes affect posterior parts of segments, some affect segment boundaries, and some affect anterior parts of segments. RNA (and protein) from one of these genes, the *engrailed* (*en*) locus, accumulates in fourteen transverse stripes (each about one cell in width) by late division cycle 14, after completion of cellularization (Kornberg et al 1985; Fjose et al 1985; DiNardo et al 1985). Later, during germ-band elongation, a fifteenth *en* stripe develops near the posterior end of the embryo. The positions of these stripes correspond to the anlagen of the posterior part of each segment (Figure 2). There are therefore about twice as many *en* stripes as *h* or *ftz* stripes. *en* has been shown to act in posterior compartment cells to distinguish them from anterior compartment cells, both during embryogenesis and during metamorphosis (Garcia-Bellido & Santamaria 1972; Lawrence & Morata 1976; Kornberg 1981a,b). The *engrailed* (*en*) gene has some attributes of a pair-rule locus (Nüsslein-Volhard & Wieschaus 1980; Kornberg 1981b), but it seems to belong in the segment polarity class since it affects all of the segments. It remains to be seen whether *en* acts in a way that is fully representative of the genes in the segment polarity class. The *en* gene also appears to function in the precellular embryo (Karr et al 1985).

The experiments employing in situ hybridization of cloned gene probes to tissues have demonstrated that the expression of many of the regulatory genes is controlled at a level that affects the accumulation of transcripts. They have also shown that different cells express different combinations of segmentation genes. A key question is the extent to which the data support the intuitive idea that the embryo is progressively subdivided by successive actions of the gap, pair-rule, and segment polarity loci. As would be expected if this idea is correct, the *localized* patterns of expression appear in a temporal sequence of successively finer subdivisions, as described above.

More direct evidence for the proposed hierarchy comes from examining the expression of genes in one class in embryos carrying mutations in genes in other classes. Only a few experiments have been reported to date, but all of them are in agreement with the gap → pair-rule → segment polarity gene hierarchy. For example, *ftz* (pair-rule) expression is altered by mutations in all four gap loci but is unaffected by mutations in three segment polarity loci, including *en* (Carroll & Scott 1986). Furthermore, in the next tier of the proposed hierarchy, mutations in any of the eight pair-rule loci alter *en* expression (S. DiNardo & P. H. O'Farrell, unpublished). However, in addition to these regulatory interactions between classes, some of the genes within a class interact. Thus, for three of the seven other pair-rule loci, a mutation alters *ftz* expression (Carroll & Scott 1986;

Howard & Ingham 1986), and within the next tier, mutation in either of at least two other segment polarity genes affects *en* expression (S. DiNardo & P. H. O'Farrell, unpublished data). Whereas mutations in the pair-rule locus *h* affect *ftz* expression, the pattern of *h* RNAs is *not* affected by *ftz* mutations (Ingham et al 1985; Howard & Ingham 1986). Therefore, even within the class of pair-rule loci, hierarchical interactions can be defined.

In an embryo mutant for a pair-rule gene, *ftz*, seven even-numbered *en* stripes are missing, but the odd-numbered stripes of *en* expression are present. Reciprocally, the pair-rule mutant *paired* (*prd*) embryo lacks the odd-numbered stripes but still expresses *en* in the even-numbered stripes (S. DiNardo & P. H. O'Farrell, unpublished data). Therefore, *en* expression can be induced by at least two independent trans-acting systems, one dependent on the *prd* gene product and one dependent on the *ftz* gene product. Since expression of *engrailed* in the head region of the embryo appears unaffected by either *ftz* or *prd* mutations, there may be additional combinations of regulators that can induce *en*.

Other observations show that *en* and *ftz* are controlled by regulators acting in combination. For example, though the expression of even-numbered bands of *en* relies on *ftz*, the *en* stripes of expression are one cell wide while the *ftz* stripes are three cells wide. Since *ftz* does not induce *en* in all the cells in which it is expressed, *ftz* action upon *en* appears to be modified by at least one other spatially localized gene product. Other observations suggest that complex combinatorial regulation may be a common feature of the regulatory interactions among segmentation genes. For example, no single zygotically active locus seems to act as a general repressor or activator of *ftz* (Carroll & Scott 1986). A particular locus may be required for proper *ftz* expression in one part of the embryo but not in other parts.

The hierarchy can be extended to genes that act only during oogenesis: Eight "maternal effect" segmentation loci have been shown to alter the pattern of expression of *ftz* (Mohler & Wieschaus 1985; Degelmann et al 1986; Carroll et al 1986), and one such locus has been shown to alter *h* expression (Ingham et al 1985). Some of the maternal effect loci alter all of the *ftz* stripes, while others disrupt only part of the pattern. As is true for the zygotically active loci, no single maternally active locus seems to act as a general repressor or activator of *ftz*. Some of the maternal gene products may be needed only in certain parts of the embryo; others may be needed throughout most or all of the embryo. The potential complexities are already evident; for example, a maternally active segmentation gene could act directly on *ftz*, or its effects could be mediated by all four gap loci and at least the three pair-rule loci that are above *ftz* in the hierarchy. The results support the view that the complex morphology

of an embryo develops as the result of combinatorial actions of segmentation genes. But what do the segmentation genes act upon other than
themselves?

HOW DO SEGMENTATION GENES CONTROL MORPHOLOGY?

To alter morphology a regulator must control many features of cell
behavior, most obviously expression of differentiated gene products, internal and surface subcellular structures, patterns of proliferation, and cell
movements. At present, our only real clue to how the segmentation genes
(and the homeotic genes, see below) might achieve control of these parameters is that some of the gene products appear to be sequence-specific
DNA-binding proteins. This is based on the observation that some of the
genes have a highly conserved sequence, the homeobox sequence
(McGinnis et al 1984; Scott & Weiner 1984), which encodes a protein
structure with homology to bacterial DNA-binding proteins (Laughon &
Scott 1984) and has been shown to have sequence-specific DNA-binding
activity (Desplan et al 1985; A. Laughon & M. P. Scott, unpublished
data). In addition, the gap gene *Krüppel* encodes a protein product that
has structural homology to the 5S ribosomal RNA transcription factor
IIIA (Rosenberg et al 1986). Thus, the homeobox-containing genes and
Krüppel may encode regulators of transcription. If so, what are the target
genes regulated by these proteins? Ultimately, genes encoding tissue-specific products must be regulated. This may occur through a regulatory
cascade and not by direct action of the segmentation or homeotic genes
on the target genes. Such a regulatory cascade has been characterized in
the case of the sex-specific expression of yolk proteins (Belote et al 1985).
The control of genes encoding tissue-specific products by segmentation
genes has not yet been examined. However, there are indications of a
regulatory cascade in which segmentation genes control the expression of
other gene regulators. We have already summarized the evidence that some
segmentation genes control the expression of other segmentation genes.
Additionally, the segmentation genes appear to contribute to control of
the expression of another class of regulatory genes: the homeotic genes.

REGULATION OF SPATIAL DISTRIBUTION OF HOMEOTIC GENE EXPRESSION

Whereas mutations in segmentation genes lead to deletions of body parts,
the consequence of a defect in a homeotic gene is that a particular part of
the embryo develops the structures appropriate to another part of the

animal (Garcia-Bellido 1975, 1977). Thus, the wild-type functions of homeotic genes are thought to direct the developmental fate of a group of cells. Many of the homeotic genes are located in two clusters, the bithorax complex (BX-C) (Lewis 1978; Bender et al 1983; Karch et al 1985) and the Antennapedia complex (ANT-C) (Kaufman et al 1980; Garber et al 1983; Scott et al 1983). To achieve normal development, the correct combinations of homeotic genes must be expressed in cells in appropriate positions in the embryo. Again we are faced with the question of how this spatial regulation of gene expression is achieved.

We summarize a few pertinent features of the patterns of expression of the homeotic genes: First, there is an excellent correspondence between the known sites of a particular gene's function and the places its transcripts and proteins are observed. Transcript distribution has been studied for the ANT-C homeotic genes *Antennapedia* (*Antp*) (Levine et al 1983), *Sex combs reduced* (*Scr*), and *Deformed* (*Dfd*) (Levine et al 1985); and the BX-C homeotic genes *Ultrabithorax* (*Ubx*) (Akam 1983; Akam & Martinez-Arias 1985), *iab-2* (also called *abdA*) (Sanchez-Herrero et al 1985; Karch et al 1985), and *iab-7* (also called *AbdB*) (Levine et al 1985). Protein distributions have been observed only for *Ubx* (White & Wilcox 1984, 1985a,b; Beachy et al 1985) and *Antp* (S. B. Carroll et al, submitted). For describing homeotic gene expression, the embryo can usefully be viewed as a series of 14 parasegmental units, with parasegment 1 in the head region and parasegment 14 at the posterior end (Martinez-Arias & Lawrence 1985). *Ubx* acts in parasegments 5–13, and that is where its transcripts and proteins are observed. *Antp* functions in parasegments 4–13, and again there is good correspondence with RNA (Levine et al 1983; Martinez-Arias 1986) and protein (S. B. Carroll et al, submitted) localizations. The localization of gene products other than *Ubx* and *Antp* has not been as thoroughly examined, but even in these cases the correspondence between the positions of gene function and gene transcription appears to be excellent. The presence of homeotic gene expression boundaries that match compartment boundaries suggests that homeotic genes may respond to boundaries established earlier by the segmentation genes.

As in the case of the segmentation genes, the homeotic genes can be placed into a regulatory hierarchy. The presence of certain homeotic gene products reduces the transcription of others. The best examples of interactions come from BX-C genes and from the effects of BX-C genes on the accumulation of *Antp* products. *Antp* mutations have no effect on the pattern of *Ubx* transcripts or protein, but lack of Ubx^+ function leads to the appearance of a high level of *Antp* RNA (Hafen et al 1984b) and protein (S. B. Carroll et al, submitted) in parasegment 6. Thus, Ubx^+ directly or indirectly appears to block *Antp* RNA accumulation. The effect can be extended by removing the entire BX-C region. Then *Antp* RNA and protein are present at a high level all the way from their normal locations in parasegments 4 and 5 through parasegment 12. (Normally *Antp* is expressed at a low level in all of these more posterior regions.) BX-C genes other than *Ubx* must therefore also act to depress *Antp* function.

BX-C genes that function in the abdomen (e.g. *abdA* and *AbdB*) negatively regulate *Ubx* expression as well as *Antp* expression (Struhl & White 1985). The *abdA* (*iab-2*) and *AbdB* (*iab-7*) transcripts have not been as precisely localized as the *Ubx* or *Antp* transcripts, but the *abdA* transcripts are found in regions approximately corresponding to parasegments 7–11, and the *AbdB* transcripts are in the region of parasegments 11–13 (Harding et al 1985). In the absence of both the *abdA* gene and the *AbdB* gene, high level *Ubx* expression extends from its normal parasegment 5 and 6 locale all the way through parasegment 13 (Struhl & White 1985). (Normally *Ubx* is expressed only at a low level in these more posterior regions.) These effects may be more subtle: Not only is the level of *Ubx* (or *Antp*) affected, but the number of cells in each segment expressing either gene changes in response to the other homeotic genes. Thus, an important regulatory paradigm, which may apply to only some tissues or stages, is emerging: Homeotic genes expressed in more posterior regions reduce transcription of homeotic genes that are expressed more anteriorly, and they also affect *which* cells within each segment (or parasegment) express the more anterior gene.

Polycomb (*Pc*) and *extra sex combs* (*esc*) are two homeotic genes that, in contrast to ANT-C and BX-C genes, are *not* position-specific in their effects. The multiple homeotic transformations caused by *Pc* and *esc* mutations appear to be due to alterations in the spatial expression of ANT-C and BX-C genes (Lewis 1978; Struhl 1981b). *Pc* and *esc* have both been shown to behave as though they are repressors of ANT-C and BX-C gene function (Duncan & Lewis 1982; Struhl 1983). The effect of *Pc* mutations on *Ubx* and *Antp* RNA and protein distribution reveals that the actual situation is somewhat complex: *Ubx* expression increases in some regions and decreases in others (Cabrera et al 1985; Wedeen et al 1986), perhaps because genes that affect *Ubx* (such as *abdA* and *AbdB*) are also affected by the loss of *Pc*[+] function. For instance, the absence of *Pc*[+] could derepress *abdA*, which in turn would repress *Ubx* in some cells. *Antp* protein is also more broadly distributed (but at low levels) in the absence of *Pc*[+] (S. B. Carroll et al, submitted). Mutations in *esc* affect *Ubx* expression in much the same way as *Pc* mutations (Struhl & Akam 1985), in keeping with the earlier predictions from genetic analyses. Abnormally widespread distributions of *iab-2*, *iab-7*, and *Dfd* RNAs in *Pc*[−] embryos were also found (Wedeen et al 1986). Therefore, the prediction that *Pc*[+]

and esc^+ are negative regulators of BX-C and ANT-C genes is largely borne out by these data.

Interestingly, the *initial* pattern of *Ubx* transcripts in an esc^- or Pc^- embryo is indistinguishable from that of wild-type embryos (Struhl & Akam 1985; Wedeen et al 1986). Only later, after gastrulation, does *Ubx* RNA accumulate where it normally would not. Therefore, *Ubx* expression is properly initiated *without* a need for either Pc^+ or esc^+ gene (Ingham 1983). This is in contrast to hypotheses that both genes are involved in initiation of position-specific homeotic gene expression (e.g. Lewis 1978; Struhl & Brower 1982). The role of *Pc* and *esc* in keeping *Ubx* and other genes off where they should be off [as was suggested by Denell & Frederick (1983)] could be related to the interactions among different BX-C (and other) genes. For example, the negative effects of *abdA* and *AbdB* functions on *Ubx* expression could involve some sort of cooperation with *Pc* or *esc* products. Since many other genes superficially similar to *Pc* and *esc* have been found (Duncan 1982; Ingham 1984; Ingham & Whittle 1980; Jürgens 1985; Dura et al 1985), it will be important to ascertain their different roles. If the other genes, like *Pc* and *esc*, are not involved in the *initiation* of position-specific homeotic gene expression, then which genes are? Some of the complexities of position-specific homeotic gene expression are due to interactions among homeotic genes, but the anterior limits of *Antp* or *Ubx* expression have not been found to be prescribed by any known homeotic gene. Perhaps it is the segmentation genes that are involved in the initiation of position-specific homeotic gene expression. Support for this idea comes from the observation that the pattern of expression of homeotic genes can be altered by mutations in segmentation genes.

The effects of segmentation genes on homeotic gene expression have been shown both by genetic analysis (Weiner et al 1984; Duncan 1986) and by immunofluorescence studies (S. DiNardo & P. H. O'Farrell, S. B. Carroll & M. P. Scott, unpublished data). For example, the ftz^{Rpl} allele causes a dominant mutant phenotype in which posterior halteres are transformed into posterior wings. The haltere-to-wing transformation is also caused by certain BX-C mutations. Several other *ftz* alleles have been isolated that cause homeotic transformations similar to those that result from certain BX-C mutations (Duncan 1986). The action of *ftz* upon *Ubx* may account for the pair-rule modulation of *Ubx* RNA accumulation seen transiently during formation of the germ band (Akam 1985). Thus, certain mutations in a segmentation gene affect determination events regulated by the bithorax complex.

A second example is provided by the effects of the *hunchback* (*hb*) gap locus (Lewis 1968; Lehmann & Nüsslein-Volhard 1986; M. Bender et al, submitted) on the expression of *Ubx* and *Antp*. Certain *hb* alleles cause transformation of the head and thorax into abdomen (Lehmann & Nüss-lein-Volhard 1986). In agreement with these observations, in the absence of hb^+ function Ubx and $Antp$ are expressed far more anteriorly in the embryo than they normally are (S. B. Carroll & M. P. Scott, unpublished data). Therefore, hb is one gene involved in preventing $Antp$ and Ubx expression in anterior regions; other gap loci may also be involved in regulating homeotic gene expression.

An important issue is how directly the interacting genes are connected. Are there many intervening steps, for example, between the expression of the *Ultrabithorax* product(s) and the reduction in the transcription of *Antennapedia* in the abdominal region where both genes are expressed? The rapidity of the responses of some genes to others suggests direct interactions. For example, the *fushi tarazu* protein product is first detectable just after the cell membranes are completed to form the cellular blastoderm embryo. The *engrailed* protein product is first observed only minutes later, and the earliest *en* pattern seen is dependent on whether *ftz* has been active (S. DiNardo & P. H. O'Farrell, unpublished data). Similarly, the pattern of expression of *ftz* is responsive to the function of the *hairy* (*h*) gene even though transcription of the two genes is activated nearly simultaneously (Carroll & Scott 1986; Ingham et al 1985; Howard & Ingham 1986). The timing of these effects suggests that a long cascade of interactions intervening between the products of one gene and the transcriptional control of the other is unlikely, but of course a few fast steps could occur in the time available.

CELLS RESPOND TO THEIR NEIGHBORS

The phenomenon of developmental induction suggests that cell communication plays a major role in organizing morphogenesis. For example, in chick wing development a thicker region of ectoderm (the apical ecto-dermal ridge) induces the development of the underlying mesoderm. Although in insects we know much less about cell communication, it is clear that it plays a role in development. In grasshopper and in *Drosophila*, neuroblast cells that are precursors of the central nervous system arise by the differentiation of a minority population of cells in the ventral ectoderm. Cell ablation experiments with embryos have shown that developing neuroblast cells prevent the surrounding ectodermal cells from following a similar developmental path (Doe & Goodman 1985).

Regeneration experiments in many organisms including insects (Bryant et al 1981) have uncovered several features that suggest that local communication plays a major role in the restoration of pattern. Only cells neighboring a wound respond. The responding cells appear to recognize that they are missing neighbors, and the new cells produced during regeneration are appropriate neighbors for the responding cells. However, in a frequent type of mistake, the wrong neighbors are produced. Rather than restoring their missing neighbors, the responding cells duplicate the neighbors that were not damaged by the wound. This type of error leads to a duplication of pattern elements in reverse polarity (French et al 1976). Its prevalence in regeneration experiments argues that regeneration is guided by local interactions rather than long-range signals specifying the polarity of a structure.

One of the major conclusions derived from studies of the patterns of expression of developmental genes is that segmentation and homeotic gene products act in the cells in which they are expressed to determine the developmental fates of cells. This conclusion was suggested by the correspondence between the positions of expression of particular developmental genes and the positions of morphological alterations in mutants defective in these genes. However, this correspondence is not perfect and actually should not be expected to be perfect. In mutants, final morphology might be expected to be modified by regenerative responses provoked by the deficiency of pattern elements.

The segment polarity mutations cause defects in all of the segments. In general, part of the normal pattern is missing from each segment and the remaining structures are duplicated in mirror image. Each of the nine identified genes in the segment polarity class affects different parts of the segments, and each has a distinct plane of mirror-image symmetry, but otherwise the mutant phenotypes are similar in format. Although nothing is known about the products specified by the segment polarity genes, it is plausible that they are regulators that specify the fate of different parts of each segment. In the regeneration studies described above and others (e.g. Lawrence 1973), it was observed that cells are sensitive to whether they have appropriate cell neighbors. This suggests how the phenotypes of segment polarity mutations may arise (Nüsslein-Volhard & Wieschaus 1980). In a mutant unable to specify a particular cell type, the only way for all cells to have only normal neighbors is to produce a polarity reversal pattern. Polarity reversals are not exclusively the result of mutations in segment polarity genes; the Kr gap locus, the runt pair-rule locus (Nüsslein-Volhard & Wieschaus 1980), and the bicaudal maternal effect locus (Nüsslein-Volhard 1979) also have polarity reversal aspects of their phenotypes.

Two kinds of cell-cell interactions may be important for development of segments. First, molecules involved in forming gradients spreading through the embryo might originate from cells in one part of the embryo.

In this case, mutations in genes encoding the gradient molecules would have an effect on the whole embryo even if only the expressing cells were mutant. A second possibility is that local cell-cell interactions of the sort observed in insect neural development (Doe & Goodman 1985) are important for formation of the segmental pattern. Mutations in genes encoding molecules involved in such local interactions would be expected to be largely, but not completely, cell autonomous. In genetically mosaic embryos having mutant and wild-type cells, an abnormal phenotype should be seen in some of the wild-type cells located near mutant cells as well as in regions of mutant cells.

Many studies have demonstrated cell autonomous behavior of homeotic genes (e.g. Garcia-Bellido & Lewis 1976; Morata & Garcia-Bellido 1976; but see Struhl 1981a for a possible exception). However, only a few studies have examined the cell autonomy of segmentation genes (Morata & Lawrence 1977; Gergen & Wieschaus 1985, 1986a). The difficulties in examining segmentation gene cell autonomy are in generating genetic mosaics at a very early stage and in obtaining visible embryonic marker mutations that distinguish the different parts of the mosaic. In the studies to date, one pair rule locus (*runt*), one gap locus (*giant*), three segment polarity loci (*wingless*, *armadillo*, and *fused*), and one unclassified locus in which mutations cause segment deletions (*unpaired*) have been examined. In each case except *wingless*, the mosaics showed that the effects of the genes were primarily cell autonomous. The *wingless* mutation is *not* cell autonomous during the development of adult cuticular structures: Mutant clones in wild-type wings have a wild-type phenotype (Morata & Lawrence 1977). *wingless* has not yet been examined for cell autonomy in embryos. In the other cases, embryonic autonomy was observed: Mutant cells had mutant phenotypes, and wild-type cells did not. No long-range effects of mutant cells on wild-type cells, or vice versa, were observed. However, short-range nonautonomous effects were observed in embryo mosaics of some of the genes. For example, in a subset of the *runt* mosaics, some genotically wild type cells adjacent to mutant *runt* cells formed non-wild-type patterns (Gergen & Wieschaus 1985), which suggests that cells are responding to their neighbors *locally*. Similar possible cases of local nonautonomy were observed for *giant* and *unpaired* (Gergen & Wieschaus 1986a). These results do not necessarily imply that the *runt* (or *giant* or *unpaired*) product diffuses from cell to cell, or even that it is a cell surface molecule. The lack of a particular regulator in a mutant may lead to a cascade of effects, some of which may alter cell-cell interactions. Thus, even if a gene encodes, for example, a transcription factor, mutations in the gene could indirectly affect genetically wild-type cells located nearby by acting upon other genes involved in cell-cell communication. It will be important to determine which segmentation genes are involved, directly or indirectly, in cell-cell interactions.

A DIGITAL MODEL OF PATTERN FORMATION

There remains a conceptual challenge. How can the new observations explain emergence of biological pattern? Data summarized in this review have revealed some fundamental steps responsible for pattern formation. It seems important to attempt to make some order out of what may seem to be a complex morass of data. We therefore present a schematic model here that is intended to highlight possible general principles in the creation of pattern. Its important points are: (a) pattern formation is controlled by the localized expression of a number of regulatory gene products; (b) pattern formation involves a series of sequential subdivisions of the embryo; (c) combinatorial control by a battery of genes leads to the induction of new gene activities in a regulatory cascade (Garcia-Bellido 1977; Lewis 1978; Kauffman 1981); (d) more than one combination of regulatory genes can lead to the activation of a particular "downstream" regulator; (e) a series of induction events is not itself sufficient to refine the pattern; and (f) cell-cell communication is likely to play an important part in pattern refinement.

Pattern formation appears to occur as a sequential process. Thus, a coarse early pattern (the gap pattern) is resolved into a finer (the pair-rule pattern) and finer pattern (the segmental pattern). Spatial information along the dorsal-ventral axis may be similarly resolved through a sequence of subdivisions. The greatest amount of information is available about the pair-rule to segmental unit transition. Like the observed patterns of expression of *ftz* and *h*, most or all of the other six pair-rule gene products are likely to show localization to stripes with a periodicity of every other segment. However, since the spatially periodic defects of pair-rule mutants are offset from one another, we expect the stripes of expression of these gene products to be displaced with respect to one another (Figure 4a) (Ingham et al 1985; Ish-Horowicz et al 1985). This means that each cell along the anterior-posterior axis will express a combination of regulators that distinguishes it from neighboring cells just ahead or just behind. It is then just a small step to the idea that certain combinations of pair rule gene products may induce or repress certain "downstream" target genes (transition from Figure 4a to 4b). The observations, described above, on the induction of the segment polarity gene *engrailed* by the genes *ftz* and *prd* support the idea that regulation is combinatorial.

The pair-rule gene products regulate the expression of *engrailed* (*en*) to produce a pattern that is finer (14 stripes) than the spatial patterns of expression of pair-rule gene products (e.g. 7 stripes of *ftz*) (Figure 2). How can a coarse pattern direct the expression of a more detailed pattern of expression? As is discussed above, *ftz* appears to be required to induce only the even-numbered *en* stripes, while *prd* is required only for the odd-numbered stripes. One important observation is therefore that the same outcome (in this case *en* activation) can result from either of two different signals or combinations of signals. Together, the different signals produce two stripes of *en* expression within each pair-rule repeat interval. This principle is shown schematically in Figure 4: the expression of gene D is induced either by the combination of genes 6, 7, and 8 "on," or by genes 2, 3, and 4 "on."

In an induction step an existing group of spatially localized regulators controls expression of the next group of regulators in a step of pattern refinement. Since pattern establishment appears to proceed stepwise, it is tempting to suggest that a repetition of induction steps may be responsible for sequential pattern refinement. However, repetition of induction steps alone is inadequate because these steps do not refine pattern. An induction step, such as that in Figure 4, is merely a process of "reading out" an existing prepattern, because it relies on the fact that each cell is *already* distinguished from its neighbor by its particular combination of regulators. Thus, although an induction step can subdivide a periodic pattern once, it does not increase the detail of the pattern, and most importantly, the subdivision process cannot be repeated directly. This is illustrated by the

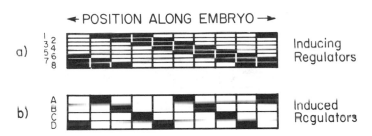

Figure 4   Combinatorial action of genes expressed in overlapping regions: Induction of a more finely divided pattern of gene expression. (*a*) In this hypothetical example, the genes 1–8 are expressed in offset, overlapping repetitive patterns along the embryo. The vertical divisions demarcate arbitrary spatial units (cells or groups of cells). Blacked-in boxes indicate places where genes are on. Each gene is expressed in three spatial units repeating once every eight units. (*b*) The genes A through D are induced by particular combinations of genes 1–8, in a spatially specific way. For example, expression of gene A is induced either by genes 1, 7, and 8 being on, or by genes 3, 4, and 5 being on. Each of the genes A–D is expressed in a single spatial unit at an interval that repeats once every four units. These genes are therefore expressed in twice as many repeats as are genes 1–8.

pattern of distribution of regulators in Figure 4b; this pattern cannot be further subdivided by another induction step. We next discuss how an "induction step" can be combined with a different sort of event (a subdivision step) to create a two-step process that creates finer pattern elements. Moreover, repetition of this two-step process of induction and subdivision can sequentially refine the pattern, whereas repetition of either step alone cannot.

To refine pattern information, a process must subdivide a region (a "spatial unit") across which regulators are initially uniformly distributed. Cell-cell interactions that cause changes in gene expression could provide the information required for this subdivision step. Although there are many ways that local interactions could subdivide spatial units, only one hypothetical format is described here.

We begin by suggesting that when a regulator is induced, a second (partner) regulator is also induced by the same signals so that two regulators are expressed in each spatial unit (Figure 5a). Although initially coincident, the distributions of a regulator and its partner are changed by cell (or, at a syncitial stage, nuclear) interactions. For example, expression of one regulator could be locally suppressed by interactions with cells to the anterior, while continued expression of is partner regulator could be

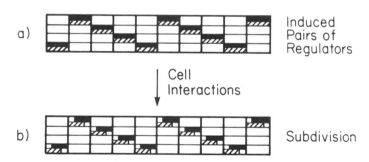

Figure 5 Schematic illustration of how cell interactions can subdivide spatial units of expression. The layout is the same as in Figure 4; gene expression is indicated by either black or shaded blocks. However, in contrast to Figure 4b, two regulators are induced in each repeating unit: a regulator (*black*) and a partner regulator (*shaded*). Although each regulator is initially expressed in the same pattern as its partner, cell interactions cause a "retreat" of each regulator into only part of the original region of expression, as shown in b. Such a "subdivision" step causes different parts of each spatial unit to have a different combination of regulators: just one, just the other, or (in the center) both. Therefore, a new round of induction events could occur, just as in Figure 4, but at a finer level of pattern. In the diagram shown, neighboring cells have a negative influence on expression of the regulator or its partner, but a formally similar model could be drawn in which the interactions are positive: A neighboring cell could induce an additional regulator rather than repressing a previously activated regulator.

locally suppressed by interactions with cells to the posterior. As a result, spatial units would be subdivided to create three zones, one expressing only one regulator, one expressing both this regulator and its partner, and one expressing only the partner regulator (Figure 5*b*). Data that suggest a model of this type include the observed refinement of expression of *hairy* and *fushi tarazu* from solid regions to stripes (Hafen et al 1984a; Ingham et al 1985) and the shrinkage of *ftz* stripes from four cells wide to three (Carroll & Scott 1985). As is illustrated in Figure 6, if steps of induction of new groups of localized regulators alternate with steps of subdivision of spatial units throughout development, this process can refine a pattern to any extent required.

The model shown in Figures 5 and 6 offers a way in which the spatial patterns of expression of homeotic genes might be regulated. The expression of homeotic genes begins at a stage when segmentation genes are expressed in a periodic pattern along the axis of the embryo. Since each homeotic gene is expressed only in particular segments (or parasegments), there must be a means of distinguishing each of the periodic repeats. The

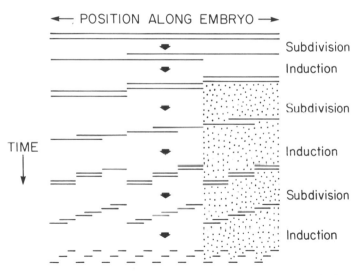

Figure 6   Reiteration of induction steps alternating with subdivision steps can lead to pattern refinement. A series of alternating induction and subdivision events using the principles diagrammed in Figures 4 and 5 is shown. For simplicity, each of the subdivision steps is shown as opposite "retreats" of a regulator and its partner in response to neighboring cells. However, other sorts of negative or positive interactions could be used for each of the subdivision steps. The stippled region indicates that some of the repeating units produced by the last induction step can be distinguished from other repeats if one of the early regulators persists and affects later events.

model suggests that if the regulators made at earlier times were to linger, the final combination of regulators could encode distinct identities for each repeat unit. For example, the broad early expression of one regulator could persist to distinguish repeating units, as indicated by the stippled region in Figure 6. Homeotic gene expression could be appropriately switched on or off in particular repeat units by such a regulatory system. The effects of the *hb* gap locus on homeotic gene expression, described above, are consistent with the idea that early acting regulators also affect later events.

This model proposes that local cell-cell communication is involved in pattern formation. Although there are a number of indications that local communication is involved in *Drosophila* development (see the previous section), the mechanism and even the properties of this communication remain largely unknown. A good deal of the pattern is laid down before there are any cells, so that cell membranes that separate nuclei cannot be required for the earliest parts of the process. Alternatively, a spatial unit could be subdivided by cell lineage mechanisms. These are particularly interesting because embryonic pattern development of some organisms is strongly dependent on lineage. A cell lineage mechanism can have substantial homology to the digital model already discussed. Thus, in an induction step, a combination of regulators in a particular cell could induce two new regulators (regulator and partner). To subdivide spatial units, mitosis could segregate two daughter cells (or, in early *Drosophila* embryos, two daughter nuclei), one producing one regulator and the other its partner regulator. Which daughter cell received which regulator could be controlled by the polarity of the parent nucleus or cell, the polarization being the result of earlier pattern-forming events. The elaborate cell cycle regulation of the yeast *HO* gene (Strathern & Herskowitz 1979; Nasmyth 1985) has many of the properties required by the proposed cell lineage paradigm.

We describe the model presented here as digital because it proposes a series of discrete subdivisions. This emphasizes its distinction from gradient models, in which a cell defines its position by interpreting the concentration of a morphogen, an analog step. The digital model is consistent with the resilience of the pattern formation process. Because interactions are local, the outcome will not easily be disturbed by experimental spatial distortions of the embryo (Vogel 1982a,b). The types of cell-cell interactions suggested here could be similar to those thought to account for regeneration phenomena (French et al 1976), as proposed by Gergen et al (1985).

It should be clear that the schematic diagram (Figure 6) is meant only as an aid to illustrate the types of processes that may be involved in creating pattern. Much more data will be required to define the actual regulatory circuitry so that such a figure could be transformed into a diagram of relationships between known genes. The number of components in the process, and whether particular interactions are positive or negative, could vary between stages of development, between tissues, and between organisms. Each step in pattern refinement could employ different sorts of interactions. Both positive and negative influences of cells on the gene expression of neighboring cells could be involved. Several developmental genes appear to have independent programs of expression in *Drosophila* epidermal ectoderm and in neural ectoderm. Most dramatically, the *ftz* gene is expressed twice, once in 7 stripes in the epidermal precursor cells and once in 15 repeating units in the developing nervous system (Carroll & Scott 1985). Thus, some of the pattern-forming machinery may be used more than once during development.

The events and mechanisms revealed by the above studies are unlikely to be unique to insect development. Hopefully, some of the *Drosophila* genes involved in pattern formation have functional analogs in other organisms, despite the extraordinary and beautiful diversity of developmental pathways. Thus, although much remains to be done, the studies described in this review may have wide-ranging implications for our understanding of the molecular basis of pattern-formation mechanisms in all organisms.

ACKNOWLEDGMENTS

We are very grateful to our many colleagues who provided copies of their published and unpublished work. The large amount of material necessitated a narrower focus than was originally planned, and we apologize to those whose research could not be discussed within the limits of the space available. We would also like to thank the members of our laboratories and Dr. William Wood for comments on the manuscript. Cathy Inouye provided superb assistance with the preparation of the manuscript, for which we thank her. Research in our laboratories is supported by grants from the National Institutes of Health and the National Science Foundation. In addition, MPS acknowledges the support of the Searle Scholars program.

Literature Cited

Akam, M. 1983. The location of *Ultrabithorax* transcripts in *Drosophila* tissue sections. *EMBO J.* 2: 2075–84

Akam, M. E. 1985. Segments, lineage boundaries and the domains of expression of homeotic genes. *Philos. Trans. R. Soc. London Ser. B.* In press

Akam, M. E., Martinez-Arias, A. 1985. The distribution of *Ultrabithorax* transcripts in *Drosophila* embryos. *EMBO J.* 4: 1689–1700

Anderson, K. V., Nüsslein-Volhard, C. 1984. Information for the dorsal-ventral pattern of the *Drosophila* embryo is stored as maternal mRNA. *Nature* 311: 223–27

Beachy, P. A., Helfand, S. L., Hogness, D. S.

1985. Segmental distribution of bithorax complex proteins during *Drosophila* development. *Nature* 313 : 545–51

Belote, J. M., Handler, A. M., Wolfner, M. F., Livak, K. J., Baker, B. S. 1985. Sex-specific regulation of yolk protein gene expression in *Drosophila. Cell* 40 : 339–48

Bender, W., Akam, M., Karch, F., Beachy, P. A., Peifer, M., et al. 1983. Molecular genetics of the bithorax complex in *Drosophila melanogaster. Science* 221 : 23–29

Blumenthal, A. B., Kriegstein, H. J., Hogness, D. S. 1973. The units of DNA replication in *Drosophila melanogaster* chromosomes. *Cold Spring Harbor Symp. Quant. Biol.* 38 : 205–23

Brower, D. 1985. The sequential compartmentalization of *Drosophila* segments revisited. *Cell* 41 : 361–64

Bryant, S. V., French, V., Bryant, P. J. 1981. Distal regeneration and symmetry. *Science* 212 : 993–1002

Cabrera, C. V., Botas, J., Garcia-Bellido, A. 1985. Distribution of *Ultrabithorax* proteins in mutants of *Drosophila* bithorax complex and its transregulatory genes. *Nature* 318 : 569–71

Carroll, S. B., Scott, M. P. 1985. Localization of the *fushi tarazu* protein during *Drosophila* embryogenesis. *Cell* 43 : 47–57

Carroll, S. B., Scott, M. P. 1986. Zygotically-active genes that affect the spatial expression of the *fushi tarazu* segmentation gene during early *Drosophila* embryogenesis. *Cell* 45 : 113–26

Carroll, S. G., Winslow, G. A., Schüpbach, T., Scott, M. P. 1986. Maternal control of *Drosophila* segmentation gene expression. *Nature*. In press

Degelmann, A., Hardy, P. A., Perrimon, N., Mahowald, A. P. 1986. Developmental analysis of the torso-like phenotype in *Drosophila* produced by a maternal effect locus. *Dev. Biol.* 115 : 479–89

Denell, R. E., Frederick, R. D. 1983. Homeosis in *Drosophila*: A description of the Polycomb lethal syndrome. *Dev. Biol.* 97 : 34–47

Desplan, C., Theis, J., O'Farrell, P. H. 1985. The *Drosophila* developmental gene, *engrailed*, encodes a sequence-specific DNA binding activity. *Nature* 318 : 630–35

DiNardo, S., Kuner, J. M., Theis, J., O'Farrell, P. H. 1985. Development of embryonic pattern in *D. melanogaster* as revealed by accumulation of the nuclear *engrailed* protein. *Cell* 43 : 59–69

Doe, C. Q., Goodman, C. S. 1985. Early events in insect neurogenesis II. The role of cell interactions and cell lineage in the determination of neuronal precursor cells. *Dev. Biol.* 111 : 206–19

Duncan, I. M. 1982. Polycomblike: A gene

that appears to be required for the normal expression of the bithorax and Antennapedia gene complexes of *Drosophila melanogaster. Genetics* 102 : 49–70

Duncan, I. M. 1986. Control of bithorax complex functions by the segmentation gene *fushi tarazu* of *Drosophila melanogaster. Cell.* In press

Duncan, I. M., Lewis, E. B. 1982. Genetic control of body segment differentiation in *Drosophila*. In *Developmental Order: Its Origin and Regulation*, ed. S. Subtelny, pp. 533–54. New York: Liss

Dura, J.-M., Brock, H. W., Santamaria, P. 1985. Polyhomeotic: A gene of *Drosophila melanogaster* required for correct expression of segmental identity. *Mol. Gen. Genet.* 198 : 213–20

Edgar, B. A., Kiehle, C. P., Schubiger, G. 1986. Cell cycle control by the nucleocytoplasmic ratio in early *Drosophila* development. *Cell* 44 : 365–72

Edgar, B. A., Schubiger, G. 1986. Parameters controlling transcriptional activation during early *Drosophila* development. *Cell* 44 : 871–77

Fjose, A., McGinnis, W. J., Gehring, W. J. 1985. Isolation of a homeobox-containing gene from the *engrailed* region of *Drosophila* and the spatial distribution of its transcript. *Nature* 313 : 284–89

Foe, V. E., Alberts, B. 1983. Studies of nuclear and cytoplasmic behavior during the five mitotic cycles that precede gastrulation in *Drosophila* embryogenesis. *J. Cell Sci.* 61 : 31–70

French, V., Bryant, P. J., Bryant, S. V. 1976. Pattern regulation in epimorphic fields. *Science* 193 : 969–81

Garber, R. L., Kuroiwa, A., Gehring, W. J. 1983. Genomic and cDNA clones of the homeotic locus *Antennapedia* in *Drosophila. EMBO J.* 2 : 2027–36

Garcia-Bellido, A. 1975. Genetic control of wing disc development in *Drosophila*. In *Cell Patterning. CIBA Found. Symp.* 29 : 161–82

Garcia-Bellido, A. 1977. Homeotic and atavic mutations in insects. *Am. Zool.* 17 : 613–29

Garcia-Bellido, A., Lewis, E. B. 1976. Autonomous cellular differentiation of homeotic bithorax mutants of *Drosophila melanogaster. Dev. Biol.* 48 : 400–10

Garcia-Bellido, A., Ripoll, P., Morata, G. 1976. Developmental compartmentalization in the dorsal mesothoracic disc of *Drosophila. Dev. Biol.* 48 : 132–47

Garcia-Bellido, A., Santamaria, P. 1972. Developmental analysis of the wing disc in the mutant *engrailed* of *Drosophila melanogaster. Genetics* 72 : 87–104

Gergen, J. P., Coulter, D., Wieschaus, E.

1985. Segmental pattern and blastoderm cell identities. *Symp. Soc. Dev. Biol.* 43: In press

Gergen, J. P., Wieschaus, E. 1985. The localized requirements for a gene affecting segmentation in *Drosophila*: Analysis of larvae mosaic for *runt*. *Dev. Biol.* 109: 321–35

Gergen, J. P., Wieschaus, E. 1986a. Localized requirements for gene activity in segmentation of *Drosophila* embryos: Analysis of *armadillo*, *fused*, *giant*, and *unpaired* mutations in mosaic embryos. *Wilhelm Roux Arch. Entwicklungsmech. Org.* 195: 49–62

Gergen, J. P., Wieschaus, E. 1986b. Dosage requirements for *runt* in the segmentation of *Drosophila* embryos. *Cell* 45: 289–99

Ghysen, A., Dambly, C., Aceves, E., Jan, L. Y., Jan, Y. N. 1986. Sensory neurons and peripheral pathways in *Drosophila* embryos. *Wilhelm Roux Arch. Entwicklungsmech. Org.* In press

Hafen, E., Kuroiwa, A., Gehring, W. J. 1984a. Spatial distribution of transcripts from the segmentation gene *fushi tarazu* of *Drosophila*. *Cell* 37: 825–31

Hafen, E., Levine, M., Gehring, W. J. 1984b. Regulation of *Antennapedia* transcript distribution by the *bithorax* complex in *Drosophila*. *Nature* 307: 287–89

Harding, K., Wedeen, C., McGinnis, W., Levine, M. 1985. Spatially regulated expression of homeotic genes in *Drosophila*. *Science* 229: 1236–42

Hartenstein, V., Campos-Ortega, J. A. 1984. Early neurogenesis in wild-type *Drosophila melanogaster*. *Wilhelm Roux Arch. Entwicklungsmech. Org.* 193: 308–25

Hartenstein, V., Campos-Ortega, J. A. 1985. Fate-mapping in wild-type *Drosophila melanogaster* I. The spatio-temporal pattern of eukaryotic cell division. *Wilhelm Roux Arch. Entwicklungsmech. Org.* 194: 181–95

Hartenstein, V., Technau, G. M., Campos-Ortega, J. A. 1985. Fate mapping in wild-type *Drosophila melanogaster*. III. A fate map of the blastoderm. *Wilhelm Roux Arch. Entwicklungsmech. Org.* 194: 213–16

Hegner, R. W. 1911. Experiments with Chrysomelid beetles. III. The effect of killing parts of the eggs of *Leptinotarsa decemlineata*. *Biol. Bull.* 10: 237–51

Hough-Evans, B. R., Jacobs-Lorena, M., Cummings, M. R., Britten, R. J., Davidson, E. H. 1980. Complexity of RNA in eggs of *Drosophila melanogaster* and *Musca domestica*. *Genetics* 95: 81–94

Howard, K., Ingham, P. W. 1986. Regulatory interactions between the segmentation genes *fushi tarazu*, *hairy*, and

engrailed in the *Drosophila* blastoderm. *Cell* 44: 949–57

Illmensee, K. 1976. Nuclear and cytoplasmic transplantation in *Drosophila*. In *Insect Development*, ed. P. A. Lawrence, pp. 76–96. London: Halsted

Illmensee, K., Mahowald, A. P. 1974. Transplantation of posterior pole plasm in *Drosophila*. Induction of germ cells at the anterior pole of the egg. *Proc. Natl. Acad. Sci. USA* 71: 1016–20

Ingham, P. W. 1983. Differential expression of bithorax complex genes in the absence of the *extra sex combs* and *trithorax* genes. *Nature* 306: 591–93

Ingham, P. W. 1984. A gene that regulates the bithorax complex differentially in larval and adult cells of *Drosophila*. *Cell* 37: 815–23

Ingham, P. W., Howard, K. R., Ish-Horowicz, D. 1985. Transcription pattern of the *Drosophila* segmentation gene *hairy*. *Nature* 318: 439–45

Ingham, P., Whittle, R. 1980. Trithorax: A new homeotic mutation of *Drosophila melanogaster* causing transformations of abdominal and thoracic segments. *Mol. Gen. Genet.* 179: 607–14

Ish-Horowicz, D., Howard, K. R., Pinchin, S. M., Ingham, P. W. 1985. Molecular and genetic analysis of the *hairy* locus in *Drosophila*. *Cold Spring Harbor Symp. Quant. Biol.* 50: 135–44

Jacobs-Lorena, M., Hough-Evans, B. R., Britten, R. J., Davidson, E. H. 1980. Complexity of RNA in developing oocytes of *Drosophila melanogaster*. *Dev. Biol.* 76: 509–13

Jan, Y. N., Ghysen, A., Christoph, I., Barbel, S., Jan, L. Y. 1985. Formation of neuronal pathways in the imaginal discs of *Drosophila melanogaster*. *J. Neurosci.* 5: 2453–64

Jürgens, G. 1985. A group of genes controlling the spatial expression of the bithorax complex in *Drosophila*. *Nature* 316: 153–55

Jürgens, G., Wieschaus, E., Nüsslein-Volhard, C., Kluding, H. 1984. Mutations affecting the pattern of the larval cuticle in *Drosophila melanogaster*. II. Zygotic loci on the third chromosome. *Wilhelm Roux Arch. Entwicklungsmech. Org.* 193: 283–95

Kalthoff, K. 1976. Specification of the anterio-posterior body pattern in insect eggs. In *Insect Development*, ed. P. A. Lawrence. New York: Wiley. 230 pp.

Kalthoff, K. 1983. Cytoplasmic determinants in dipteran eggs. In *Time, Space and Pattern in Embryonic Development*, ed. W. R. Jeffery, R. A. Raff, pp. 313–48. New York: Liss

78   SCOTT & O'FARRELL

Kalthoff, K., Rau, K.-G., Edmond, J. C. 1982. Modifying effects of ultraviolet irradiation on the development of abnormal body patterns in centrifuged insect embryos. (*Smittia* sp., Chironomidae, Diptera). *Dev. Biol.* 91 : 413–22

Karch, F., Weiffenbach, B., Piefer, M., Bender, W., Duncan, I., et al. 1985. The abdominal region of the bithorax complex. *Cell* 43 : 81–96

Karr, T. L., Ali, Z., Drees, B., Kornberg, T. 1985. The *engrailed* locus of *D. melanogaster* provides an essential zygotic function in precellular embryos. *Cell* 43 : 591–601

Kauffman, S. A. 1981. Pattern formation in the *Drosophila* embryo. *Philos. Trans. R. Soc. London Ser. B* 295 : 567–94

Kaufman, T. C., Lewis, R., Wakimoto, B. 1980. Cytogenetic analysis of chromosome 3 in *Drosophila melanogaster*: The homeotic gene complex in polytene chromosome interval 84A,B. *Genetics* 94 : 115–33

Knipple, D. C., Seifert, E., Rosenberg, U. B., Preiss, A., Jäckle, H. 1985. Spatial and temporal patterns of *Krüppel* gene expression in early *Drosophila* embryos. *Nature* 317 : 40–44

Kornberg, T. 1981a. Compartments in the abdomen of *Drosophila* and the role of the *engrailed* locus. *Dev. Biol.* 86 : 363–81

Kornberg, T. 1981b. *engrailed*: A gene controlling compartment and segment formation in *Drosophila*. *Proc. Natl. Acad. Sci. USA* 78 : 1095–99

Kornberg, T., Siden, I., O'Farrell, P., Simon, F. 1985. The *engrailed* locus of *Drosophila*: In situ localization of transcripts reveals compartment-specific expression. *Cell* 40 : 45–53

Kuwada, J. Y., Goodman, C. S. 1985. Neuronal determination during embryonic development of the grasshopper nervous system. *Dev. Biol.* 110 : 114–26

Laughon, A., Scott, M. P. 1984. Sequence of a *Drosophila* segmentation gene: Protein structure homology with DNA-binding proteins. *Nature* 310 : 25–31

Lawrence, P. A. 1973. The development of spatial patterns in the integument of insects. In *Developmental Systems: Insects*, eds. S. J. Counce, C. H. Waddington, 1 : 157–209. New York/London : Academic

Lawrence, P. A., ed. 1976. *Insect Development*. London : Halsted

Lawrence, P. A., Morata, G. 1976. Compartments in the wing of *Drosophila*: A study of the *engrailed* gene. *Dev. Biol.* 50 : 321–37

Lehmann, R. 1985. Regionsspezifische Segmentierungsmutanten bei *Drosophila*

melanogaster meigen. PhD dissertation, Univ. Tübingen

Lehmann, R., Nüsslein-Volhard, C. 1986. *hunchback*, a gene required for segmentation of an anterior and posterior region of the *Drosophila* embryo. *Dev. Biol.* In press

Levine, M., Hafen, E., Garber, R. L., Gehring, W. J. 1983. Spatial distribution of *Antennapedia* transcripts during *Drosophila* development. *EMBO J.* 2 : 2037–46

Levine, M., Harding, K., Wedeen, C., Doyle, H., Hoey, T., Radomska, H. 1985. Expression of the homeobox gene family in *Drosophila.* *Cold Spring Harbor Symp. Quant. Biol.* 50 : 209–22

Lewis, E. B. 1968. Genetic control of developmental pathways in *Drosophila melanogaster*. *Proc. 12th Int. Congr. Genet.* 2 : 96–97

Lewis, E. B. 1978. A gene complex controlling segmentation in *Drosophila*. *Nature* 276 : 565–70

Lohs-Schardin, M., Cremer, C., Nüsslein-Volhard, C. 1979. A fate map for the larval epidermis of *Drosophila melanogaster*: Localized defects following irradiation of the blastoderm with an ultraviolet laser microbeam. *Dev. Biol.* 73 : 239–55

Mahowald, A. P., Hardy, P. A. 1985. Genetics of *Drosophila* embryogenesis. *Ann. Rev. Genet.* 19 : 149–77

Martinez-Arias, A. 1986. The *Antennapedia* gene is required and expressed in parasegments 4 and 5 of the *Drosophila* embryo. *EMBO J.* 5 : 135–41

Martinez-Arias, A., Lawrence, P. A. 1985. Parasegments and compartments in the *Drosophila* embryo. *Nature* 313 : 639–42

McGinnis, W., Garber, R. L., Wirz, J., Kuroiwa, A., Gehring, W. J. 1984. A homologous protein-coding sequence in *Drosophila* homeotic genes and its conservation in other metazoans. *Cell* 37 : 403–8

McKnight, S. L., Miller, O. L. Jr. 1976. Ultrastructural patterns of RNA synthesis during early embryogenesis of *Drosophila melanogaster*. *Cell* 8 : 305–19

Meinhardt, H. 1982. *Models of Biological Pattern Formation*. New York : Academic. 230 pp.

Meinhardt, H., Gierer, A. 1980. Generation and regeneration of sequences of structures during morphogenesis. *J. Theor. Biol.* 85 : 429–50

Mohler, J., Wieschaus, E. 1985. *Bicaudal* mutations of *Drosophila melanogaster*: Alteration of blastoderm cell fate. *Cold Spring Harbor Symp. Quant. Biol.* 50 : 105–12

Morata, G., Garcia-Bellido, A. 1976. Developmental analysis of some mutants of the bithorax system of *Drosophila*. *Wilhelm Roux Arch. Entwicklungsmech. Org.* 179: 125–43

Morata, G., Lawrence, P. A. 1977. The development of *wingless*, a homeotic mutation of *Drosophila*. *Dev. Biol.* 56: 227–40

Nasmyth, K. 1985. At least 1400 base pairs of 5' flanking DNA is required for the correct expression of the HO gene in yeast. *Cell* 42: 213–23

Newport, J., Kirschner, M. 1982. A major developmental transition in early *Xenopus* embryos: I. Control of the onset of transcription. *Cell* 30: 687–96

Nüsslein-Volhard, C. 1979. Maternal effect mutations that alter the spatial coordinates of the embryo of *Drosophila melanogaster*. In *Determinants of Spatial Organization*, ed. S. Subtelny, I. R. Konigsberg, pp. 185–211. New York: Academic

Nüsslein-Volhard, C., Wieschaus, E. 1980. Mutations affecting segment number and polarity in *Drosophila*. *Nature* 287: 795–801

Nüsslein-Volhard, C., Wieschaus, E., Kluding, H. 1984. Mutations affecting the pattern of the larval cuticle in *Drosophila melanogaster*. I. Zygotic loci on the second chromosome. *Wilhelm Roux Arch. Entwicklungsmech. Org.* 193: 267–82

Okada, M., Kleinman, I. A., Schneiderman, H. A. 1974. Restoration of fertility in sterilized *Drosophila* eggs by transplantation of polar cytoplasm. *Dev. Biol.* 37: 43–54

Poulson, D. F. 1950. Histogenesis, organogenesis, and differentiation in the embryo of *Drosophila melanogaster* meigen. In *The Biology of* Drosophila, ed. M. Demerec, pp. 168–274. New York: Wiley

Raff, R. A., Kaufman, T. C. 1983. *Embryos, Genes, and Evolution*. New York: Macmillan. 395 pp.

Rau, K.-G., Kalthoff, K. 1980. Complete reversal of anterio-posterior polarity in a centrifuged insect embryo. *Nature* 287: 635–37

Regulski, A., Harding, K., Kostriken, R., Karch, F., Levine, M., et al. 1985. Homeobox genes of the Antennapedia and bithorax complexes of *Drosophila*. *Cell* 43: 71–80

Rosenberg, U. B., Schröder, C., Preiss, A., Kienlin, A., Cote, S., et al. 1986. Structural homology of the product of the *Drosophila Krüppel* gene with *Xenopus* transcription factor IIIA. *Nature* 319: 336–39

Russell, M. A. 1985. Positional information in insect segments. *Dev. Biol.* 108: 269–83

Sanchez-Herrero, E., Vernos, I., Marco, R., Morata, G. 1985. Genetic organization of *Drosophila* bithorax complex. *Nature* 313: 108–13

Sander, K. 1984. Embryonic pattern formation in insects: Basic concepts and their experimental foundations. In *Pattern Formation*, ed. G. Malacinski, P. Bryant, Ch. 10, pp. 245–68

Schubiger, G., Newman, S. M. 1982. Determination in *Drosophila* embryos. *Am. Zool.* 22: 47–55

Schüpbach, T., Wieschaus, E. 1986. Maternal-effect mutations altering the anterior-posterior pattern of the *Drosophila* embryo. *Wilhelm Roux Arch. Entwicklungsmech. Org.* In press

Scott, M. P., Weiner, A. J. 1984. Structural relationships among genes that control development: Sequence homology between the *Antennapedia*, *Ultrabithorax*, and *fushi tarazu* loci of *Drosophila*. *Proc. Natl. Acad. Sci. USA* 81: 4115–19

Scott, M. P., Weiner, A. J., Hazelrigg, T. I., Polisky, B. A., Pirrotta, V., et al. 1983. The molecular organization of the *Antennapedia* locus of *Drosophila*. *Cell* 35: 763–76

Simcox, A. A., Sang, J. M. 1983. When does determination occur in *Drosophila* embryos? *Dev. Biol.* 97: 212–21

Slack, J. M. W. 1983. *From Egg to Embryo*. Cambridge, England: Cambridge Univ. Press. 241 pp.

Strathern, J. N., Herskowitz, I. 1979. Asymmetry and directionality in production of new cell types during clonal growth; the switching pattern of homothallic yeast. *Cell* 17: 371–81

Struhl, G. 1981a. A homeotic mutation transforming leg to antenna in *Drosophila*. *Nature* 292: 635–38

Struhl, G. 1981b. A gene product required for correct initiation of segmental determination in *Drosophila*. *Nature* 293: 36–41

Struhl, G. 1983. Role of the *esc+* gene product in ensuring the selective expression of segment specific homeotic genes in *Drosophila*. *J. Embryol. Exp. Morphol.* 76: 297–331

Struhl, G. 1985. Near-reciprocal phenotypes caused by inactivation or indiscriminate expression of the *Drosophila* segmentation gene *ftz*. *Nature* 318: 677–80

Struhl, G., Akam, M. 1985. Altered distributions of *Ultrabithorax* transcripts in *extra sex combs* mutant embryos of *Drosophila*. *EMBO J.* 4: 3259–64

Struhl, G., Brower, D. 1982. Early role of the *esc+* gene product in the determination of segments in *Drosophila*. *Cell* 31: 285–92

Struhl, G., White, R. A. H. 1985. Regulation of the *Ultrabithorax* gene of *Drosophila* by other bithorax complex genes. *Cell* 43: 507–19

Turner, F. R., Mahowald, A. P. 1976. Scanning electron microscopy of *Drosophila* embryogenesis. I. The structure of the egg envelopes and the formation of the cellular blastoderm. *Dev. Biol.* 50: 95–108

Turner, F. R., Mahowald, A. P. 1977. Scanning electron microscopy of *Drosophila melanogaster* embryogenesis. II. Gastrulation and segmentation. *Dev. Biol.* 57: 403–16

Underwood, E. M., Turner, F. R., Mahowald, A. P. 1980. Analysis of cell movements and fate mapping during early embryogenesis in *Drosophila melanogaster*. *Dev. Biol.* 74: 286–301

Vogel, O. 1982a. Experimental test fails to confirm gradient interpretation of embryonic patterning in leafhopper eggs. *Dev. Biol.* 90: 160–64

Vogel, O. 1982b. Development of complete embryos in drastically deformed leafhopper eggs. *Wilhelm Roux Arch. Entwicklungsmech. Org.* 191: 134–36

Wakimoto, B. T., Kaufman, T. C. 1981. Analysis of larval segmentation in lethal genotypes associated with the Antennapedia gene complex in *Drosophila melanogaster*. *Dev. Biol.* 81: 51–64

Wedeen, C., Harding, K., Levine, M. 1986. Spatial regulation of Antennapedia and bithorax gene expression by the *Polycomb* locus in *Drosophila*. *Cell* 44: 739–48

Weiner, A. J., Scott, M. P., Kaufman, T. C.

1984. A molecular analysis of *fushi tarazu*, a gene in *Drosophila melanogaster* that encodes a product affecting embryonic segment number and cell fate. *Cell* 37: 843–51

Weir, M. P., Kornberg, T. 1985. Patterns of *engrailed* and *fushi tarazu* transcripts reveal novel intermediate stages in *Drosophila* segmentation. *Nature* 318: 433–39

White, R. A. H., Wilcox, M. 1984. Protein products of the bithorax complex in *Drosophila*. *Cell* 39: 163–71

White, R. A. H., Wilcox, M. 1985a. Distribution of *Ultrabithorax* proteins in *Drosophila*. *EMBO J.* 4: 2035–43

White, R. A. H., Wilcox, M. 1985b. Regulation of the distribution of *Ultrabithorax* proteins in *Drosophila*. *Nature* 318: 563–67

Wieschaus, E., Gehring, W. J. 1976. Clonal analysis of primordial disc cells in the early embryo of *Drosophila melanogaster*. *Dev. Biol.* 50: 249–63

Wieschaus, E., Nüsslein-Volhard, C., Jürgens, G. 1984a. Mutations affecting the pattern of the larval cuticle in *Drosophila melanogaster*. III. Zygotic loci on the X chromosome and fourth chromosome. *Wilhelm Roux Arch. Entwicklungsmech. Org.* 193: 296–307

Wieschaus, E., Nüsslein-Volhard, C., Kluding, H. 1984b. *Krüppel*, a gene whose activity is required early in the zygotic genome for normal embryonic segmentation. *Dev. Biol.* 104: 172–86

Ann. Rev. Cell Biol. 1986. 2 : 81–116

CELL ADHESION MOLECULES IN THE REGULATION OF ANIMAL FORM AND TISSUE PATTERN

Gerald M. Edelman

Laboratory of Developmental and Molecular Biology, The Rockefeller University, New York, New York 10021

CONTENTS

Historical Introduction

The main purpose of this review is to bring to the reader unfamiliar with the details of cell-cell adhesion a reasonably up-to-date description of facts and problems in the field, with a critical emphasis on recent developments. In reference to cells, "adhesion" has been a portmanteau word applied to the interaction of cells with each other and with surfaces, to early cell-binding events and late, and to specific and nonspecific phenomena. Only with the advent of concepts and methodologies leading to the characterization of molecules of defined function has it been possible to qualify precisely the nature and regulation of a particular kind of adhesion. Before

81

0743–4634/86/1115–0081$02.00

describing these developments, a brief mention of past approaches in the field may provide a useful perspective.

Adhesion phenomena can be seen in a wide range of taxa, from pro-karyotes to eukaryotes and from unicellular organisms to metazoans. The study of the differential binding and sorting out of cells in multicellular organisms began in 1907 when Wilson (1907) showed that mechanically dissociated and mixed cells of two different species of marine sponges sorted out to produce two types of organisms, each consisting of the cells of only one species. Comparable results were obtained in coelenterates (DeMorgan & Drew 1914; Chalkey 1945). A similar approach was later applied by Holtfreter (1939, 1948a,b), who showed that embryonic cells dissociated from different tissues could sort out in a mixture to form regions with structures characteristic of parent tissues—so-called histo-typic aggregates. These experiments showed that there is selective adher-ence among cells of various types. This paradigm was extensively employed by Moscona (1952, 1962) and his colleagues to show similar phenomena in cells from chicken and mice.

Various attempts were subsequently made to develop more direct short-term assays of cell-cell adhesion (see Frazier & Glaser 1979 for a review) and to isolate molecular fractions that might be responsible for differential selectivity or specificity (Balsamo & Lilien 1974; Merrell et al 1975; Oppenheimer 1975; Shur & Roth 1975; Hausman & Moscona 1976). Only limited success was achieved, however, and various interpretations of the results reflected different views about the nature of adhesion. These interpretations included ideas that adhesion differences were due to differ-ences in generalized interactions at the cell surface (such as electrostatic or van der Waals interaction) (Curtis 1967); that differential adhesion was based on thermodynamic considerations (Steinberg 1970); and that cell recognition mediated by intermolecular specificity was involved (Moscona 1962; Hausman & Moscona 1976). The most detailed proposal of the latter type was the chemoaffinity hypothesis of Sperry (1963), which pro-posed that the exquisite order of neural maps and projections was derived from cell recognition markers that were specific down to the level of individual neurons. It is important to note that these ideas were concurrent with a number of proposals (Turing 1952; Wolpert 1971; see Harrison 1982 for review) concerned with the nature of the specification of tissue pattern or positional information in morphogenesis.

More recently, these issues have been greatly clarified as a result of two developments leading to the isolation and characterization of cell adhesion molecules (CAMs) (see Edelman 1983, 1984a, 1985a; Damsky et al 1984, for extensive reviews). The first development was methodological: Short-term, immunologically based cell adhesion assays were devised in which antibodies to putative CAMs were used to block cell adhesion and sub-sequently to isolate candidate molecules (Brackenbury et al 1977; Thiery et al 1977). The second development was conceptual: Cell adhesion was viewed as a cell-regulatory phenomenon as well as one of molecular speci-ficity, in accord with the suggestion that cell surface modulation is a major mechanism in pattern formation (Edelman 1976). These two developments implied that cell adhesion molecules (CAMs) must be defined by a multiplicity of criteria, including (1) demonstration of specific adhesive behavior in an immunologically based assay; (2) sufficient characterization of the molecular structure, binding mechanism, and specificity; (3) dem-onstration of CAMs on the membrane of cells they ligate and CAM synthesis by these cells; (4) the appearance of CAMs in definite sequences of expression during embryogenesis and histogenesis in a way consistent with the initial formation of cell collectives and boundaries having morphological and functional significance; (5) direct evidence for mor-phogenetic function as shown by alteration of tissue structure after per-turbation of CAM binding function and by alteration of CAM expression after mechanical or chemical disruption of morphology.

Cell Function and Molecular Categories of Adhesion

Application of these criteria has provided a clearer discrimination between molecules carrying out cell-cell interaction (CAMs), those carrying out cell-substrate interaction (SAMs, substrate adhesion molecules), and those appearing at junctions (CJMs, cell junctional molecules). There is an emerging realization that the regulation of the time and place of the appearance of these separate molecular families is one of the main bases of morphogenesis and histogenesis (Edelman & Thiery 1985). The different families of molecules comprising CAMs, SAMs, and CJMs clearly play different roles in the regulation of morphogenetic sequences, as indicated by differences in their chemical structures, locations, and schedules of appearances. CAM regulation is intimately tied to *initial* boundary for-mation, embryonic induction and migration, tissue stabilization, and regeneration. The known functions of SAMs are related to migration, stabilization of epithelia, and the development of hard tissues. The known functions of CJMs are formation of specialized cell connections, cell com-munication (as seen in gap junctions), and the sealing of the surfaces of epithelial sheets (see Edelman & Thiery 1985 for discussions by various authors).

The distinctions above must be made clear in order to avoid confusion about the criteria for adhesion at different times and in different cir-cumstances, particularly because cells may "adhere" by CAMs, SAMs, or CJMs in any combination. In working out the conjugate functions of these structurally different molecular families, it is important to establish the temporal order of their appearance at a given site; the contingent or independent appearance of molecules in each family; their different effects on the driving-force processes of cell division, motion, and death; and their respective roles in the regulation of local differentiation events. Unlike many intracellular proteins, all of these morphoregulatory molecules carry out mechanochemical functions (Edelman 1985a,b), which serve to relate developmental genetics to the acquisition of form in tissues and in the whole animal.

With this background, we now turn to the specialized functions of CAMs. I briefly survey work on the CAMs, proceeding from molecular structural and molecular genetic data to embryonic and histogenetic functions, and finally to the challenging and largely unexplored problems of CAM signaling and regulation. When findings conflict or are deficient, my approach is critical and normative; other reviews (Moscona 1974; Grinnell 1978; Frazier & Glaser 1979; Lilien et al 1979; Letourneau et al 1980; Garrod & Nicol 1981; Damsky et al 1984) may be consulted for different views and additional balance. Because a sufficient number of publications now exist in the field, a more or less comprehensive CAM bibliography may be useful for workers in related fields; for this reason certain references are included, often without further comment.

As it has become clearer that CAM-mediated adhesion is a cell regulatory mechanism and not a matter only of "cell recognition," a number of related aspects of the subject have come increasingly to the fore and, accordingly, are emphasized below. These include: (a) the subtlety and multiplicity of cell surface modulation mechanisms affecting CAM distribution, binding state, and prevalence; (b) the ubiquitousness of CAMs in development, histogenesis, stabilization, and regeneration of tissues; and (c) the emergence of a general set of definite differentiation rules (Crossin et al 1985; Edelman 1985b, 1986) relating CAM expression sequences during development to the establishment of the borders between cells and cell collectives involved in embryonic induction.

CAM Structure and Binding

CAMs were first definitively identified by means of immunologically based adhesion assays in which specific antibodies capable of blocking cell adhesion in vivo were used to purify cell surface molecules as putative CAMs (Brackenbury et al 1977). Their provisional acceptance as CAMs was based upon their appearance at the cell surface at appropriate developmental stages, the perturbation of various tissue functions by the specific antibodies, and the chemical definition of their binding mechanisms. A number of proteins studied in various laboratories and given different names were subsequently identified as CAMs that had already been described, and several new candidate CAMs have now been discovered (see Table 1).

Although a variety of candidate CAMs exist, only three have been purified in quantity and chemically characterized; they are named after the tissues from which they were initially isolated, but they can have much wider tissue distributions. These are N-CAM (neural cell adhesion molecule) (Thiery et al 1977; Hoffman et al 1982), L-CAM (liver cell adhesion molecule) (Bertolotti et al 1980; Gallin et al 1983), and Ng-CAM (neuron-glia cell adhesion molecule) (Grumet & Edelman 1984; Grumet et al 1984a,b). Only for N-CAM has a definite binding mechanism been determined; for this reason N-CAM is discussed more extensively than the others.

All of the well-characterized CAMs are large cell surface glycoproteins; the evidence suggests that they are also intrinsic membrane proteins (Hoffman et al 1982; Gennarini et al 1984a,b). Their known structures are summarized in Figure 1. Although N-CAM is specified by a single gene located on chromosome 9 in the mouse (D'Eustachio et al 1985), N-CAM fractions contain three related polypeptide chains (Hoffman et al 1982; Cunningham et al 1983; Hansen et al 1985), each of which has three domains: an N-terminal binding domain, a polysialic acid–rich middle domain, and a cell-associated or cytoplasmic domain (Figure 1). The large domain (ld) polypeptide (Hemperly et al 1986) differs from the small domain (sd) and smallest domain (ssd) polypeptides in the size of its

Table 1   CAMs and candidate CAMs

CAM	Primary or secondary	Proteins identified as same	Binding mechanism	Ion dependence	Refs.[a]
Characterized					
N-CAM	1°	D_2, BSP-2	Homophilic	Ca^{2+} independent	1–4
Ng-CAM	2°	L1, NILE	Heterophilic	Ca^{2+} independent	5–10
L-CAM	1°	E-cadherin, cell CAM 120/80, uvomorulin	Homophilic(?)	Ca^{2+} dependent	11–16
Candidate					
Cell-CAM 105	2°	—	?	Ca^{2+} independent	17
N-cadherin	?	—	?	Ca^{2+} dependent	16
P-cadherin	?	—	?	Ca^{2+} dependent	16

[a] (1) Brackenbury et al 1977; (2) Jorgensen et al 1980, Jorgensen & Haller 1980; (3) Hirn et al 1983; (4) Rieger et al 1985; (5) Grumet & Edelman 1984; (6) Grumet et al 1984a; (7) Faissner et al 1984, 1985; (8) Salton et al 1983a,b; (9) Bock et al 1985; (10) Friedlander et al 1986; (11) Bertolotti et al 1980; (12) Gallin et al 1983; (13) Hyafil et al 1981; (14) Yoshida & Takeichi 1982; (15) Damsky et al 1983; (16) Hatta et al 1985; (17) Ocklind & Obrink 1982, Ocklind et al 1983.

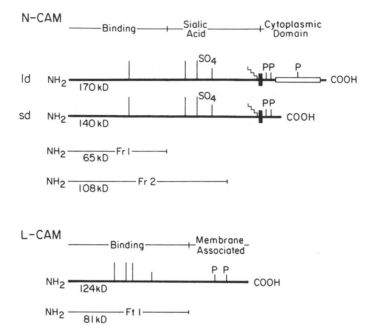

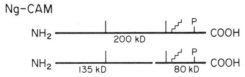

Figure 1 Diagrams of the linear chain structure of two primary CAMs (N-CAM and L-CAM) and of the secondary Ng-CAM. N-CAM is comprised of two chains that differ in the size of their cytoplasmic domains. As indicated by the open bar at the COOH terminus, the ld (large domain) polypeptide contains approximately 250 more amino acid residues in this region than does the sd (small domain) polypeptide. A third, and the smallest (ssd), polypeptide is not shown. The thick vertical bar indicates the membrane-spanning region. Below the chains are the fragments Fr1 and Fr2 derived by limited proteolysis. As indicated by vertical lines, most of the carbohydrate is covalently attached in the middle domain at three sites and it is sulfated, although the exact sulfation site is unknown. Attached to these carbohydrates is polysialic acid. There are phosphorylation (P) sites as well in the COOH terminal domains. The diagonal staircases refer to covalent attachment of palmitate. L-CAM yields one major proteolytic fragment (Ft1) and has four attachment sites for carbohydrate (vertical lines) but lacks polysialic acid. It is also phosphorylated in the COOH terminal region. Ng-CAM is shown as a major 200-kDa chain. There are two components (135 and 80 kDa) that are probably derived from a posttranslationally cleaved precursor. Each is related to the major 200-kDa chain (which may be this precursor), and the smaller is arranged as shown on the basis of a known phosphorylation site.

cytoplasmic domain. Present evidence suggests that, proceeding outward from the position of the molecule at the cell membrane to the amino terminus, these polypeptides are very similar or identical.

In the chicken each N-CAM polypeptide contains at least four asparagine-linked carbohydrate groups (Cunningham et al 1983; Crossin et al 1984). The carbohydrates that are initially added are of the high-mannose type and are converted to complex-type carbohydrate moieties within 30 min of their addition (Lyles et al 1984a,b). Polysialic acid is present on some of these groups (Hoffman et al 1982; Rougon et al 1982; Finne et al 1983), and it has been shown that there are three attachment sites in the middle domain (Crossin et al 1984). The sialic acid content of N-CAM in the adult is less than in the embryo; the conversion from embryonic to adult forms (E-A conversion) is accomplished by de novo synthesis of the adult forms (Friedlander et al 1985). Unlike N-CAM, the other two well-characterized CAMs lack polysialic acid. N-CAM is sulfated: Sorkin et al (1984) found that in the chicken the sulfate is on asparagine-linked carbohydrate, while Lyles et al (1984a) found that in the rat the sulfate may be on polypeptide. In the chicken both N-CAM polypeptides are phosphorylated on serine and threonine residues (Sorkin et al 1984); in the mouse only phosphoserine has been detected (Gennarini et al 1984a). Phosphorylation occurs after transport to the cell surface (Lyles et al 1984a,b). N-CAM is also acylated (Sorkin et al 1985).

L-CAM (Figure 1) also consists of an amino-terminal binding region and a cell-associated region and contains several sites susceptible to proteolytic cleavage in the absence of Ca^{2+} ions (Cunningham et al 1984). After detection of L-CAM (Bertolotti et al 1980) and description of Ca^{2+}-dependent adhesion in fibroblasts (Urushihara & Takeichi 1980), various studies on Ca^{2+} dependence were carried out (Brackenbury et al 1981; Hyafil et al 1981; Grunwald et al 1980, 1982; Magnani et al 1981; Nielson et al 1981; Thomas & Steinberg 1981; Thomas et al 1981; Damsky et al 1983). The most extensive were those of Takeichi and coworkers (Takeichi et al 1981; Yoshida & Takeichi 1982; Ogou et al 1982, 1983, Shirayoshi et al 1983; Yoshida-Noro et al 1984). L-CAM is initially synthesized as a 135-kDa precursor (Peyrieras et al 1983; Vestweber & Kemler 1984), which is converted to a stable 124-kDa form in about 30 min. In the chicken the 124-kDa protein contains one high-mannose and three complex carbohydrate groups and is phosphorylated on serine and threonine residues (Cunningham et al 1984). Its precursor differs from the mature molecule in both polypeptide component and the nature of the carbohydrate groups (B. Sorkin, G. M. Edelman, B. A. Cunningham, unpublished observations).

Both N-CAM and L-CAM are primary adhesion molecules that appear early in development. Ng-CAM, a secondary CAM, consists of a large glycoprotein ($M_r = 200,000$) that appears to be posttranslationally cleaved into two components of 135,000 and 80,000 M_r. The evidence suggests that the amino-terminal portions of the two larger chains (Figure 1) are similar and that the M_r 80,000 chain corresponds to the more carboxy-terminal portion of the M_r 200,000 chain (Grumet et al 1984a,b; Faissner et al 1985). The neuronal antigen L1 appears to be murine Ng-CAM (Grumet et al 1984a; Rathjen & Schachner 1984; Faissner et al 1984, 1985). Recent studies have revealed immunological (Bock et al 1985) and structural identity (Friedlander et al 1986) between the nerve growth factor–inducible large external (NILE) glycoprotein and Ng-CAM (Salton et al 1982a,b). As shown by studies on NILE, Ng-CAM contains aspar-agine-linked carbohydrate groups (Salton et al 1983b) and sialic acid (Margolis et al 1983; Grumet et al 1984a). It is also phosphorylated (Salton et al 1983b). In the chicken, phosphate is present only in the 200-kDa and 80-kDa polypeptide components. Fatty acid is also incorporated into these species (Sorkin et al 1984).

Besides the three well-characterized CAMs, a number of other molecules (Table 1) may be candidates for new secondary CAMs. These include the so-called N- and P-cadherins (Hatta et al 1985). The protein called E-cadherin appears to be L-CAM. N-cadherin (M_r 127,000) is a Ca^{2+}-dependent CAM found in a distribution similar to that of N-CAM (Hatta et al 1985; Thiery et al 1984); it may be the basis of the residual Ca^{2+}-dependent activity seen after transformation by tumor viruses down-regulates N-CAM from the surface of neurons (Greenberg et al 1984; Brackenbury et al 1984). An additional putative CAM of M_r 105,000, cell-CAM 105 (Ocklind & Obrink 1982; Ocklind et al 1983, 1984; Vestweber et al 1985), has been isolated from hepatocytes. This molecule may be a secondary CAM and is definitely distinct from the Ca^{2+}-dependent L-CAM. All of these molecules originate in solid tissues; the possibility that cells from fluid tissues, e.g. lymphocytes, may also have CAMs has been adumbrated (Gallatin et al 1983).

Although in most of these cases the binding specificities are unknown, all three of the well-characterized CAMs appear to have different binding specificities. N-CAM is the first CAM whose binding mechanism has been worked out. Its binding is second order homophilic, i.e. N-CAM-to-N-CAM on an apposed cell. This *trans* binding is Ca^{2+}-independent. Small changes in surface density (prevalence modulation) lead to large changes in the binding rates of N-CAM-bearing lipid vesicles (Hoffman & Edelman 1983), which indicates the likelihood of formation of multivalent structures on the same cell (*cis* interaction) at higher CAM surface densities. It has not been proven that L-CAM binding, which is Ca^{2+}-dependent, is also homophilic, although preliminary experiments support this idea. In contrast, if Ng-CAM, which binds neurons to neurons and neurons to glia, has a ligand on an apposed cell (either neuron or glia), the binding must be heterophilic, as shown by its presence on neurons and its absence on glia in the central nervous system (Grumet et al 1984a,b), as well as by direct neuron-to-neuron vesicle-binding experiments (Hoffman et al 1986). The possibility that N-CAM is a *trans* ligand for Ng-CAM has definitely been excluded.

A significant ambiguity centers around Ng-CAM. A molecule, L1, with the chemical and immunological properties of Ng-CAM has been described in the mouse (Faissner et al 1985; Grumet et al 1985). Although Ng-CAM has been implicated in neuron-neuron and neuron-glia binding (Grumet et al 1985), attempts to show a role for L1 in the latter type of binding have failed (Keilhauer et al 1985). This difference may be related to differences in cell preparations and assay conditions. It is significant that antibodies to Ng-CAM block cross-species binding of chick neurons to mouse glia which, unlike chick glia, can be definitively identified by biochemical and differentiation markers (M. Grumet, G. M. Edelman, unpublished observations). In view of the occurrence of cell surface modulation, this issue cannot be resolved until a heterophilic ligand for Ng-CAM is found or it is shown that the molecule acts indirectly, i.e. without a *trans* ligand, by *cis* interaction with N-CAM.

This discussion points up a fundamental gap in our understanding: the detailed binding mechanism has only been elucidated for N-CAM, not for the other CAMs. Data on binding thermodynamics and kinetics in solution (Bell 1978) are difficult to obtain because of the small amounts of available protein, preparation instability, and unfavorable aggregation properties related to the presence of membrane spanning regions. Such physico-chemical information will obviously be necessary for a complete description of cellular mechanisms of CAM binding. The obstacles may finally be overcome by synthesis of binding region peptides, which will be possible once extensive molecular genetic analyses reveal the relevant amino acid sequences.

CAM Genes

cDNA clones for the two primary CAMs and for Ng-CAM have been obtained. Initially, two independent cDNA clones were derived from enriched mRNA for chick N-CAM that had been prepared by immuno-precipitation of polysomes with antibodies to N-CAM (Murray et al

1984). The plasmids hybridized to two discrete 6–7 kilobase (kb) RNA species in poly(A)$^+$ mRNA from embryonic chick brain but not to comparably prepared RNA from liver. A probe for N-CAM has been obtained independently in the mouse (Goridis et al 1985). A cDNA clone for L-CAM has been obtained from poly(A)$^+$ RNA of embryonic liver using the λgt11 expression vector (Gallin et al 1985). The clone was complementary to a single 4-kb mRNA found in tissues from all organs expressing L-CAM but not from those lacking this CAM.

These probes have been used to explore the multiplicity of genes coding for primary CAMs. For neither of the two primary CAMs is there evidence for a large gene family. Southern blotting analysis with one of the probes for N-CAM detected only one fragment in chicken genomic DNA digested with several restriction enzymes (Murray et al 1984), which suggests that sequences corresponding to those of the probe are present at most a few times and probably only once in the chicken genome. Similar results have been obtained for the probe for Ng-CAM (M. Burgoon, R. Brackenbury, G. M. Edelman, B. A. Cunningham, unpublished observations). In mouse, in which three N-CAM polypeptides are detected, three primary translation products are found in in vitro translations of N-CAM mRNA (Hansen et al 1985), and five N-CAM mRNAs can be detected (Goridis et al 1985). Southern blotting analysis of the cDNA probe for L-CAM has shown components consistent with the presence of one to three L-CAM genes (Gallin et al 1985). Although there may be several genes (or pseudogenes) for L-CAM, L-CAM mRNAs from all tissues studied are of the same apparent size (Gallin et al 1985).

Somatic hybridization methods have been used to localize the N-CAM gene in mouse (D'Eustachio et al 1985) and human (Nguyen et al 1985) chromosomes. A single locus, *n-cam*, has been found on mouse chromosome 9, and its counterpart has been tentatively identified on human chromosome 11. These data are in accord with the results of Southern blotting experiments and confirm that N-CAM is specified by a single gene.

In more recent experiments (Murray et al 1986), rabbit antibodies to the chicken N-CAM were used to isolate four overlapping cDNA clones from a chicken cDNA expression library in bacteriophage λgt11. These clones collectively accounted for 3.8 kb of N-CAM mRNA sequence and hybridized specifically to two 6–7 kb brain poly(A)$^+$ RNA species that comigrated with previously identified N-CAM mRNAs. DNA fragments derived from an internal region of the cloned cDNA sequences hybridized to the larger but not to the smaller N-CAM mRNA species, while fragments on either side of this region hybridized to both mRNAs. A cDNA fragment that recognized only the larger mRNA was subcloned into λgt11, and an expressed fusion protein made with a portion of the β-galactosidase gene was used to affinity-purify rabbit polyclonal antibodies. The antibodies recognized only the larger of the two structurally related N-CAM polypeptides (which we called ld and sd polypeptides on the basis of cDNA sequences; see Hemperly et al 1986). In contrast, when several cDNA clones that recognized both mRNAs were used to purify antibodies, the antibodies recognized both polypeptides. The results, in conjunction with the data indicating that one gene specifies N-CAM, suggest that different N-CAM polypeptides are synthesized from multiple N-CAM messages generated by alternative splicing of transcripts from a single N-CAM gene. Subsequent to this work, complete sequence analysis of cDNA probes corresponding to the polysialic acid–rich and cytoplasmic domains (Hemperly et al 1986) showed that the ld polypeptide has a cytoplasmic domain approximately 250 amino acids larger than that of the sd polypeptide.

The data on known CAM genes suggest that a given CAM is not diversified in its action as a result of the presence of multigene families each corresponding to a different binding specificity or cellular address. Instead, it appears that the initial control of CAM expression is seated in early steps in the control of the expression of the respective structural gene and is accomplished by means of RNA splicing events and possibly by regulation of the rate of mRNA turnover. A system for perturbation of CAM differentiation using temperature-sensitive Rous sarcoma virus (RSV) mutants to transform rat cerebellar cell lines (Greenberg et al 1984; Brackenbury et al 1984) has been used to analyze the level of control of the expression of N-CAM. The data support the notion that a major element of control occurs at the transcription level (R. Brackenbury, G. M. Edelman, unpublished observations). These conclusions are consistent with the idea that CAM functions are diversified by cell surface modulation mechanisms; in some cases local modulation can occur via alternative RNA splicing (Murray et al 1986b).

Cell Surface Modulation of CAMs

Local cell surface modulation is alteration over time of the amount, distribution, or chemical properties of a particular kind of molecule at the cell surface (Edelman 1976). CAMs have been shown to change in amount (prevalence modulation), position or distribution (polarity modulation), or in molecular structure (chemical modulation) during development. The known nonlinear binding properties of CAMs at the cell surface that result from cell surface modulation are consistent with the cell-regulatory nature of cell-cell adhesion. Extensive examples of the first two forms of modulation are given below in our description of the expression sequences of CAMs. An example of chemical modulation is seen in the so-called embryonic to adult (E-A) conversion of N-CAM (Rothbard et al 1982; Hoffman et al 1982). This is a gradual but large decrease during development in polysialic acid content from 30 per 100 g polypeptide to 10 per 100 g polypeptide. E-A conversion is seen in brain, muscle (Rieger et al 1985), and skin (Chuong & Edelman 1985a,b). The three major attachment sites of sialic acid in embryonic and adult forms are alike or identical (Crossin et al 1984); thus it appears that the E-A conversion results from altered sialyl transferase or neuraminidase (Rutishauser et al 1985) activity. Experiments on the E-A conversion in vitro (Friedlander et al 1985) suggest cellular turnover of embryonic forms and replacement with newly synthesized adult forms, however, which supports the notion that the enzymatic activity is intracellular. Binding experiments (Hoffman & Edelman 1983) indicate that the E-A conversion is accompanied by a fourfold increase in rates of binding, as tested by vesicle assays in vitro.

Another form of modulation is seen in the differential cellular expression (Pollerberg et al 1985) of the ld polypeptide (Murray et al 1986b) of N-CAM, which may be related in some cases to its possible pairing with Ng-CAM on the same cell surface (*cis* interaction). While evidence for this is circumstantial, it raises the possibility that differential cytoskeletal binding or *cis* interactions of the same or different cell surface CAMs could alter their *trans* binding to CAMs on apposed cells and their polarity modulation in the same cell.

The evidence reviewed above makes it clear that cell surface modulation is a key mechanism in CAM action at the cellular level and that a large variety of actual and potential modulation mechanisms can operate (Table 2). The existence of such mechanisms shifts attention from the binding sites of different CAMs (each undoubtedly with a different specificity) to the functions of those parts of each molecule related to secretion, membrane insertion, cytoskeletal interaction, *cis* interaction, and alternative RNA splicing that yields different forms of the same CAM. Each of these processes may have a significant role in the signaling events discussed at the end of this article, and a rich set of biochemical problems related to the relevant CAM properties remains to be worked out.

The existence of cell surface modulation in various forms indicates that individual cell shape and specialization will constrain and alter the effects of various modulatory modes of cell adhesion in different contexts. For example, N-CAM appears to be enriched in neural growth cones (Chuong et al 1982; Wallis et al 1985; Ellis et al 1985); the evidence is consistent with the presence of the ld chain at this location. Direct evidence for the context-dependent nature of adhesion mechanisms is provided by a recent study (Hoffman et al 1986), which has shown that changes in the prevalence modulation of Ng-CAM and N-CAM in chick dorsal root ganglion cells and retinas in culture alter the relative binding roles of the two molecules. In the ganglia the Ng/N ratio at the cell surface was higher than in the retinas and increased with time. Fasciculation of out-growing neurites was

Table 2 CAM modulation mechanisms

Mechanism	Effect	Examples	Reference
Prevalence (Change in synthesis or expression on cell surface)	Increase or decrease in binding rate	N-CAM homophilic binding	Edelman et al 1983
		twofold increase → thirtyfold rate increase	Hoffman & Edelman 1983
Differential prevalence (Change in relative expression of CAMs of different specificity on same cell)	Border formation for cell collectives	CAM expression rules (see Table 3) N-CAM/Ng-CAM (see Figure 3)	Chuong & Edelman 1985a,b; Crossin et al 1985 Daniloff et al 1986a; Hoffman et al 1986
Cytoplasmic domain switch (Alteration in size and structure of cytoplasmic domain by RNA splicing)	Selective expression of ld or sd polypeptides; altered cytoskeletal interaction (?)	Selective expression of N-CAM ld chain in cerebellar layers	Murray et al 1986b Pollerberg et al 1985
Polarity (Change in location on selected cell regions)	Localization of binding on cell	Motor endplate (N-CAM)	Rieger et al 1985
		neurite extension and leading edge migrating cells (Ng-CAM)	Daniloff et al 1986a
		exocrine pancreatic cells	Thiery et al 1984
Chemical (Posttranslational change in structure)	Altered binding rates abrogation of binding (?)	E-A conversion of N-CAM	Rothbard et al 1982
		loss of 2/3 polysialic acid → fourfold increase in binding	Chuong & Edelman 1984; Hoffman et al 1982

mainly perturbed by anti-Ng-CAM antibodies. In contrast, in retina, in which the Ng/N ratio was low, retinal layering was perturbed mainly by anti-N-CAM antibodies. In both cases adherent structures displaying both molecules (such as certain cell somata) may not have changed their binding despite dynamic changes in the Ng/N ratio at the cell surface; on the other hand, such changes would be expected to have large local mechanistic effects on dynamic structures such as extending neurites. In view of the identity of Ng-CAM and NILE, these findings are consistent with those of experiments in which anti-NILE antibodies inhibited fasciculation (Stallcup & Beasley 1985; Stallcup et al 1985) in a different fashion than anti-N-CAM antibodies.

The facts reviewed here suggest that in interpreting CAM function one must consider where, when, and on what cells the CAMs are expressed. As shown in Table 2, these constraints are as important as binding specificities. The problem of CAM-mediated adhesion must therefore be solved in a context-dependent fashion that takes into account the actual tissue and cell morphology. This dependency will concomitantly increase analytical difficulties. However, the combinations of mechanisms having important effects on morphogenesis that result from such context dependency are virtually limitless, and the researcher is freed from having to specify a very large number of definite molecular cell addresses (Sperry 1963; Edelman 1984c).

CAM Expression Sequences in Development: Map Restriction

N-CAM and L-CAM appear early in development on derivatives of all three germ layers, although with time they tend to be distributed in a characteristic fashion that limits their distributions (Thiery et al 1982; Edelman et al 1983; Thiery et al 1984; Crossin et al 1985). Because of these properties, they are called primary CAMs. In contrast, Ng-CAM, as a secondary CAM, appears only on postmitotic cells in derivatives of the neuroectoderm.

A systematic examination of the spatiotemporal appearance of each of the CAMs during development (Edelman et al 1983) shows a characteristic expression sequence (Figure 2) and definite restriction in fate maps. In the chick the two primary CAMs appear together on all cells; at gastrulation, however, ingressing cells stain much less with anti-N-CAM and anti-L-CAM. Cells in the chordamesoderm then reexpress N-CAM at neural induction, and endodermal cells reexpress L-CAM. During neural induction a remarkable transition occurs in the epiblast: After completion of induction the presumptive neural plate shows only N-CAM, and the surrounding somatic ectoderm shows mainly L-CAM. Thereafter, at all

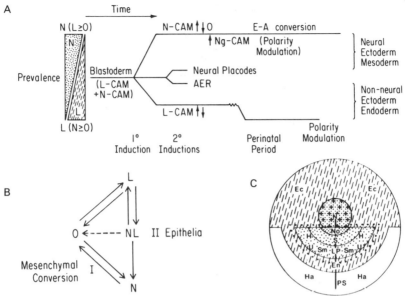

Figure 2 Major CAM expression sequence, epigenetic expression rules, and composite CAM fate map in the chick. (A) Schematic diagram showing the temporal sequence of expression of CAMs during embryogenesis starting from the stage of the blastoderm and proceeding through neural and secondary inductions. Germ layer derivatives are indicated at right. The vertical wedges at the left refer to the approximate relative amounts of each CAM in the different layers or parts of the embryo, e.g. the line referring to blastoderm has relatively large amounts of each CAM, whereas that for neural ectoderm has major amounts of N-CAM. After they diverge in cellular distribution, the CAMs are then modulated in prevalence within various regions of inductions or actually decrease greatly when mesenchyme appears or cell migration occurs. Note that placodes that have both CAMs echo the events seen for neural induction. Just before the appearance of glia, a secondary CAM (Ng-CAM) emerges; unlike the other two CAMs, this CAM would not be found in the fate map shown in (C) before 3.5 days. In the perinatal period, a series of modulations occurs: E-A conversion for N-CAM and polar redistribution for L-CAM. (B) Diagram summarizing epigenetic rules for CAM expression (see Table 3). Arrows refer to transitions. Mesenchyme cells derived from blastoderm express L-CAM and never reexpress L-CAM; they can lose or reduce N-CAM or reexpress it (rule I). Epithelia express N-CAM and L-CAM and can lose one or the other (rule II); tissues showing N-CAM do not reexpress L-CAM. (C) Composite CAM fate map in the chick. The distribution of N-CAM (*stipple*), L-CAM (*slashes*), and Ng-CAM (*crosses*) on tissues of 5- to 14-day-old embryos is mapped back onto the tissue precursor cells in the blastoderm. Additional regions of N-CAM staining in the early embryo (5 days) are shown by larger dots. In the early embryo the borders of CAM expression overlap the borders of the germ layers, i.e. derivatives of all three germ layers express both CAMs. At later times overlap is more restricted: N-CAM disappears from somatic ectoderm and from endoderm, except for a population of cells in the lung. L-CAM is expressed on all ectodermal and endodermal epithelia but remains restricted in the mesoderm to epithelial derivatives of the urogenital system. The vertical bar represents the primitive streak (PS); Ec = intraembryonic and extraembryonic ectoderm; En = endoderm; N = nervous system; No = prechordal and chordamesoderm; S = somite; Sm = smooth muscle; Ha = hemangioblastic area; H = heart; LP = lateral plate mesoderm; U = urogenital system.

sites of secondary embryonic induction, N-CAM or L-CAM or both undergo a series of prevalence modulations that follow two main rules (Crossin et al 1985; Edelman 1985c,d): All epithelial-mesenchymal transformations show a modal transition of $N \rightarrow 0 \rightarrow N$, where 0 means either low or undetectable levels of the CAM (rule I), and epithelial cells show other modal transitions, i.e. either $NL \rightarrow L$ or $NL \rightarrow N$ (rule II). The modes of modulation and their corresponding rules are listed in Table 3. The expression of CAMs in epigenetic sequences expressing these rules is map restricted: Particular CAMs appear in an ordered pattern in a fate map of chick blastoderm (Figure 2).

One of the striking aspects of these epigenetic sequences is the finding that at many sites of induction cell collectives consisting of mesenchymal condensations that follow rule I are found in proximity to epithelial cell collectives that follow rule II. This reflects two facts: (*a*) With the possible exception of urogenital structures, once mesodermal mesenchyme is formed it never reexpresses L-CAM; (*b*) this mesenchyme plays an important inductive role at many sites. Borders at such sites are between mesenchymally derived collectives expressing N-CAM (rule I) and epithelial collectives expressing either N-CAM, or N-CAM and L-CAM together (rule II, see Figure 2).

Several striking examples illustrating the recursive expression of these rules have been worked out. Placodes such as the otic placode show a remarkable relationship between the prevalence modulation of the two primary CAMs and the borders of prospective structures in the cochlea (Richardson et al 1986). Feathers show a series of morphogenetic events in which variant sequences of N-CAM and L-CAM expression are seen after repeated recursive application of each of the rules (Chuong & Edelman 1985a,b). Before considering some of this evidence, it is illuminating first to review data on the nervous system, as it is the one known case in which the sequence of expression of a secondary CAM (Ng-CAM) plays an important role together with only one of the two well-characterized primary CAMs (N-CAM).

Histogenesis: Role of CAMs in Border and Pattern Formation

After neural induction (Figure 2) and the exclusion of L-CAM from neural derivatives (rule II), the secondary Ng-CAM is expressed on postmitotic neurons. In the central nervous system (CNS) this CAM is seen on extending neurites and in very slight amounts on cell somata. At just those sites and times at which neuronal migrations on guide glia take place, however, Ng-CAM is strongly expressed on somata and leading processes in addition to its expression on neurites (Thiery et al 1985b; Daniloff et al 1986). Ng-CAM is not seen on glia in the CNS but is found in the peripheral

Table 3   Sites showing epigenetic rules for CAM expression during chicken embryogenesis

Rule I: Mesenchymal conversion[a]	Rule II: Epithelia[b]
Ectodermal	Ectodermal
$N \to 0 \to N$	$NL \to N$
Neural crest	Neural tube
—Peripheral nerve	Placode-derived ganglia
—Ganglia	$NL \to L$
	Somatic ectoderm
Mesodermal	Stratum germinativum
$N \to 0 \to N$	Apical epidermal ridge
Somite	Branchial ectoderm
—Skeletal muscle (end plate only)	$NL \to N \to *$
—Dermal papilla (feather)	Lens
Nephrotome	Marginal and axial plate
—Germinal epithelium of gonad	of feather
—Gonadal stroma	$NL \to L \to *$
Splanchnopleure	Stratum corneum
—Spleen stroma	Feather barbule, rachis
—Lamina propria of gut	
—Some mesenteries	Mesodermal
$N \to 0 \to N \to *$	$N \to NL \to L$
Somite	Wolffian duct
—Chondrocytes	Mesonephric tubules
Lateral plate	Müllerian duct
—Smooth muscle	
	Endodermal
	$NL \to L$
	Epithelium of:
	Trachea
	Gastrointestinal tract
	Hepatic duct
	Gall bladder
	Thyroid
	Pharyngeal derivatives
	NL
	Parabronchi (lung epithelia)

[a] Rule I shows cyclic changes in, or disappearance of, N-CAM. Some of these transitions occur with movement; 0 represents low levels of CAM. The original tissues are listed at the left margin. Tissues containing high levels of N-CAM are preceded by a dash. Where * appears the CAM can be replaced by a differentiation product.

[b] Rule II shows replacement of one CAM by another or the disappearance of CAM. * represents differentiation products (e.g. keratin, crystallin) with disappearance of the CAM.

nervous system (PNS); Ng-CAM and N-CAM are both present on Schwann cells and neurons. It is notable that by immunocytochemical techniques PNS neurons do not exhibit the same kind of polarity modulation of Ng-CAM seen in the CNS.

As shown in Figure 3, there is a remarkable site-specific microsequence

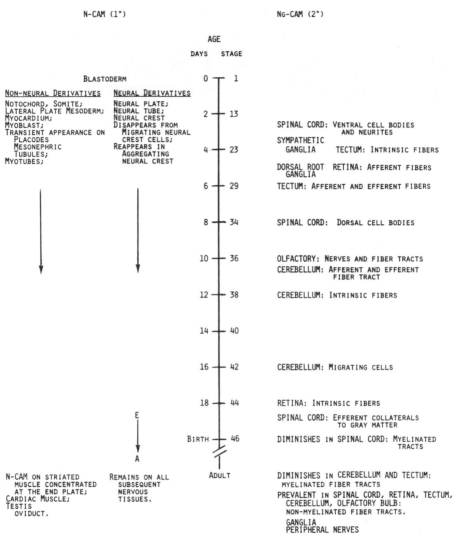

Figure 3 Expression sequence of two neuronal CAMs in the developing chick nervous system. N-CAM, a primary cell adhesion molecule, appears early (blastoderm stage) in both neural and nonneural derivatives. After birth it remains on all subsequent nervous tissues. Ng-CAM appears later during embryogenesis and is first observed on the developing spinal cord (Thiery et al 1985b; Daniloff et al 1986). The sequential appearance of Ng-CAM in CNS and PNS region is summarized. In the adult, Ng-CAM is limited in the unmyelinated CNS regions (tracts or laminae) and PNS.

(Daniloff et al 1986) of CAM expression during CNS and PNS development that results in grossly altered distribution patterns of N-CAM and Ng-CAM with time. Two particularly striking events are the downregulation of Ng-CAM in presumptive myelinating regions (prevalence modulation) and the perinatal E-A conversion of N-CAM in tracts. Neural crest cells that form the PNS (among other structures) show a remarkable expression of rule I as applied to ectomesenchyme: N-CAM is lost from, or lowered on, the surface of migrating crest cells and reappears at sites where ganglion formation occurs (Thiery et al 1982, 1985c). All of these sequences reveal coordinated cell surface modulation events during the formation of particular neural structures, including prevalence modulation, polarity modulation, and chemical modulation (E-A conversion).

Recent studies of the ld polypeptide of N-CAM indicate that it is differentially expressed on various cells (Williams et al 1985) and that it appears differentially in certain layers during development of the cerebellum and retina (Pollerberg et al 1985; Murray et al 1986b). The data on the existence of alternative RNA splicing (Murray et al 1986a) and the finding of a large cytoplasmic domain on the ld polypeptide that may interact differentially with the cytoskeleton (Hemperly et al 1986) suggest that differential cell surface modulation of the ld and sd chains could lead to altered patterns of cell interaction, migration, and layering. The adequate signal may be one that controls alternative splicing in a local tissue region.

One of the key structural events in embryogenesis and histogenesis is boundary formation, an event that appears to be strongly tied to the early differentiation rules (Edelman 1985c, 1986; Crossin et al 1985) found for the two well-characterized primary CAMs. A specialized structure with neural components, the otic placode, shows the expression of these epigenetic rules (Table 3, Figure 2) particularly well (Richardson et al 1986). An inducing collective reexpressing N-CAM (rule I) induces an epithelial placode showing N-CAM and L-CAM (rule II). Then site-specific borders of various presumptive structures form within the developing placode by a sequence of prevalence modulations of the two primary CAMs following rule II. Neural structures related to the cochlea express N-CAM; non-neural structures express L-CAM and N-CAM in various sequences.

Perhaps the most striking example of the recursive application of the epigenetic rules for primary CAM expression is the feather. This periodic and hierarchically organized structure provides an opportunity to analyze the coupling of cell collectives in detail and to relate their interactions to cytodifferentiation events within a dimensionally well-organized appendage. Feathers are induced through the formation of dermal condensations of mesodermally derived mesenchyme, which act upon ectodermal cells to form placodes (Sengel 1976). Such placodes and condensations are eventually hexagonally close-packed as feather induction proceeds in rows from medial to lateral aspects of the chicken skin. Within each induced placode a dermal papilla is subsequently formed by repeated inductive interactions between mesoderm and ectoderm. Subsequently, the cellular proliferation of barb ridges (with fusion to form a rachis), followed by barbule plate formation, yields the basis for three levels of branching: rachis, ramis, and barbule.

An extraordinary series of events involving L-linked collectives of cells coupled to N-linked collectives either by movement and adhesion or by cell division and adhesion is seen at each of these levels (Chuong & Edelman 1985a,b). Initially, L-linked ectodermal cells are approached by CAM-negative mesenchyme cells, which become N-CAM positive in the ectodermal vicinity (rule I). As these cells accumulate to form condensations, placodes are induced in the L-linked cells. A similar couple consisting of L-linked ectodermal cell collectives adjacent to N-linked mesenchymal collectives is seen in the papilla, which forms subsequently. After N-CAM-positive mesenchyme cells are excluded by a basement membrane, the collar cells derived from the L-CAM-positive papillar ectoderm express both N-CAM and L-CAM (rule II).

After this event, site-restricted applications of rule II lead to a remarkable periodicity of borders: Cells derived from papillar ectoderm by division form barb ridges and express L-CAM. In the valleys between the neighboring ridges the basilar cells lose L-CAM and express N-CAM to form the marginal plate. A similar process occurs as the ridge cells organize into L-CAM-positive barbule plates—N-CAM is expressed in cells between the barbule plate cells to form the axial plate. The end result is a series of cellular patterns that follows rule II, in which cell collectives linked by L-CAM alternate with those expressing N-CAM at both the secondary barb level and the tertiary barbule level. After further extension of the barb ridges into rami, the L-CAM-positive cells keratinize, and the N-CAM-positive cells die without keratinization, leaving spaces between barbules and yielding the characteristic feather morphology.

In this histogenetic CAM expression sequence, one observes periodic CAM modulation, periodic and successive formation of L-CAM-linked and N-CAM-linked cell collectives (CAM couples), and the definite association of gene expression events during cytodifferentiation with particular kinds of CAMs. The most dramatic example is the expression of keratins only in L-containing cells. Throughout this histogenetic process there is an intimate connection between the regulatory process of adhesion and the epigenetic sequences, which consist of those different primary processes

that act as driving forces. These processes include morphogenetic movement for the original mesenchymal induction, mitosis for the formation
of papillar ectoderm and barb ridges, and death for the N-linked collectives
in the final period of feather formation. These findings have several important implications: They raise the possibility that CAM function is important in inductive sequences, and they suggest that a series of local signals
is responsible for particular sequences of CAM expression. As discussed
below, direct evidence for this conclusion comes from experiments on
perturbation of CAM function and from experiments on altered CAM
expression after disruption of morphology.

CAM Perturbation and Morphology-Dependent Regulation

A positive result in an in vitro adhesion assay is insufficient evidence for
acceptance of a molecule as a CAM. Several other criteria must be met:
(1) synthesis and control by the cells that a putative CAM ligates; (2)
expression in definite spatiotemporal sequences related to function during
morphogenesis, stabilization, or regeneration; (3) perturbation of morphology after abrogation or blockade of CAM binding function; (4)
alteration of CAM expression after destruction of morphology, implying
the existence of sustained structure-dependent regulatory signals for CAM
expression and modulation. We have considered some evidence for the
first two criteria; evidence fulfilling the third and fourth is accumulating
in a variety of systems.

Specific antibodies to CAMs can block their function and lead to altered
morphology. Addition of anti-Ng-CAM antibodies disrupts neural fasciculation of ganglion cells in vitro (Hoffman et al 1986; Friedlander et al
1986; Stallcup et al 1985), and addition of anti-N-CAM antibodies greatly
disrupts layer formation in the chick retina during organ culture (Buskirk
et al 1980). The study by Hoffman et al mentioned in the section on
cell surface modulation provided evidence that the early observations
(Rutishauser et al 1978) of anti-N-CAM effects on fasciculation were
actually the result of small amounts of cross-reactive anti-Ng-CAM antibodies. Implanted anti-N-CAM antibodies disrupt retinotectal map formation in the frog (Fraser et al 1984) and optic nerve fiber migration in
the chicken (Thanos et al 1984), and anti-Ng-CAM antibodies inhibit the
characteristic migration of external granule cells in cerebellar tissue slices
(Lindner et al 1983; Grumet et al 1984a,b; Hoffman et al 1986).

Perhaps the most striking of the perturbation experiments are those
recently performed on an in vitro feather induction system (Gallin et al
1986). Anti-L-CAM antibodies added to chick skin explants altered the
pattern of N-CAM-linked dermal condensations from a sixfold symmetry

pattern to one of fourfold symmetry, and the feather placodes tended to fuse mediolaterally into stripes. Longer term cultures with anti-L-CAM present showed formation of scale-like plates rather than the feather-like patterns seen in unperturbed controls. Thus, antibodies to one member of a CAM couple in an inducing collective can alter the patterns formed by cells in the other member. While these results are preliminary, they suggest that CAMs play a key role in causal sequences during induction and histogenesis.

In view of the epigenetic nature of tissue formation and the existence of defined CAM expression sequences, it is a reasonable conclusion that CAM expression and modulation depend upon local signals that vary according to the state, composition, and integrity of particular interacting cell collectives. Thus, while CAMs may serve to stabilize tissue structures (along with SAMs and CJMs), their modulation should depend in turn upon local cellular interactions in the stabilized structure. A number of studies have shown that disrupted morphology or altered morphogenesis can actually lead to changes in CAM modulation patterns. For example, perturbation in vivo of normal cell-cell interactions during degeneration and regeneration has been shown to result in alteration of CAM expression and distribution. N-CAM is present at the neuromuscular junction of striated muscles but is absent from the rest of the surface of the myofibril (Rieger et al 1985). After the sciatic nerve is cut, the molecule appears diffusely at the cell surface and in the cytoplasm of the muscle cell (Rieger et al 1985; Covault & Sanes 1985). These experiments indicate that early events related to regeneration can be accompanied by altered CAM modulation. More recent experiments (Daniloff et al 1986b) show that the crushing or cutting of a nerve has widespread effects that range from altered anti-CAM staining in motor neurons of the spinal cord (on the affected side) to modulatory changes in N-CAM and Ng-CAM within Schwann cells that are local to the lesion.

These observations have a direct bearing upon the possible role of CAMs in disease. For example, in fetal neural tube defects, N-CAM levels are elevated (Jorgensen & Norgaardpedersen 1981). A recent study of various muscle diseases indicates alterations in the amount of N-CAM expression (Walsh & Moore 1985). This is in accord with the observations on degeneration and regeneration in nerve-muscle interactions. Certain viral transformations result in definite alteration of CAM expression (Greenberg et al 1984; Brackenbury et al 1984; Hixson et al 1984; Hixson & McIntire 1985), and a potential role for CAMs in metastasis is suggested. When neural cells were transformed by a temperature-sensitive mutant of RSV, they retained normal morphology, adult N-CAM levels, and aggregation behavior at the nonpermissive temperature. At the permissive temperature, however, they transformed and within hours down-regulated N-CAM and became more mobile; reversal of these events occurred when transformed cells grown at the permissive temperature were raised to the nonpermissive temperature (Greenberg et al 1984). Recent double-blind analyses of a variety of central nervous system tumors (C.-M. Chuong, J. Zabriskie, G. M. Edelman, unpublished observations) suggest definite patterns of CAM expression that differ in tumors of different origin; this finding suggests that CAM characterization in patients with cancer may have diagnostic value.

CAM modulation has also been found to be perturbed in genetic diseases with altered morphology. A recent study (Rieger et al 1986) has indicated that N-CAM and Ng-CAM are colocalized at nodes of Ranvier in peripheral myelinated nerves. In contrast, nonmyelinated fibers showed uniform distribution of the molecules. In two dysmyelinating mouse mutants, *trembler* $(tr/+)$ and motor endplate disease *med/med*, the distribution pattern of these molecules was found to be disrupted in myelinated fibers. In the mouse mutant *staggerer*, which shows connectional defects between Purkinje cells and parallel fibers in the cerebellum and extensive granule cell death, E-A conversion of N-CAM is greatly delayed in the cerebellum (Edelman & Chuong 1982). A variety of preliminary phenomenological observations of the protein D2 (which is N-CAM, see Jorgensen et al 1980) showed that it was present in the amniotic fluid of mothers with fetuses having neural tube defects (Jorgensen 1981; Ibsen et al 1983). Further understanding of the role of CAMs in disease may have important diagnostic and therapeutic consequences in dealing with congenital defects, bone repair, nerve regeneration, degenerative disorders, and neoplasia.

All of the findings described in this section argue for the existence of an elaborate series of local morphology-dependent signals regulating CAM expression in various tissues. To confirm this conclusion better data are needed on the exact sequences of expression of CAM messengers. These data can be obtained using methods such as in situ hybridization, direct identification of the source and nature of the chemical signals involved, and a clearer definition of the conjugate roles of SAMs and CJMs in the processes of morphogenesis and regeneration.

Comparison of Map Restriction in CAMs and SAMs

The finding that CAM expressions are map restricted (Figure 2) and the evidence of modulation of N-CAM and fibronectin appearance during neural crest migration (Thiery et al 1982, 1985c) raise the possibility that some extracellular matrix proteins may, like CAMs, also be map restricted and thus play an important role in pattern formation. This role might be clarified by searching for SAMs that appear sequentially in a site-restricted fashion during development.

Recently, this search has been successfully carried out, aided by the identification by Thiery and colleagues (Thiery et al 1985a,c; Vincent & Thiery 1984; Vincent et al 1983) of a carbohydrate antigen (NC-1), first detected by immunizing rabbits wih ciliary ganglion cells. This epitope was found to appear on migrating neural crest cells (at a time when they lacked N-CAM) but was nevertheless subsequently found to be present on the carbohydrate moieties of N-CAM and Ng-CAM. Extensive search showed that the epitope was related to the so-called HNK antigen of lymphocytes (Abo & Balch 1981); it therefore can appear on a variety of different proteins found in different tissues and presumably involved in different functions.

Immunocytochemical staining for the NC-1 antigen and the presence of antibodies to the antigen in sera that blocked neuron-glia adhesion led to the identification and isolation of an extracellular matrix protein (cytotactin) with widespread tissue distribution (Grumet et al 1985). Characterization of this protein suggests that, while it is a SAM, it appears in a sequential site-restricted fashion during development. Cytotactin mediates glia-neuron adhesion in vitro, but unlike Ng-CAM, it is absent from neurons. All available evidence indicates that it is not the Ng-CAM receptor. Cytotactin was isolated from 14-day-old embryonic chicken brains as structurally related polypeptides of M_r 220,000, 200,000, and 190,000. These polypeptides were efficiently extracted in the absence of detergent and appeared to be disulfide-linked into higher polymers. Cytotactin is immunologically distinct from both laminin and fibronectin. Immunofluorescence staining with specific antibodies indicated that cytotactin is present in extracellular spaces and in basement membranes of a variety of nonneural tissues, including smooth muscle, lung, and kidney. In the cerebellum it appears on glial end-feet and on Bergmann glial fibers as well as in extracellular spaces. The molecule is synthesized by glia and cells from smooth muscle, lung, and kidney, and it is found at the surface of glia in culture in a cell-associated fibrillar pattern.

A survey of the expression sequence of cytotactin in chick embryos (Crossin et al 1986) confirmed that this protein is distinct from the SAMs mentioned above. Cytotactin is first present in the gastrulating chicken embryo. It appears later in the basement membrane of the developing neural tube and notochord, beginning in the cephalic regions and proceeding caudally. Between 2 and 3 days of development the molecule is present at high levels in the early neural crest pathways (surrounding the neural tube and somites), but in contrast to fibronectin and laminin, it is restricted to these pathways and is not found in the lateral plate mesoderm or ectoderm. At later times cytotactin is expressed extensively in the central nervous system, in lesser amounts in the peripheral nervous system, and in a number of nonneural sites, most prominently in all smooth muscle and in basement membranes of lung and kidney. A striking feature of its expression sequence is that at each of its major sites of appearance—gastrula, somites, and neural crest pathways—it is first seen in anterior parts of the embryo and then is expressed posteriorly. Cytotactin is never seen in voluntary or cardiac muscle or their precursors, and a comparison of composite fate maps (Figure 4) shows that in its map-restriction it is distributed differently from CAMs or other SAMs such as fibronectin.

A murine molecule called the J1 glycoprotein has been described (Kruse et al 1985); it shows properties very similar to those of cytotactin. J1 was described as a CAM restricted to the nervous system. Recent experiments (K. L. Crossin, G. M. Edelman, unpublished observations) using anti-cytotactin antibodies that are cross-reactive in a variety of species, however, strongly support the suggestion (Grumet et al 1985; Crossin et al 1986) that the J1 glycoprotein is a SAM similar or identical to cytotactin and that further study will reveal it to be present in a variety of tissues.

It is definite that cytotactin, like J1 (Kruse et al 1985), mediates glia-neuron interaction. Recent experiments (Grumet et al 1985; C.-M. Chuong, G. M. Edelman, unpublished observations) indicate that anti-cytotactin antibodies inhibit extension of external granule cells into the internal granular layer of cerebellar tissue slices. While anti-Ng-CAM completely blocks movement of external granule cells into the molecular layer, anticytotactin results in a pile-up of cells on guide glia in this layer. The function of cytotactin thus may be to regulate cell motion, possibly by retarding migration in the pathways at which it appears.

These findings raise the possibility that additional extracellular matrix proteins and SAMs may contribute to pattern formation in embryogenesis as a result of their restricted expression in a spatiotemporally regulated fashion at some sites but not at others. Clearly, regulation of CAMs and SAMs cannot follow identical rules: CAMs are intrinsic to the cells that synthesize them and can be directly modulated by these cells, but SAMs and ECMs, once released from cells, must be regulated, destroyed, or rebound to other cells in a different fashion (see Edelman & Thiery 1985 for reviews). This contrast focuses attention on the key issues raised by the experimental findings surveyed in this review: What signals govern CAM expression and what is the relationship of CAM gene expression to the subsequent expression of genes mediating specific cytodifferentiation?

CAM Regulation and Signals for CAM Expression

The observations indicating that there are epigenetic rules for CAM expression (Edelman 1985b, 1986; Crossin et al 1985) and the results of various perturbation experiments are consistent with the so-called regu-

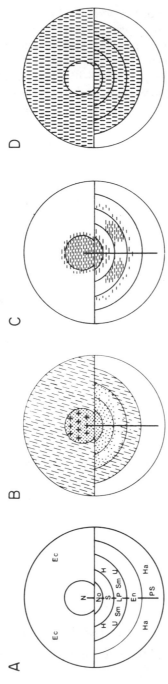

Figure 4   Fate maps comparing the distributions of CAMs, cytotactin, and fibronectin (Crossin et al 1986). (A) The presumptive areas of the blastoderm that will give rise to the different tissues have been established by various marking techniques (for review, see Vaka et 1985). Abbreviations are the same as those used in Figure 2C. (B) The distributions of the three known CAMs (N-CAM = ·, L-CAM = /, and Ng-CAM = +) are superimposed upon a classical fate map. N-CAM- and L-CAM-containing tissues map to distinct and contiguous regions within the blastoderm, and Ng-CAM appears later and only within neuroectodermal derivatives. The distribution of tissues expressing cytotactin is shown in C (cytotactin = −), and that of fibronectin is shown in D (fibronectin = |). Cellular fibronectin occurs in most tissues, exclusive of the neuroectoderm, in contrast to the restricted distribution of cytotactin, which is present in neuroectoderm and notochord, smooth muscle, the urogenital system, in certain somite derivatives (sclerotome), and at low levels and at restricted times in ectoderm and endoderm.

lator hypothesis (Edelman 1984b). This hypothesis states that *early* control of CAM expression by regulatory genes (and thus the realization of the epigenetic rules) is separate from control of expression of the molecules concerned with other cytodifferentiation events, such as those governing intracellular economy (Han et al 1986) or regulation of cell shape. According to this hypothesis, one may usefully distinguish "morphoregulatory" genes, which regulate CAM and SAM expression, from "historegulatory" genes, which regulate specific cytodifferentiation events. With this distinction in mind, CAM expression (Edelman 1985b) may be viewed as taking place in a cycle (Figure 5). Traversal of the outer loop of the cycle may lead to expression or down-regulation of N-CAM genes (rule I, see

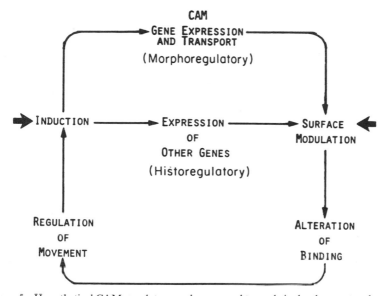

Figure 5 Hypothetical CAM-regulatory cycle proposed to apply in development and regeneration. Early induction signals (heavy arrow at left) lead to CAM gene expression. Surface modulation (by prevalence changes, polar redistribution on the cell, or chemical changes such as E-A conversion) alters the binding rates of cells. This regulates morphogenetic movements, which in turn affect embryonic induction or milieu-dependent differentiation. The inductive changes can again affect expression of CAM genes and that of other genes for specific tissues. It is assumed that morphoregulatory genes affecting CAM expression are separate from historegulatory genes affecting other cytodifferentiation events. The heavy arrows at left and right refer to candidate signals for initiation of induction, either through surface modulation as a result of CAM binding (*right*) or through the release of morphogens affecting induction (*left*). The evidence from work on the feather suggests that mechanochemical signals and morphogens may act together at key local sites to change CAM expression, as seen at the origins of the marginal and axial plates (see Chuong & Edelman 1985a,b).

Table 3), as is seen in the case of mesenchyme cells contributing to the dermal condensation (Chuong & Edelman 1985a,b) and in the case of neural crest cells (Thiery et al 1982). Alternatively, switching off of one or the other primary CAM gene may occur in an epithelium expressing both CAMs (rule II, Table 3).

The action of historegulatory genes (the inner loop in Figure 5) responding to signals from the new milieux that result from CAM-mediated cell aggregation and cell motion, as well as from tissue folding and tension, may lead to the entry of altered cells into the cycle. Among other changes, this could result in the expression of gene products that alter cell shape or that govern enzymes affecting CAM action, as in E-A conversion. The combined effects of the inner and outer loops of the cycle and the linking of two such cycles by formation of CAM couples (Crossin et al 1985) would lead to a rich set of morphogenetic structures. If proven, the CAM regulatory cycle would provide a particularly elegant solution to the problem of mechanochemical control of pattern through several levels of organization (from the gene and gene product to the cell, to tissues and organs, and back to the gene) that is posed by the occurrence of epigenetic sequences both in regulative development and in regeneration.

The fact that CAMs are specified by one or a few genes and the availability of CAM cDNA probes make it feasible to test this hypothesis. The level at which CAM expression is controlled may be checked, and in situ hybridization experiments in a given tissue should allow comparison of the temporal sequence of CAM mRNA expression with that for particular tissue-specific gene products such as keratin. If the outcome of these experiments is that CAM gene expression is the prior event, there would be a satisfying interpetation of the crossing of histotypic boundaries by CAM distributions, as seen in composite fate maps (Figures 2 and 4). In this view, CAMs would first regulate cell movement by modulation, leading to the formation of a milieux capable of generating subsequent signals for historegulatory genes, and leading in turn to the fixation of histotypic boundaries.

What of the actual nature of the signals for CAM expression? Recently, a possible candidate has been found: As discussed above, nerve growth factor (NGF) has been shown to enhance the expression of Ng-CAM in PC12 pheochromocytoma cells (Friedlander et al 1986), and this CAM has been shown to be identical to the NGF-inducible large external glycoprotein (NILE) (Bock et al 1985; Friedlander et al 1986). This finding raises the possibility that the in vivo morphogens postulated in the CAM cycle may be common hormones, neurotransmitters, or growth factors.

Even for one kind of CAM, cell surface modulation (which is critical to the proposed cycle) is under the control of a variety of factors, from RNA splicing to signal sequences that determine the insertion of CAMs into the cell membrane. For example, recent experiments on the base sequence of an N-CAM cDNA clone (Hemperly et al 1986) indicate that the ld chain of N-CAM differs from the sd chain in possessing a much larger cytoplasmic domain (see Figure 1). The finding that the ld polypeptide is modulated and differentially expressed in particular neural populations (Pollerberg et al 1985) and the observations (Murray et al 1984; Goridis et al 1985; Murray et al 1986b) on various types of RNA species specifying different forms of N-CAM (despite the existence of a single N-CAM gene) are consistent with the proposal that the sd chain arises by alternative splicing (Murray et al 1986a) of transcripts from a single gene. Clearly at least one form of cell surface modulation can be regulated by such splicing events.

One of the most challenging tasks in developing a molecular histology will be to define both the exact signal pathways and the modulation mechanisms relating such control events to the development and evolution of animal form. It may not be amiss, therefore, to mention the possible bearing of studies of cell adhesion on the understanding of morphologic evolution. Although CAMs in the forms described here appear to be restricted to vertebrate species (Hoffman et al 1984; Hall & Rutishauser 1985), and hybridization studies have so far failed to detect homologies between mouse, chick, and *Drosophila* (R. Brackenbury, G. M. Edelman, unpublished observation), molecules with functions similar to the vertebrate CAMs will probably be found in various invertebrate species.

The developmental genetic analysis of even the known CAMs promises to open up new understanding of the evolution of form. A major basis of morphologic evolution is heterochrony, which consists of changes during embryonic development in the relative rate of appearance of characters present in ancestors. It has been suggested that heterochrony results from mutations in regulatory genes that affect the timing of biochemical events, the integration of structural gene expression, and the fates of embryonic cells (see Raff & Kaufman 1983). One of the outstanding problems in modern evolutionary theory is to provide a molecular basis for heterochrony by determining the nature of the key regulatory genes and by establishing the mechanism of action of the proteins they regulate. It is an attractive hypothesis (Edelman 1984c, 1985d, 1986) that morpho-regulatory genes controlling the types and sequences of expression of CAMs, SAMs, and possibly CJMs, during animal development play a central role in mediating heterochronic effects. The data on the structure, function, genes, sequences of expression, and regulation of cell adhesion molecules during development that have been reviewed here are consistent with this hypothesis. Evolutionary alterations in the control of molecules such as the CAMs may provide a main link between the genetic code

110   EDELMAN

and the mechanochemical events that lead to the specification of three-dimensional form in the phenotype. Unraveling the intricacies of this link may be aided by constructing transgenic mice containing new or misplaced CAM genes and by experiments in which the CAM regulation of developing endocrine organs in metamorphic animals is perturbed, leading to major heterochronic changes (Alberch 1986).

ACKNOWLEDGEMENT

The work of the author cited here was supported by National Institutes of Health grants HD-16550, HD-09635, and AM-04256, and by a Senator Jacob Javits Center for Excellence in Neuroscience Grant (NS-22789).

Literature Cited

Abo, T., Balch, C. M. 1981. A differentiation antigen of human NK and K cells identified by a monoclonal antibody (HNK-1). *J. Immunol.* 127: 1024–29

Alberch, P. 1986. The evolution of a developmental process. In *Marine Biological Laboratories Lectures in Biology*, ed. R. A. Raff, E. Raff. In press

Balsamo, J., Lilien, J. 1974. Functional identification of three components which mediate tissue-type specific embryonic cell adhesion. *Nature* 251: 522–24

Bell, G. I. 1978. Models for the specific adhesion of cells to cells. *Science* 200: 618–27

Bertolotti, R., Rutishauser, U., Edelman, G. M. 1980. A cell surface molecule involved in aggregation of embryonic liver cells. *Proc. Natl. Acad. Sci. USA* 77: 4831–35

Bock, E., Richter-Landsberg, C., Faissner, A., Schachner, M. 1985. Demonstration of immunochemical identity between the nerve growth factor-inducible large external (NILE) glycoprotein and the cell adhesion molecule-L1. *EMBO J.* 4: 2765–68

Brackenbury, R., Greenberg, M. E. Edelman, G. M. 1984. Phenotypic changes and loss of N-CAM mediated adhesion in transformed chicken retinal cells. *J. Cell Biol.* 99: 1944–54

Brackenbury, R., Rutishauser, U., Edelman, G. M. 1981. Distinct calcium-independent and calcium-dependent adhesion systems of chicken embryo cells. *Proc. Natl. Acad. Sci. USA* 78: 387–91

Brackenbury, R., Thiery, J.-P., Rutishauser, U., Edelman, G. M. 1977. Adhesion among neural cells of the chick embryo. I. An immunological assay for molecules involved in cell-cell binding. *J. Biol. Chem.* 252: 6835–40

Buskirk, D. R., Thiery, J.-P., Rutishauser, U., Edelman, G. M. 1980. Antibodies to a neural cell adhesion molecule disrupt histogenesis in cultured chick retinae. *Nature* 285: 488–89

Chalkey, H. W. 1945. Quantitative relation between the number of organized centers and tissue volume in regenerating masses of minced body sections of *Hydra. J. Natl. Cancer Inst.* 6: 191–95

Chuong, C.-M., Edelman, G. M. 1984. Alterations in neural cell adhesion molecules during development of different regions of the nervous system. *J. Neurosci.* 4: 2354–68

Chuong, C.-M., Edelman, G. M. 1985a. Expression of cell adhesion molecules in embryonic induction: I. Morphogenesis of nestling feathers. *J. Cell Biol.* 101: 1009–26

Chuong, C.-M., Edelman, G. M. 1985b. Expression of cell adhesion molecules in embryonic induction: II. Morphogenesis of adult feathers. *J. Cell Biol.* 101: 1027–43

Chuong, C.-M., McClain, D. A., Streit, P., Edelman, G. M. 1982. Neural cell adhesion molecules in rodent brains isolated by monoclonal antibodies with cross-species reactivity. *Proc. Natl. Acad. Sci. USA* 79: 4234–38

Covault, J., Sanes, J. R. 1985. Neural cell-adhesion molecule (N-CAM) accumulates in denervated and paralyzed skeletal-muscles. *Proc. Natl. Acad. Sci. USA* 82: 4544–48

Crossin, K. L., Chuong, C.-M., Edelman, G. M. 1985. Expression sequences of cell adhesion molecules. *Proc. Natl. Acad. Sci. USA* 82: 6942–46

Crossin, K. L., Edelman, G. M., Cunningham, B. A. 1984. Mapping of three carbohydrate attachment sites in embryonic and adult forms of the neural cell adhesion molecule. *J. Cell Biol.* 99: 1848–55

Crossin, K. L., Hoffman, S., Grumet, M., Thiery, J.-P., Edelman, G. M. 1986. Site-restricted expression of cytotactin during development of the chicken embryo. *J. Cell Biol.* 102: 1917–30

Cunningham, B. A., Hoffman, S., Rutis-hauser, U., Hemperly, J. J., Edelman, G. M. 1983. Molecular topography of the neural cell adhesion molecule N-CAM: Surface orientation and location of sialic acid-rich and binding regions. *Proc. Natl. Acad. Sci. USA* 80: 3116–20

Cunningham, B. A., Leutzinger, Y., Gallin, W. J., Sorkin, B. C., Edelman, G. M. 1984. Linear organization of the liver cell adhesion molecule L-CAM. *Proc. Natl. Acad. Sci. USA* 81: 5787–91

Curtis, A. S. G. 1967. *The Cell Surface: Its Molecular Role in Morphogenesis.* New York: Academic

Damsky, C. H., Knudsen, K. A., Buck, C. A. 1984. Integral membrane proteins in cell-cell and cell-substratum adhesion. In *The Biology of Glycoproteins*, ed. R. J. Ivatt, pp. 1–64. New York/London: Plenum

Damsky, C. H., Richa, J., Knudsen, K., Solter, D., Buck, C. A. 1982. Identification of a cell-cell adhesion glycoprotein from mammary tumor epithelium. *J. Cell Biol.* 95: 22 (Abstr.)

Damsky, C. H., Richa, J., Solter, D., Knudsen, K., Buck, C. A. 1983. Identification and purification of a cell surface glycoprotein involved in cell-cell interactions. *Cell* 34: 455–66

Daniloff, J. K., Chuong, C.-M., Levi, G., Edelman, G. M. 1986a. Differential distribution of cell adhesion molecules during histogenesis of the chick nervous system. *J. Neurosci.* 6: 739–58

Daniloff, J. K., Levi, G., Grumet, M., Rieger, F., Edelman, G. M. 1986b. Altered expression of neuronal cell adhesion molecules induced by nerve injury and repair. *J. Cell Biol.* In press

DeMorgan, W., Drew, H. 1914. A study of the restitution masses formed by the dissociated cells of the hybrids *Antennularia ramosa* and *A. antennine. J. Mar. Biol. Assoc. UK* 10: 440–63

D'Eustachio, P., Owens, G., Edelman, G. M., Cunningham, B. A. 1985. Chromosomal location of the gene encoding to the neural cell adhesion molecule (N-CAM) in the mouse. *Proc. Natl. Acad. Sci. USA* 82: 7631–35

Edelman, G. M. 1976. Surface modulation in cell recognition and cell growth. *Science* 192: 218–26

Edelman, G. M. 1983. Cell adhesion mole-cules. *Science* 219: 450–57

Edelman, G. M. 1984a. Modulation of cell adhesion during induction, histogenesis, and perinatal development of the nervous system. *Ann. Rev. Neurosci.* 7: 339–77

Edelman, G. M. 1984b. Cell adhesion and morphogenesis: The regulator hypothesis. *Proc. Natl. Acad. Sci. USA* 81: 1460–64

Edelman, G. M. 1984c. Cell surface modulation and marker multiplicity in neural patterning. *Trends Neurosci.* 7: 78–84

Edelman, G. M. 1985a. Cell adhesion and the molecular processes of morphogenesis. *Ann. Rev. Biochem.* 54: 135–69

Edelman, G. M. 1985b. Expression of cell adhesion molecules during embryogenesis and regeneration. *Exp. Cell Res.* 161: 1–16

Edelman, G. M. 1985c. Evolution and morphogenesis: The regulator hypothesis. In *17th. Stadler Genetics Symposium on Genetics, Development and Evolution*, pp. 1–28. New York: Plenum

Edelman, G. M. 1986b. Molecular mechanisms of morphogenetic evolution. In *Chemica Scripta of the Royal Academy.* Cambridge: Cambridge Univ. Press. In press

Edelman, G. M. 1986a. Epigenetic rules for expression of cell adhesion molecules during morphogenesis. *Junctional Complexes of Epithelial Cells. CIBA Found. Symp.* No. 125. London: Wiley

Edelman, G. M., Chuong, C.-M. 1982. Embryonic to adult conversion of neural cell-adhesion molecules in normal and *staggerer* mice. *Proc. Natl. Acad. Sci. USA* 79: 7036–40

Edelman, G. M., Gallin, W. J., Delouvée, A., Cunningham, B. A., Thiery, J.-P. 1983. Early epochal maps of two different cell adhesion molecules. *Proc. Natl. Acad. Sci. USA* 80: 4384–88

Edelman, G. M., Thiery, J.-P., eds. 1985. *The Cell in Contact: Adhesions and Junctions as Morphogenetic Determinants.* New York: Wiley.

Ellis, L., Wallis, I., Abreu, E., Pfenninger, K. H. 1985. Nerve growth cones isolated from fetal rat brain. IV. Preparation of a membrane subfraction and identification of a membrane glycoprotein expressed on sprouting neurons. *J. Cell Biol.* 101: 1977–89

Faissner, A., Kruse, J., Goridis, C., Bock, E., Schachner, M. 1984. The neural cell adhesion molecule L1 is distinct from the N-CAM related group of surface antigens BSP-2 and D2. *EMBO J.* 3: 733–37

Faissner, A., Teplow, D. B., Kubler, D., Keilhauer, G., Kinzel, V., Schachner, M. 1985. Biosynthesis and membrane topography of the neural cell adhesion molecule L1. *EMBO J.* 4: 3105–13

Finne, J., Finne, U., Deagostini-Bazin, H., Goridis, C. 1983. Occurrence of α2-8 linked polysialosyl units in a neural cell adhesion molecule. *Biochem. Biophys. Res. Commun.* 112: 482–87

Fraser, S. E., Murray, B. A., Chuong, C.-M., Edelman, G. M. 1984. Alteration of the retinotectal map in *Xenopus* by antibodies to neural cell adhesion molecules. *Proc. Natl. Acad. Sci. USA* 81: 4222–26

Frazier, W., Glaser, L. 1979. Surface components and cell recognition. *Ann. Rev. Biochem.* 48: 491–523

Friedlander, D. R., Brackenbury, R., Edelman, G. M. 1985. Conversion of embryonic forms of N-CAM *in vitro* results from *de novo* synthesis of adult forms. *J. Cell Biol.* 101: 412–19

Friedlander, D. R., Grumet, M., Edelman, G. M. 1986. Nerve growth factor enhances expression of neuron-glia cell adhesion molecule in PC12 cells. *J. Cell Biol.* 102: 413–19

Gallatin, W. M., Weissman, I. L., Butcher, E. C. 1983. A cell surface molecule involved in organ-specific homing of lymphocytes. *Nature* 304: 30–34

Gallin, W. J., Chuong, C. M., Finkel, L. H., Edelman, G. M. 1986. Antibodies to L-CAM perturb inductive interactions and alter feather pattern and structure. *Proc. Natl. Acad. Sci. USA.* In press

Gallin, W. J., Edelman, G. M., Cunningham, B. A. 1983. Characterization of L-CAM, a major cell adhesion molecule from embryonic liver cells. *Proc. Natl. Acad. Sci. USA* 80: 1038–42

Gallin, W. J., Prediger, E. A., Edelman, G. M., Cunningham, B. A. 1985. Isolation of a cDNA clone for the liver cell adhesion molecule (L-CAM). *Proc. Natl. Acad. Sci. USA* 82: 2809–13

Garrod, D. R., Nicol, A. 1981. Cell behavior and molecular mechanisms of cell-cell adhesion. *Biol. Rev. Cambridge Philos. Soc.* 56: 199–242

Gennarini, G., Hirn, M., Deagostini-Bazin, H., Goridis, C. 1984a. Studies on the transmembrane disposition of the neural cell adhesion molecule N-CAM. A monoclonal antibody recognizing a cytoplasmic domain and evidence for the presence of phosphoserine residues. *Eur. J. Biochem.* 142: 65–73

Gennarini, G., Rougon, G., Deagostini-Bazin, H., Hirn, M., Goridis, C. 1984b. Studies on the transmembrane disposition of the neural cell adhesion molecule N-CAM. The use of liposome-inserted radioiodinated N-CAM to study its transbilayer orientation. *Eur. J. Biochem.* 142: 57–64

Goridis, C., Hirn, M., Santoni, M. J., Gennarini, G., Deagostini-Bazin, H., Jordan, B. R., Kiefer, M., Steinmetz, M. 1985. Isolation of mouse N-CAM-related cDNA—Detection and cloning using monoclonal-antibodies. *EMBO J.* 4: 631–35

Greenberg, M. E., Brackenbury, R., Edelman, G. M. 1984. Alteration of the neural cell adhesion molecule (N-CAM) expression after neuronal cell transformation by Rous sarcoma virus. *Proc. Natl. Acad. Sci. USA* 81: 969–73

Grinnell, A. 1978. Cellular adhesiveness and extracellular substrata. *Int. Rev. Cytol.* 53: 63–144

Grumet, M., Edelman, G. M. 1984. Heterotypic binding between neuronal membrane vesicles and glial cells is mediated by a specific neuron-glial cell adhesion molecule. *J. Cell Biol.* 98: 1746–56

Grumet, M., Hoffman, S., Chuong, C.-M., Edelman, G. M. 1984a. Polypeptide components and binding functions of neuron-glia cell adhesion molecules. *Proc. Natl. Acad. Sci. USA* 81: 7989–93

Grumet, M., Hoffman, S., Crossin, K. L., Edelman, G. M. 1985. Cytotactin, an extracellular matrix protein of neural and non-neural tissues that mediates glia-neuron interaction. *Proc. Natl. Acad. Sci. USA* 82: 8075–79

Grumet, M., Hoffman, S., Edelman, G. M. 1984b. Two antigenically related neuronal CAMs of different specificities mediate neuron-neuron and neuron-glia adhesion. *Proc. Natl. Acad. Sci. USA* 81: 267–71

Grunwald, G. B., Geller, R. L., Lilien, J. 1980. Enzymatic dissection of embryonic cell adhesive mechanisms. *J. Cell Biol.* 85: 766–76

Grunwald, G. B., Pratt, R. S., Lilien, J. 1982. Enzymic dissection of embryonic cell adhesive mechanisms. III. Immunological identification of a component of the calcium-dependent adhesive system of embryonic chick neural retina cells. *J. Cell Sci.* 55: 69–83

Hall, A. K., Rutishauser, U. 1985. Phylogeny of a neural cell-adhesion molecule. *Dev. Biol.* 110: 39–46

Han, J. H., Rall, L., Rutter, W. J. 1986. Selective expression of rat pancreatic genes during embryonic development. *Proc. Natl. Acad. Sci. USA* 83: 110–14

Hansen, O. C., Nybroe, O., Bock, E. 1985. Cell-free synthesis of the D2-cell adhesion molecule—evidence for three primary translation products. *J. Neurochem.* 44: 712–17

Harrison, L. G. 1982. An overview of kinetic theory. In *Developmental Modelling in Developmental Order: Its Origin and Regulation*, pp. 3–33. New York: Liss

Hatta, K., Okada, T. S., Takeichi, M. 1985. A monoclonal antibody disrupting calcium-dependent cell-cell adhesion of brain tissues: Possible role of its target antigen in animal pattern formation. *Proc. Natl. Acad. Sci. USA* 82: 2789–93

Hausman, R. E., Moscona, A. A. 1976. Isolation of a retina-specific cell-aggregating factor from membranes of embryonic neural retina tissue. *Proc. Natl. Acad. Sci. USA* 73: 3594–98

Hemperly, J. J., Murray, B. A., Edelman, G. M., Cunningham, B. A. 1986. Sequence of a cDNA clone encoding the polysialic acid-rich and cytoplasmic domains of the neural cell adhesion molecule N-CAM. *Proc. Natl. Acad. Sci. USA* 83: 3037–41

Hirn, M., Ghandour, M. S., Deagostini-Bazin, H., Goridis, C. 1983. Molecular heterogeneity and structural evolution during cerebellar ontogeny detected by monoclonal antibody of the mouse cell surface antigen BSP-2. *Brain Res.* 265: 87–100

Hixson, D. C., McIntire, K. D. 1985. Detection using monoclonal antibodies of a structurally altered form of cell CAM 105 on rat hepatocellular carcinomas. In *Molecular Determinants of Animal Form*, ed. G. M. Edelman, pp. 253–70. New York: Liss

Hixson, D. C., McIntire, K. D., Obrink, B. 1984. Altered expression of cell-adhesion molecules on transplantable rat hepatocellular carcinomas. *J. Cell Biol.* 99: 164

Hoffman, S., Chuong, C.-M., Edelman, G. M. 1984. Evolutionary conservation of key structures and binding functions of neural cell adhesion molecules. *Proc. Natl. Acad. Aci. USA* 81: 6881–85

Hoffman, S., Edelman, G. M. 1983. Kinetics of homophilic binding by E and A forms of the neural cell adhesion molecule. *Proc. Natl. Acad. Sci USA* 80: 5762–66

Hoffman, S., Friedlander, D. R., Chuong, C.-M., Grumet, M., Edelman, G. M. 1986. Differential contributions of Ng-CAM and N-CAM to cell adhesion in different neural regions. *J. Cell Biol.* In press

Hoffman, S., Sorkin, B. C., White, P. C., Brackenbury, R., Mailhammer, R., et al. 1982. Chemical characterization of a neural cell adhesion molecule purified from embryonic brain membranes. *J. Biol. Chem.* 257: 7720–29

Holtfreter, J. 1939. Gewebeaffinität, ein Mittel der embryonalen Formbildung. *Arch. Exp. Zellforsch. Besonders Gewebezuecht.* 23: 169–209

Holtfreter, J. 1948a. The mechanism of embryonic induction and its relation to parthenogenesis and malignancy. *Symp. Soc. Exp. Biol.*, No. 11, p. 17. Cambridge:

Cambridge Univ.

Holtfreter, J. 1948b. Significance of the cell membrane in embryonic processes. *Ann. NY Acad. Sci.* 49: 709–60

Hyafil, F., Babinet, C., Jacob, F. 1981. Cell-cell interactions in early embryogenesis: A molecular approach to the role of calcium. *Cell* 26: 447–54

Ibsen, S., Berezin, V., Norgaardpedersen, B., Bock, E. 1983. Quantification of the D2-glycoprotein in amniotic-fluid and serum from pregnancies with fetal neural-tube defects. *J. Neurochem.* 41: 363–66

Jorgensen, O. S. 1981. Neuronal membrane D2-protein during rat brain ontogeny. *J. Neurochem.* 37: 939–46

Jorgensen, O. S., Delouvée, A., Thiery, J.-P., Edelman, G. M. 1980. The nervous system specific protein D2 is involved in adhesion among neurites from cultured rat ganglia. *FEBS Lett.* 111: 39–42

Jorgensen, O. S., Moller, M. 1980. Immunocytochemical demonstration of the D2 protein in the presynaptic complex. *Brain Res.* 194: 419–29

Jorgensen, O. S., Norgaardpedersen, B. 1981. The synaptic membrane D2-protein in amniotic-fluid from pregnancies with fetal neural tube defects. *Prenatal Diag.* 1: 3–6

Keilhauer, G., Faissner, A., Schachner, M. 1985. Differential inhibition of neuron neuron, neuron astrocyte and astrocyte astrocyte adhesion by L1, L2 and N-CAM antibodies. *Nature* 316: 728–80

Kruse, J., Keilhauer, G., Faissner, A., Timpl, R., Schachner, M. 1985. The J1-glycoprotein—a novel nervous-system cell-adhesion molecule of the L2/HNK-1 family. *Nature* 316: 146 48

Letourneau, P. C., Ray, P. N., Bernfield, M. R. 1980. The regulation of cell behavior by cell adhesion. In *Biological Regulation and Development*, ed. R. Goldberger, 2: 339–76. New York: Plenum. 655 pp.

Lilien, J., Balsamo, J., McDonough, J., Hermolin, J., Cook, J., Rutz, R. 1979. Adhesive specificity among embryonic cells. In *Surfaces of Normal and Malignant Cells*, ed. R. O. Hynes, pp. 389–418. New York: Wiley. 471 pp.

Lindner, J., Rathjen, F. G., Schachner, M. 1983. Monoclonal and polyclonal antibodies modify cell-migration in early postnatal mouse cerebellum. *Nature* 305: 427–30

Lyles, J. M., Linneman, D., Bock, E. 1984a. Biosynthesis of the D2-cell adhesion molecule: Post-translational modifications, intracellular transport, and developmental changes. *J. Cell Biol.* 99: 2082–91

Lyles, J. M., Norrild, B., Bock, E. 1984b. Biosynthesis of the D2-cell adhesion molecule: Pulse-chase studies in cultured fetal rat neuronal cells. *J. Cell Biol.* 98: 2077–81

Magnani, J. L., Thomas, W. A., Steinberg, M. S. 1981. Two distinct adhesion mechanisms in embryonic neural retina cells. I. A kinetic analysis. *Dev. Biol.* 81: 96–105

Margolis, R. K., Salton, S. R. J., Margolis, R. U. 1983. Structural features of the nerve growth factor inducible large external glycoprotein of PC12 pheochromocytoma cells and brain. *J. Neurochem.* 41: 1635–40

Merrell, R., Gottlieb, D. I., Glaser, L. 1975. Embryonal cell surface recognition. Extraction of an active plasma membrane component. *J. Biol. Chem.* 250: 5655–59

Moscona, A. A. 1952. Cell suspensions from organ rudiments of chick embryos. *Exp. Cell Res.* 3: 536–39

Moscona, A. A. 1962. Analysis of cell recombinations in experimental synthesis of tissues *in vitro. J. Cell Comp. Physiol.* 60: 65–80 (Suppl. 1)

Moscona, A. A. 1974. Surface specification of embryonic cells: Lectin receptors, cell recognition, and specific ligands. In *The Cell Surface in Development,* ed. A. A. Moscona, pp. 67–99. New York: Wiley. 334 pp.

Murray, B. A., Hemperly, J. J., Gallin, W. J., MacGregor, J. S., Edelman, G. M., Cunningham, B. A. 1984. Isolation of cDNA clones for the chicken neural cell adhesion molecule (N-CAM). *Proc. Natl. Acad. Sci. USA* 81: 5584–88

Murray, B. A., Hemperly, J. J., Prediger, E. A., Edelman, G. M., Cunningham, B. A. 1986a. Alternatively spliced mRNAs code for different polypeptide chains of the chicken neural cell adhesion molecule (N-CAM). *J. Cell Biol.* 102: 189–93

Murray, B. A., Owens, G. C., Prediger, E. A., Crossin, K. L., Cunningham, B. A., Edelman, G. M. 1986b. Cell surface modulation of the neuronal cell adhesion molecule resulting from alternative mRNA splicing in a tissue-specific developmental sequence. *J. Cell Biol.* In press

Nguyen, C., Mattei, M. G., Goridis, C., Mattei, J. F., Jordan, B. R. 1985. Localization of the human N-CAM gene to chromosome 11 by *in situ* hybridization with a murine N-CAM cDNA probe. *Cytogenet. Cell Genet.* 40: 713

Nielson, L. D., Pitts, M., Grady, S. R., McGuire, E. J. 1981. Cell-cell adhesion in the embryonic chick: Partial purification of liver adhesion molecules from liver membranes. *Dev. Biol.* 86: 315–26

Ocklind, C., Forsum, U., Obrink, B. 1983. Cell surface localization and tissue distribution of a hepatocyte cell-cell adhesion glycoprotein (Cell-CAM 105). *J. Cell Biol.* 96(4): 1168–71

Ocklind, C., Obrink, B. 1982. Intercellular adhesion of rat hepatocytes. Identification of a cell surface glycoprotein involved in the initial adhesion process. *J. Biol. Chem.* 257: 6788–95

Ocklind, C., Odin, P., Obrink, B. 1984. Two different cell adhesion molecules—cell-CAM 105 and a calcium-dependent protein—occur on the surface of rat hepatocytes. *Exp. Cell Res.* 151: 29–45

Ogou, S.-I., Okada, T. S., Takeichi, M. 1982. Cleavage stage mouse embryos share a common cell-adhesion system with teratocarcinoma cells. *Dev. Biol.* 92: 521–28

Ogou, S.-I., Yoshida-Noro, C., Takeichi, M. 1983. Calcium-dependent cell-cell adhesion molecules common to hepatocytes and teratocarcinoma stem cells. *J. Cell Biol.* 97: 944–48

Oppenheimer, S. B. 1975. Functional involvement of specific carbohydrate in teratoma cell adhesion factor. *Exp. Cell Res.* 92: 122–26

Peyrieras, N., Hyafil, F., Louvard, D., Ploegh, H. L., Jacob, F. 1983. Uvomorulin: A nonintegral membrane protein of early mouse embryo. *Proc. Natl. Acad. Sci. USA* 80: 6274–77

Pollerberg, E. G., Sadoul, R., Goridis, C., Schachner, M. 1985. Selective expression of the 180-KD component of the neural cell-adhesion molecule N-CAM during development. *J. Cell Biol.* 101: 1921–29

Raff, R. A., Kaufman, T. C. 1983. *Embryos, Genes, and Evolution.* New York: Macmillan

Rathjen, F. G., Schachner, M. 1984. Immunocytological and biochemical characterization of a new neuronal cell surface component (L1 antigen) which is involved in cell adhesion. *EMBO J.* 3: 1–10

Richardson, G., Crossin, K. L., Chuong, C.-M., Edelman, G. M. 1986. Expression of cell adhesion molecules in embryonic induction. III. Development of the otic placode. In *Functions of the Auditory System,* ed. G. M. Edelman, W. E. Gall, W. M. Cowan. New York: Wiley. In press

Rieger, F., Daniloff, J. K., Pincon-Raymond, M., Crossin, K. L., Grumet, M., Edelman, G. M. 1986. Neuronal cell adhesion molecules and cytotactin colocalize at the node of Ranvier. *J. Cell Biol.* In press

Rieger, F., Grumet, M., Edelman, G. M. 1985. N-CAM at the vertebrate neuromuscular junction. *J. Cell Biol.* 101: 285–93

Rothbard, J. B., Brackenbury, R., Cunningham, B. A., Edelman, G. M. 1982. Differences in the carbohydrate structures of neural cell adhesion molecules from adult and embryonic chicken brains. *J. Biol. Chem.* 257: 11064–69

Rougon, G., Deagostini-Bazin, H., Hirn, M., Goridis, C. 1982. Tissue and developmental stage-specific forms of a neural cell surface antigen linked to differences in glycosylation of a common polypeptide. *EMBO J.* 1: 1239–44

Rutishauser, U., Gall, W. E., Edelman, G. M. 1978. Adhesion among neural cells of the chick embryo. III. Role of the cell surface molecule CAM in the formation of neurite bundles in cultures of spinal ganglia. *J. Cell Biol.* 79: 371–93

Rutishauser, U., Watanabe, M., Silver, J., Troy, F. A., Vimr, E. R. 1985. Specific alteration of NCAM-mediated cell adhesion by an endoneuraminidase. *J. Cell Biol.* 101: 1842–49

Salton, S. R. J., Richter-Landsberg, C., Green, L. A., Shelanski, M. L. 1983a. Nerve growth factor–inducible large external (NILE) glycoprotein: Studies of a central and peripheral neuronal marker. *J. Neurosci.* 3: 441–54

Salton, S. R. J., Shelanski, M. L., Greene, L. A. 1983b. Biochemical properties of the nerve growth factor–inducible large external (NILE) glycoprotein. *J. Neurosci.* 3: 2420–30

Sengel, P. 1976. *Morphogenesis of Skin,* Cambridge: Cambridge Univ. Press. 277 pp.

Shirayoshi, Y., Okada, T. S., Takeichi, M. 1983. The calcium-dependent cell-cell adhesion system regulates inner cell mass formation and cell surface polarization in early mouse development. *Cell* 35: 631–38

Shur, B. D., Roth, S. 1975. Cell surface glycosyltransferases. *Biochim. Biophys. Acta* 415: 473–512

Sorkin, B. C., Grumet, M., Cunningham, B. A., Edelman, G. M. 1985. Structures of two neuronal cell adhesion molecules. *Soc. Neurosci.* 2: 1138 (Abstr.)

Sorkin, B. C., Hoffman, S., Edelman, G. M., Cunningham, B. A. 1984. Sulfation and phosphorylation of the neural cell adhesion molecule N-CAM. *Science* 225: 1476–78

Sperry, R. W. 1963. Chemoaffinity in the orderly growth of nerve fiber patterns and connections. *Proc. Natl. Acad. Sci. USA* 50: 703–10

Stallcup, W. B., Beasley, L. L. 1985. Involvement of the nerve growth factor–inducible large external glycoprotein (NILE) in neurite fasciculation in primary cultures of rat brain. *Proc. Natl. Acad. Sci. USA*
82: 1276–80

Stallcup, W. B., Beasley, L. L., Levine, J. M. 1985. Antibody against nerve growth factor–inducible external (NILE) glycoprotein labels nerve fiber tracts in developing rat nervous system. *J. Neurosci.* 5: 1090–1101

Steinberg, M. S. 1970. Does differential adhesion govern self-assembly processes in histogenesis? Equilibrium configurations and the emergence of a hierarchy among populations of embryonic cells. *J. Exp. Zool.* 173: 395–433

Takeichi, M., Atsumi, T., Yoshida, C., Uno, K., Okada, T. S. 1981. Selective adhesion of embryonal carcinoma cells and differentiated cells by Ca^{2+}-dependent sites. *Dev. Biol.* 87: 340–50

Thanos, S., Bonhoeffer, F., Rutishauser, U. 1984. Development of the chicken retinotectal projection. *Proc. Natl. Acad. Sci. USA* 81: 1906–10

Thiery, J.-P., Boucaut, J. C., Yamada, K. M. 1985a. Cell migration in the vertebrate embryo. In *Molecular Determinants of Animal Form,* ed. G. M. Edelman, pp. 167–93. New York: Liss

Thiery, J.-P., Brackenbury, R., Rutishauser, U., Edelman, G. M. 1977. Adhesion among neural cells of the chick embryo. II. Purification and characterization of a cell adhesion molecule from neural retina. *J. Biol. Chem.* 252: 6841–45

Thiery, J.-P., Delouvée, A., Gallin, W. J., Cunningham, B. A., Edelman, G. M. 1984. Ontogenetic expression of cell adhesion molecules: L-CAM is found in epithelia derived from the three primary germ layers. *Dev. Biol.* 102: 61–78

Thiery, J.-P., Delouvée, A., Grumet, M., Edelman, G. M. 1985b. Initial appearance and regional distribution of the neuron-glia cell adhesion molecule in the chick embryo. *J. Cell Biol.* 100: 442–56

Thiery, J.-P., Duband, J.-L., Delouvée, A. 1985c. The role of cell adhesion in morphogenetic movements during early embryogenesis. In *The Cell in Contact; Adhesions and Junctions as Morphogenetic Determinants,* ed. G. M. Edelman, J.-P. Thiery, pp. 169–96. New York: Wiley

Thiery, J.-P., Duband, J.-L., Rutishauser, U., Edelman, G. M. 1982. Cell adhesion molecules in early chick embryogenesis. *Proc. Natl. Acad. Sci. USA* 79: 6737–41

Thomas, W. A., Steinberg, M. S. 1981. Two distinct adhesion mechanisms in embryonic neural retina cells. II. An immunological analysis. *Dev. Biol.* 81: 106–14

Thomas, W. A., Thomson, J., Magnani, J. L., Steinberg, M. S. 1981. Two distinct adhesion mechanisms in embryonic neural retina cells. III. Functional specificity. *Dev. Biol.* 81 : 379–85

Turing, A. M. 1952. The chemical basis of morphogenesis. *Philos. Trans. R. Soc. London Ser. B* 237 : 37–72

Urushihara, H., Takeichi, M. 1980. Cell-cell adhesion molecule: Identification of a glycoprotein relevant to the Ca^{2+}-independent aggregation of chinese hamster fibroblasts. *Cell* 20 : 363–71

Vakaet, L. 1985. Morphogenetic movements and fate maps in the avian blastoderm. In *Molecular Determinants of Animal Form*, ed. G. M. Edelman. New York: Liss

Vestweber, D., Kemler, R. 1984. Some structural and functional aspects of the cell adhesion molecule uvomorulin. *Cell Differentiation* 15 : 269–73

Vestweber, D., Ocklind, C., Gossler, A., Odin, P., Obrink, B., Kemler, R. 1985. Comparison of two cell-adhesion molecules uvomorulin and cell-CAM 105. *Exp. Cell Res.* 157 : 451–61

Vincent, M., Duband, J.-L., Thiery, J.-P. 1983. A cell surface determinant expressed early in migrating avian neural crest cells. *Dev. Brain Res.* 9 : 235–38

Vincent, M., Thiery, J.-P. 1984. A cell surface marker for neural crest and placodal cells: Further evolution in peripheral and central nervous systems. *Dev. Brain Res.*

103 : 468–81

Wallis, I., Ellis, L., Suh, K., Pfenninger, K. H. 1985. Immunolocalization of a neuronal growth-dependent membrane glycoprotein. *J. Cell Biol.* 101 : 1990–98

Walsh, F. S., Moore, S. E. 1985. Expression of cell-adhesion molecule, N-CAM, in diseases of adult human skeletal-muscle. *Neurosci. Lett.* 59 : 73–78

Williams, R. K., Goridis, C., Akeson, R. 1985. Individual neural cell-types express immunologically distinct N-CAM forms. *J. Cell Biol.* 101 : 36–42

Wilson, H. V. 1907. On some phenomena of coalescence and regeneration in sponges. *J. Exp. Zool.* 5 : 245–58

Wolpert, L. 1971. Positional information and pattern formation. *Curr. Top. Dev. Biol.* 6 : 183–224

Yoshida, C., Takeichi, M. 1982. Teratocarcinoma cell adhesion: Identification of a cell-surface protein involved in calcium-dependent cell aggregation. *Cell* 28 : 217–24

Yoshida-Noro, C., Suzuki, N., Takeichi, M. 1984. Molecular nature of the calcium-dependent cell-cell adhesion system in mouse teratocarcinoma and embryonic cells studied with a monoclonal antibody. *Dev. Biol.* 101 : 19–27

Ann. Rev. Cell Biol. 1986. 2 : 117–47

CORE PARTICLE, FIBER, AND TRANSCRIPTIONALLY ACTIVE CHROMATIN STRUCTURE[1]

D. S. Pederson, F. Thoma,[2] and R. T. Simpson

Laboratory of Cellular and Developmental Biology, National Institute of Diabetes and Digestive and Kidney Diseases, National Institutes of Health, Bethesda, Maryland 20892

CONTENTS

INTRODUCTION

The last decade has seen a rapid advance in our knowledge of the structure of chromatin. Observations early in the 1970s of specific oligomeric histone complexes and a tandemly repeated beaded morphology for chromatin

---

[1] The US Government has the right to retain a nonexclusive, royalty-free license in and to any copyright covering this paper.

[2] Current address: Eidgenossische Technische Hochschule, Institut fur Zellbiologie, Honggerberg, CH-8093 Zurich, Switzerland.

rapidly led to development of the nucleosome paradigm. The ability to isolate large quantities of homogeneous nucleosome core particles allowed the armamentarium of the physical chemist to be applied to the determination of the structure of this fundamental chromatin subunit, culminating in the recent descriptions of the crystal structures of the core particle and the histone octamer. The development of recombinant DNA techniques has permitted questions about the chromatin structure of single-copy genes to be posed, particularly how this structure is related to the functional roles of the gene.

It is well established that selective transcription of genomic DNA is important to differentiation and development and that this transcription is affected by the way in which genes are packaged as chromatin (recently reviewed by Eissenberg et al 1985; Reeves 1984; Weintraub 1985; Yamamoto 1985). Of current importance is understanding how the events of DNA metabolism—transcription, replication, recombination, and repair—occur in a chromatin (as opposed to free DNA) milieu. To what extent does the packaging of DNA in chromatin, per se, play an active role in differential gene activity? For instance, how does the position of a promotor in the nucleus, its location in condensed chromatin, or its exact orientation on the face of a nucleosome affect the binding of transcription factors? Is chromatin remodeling regulated to alter binding of transcription factors in different cell states? How do nucleosomes in transcribed regions differ from nucleosomes of inactive chromatin, and how are those differences regulated? In this review we describe some of the most promising approaches to answering these questions.

Work on the nuclear matrix, DNA conformations, the enzymology of transcription, and chromatin structure has been converging in a way that deepens our understanding of nuclear function. We concentrate on three areas of chromatin research: (a) core particle structure, (b) higher order organization of the chromatin fiber, and (c) the structure of transcriptionally active chromatin. This review draws principally on work published between 1982 and 1985 and is intended to be heuristic rather than exhaustive. Because of space limitations, we frequently cite only the most recent publication in a series from a single laboratory. Statements that we do not explicitly document are supported by work that has been frequently reviewed.

CORE PARTICLE STRUCTURE

Core Particle

A decade of effort culminated in 1984 with the description of a crystal structure for the nucleosome core particle at 7 Å resolution by Richmond et al (1984). This particle consists of an octamer of histones—two each of H2A, H2B, H3, and H4—and 146 base pairs (bp) of DNA. The overall size and shape of the particle agrees with earlier diffraction and electron microscopic measurements; it is a disk about 57 Å in height and 110 Å in diameter. The path of DNA in the particle is clearly defined through at least 12 helical turns with the major and minor grooves clearly visible in places. Right-handed B-form DNA is folded into a left-handed superhelix around the histones, as expected from earlier biochemical studies of the core particle. Each superhelical turn contains 7.6 turns of double helix; the number of nucleotides per double helical turn can not be determined at the current level of resolution. The pitch of the superhelix varies; it is about 30 Å for the central turn and decreases to 25 Å for the flanking regions. The minor groove of DNA faces outward from the particle at the dyad axis.

Surprisingly, the superhelical DNA path is not smooth; rather, there are four loci where the nucleic acid bends or kinks. These bends are located about 10 and 40 bp to either side of the midpoint of the core particle DNA. The exact nature of the bends is not apparent; they are, however, spread over several nucleotides as opposed to previously proposed kinks. Bending DNA into a circular path alters the widths of the grooves. At the bends about 10 bp from the dyad, the minor groove width is 7 Å where it faces the octamer and 13 Å facing the solvent; corresponding values for the major groove are 11 Å and 20 Å, respectively. There are substantial contacts of DNA with internal protein on one side of each bend, a half turn of the DNA helix away, and lesser contacts on the other side. Most of the protein contacts are in the minor groove of the DNA helix.

Within the DNA superhelix, clustered densities arise from the histone octamer. In general the disposition of the protein agrees with the lower-resolution neutron diffraction structure determined for the particle by Bentley et al (1984). No protein is found either outside the DNA or between the gyres of the superhelix, although an extended chain or disordered protein region would escape detection at the current level of resolution. Some penetration of protein between DNA gyres at positions about 30 and 50 bp to either side of the dyad was seen in the neutron study. The density corresponding to H3 is identified by the binding site for the mercurial heavy atom; H3 is the only histone with a cysteinyl residue. Presumptive identification of the other densities as specific histones is based on previous studies that showed that the octamer is made up of a tetramer of H3-H4 and two dimers of H2A-H2B, image reconstruction of negatively stained ordered aggregates of octamers (Klug et al 1980), and DNA-protein cross-linking experiments (Shick et al 1980). A number of rodlike densities are present; these are presumed to be α-helices. The densities identified as the H3-H4 tetramer are related by a dyad axis, which also fits the central turn of the DNA superhelix. The putative H2A-H2B dimers depart from dyad symmetry in the crystal: one dimer is rotated and shifted in position to allow H2A to bind to the DNA of an adjacent core particle. This change is likely forced by the packing in the crystal. The structure of the particle is such that the trajectory of linker DNA can not be a smooth continuation of the superhelix. Richmond et al (1984) postulated a 50–60° bend for the linker–core particle DNA junction; this affords a wedge-shaped cavity, which might accommodate the globular portion of histone H1 in chromatin.

The structure of the core particle at 15 Å resolution (Uberbacher & Bunick 1985a) was solved from X-ray diffraction data using molecular replacement and model building, as opposed to determining phases with heavy atom derivatives. This structure is similar to that reported by Richmond et al (1984); the differences are a more uniform DNA curvature, some protein extending between the DNA gyres, H2A near the ends of the DNA and H4 close to the dyad axis, and maintenance of symmetry in the positions of all histone densities.

Histone Octamer

One would hope that the solution of the core particle structure would be facilitated by determining the structure of the histone octamer crystallized at high ionic strength. Such a structure was recently determined at 3.3 Å resolution by Burlingame et al (1985), with the surprising result that it differs strikingly from the histones in the core particle structure (Richmond et al 1984). The overall shape of the octamer is a prolate ellipsoid 110 Å in length and 65–70 Å in maximum diameter; this is larger than the core particle, which has DNA wound on the outside of the histones. Like the core particle structure, the organization of the octamer is tripartite with a central H3-H4 tetramer flanked by H2A-H2B dimers. The electron density contiguous to the binding site for the mercury heavy atom, which includes a long α-helical segment, was identified as H3. Two domains within the central region of the octamer, separate from H3 but continuous with each other, are H4. Unlike the protein in the core particle structure, the octamer has large solvent channels between the central tetramer and the symmetrically related flanking dimers. It was reported that two polypeptide chains could be traced nearly end to end in the dimer. H2B was tentatively identified by characteristic amino acid sequence features, and therefore it was concluded that the other is H2A. These two polypeptides are rich in α-helical regions. At the current level of analysis the positions of the proteins are consistent with most cross-linking results obtained using formaldehyde. Since the orientations of the polypeptide chains in the octamer have been suggested only for H3 and H2B, it is not known whether this structure is consistent with the positions of cross-links between histones in chromatin (summarized in McGhee & Felsenfeld 1980).

A series of grooves on the outside of the octamer suggest a ramp that could bind DNA in the nucleosome; the path is of sufficient length to accommodate about 160 base pairs of DNA. The histones that contact DNA from end to end along this path are largely consistent with the results of DNA-protein cross-linking studies (Shick et al 1980). If this is indeed the structure of the nucleosome, its dimensions are strikingly different from those of the core particle structure. A vigorous discussion of these structures has ensued; the topics brought into question include the correctness of the two structures, the possibility of conformational changes in core particles vis-à-vis the nucleosome, hydrodynamic properties of various chromatin subunits and components and more (Klug et al 1985; Moudrianakis et al 1985a,b; Uberbacher & Bunick 1985b).

The crystallographic quality that allows definition of major and minor DNA grooves in the core particle and resolution of α-helices and some side chains in the octamer suggests that both structures may well be correct. The apparent paradox could be resolved if it is found that the octamer undergoes a major conformational collapse when complexed with DNA. This leads us to examine the evidence that has led most workers to believe that the structure of the histones is the same in solution and in chromatin. Cross-linking of histone octamers with dimethyl suberimidate free of DNA in high-salt solutions or in chromatin leads to the same pattern of dimers at the initial stage of reaction and a final octameric product (Thomas & Kornberg 1975), which suggests that the histones are similarly related in structure in each environment. However, dimethyl suberimidate is a long cross-linking agent, thus it does not detect interactions with the precision possible with zero-length cross-linkers. Note, however, that the H2A-H2B contact that leads to a UV-induced cross-link is the same in nuclei and the isolated dimer (Callaway et al 1985). On the other hand, a recent report indicates that H3 containing a modified cysteinyl residue fails to form octamers in 2-M NaCl (Lindsey et al 1985), although it does appear to form nucleosomes, even in vivo (Prior et al 1980). The infrared spectrum of the core particle, when the DNA contribution is subtracted out, is similar to that of the octamer (Cotter & Lilley 1977), although the spectrum consists of a single broad peak with a slight shoulder. Laser Raman spectra of octamers plus DNA are quite similar to the spectrum of the core particle in either low salt or 2-M NaCl (Thomas et al 1977). Both these methods detect largely the secondary structure of proteins, not their tertiary structure or the details of protein-protein interactions. X-ray scattering results for the octamer at high ionic strength were interpreted to indicate a disk-shaped particle about 12 nm in diameter and 3 nm in height (Damaschun et al 1983). A study of the accessibility of the amino and carboxyl terminal regions of H2A to tryptic digestion showed clear differences in the exposure of the carboxyl terminal region for different complexes; the region was trypsin resistant in octamers, somewhat susceptible in chromatin, and fully susceptible in the H2A-H2B dimer (Hatch et al 1984).

An octamer model based on three-dimensional reconstruction from electron micrographs of negatively stained helical stacks of octamers (Klug et al 1980) has been claimed to support the Richmond et al (1984) structure and to be at odds with the Burlingame et al (1985) version of the octamer (Klug et al 1985). However, this argument appears less than conclusive on the grounds of (a) the relatively low resolution of the electron microscopic study and (b) the uncertainty as to whether molecular envelopes deduced from negatively stained specimens convey, in their finest details, the native structure. The reconstruction was assigned a nominal resolution of 22 Å, although only one meridional reflection was present at spacings finer than 30 Å. There are now indications from several sources that while negative staining can yield models that are generally sound, literal interpretation of their finest details can be misleading. For instance, a comparison of reconstructions of crystalline ribosomes in the frozen-hydrated (presumably native) state and after negative staining documented significant discrepancies (Milligan 1985). In view of the abundance of basic residues in histones, some coincident positive staining would seem a real possibility for negatively stained particles. Additionally, although both studies were done at high ionic strength, the salts used differed. In summary, the evidence that the octamer assumes precisely the same structure with and without DNA is somewhat sparse.

Generally, solution chemical data on protein structures have been consistent with those on crystal structures, and there is no precedent for a conformational change of the magnitude necessary to allow the octamer free of DNA to fit into the center of the core particle. It has been assumed that the high ionic strength conditions, which are required to keep the octamer intact, provide charge neutralization equivalent to that in the core particle. The cooperative nature of protein interactions with a large polyanion such as the 96,000 dalton core particle DNA may, however, lead to structural differences hitherto unobserved for a protein. While the structures of some DNA-binding proteins have been determined alone and others as complexes with DNA, no protein other than the histone octamer has been studied both in its free state and in association with the nucleic acid.

Other Features

A dynamic feature of DNA structure in the core particle that potentially is of functional importance is the degree to which the nucleic acid is constrained from free rotation on the surface of the nucleosome. An investigation of local phosphodiester backbone mobility by triplet-state anisotropy decay concluded that rapid internal motions could occur for nucleosomal DNA about as readily as for DNA in solution (Wang et al 1982). The more global flexibility of DNA can be investigated by study of the thermal dependence of DNA twisting when in a nucleosome. Such a study using plasmid DNA reassociated with chicken erythrocyte inner histones led Morse & Cantor (1985) to conclude that more than 200 bp of DNA were constrained from twisting by each octamer of histones. This rather surprising result suggests that portions of linker DNA may interact with histones, a finding not suspected from previous work. In contrast, similar experiments performed in vivo in yeast suggest that DNA in yeast plasmid chromatin maintains 70% of the ability of protein-free DNA to twist in response to temperature changes (Saavedra & Huberman 1986). Possibly, the large fraction of 2 μm DNA that is transcribed is free to twist. If the constraints were spread throughout the plasmid chromatin, this would correspond to a limitation of twisting of only about 50 bp per core particle. Whether the differences in these results arise from the fact that the histones derived from the avian erythrocytes were from transcriptionally inert cells, while yeast cells are highly active, is a question of high interest.

It has long been suspected that the highly basic amino terminal regions of the histones are the primary sites of interaction with DNA and that their central, and likely more globular, portions are the sites for interactions with other histones. Grunstein and collaborators have recently used yeast genetics to address this question and found an unexpected result. There are two genes for H2A and two genes for H2B in yeast. However, either of the yeast H2B or H2A subtypes can provide proteins for a normal or nearly normal phenotype (Rybowski et al 1981; Kolodrubetz et al 1982), thus the amino acid differences between the subtypes for each histone seem to have no detectable effect on protein function. Furthermore, using in vitro mutagenesis of H2B genes on plasmids in yeast lacking both H2B subtypes, they found that deletion of up to 20 amino acid residues (positions 3–22) of the amino terminus of H2B has no detectable effect on phenotype (Wallis et al 1983). Histones lacking amino terminal regions (removed by tryptic proteolysis) had previously been shown to be capable of refolding core particle–length DNA (Whitlock & Stein 1977), but the finding that a protein lacking a highly conserved region of distinctive character could function throughout the cell cycle and through meiosis was unexpected.

HIGHER ORDER CHROMATIN STRUCTURE

The next chromatin structure to consider is the chromatosome. The chromatosome is similar to the core particle, but it contains about 165 bp of DNA and a molecule of histone H1. It is thought to contain two full superhelical turns of DNA, and the H1 is thought to bind at the points where the DNA enters and exits from its path around the inner histones. Chromatosomes are connected in tandem arrays by linker DNA, which we define here as the DNA between chromatosomes. Linker length varies from 0 bp in neurons to 80 bp in sea urchin sperm; most tissues have linkers of about 30–40 bp. There are also variations in linker DNA length within a single tissue. Tandem arrays of chromatosomes have a zigzag structure at very low ionic strength (Thoma et al 1979; Worcel et al 1981). Within interphase nuclei or metaphase chromosomes, this nucleofilament is further folded into a "compact" or "thick" fiber with a knobby, irregular appearance and an approximate diameter of 30 nm. The structure of this fiber has important implications for the mechanism of regulation of chromatin condensation and transcription. For instance, accessibility of the linker DNA to other proteins may be different in different regions of chromatin, depending on the exact structure of the compact fiber.

Because of solubility problems, solution studies on the thick fiber require use of low ionic strength buffers and relatively short segments of chromatin. It is encouraging to note that recent X-ray studies of whole cells, nuclei, and chromatin fibers indicate a similar structure for the fiber in solution and in situ (Langmore & Paulson 1983; Paulson & Langmore 1983; Widom & Klug 1985). Here we discuss studies of compaction of the thick fiber, nucleosome and linker orientation in the fiber, fiber structure and linker length, and the role and location of histone H1 in the fiber. We then use the results from these studies to evaluate the models that have been proposed to explain the organization of the thick fiber. These models are illustrated in Figure 1. Reference to the solenoid (Figure 1) will be of assistance in reading the next two sections. Our summaries of biophysical studies and features of H1 structure are brief; these have been reviewed in detail by others (Butler 1983; Cole 1984).

Fiber Structure

With increasing ionic strength, the zigzag chain observed at very low ionic strengths compacts through intermediate states with ill-defined beaded morphology. At ionic strengths ≥60 mM, compact fibers of about

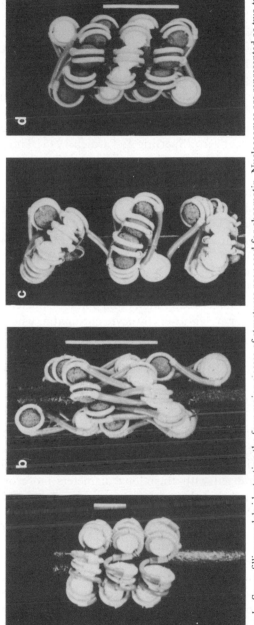

Figure 1   Space-filling models illustrating the four major classes of structures proposed for chromatin. Nucleosomes are represented as two turns of DNA wrapped around a central core of histones. H1 is not represented. Core particle DNA is light gray; linker DNA is dark gray. The white bar to the right of the structures in *a*, *b*, and *d* indicates the pitch of the helical structure. In *b*, *c*, and *d* alternate nucleosomes have histones colored white and medium gray. The tilt angle of the nucleosome is the angle made by the plane of the nucleosome disk and the long axis of the chromatin fiber (*a*) A single-start contact solenoidal model. This structure has a helical repeat of 6 nucleosomes, a diameter of 30 nm, and a pitch of 11 nm. (*b*) A twisted ribbon model. Note the single ribbon of odd (white) and even (gray) nucleosome cores. This structure has a helical repeat of 18 nucleosomes, a diameter of 30 nm, and a pitch of 32 nm. (*c*) A superbead model. This structure was formed by dislocation of a model such as shown in *d*. (*d*) A cross-linker model. Note the alternating ramps of odd and even nucleosomes. This structure has a helical repeat of 18 nucleosomes, a diameter of 31 nm, and a pitch of 26 nm. (Adapted from Williams et al 1986.)

30 nm in diameter with poorly resolved nucleosomes are observed. Both morphological and biophysical approaches have been used to evaluate the density of nucleosomes along the fiber. Many types of evidence indicate that nucleosome disks are arranged with the flat sides of the disk roughly parallel to the long axis of the 30-nm fiber (see below). Thus the diameter of the nucleosome disk (10–11 nm) equals the pitch of a simple contact solenoid of nucleosomes. Hence, the nucleosome density is usually described in terms of number of nucleosomes per 10–11 nm fiber length. The number of nucleosomes per 10–11-nm fiber length (the length corresponding to the diameter of the core particle disk; see below) increases from 1 at ionic strengths near zero (Sperling & Tardieu 1976; Thoma et al 1979) to 6–8 in 60–100-mM NaCl (Campbell et al 1978; Suau et al 1979; Engel et al 1980; Ausio et al 1984; Woodcock et al 1984; Williams et al 1986) or in the presence of millimolar concentrations of Mg^{2+} (McGhee et al 1983). Two much higher values for mass per unit length have been reported. Woodcock et al (1984) calculated 11.6 nucleosomes per 11 nm for chicken erythrocyte chromatin in 150-mM NaCl; their values at lower ionic strengths agreed with those of others. Williams et al (1986) calculated 12 nucleosomes per 10 nm by scanning transmission electron microscopy of *Thyone* (sea cucumber) sperm chromatin fixed at 60-mM NaCl; this chromatin has a long linker and a fiber diameter of 39 nm. Mass per unit length for *Necturus* erythrocyte chromatin, which has a 31-nm diameter fiber, was similar to that reported by others for 30-nm fibers. Most observations seem to indicate that the isolated 30-nm thick fiber has 6–8 nucleosomes per 10–11 nm axial length under conditions of moderate ionic strength; however, the actual number of nucleosomes per unit length in situ is not known.

In some electron micrographs, 11-nm oblique lines or striations are seen along the 30-nm fiber, which suggests that nucleosomes are not oriented with their flat faces in the axial direction (Finch & Klug 1976; Thoma et al 1979). Studies employing several different physical methods indicate that nucleosomes are arranged with their long diameters more or less parallel to the fiber axis and that the nucleosome disk is placed radially rather than tangentially (Suau et al 1979; McGhee et al 1980). Recently, Widom & Klug (1985) have obtained X-ray diffraction patterns from partially oriented compact fibers. Importantly, they found preferential orientations of several reflections. The major meridional reflection at 11 nm correlates with the outer dimension of the disk and the 34-nm equatorial reflection correlates with interfiber spacing. An equatorial 5.7-nm reflection is thought to arise from the disk thickness. Computation of expected fiber diffraction of the nucleosome core particles (from the known crystal structure) in various possible orientations gave a best fit with a radial orientation of the flat faces of the disk.

Dichroism methods have been used to assess the orientation of the core particle and linker in the thick fiber; electric dichroism requires making assumptions about the path of DNA to fit the data to a model. McGhee et al (1980; 1983) calculated tilt angles of 20–33° for the core particle relative to the fiber axis, based on the assumption that linker DNA continued smoothly along the superhelical path of the core particle. Fibers with linker lengths from 10 to 80 bp could be fit equally well by this model. Crothers and coworkers have developed photochemical dichroism, a method that allows separation of the dichroism of linker and core particle DNA and inferences about their relative orientations within the fiber (Mitra et al 1984; Sen et al 1986). They conclude that the angle between the nucleosome plane and the fiber axis is about 12° for long linker (80 bp) and 30° for short linker (10 bp) chromatin. Linker dichroism is interpreted as being incompatible with a continuously supercoiled linker. Thus, the plane of the nucleosome disk appears to be directed radially and to be tilted with respect to the fiber axis; the tilt may be a rather flexible feature of chromatin or depend on the linker length.

What other features of fiber structure vary as a function of linker length? This is an important question for evaluation of the model structures that have been proposed for the fiber. Hydrodynamic (see Butler 1983) and dichroism (McGhee et al 1983) measurements are consistent with fiber widths of about 30 nm for fibers from several tissues with widely varying linker lengths. Moreover, fiber diffraction patterns and fiber diameters estimated from electron micrographs were similar for chromatin from chicken erythrocytes ($\sim$40-bp linker) and from sea urchin sperm ($\sim$80-bp linker) (Widom et al 1985).

In contrast, other electron microscope measurements suggested that linker length does affect fiber dimensions. Fiber (or granule, see below) diameters in thin sections of nuclei from rat or chicken liver (30-bp linker), chicken erythrocytes, and sea urchin sperm were 26, 28, and 37 nm, respectively (Zentgraf & Franke 1984). Negatively stained chromatin fibers prepared from *Necturus* erythrocytes (30-bp linker) and *Thyone* sperm (65-bp linker) had diameters of 31 and 39 nm, respectively (Williams et al 1986).

To what might these differences in assessment of fiber diameter as a function of linker length be attributed? Hydrodynamic measurements average over the entire population of molecules and may not be sensitive enough to detect small differences, in particular since chromatin fragments are often irregular and knobby. Electron microscopy allows selection of straight fibers and selection against irregular features; however, it is not known whether the actual linker length of the particular segment of fiber measured is identical to that determined for the tissue as a whole. Variations in linker length might be expected to lead to local irregularities of higher order structure. The question is not closed, but evidence is building that there may indeed be a correlation between the diameter of the thick fiber and the chromatin linker length, with increased linker lengths producing fibers of greater diameter.

H1 Location and Role

The proteins of the histone H1 class (simply termed H1 for this review) are polymorphic and frequently modified; they also exhibit tissue and organism specificity. A common structure for all suggests a conserved function; each H1 contains a 40–amino acid random coil basic amino-terminal domain, a conserved central 80–amino acid globular domain, and a highly basic 100–amino acid carboxyl terminal domain (Hartman et al 1977). There is on average about one H1 per nucleosome, although the stoichiometry varies from about 0.8 to 1.3 (Bates & Thomas 1981).

As noted above, an H1 molecule is thought to tie together the two DNA turns in the chromatosome. Only chromatin containing H1, either native or H1 repleted after depletion, forms fibers (Finch & Klug 1976; Renz et al 1977; Thoma et al 1979; Allan et al 1981), which clearly demonstrates the pivotal role of this histone in formation of chromatin higher order structure. The exact location of H1 in the fiber is, unfortunately, not known; this information would contribute significantly to the understanding of fiber structure. A location inside the fiber is suggested by the observations of Losa et al (1984). They found that chymotrypsin can cut the globular region of H1 only under solvent conditions in which the fiber does not form, either after dissociation of H1 or unfolding of the fiber without histone dissociation. It seems likely that H1-H1 interactions may be necessary for formation of the fiber, so the results of cross-linking studies are germane to the solution of this structure. Ring & Cole (1983) found mainly carboxyl terminal cross-links in nuclei and Nicolaev et al (1983) found all possible amino and carboxyl terminal interactions in extended chromatin; however, more quantitative studies on purified chromatin fragments indicate that the major contacts are between amino and carboxyl terminal domains in extended chromatin (Lennard & Thomas 1985). This would suggest a head-to-tail arrangement for H1 along the length of the extended fiber. When compact fibers were studied, an increase in the number of cross-links between two carboxyl terminal regions was noted.

A number of studies have attempted to dissect the roles played by the various domains of H1 in the higher order structure of chromatin. The globular domain restores the chromatosome nuclease digestion pause, which suggests that it links the two DNA turns in the particle (Allan et al 1980). The carboxyl terminal domain of H1 appears to be its major site of binding to DNA at physiological ionic strength (Allan et al 1981; Thoma et al 1983). A major problem in reassociation studies with H1 or its domains is lack of a definitive means of confirming correct binding. Additionally, H1 in excess of that present in native chromatin can bind to either native or H1-depleted chromatin, inducing its condensation and eventual precipitation (Thoma & Koller 1981).

Models for Higher Order Structure

Based on the assumption that nucleosomes are 11-nm-diameter spherical particles, Finch & Klug (1976) postulated that a nucleosome chain can fold into a solenoid-like shape; i.e. a single-start contact helix, with a pitch of 11 nm and a diameter of about 30 nm, containing six nucleosomes per turn. The observation of a zigzag fiber at low ionic strength makes simple helical coiling unlikely, and an alternative hypothesis concerning the condensation of chromatin fibers has been developed by Thoma et al (1979). The zigzag fiber was interpreted as a contact helix with two nucleosomes per turn stabilized by H1-H1 interactions in the center. Condensation of the zigzag fiber into a compact fiber was postulated to occur by twisting the helix around its axis, increasing the number of nucleosomes per turn to six to eight, and decreasing the angle between nucleosome disks while keeping the pitch constant at about 11 nm and maintaining the radial orientation of the nucleosomes. Solenoidal models can accommodate variable length and internally located linkers provided the linker is allowed to form reversed loops, kinks, and/or sharp bends (see Butler 1984 for a detailed discussion of linker accommodation). A constant, internal location of the H1 binding site, suggested by data cited previously, is a major feature of the fiber structure suggested by this model.

In a variant of this model, McGhee et al (1980; 1983) proposed that the linker DNA continues the superhelical path of the nucleosomal DNA with about 80 bp per turn. This leads to variable orientation of the entry and exit sites of the nucleosomes in the fiber; nucleosomes connected by linkers of 40 bp would have the entry and exit sites alternating inside and outside, whereas H1 associated with 20- or 60-bp linkers would disrupt spacing between gyres of the solenoid. Recent photochemical dichroism results do not support such a continuous superhelical model for location of the linker DNA (Sen et al 1986), although these workers could not unambiguously define the linker path. The inaccessibility of H1 to chymotrypsin suggests that this histone is in the center of the fiber and does not support models in which H1 location is a function of linker length. If linker DNA is continuously supercoiled and H1 is in the center of the fiber, H1 must be able to bind at many sites on the nucleosome, not only at the entry and exit points.

Worcel et al (1981) and Woodcock et al (1984) proposed that the zigzag fiber, as a ribbon, winds helically to form thick fibers. Nucleosomes would be arranged radially. Linker DNA would run more or less parallel to the fiber axis; for long linkers this is difficult to reconcile with dichroism data (McGhee et al 1983). Since the width of the ribbon depends on linker length, these models predict linker length–dependent variations in spacing along the fiber. This has not been observed in X-ray diffraction studies (Widom et al 1985). Evidence supporting the coiled ribbon model (Woodcock et al 1984) is derived from electron micrographs that suggest face-to-face packaging of nucleosomes in the ribbon. However, few of the observed nucleosomes were closely spaced. A pitch angle of $23°$, steeper than that predicted by solenoidal models, was inferred from oblique striations of the fiber, but the measurements scattered over a rather broad range. Finally, it is not certain that these oblique striations reflect the helical ramps (see below). The models differ in the relative orientation of nucleosomes within the zigzag fiber in order to cope with the "linking number paradox," a problem which has been discussed in detail by others (Worcel et al 1981; Wang 1982; Butler 1983; Woodcock et al 1984).

Langmore and coworkers have reported a linker length dependence of fiber diameter and mass per unit length. A left-handed helical symmetry with a pitch of about 26 nm and a pitch angle of about $32°$ (much steeper than predicted by solenoidal models) was calculated from three (predominantly one-sided) optical diffraction patterns of electron micrographs of negatively stained fibers (Williams et al 1986). Their model proposes that the zigzag fibers form a two-start contact helix with the zig nucleosomes starting on one side and forming a ramp and the zag nucleosomes starting on the opposite side to form the second ramp. Nucleosomes are arranged radially in the cross-linker model, as in the solenoidal model, and the model structure can fit the X-ray data. Linker DNA would cross the center of the fiber. For very long linker DNA, the model structure may conflict with the electric dichroism data. For chromatin with no linker DNA, the model appears to be sterically impossible. The model fits well with a constant, internal location of H1 and is supported by studies that suggest linker length–dependent changes in fiber diameter. A caveat to this and the ribbon models is the possibility that the pitch angle measured in negatively stained fibers is not the pitch angle of the helical ramp of nucleosomes. If nucleosomes are arranged radially, as seems likely, there is a cleft between adjacent nucleosomes that might be penetrated by the stain. If nucleosomes of adjacent gyres are in close edge-to-edge contact, the internucleosomal grooves running from turn to turn might be the dominant stained feature.

Certain electron microscopic and biochemical observations led to a proposal that the thick fiber is organized as a repeating series of granular 30-nm "superbeads" (see references in Zentgraf & Franke 1984). Subsequent observations, reviewed in detail by Butler (1983) and Reeves (1984), suggest that this is unlikely. The knobby appearance of the fiber more likely reflects local variation in linker length, variation in DNA structure or protein content, and/or alterations in structure occurring as a consequence of the preparation of the material for morphologic analysis.

Despite the intrinsic irregularities of chromatin fibers, certain features of the structure of this entity are emerging. Fiber diameters may well vary with linker length. Nucleosomal disks appear to be oriented radially with a tilt between 10 and 30°, which may depend on linker length. The fibers are stabilized by interactions between H1 histones. Although most workers would prefer to place H1 and the linker DNA in the center of the fiber, the actual location of both species remains to be determined. The superbead and ribbon models are less consistent with current data than are the solenoidal and crossed-linker structural proposals, but it seems too early to draw a general picture of chromatin folding.

The microheterogeneity of chromatin fibers is a property that may reflect variable modes of packaging along the fiber. Certain regions may be compact and regular whereas others may be less well organized in order to adapt to functional requirements or packaging. One approach to determining the structures nucleosome chains can adopt would be to reconstitute chromatin with constant linkers and precisely positioned nucleosomes (Simpson et al 1985), which would allow high-resolution solution and, perhaps, crystallographic studies of this important level of organization of the eukaryotic genome.

STRUCTURE OF ACTIVE CHROMATIN

At present, few explicit statements can be made about the structure of the small fraction of chromatin that is transcriptionally active. Nonetheless, investigations of the structure of nuclease-hypersensitive regions (HSRs) and nucleosomes in transcribed regions (reviewed below) and of the existence and mechanisms of nucleosome positioning (reviewed by Simpson, 1986) suggest that the following general concepts can be applied to relating transcription to chromatin structure : (a) Alterations of canonical B-form DNA structure occur because of DNA sequence variation and modification and interactions of DNA with proteins, both histones and non-histones. These variations may function in aspects of DNA metabolism. For instance, propagation of an altered DNA structure might confer a readily dissociable character to nucleosomes in transcribed regions. (*b*) Some variations in DNA structure can be accommodated in nucleosomes, while other sequences or interactions with factors may occur only in nucleosome-free regions. Yet other factors may bind only to nucleosomal DNA. We elaborate on these concepts by considering DNA structure and protein binding in nonnucleosomal regions, especially HSRs, and then describe what is known concerning the nucleosomes of transcriptionally active regions.

Because HSRs are exceptionally sensitive to different nucleases that have varying structural specificities, they most likely lie between and not within nucleosomes. Their locations suggest that HSRs are important to DNA function. Thus, HSRs are associated with promotors and enhancers of transcriptionally competent genes. HSRs also appear to mark origins of replication (Palen & Cech 1984; Thoma 1986) and regions at which site-specific recombination occurs (Nasmyth 1982). HSRs therefore represent landmarks formed by and reflecting a variety of different chromatin structures. Although HSRs are commonly defined as regions with an enhanced nuclease sensitivity relative to that of neighboring regions during a brief digestion, a site could be viewed as truly hypersensitive only if its sensitivity is enhanced relative to free DNA. However, such an enhancement is difficult to determine, both since the target size and accessibility of a sequence in chromatin differ from those in free DNA. Comparing a DNA sequence in different tissues or physiological states may provide a more reliable indication of a locus at which biologically important chromatin structure changes occur.

A common feature of HSRs, so defined, must be an alteration of the DNA structure at the sensitive site. For this reason a discussion of possible DNA structures at an HSR is warranted. Possibilities include bent DNA, the unwound and A-form DNA associated with transcription, "alternate" (including underwound and supercoiled) B-form DNA structures, the left-handed Z-form DNA, and DNA modified by cytosine methylation.

DNA Structure at Hypersensitive Regions

Bending of DNA in some protein-DNA interactions is supported by X-ray crystallographic and electrophoretic studies. It has been invoked to explain the finding that DNA-binding proteins can sometimes interact even when separated by variable lengths of DNA so long as the insertions do not alter the relative helical orientation of the binding sites (Dunn et al 1984). Nucleases sensitive to bending-induced variation in the width of major and minor grooves, like deoxyribonuclease I (Drew 1984), could detect bent DNA at HSRs. However, it is also likely that alterations in groove geometry induced by under- or overwound DNA, for example, will affect cutting by DNAse I. Thus, in the absence of corroborating physical studies on the proteins bound at HSRs, it is risky to make inferences from nuclease digestion data about whether DNA is bent at an HSR.

During transcription, DNA is partially unwound and a short RNA-DNA hybrid region is formed. Some structure associated with these changes may be nuclease hypersensitive. If a gene is transcribed at extremely high frequency, RNA polymerase–DNA complexes, stalled at various places along a gene in isolated nuclei, may be detected in aggregate as a large HSR. This proposition would explain the observation that the entire transcribed region of highly active genes is nuclease hypersensitive.

Underwound or torsionally stressed DNA may be present at HSRs, as suggested by the observation that some HSRs are sensitive to S1 nuclease (S1), which is usually thought to be a single-strand specific nuclease. The same or nearby sequences are also S1 sensitive in supercoiled but not relaxed or linear plasmids (Larsen & Weintraub 1982). In investigating the DNA structure of S1-sensitive chromatin, Weintraub (1983) found that the S1 sensitivity of in vitro reconstituted chromatin was lost when reconstitutes were digested with restriction enzymes, which supports the notion that S1-sensitive HSRs are under torsional stress. However, the S1 sensitivity of native β-globin chromatin was not eliminated by treatment of nuclei with restriction nucleases, DNA topoisomerase I, or ethidium bromide (Larsen & Weintraub 1982).

In naked DNA plasmids, including those containing β-globin gene sequences, S1-sensitive sites are often found in homopurine–homopyrimidine stretches (Evans et al 1984). Kohwi-Shigematsu et al (1983) found that the S1-sensitive region of the chicken β^A-globin gene was reactive with bromoacetaldehyde (which leaves DNA sensitive to S1 after linearization by restriction digestion) and concluded that the S1-sensitive region contains unpaired nucleotides. Naked DNA reacted with bromoacetaldehyde only when supercoiled (Kohwi-Shigematsu & Kohwi 1985). In contrast, Schon et al (1983) studied the S1 sensitivity of the chicken β-globin gene in supercoiled plasmids and found that S1 nicks only one of two strands in some instances and that, where testable, S1-sensitive regions could be cut by restriction enzymes. This, together with examination of cutting patterns at varying temperatures, led them to conclude that S1 sensitivity was not a reflection of single strandedness or DNA breathing. In addition, Cantor & Efstratiadis (1984) found that the N-1 position of adenines and the N-3 position of cytosines of this S1-sensitive sequence were not methylated by dimethylsulfate, as they would be were the region single stranded. They suggested that the region may be in a left-handed DNA conformation. Experiments with dimethylsulfate and examination of the linking number change required to generate S1 sensitivity as a function of pH led Pulleyblank et al (1985) to suggest that poly d(TC)-poly d(GA) is right-handed with Watson-Crick base pairs alternating with protonated Hoogsteen base pairs. Importantly, Pulleyblank et al found that this sequence reacts with an anti-Z DNA antibody raised against brominated poly d(GC), which suggests that caution should be taken in studies on Z DNA that employ these antibodies.

Whether Z DNA exists in vivo is still an open question, given the concern over anti-Z DNA antibody specificities and the observation that the extent of antibody reactivity in situ varies with the pH of the chromatin fixation (Hill & Stollar 1983). The best evidence for its in vivo existence lies in the isolation of Z DNA binding proteins (Nordheim et al 1982; Azorin & Rich 1985). These proteins preferentially bind Z over B DNA, but histones [which apparently do not form nucleosomes with long stretches of Z DNA (Nickol et al 1982)] bind Z DNA as well as they do B DNA (Azorin & Rich 1985). Thus, it is difficult to rule out the possibility that Z DNA binding proteins may function in vivo through interactions with B or "alternate B" DNA. Development of chemical probes for Z DNA and B-Z boundaries should help clarify the existence and role of this structure in vivo (Johnston & Rich 1985). Z DNA itself is S1 resistant but B-Z junctions are sensitive (Kang & Wells 1985), which suggests that Z DNA could be found at HSRs. Poly d(TG)-poly d(CA) is a sequence widely present in lengths of about 50 bp in eukaryotic genomes, and it is capable of forming the Z structure in vitro. However, Gross et al (1985) did not find these sequences to be preferentially located at or near S1-sensitive HSRs. Rather, they were in nucleosome core particles with an apparent helical repeat of 10.5 bp rather than the 12 bp repeat that would be obtained for Z DNA.

While an inverse correlation between cytosine methylation and gene activity exists (reviewed by Doerfler 1983), methylation per se does not appear to create or inhibit nuclease hypersensitivity. For instance, Nickol & Felsenfeld (1983) found that methylation at Hpa II sites has no effect on the S1 sensitivity of supercoiled plasmids containing the 5′ HSR sequences of the chicken β-globin gene. Nonetheless, cytosine methylation can affect protein interactions, as evidenced by the existence of restriction enzymes whose action is inhibited or enhanced by DNA methylation and by the association of histone H1 with methylated DNA sequences (Ball et al 1983).

Protein Content of HSRs

Proteins at or near HSRs probably include transcription factors (reviewed by Dynan & Tjian 1985). We limit our review to examples of work explicitly linking chromatin structure and binding of transcription factors or other proteins. Such links have developed through investigations on the fine structure of HSRs and from functional studies on purified transcription factors. For instance, a subset of SV40 molecules are nuclease hypersensitive in a region containing transcription promotors and the replication origin. Choder et al (1984) labeled transcriptionally active SV40 molecules by elongating preinitiated transcripts in vitro and separated the polymerase-RNA-DNA ternary complexes by electrophoresis or sedimentation. When they briefly digested molecules with DNAse I prior to labeling, they found that the transcriptionally active fraction had been preferentially linearized, which indicates that the transcriptionally active molecules contain an HSR. In what way then does the SV40 HSR facilitate or reflect binding of transcription factors?

Sp1, which is involved in the transcription of many genes (Gidoni et al 1985), binds within the 300-bp SV40 HSR; RNA polymerase II and the general transcription factor Sp2 are also required for in vitro transcription of SV40 (Dynan & Tjian 1983). Sp1 can bind all six GGGCGG boxes in the imperfect 21-bp tandem repeat region with differing affinities. In one instance, binding to one box (V) prevents efficient binding to an adjacent box (IV). Binding to different subsets of the six boxes alternately enhances early and late transcription. Thus Sp1's action is bidirectional. Jongstra et al (1984) found that some features of the 300-bp HSR depend solely on the 72-bp repeats and presumably not on Sp1 binding. However, hypersensitivity of "region I," adjacent to the 21-bp repeats, depends on the presence of the 21-bp repeats, perhaps due to binding by Sp1. Thus, Sp1 may act by propagating an altered DNA structure, which is then detected as an HSR. Alternatively, region I hypersensitivity may reflect binding of another factor stabilized by interaction with bound Sp1. Support for the latter notion comes from HeLa cell transfection experiments by Takahashi et al (1986). They found that insertion of 1/2, 3/2, or 5/2 helical turns of DNA between the 21-bp repeats and the 72-bp repeat enhancer region reduced early and late transcription more (five-to tenfold) than did insertions of 1 or 2 helical turns (about twofold). This suggests an interaction between different proteins binding on the same face of the DNA helix in the 21-bp repeat and enhancer regions. Similarly, insertions of 10 and 21 bp between the early gene TATA box and the 21-bp repeats were less detrimental to early gene transcription than were 4- and 15-bp inserts. Since Sp1 binding is not affected by the absence of the TATA sequences, Sp1 may facilitate secondary binding by the TATA factor through protein-protein interactions. The helical dependence seen with insertions of varying length is consistent with DNA bending in these interactions.

Yang et al (1985) formed topoisomerase II–DNA cross-links by treating SV40 infected cells with an enzyme inhibitor and then with sodium dodecyl sulfate. They mapped the site of these cross-links to the SV40 HSR by indirect end labeling. Importantly, the abundance of these cross-links was highest late in infection, which suggests that association with topo-isomerase II may be related to replication.

Zaret & Yamamoto (1984) found that an HSR developed in parallel with transcription induction by dexamethasone and declined after hormone withdrawal. The location of the HSR in the enhancer-like glucocorticoid response element and studies that demonstrate hormone receptor binding to this element strongly suggest that the HSR forms as a result of receptor protein binding.

In characterizing the fine structure of HSRs 5′ to *Drosophila* heat-shock genes, Wu (1984a) found discrete exonuclease III barriers near DNAse I or restriction enzyme cuts made in HSRs and proposed that the TATA box region was associated with a protein. Changes in the exonuclease III pattern upon heat shock suggested that heat shock causes a second protein ("heat-shock activating protein") to associate with a promotor element 5′ to the TATA box. Costlow et al (1985) found that deletions in sequences bound by the heat-shock activating protein did not eliminate the DNAse I hypersensitivity of that region, whereas deletions in the TATA region did. Thus, the TATA box–associated protein may establish a hyper-sensitive region and facilitate binding of the heat-shock activating protein. By adding 0.4-M NaCl nuclear extracts from heat-shocked cells to nuclei from normal cells (Wu 1984b) or to naked DNA (Wu 1985), Wu was able to reproduce the exonuclease III resistance pattern characteristic of heat-shocked cells. The sequences bound by the TATA box and heat-shock activating proteins appear to associate with, respectively, the B factor (Parker & Topol 1984a) and the transcription factor specific to the heat-shock gene (Parker & Topol 1984b; Topol et al 1985), which were exten-sively purified and characterized by Parker and coworkers. This suggests that these transcription factors cause the observed effects on chromatin structure.

Emerson & Felsenfeld (1984) recreated an HSR 5′ to chick β-globin sequences by adding 0.15-M NaCl nuclear extracts from tissues in which that gene is expressed to DNA prior to, or along with, the addition of histones. In the presence of competitor DNA, they found preferential retention of 5′ gene sequences on nitrocellulose after incubation with extracts purified by DNA cellulose chromatography. Assuming that one molecule of factor binds each molecule of specific DNA, the globin gene factor has a specific binding constant of roughly 1.5×10^{12} liters/mol and a specific/nonspecific DNA binding ratio of 3.5×10^4 (Emerson et al 1985). Footprinting experiments showed protection of two regions, each about 25-bp long and separated by about 15 bp. When tested separately each region still gave a footprint, which indicates that binding to the two regions is independent. The fact that there are different sequences in the two regions suggests that two factors are being detected; the binding constants measured by nitrocellulose filter binding would apply to the tighter binding protein of the pair. Importantly, the in vitro footprint corresponds well with a high-resolution in vivo footprint (Jackson & Felsenfeld 1985). Characterization of globin gene transcription factors (Bazett-Jones et al 1985) may soon permit us to relate binding by the HSR factors to transcription.

Westergaard and colleagues found that the SDS-promoted cleavage at HSRs in *Tetrahymena* large ribosomal chromatin was due to a topo-isomerase I, which they have partially purified (Bonven et al 1985). They determined that the enzyme has a sequence specificity that is largely dependent on Ca^{2+} in the reaction, which indicates that its location in HSRs is not simply a reflection of ease of access. This finding has important implications for studies on topoisomerase I binding. Cytological studies on *Drosophila* by Fleischman et al (1984) indicate that topoisomerase I is associated with heat-shock loci after, but not before, induction of transcription by heat shock. One level at which topoisomerase I may play a regulatory role in transcription is suggested by the finding of Constantinou et al (1985) that the regulatory subunit of a cyclic AMP–dependent protein kinase, which upon binding cyclic AMP dissociates from the kinase subunit and is phosphorylated, acts as a topoisomerase I in vitro.

Assembly and Transcription of Active Chromatin

The preceding section concentrated on genes for which clear connections exist between binding by nonhistone proteins associated with transcription and disruption of chromatin structure. Of interest is how nucleosome structure might be locally dismantled either prior to, or by, the binding of transcription factors. In principle, this could be initiated by the binding of a nonhistone protein to nucleosomal DNA. Experiments by Schlissel & Brown (1984) suggest that transcription factor IIIA (TF IIIA) may initiate formation of a 5S gene transcription complex in such a manner (Bieker et al 1985; Setzer & Brown 1985). They found that after extraction of *Xenopus* erythrocyte chromatin with 0.6-M NaCl or the cationic resin BioRex-70, two treatments that remove histone H1, the normally inactive oocyte-type 5S genes could be transcribed by the addition of transcription factors and RNA polymerase III. Such extracted oocyte 5S gene chromatin became refractile to transcriptional activation after readdition of histone H1. Although earlier work suggested that TF IIIA cannot form a functional complex in the presence of already-associated histones (Bogenhagen et al 1982; Gottesfeld & Bloomer 1982; Gargiulo et al 1984), this result suggests that TF IIIA may bind to H1-depleted nucleosomes. It seems likely that such binding would be affected by whether essential DNA contacts are turned inward or outward from the histone core, which would in turn depend on nucleosome position with respect to DNA sequence. In addition, as suggested previously, factors that require or induce an altered DNA twist may not be able to bind nucleosomal DNA. TF IIIA binding to the 5S RNA gene results in only a small unwinding of one bp (Hanas et al 1984) or two to four bp (Reynolds & Gottesfeld 1983) per gene. Regardless of exactly how TF IIIA initiates binding in chromatin, we presume that complete formation of a transcription complex, which involves subsequent binding by transcription factors IIIC and IIIB and RNA polymerase III, results in nucleosome displacement, since recent studies on 5S gene chromatin suggest that the coding region is nucleosome free (Pederson et al 1984; Cartwright & Elgin 1984).

To test the notion that TF IIIA binds to H1-depleted nucleosomes, Rhodes (1985) reconstituted core histones and *Xenopus* somatic 5S gene DNA and found a nonrandomly positioned nucleosome containing the 5′ portion of the TF IIIA binding sequence (nucleotides +45–97, where nucleotide +1 is the 5S gene transcription start site). A comparison of prominent DNAse I cuts near position +60 bp within both the nucleosome and TF IIIA footprints suggests that the core histones and TF IIIA bind to the same face of the DNA helix. TF IIIA added to the reconstituted nucleosome changed the nucleosome's electrophoretic mobility in a nucleoprotein gel, which indicates formation of a ternary complex. The nature of this complex was studied by footprint analysis using DNA end-labeled at the 3′ side of the gene. The footprint of the nucleosome alone suggests that its 3′ border lies at about +70 bp (the 5′ border lies at about −77 bp). The ternary-complex footprint is largely nucleosome-like to about +60 bp and largely TF IIIA–like from +60 bp through the end of the TF IIIA binding domain. Similar results have been obtained in our laboratory (P. C. FitzGerald, unpublished observations). Thus TF IIIA appears to be incompletely bound, although it may displace nucleosomal DNA over a short region where the footprint is neither fully nucleosomal- nor TF IIIA–like. Rhodes suggested that this footprint in the overlap region may reflect a competition that causes alternate binding of the DNA in the region to histones or TF IIIA. Whether TF IIIA interacts with nucleosomal DNA in the same fashion in vivo is not known.

Even for a gene whose promotor lies outside the transcribed region, topological considerations suggest that transcription (and replication) cannot occur through the intact nucleosome. This is difficult to test experimentally because most sequences are only infrequently associated with polymerases. Most active genes appear to exist as nucleosomal chromatin; this finding suggests that nucleosome disruption occurs only long enough to permit passage of the polymerase, which would not be detectable, except possibly in the case of very rapidly transcribed sequences. In the following section we review recent work on the structure and behavior of nucleosomes of transcribed sequences.

The micrococcal nuclease-generated nucleosome ladder associated with heavily transcribed genes often appears less distinct than it does for the same genes in low activity or inactive states (Ness et al 1983; Rose & Garrard 1984; Cohen & Sheffery 1985; Wu & Simpson 1985). This observation is consistent with the notion that nucleosomes on transcribed sequences are present only part of the time or are disrupted in some other way. Similarly, some indirect end-label studies find that sequence-preferred cuts found in the digestion of naked DNA are also made in transcribed regions of chromatin more often than in nontranscribed chromatin (Thoma et al 1984). While such observations have sometimes been interpreted as indicating that nucleosomes of transcribed regions are not positioned, they are equally consistent with the possibility that nucleosomes of transcribed regions are less stable than nucleosomes of inactive regions and that polymerase activity prior to nucleus isolation, or nuclease digestion per se, result in occasional nucleosome dissociation or disruption.

Karpov et al (1984) cross-linked histones to DNA in nuclei and analyzed the products by two-dimensional gel electrophoresis and hybridization. Possibly because the cross-linking is not quantitative, only some of the nontranscribed sequences migrate with a mobility characteristic of H1-containing nucleosomes and the remainder show mobility characteristic of H1-depleted nucleosomes or free DNA. However, sequences 5′ to the *hsp 70* gene (both before and after heat shock) and its coding sequences (after heat shock) could not be cross-linked to histones. This suggests that octamer histones are fully displaced from nucleosomal DNA (rather than, say, unfolded) during transcription of activated heat-shock genes. Work by Stein et al (1985) indicates that such displacement may occur with only a small expenditure of energy. Their finding that core histone octamers can bind to nucleosomes suggests that a core octamer may associate transiently with another nucleosome. In this view, the protein components of a nucleosome remain confined to the neighborhood and available to reassociate with DNA after passage of RNA polymerase. Their finding that histone octamers bound to nucleosomes transfer to DNA more readily than do histones in nucleosomes suggests that, if a dissociation-reassociation cycle occurs, the octamer displacement from nucleosomes is rate limiting.

We have summarized the evidence for stressed DNA at HSRs and for topoisomerases that create or reside at, HSRs. Stressed DNA may be distributed throughout transcribed chromatin. This possibility was suggested by the experiments of Ryoji & Worcel (1984), who found that 5S RNA synthesis from circular plasmid molecules injected into *Xenopus* oocytes increased in parallel with supercoiling of the injected substrate and in parallel with nucleosome assembly (judged by the progressive increase in definition of a nuclease-generated oligomeric ladder). Injection of topoisomerase I, DNAse I, or the topoisomerase II inhibitor novobiocin into oocytes containing assembled minichromosomes led to complete relaxation of a fraction of the plasmid molecules carrying the 5S gene. Injection of novobiocin also resulted in a virtually complete shutdown of 5S RNA synthesis, which suggests that the transcriptionally active molecules are those that contain unrestrained superhelical density. This is reminiscent of observations by Luchnik et al (1985), who found that a small fraction of SV40 chromatin could be completely relaxed by topoisomerase I and that this fraction cosedimented with RNA polymerase–containing DNA molecules. The inhibition of 5S RNA synthesis by novobiocin suggested to Ryoji & Worcel that a previously unknown eukaryotic gyrase was involved in transcription. In apparent conflict with this inference is the observation that yeast with a temperature-sensitive topoisomerase II mutation are defective only in mitosis, apparently due to their inability to decatenate replicated DNA (DiNardo et al 1984; Holm et al 1985). Other ATP-requiring enzymes besides topoisomerase II may be inhibited by novobiocin.

The fraction of plasmid molecules that are completely relaxed may reflect assembly of the transcriptionally active DNA into unstable nucleosomes. As suggested above, such nucleosomes may be prone to rapid dissociation, transiently creating unrestrained superhelical density prior to their reassociation. If one assumes that reassociation is promoted by local superhelical density, the relaxation of transiently formed supercoils by injection of topoisomerase I or DNAse I would be inhibitory, as observed.

Ryoji & Worcel (1984) found that a fraction of the plasmids containing only bacterial sequences were also assembled into this so-called dynamic chromatin. This observation indicates that the pathway of chromatin assembly is not the sole component determining transcriptional specificity. Nonetheless, transcription factors may be involved in specifying pathways of chromatin assembly. Kmiec & Worcel (1985) found that TF IIIA enhances or accelerates a supercoiling reaction in a S-150 extract from *Xenopus* oocytes, that this reaction has a specificity for 5S gene–containing sequences on the same order as TF IIIA has for binding 5S promotor over vector sequences, and that this reaction is novobiocin sensitive. Assembly of an active 5S gene transcription complex, which requires ATP (Bieker et al 1985), may be accompanied by association with topoisomerases, which would accelerate nucleosome assembly. A similar mechanism might explain the observation of Glikin et al (1984) that assembly of dynamic chromatin on a number of DNA's in oocyte extracts is enhanced by ATP and Mg^{2+}.

Progress in determining compositional differences between active and inactive nucleosomes has come mainly from indirect methods (reviewed by Reeves 1984). Recently, doubt has been cast on the suggestion that nucleosomes of active sequences are associated with ubiquitinated histones (Barsoum & Varshavsky 1985). Huang et al (1986) found that the electrophoretic mobility of active κ immunoglobulin gene nucleosomes was not altered by treatment of the nucleosomes with isopeptidase to remove ubiquitin moieties. Development of methods to purify unique genes as chromatin (Workman & Langmore 1985; Pederson et al 1986) should soon make direct compositional comparisons possible.

Our review of HSR structure suggests the existence of at least two families of HSR-associated proteins. One family consists of proteins like the glucocorticoid receptor and globin gene factor(s) that create an HSR upon binding. A second family consists of proteins, like the heat-shock transcription factor, that bind only at preexisting HSRs. Although it has not been demonstrated directly, present observations suggest that TATA binding protein(s) and the transcription factor Sp1 belong to the first family and play a role in establishing HSRs, while topoisomerase I belongs to the second family. For those HSRs for which compositional features are known, nonhistone proteins appear to be associated with the HSR; no HSRs have been shown to be due to sequence-related interactions with histones. We have reviewed evidence concerning possible disruption of nucleosomes in transcribed chromatin. The structural basis for such instability may lie in compositional differences between nucleosomes of transcriptionally active and inactive regions; it is unclear how such differences might come about. The possibility that unstable nucleosomes assemble in response to propagation of an alternate DNA structure from an HSR has not been demonstrated, but it is consistent with findings that HSR formation and transcriptional activity are related and that at least some HSR-forming factor(s) must bind prior to, or concurrent with, nucleosome assembly.

SUMMARY AND PROSPECTS

This brief review of three selected areas of chromatin structure documents the progress alluded to in the introduction. However, many problems remain in each of the areas examined. High-resolution studies of core particle structure, together with localization of side chains in the octamer structure, should resolve the confusing differences between the currently proposed structures for the histones in these two situations. Crystallographic analysis of the chromatosome would provide details that would elucidate exactly how H1 makes its primary association with the remainder of chromatin. Experiments designed to test the solenoidal and crossed-linker models for higher order chromatin structure should allow critical evaluation of these proposals. Hopefully, a unique fragment of higher order chromatin containing tandemly repeated and exactly positioned nucleosomes can be made and crystallized, making the linker path and H1-H1 interactions accessible to detailed experimental analysis.

The road to understanding the structure of active chromatin is likely to be somewhat longer. Isolation of gene regulatory factors and the study of their interactions with DNA is a good beginning. These studies must be extended to an evaluation of their interactions with chromatin, and perhaps also with nuclear matrices and polymerases. Even in prokaryotic systems, transcription and replication are not fully understood. In eukaryotic systems, the cast of actors is larger, and with the exception of studies on yeast, the contributions of genetics to the analysis are much less. The possibility of isolating unique genes as chromatin in different functional states is an exciting one for the study of DNA metabolism in eukaryotic cells; its potential remains to be realized.

ACKNOWLEDGMENTS

We recognize discussions with and criticism from many members of our laboratory as contributing to this review. We specifically thank Drs. Alasdair Steven, Alan Kimmel, and Jurrien Dean for critiques of the manuscript. We are grateful to Dr. John Langmore for photographs of his models of proposed higher order chromatin structures.

Literature Cited

Allan, J., Cowling, G. J., Harborne, N., Cattini, P., Craigie, R., et al. 1981. Regulation of the higher order structure of chromatin by histones H1 and H5. *J. Cell Biol.* 90: 279–88

Allan, J., Hartmann, P. G., Crane-Robinson, C., Aviles, F. X. 1980. The structure of histone H1 and its location in chromatin. *Nature* 288: 675–79

Ausio, J., Borochov, N., Seger, D., Eisenberg, H. 1984. Interaction of chromatin with NaCl and MgCl₂. *J. Mol. Biol.* 177: 373–98

Azorin, F., Rich, A. 1985. Isolation of Z-DNA binding proteins from SV40 minichromosomes: Evidence for binding to the viral control region. *Cell* 41: 365–74

Ball, D., Gross, D., Garrard, W. 1983. 5-methyl cytosine is localized to nucleosomes that contain histone H1. *Proc. Natl. Acad. Sci. USA* 80: 5490–94

Barsoum, J., Varshavsky, A. 1985. Preferential localization of variant nucleosomes near the 5′ end of the mouse dihydrofolate reductase gene. *J. Biol. Chem.* 260: 7688–97

Bates, D. L., Thomas, J. O. 1981. Histones H1 and H5: One or two molecules per nucleosome? *Nucleic Acids Res.* 9: 5883–94

Bazett-Jones, D. P., Yeckel, M., Gottessfeld, J. M. 1985. Nuclear extracts from globin synthesizing cells enhance globin transcription in vitro. *Nature* 317: 824–28

Bentley, G. A., Lewit-Bentley, A., Finch, J. T., Podjarny, A. D., Roth, M. 1984. Crystal structure of the nucleosome core particle at 16 Å resolution. *J. Mol. Biol.* 176: 55–75

Bieker, J. J., Martin, P. L., Roeder, R. G. 1985. Formation of a rate limiting intermediate in 5S RNA gene transcription. *Cell* 40: 119–27

Bogenhagen, D. F., Wormington, W. M., Brown, D. D. 1982. Stable transcription complexes of Xenopus 5S RNA genes: A means to maintain the differentiated state. *Cell* 28: 413–21

Bonven, B., Gocke, E., Westergaard, O. 1985. A high affinity topoisomerase I binding sequence is clustered at DNAse I hypersensitive sites in *Tetrahymena* r-chromatin. *Cell* 41: 541–51

Burlingame, R. W., Love, W. E., Wang, B. C., Hamlin, R., Xuong, N.-H., et al. 1985. Crystallographic structure of the octameric histone core of the nucleosome at a resolution of 3.3 Å. *Science* 228: 546–53

Butler, P. J. G. 1983. The folding of chromatin. *CRC Crit. Rev. Biochem.* 15: 57–91

Butler, P. J. G. 1984. A defined structure of the 30 nm chromatin fiber which accommodates different nucleosomal repeat lengths. *EMBO J.* 3: 2599–2604

Callaway, J. E., DeLange, R. J., Martinson, H. G. 1985. Contact site of histones 2A and 2B in chromatin and in solution. *Biochemistry* 24: 2686–92

Campbell, A. M., Cotter, R. I., Pardon, J. F. 1978. Light scattering measurements supporting helical structures for chromatin in solution. *Nucleic Acids Res.* 5: 1571–80

Cantor, C., Efstratiadis, A. 1984. Possible structures of homopurine-homopyrimidine S1 hypersensitive sites. *Nucleic Acids Res.* 12: 8059–72

Cartwright, I. L., Elgin, S. C. R. 1984. Chemical footprinting of 5S RNA chromatin in embryos of *Drosophila melanogaster*. *EMBO J.* 3: 3101–8

Choder, M., Bratosin, S., Aloni, Y. 1984. A direct analysis of transcribed minichromosomes: All transcribed SV40 minichromosomes have a nuclease hypersensitive region within a nucleosome free domain. *EMBO J.* 3: 2929–36

Cohen, R. B., Sheffery, M. 1985. Nucleosome disruption precedes transcription and is largely limited to the transcribed domain of globin genes in murine erythroleukemia cells. *J. Mol. Biol.* 182: 109–29

Cole, R. D. 1984. A minireview of microheterogeneity in H1 histone and its possible significance. *Anal. Biochem.* 136: 24–30

Constantinou, A., Squinto, S., Jungman, R. 1985. The phosphoform of the regulatory subunit RII of the cyclic AMP dependent protein kinase possesses intrinsic topoisomerase activity. *Cell* 42: 429–37

Costlow, N. A., Simon, J. A., Lis, J. T. 1985. A hypersensitive site in *hsp70* chromatin requires adjacent not internal DNA sequence. *Nature* 313: 147–49

Cotter, R. I., Lilley, D. M. 1977. The conformation of DNA and protein within chromatin subunits. *FEBS Lett.* 82: 63–68

Damaschun, H., Zalenskaya, I. A., Damaschun, G., Vovob'ev, V. I., Misselwitz, R., et al. 1983. Shape and α-helix content of chromatin core histone particles in solution at high ionic strength. *Stud. Biophys.* 97: 105–12

DiNardo, S., Voelkel, K., Sternglanz, R. 1984. DNA topoisomerase II mutants of *Saccharomyces cerevisiae*: Topoisomerase II is required for segregation of daughter molecules at the termination of DNA replication. *Proc. Natl. Acad. Sci. USA* 81: 2616–20

Doerfler, W. 1983. DNA methylation and gene activity. *Ann. Rev. Biochem.* 52: 93–124

Drew, H. R. 1984. Structural specificities of five commonly used DNA nucleases. *J. Mol. Biol.* 176: 5350–57

Dunn, T. M., Hahn, S., Ogden, S., Schlief, R. F. 1984. An operator at 280 base pairs that is required for repression of *araBAD* operon promoter: Addition of DNA helical turns between the operator and promoter cyclically hinders repression. *Proc. Natl. Acad. Sci. USA* 81. 5017–20

Dynan, W. S., Tjian, R. 1985. Control of eukaryotic messenger RNA synthesis by sequence specific DNA binding proteins. *Nature* 316: 774–78

Dynan, W. S., Tjian, R. 1983. Isolation of transcription factors that discriminate between different promoters recognized by RNA polymerase II. *Cell* 32: 669–80

Eissenberg, J. C., Cartwright, I. L., Thomas, G. H., Elgin, S. C. R. 1985. Selected topics in chromatin structure. *Ann. Rev. Genet.* 19: 485–536

Emerson, B. M., Felsenfeld, G. 1984. Specific factor conferring nuclease hypersensitivity at the 5' end of the chicken adult β-globin gene. *Proc. Natl. Acad. Sci. USA* 81: 95–99

Emerson, B. M., Lewis, C. D., Felsenfeld,

G. 1985. Interaction of specific nuclear factors with the nuclease hypersensitive region of the chicken adult β-globin gene: Nature of the binding domain. *Cell* 41: 21–30

Engel, A., Supterlin, S., Koller, T. 1980. Estimation of the mass per unit length of soluble chromatin using the STEM. In *Proc. VII Eur. Congr. Electron Microsc.*, ed. P. Brederoo, W. E. E. Priester, Vol. 2, pp. 548–49. The Hague: Dr. W. Junk

Evans, T., Schon, E., Gora-Maslak, G., Patterson, J., Efstratiadis, A. 1984. S1-hypersensitive sites in eukaryotic promoter regions. *Nucleic Acids Res.* 12: 8043–58

Finch, J. T., Klug, A. 1976. Solenoidal model for superstructure of chromatin. *Proc. Natl. Acad. Sci. USA* 73: 1897–1901

Fleischman, G., Pflügfelder, G., Steiner, E. K., Javaherian, K., Howard, G. C., et al. 1984. *Drosophila* DNA topoisomerase I is associated with transcriptionally active regions of the genome. *Proc. Natl. Acad. Sci. USA* 81: 6958–62

Gargiulo, G., Razvi, F., Worcel, A. 1984. Assembly of transcriptionally active chromatin in *Xenopus* oocytes requires specific DNA binding factors. *Cell* 38: 511–21

Gidoni, D., Kadonaga, J., Barrera-Saldana, H., Takahashi, K., Chambon, P., et al. 1985. Bidirectional SV40 transcription mediated by tandem Sp1 binding interactions. *Science* 230: 511–17

Glikin, G. C., Ruberti, I., Worcel, A. 1984. Chromatin assembly in *Xenopus* oocytes: In vitro studies. *Cell* 37: 33–41

Gottesfeld, J., Bloomer, L. S. 1982. Assembly of transcriptionally active 5S RNA gene chromatin in vitro. *Cell* 28: 781–91

Gross, D., Huang, S.-Y., Garrard, W. 1985. Chromatin structure of the potential Z-forming sequence $(dT-dG)_n \cdot (dT-dC)_n$. *J. Mol. Biol.* 183: 251–65

Hanas, J., Bogenhagen, D., Wu, C.-W. 1984. DNA unwinding ability of *Xenopus* transcription factor A. *Nucleic Acids Res.* 12: 1265–76

Hartman, P. G., Chapman, G. E., Moss, T., Bradbury, E. M. 1977. Studies on the role and mode of operation of the very lysine rich histone H1 in eukaryotic chromatin. *Eur. J. Biochem.* 77: 45–51

Hatch, C. L., Bonner, W. M., Moudrianakis, E. N. 1984. Differential accessibility of the amino and carboxy termini of histone H2A in the nucleosome and its histone subunits. *Biochemistry* 22: 3016–23

Hill, R. J., Stollar, B. D. 1983. Dependence of Z DNA antibody to polytene chromosomes on acid fixation and DNA torsional

strain. *Nature* 305: 338–40

Holm, C., Goto, T., Wang, J., Botstein, D. 1985. DNA topoisomerase II is required at the time of mitosis in yeast. *Cell* 41: 553–63

Huang, S.-Y., Barnard, M. B., Xu, M., Matsui, S.-I., Rose, S. M., et al. 1986. The active κ immunoglobulin gene is packaged by non-ubiquitin-conjugated nucleosomes. *Proc. Natl. Acad. Sci. USA* 83: 3738–42

Jackson, P. D., Felsenfeld, G. 1985. A method for mapping intranuclear protein interactions and its application to a nuclease hypersensitive site. *Proc. Natl. Acad. Sci. USA* 82: 2296–2300

Johnston, B., Rich, A. 1985. Chemical probes of DNA conformation: Detection of Z-DNA at nucleotide resolution. *Cell* 42: 713–24

Jongstra, J., Reudelhuber, T., Oudet, P., Benoist, C., Chae, C.-B., et al. 1984. Induction of altered chromatin structure by simian virus 40 enhancer and promoter elements. *Nature* 307: 708–14

Kang, D., Wells, R. 1985. B-Z junctions contain few, if any, nonpaired bases at physiological superhelical densities. *J. Biol. Chem.* 260: 7783–90

Karpov, V. L., Preobrazhenskaya, O. V., Mirzabekov, A. D. 1984. Chromatin structure of *hsp 70* genes, activated by heat shock: Selective removal of histones from the coding region and their absence from the 5' region. *Cell* 36: 423–31

Klug, A., Finch, J. T., Richmond, T. J. 1985. Crystallographic structure of the octamer histone cone of the nucleosome. *Science* 229: 1109–10

Klug, A., Rhodes, D., Smith, J., Finch, J. T., Thomas, J. O. 1980. A low resolution structure for the histone core of the nucleosome. *Nature* 287: 509–16

Kmiec, E., Worcel, A. 1985. The positive transcription factor of the 5S RNA gene induces a 5S DNA specific gyration in *Xenopus* oocyte extracts. *Cell* 41: 945–53

Kohwi-Shigematsu, T., Gelinas, R., Weintraub, H. 1983. Detection of an altered DNA conformation at specific sites in chromatin and supercoiled DNA. *Proc. Natl. Acad. Sci. USA* 80: 4389–93

Kohwi-Shigematsu, T., Kohwi, Y. 1985. Poly(dG)-poly(dC) sequences, under torsional stress, induce an altered DNA conformation upon neighboring DNA sequences. *Cell* 43: 199–206

Kolodrubetz, D., Rybowski, M. C., Grunstein, M. 1982. Histone H2A subtypes associate interchangeably in vivo with histone H2B subtypes. *Proc. Natl. Acad. Sci. USA* 79: 7814–18

Langmore, J. P., Paulson, J. R. 1983. Low

angle x-ray diffraction studies of chromatin structure in vivo and in isolated nuclei and metaphase chromosomes. *J. Cell Biol.* 96 : 1120–31

Larsen, A., Weintraub, H. 1982. An altered DNA conformation detected by S1 nuclease occurs at specific regions in active chicken globin chromatin. *Cell* 29 : 609–22

Lennard, A. C., Thomas, J. O. 1985. The arrangement of H5 molecules in extended and condensed chicken erythrocyte chromatin. *EMBO J.* 4 : 3455–62

Lindsey, G. G., Thompson, P., Pretorius, L., von Holt, C. 1985. Fluorescent labelling of histone H3 : Effect on histone-histone interaction and core particle assembly. *FEBS Lett.* 192 : 230–34

Losa, R., Thoma, F., Koller, T. 1984. Involvement of the globular domain of histone H1 in the higher order structures of chromatin. *J. Mol. Biol.* 175 : 529–51

Luchnik, A., Bakayev, V., Yugai, A., Zbarsky, I., Georgiev, G. 1985. DNAse I hypersensitive minichromosomes of SV40 possess an elastic torsional strain in DNA. *Nucleic Acids Res.* 13 : 1135–49

McGhee, J. D., Felsenfeld, G. 1980. Nucleosome structure. *Ann. Rev. Biochem.* 49 : 1115–56

McGhee, J. D., Nickol, J. M., Felsenfeld, G., Rau, D. C. 1983. Higher order structure of chromatin : Orientation of nucleosomes within the 30 nm chromatin solenoid is independent of species and spacer length. *Cell* 33 : 831–41

McGhee, J. D., Rau, D. C., Charney, E., Felsenfeld, G. 1980. Orientation of the nucleosome within the higher structure of chromatin. *Cell* 22 : 87–96

Milligan, R. D. 1985. The native structure of the ribosome : Electron crystallography of 2-D crystals in frozen solution. PhD thesis. Stanford Univ., Stanford, Calif.

Mitra, S., Sen, D., Crothers, D. M. 1984. Orientation of nucleosomes and linker DNA in calf thymus chromatin determined by photochemical dichroism. *Nature* 308 : 247–50

Morse, R. II., Cantor, C. R. 1985. Nucleosome core particles suppress the thermal untwisting of core DNA and adjacent linker DNA. *Proc. Natl. Acad. Sci. USA* 82 : 4653–57

Moudrianakis, E. N., Love, W. E., Wang, B. C., Xuong, N. G., Burlingame, R. W. 1985a. Crystallographic structure of the octamer histone core of the nucleosome. *Science* 229 : 1110–12

Moudrianakis, E. N., Love, W. E., Burlingame, R. W. 1985b. Crystallographic structure of the octamer histone core of the nucleosome. *Science* 229 : 1113

Nasmyth, K. A. 1982. The regulation of yeast mating type chromatin by SIR : An action at a distance affecting both transcription and transposition. *Cell* 30 : 567–78

Ness, P. J., Labhart, P., Banz, E., Koller, T., Parish, R. W. 1983. Chromatin structure along the ribosomal DNA of *Dictyostelium*. Regional differences and changes accompanying cell differentiation. *J. Mol. Biol.* 166 : 361–81

Nickol, J., Behe, M., Felsenfeld, G. 1982. Effect of the B-Z transition in poly(dG-m⁵dC)-poly(dG-m⁵dC) on nucleosome formation. *Proc. Natl. Acad. Sci. USA* 79 : 1771–75

Nickol, J. M., Felsenfeld, G. 1983. DNA conformation at the 5′ end of the chicken adult β-globin gene. *Cell* 35 : 467–77

Nicolaev, L. G., Glotov, B. O., Dashkevich, V. K., Barbashav, S. F., Severin, E. S. 1983. Mutual arrangement of histone H1 molecules in extended chromatin. *FEBS Lett.* 163 : 66–68

Nordheim, A., Tesser, P., Azorin, F., Kwon, Y., Moller, A., et al. 1982. Isolation of *Drosophila* proteins that bind selectively to left-handed Z-DNA. *Proc. Natl. Acad. Sci. USA* 79 : 7729–33

Palen, T. E., Cech, T. R. 1984. Chromatin structure at the replication origins and transcription initiation regions of the ribosomal genes of *Tetrahymena*. *Cell* 36 : 933–42

Parker, C. S., Topol, J. 1984a. A *Drosophila* RNA polymerase II transcription factor contains a promoter region specific DNA binding activity. *Cell* 37 : 357–69

Parker, C. S., Topol, J. 1984b. A *Drosophila* RNA polymerase II transcription factor binds to the regulatory site of an *hsp 70* gene. *Cell* 37 : 272–83

Paulson, J. R., Langmore, J. P. 1983. Low angle x-ray diffraction studies of HeLa metaphase chromosomes : Effects of histone phosphorylation and chromosome isolation procedure. *J. Cell Biol.* 96 : 1132–37

Pederson, D., Shupe, K., Gorovsky, M. 1984. Changes in chromatin structure accompany modulation of the rate of transcription of 5S ribosomal genes in *Tetrahymena*. *Nucleic Acids Res.* 12 : 8489–8507

Pederson, D. S., Venkatesan, M., Thoma, F., Simpson, R. T. 1986. Isolation of a yeast gene and replication origin as chromatin. *Proc. Natl. Acad. Sci. USA*. In press

Prior, C. P., Cantor, C. R., Johnson, E. M., Allfrey, V. G. 1980. Incorporation of exogenous pyrene-labeled histone into *Physarum* chromatin : A system for studying changes in nucleosomes assembled in vivo. *Cell* 20 : 597–608

Pulleyblank, D., Haniford, D., Morgan, A. 1985. A structural basis for S1 nuclease sensitivity of double stranded DNA. *Cell* 42 : 271–80

Reeves, R. 1984. Transcriptionally active chromatin. *Biochim. Biophys. Acta* 782 : 343–93

Renz, M., Nehls, P., Hosier, J. 1977. Involvement of histone H1 in the organization of the chromatin fiber. *Proc. Natl. Acad. Sci. USA* 74 : 1879–83

Reynolds, W. F., Gottesfeld, J. M. 1983. 5S rRNA gene transcription factor IIIA alters the helical configuration of DNA. *Proc. Natl. Acad. Sci. USA* 80 : 1262–66

Rhodes, D. 1985. Structural analysis of a triple complex between the histone octamer, a *Xenopus* gene for 5S RNA and transcription factor TF IIIA. *EMBO J.* 4 : 3473–82

Richmond, T. J., Finch, J. T., Rushton, B., Rhodes, D., Klug, A. 1984. Structure of the nucleosome core particle of 7 Å resolution. *Nature* 311 : 532–37

Ring, D., Cole, R. D. 1983. Close contacts between H1 histone molecules in nuclei. *J. Biol. Chem.* 258 : 15361–64

Rose, S. M., Garrard, W. T. 1984. Differentiation dependent chromatin alterations precede and accompany transcription of immunoglobulin light chain genes. *J. Biol. Chem.* 259 : 8534–44

Rybowski, M., Wallis, J., Choe, J., Grunstein, M. 1981. Histone H2B subtypes are dispensable during the yeast cell cycle. *Cell* 25 : 477–87

Ryoji, M., Worcel, A. 1984. Chromatin assembly in *Xenopus* oocytes: In vivo studies. *Cell* 37 : 21–32

Saavedra, R. A., Huberman, J. A. 1986. Both DNA topoisomerases I and II relax 2 μm DNA in living yeast cells. *Cell* 45 : 65–70

Schlissel, M. S., Brown, D. D. 1984. The transcriptional regulation of *Xenopus* 5S RNA genes in chromatin: The roles of active stable transcription complexes and histone H1. *Cell* 378 : 903–13

Schon, E., Evan, T., Welsh, J., Effstratiadis, A. 1983. Conformation of promoter DNA: Fine structure mapping of S1 hypersensitive sites. *Cell* 35 : 837–48

Sen, D., Mitra, S., Crothers, D. M. 1986. Higher order structure of chromatin: Evidence from photochemically detected linear dichroism. *Biochemistry* 25 : 3441–47

Setzer, D. R., Brown, D. D. 1985. Formation and stability of the 5S RNA transcription complex. *J. Biol. Chem.* 260 : 2483–92

Shick, V. V., Belyavsky, A. G., Bavykin, S. G., Mirzabekov, A. D. 1980. Primary

organization of the nucleosome core particles. Sequential arrangement of histones along DNA. *J. Mol. Biol.* 139 : 491–517

Simpson, R. T. 1986. Nucleosome positioning in vivo and in vitro. *BioEssays* 4 : 172–76

Simpson, R. T., Thoma, F., Brubaker, J. M. 1985. Chromatin reconstituted from tandemly repeated cloned DNA fragments and core histones: A model system for study of higher order structure. *Cell* 42 : 799–808

Sperling, L., Tardieu, A. 1976. The mass per unit length of chromatin by low angle x-ray scattering. *FEBS Lett.* 64 : 89–91

Stein, A., Holley, K., Zellif, J., Townsend, T. 1985. Interactions between core histones and chromatin at physiological ionic strength. *Biochemistry* 24 : 1783–90

Suau, P., Bradbury, E. M., Baldwin, J. P. 1979. Higher order structures of chromatin in solution. *Eur. J. Biochem.* 97 : 593–602

Takahashi, K., Vigneron, M., Matthes, H., Wildeman, A., Zenke, M., et al. 1986. Requirement of stereospecific alignments for initiation from the simian virus 40 early promoter. *Nature* 319 : 121–26

Thoma, F. 1986. Protein-DNA interactions and nuclease sensitive regions determine nucleosome positions on yeast plasmid chromatin. *J. Mol. Biol.* In press

Thoma, F., Bergman, L. W., Simpson, R. T. 1984. Nuclease digestion of circular TRP1ARS1 chromatin reveals positioned nucleosomes separated by nuclease sensitive regions. *J. Mol. Biol.* 177 : 715–33

Thoma, F., Koller, T. 1981. Unravelled nucleosomes, nucleosome beads and higher order structures of chromatin: Influence of nonhistone components and histone H1. *J. Mol. Biol.* 149 : 709–33

Thoma, F., Koller, T., Klug, A. 1979. Involvement of histone H1 in the organization of the nucleosome and of the salt dependent superstructures of chromatin. *J. Cell Biol.* 83 : 403–27

Thoma, F., Losa, R., Koller, T. 1983. Involvement of the domains of histones H1 and H5 in the structural organization of soluble chromatin. *J. Mol. Biol.* 167 : 619–40

Thomas, G. J. Jr., Prescott, B., Olins, D. E. 1977. Secondary structure of histones and DNA in chromatin. *Science* 197 : 385–88

Thomas, J. O., Kornberg, R. D. 1975. An octamer of histones in chromatin and free in solution. *Proc. Natl. Acad. Sci. USA* 72 : 2626–30

Topol, J., Ruden, D. M., Parker, C. S. 1985. Sequences required for in vitro transcriptional activation of a *Drosophila hsp* 70 gene. *Cell* 42 : 527–37

Uberbacher, E. C., Bunick, G. J. 1985a. X-ray structure of the nucleosome core particle. *J. Biomol. Struct. Dynam.* 2: 1033–55

Uberbacher, E. C., Bunick, G. J. 1985b. Crystallographic structure of the octamer histone core of the nucleosome. *Science* 229: 1112–13

Wallis, J. W., Rykowski, M., Grunstein, M. 1983. Yeast histone H2B containing large amino terminus deletions can function in vivo. *Cell* 35: 711–19

Wang, J., Hogan, M., Austin, R. H. 1982. DNA motions in the nucleosome core particle. *Proc. Natl. Acad. Sci. USA* 79: 5896–5900

Wang, J. C. 1982. The path of DNA in the nucleosome. *Cell* 29: 724–26

Weintraub, H. 1983. A dominant role for DNA secondary structure in forming hypersensitive structures in chromatin. *Cell* 32: 1191–1203

Weintraub, H. 1985. Assembly and propagation of repressed and derepressed chromosomal states. *Cell* 42: 705–11

Whitlock, J. P. Jr., Stein, A. 1977. Folding of DNA by histones which lack their NH$_2$-terminal regions. *J. Biol. Chem.* 253: 3857–61

Widom, J., Finch, J. T., Thomas, J. O. 1985. Higher order structure of long repeat chromatin. *EMBO J.* 4: 3189–94

Widom, J., Klug, A. 1985. Structure of the 300 Å chromatin filament: X-ray diffraction from oriented samples. *Cell* 43: 207–13

Williams, S. P., Athey, B. D., Muglia, L. J., Schappe, R. S., Gough, A. H., et al. 1986. Chromatin fibers are left-handed double helices with diameter and mass per unit length that depend on linker length. *Biophys. J.* 49: 233–48

Woodcock, C. L. F., Frado, L. Y., Rattner, J. B. 1984. The higher order structure of chromatin. Evidence for a helical ribbon arrangement. *J. Cell Biol.* 99: 42–52

Worcel, A., Strogatz, S., Riley, D. 1981. Structure of chromatin and the linking number of DNA. *Proc. Natl. Acad. Sci. USA* 78: 1461–65

Workman, J. L., Langmore, J. P. 1985. Nucleoprotein hybridization: A method for isolating specific genes as high molecular weight chromatin. *Biochemistry* 24: 7486–97

Wu, C. 1984a. Two protein binding sites in chromatin implicated in the activation of heat shock genes. *Nature* 309: 229–34

Wu, C. 1984b. Activating protein factor binds in vitro to upstream control sequences in heat shock gene chromatin. *Nature* 311: 81–84

Wu, C. 1985. An exonuclease protection assay reveals heat shock element and TATA box DNA binding proteins in crude nuclear extracts. *Nature* 317: 84–87

Wu, T.-C., Simpson, R. T. 1985. Transient alterations of the chromatin structure of sea urchin early histone genes during embryogenesis. *Nucleic Acids Res.* 13: 6185–6203

Yamamoto, K. R. 1985. Steroid receptor regulated transcription of specific genes and gene networks. *Ann. Rev. Genet.* 19: 209–52

Yang, L., Rowe, T. C., Nelson, E. M., Liu, L. F. 1985. In vivo mapping of DNA topoisomerase II specific cleavage sites on SV40 chromatin. *Cell* 41: 127–32

Zaret, K. S., Yamamoto, K. R. 1984. Reversible and persistent changes in chromatin structure accompany activation of a glucocorticoid dependent enhancer element. *Cell* 38: 29–38

Zentgraf, H., Franke, W. W. 1984. Differences of supranucleosomal organization in different kinds of chromatin: Cell type specific globular subunits containing different numbers of nucleosomes. *J. Cell Biol.* 99: 272–85

Ann. Rev. Cell Biol. 1986. 2 : 149–78

THE ROLE OF PROTEIN KINASE C IN TRANSMEMBRANE SIGNALLING

Ushio Kikkawa and Yasutomi Nishizuka

Department of Biochemistry, Kobe University School of Medicine, Kobe 650, Japan

CONTENTS

INTRODUCTION

Inositol phospholipids have attracted great attention from researchers studying the activation of cellular functions and proliferation. In the action of a group of hormones, neurotransmitters, or other biologically active substances, including some growth factors, a signal-induced degradation of the phospholipids may generate important intracellular second messengers that function differently from cyclic AMP. The response of inositol phospholipids to the activation of cell surface receptors was first recognized by Hokin & Hokin (1953), who showed with pancreatic slices that acetylcholine induces a rapid incorporation of ^{32}P into phosphatidylinositol (PI) and phosphatidic acid. It later became evident that breakdown and resynthesis of inositol phospholipids occur in many cell types in response

149

0743–4634/86/1115–0149$00.00

to a wide variety of external signals (for reviews see Hokin & Hokin 1964; Michell 1975; Hawthorne & Pickard 1979; Irvine et al 1982). Durell et al (1969) suggested a potential role of this inositol phospholipid turnover in the mechanism of receptor function and Michell (1975) subsequently proposed that this phospholipid breakdown may open a Ca^{2+} gate.

Although there are some exceptions, such as bovine adrenal medullary cells (Hawthorne 1986) and some presynaptic muscarinic receptors (Starke 1980), it seems clear that receptor-mediated calcium signalling depends upon the breakdown of inositol phospholipids. Berridge and his coworkers (Streb et al 1983; Berridge & Irvine 1984) have provided evidence that inositol 1,4,5-trisphosphate (I-1,4,5-P_3), one of the earliest products of phosphatidylinositol 4,5-bisphosphate (PIP$_2$) hydrolysis, serves as a mediator of Ca^{2+} release from its internal store. A series of studies in this laboratory suggests that 1,2-diacylglycerol, the other product that remains in membranes, initiates the activation of a specialized protein kinase, protein kinase C, and that the information from extracellular signals is transduced across the membrane to result in protein phosphorylation (Nishizuka 1984a,b). The present article briefly summarizes the current status of research on the possible role of protein kinase C in stimulus-response coupling. Research on this subject is moving so fast that this article cannot be comprehensive. Several aspects of this protein kinase C pathway have been reviewed previously (Nishizuka 1984a,b; Rasmussen & Barrett 1984; Williamson et al 1985; Nishizuka 1986; Kikkawa & Nishizuka 1986).

PATHWAYS OF SIGNAL TRANSDUCTION

Inositol phospholipids most frequently contain the 1-stearoyl-2-arachi-donyl-glycerol backbone, and a small portion of PI contains an additional phosphate at position 4 (phosphatidylinositol 4-phosphate, PIP) or two phosphates at positions 4 and 5 (PIP$_2$) of the inositol moiety. These minor phospholipids, each of which represents 1–2% of the total inositol phospholipids, are produced through sequential phosphorylation of PI. In various tissues, signal-induced breakdown of PIP$_2$ appears to precede the previously known PI response, and PIP$_2$ has recently been regarded as a prime breakdown target (for a review see Fisher et al 1984). Thus, the information of external signals may flow from the cell surface into the cell interior through two pathways, Ca^{2+} mobilization and protein kinase C activation. However, in most tissues PI and PIP also disappear when cell surface receptors are stimulated, and it is not absolutely clear whether they are hydrolyzed directly or are phosphorylated to form PIP$_2$ before being hydrolyzed. Plausible evidence at present suggests that in many cell types, such as platelets, the three inositol phospholipids are broken down, probably at different rates, starting at different times, and resulting in the formation of diacylglycerol and inositol mono-, bis-, and trisphosphate (Fisher et al 1984; Majerus et al 1985). Phospholipases C purified to near homogeneity from sheep seminal vesicular glands (Wilson et al 1984) and partially purified from rat brain and liver (Nakanishi et al 1985) hydrolyze PIP_2 and PIP more rapidly than PI, particularly at low Ca^{2+} concentrations.

The phospholipases C in mammalian tissues, except for the enzyme of lysosomal origin, require Ca^{2+}. These enzymes are usually specific for inositol phospholipids, although a phospholipase C that acts on other phospholipids has been reported (Wolf & Gross 1985). The receptor mechanism of inositol phospholipid hydrolysis is unclear, and various results reported so far suggest that this process is Ca^{2+}-dependent but may not be regulated by Ca^{2+} (for reviews see Fisher et al 1984; Hirasawa & Nishizuka 1985). It is highly possible that GTP and its binding protein are involved in this process, as first suggested by Gomperts (1983) and Haslam & Davidson (1984). However, the signal-induced hydrolysis of inositol phospholipids in various tissues may not be explained by a unified single mechanism, since there appears to be considerable tissue variation in the mode of interaction of cell surface signalling systems, as discussed below.

During the early phase of cellular response to relevant ligands, the two signal pathways mentioned above begin to exert their effects within seconds. Under normal conditions intracellular Ca^{2+} concentration is maintained between 0.1 and 0.4 μM. The plasma membrane is relatively impermeable to Ca^{2+}, which is pumped back out of the cell by both Ca^{2+} transport ATPase and the Na^+-Ca^{2+} exchange system to sustain cellular Ca^{2+} homeostasis (Carafoli & Zurini 1982). Evidence for I-1,4,5-P_3 as an intracellular messenger for Ca^{2+} mobilization was obtained from studies of the effect of this compound on various permeabilized cells and also from the effects of microinjecting it into intact cells (Berridge & Irvine 1984). I-1,4,5-P_3 appears to act on internal Ca^{2+} stores, probably in the endoplasmic reticulum, through interaction with its receptor. This receptor has now been detected by direct binding studies with liver microsomes (Spät et al 1986). Inositol trisphosphate, once produced disappears very rapidly, and the major mechanism for terminating this signal pathway is thought to be removal of the 5-phosphate by the action of a specific phosphatase (Berridge & Irvine 1984). However, there may be more than one mechanism eventually leading to the intracellular concentration of Ca^{2+} required for full activation of cellular function, since the presence of extracellular Ca^{2+} is often essential for evoking physiological responses. In any case, the Ca^{2+} signal in most tissues is transient and returns quickly to basal or even below basal levels. In experiments with the photoprotein aequorin this transient Ca^{2+} spike has been successfully observed (Morgan & Morgan 1982; Johnson et al 1985b; Kikkawa et al 1986c). However, the role played by the rapid upstroke of the Ca^{2+} spike in sequential events underlying cellular responses is not clear.

Diacylglycerol is normally absent from membranes. When receptors are stimulated, this neutral lipid appears transiently and disappears within a few seconds or at most a minute. Biochemical studies indicate that this rapid disappearance is due both to its conversion back to inositol phospholipids by way of phosphatidic acid (PI turnover) and to its further degradation to arachidonic acid, which in turn can generate other messengers such as prostaglandins. Thus protein kinase C reveals its activity for only a short time after the stimulation of the receptor and is inactivated rapidly with the disappearance of diacylglycerol and presumably also with the proteolytic degradation of the kinase itself. However, as we briefly discuss later, the consequence of this enzyme activation may persist for a long period, depending upon the biological stability of the phosphate that is covalently attached to each substrate protein molecule. Although both Ca^{2+} and diacylglycerol signals are transient, the two pathways are essential and often synergistic for evoking subsequent cellular responses.

Recently, inositol-1,3,4-trisphosphate, an isomer of I-1,4,5-P_3 (Batty et al 1985), and other inositol polyphosphates such as tetrakisphosphate, pentaphosphate, and hexaphosphate were shown to be present for several minutes after stimulation of cell surface receptors (Berridge & Irvine 1986). However, the functions of these compounds remain to be explored (for a review see Michell 1986).

BIOCHEMICAL AND PHYSIOLOGICAL ACTIVATION

Protein kinase C is widely distributed in tissues and organs, with brain having the highest activity (Kuo et al 1980; Minakuchi et al 1981). In brain tissues a large quantity of the enzyme is associated with synaptic membranes, whereas in most other tissues the enzyme is present mainly in the soluble fraction as an inactive form (Kikkawa et al 1982). Upon cell stimulation the enzyme is apparently translocated to membrane fractions (Kraft & Anderson 1983). However, the exact intracellular topography of this protein kinase is not known because the enzyme has always been extracted for assay in the presence of high concentrations of a Ca^{2+} chelator to prevent its Ca^{2+}-dependent proteolysis. Under resting conditions protein kinase C appears to be loosely bound to membranes, and it is likely that the enzyme is dissociated from membranes by removing

Ca^{2+}. Recent analysis in this laboratory using the immunocytochemical procedure with monoclonal antibodies against protein kinase C indicates that this enzyme is seemingly absent or poorly represented in the nucleus (Kikkawa et al 1986b).

Protein kinases C so far obtained from various tissues appear to be practically indistinguishable from one another, at least in their kinetic and catalytic properties. However, the enzyme isolated from brain tissues occasionally reveals a duplex band upon sodium dodecylsulfate poly-acrylamide gel electrophoresis, but the precise nature of the doublet is not known at present (Kikkawa et al 1986a). The activation of protein kinase C normally depends on Ca^{2+} as well as phospholipids and diacylglycerol. However, diacylglycerol dramatically increases the affinity of this enzyme for Ca^{2+}, and thereby renders it fully active without a net increase in the Ca^{2+} concentration (Kishimoto et al 1980; Kaibuchi et al 1981). Thus, the activation of this protein kinase is biochemically dependent on Ca^{2+}, but physiologically independent of Ca^{2+} concentration. Obviously, the enzyme can be activated effectively and synergistically by the simultaneous increase of diacylglycerol and Ca^{2+}. Phosphatidylserine is absolutely required for the enzyme activation; other phospholipids, such as phosphatidylethanolamine, phosphatidylinositol, phosphatidylcholine, and sphingomyelin, are all inert.

Diacylglycerols having various fatty acyl moieties are capable of activating protein kinase C, and apparently no specificity is observed for the 1-stearoyl-2-arachidonyl structure. In in vitro assay diacylglycerols that contain at least one unsaturated fatty acyl moiety at either position 1 or 2 are active (Kishimoto et al 1980). When one fatty acyl moiety is replaced by a short chain, the resulting diacylglycerols, such as 1-oleoyl-2-acetyl-glycerol (Mori et al 1982) and 1-palmitoyl-2-acetylglycerol (Kajikawa et al 1986), also support enzyme activation. However, all active diacyl-glycerols have a 1,2-*sn* configuration, and other stereoisomers are inactive, which suggests that a highly specific lipid-protein interaction is needed for this enzyme activation (Rando & Young 1984). The detailed biochemical mechanism of this enzyme activation is largely unknown.

Protein kinase C can also be activated by proteolysis with Ca^{2+}-dependent protease. A smaller, enzymatically active component is produced (Inoue et al 1977; Kishimoto et al 1983). This active component is totally independent of Ca^{2+}, phospholipid, and diacylglycerol. Protein kinase C, which is attached to the membrane, is more susceptible to this limited proteolysis (Kishimoto et al 1983), and such proteolysis of protein kinase C may occur in intact platelets (Tapley & Murray 1984). The physiological significance of this reaction remains unknown, but it may take a part in the down-regulation of protein kinase C itself.

PERMEANT DIACYLGLYCEROL AND TUMOR PROMOTER

To explore the link between inositol phospholipid breakdown and protein kinase C activation and the role of the latter in stimulus-response coupling, a synthetic diacylglycerol such as 1-oleoyl-2-acetylglycerol has often been used. It can be readily intercalated into intact cell membranes and dispersed in the phospholipid bilayer, and it activates protein kinase C directly (Kaibuchi et al 1983). Recently, 1,2-dioctanoylglycerol and 1,2-didecano-ylglycerol were shown to be effective, permeant diacylglycerols (Lapetina et al 1985). Synthetic diacylglycerols having a 2,3-*sn* configuration are not active for intact cell systems (Nomura et al 1986). It is worth noting that the diacylglycerols derived from triacylglycerol by the action of lipoprotein lipase (Morley & Kuksis 1972) or a heparin-releasable hepatic lipase (Åkesson et al 1976) have a 2,3-*sn* configuration. When added to intact cells, the active permeant diacylglycerols appear to be metabolized rapidly and converted to the corresponding phosphatidic acids and probably to inositol phospholipids (Kaibuchi et al 1983; Lapetina 1986).

Tumor-promoting phorbol esters, such as 12-*O*-tetradecanoylphorbol-13-acetate (TPA), have a molecular structure that in part is very similar to that of diacylglycerol and activate protein kinase C directly both in vitro and in vivo (Castagna et al 1982; Yamanishi et al 1983). Like diacylglycerol, phorbol esters dramatically increase the affinity of this enzyme for Ca^{2+}, resulting in its full activation at physiological concentrations of this divalent cation. Several lines of evidence provided by many laboratories seem to suggest that protein kinase C is probably a prime target of tumor promoters (Castagna et al 1982; Yamanishi et al 1983; Niedel et al 1983; Sando & Young 1983; Leach et al 1983; Ashendel et al 1983; Kikkawa et al 1983). Kinetic analysis with purified protein kinase C (Kikkawa et al 1983) or with intact platelets (Uratsuji et al 1985) suggests that roughly one molecule of tumor promoter can activate one molecule of this protein kinase. Mezerein (Couturier et al 1984; Miyake et al 1984), teleocidin, and *Aplysia* toxin (tumor promoters structurally unrelated to phorbol esters) (Fujiki et al 1984; Arcoleo & Weinstein 1985) also activate protein kinase C. This finding suggests that a diacylglycerol-like structure is not always essential and that many tumor promoters so far identified induce a membrane perturbation analogous to that caused by diacylglycerol. Thus, the traditional concept of tumor promotion has been replaced by an explicit biochemical explanation, which will make for a better understanding of the role of protein kinase C (Nishizuka 1984a).

In experiments with phorbol esters, the resulting cellular responses sometimes appear to cast doubt as to the suitability of these esters for cell biology studies. Diacylglycerol, the physiological activator of protein kinase C, is present only transiently in membranes, as noted above, whereas TPA is hardly metabolizable and persists for longer periods of time. Therefore, TPA may extend a usually limited phase of cellular response, thereby distorting the normal sequence of events. As will become evident later, protein kinase C presumably has both positive and negative actions that depend upon the function of the target proteins. At an early phase of cellular responses, this enzyme presumably acts synergistically with Ca^{2+} as part of a positive response, but more frequently it acts as a negative feedback control, as in the down-regulation of some receptors, which comes into play immediately upon stimulation (Nishizuka 1986). Therefore, it is essential to understand the temporal sequence of events occurring in different phases of a cellular response.

In addition, the concentration of phorbol ester employed should be given special consideration when attempting to evaluate the exact contribution of the ester to physiological responses. As emphasized repeatedly elsewhere (Nishizuka 1984a,b), at higher concentrations tumor promoters per se can induce significant biological effects. There is no proof at present that protein kinase C is the sole target of tumor promoters, and the possibility exists that these compounds act as membrane perturbers or fusigens, particularly at higher concentrations. To interpret any experimental result for phorbol ester correctly, the parallel demonstration of other parameters such as phosphorylation of some endogenous proteins is obviously desirable, as described in our earlier experiments with platelets (Yamanishi et al 1983) and lymphocytes (Kaibuchi et al 1985). At higher concentrations, permeant diacylglycerols appear to show many nonspecific actions similar to those of phorbol ester. However, it should not be surprising that slightly different effects are obtained from stimulation of living cells with diacylglycerol as opposed to phorbol ester (Kreutter et al 1985).

POSSIBLE ROLES IN CELLULAR RESPONSES

Under appropriate conditions, protein kinase C activation and Ca^{2+} mobilization can be induced selectively and independently by the application of a permeant diacylglycerol or phorbol ester for the former and a Ca^{2+} ionophore such as A23187 or ionomycin for the latter. By using this procedure it has been shown that the two signal pathways are both essential to elicit full cellular responses. The role of protein kinase C in stimulus-response coupling was first demonstrated for the release of serotonin from platelets (Kaibuchi et al 1983) and was subsequently shown for release reactions, secretion, and exocytosis from a wide variety of cell types and for many other cellular processes, as shown in Table 1. The potential role

Table 1   Potential roles of protein kinase C in cellular responses

Tissues and cells	Responses	References
Blood cell systems		
Platelets	Serotonin release	Kaibuchi et al 1983; Rink et al 1983; Knight et al 1984
	Lysosomal enzyme release	Kajikawa et al 1983; Knight et al 1984
	Arachidonate release	Halenda et al 1985
	Thromboxane synthesis	Mobley & Tai 1985
Neutrophils	Superoxide generation	Serhan et al 1983; Fujita et al 1984; Dale & Penfield 1984; Di Virgilio et al 1984; Robinson et al 1985; Gennaro et al 1985
	Lysosomal enzyme release	Kajikawa et al 1983; O'Flaherty et al 1984; White et al 1984
	Hexose transport	McCall et al 1985
Basophils	Histamine release	Sagi-Eisenberg et al 1985
Mast cells	Histamine release	Katakami et al 1984
Endocrine systems		
Adrenal medulla	Catecholamine secretion	Knight & Baker 1983; Brocklehurst et al 1985; Houchi et al 1985; Pocotte et al 1985
Adrenal cortex	Aldosterone secretion	Kojima et al 1983
	Steroidogenesis	Culty et al 1984; Vilgrain et al 1984
Pancreatic islets	Insulin release	Tanigawa et al 1982; Zawalich et al 1983; Hubinont et al 1984; Tamagawa et al 1985
Insulinoma cells	Insulin release	Hutton et al 1984
Pituitary cells	Anterior pituitary hormone release	Negro-Vilar & Lapetina 1985
	Growth hormone release	Ohmura & Friesen 1985
	Luteinizing hormone release	Conn et al 1985; Naor & Eli 1985
	Prolactin release	Delbeke et al 1984; Ronning & Martin 1985
	Thyrotropin release	Martin & Kowalchyk 1984
Parathyroid cells	Parathyroid hormone release	Brown et al 1984a
Thyroid C cells	Calcitonin release	Hishikawa et al 1985
Leydig cells	Steroidogenesis	Mukhopadhyay et al 1984; Papadojoulos et al 1985; Moger 1985

Table 1   (*continued*)

Tissues and cells	Responses	References
Exocrine systems		
Pancreas	Amylase secretion	du Pont & Fleuren-Jakobs 1984; Merritt & Rubin 1985
Parotid gland	Protein secretion	Putney et al 1984
Gastric gland	Pepsinogen secretion	Sakamoto et al 1985
Alveolar cells	Surfactant secretion	Sano et al 1985
Nervous systems		
Ileal nerve endings	Acetylcholine release	Tanaka et al 1984
Neuromuscular preparation	Transmitter release	Publicover 1985
Caudate nucleus	Acetylcholine release	Tanaka et al 1986
PC12 cells	Dopamine release	Pozzan et al 1984
Neurons	Dopamine release	Zurgil & Zisapel 1985
Muscular systems		
Smooth muscle	Contraction	Rasmussen et al 1984; Danthuluri & Deth 1984; Nakaki et al 1985; Dale & Obianime 1985; Baraban et al 1985a
Other systems		
Adipocytes	Lipogenesis	van de Werve et al 1985a
	Glucose transport	Kirsch et al 1985
Hepatocytes	Glycogenolysis	Roach & Goldman 1983; Garrison et al 1984; Fain et al 1984; van de Werve et al 1985b
Epidermal cells	Inhibition of gap junction	Gainer & Murray 1985
Fibroblasts	Inhibition of gap junction	Enomoto & Yamasaki 1985

of protein kinase C in signal transduction at the cell surface has also been extrapolated to neural tissues, particularly in relation to neuropeptide and transmitter release in both peripheral and central nervous systems. The activation of cellular responses by the protein kinase C pathway appears to be separate from, but often synergistic with, activation via an increase of intracellular Ca^{2+}. The unique role of protein kinase C is to provide a logical basis for understanding completely new aspects of signal-dependent processes and to add a new dimension essential to our knowledge of cell biology.

The role of the two signal pathways mentioned above may also be extended to long-term responses such as gene expression and cell proliferation. It has been shown with macrophage-depleted cultures of human peripheral lymphocytes that the two pathways are essential and act synergistically to promote DNA synthesis. However, some growth factor is still necessary to stimulate rapid cell proliferation, although a low concentration may be sufficient. Thus an additional signal pathway is involved in eliciting the full activation of cell proliferation (Kaibuchi et al 1985). Interleukin 1 presumably utilizes the two signal pathways described above; it promotes T lymphocyte proliferation by initiating the induction of the receptor of interleukin 2 (IL2) (Shackelford & Trowbridge 1984; Depper et al 1984; Truneh et al 1985) and by facilitating the release of IL2 (Isakov et al 1985; Truneh et al 1985). The latter functions much like a progression factor that activates this additional signal pathway. Synergism between a permeant diacylglycerol and a Ca^{2+} ionophore also leads to proliferation of human B lymphocytes, and the mechanism of growth promotion for these cells may be under similar control (Guy et al 1985; Nel et al 1985).

Tumor promoters and growth factors have been known for many years to act in concert to stimulate cell proliferation. For instance, insulin is needed in addition to TPA or permeant diacylglycerol to stimulate the growth response of Swiss 3T3 cell lines (Rozengurt et al 1984). Epidermal growth factor (EGF) does not appear to provoke inositol phospholipid breakdown in most tissues, except for a few cell types such as human A431 epidermoid carcinoma cells (Sawyer & Cohen 1981), whereas some other growth factors such as platelet-derived growth factor (Habenicht et al 1981) do so, particularly at higher concentrations. Tumor promoters and permeant diacylglycerols both induce expression of some genes, such as those of ornithine decarboxylase (Verma et al 1980; Weeks et al 1982; Otani et al 1985; Jetten et al 1985), histidine decarboxylase (Watanabe et al 1981), serotonin N-acetyltransferase (Zatz 1985), prolactin (Murdoch et al 1985), and probably γ-interferon (Wilkinson & Morris 1983; Johnson et al 1985a) and plasminogen activator (Loskutoff & Edgington 1977; Lee & Weinstein 1978; Wilson & Reich 1978; Crutchley et al 1980; Degen et al 1985). Some proto-oncogenes such as c-fos (Greenberg & Ziff 1984; Kruijer et al 1984; Muller et al 1984; Mitchell et al 1985; Bravo et al 1985; Greenberg et al 1985) and c-myc (Kelly et al 1983; Muller et al 1984; Coughlin et al 1985; Bravo et al 1985; Greenberg et al 1985; de Bustros et al 1985; Faletto et al 1985) are potential targets of protein kinase C action.

By analogy to lymphocyte activation, discussed above, it is tempting to assume that protein kinase C may induce expression of genes related to the action of growth factors, thereby promoting cell proliferation in the long term. However, as noted above, protein kinase C seems to be absent or to have only a weak presence in the nucleus. Presumably, an additional step of signal transduction is needed prior to the ultimate activation of nuclear events. It is well known that some of the receptors for growth factors have, or are associated with, a tyrosine-specific protein kinase activity (Hunter & Cooper 1985). However, the biological role of the tyrosine phosphorylation reactions is still unknown.

FEEDBACK CONTROL AND DOWN-REGULATION

In biological systems positive signals are usually followed by immediate feedback control to prevent overresponse. A major function of protein kinase C appears to be intimately related to such feedback control or down-regulation. The stimulation of α_1-adrenergic receptors is well known to induce inositol phospholipid breakdown, which initiates signal transduction to regulate various cellular functions (Exton 1985; Williamson et al 1985). Recent evidence obtained for hepatocytes (Corvera & Garcia-Sainz 1984; Lynch et al 1985; Cooper et al 1985; Garcia-Sainz et al 1985) and smooth muscle cells (Leeb-Lundberg et al 1985) strongly suggests that protein kinase C exerts negative feedback control on α_1-adrenergic receptors and inhibits the Ca^{2+} signal pathway. Similar feedback control by protein kinase C of its own receptors[1] has been suggested for many other signalling systems in various cell types, including platelets (Rittenhouse & Sasson 1985; MacIntyre et al 1985; Watson & Lapetina 1985; Zavoico et al 1985), neutrophils (Naccache et al 1985), pituitary cells (Drummond 1985), hippocampal slices (Labarca et al 1984), PC12 cells (Vicentini et al 1985), and astrocytoma cells (Orellana et al 1985). The biochemical basis of this negative feedback control remains to be clarified.

The feedback control or down-regulation by protein kinase C is not confined to its own receptors but appears to extend to growth factor receptors. For instance, this enzyme was shown to phosphorylate the EGF receptor with a concomitant decrease in both its tyrosine-specific protein kinase and growth factor binding activities (Cochet et al 1984; Iwashita & Fox 1984; McCaffrey et al 1984; Davis et al 1985). In the EGF receptor, threonine-654, the phosphorylated residue, is located on the N-terminal domain of the molecule between the membrane-spanning and tyrosine-specific protein kinase active site (Hunter et al 1984). In fact, TPA was previously shown to reduce markedly the EGF binding in many mitogenically responsive cell types (Lee & Weinstein 1978; Shoyab et al 1979; King & Cuatrecasas 1982; Krupp et al 1982). The receptors for other

---

[1] The receptors are those coupled to inositol phospholipid hydrolysis.

growth factors, such as insulin (Jacobs et al 1983), somatomedin C (Jacobs et al 1983), transferrin (May et al 1984; Klausner et al 1984), and interleukin 2 (Shackelford & Trowbridge 1984), may also be targets of protein kinase C. However, the sites of phosphorylation and the effects of these receptor modifications are not known. It has been suggested that protein kinase C and Ca^{2+} synergistically inhibit the internalization of EGF (Logsdon & Williams 1984). Negative feedback control by protein kinase C may also be anticipated in many other membrane functions as well as intracellular processes as discussed below.

MODULATION OF MEMBRANE FUNCTIONS

Protein kinase C phosphorylates many proteins associated with membranes. It has been postulated that the enzyme plays a role in Ca^{2+} extrusion immediately after Ca^{2+} mobilization into the cytosol and that Ca^{2+}-ATPase is a possible target of protein kinase C. Studies with vascular smooth muscle (Morgan & Morgan 1982) and platelets (Johnson et al 1985b; Kikkawa et al 1986c) have shown that when receptors are stimulated, the increase in intracellular Ca^{2+} concentration is transient, as noted above. Protein kinase C is proposed to be involved in the Ca^{2+} extrusion process, since in various cell types the cytosolic Ca^{2+} concentration is often decreased by the addition of phorbol ester (Tsien et al 1982; Moolenaar et al 1984; Lagast et al 1984; Sagi-Eisenberg et al 1985; Drummond 1985; Albert & Tashjian 1985). In fact, it has been proposed that Ca^{2+}-ATPase is activated in sarcoplasmic reticulum preparations by the addition of protein kinase C (Limas 1980; Movsesian et al 1984). It has also been proposed that protein kinase C may play a role in enhancing Ca^{2+} entry, since the microinjection of phorbol ester or protein kinase C itself into bag cell neurons of *Aplysia* enhances the voltage-sensitive Ca^{2+} current (DeRiemer et al 1985). However, more information is needed to clarify the role, if any, of this protein kinase in maintaining Ca^{2+} homeostasis under physiological conditions.

Similarly, a possible role of protein kinase C in activating Na,K-ATPase in peripheral nerve has been proposed (Greene & Lattimer 1986). Moreover, in the photoreceptor cells of *Hermissenda* (Farley & Auerbach 1986; Alkon et al 1986) and rat hippocampal pyramidal neurons (Baraban et al 1985b) treatment with phorbol ester or permeant diacylglycerol, or intracellular injections of protein kinase C modulates Ca^{2+}-dependent K^+ conductance. Protein kinase C has previously been proposed to play a role in the expression of functional plasticity, particularly in hippocampal neural activity (Routtenberg & Lovinger 1985). The Na^+-H^+ exchange system also appears to be activated by phorbol ester and by permeant diacylglycerols; thus protein kinase C may have a function in increasing cytoplasmic pH (Burns & Rozengurt 1983; Rosoff et al 1984; Moolenaar et al 1984; Besterman et al 1985). Several lines of evidence so far described all suggest that protein kinase C modulates ion conductance by phosphorylating membrane proteins such as channels, pumps, and ion exchangers. It is equally possible that other active membrane transport systems are modulated by protein kinase C. Research with monoclonal antibodies has confirmed the presence of a protein kinase C immunoreactive material in many regions of the cytoplasm, including axons and dendrites of neuronal cells (Kikkawa et al 1986b). However, establishing a definitive biochemical basis for a protein kinase C role in modulating various membrane functions will require further investigations. Protein kinase C is presumably identical with an enzyme, B50 protein kinase, found in the presynaptic region of brain tissues (Aloyo et al 1983).

It is particularly worth noting that the phosphates attached to the substrate proteins of protein kinase C are frequently resistant to the action of phosphatases, and therefore the consequences of these phosphorylation reactions may persist for long periods. This may provide a basis for the dual action of protein kinase C: negative feedback control or down-regulation and positive forward or sustained control of cellular functions, depending on the functional consequence of the phosphorylation of each substrate protein molecule. Ca^{2+} may initiate the physiological responses, and protein kinase C may help to ensure that the resulting biological events persist. However, crucial information on the target proteins that might mediate such persistent effects is still limited.

INTERACTION WITH OTHER SIGNALLING SYSTEMS

The signalling systems used by cells often display extensive heterogeneity, and many variations exist from tissue to tissue. Most tissues seem to have at least two major receptor classes for transducing information across the cell membrane. One class depends on the generation of cyclic AMP as a secondary messenger, while the other class of receptors induces rapid turnover of inositol phospholipids as well as mobilization of Ca^{2+}. Stimulation of the latter class normally leads to the release of arachidonate and often increases cyclic GMP. Thus, protein kinase C activation, Ca^{2+} mobilization, arachidonate release, and cyclic GMP formation appear to be integrated into a single receptor cascade.

Cellular responses may be tentatively divided into several modes, as shown schematically in Figure 1. In *bidirectional* control systems the two classes of receptors appear to counteract each other, whereas in *mono-*

directional control systems one receptor class may potentiate the other one. For instance, in some cell types including platelets, lymphocytes, mast cells, and many others, the signals that induce inositol phospholipid breakdown promote the activation of cellular functions, but the signals that produce cyclic AMP usually antagonize such activation. In such tissues the signal-induced breakdown of inositol phospholipids and the subsequent events eventually leading to cellular responses are all profoundly blocked by cyclic AMP (Nishizuka 1984a,b). Conversely, in another group of cell types, which includes erythrocytes (Kellehers et al 1984; Sibley et al 1984), Leydig cells (Mukhopadyay & Schumacher 1985; Papadojoulos et al 1985; Rebois & Patel 1985), hepatocytes (Heyworth et al 1984; Garcia-Sainz et al 1985), cardiac muscle (Limas & Limas 1985), glioma C6 cells (Kassis et al 1985), ovarian glanulosa cells (Shinohara et al 1985), and *Xenopus* oocytes (Dascal et al 1985), protein kinase C inhibits and desensitizes the adenylate cyclase system. In several other cell types, including pinealocytes (Sugden et al 1985; Zatz 1985), pituitary cells (Cronin & Canonico 1985; Quilliam et al 1985), lymphoma S49 cells (Bell et al 1985; Katada et al 1985; Kiss & Steinberg 1985), embryonal myoblasts (Sulakhe et al 1985), vascular smooth muscle (Nabika et al 1985), and cells in cerebral cortex

Bidirectional control systems

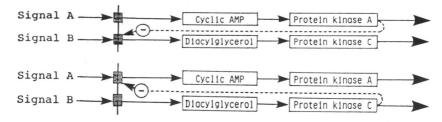

Monodirectional control systems

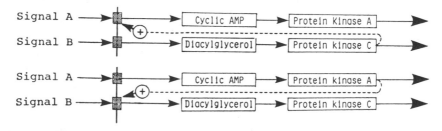

Figure 1 Several modes of interaction of two major signalling systems: Signal A, protein kinase A (cyclic AMP–dependent protein kinase); Signal B, protein kinase C (diacylglycerol-activated protein kinase).

slices (Hollingworth et al 1985), protein kinase C greatly potentiates cyclic AMP production. However, there is as yet no example of a tissue in which cyclic AMP potentiates signal-induced turnover of inositol phospholipids. Yet, these two signal transduction pathways frequently act in concert in many endocrine cells such as pancreatic islets (Tamagawa et al 1985). The evidence presented thus far is still incomplete, but it is reasonable to assume that various combinations of the two receptor signalling systems may cooperate positively and thereby intensify responses in many physiological processes. Further exploration of such receptor interactions in individual cell types will be extremely important for understanding the molecular basis of signal-induced cell growth and differentiation. It is well known that tumor promoters often enhance cell differentiation rather than cell proliferation.

CATALYTIC PROPERTIES AND TARGET PROTEINS

Protein kinase C shows a broad substrate specificity when tested in vitro. The enzyme phosphorylates seryl and threonyl residues but not tyrosyl residues of endogenous proteins. The enzyme does not utilize GTP as a phosphate donor. Like many other protein kinases the enzyme phosphorylates itself in the presence of Ca^{2+}, phospholipid, and diacylglycerol, but the significance of this autophosphorylation is not known.

Protein kinase C and cyclic AMP–dependent protein kinase (protein kinase A) relay information along different signal pathways within the cell, as discussed above, but these two enzymes sometimes cause apparently similar cellular responses. Moreover, recent analysis indicates that protein kinases C and A often share the same phosphate acceptor proteins and even the same seryl and threonyl residues in a single protein molecule for phosphorylation (Kishimoto et al 1985). Extensive work by Krebs and many other investigators (Krebs & Beavo 1979) has shown that the primary structure in the vicinity of the phosphorylation site is one of the determinant factors for substrate recognition and that protein kinase A reacts with the seryl and threonyl residues located near the carboxyl side of a lysyl or arginyl residue. With some model substrate proteins, such as bovine myelin basic protein, it has been shown that, unlike protein kinase A, protein kinase C appears to favor the hydroxy amino acids located near the amino side close to the basic amino acyl residue. All seryl and threonyl residues that are commonly phosphorylated by these two enzymes have basic amino acyl residues at both the amino and carboxyl side of phosphorylatable residues (Kishimoto et al 1985). However, it is not known whether the above findings are applicable to many other protein substrates.

164   KIKKAWA & NISHIZUKA

Analyses have been made with some synthetic peptides (O'Brian et al 1984; Turner et al 1985; Ferrari et al 1985), but relative reaction velocities of protein kinases C and A for the peptides have not been well investigated. The topographical arrangement or subcellular localization of the enzyme and its target proteins is another crucial factor for determining substrate recognition.

A potential role of protein kinase C in the control or modulation of many metabolic processes has been postulated. Table 2 is a list of some functionally defined proteins that have been proposed as substrates of protein kinase C mainly on the basis of in vitro experiments. Although the list of such phosphate-acceptor proteins is expanding rapidly, the physiological significance of many of these phosphorylation reactions remains to be explored.

Table 2   Possible phosphate acceptor proteins of protein kinase C

Phosphate acceptor proteins	References
Receptor proteins	
Epidermal growth factor receptor	Cochet et al 1984; Iwashita & Fox 1984; McCaffrey et al 1984; Davis & Czech 1984; Hunter et al 1984; Brown et al 1984b; Moon et al 1984
Insulin receptor	Jacobs et al 1983; Takayama et al 1984
Somatomedin C receptor	Jacobs et al 1983
Transferrin receptor	Klausner et al 1984; May et al 1984
Interleukin 2 receptor	Shackelford & Trowbridge 1984
Nicotinic acetylcholine receptor	Huganir et al 1984
β-Adrenergic receptor	Kelleher et al 1984; Sibley et al 1984
IgE receptor	Teshima et al 1984
Membrane proteins	
Ca^{2+} transport ATPase	Limas 1980; Movsesian et al 1984; Iwasa & Hosey 1984; Lagast et al 1984
Na,K-ATPase	Greene & Lattimer 1986
Na^+ channel protein	Costa & Catterall 1985
Na^+-H^+ exchanger	Burns & Rozengurt 1983; Moolenaar et al 1984; Rosoff et al 1984; Besterman et al 1984; Vara et al 1985
Glucose transporter	Witters et al 1985
GTP-binding protein	Katada et al 1985
HLA antigen	Feuerstein et al 1985
Chromaffin granule–binding protein	Summers & Creutz 1985
Synaptic B50 (F1) protein	Aloyo et al 1983; Akers & Routtenberg 1985

Table 2   (*continued*)

Phosphate acceptor proteins	References
Contractile and cytoskeletal proteins	
Myosin light chain	Naka et al 1983
Troponin T and I	Mazzei & Kuo 1984
Vinculin	Werth et al 1983; Kawamoto & Hidaka 1984
Filamin	Kawamoto & Hidaka 1984
Caldesmon	Umekawa & Hidaka 1985
Cardiac C protein	Lim et al 1985
Microtubule-associated proteins	Takai et al 1984
Enzymes	
Glycogen phosphorylase kinase	Kishimoto et al 1977
Glycogen synthase	Ahmed et al 1984
Phosphofructokinase	Hofer et al 1985
3-Hydroxy-3-methylglutaryl-CoA reductase	Beg et al 1985
Tyrosine hydroxylase	Albert et al 1984; Vuilliet et al 1985
NADPH oxidase	Papini et al 1985
Cytochrome P450	Vilgrain et al 1984
Guanylate cyclase	Zwiller et al 1985
DNA methylase	DePaoli-Roach et al 1986
Myosin light chain kinase	Ikebe et al 1985; Nishikawa et al 1985
Initiation factor 2	Schatzman et al 1983
Other proteins	
Fibrinogen	Humble et al 1984
Retinoid-binding proteins	Cope et al 1984
Vitamin D binding protein	Wooten et al 1985
Ribosomal S6 protein	Le Peuch et al 1983; Padel & Söling 1985; Parker et al 1985; Trevillyan et al 1985
GABA modulin	Wise et al 1983
Stress proteins	Welch 1985
Myelin basic protein	Sulakhe et al 1980; Wise et al 1982; Kishimoto et al 1985
High-mobility group proteins	Ramachandron et al 1984; Wolf et al 1984; Kimura et al 1985
Middle T antigen	Hirata et al 1984
pp60^{src} protein	Gould et al 1985

CONCLUSION

This article summarizes some of our current knowledge of protein kinase C. The evidence available to date suggests that it plays a crucial role in signal transduction for the activation of many cellular functions, including proliferation, particularly during the early phase of response. Perhaps the signal-induced breakdown of inositol phospholipids initiates a cascade of events starting with Ca^{2+} mobilization and protein kinase C activation and ending with alterations of a variety of biological processes such as cell growth and differentiation. However, it seems premature to discuss the details of the relationship between the role of Ca^{2+} and that of protein kinase C. Each of these signal pathways may play diverse roles in controlling biochemical reactions. Most probably, the protein phosphorylation catalyzed by protein kinase C profoundly modulates various Ca^{2+}-mediated processes, such as release reactions and exocytosis, cell proliferation and differentiation, membrane conductance and transport, smooth muscle contraction, potentiation and desensitization of other receptor systems, and many metabolic processes, apparently by markedly increasing the sensitivity of the corresponding reactions to internal Ca^{2+} concentrations.

ACKNOWLEDGMENTS   The author's research has been supported by grants from the Ministry of Education, Science and Culture, Japan; the Muscular Dystrophy Association; the Yamanouchi Foundation; the Merck Sharp and Dohme Research Laboratories; and the Biotechnology Laboratories of Takeda Chemical Industries.

NOTE ADDED IN PROOF   The complete amino acid sequences of several forms of protein kinase C in bovine brain (Parker, P. J., et al. 1986. *Science* 233: 853–59) and rat brain (Ono, Y., et al. 1986. *FEBS Lett.* In press) have recently been deduced from cDNA clones. These enzymes are similar and show significant homology to other serine/threonine protein kinases (Ono, Y., et al 1986. *FEBS Lett.* 203: 111–15). Although the gene appears to be multiple and located on distinct chromosomes (Coussens, L., et al. 1986. *Science* 233: 859–66), some of the heterogeneity is apparently derived from alternative splicing (Ono, Y., et al. 1986. *FEBS Lett.* In press).

Literature Cited

Ahmed, Z., Lee, F. T., DePaoli-Roach, A., Roach, P. J. 1984. Phosphorylation of glycogen synthase by Ca^{2+}- and phospholipid-activated protein kinase (protein kinase C). *J. Biol. Chem.* 259: 8743–47

Akers, R. F., Routtenberg, A. 1985. Protein kinase C phosphorylates a 47 M_r protein (F1) directly related to synaptic plasticity. *Brain Res.* 334: 147–51

Åkesson, B., Gronowitz, S., Hersloef, B. 1976. Stereospecificity of hepatic lipases. *FEBS Lett.* 71: 241–44

Albert, K. A., Helmer-Matyjek, E., Nairn, A. C., Müller, T. H., Haycock, J. W., et al. 1984. Calcium/phospholipid-dependent protein kinase (protein kinase C) phosphorylates and activates tyrosine hydroxylase. *Proc. Natl. Acad. Sci. USA* 81: 7713–17

Albert, P. R., Tashjian, A. H. 1985. Dual

actions of phorbol esters on cytosolic free Ca^{2+} concentrations and reconstitution with iononycin of acute thyrotropin-releasing hormone responses. *J. Biol. Chem.* 260: 8746–59

Alkon, D. L., Kubota, M., Neary, J. T., Naito, S., Coulter, D., et al. 1986. C-Kinase activation prolongs Ca^{2+}-dependent inactivation of K^+ currents. *Biochim. Biophys. Res. Commun.* 134: 1245–53

Aloyo, V. J., Zwiers, H., Gispen, W. H. 1983. Phosphorylation of B-50 protein by calcium-activated, phospholipid-dependent protein kinase and B-50 protein kinase. *J. Neurochem.* 41: 649–53

Arcoleo, J. P., Weinstein, I. B. 1985. Activation of protein kinase C by tumor promoting phorbol esters, teleocidin and aplysiatoxin in the absence of added calcium. *Carcinogenesis* 6: 213–17

Ashendel, C. L., Staller, J. M, Boutwell, R. K. 1983. Identification of a calcium- and phospholipid-dependent phorbol ester binding activity in the soluble fraction of mouse tissues. *Biochem. Biophys. Res. Commun.* 111: 340–45

Baraban, J. M., Gould, R. J., Peroutka, S. J., Snyder, S. H. 1985a. Phorbol ester effects on neurotransmission: Interaction with neurotransmitters and calcium in smooth muscle. *Proc. Natl. Acad. Sci. USA* 82: 604–7

Baraban, J. M., Snyder, S. H., Alger, B. E. 1985b. Protein kinase C regulates ionic conductance in hippocampal pyramidal neurons: Electrophysiological effects of phorbol esters. *Proc. Natl. Acad. Sci. USA* 82: 2538–42

Batty, I. R., Nahorski, S. R., Irvine, R. F. 1985. Rapid formation of inositol 1,3,4,5-tetrakisphosphate following muscarinic receptor stimulation of rat cerebral cortical slices. *Biochem. J.* 232. 211–15

Beg, Z. H., Stonik, J. A., Brewer, H. B. Jr. 1985. Phosphorylation of hepatic 3-hydroxy-3-methylglutaryl coenzyme A reductase and modulation of its enzymatic activity by calcium-activated and phospholipid-dependent protein kinase. *J. Biol. Chem.* 260: 1682–87

Bell, J. D., Buxton, I. L. O., Brunton, L. L. 1985. Enhancement of adenylate cyclase activity in S49 lymphoma cells by phorbol esters. *J. Biol. Chem.* 260: 2625–28

Berridge, M. J., Irvine, R. F. 1984. Inositol trisphosphate, a novel second messenger in cellular signal transduction. *Nature* 312: 315–21

Berridge, M. J., Irvine, R. F. 1986. Inositol lipids and calcium mobilization. In *Calcium and the Cell. Ciba Found. Symp.* 122. In press

Besterman, J. M., May, W. S. Jr., LeVine,
H. III, Cragoe, E. J. Jr., Cuatrecasas, P. 1985. Amiloride inhibits phorbol ester–stimulated Na^+/H^+ exchange and protein kinase C. An amiloride analog selectively inhibits Na^+/H^+ exchange. *J. Biol. Chem.* 260: 1155–59

Bravo, R., Burckhardt, J., Curran, T., Muller, R. 1985. Stimulation and inhibition of growth by EGF in different A431 cell clones is accompanied by the rapid induction of c-fos and c-myc protooncogenes. *EMBO J.* 4: 1193–97

Brocklehurst, K. W., Morita, K., Pollard, H. B. 1985. Characterization of protein kinase C and its role in catecholamine secretion from bovine adrenal-medullary cells. *Biochem. J.* 228: 35–42

Brown, E. M., Redgrave, J., Thatcher, J. 1984a. Effect of the phorbol ester TPA on PTH secretion. Evidence for a role for protein kinase C in the control of PTH release. *FEBS Lett.* 175: 72–75

Brown, K. D., Blay, J., Irvine, R. F., Heslop, J. P., Berridge, M. J. 1984b. Reduction of epidermal growth factor receptor affinity by heterologous ligands: Evidence for mechanism involving the breakdown of phosphoinositides and the activation of protein kinase C. *Biochem. Biophys. Res. Commun.* 123: 377–84

Burns, C. P., Rozengurt, E. 1983. Serum, platelet-derived growth factor, vasopression and phorbol esters increase intracellular pH in Swiss 3T3 cells. *Biochem. Biophys. Res. Commun.* 116: 931–38

Carafoli, E., Zurini, M. 1982. The Ca^{2+}-pumping ATPase of plasma membranes: Purification, reconstitution and properties. *Biochim. Biophys. Acta* 683: 279–301

Castagna, M., Takai, Y., Kaibuchi, K., Sano, K., Kikkawa, U., et al. 1982. Direct activation of calcium-activated, phospholipid-dependent protein kinase by tumor-promoting phorbol esters. *J. Biol. Chem.* 257: 7847 51

Cochet, C., Gill, G. N., Meisenhelder, J., Cooper, J. A., Hunter, T. 1984. C-kinase phosphorylates the epidermal growth factor receptor and reduces its epidermal growth factor–stimulated tyrosine protein kinase activity. *J. Biol. Chem.* 259: 2553–58

Conn, P. M., Ganong, B. R., Ebeling, J., Staley, D., Niedel, J. E., et al. 1985. Diacylglycerols release LH: Structure-activity relations reveal a role for protein kinase C. *Biochem. Biophys. Res. Commun.* 126: 532–39

Cooper, R. H., Coll, K. E., Williamson, J. R. 1985. Differential effects of phorbol ester on phenylephrine and vasopressin-induced Ca^{2+} mobilization in isolated hepatocytes. *J. Biol. Chem.* 260: 3281–88

Cope, F. O., Staller, J. M., Mahsem, R. A., Boutwell, R. K. 1984. Retinoid-binding proteins are phosphorylated *in vitro* by soluble Ca^{2+}- and phosphatidylserine-dependent protein kinase from mouse brain. *Biochem. Biophys. Res. Commun.* 120: 593–601

Corvera, S., Garcia-Sainz, J. A. 1984. Phorbol esters inhibit alpha 1 adrenergic stimulation of glycogenolysis in isolated rat hepatocytes. *Biochem. Biophys. Res. Commun.* 119: 1128–33

Costa, M. R., Catterall, W. A. 1985. Phosphorylation of the alpha subunit of the sodium channel by protein kinase C. *Cell. Mol. Neurobiol.* 4: 291–97

Coughlin, S. R., Lee, W. M. F., Williams, P. W., Giels, G. M., Williams, L. T. 1985. *c-myc* gene expression is stimulated by agents that activate protein kinase C and does not account for the mitogenic effect of PDGF. *Cell* 43: 243–51

Couturier, A., Bazgar, S., Castagna, M. 1984. Further characterization of tumor-promoter-mediated activation of protein kinase C. *Biochem. Biophys. Res. Commun.* 121: 448–55

Cronin, M. J., Canonico, P. L. 1985. Tumor promoters enhance basal and growth hormone releasing factor stimulated cyclic AMP levels in anterior pituitary cells. *Biochem. Biophys. Res. Commun.* 129: 404–10

Crutchley, D., Conanan, L., Maynard, J. 1980. Induction of plasminogen activator and prostaglandin biosynthesis in HeLa cells by 12-O-tetradecanoyl-phorbol-13-acetate. *Cancer Res.* 40: 849–52

Culty, M., Vilgrain, I., Chambaz, E. M. 1984. Steroidogenic properties of phorbol ester and a Ca^{2+} ionophore in bovine adrenocortical cell suspensions. *Biochem. Biophys. Res. Commun.* 121: 499–506

Dale, M. M., Obianime, W. 1985. Phorbol myristate acetate causes in guinea-pig lung parenchymal strip a maintained spasm which is relatively resistant to isoprenaline. *FEBS Lett.* 190: 6–10

Dale, M. M., Penfield, A. 1984. Synergism between phorbol ester and A23187 in superoxide production by neutrophils. *FEBS Lett.* 175: 170–72

Danthuluri, N. R., Deth, R. C. 1984. Phorbol ester-induced contraction of arterial smooth muscle and inhibition of α-adrenergic response. *Biochem. Biophys. Res. Commun.* 125: 1103–9

Dascal, N., Lotan, I., Gillo, B., Lester, H. A., Lass, Y. 1985. Acetylcholine and phorbol esters inhibit potassium currents evoked by adenosine and cAMP in *Xenopus* oocytes. *Proc. Natl. Acad. Sci. USA* 82: 6001–5

Davis, R. J., Czech, M. P. 1984. Tumor-promoting phorbol diesters mediate phosphorylation of the epidermal growth factor receptor. *J. Biol. Chem.* 259: 8545–49

Davis, R. J., Ganong, B. R., Bell, R. M., Czech, M. P. 1985. *sn*-1,2-Dioctanolyglycerol. A cell-permeable diacylglycerol that mimics phorbol diester action on the epidermal growth factor receptor and mitogenesis. *J. Biol. Chem.* 260: 1562–66

de Bustros, A., Baylin, S. B., Berger, C. L., Ross, B. A., Leong, S. S., et al. 1985. Phorbol esters increase calcitonin gene transcription and decrease *c-myc* mRNA levels in cultured human medullary thyroid carcinoma. *J. Biol. Chem.* 260: 98–104

Degen, J. L., Estensen, R. D., Nagamine, Y., Reich, E. 1985. Induction and desensitization of plasminogen activator gene expression by tumor promoters. *J. Biol. Chem.* 260: 12426–33

Delbeke, D., Kojima, I., Dannies, P. S., Rasmussen, H. 1984. Synergistic stimulation of prolactin release by phorbol ester, A23187 and forskolin. *Biochem. Biophys. Res. Commun.* 123: 735–41

DePaoli-Roach, A., Roach, P. J., Zucker, K. E., Smith, S. S. 1986. Selective phosphorylation of human DNA methyltransferase by protein kinase C. *FEBS Lett.* 197: 149–53

de Pont, J. J. H. H. M., Fleuren-Jakobs, A. M. M. 1984. Synergistic effect of A23187 and a phorbol ester on amylase secretion from rabbit pancreatic acini. *FEBS Lett.* 170: 64–68

Depper, J. M., Leonard, W. J., Kronke, M., Noguchi, P. D., Cunningham, R. E., et al. 1984. Regulation of interleukin 2 receptor expression: Effects of phorbol ester, phospholipase C, and reexposure to leutin or antigen. *J. Immunol.* 133: 3054–61

DeRiemer, S. A., Strong, J. A., Albert, K. A., Greengard, P., Kaczmarek, L. K. 1985. Enhancement of calcium current in *Aplysia* neurones by phorbol ester and protein kinase C. *Nature* 313: 313–16

Di Virgilio, F., Lew, D. P., Pozzan, T. 1984. Protein kinase C activation of physiological processes in human neutrophils at vanishing small cytosolic Ca^{2+} levels. *Nature* 310: 691–93

Drummond, A. H. 1985. Bidirectional control of cytosolic free calcium by thyrotropin-releasing hormone in pituitary cells. *Nature* 315: 752–55

Durell, J., Garland, J. T., Friedel, R. O. 1969. Acetylcholine action: biochemical aspects. *Science* 165: 862–66

Enomoto, T., Yamasaki, H. 1985. Rapid inhibition of intracellular communication between BALB/c 3T3 cells by diacylglycerol, a possible endogeneous functional analogue of phorbol esters. *Cancer Res.* 45: 3706–10

Exton, J. H. 1985. Role of calcium and phosphoinositides in the actions of certain hormones and neurotransmitters. *J. Clin. Invest.* 75: 1753–57

Fain, J. N., Li, S.-Y., Litosch, I., Wallace, M. 1984. Synergistic activation of rat hepatocyte glycogen phosphorylase by A23187 and phorbol ester. *Biochem. Biophys. Res. Commun.* 119: 88–94

Faletto, D. L., Arrow, A. S., Macara, I. G. 1985. An early decrease in phosphatidylinositol turnover occurs on induction of Friend cell differentiation and produces the decrease in *c-myc* expression. *Cell* 43: 315–25

Farley, J., Auerbach, S. 1986. Protein kinase C activation induces conductance changes in *Hermissenda* photoreceptors like those seen in associative learning. *Nature* 319: 220–23

Ferrari, S., Marchiori, F., Borin, G., Pinna, L. A. 1985. Distinct structural requirement of Ca^{2+}/phospholipid-dependent protein kinase (protein kinase C) and cAMP-dependent protein kinase as evidenced by synthetic peptide substrates. *FEBS Lett.* 184: 72–77

Feuerstein, N., Monos, D. S., Cooper, H. L. 1985. Phorbol ester effect in platelets, lymphocytes, and leukemic cells (HL-60) is associated with enhanced phosphorylation of class I HLA antigens. Coprecipitation of myosin light chain. *Biochem. Biophys. Res. Commun.* 126: 206–13

Fisher, S. K., Van Rooijen, L. A. A., Agranoff, B. W. 1984. Renewed interest in the polyphosphoinositides. *Trends Biochem. Sci.* 9: 53–56

Fujiki, H., Tanaka, Y., Miyake, R., Kikkawa, U., Nishizuka, Y., et al. 1984. Activation of calcium-activated, phospholipid-dependent protein kinase (protein kinase C) by new classes of tumor promoters: Teleocidin and debromoaplysiatoxin. *Biochem. Biophys. Res. Commun.* 120: 339–43

Fujita, I., Irita, K., Takeshige, K., Minakami, S. 1984. Diacylglycerol, 1-oleoyl-2-acetyl-glycerol, stimulates superoxide generation from human neutrophils. *Biochem. Biophys. Res. Commun.* 120: 318–24

Gainer, H. S., Murry, A. W. 1985. Diacylglycerol inhibits gap junctional communication in cultured epidermal cells: Evidence for a role of protein kinase C. *Biochem. Biophys. Res. Commun.* 126: 1109–13

Garcia-Sainz, A. J., Mendlovic, F., Martinez-Olmedo, A. 1985. Effects of phorbol esters on α_1-adrenergic-mediated and glucagon-mediated actions in isolated rat hepatocytes. *Biochem. J.* 228: 277–80

Garrison, J. C., Johnsen, D. E., Campanile, C. P. 1984. Evidence for the role of phosphorylase kinase, protein kinase C, and other Ca^{2+}-sensitive protein kinases in the response of hepatocytes to angiotensin II and vasopressin. *J. Biol. Chem.* 259: 3283–92

Gennaro, R., Florio, C., Romeo, D. 1985. Activation of protein kinase C in neutrophil cytoplasts. Localization of protein substrates and possible relationship with stimulus-response coupling. *FEBS Lett.* 180: 185–90

Gomperts, B. D. 1983. Involvement of guanine nucleotide-binding protein in the gating of Ca^{2+} by receptors. *Nature* 306: 64–66

Gould, K. L., Woodgett, J. R., Cooper, J. A., Buss, J. E., Shalloway, D., et al. 1985. Protein kinase C phosphorylates pp60^{src} at a novel site. *Cell* 42: 849–57

Greenberg, M. E., Greene, L. A., Ziff, E. B. 1985. Nerve growth factor and epidermal growth factor induce rapid transient changes in proto-oncogene transcription in PC12 cells. *J. Biol. Chem.* 260: 14101–10

Greenberg, M. E., Ziff, E. B. 1984. Stimulation of 3T3 cells induces transcription of the *c-fos* proto-oncogene. *Nature* 311: 433–38

Greene, D. A., Lattimer, S. A. 1986. Protein kinase C agonists acutely normalize decreased ouabain-inhibitable respiration in diabetic rabbit nerve: Implications for (Na,K)-ATPase regulation and diabetic complications. *Diabetes* 35: 242–45

Guy, G. R., Gordon, J., Michell, R. H., Brown, G. 1985. Synergism between diacylglycerols and calcium ionophore in the induction of human B cell proliferation mimics the inositol lipid polyphosphate breakdown signals induced by crosslinking surface immunoglobulin. *Biochem. Biophys. Res. Commun.* 131: 484–91

Habenicht, A. J. R., Glomset, J. A., King, W. C., Nist, C., Mitchell, C. D., et al. 1981. Early changes in phosphatidylinositol and arachidonic acid metabolism in quiescent Swiss 3T3 cells stimulated to divide by platelet-derived growth factor. *J. Biol. Chem.* 256: 12329–35

Halenda, S. P., Zavoico, G. B., Feinstein, M. B. 1985. Phorbol esters and oleoyl acetyl glycerol enhance release of arachidonic acid in platelets stimulated by Ca^{2+} ionophore A23187. *J. Biol. Chem.* 260: 12484–89

Haslam, R. J., Davidson, M. M. L. 1984. Guanine nucleotides decrease the free $[Ca^{2+}]$ required for secretion of serotonin from permeabilized blood platelets: Evidence of a role for a GTP-binding protein in platelet activation. *FEBS Lett.* 174: 90–95

Hawthorne, J. N. 1986. Does receptor-linked phosphoinositide metabolism provide messengers mobilizing calcium in nervous tissues? *Int. Rev. Neurobiol.* In press

Hawthorne, J. N., Pickard, M. R. 1979. Phospholipid in synaptic function. *J. Neurochem.* 32: 5–14

Heyworth, C. M., Whetton, A. D., Kinsella, A. R., Houslay, M. D. 1984. The phorbol ester, TPA inhibits glucagon-stimulated adenylate cyclase activity. *FEBS Lett.* 170: 38–42

Hirasawa, K., Nishizuka, Y. 1985. Phosphatidylinositol turnover in receptor mechanism and signal transduction. *Ann. Rev. Pharmacol. Toxicol.* 25: 147–70

Hirata, F., Matsuda, K., Notsu, Y., Hattori, T., Del Carmie, R. 1984. Phosphorylation at a tyrosine residue of lipomodulin in mitogen-stimulated murine thymocytes. *Proc. Natl. Acad. Sci. USA* 81: 4717–21

Hishikawa, R., Fukase, M., Yamatani, T., Kadowaki, S., Fujita, T. 1985. Phorbol ester stimulates calcitonin secretion synergistically with A23187, and additively with dibutyryl cyclic AMP in a rat C-cell line. *Biochem. Biophys. Res. Commun.* 132: 424–29

Hofer, H. W., Schlatter, S., Graefe, M. 1985. Phosphorylation of phosphofructokinase by protein kinase C changes the allosteric properties of the enzyme. *Biochem. Biophys. Res. Commun.* 129: 892–97

Hokin, M. R., Hokin, L. E. 1953. Enzyme secretion and the incorporation of ^{32}P into phospholipids of pancreatic slices. *J. Biol. Chem.* 203: 967–77

Hokin, M. R., Hokin, L. E. 1964. Interconversion of phosphatidylinositol and phosphatidic acid involved in the response to acetylcholine in the salt gland. In *The Metabolism and Physiological Significance of Lipids*, ed. R. M. C. Dawson, D. N. Rhodes, pp. 423–34. London/New York/Sydney: Wiley

Hollingsworth, E. B., Sears, E. B., Daly, J. W. 1985. An activator of protein kinase C (phorbol-12-myristate-13-acetate) augments 2-chloroadenosine-elicited accumulation of cyclic AMP in guinea pig cerebral cortical particulate preparations. *FEBS Lett.* 184: 339–42

Houchi, H., Nakanishi, A., Uddin, M. M., Ohuchi, T., Oka, M. 1985. Phorbol ester stimulates catecholamine synthesis in isolated bovine adrenal medullary cells.

FEBS Lett. 188: 205–8

Hubinont, C. J., Best, L., Sener, A., Malaisse, W. J. 1984. Activation of protein kinase C by a tumor-promoting phorbol ester in pancreatic islets. *FEBS Lett.* 170: 247–53

Huganir, R. L., Miles, K., Greengard, P. 1984. Phosphorylation of the nicotinic acetylcholine receptor by an endogenous tyrosine-specific protein kinase. *Proc. Natl. Acad. Sci. USA* 81: 6968–72

Humble, E., Heldin, P., Forsberg, P.-O., Engström, L. 1984. Phosphorylation of human fibrinogen *in vitro* by calcium-activated phospholipid-dependent protein kinase from pig spleen. *J. Biochem.* 95: 1435–43

Hunter, T., Cooper, J. A. 1985. Protein-tyrosine kinases. *Ann. Rev. Biochem.* 54: 897–930

Hunter, T., Ling, N., Cooper, J. A. 1984. Protein kinase C phosphorylation of the EGF receptor at a threonine residue close to the cytoplasmic face of the plasma membrane. *Nature* 311: 480–83

Hutton, J. C., Peshavaria, M., Brocklehurst, K. W. 1984. Phorbol ester stimulation of insulin release and secretory-granule protein phosphorylation in a transplantable rat insulinoma. *Biochem. J.* 224: 483–90

Ikebe, M., Inagaki, M., Kanamaru, K., Hidaka, H. 1985. Phosphorylation of smooth muscle myosin light chain kinase by Ca^{2+}-activated, phospholipid-dependent protein kinase. *J. Biol. Chem.* 260: 4547–50

Inoue, M., Kishimoto, A., Takai, Y., Nishizuka, Y. 1977. Studies on a cyclic nucleotide-independent protein kinase and its proenzyme in mammalian tissues. II. Proenzyme and its activation by calcium-dependent protease from rat brain. *J. Biol. Chem.* 252: 7610–16

Irvine, R. F., Dawson, R. M. C., Freinkel, N. 1982. Stimulated phosphatidylinositol turnover: A brief proposal. In *Contemporary Metabolism*, ed. N. Freinkel, 2: 301–42. New York: Plenum

Isakov, N., Bleackly, R. C., Shaw, J., Altman, A. 1985. Teleocidin and phorbol ester tumor promoters exert similar mitogenic effects on human lymphocytes. *Biochem. Biophys. Res. Commun.* 130: 724–31

Iwasa, Y., Hosey, M. M. 1984. Phosphorylation of cardiac sarcolemma proteins by the calcium-activated, phospholipid-dependent protein kinase. *J. Biol. Chem.* 259: 534–40

Iwashita, S., Fox, C. F. 1984. Epidermal growth factor and potent phorbol tumor promoters induce epidermal growth factor receptor phosphorylation in a similar but distinctly different manner in human epidermoid carcinoma A431 cells. *J. Biol. Chem.* 259 : 2559–67

Jacobs, S., Sahyoun, N. E., Saltiel, A. R., Cuatrecasas, P. 1983. Phorbol esters stimulate the phosphorylation of receptors for insulin and somatomedin C. *Proc. Natl. Acad. Sci. USA* 80 : 6211–13

Jetten, A. M., Ganong, B. R., Vandenbark, G. R., Shirley, J. E., Bell, R. M. 1985. Role of protein kinase C in diacylglycerol-mediated induction of ornithine decarboxylase and reduction of epidermal growth factor binding. *Proc. Natl. Acad. Sci. USA* 82 : 1941–45

Johnson, H. M., Vassallo, T., Torres, B. A. 1985a. Interleukin 2–mediated events in γ-interferon production are calcium dependent at more than one site. *J. Immunol.* 134 : 967–70

Johnson, P. C., Ware, J. A., Cliveden, P. B., Smith, M., Dvorak, A. M., et al. 1985b. Measurement of ionized calcium in blood platelets with the photoprotein aequorin. *J. Biol. Chem.* 260 : 2069–76

Kaibuchi, K., Takai, Y., Nishizuka, Y. 1981. Cooperative roles of various membrane phospholipids in the activation of calcium-activated, phospholipid-dependent protein kinase. *J. Biol. Chem.* 256 : 7146–49

Kaibuchi, K., Takai, Y., Nishizuka, Y. 1985. Protein kinase C and calcium ion in mitogenic response of macrophage-depleted human peripheral lymphocytes. *J. Biol. Chem.* 260 : 1366–69

Kaibuchi, K., Takai, Y., Sawamura, M., Hoshijima, M., Fujikura, T., et al. 1983. Synergistic functions of protein phosphorylation and calcium mobilization in platelet activation. *J. Biol. Chem.* 258 : 6701–4

Kajikawa, N., Kaibuchi, K., Matsubara, T., Kikkawa, U., Takai, Y., et al. 1983. A possible role of protein kinase C in signal-induced lysosomal enzyme release. *Biochem. Biophys. Res. Commun.* 116 : 743–50

Kajikawa, N., Kikkawa, U., Itoh, K., Nishizuka, Y. 1986. Permeable diacylglycerol, its application to release reactions, and partial purification and properties of protein kinase C from platelets. *Methods Enzymol.* In press

Kassis, S., Zaremba, T., Patel, J., Fishman, P. H. 1985. Phorbol esters and β-adrenergic agonists mediate desensitization of adenylate cyclase in rat glioma C6 cells by distinct mechanism. *J. Biol. Chem.* 260 : 8911–17

Katada, T., Gilman, A. G., Watanabe, Y., Bauer, S., Jakobs, K. H. 1985. Protein kinase C phosphorylates the inhibitory guanosine-nucleotide-binding regulatory component and apparently suppresses its function in hormonal inhibition of adenylate cyclase. *Eur. J. Biochem.* 151 : 431–37

Katakami, Y., Kaibuchi, K., Sawamura, M., Takai, Y., Nishizuka, Y. 1984. Synergistic action of protein kinase C and calcium for histamine release from rat peritoneal mast cells. *Biochem. Biophys. Res. Commun.* 121 : 573–78

Kawamoto, S., Hidaka, H. 1984. Ca^{2+}-activated, phospholipid-dependent protein kinase catalyses the phosphorylation of actin-binding proteins. *Biochem. Biophys. Res. Commun.* 118 : 736–42

Kelleher, D. J., Pessin, J. E., Ruoho, A. E., Johnson, G. L. 1984. Phorbol ester induces desensitization of adenylate cyclase and phosphorylation of the β-adrenergic receptor in turkey erythrocytes. *Proc. Natl. Acad. Sci. USA* 81 : 4316–20

Kelly, K., Cochran, B. H., Stiles, C. D., Leder, P. 1983. Cell-specific regulation of the *c-myc* gene by lymphocyte mitogens and platelet-derived growth factor. *Cell* 35 : 603–10

Kikkawa, U., Go, M., Koumoto, J., Nishizuka, Y. 1986a. Rapid purification of protein kinase C by high performance liquid chromatography. *Biochem. Biophys. Res. Commun.* 135 : 636–43

Kikkawa, U., Kitano, T., Saito, N., Fujiwara, H., Nakanishi, H., et al. 1986b. Possible roles of protein kinase C in signal transduction in nervous tissues. *Prog. Brain Res.* In press

Kikkawa, U., Kitano, T., Saito, N., Kishimoto, A., Taniyama, K., et al. 1986c. Role of protein kinase C in calcium-mediated signal transduction. In *Calcium and the Cell. Ciba Found. Symp.* 122. In press

Kikkawa, U., Nishizuka, Y. 1986. Protein kinase C. *Enzymes* 17 : 167–89

Kikkawa, U., Takai, Y., Minakuchi, R., Inohara, S., Nishizuka, Y. 1982. Calcium-activated, phospholipid-dependent protein kinase from rat brain. Subcellular distribution, purification, and properties. *J. Biol. Chem.* 257 : 13341–48

Kikkawa, U., Takai, Y., Tanaka, Y., Miyake, R., Nishizuka, Y. 1983. Protein kinase C as a possible receptor protein of tumor-promoting phorbol esters. *J. Biol. Chem.* 258 : 11442–45

Kimura, K., Katoh, N., Sakurada, K., Kubo, S. 1985. Phosphorylation of high mobility group 1 protein by phospholipid-sensitive Ca^{2+}-dependent protein kinase from pig testis. *Biochem. J.* 227 : 271–76

King, A. C., Cuatrecasas, P. 1982. Resolution of high and low affinity epidermal growth factor receptors. *J. Biol. Chem.* 257 : 3053–60

Kirsch, D., Obermaier, B., Haring, H. U. 1985. Phorbolesters enhance basal D-glucose transport but inhibit insulin stimulation of D-glucose transport and insulin binding in isolated rat adipocytes. *Biochem. Biophys. Res Commun.* 128: 824–32

Kishimoto, A., Kajikawa, N., Shiota, M., Nishizuka, Y. 1983. Proteolytic activation of calcium-activated, phospholipid-dependent protein kinase by calcium-dependent neutral protease. *J. Biol. Chem.* 258: 1156–64

Kishimoto, A., Nishiyama, K., Nakanishi, H., Uratsuji, Y., Nomura, H., et al. 1985. Studies on the phosphorylation of myelin basic protein by protein kinase C and adenosine - 3′,5′ - monophosphate - dependent protein kinase. *J. Biol. Chem.* 260: 12492–99

Kishimoto, A., Takai, Y., Mori, T., Kikkawa, U., Nishizuka, Y. 1980. Activation of calcium and phospholipid-dependent protein kinase by diacylglycerol, its possible relation to phosphatidylinositol turnover. *J. Biol. Chem.* 255: 2273–76

Kishimoto, A., Takai, Y., Nishizuka, Y. 1977. Activation of glycogen phosphorylase kinase by a calcium-activated, cyclic nucleotide-independent protein kinase system. *J. Biol. Chem.* 252: 7449–52

Kiss, Z., Steinberg, R. A. 1985. Interaction between cyclic AMP- and phorbol ester-dependent phosphorylation systems in S49 mouse lymphoma cells. *J. Cell. Physiol.* 125: 200–6

Klausner, R. D., Harford, J., van Renswoude, J. 1984. Rapid internalization of the transferrin receptor in K562 cells is triggered by ligand binding or treatment with a phorbol ester. *Proc. Natl. Acad. Sci. USA* 81: 3005–9

Knight, D. E., Baker, P. F. 1983. The phorbol ester TPA increases the affinity of exocytosis for calcium in "leaky" adrenal medullary cells. *FEBS Lett.* 160: 98–100

Knight, D. E., Niggli, V., Scrutton, M. C. 1984. Thrombin and activators of protein kinase C modulate secretory responses of permeabilised human platelets induced by Ca^{2+}. *Eur. J. Biochem.* 143: 437–46

Kojima, I., Lippes, H., Kojima, K., Rasmussen, H. 1983. Aldosterone secretion: Effect of phorbolester and A23187. *Biochem. Biophys. Res. Commun.* 116: 555–62

Kraft, A. S., Anderson, W. B. 1983. Phorbol esters increase the amount of Ca^{2+}, phospholipid-dependent protein kinase associated with plasma membrane. *Nature* 301: 621–23

Krebs, E. G., Beavo, J. A. 1979. Phospho-rylation-dephosphorylation of enzymes. *Ann. Rev. Biochem.* 48: 923–59

Kreutter, D., Caldwell, A. B., Morin, M. J. 1985. Dissociation of protein kinase C activation from phorbol ester–induced maturation of HL-60 leukemia cells. *J. Biol. Chem.* 260: 5979–84

Kruijer, W., Cooper, J. A., Hunter, T., Verma, I. M. 1984. Platelet-derived growth factor induces rapid but transient expression of the *c-fos* gene and protein. *Nature* 312: 711–16

Krupp, M. N., Connolly, D. T., Lane, M. D. 1982. Synthesis, turnover, and down-regulation of epidermal growth factor receptors in human A431 epidermoid carcinoma cells and skin fibroblasts. *J. Biol. Chem.* 257: 11489–96

Kuo, J. F., Andersson, R. G. G., Wise, B. C., Mackerlova, L., Solomonsson, I., et al. 1980. Calcium-dependent protein kinase: Widespread occurrence in various tissues and phyla of the animal kingdom and comparison of effects of phospholipid, calmodulin, and trifluoperazine. *Proc. Natl. Acad. Sci. USA* 77: 7039–43

Labarca, R., Janowsky, A., Patel, J., Paul, S. M. 1984. Phorbol ester inhibit agonist-induced [^{3}H]inositol-1-phosphate accumulation in rat hippocampal slices. *Biochem. Biophys. Res. Commun.* 123: 703–9

Lagast, H., Pozzan, T., Waldvogel, F. A., Lew, P. D. 1984. Phorbol myristate acetate stimulates ATP-dependent calcium transport by the plasma membrane of neutrophils. *J. Clin. Invest.* 73: 878–83

Lapetina, E. G. 1986. Incorporation of synthetic 1,2-diacylglycerol into platelet phosphatidylinositol is increased by cyclic AMP. *FEBS Lett.* 195: 111–14

Lapetina, E. G., Reep, B., Ganong, B. R., Bell, R. M. 1985. Exogenous *sn*-1,2-diacylglycerols containing saturated fatty acids function as bioregulators of protein kinase C in human platelets. *J. Biol. Chem.* 260: 1358–61

Leach, K. L., James, M. L., Blumberg, P. M. 1983. Characterization of a specific phorbol ester aporeceptor in mouse brain cytosol. *Proc. Natl. Acad. Sci. USA* 80: 4208–12

Lee, L. S., Weinstein, I. B. 1978. Epidermal growth factor, like phorbol esters, induces plasminogen activator in HeLa cells. *Nature* 274: 696–97

Leeb-Lundberg, L. M. F., Cotechia, S., Lomasney, J. W., DeBernardis, J. F., Lefkowitz, R. J., et al. 1985. Phorbol esters promote α_1-adrenergic receptor phosphorylation and receptor uncoupling from inositol phospholipid metabolism. *Proc. Natl. Acad. Sci. USA* 82: 5651–55

Le Peuch, C. J., Ballester, R., Rosen, O. M.

1983. Purified rat brain calcium- and phospholipid-dependent protein kinase phosphorylates ribosomal protein S6. *Proc. Natl. Acad. Sci. USA* 80: 6858–62

Lim, M. S., Sutherland, C., Walsh, M. P. 1985. Phosphorylation of bovine cardiac C-protein by protein kinase C. *Biochem. Biophys. Res. Commun.* 132: 1187–95

Limas, C. J. 1980. Phosphorylation of cardiac sarcoplasmic reticulum by a calcium-activated, phospholipid-dependent protein kinase. *Biochem. Biophys. Res. Commun.* 96: 1378–83

Limas, C. J., Limas, C. 1985. Phorbol ester- and diacylglycerol-mediated desensitization of cardiac beta-adrenergic receptors. *Circ. Res.* 57: 443–49

Logsdon, C. D., Williams, J. A. 1984. Intracellular Ca^{2+} and phorbol esters synergistically inhibit internalization of epidermal growth factor in pancreatic acini. *Biochem. J.* 223: 893–900

Loskutoff, D. J., Edgington, T. S. 1977. Synthesis of a fibrinolytic activator and inhibitor by endothelial cells. *Proc. Natl. Acad. Sci. USA* 74: 3903–7

Lynch, C. J., Charest, R., Bocckino, S. B., Exton, J. H., Blackmore, P. F. 1985. Inhibition of hepatic α_1-adrenergic effects and binding by phorbol myristate acetate. *J. Biol. Chem.* 260: 2844–51

MacIntyre, D. E., McNicol, A., Drummond, A. H. 1985. Tumour-promoting phorbol esters inhibit agonist-induced phosphatidate formation and Ca^{2+} flux in human platelets. *FEBS Lett.* 180: 160–64

Majerus, P. W., Wilson, D. B., Connolly, T. E., Bross, T. E., Neufeld, E. J. 1985. Phosphoinositide turnover provides a link in stimulus-response coupling. *Trends Biochem. Sci.* 10: 168–71

Martin, T. F., Kowalchyk, J. A. 1984. Evidence for the role of calcium and diacylglycerol as dual second messengers in thyrotropin-releasing hormone action: Involvement of diacylglycerol. *Endocrinology* 115: 1517–26

May, W. S., Jacobs, S., Cuatrecasas, P. 1984. Association of phorbol ester-induced hyperphosphorylation and reversible regulation of transferrin membrane recoptors in HL-60 cells. *Proc. Natl. Acad. Sci. USA* 81: 2016–20

Mazzei, G. J., Kuo, J. F. 1984. Phosphorylation of skeletal-muscle troponin I and troponin T by phospholipid-sensitive Ca^{2+}-dependent protein kinase and its inhibition by troponin C and tropomyosin. *Biochem. J.* 218: 361–69

McCaffrey, P. G., Friedman, B., Rosner, M. R. 1984. Diacylglycerol modulates binding and phosphorylation of the epidermal growth factor receptor. *J. Biol. Chem.* 259: 12502–7

McCall, C., Schmitt, J., Cousart, S., O'Flaherty, J., Bass, D., et al. 1985. Stimulation of hexose transport by human polymorphonuclear leukocytes: A possible role for protein kinase C. *Biochem. Biophys. Res. Commun.* 126: 450–56

Merritt, J. E., Rubin, R. P. 1985. Pancreatic amylase secretion and cytoplasmic free calcium. Effect of ionomycin, phorbol dibutyrate and diacylglycerols alone and in combination. *Biochem. J.* 230: 151–59

Michell, R. H. 1975. Inositol phospholipids and cell surface receptor function. *Biochim. Biophys. Acta* 415: 81–147

Michell, R. H. 1986. Profusion and confusion. *Nature* 319: 176–77

Minakuchi, R., Takai, Y., Yu, B., Nishizuka, Y. 1981. Widespread occurrence of calcium-activated, phospholipid-dependent protein kinase in mammalian tissues. *J. Biochem.* 89: 1651–54

Mitchell, R. L., Zokas, L., Schreiber, R. D., Verma, I. M. 1985. Rapid induction of the expression of proto-oncogene *fos* during human monocytic differentiation. *Cell* 40: 209–17

Miyake, R., Tanaka, Y., Tsuda, T., Kaibuchi, K., Kikkawa, U., et al. 1984. Activation of protein kinase C by non-phorbol tumor promoter, mezerein. *Biochem. Biophys. Res. Commun.* 121: 649–56

Mobley, A., Tai, H. H. 1985. Synergistic stimulation of thromboxane biosynthesis by calcium ionophore and phorbol ester or thrombin in human platelets. *Biochem. Biophys. Res. Commun.* 130: 717–23

Moger, W. H. 1985. Stimulation of Leydig cell steroidogenesis by the phorbol ester 12 - O - tetradecanoylphorbol - 13 - acetate: Similarity to the effects of gonadotropin-releasing hormone. *Life Sci.* 37: 869–73

Moolenaar, W. H., Tertoolen, L. G. J., de Laat, S. W. 1984. Phorbol ester and diacylglycerol mimic growth factors in raising cytoplasmic pH. *Nature* 312: 371–74

Moon, S. K., Palfrey, H. C., King, A. C. 1984. Phorbol esters potentiate tyrosine phosphorylation of epidermal growth factor receptors in A431 membranes by a calcium-independent mechanism. *Proc. Natl. Acad. Sci. USA* 81: 2298–2302

Morgan, J. P., Morgan, K. G. 1982. Vascular smooth muscle: The first recorded Ca^{2+} transients. *Pflügers Arch.* 395: 75–77

Mori, T., Takai, Y., Yu, B., Takahashi, J., Nishizuka, Y., et al. 1982. Specificity of the fatty acyl moieties of diacylglycerol for the activation of calcium-activated, phospholipid-dependent protein kinase. *J. Biochem.* 91: 427–31

Morley, N., Kuksis, A. 1972. Positional specificity of lipoprotein lipase. *J. Biol. Chem.* 247: 6389–93

Movsesian, M. A., Nishikawa, M., Adelstein, R. S. 1984. Phosphorylation of phospholamban by calcium-activated, phospholipid-dependent protein kinase. *J. Biol. Chem.* 259: 8029–32

Mukhopadhyay, A. K., Bohnet, H. G., Leidenberger, F. A. 1984. Phorbol esters inhibit LH stimulated steroidogenesis by mouse Leydig cells *in vitro*. *Biochem. Biophys. Res. Commun.* 119: 1062–67

Mukhopadhyay, A. K., Schumacher, M. 1985. Inhibition of hCG-stimulated adenylate cyclase in purified mouse Leydig cells by the phorbol ester PMA. *FEBS Lett.* 187: 56–60

Muller, R., Bravo, R., Burckhardt, J., Curran, T. 1984. Induction of *c-fos* gene and protein by growth factors precedes activation of *c-myc*. *Nature* 312: 716–20

Murdoch, G. H., Waterman, M., Evans, R. M., Rosenfeld, M. G. 1985. Molecular mechanism of phorbol ester, thyrotropin-releasing hormone, and growth factor stimulation of prolactin gene transcription. *J. Biol. Chem.* 260: 11852–58

Nabika, T., Nara, Y., Yamori, Y., Lovenberg, W., Endo, J. 1985. Angiotensin II and phorbol ester enhance isoproterenol- and vasoactive intestinal peptide (VIP)-induced cyclic AMP accumulation in vascular smooth muscle cells. *Biochem. Biophys. Res. Commun.* 131: 30–36

Naccache, P. H., Molski, T. F. P., Borgeat, P., White, J. R., Sha'afi, R. I. 1985. Phorbol esters inhibit the fMet-Leu-Phe- and leukotriene B$_4$-stimulated calcium mobilization and enzyme secretion in rabbit neutrophils. *J. Biol. Chem.* 260: 2125–31

Naka, M., Nishikawa, M., Adelstein, R. S., Hidaka, H. 1983. Phorbol ester-induced activation of human platelets is associated with protein kinase C phosphorylation of myosin light chains. *Nature* 306: 490–92

Nakaki, T., Roth, B. L., Chuang, D. M., Costa, E. 1985. Phasic and tonic components in 5-HT$_2$ receptor-mediated rat aorta contraction: Participation of Ca^{++} channels and protein kinase C. *J. Pharmacol. Exp. Ther.* 234: 442–46

Nakanishi, H., Nomura, H., Kikkawa, U., Kishimoto, A., Nishizuka, Y. 1985. Rat brain and liver soluble phospholipase C: Resolution of two forms with different requirements for calcium. *Biochem. Biophys. Res. Commun.* 132: 582–90

Naor, Z., Eli, Y. 1985. Synergistic stimulation of luteinizing hormone (LH) release by protein kinase C activators and Ca^{2+}-ionophore. *Biochem. Biophys. Res. Commun.* 130: 848–53

Negro-Vilar, A., Lapetina, E. G. 1985. 1,2-Didecanoylglycerol and phorbol 12,13-dibutyrate enhance anterior pituitary hormone secretion *in vitro*. *Endocrinology* 117: 1559–64

Nel, A. E., Wooten, M. W., Goldschmidt-Clermont, P. J., Miller, P. J., Stevenson, H. C., et al. 1985. Polymixin B causes coordinate inhibition of phorbol ester-induced C-kinase activity and proliferation of B lymphocytes. *Biochem. Biophys. Res. Commun.* 128: 1364–72

Niedel, J. E., Kuhn, L. J., Vandenbark, G. R. 1983. Phorbol dieter receptor copurifies with protein kinase C. *Proc. Natl. Acad. Sci. USA* 80: 36–40

Nishikawa, M., Shirakawa, S., Adelstein, R. S. 1985. Phosphorylation of smooth muscle myosin light chain kinase by protein kinase C. Comparative studies of the phosphorylated sites. *J. Biol. Chem.* 260: 8979–88

Nishizuka, Y. 1984a. The role of protein kinase C in cell surface signal transduction and tumor promotion. *Nature* 308: 693–98

Nishizuka, Y. 1984b. Turnover of inositol phospholipids and signal transduction. *Science* 225: 1365–70

Nishizuka, Y. 1986. Perspectives on the role of protein kinase C in stimulus-response coupling. *J. Natl. Cancer Inst.* 76: 363–70

Nomura, H., Nakanishi, H., Ase, K., Kikkawa, U., Nishizuka, Y. 1986. Inositol phospholipid turnover in stimulus-response coupling. *Prog. Hemostasis Thromb.* In press

O'Brian, C. A., Lawrence, D. S., Kaiser, E. T., Weinstein, I. B. 1984. Protein kinase C phosphorylates the synthetic peptide Arg-Arg-Lys-Ala-Ser-Gly-Pro-Pro-Val in the presence of phospholipid plus either Ca^{2+} or a phorbol ester tumor promoter. *Biochem. Biophys. Res. Commun.* 124: 296–302

O'Flaherty, J. T., Schmitt, J. D., McCall, C. E., Wykle, R. L. 1984. Diacylglycerols enhance human neutrophil degranulation responses: Relevancy to a multiple mediator hypothesis of cell function. *Biochem. Biophys. Res. Commun.* 123: 64–70

Ohmura, E., Friesen, H. G. 1985. 12-O-Tetradecanoyl phorbol-13-acetate stimulate rat growth hormone (GH) release through different pathways from that of human pancreatic GH-releasing factor. *Endocrinology* 116: 728–33

Orellana, S. A., Solski, P. A., Brown, J. H. 1985. Phorbol ester inhibits phosphoinositide hydrolysis and calcium mobilization in cultured astrocytoma cells. *J. Biol. Chem.* 260: 5236–39

Otani, S., Matsui, I., Kuramoto, A., Morisawa, S. 1985. Induction of ornithine decarboxylase in guinea-pig lymphocytes. Synergistic effect of diacylglycerol and calcium. *Eur. J. Biochem.* 147: 27–31

Padel, U., Söling, H. D. 1985. Phosphorylation of the ribosomal protein S6 during agonist-induced exocytosis in exocrine glands is catalysed by calcium-activated, phospholipid-dependent protein kinase (protein kinase C). Experiments with guinea pig parotid glands. *Eur. J. Biochem.* 151: 1–10

Papadojoulos, V., Carreau, S., Drosdowsky, M. A. 1985. Effect of phorbol ester and phospholipase C on LH-stimulated steroidogenesis in purified rat Leydig cells. *FEBS Lett.* 188: 312–16

Papini, E., Grzeskowiak, M., Bellavite, P., Rossi, F. 1985. Protein kinase C phosphorylates a component of NADPH oxidase of neutrophils. *FEBS Lett.* 190: 204–8

Parker, P. J., Katan, M., Waterfield, M. D., Leader, D. P. 1985. The phosphorylation of eukaryotic ribosomal protein S6 by protein kinase C. *Eur. J. Biochem.* 148: 579–86

Pocotte, S. L., Frye, R. A., Senter, R. A., TerBush, D. R., Lee, S. A., et al. 1985. Effects of phorbol ester on catecholamine secretion and protein phosphorylation in adrenal medullary cell cultures. *Proc. Natl. Acad. Sci. USA* 82: 930–34

Pozzan, T., Gatti, G., Dozio, N., Vicentini, L. M., Meldolesi, J. 1984. Ca^{2+}-dependent and -independent release of neurotransmitters from PC12 cells: A role for protein kinase C activation? *J. Cell Biol.* 99: 628–38

Publicover, S. J. 1985. Stimulation of spontaneous transmitter release by the phorbol ester, 12-O-tetradecanoylphorbol-13-acetate, an activator of protein kinase C. *Brain Res.* 333: 185–87

Putney, J. W. Jr., McKinney, J. S., Aub, D. L., Leslie, B. A. 1984. Phorbol ester-induced protein secretion in rat parotid gland. Relationship to the role of inositol lipid breakdown and protein kinase C activation in stimulus-secretion coupling. *Mol. Pharmacol.* 26: 261–66

Quilliam, L. A., Dobson, P. R., Brown, B. L. 1985. Modulation of cyclic AMP accumulation in GH3 cells by a phorbol ester and thyroliberin. *Biochem. Biophys. Res. Commun.* 129: 898–903

Ramachandran, C., Yau, P., Bradbury, E. M., Shyamala, G., Yasuda, H., et al. 1984. Phosphorylation of high-mobility-group proteins by the calcium-phospholipid-dependent protein kinase and cyclic AMP-dependent protein kinase. *J. Biol. Chem.* 259: 13495–13503

Rando, R. R., Young, N. 1984. The stereospecific activation of protein kinase C. *Biochem. Biophys. Res. Commun.* 122: 818–23

Rasmussen, H., Barrett, P. Q. 1984. Calcium messenger system: An integrated view. *Physiol. Rev.* 64: 938–84

Rasmussen, H., Forder, J., Kojima, I., Scriabine, A. 1984. TPA-induced contraction of isolated rabbit vascular muscle. *Biochem. Biophys. Res. Commun.* 122: 776–84

Rebois, R. V., Patel, J. 1985. Phorbol ester causes desensitization of gonadotropin-responsive adenylate cyclase in a murine Leydig tumor cell line. *J. Biol. Chem.* 260: 8026–31

Rink, T. J., Sanchez, A., Hallam, T. J. 1983. Diacylglycerol and phorbol ester stimulate secretion without raising cytoplasmic free calcium in human platelets. *Nature* 305: 317–19

Rittenhouse, S. E., Sasson, J. P. 1985. Mass changes in myoinositol triphosphate in human platelets stimulated by thrombin. *J. Biol. Chem.* 260: 8657–60

Roach, P. J., Goldman, M. 1983. Modification of glycogen synthase activity in isolated rat hepatocytes by tumor-promoting phorbol esters: Evidence for differetial regulation of glycogen synthase and phosphorylase. *Proc. Natl. Acad. Sci. USA* 80: 7170–72

Robinson, J. M., Badway, J. A., Karnovsky, M. L., Karnovsky, M. J. 1985. Release of superoxide and change in morphology by neutrophils in response to phorbol esters: Antagonism by inhibitors of calcium binding proteins. *J. Cell Biol.* 101: 1052–58

Ronning, S. A., Martin, T. F. 1985. Prolactin secretion in permeable GH3 pituitary cells is stimulated by Ca^{2+} and protein kinase C activators. *Biochem. Biophys. Res. Commun.* 130: 524–32

Rosoff, P. M., Stein, L. F., Cantley, L. C. 1984. Phorbol esters induce differentiation in a pre-B-lymphocyte cell line by enhancing Na^+/H^+ exchange. *J. Biol. Chem.* 259: 7056–60

Routtenberg, A., Lovinger, D. M. 1985. Selective increase in phosphorylation of a 47-kDa protein (F1) directly related to long-term potentiation. *Behav. Neur. Biol.* 43: 3–11

Rozengurt, E., Rodriguez-Pena, A., Coombs, M., Sinnett-Smith, J. 1984. Diacylglycerol stimulates DNA synthesis and cell division in mouse 3T3 cells: Role of Ca^{2+}-sensitive phospholipid-dependent protein kinase. *Proc. Natl. Acad. Sci. USA* 81: 5748–52

Sagi-Eisenberg, R., Lieman, H., Pecht, I. 1985. Protein kinase C regulation of the receptor-coupled calcium signal in histamine-secreting rat basophilic leukemia cells. *Nature* 313 : 59–60

Sakamoto, C., Matozaki, T., Nagao, M., Baba, S. 1985. Combined effect of phorbol ester and A23187 or dibutyryl cyclic AMP on pepsinogen secretion from isolated gastric gland. *Biochem. Biophys. Res. Commun.* 131 : 314–19

Sando, J. J., Young, M. C. 1983. Identification of high-affinity phorbol ester receptor in cytosol of EL4 thymoma cells: Requirement for calcium, magnesium, and phospholipids. *Proc. Natl. Acad. Sci. USA* 80 : 2642–46

Sano, K., Voelker, D. R., Mason, R. J. 1985. Involvement of protein kinase C in pulmonary surfactant secretion from alveolar type II cells. *J. Biol. Chem.* 260 : 12725–29

Sawyer, S. T., Cohen, S. 1981. Enhancement of calcium uptake and phosphatidylinositol turnover by epidermal growth factor in A-431 cells. *Biochemistry* 20 : 6280–86

Schatzman, R. C., Grifo, J. A., Merrick, W. C., Kuo, J. F. 1983. Phospholid-sensitive Ca^{2+}-dependent protein kinase phosphorylates the β subunit of eukaryotic initiation factor 2 (eIF-2). *FEBS Lett.* 159 : 167–70

Serhan, C. N., Broekman, M. J., Korchak, H. M., Smolen, J. E., Marcus, A. J., et al. 1983. Changes in phosphatidylinositol and phosphatidic acid in stimulated human neutrophils. Relationship to calcium mobilization, aggregation and superoxide radical generation. *Biochim. Biophys. Acta* 762 : 420–28

Shackelford, D. A., Trowbridge, I. S. 1984. Induction of expression and phosphorylation of the human interleukin 2 receptor by a phorbol diester. *J. Biol. Chem.* 259 : 11706–12

Shinohara, O., Knecht, M., Catt, K. J. 1985. Inhibition of gonadotropin-induced granulosa cell differentiation by activation of protein kinase C. *Proc. Natl. Acad. Sci. USA* 82 : 8518–22

Shoyab, M., De Larco, J. E., Todaro, G. J. 1979. Biologically active phorbol esters specifically alter affinity of epidermal growth factor membrane receptors. *Nature* 279 : 387–91

Sibley, D. R., Nambi, P., Peter, J. R., Lefkowitz, R. J. 1984. Phorbol diesters promote β-adrenergic receptor phosphorylation and adenylate cyclase desensitization in duck erythrocytes. *Biochem. Biophys. Res. Commun.* 121 : 973–79

Spät, A., Fabiato, A., Rubin, R. P. 1986. Binding of inositol trisphosphate by a liver microsomal fraction. *Biochem. J.* 233 :

929–32

Starke, K. 1980. Presynaptic receptors and the control of noradrenaline release. *Trends Pharmacol. Sci.* 1 : 268–71

Streb, H., Irvine, R. F., Berridge, M. J., Schulz, I. 1983. Release of Ca^{2+} form a nonmitochondrial intracellular store in pancreatic acinar cells by inositol-1,4,5-trisphosphate. *Nature* 306 : 67–69

Sugden, D., Vanecek, J., Klein, D. C., Thomas, T. P., Anderson, W. B. 1985. Activation of protein kinase C potentiates isoprenaline-induced cyclic AMP accumulation in rat pinealocytes. *Nature* 314 : 359–61

Sulakhe, P. V., Johnson, D. D., Phan, N. T., Wilcox, R. 1985. Phorbol ester inhibits myoblast fusion and activates β-adrenergic receptor coupled adenylate cyclase. *FEBS Lett.* 186 : 281–85

Sulakhe, P. V., Petrali, E. H., Davis, E. R., Thiessen, B. J. 1980. Calcium ion stimulated endogenous protein kinase catalyzed phosphorylation of basic proteins in myelin subfractions and myelin-like membrane fraction from rat brain. *Biochemistry* 19 : 5363–71

Summers, T. A., Creutz, C. E. 1985. Phosphorylation of a chromaffin granule-binding protein by protein kinase C. *J. Biol. Chem.* 260 : 2437–43

Takai, Y., Kikkawa, U., Kaibuchi, K., Nishizuka, Y. 1984. Membrane phospholipid metabolism and signal transduction for protein phosphorylation. *Adv. Cyclic Nucleotide Protein Phosphorylation Res.* 18 : 119–58

Takayama, S., White, M. F., Lauris, V., Kahn, C. R. 1984. Phorbol esters modulate insulin receptor phosphorylation and insulin action in cultured hepatoma cells. *Proc. Natl. Acad. Sci. USA* 81 : 7797–7801

Tamagawa, T., Niki, H., Niki, A. 1985. Insulin release independent of a rise in cytosolic free Ca^{2+} by forskolin and phorbol ester. *FEBS Lett.* 183 : 430–32

Tanaka, C., Fujiwara, H., Fujii, Y. 1986. Acetylcholine release from guinea pig caudate slices evoked by phorbol ester and calcium. *FEBS Lett.* 195 : 129–34

Tanaka, C., Taniyama, K., Kusunoki, M. 1984. A phorbol ester and A23187 act synergistically to release acetylcholine from the guinea pig ileum. *FEBS Lett.* 175 : 165–69

Tanigawa, K., Kuzuya, H., Imura, H., Taniguchi, H., Baba, S., et al. 1982. Calcium-activated, phospholipid-dependent protein kinase in rat pancreatic islets of Langerhans. Its possible role in glucose-induced insulin release. *FEBS Lett.* 138 : 183–86

Tapley, P. M., Murray, A. W. 1984. Platelet

Ca^{2+}-activated, phospholipid-dependent protein kinase: Evidence for proteolytic activation of the enzyme in cells treated with phospholipase C. *Biochem. Biophys. Res. Commun.* 118: 835–41

Teshima, R., Ikeuchi, H., Terao, T. 1984. Ca^{2+}-dependent and phorbol ester activating phosphorylation of a 36K-dalton protein of rat basophilic leukemia cell membranes and immunoprecipitation of the phosphorylated protein with IgE-anti IgE system. *Biochem. Biophys. Res. Commun.* 125: 867–74

Trevillyan, J. M., Perisic, O., Traugh, J. A., Byus, C. V. 1985. Insulin- and phorbol ester–stimulated phosphorylation of ribosomal protein S6. *J. Biol. Chem.* 260: 3041–44

Truneh, A., Albert, F., Golstein, P., Schmitt-Verhulst, A.-M. 1985. Early steps of lymphocyte activation bypassed by synergy between calcium ionophores and phorbol ester. *Nature* 313: 318–20

Tsien, R. Y., Pozzan, T., Rink, T. J. 1982. T-cell mitogens cause early changes in cytoplasmic free Ca^{2+} and membrane potential in lymphocytes. *Nature* 295: 68–71

Turner, R. S., Kempf, B. E., de Su, H., Kuo, J. F. 1985. Substrate specificity of phospholipid/Ca^{2+}-dependent protein kinase as probed with synthetic peptide fragments of the bovine myelin basic protein. *J. Biol. Chem.* 260: 11503–7

Umekawa, H., Hidaka, H. 1985. Phosphorylation of caldesmon by protein kinase C. *Biochem. Biophys. Res. Commun.* 132: 56–62

Uratsuji, Y., Nakanishi, H., Takeyama, Y., Kishimoto, A., Nishizuka, Y. 1985. Activation of cellular protein kinase C and mode of inhibitory action of phospholipid-interacting compounds. *Biochem. Biophys. Res. Commun.* 130: 654–61

van de Werve, G., Proietto, J., Jeanrenaud, B. 1985a. Tumor-promoting phorbol esters increase basal and inhibit insulin-stimulated lipogenesis in rat adipocytes without decreasing insulin binding. *Biochem. J.* 225: 523–27

van de Werve, G., Proietto, J., Jeanrenaud, B. 1985b. Control of glycogen phosphorylase interconversion by phorbol esters, diacylglycerol, Ca^{2+} and hormones in isolated rat hepatocytes. *Biochem. J.* 231: 511–16

Vara, F., Schneider, J. A., Rozengurt, E. 1985. Ionic responses rapidly elicited by activation of protein kinase C in quiescent Swiss 3T3 cells. *Proc. Natl. Acad. Sci. USA* 82: 2384–88

Verma, A. K., Ashendel, C., Boutwell, R. K. 1980. Inhibition by prostaglandin synthesis inhibitors of the induction of epidermal ornithine decarboxylase activity, the accumulation of prostaglandins, and tumor promotion caused by 12-O-tetradecanoyl-phorbol-13-acetate. *Cancer Res.* 40: 308–15

Vicentini, L. M., di Virgilio, F., Ambrosini, A., Pozzan, T., Meldolesi, J. 1985. Tumor promoter phorbol 12-myristate, 13-acetate inhibits phosphoinositide hydrolysis and cytosolic Ca^{2+} rise induced by the activation of muscarinic receptor in PC12 cells. *Biochem. Biophys. Res. Commun.* 127: 310–17

Vilgrain, I., Defaye, G., Chambaz, E. M. 1984. Adrenocortical cytochrome P-450 responsible for cholesterol side chain cleavage (P-450 scc) is phosphorylated by the calcium-activated, phospholipid-sensitive protein kinase (protein kinase C). *Biochem. Biophys. Res. Commun.* 125: 554–61

Vuilliet, P. R., Woodgett, J. R., Ferrari, S. Hardie, D. G. 1985. Characterization of the sites phosphorylated on tyrosine hydroxylase by Ca^{2+} and phospholipid-dependent protein kinase, calmodulin-dependent multiprotein kinase and cyclic AMP-dependent protein kinase. *FEBS Lett.* 182: 335–39

Watanabe, T., Taguchi, Y., Sasaki, K., Tsuyama, K., Kitamura, Y. 1981. Increase in histidine decarboxylase activity in mouse skin after application of the tumor promoter tetradecanoylphorbol acetate. *Biochem. Biophys. Res. Commun.* 100: 427–32

Watson, S. P., Lapetina, E. G. 1985. 1,2-Diacylglycerol and phorbol ester inhibit agonist-induced formation of inositol phosphate in human platelets: Possible implications for negative feedback regulation of inositol phospholipid hydrolysis. *Proc. Natl. Acad. Sci. USA* 82: 2623–26

Weeks, C. E., Herrmann, A. L., Nelson, F. R., Slaga, T. J. 1982. α-Difluoromethylornithine, an irreversible inhibitor of ornithine decarboxylase, inhibits tumor promoter–induced polyamine accumulation and carcinogenesis in mouse skin. *Proc. Natl. Acad. Sci. USA* 79: 6028–32

Welch, W. J. 1985. Phorbol ester, calcium ionophore, or serum added to quiescent rat embryo fibroblast cells all result in the elevated phosphorylation of two 28,000-dalton mammalian stress proteins. *J. Biol. Chem.* 260: 3058–62

Werth, D. K., Niedel, J. E., Pastan, I. 1983. Vinculin, a cytoskeletal substrate of protein kinase C. *J. Biol. Chem.* 258: 11423–26

White, J. R., Huang, C.-K., Hill, J. M. Jr., Naccache, P. H., Becker, E. L., et al. 1984.

Effect of phorbol 12-myristate 13-acetate and its analogue 4α-phorbol 12,13-didecanoate on protein phosphorylation and lysosomal enzyme release in rabbit neutrophil. *J. Biol. Chem.* 259: 8605–11

Wilkinson, M., Morris, A. 1983. Interferon with novel characteristics produced by human mononuclear leukocytes. *Biochem. Biophys. Res. Commun.* 111: 498–503

Williamson, J. R., Cooper, R. H., Joseph, S. K., Thomas, A. P. 1985. Inositol trisphosphate and diacylglycerol as intracellular second messengers in liver. *Am. J. Physiol.* 248: C203–16

Wilson, D. B., Bross, T. E., Hofmann, S. L., Majerus, P. W. 1984. Hydrolysis of polyphosphoinositides by purified sheep seminal vesicle phospholipase C enzymes. *J. Biol. Chem.* 259: 11718–24

Wilson, E. L., Reich, E. 1978. Plasminogen activator in chick fibroblasts: Induction of synthesis by retinoic acid; synergism with viral transformation and phorbol ester. *Cell* 15: 385–92

Wise, B. C., Glass, D. B., Chou, C. H. J., Raynor, R. L., Katoh, N., et al. 1982. Phospholipid-sensitive Ca^{2+}-dependent protein kinase from heart. II. Substrate specificity and inhibition by various agents. *J. Biol. Chem.* 257: 8489–95

Wise, B. C., Guidotti, A., Costa, E. 1983. Phosphorylation induces a decrease in the biological activity of the protein inhibitor (GABA-modulin) of γ-aminobutyric acid binding sites. *Proc. Natl. Acad. Sci. USA* 80: 886–90

Witters, L. A., Vacter, C. A., Lienhard, G. E. 1985. Phosphorylation of the glucose transporter in vitro and in vivo by protein kinase C. *Nature* 315: 777–78

Wolf, M., Sahyoun, N., LeVine, H. III, Cuatrecasas, P. 1984. Protein kinase C: Rapid enzyme purification and substrate-dependence of the diacylglycerol effect.

Biochem. Biophys. Res. Commun. 122: 1268–75

Wolf, R. A., Gross, R. W. 1985. Identification of neutral active phospholipase C which hydrolyzes choline glycerophospholipids and plasmalogen selective phospholipase A_2 in canine myocardium. *J. Biol. Chem.* 260: 7295–7303

Wooten, M. W., Nel, A. E., Goldschmidt-Clermont, P. J., Galbraith, R. M., Wrenn, R. W. 1985. Identification of a major endogenous substrate for phospholipid/Ca^{2+}-dependent kinase in pancreatic acini as Gc (vitamin D–binding protein). *FEBS Lett.* 191: 97–101

Yamanishi, J., Takai, Y., Kaibuchi, K., Sano, K., Castagna, M., et al. 1983. Synergistic functions of phorbol ester and calcium in serotonin release from human platelets. *Biochem. Biophys. Res. Commun.* 112: 778–86

Zatz, M. 1985. Phorbol esters mimic α-adrenergic potentiation of serotonin N-acetyltransferase induction in the rat pineal. *J. Neurochem.* 45: 637–39

Zavoico, G. B., Halenda, S. P., Sha'afi, R. I., Feinstein, M. B. 1985. Phorbol myristate acetate inhibits thrombin-stimulated Ca^{2+} mobilization and phosphatidylinositol 4,5-bisphosphate hydrolysis in human platelets. *Proc. Natl. Acad. Sci. USA* 82: 3859–62

Zawalich, W., Brown, C., Rasmussen, H. 1983. Insulin secretion: Combined effect of phorbol ester and A23187. *Biochem. Biophys. Res. Commun.* 117: 448–55

Zurgil, N., Zisapel, N. 1985. Phorbol ester and calcium act synergistically to enhance neurotransmitter release by brain neurons in culture. *FEBS Lett.* 185: 257–61

Zwiller, J., Revel, M., Malviya, A. N. 1985. Protein kinase C catalyzes phosphorylation of guanylate cyclase *in vitro*. *J. Biol. Chem.* 260: 1350–53

Ann. Rev. Cell Biol. 1986. 2 : 179–99

PROTON-TRANSLOCATING ATPASES

Qais Al-Awqati

Departments of Medicine and Physiology, Columbia University, College of Physicians and Surgeons, 630 West 168th Street, New York, New York 10032

CONTENTS

INTRODUCTION

Proton transport across biological membranes is a central process in many energy conversion reactions in the cell. It is mediated by a variety of proteins that include electron-transporting chains, photosynthetic centers, photoactivatable proteins, proton-solute cotransporters, and proton-translocating ATPases. As a rule, each cell operates several of these mechanisms. I will only discuss the proton ATPases and will further limit myself to a survey of recent progress in the analysis of proton pumps in eukaryotic cells.

CHARACTERISTICS OF PROTON ATPASES

All ion-transporting ATPases translocate cations and, despite much noise in the literature, no ATPase has been discovered that translocates anions by direct coupling to ATP hydrolysis. As their name suggests, proton ATPases transport H^+ rather than OH^- (in the opposite direction); however, this conclusion (except in one instance) is not based on any firm

179

0743–4634/86/1115–0179$02.00

experimental evidence. Here I first discuss what constitutes good evidence for the presence of a H^+-translocating ATPase.

ATP-Driven H^+ Transport

The most fundamental criterion for the presence of active proton ATPases is the demonstration of direct coupling between ion transport and ATP hydrolysis. Because most of the H^+-transporting ATPases can function in reverse, demonstration of ATP synthesis (or ATP-P_i exchange) following the imposition of proton gradients constitutes good evidence for the existence of a proton pump. Ideally, one would like to demonstrate that addition of ATP to a purified protein reconstituted into liposomes can induce vectorial proton movement across the membrane. Stimulation of ATP hydrolysis by changes in transmembrane pH or potential gradients can also be useful. The addition of proton ionophores to ATPase-containing vesicles should also stimulate ATP hydrolysis since the ionophores will cause the proton electrochemical gradient generated by the ATPase to collapse, thereby accelerating ATP turnover.

ATP-Induced Membrane Potential

Proton-translocating ATPases can be electrogenic or neutral. Neutral ATPases, like the gastric proton pump, exchange H^+ for K^+ with a 1:1 stoichiometry (Sachs et al 1982). Electrogenic ATPases are able to translocate net charge across a membrane and hence can generate a membrane potential. However, the generation of a membrane potential by an ATPase should not be construed as rigorous evidence for the presence of an electrogenic proton pump since it could be produced by a secondary effect of a pH gradient generated by a neutral ATPase. Such pH gradients could either produce diffusion potentials or they could change the ion conductance of the membrane. Further, the lack of a membrane potential need not indicate that the ATPase is neutral since the membrane in which it is located could be electrically leaky, as was found in the case of lysosomes (Harikumar & Reeves 1983). An important characteristic of electrogenic pumps is that their transport rate can be affected by a transmembrane electrical potential difference. However, because their turnover number is low, well below 1000 ions/sec, it is not possible with present electrophysiological methods to reconstitute the ATPases into planar black lipid films to measure the currents they generate as a function of ATP hydrolysis.

Presence of Proton Gradients

The demonstration of H^+ concentration gradients is not sufficient evidence for the presence of an ion pump since Donnan equilibria can generate pH gradients. Further, some ion pumps that do not require ATP, such as redox pumps, can generate pH gradients.

Inhibitors

The use of inhibitors can give valuable information on characterizing relatively purified proton ATPases. However, their use in cruder systems to provide evidence for the existence of a proton ATPase should be avoided. In intact cells, which frequently contain many types of proton ATPases, the use of inhibitors to prove the presence of one kind or another of proton ATPase is fraught with such ambiguity as to render the information meaningless. I have included in Table 1 many inhibitors that have been used to study the various H^+ ATPases. As shown in that table, there is a paucity of specific inhibitors. One is frequently reduced to studying the response of the ATPase to various concentrations of inhibitor and finding a range in which a drug inhibits one but not another ATPase.

Antibodies

Another approach to the identification of ATPases is the use of antibodies that were prepared against subcellular compartments and were found later to cross-react with a known proton ATPase. Since cation-translocating ATPases such as the Na,K-ATPase, Ca-ATPase, and yeast plasma membrane ATPase have sequence homologies, antibodies to one of them may cross-react with the others. Since intracellular organelles may contain more than one of these ATPases, the finding that an antibody raised against the organelle cross-reacts with a proton ATPase is not rigorous evidence for the presence of a proton ATPase in that vesicle. Rather, it may signify the presence of a Ca-ATPase or a Na,K-ATPase. Such results are interesting only *after* a H^+-ATPase has been identified. At that stage immunological methods can provide powerful insights into the biosynthesis and deployment of the ATPase in various cellular compartments. However, demonstrating an ATPase by immunological methods does not imply that the pump is functional.

TYPES OF PROTON ATPASES

Proton-translocating ATPases are membrane proteins that vectorially translocate H^+ from one surface to the other. Studies on a variety of proton-translocating ATPases (as well as other ion-translocating ATPases) have uncovered sufficient detail to allow a broad classification of these ion pumps into two classes, an E_1-E_2 type and an F_o-F_1 type (Table 1). A third type of ATPase occurs in eukaryotic cells and is located in plasma membranes and a variety of intracellular organelles, including endosomes,

Table 1 Classification and properties of proton translocating ATPases

Type	Enzyme	Source	DCCD	DES	DIDS	Efr	N₃	NEM	Oli	Oss	Oub	Tri	VO₄	Ven	References
E_1-E_2	*Neurospora*	Plasma membrane	S*	S	ND	ND	R	S*	R	ND	ND	ND	S	R	Goffeau & Slayman 1981; Perlin et al 1984
	Yeast	Plasma membrane	S*	S	ND	ND	R	ND	R	ND	R	S*	S	R	Serrano 1984
	H,K-ATPase	Stomach plasma membrane and microsome	S*	ND	ND	ND	R	R	R	ND	R	ND	S	ND	Sachs et al 1982
F_o-F_1	Mitochondria	Inner membrane	S	S*	S	S	S	S	S	S	R	S	R	S	Pedersen & Hullihen 1984
	Bacteria	Plasma membrane	S	ND	ND	S	S	R	S*	S	ND	ND	R	S	Futai & Kanazawa 1983
	Chloroplast	Thykaloid	S	ND	ND	S	S	S*	R	ND	ND	ND	R	ND	Selman-Reimer et al 1985
Microsomal		Endosomes	S*	ND	S*	R	R	S	R	ND	ND	ND	R	ND	Galloway et al 1983
		Lysosomes	S	ND	S*	R	R	S	R	ND	ND	ND	R	ND	Ohkuma et al 1982; Schneider 1981
		Clathrin-coated vesicles	S*	ND	ND	R	R	S	R	ND	ND	R	R	R	Forgac et al 1983; Stone et al 1983
		Golgi vesicles	S	ND	ND	ND	R	S	R	ND	ND	ND	R	ND	Glickman et al 1983
		Chromaffin granules and neurosecretory vesicles	S	ND	ND	R	R	S	R	ND	R	ND	R	ND	Flatmark et al 1982; Russell 1984
		Endoplasmic reticulum	S	ND	ND	R	R	S	R	ND	ND	ND	R	ND	Rees-Jones & Al-Awqati 1984
		Serotonin granules	S	ND	ND	ND	R	S	R	ND	ND	ND	R	ND	Dean et al 1984
		Renal epithelia	S	ND	ND	R	R	S	R	ND	R	ND	R	ND	Gluck et al 1982; Gluck & Al-Awqati 1984; Kaunitz et al 1985
		Synaptic vesicles	S	ND	ND	ND	R	S	R	ND	R	ND	R	ND	Stadler & Tsukita 1984; Harlos et al 1984
		Yeast vacuoles	S	ND	ND	ND	R	S	R	ND	ND	ND	R	ND	Ohsumi & Anraku 1981; Bowman & Bowman 1982
		Plant vacuoles	S	S	S	ND	R	S	R	ND	R	ND	R	ND	Bennet & Spanswick 1983; Sze 1985
		Multivesicular bodies	S	ND	ND	ND	ND	S	R	ND	R	ND	R	ND	Van Dyke et al 1985

[a] Concentration of inhibitor required for 50% inhibition (IC$_{50}$) of either ATPase or pumping activity.
S* = sensitive (IC$_{50}$ > 100 μM); S = highly sensitive (IC$_{50}$ < 100 μM); R = resistant; ND = not determined; DCCD = N,N'-dicyclohexylcarbodiimide; DES = diethylstilbestrol; DIDS = 4,4'-diisothiocyanostilbene-2,2'-disulfonate; Efr = efrapeptin; N$_3$ = azide; NEM = N-ethylmaleimide; Oli = oligomycin; Oss = ossamycin; Oub = ouabain; Tri = triethyltin; VO$_4$ = vanadate; Ven = venturicidin.

the Golgi complex, and endoplasmic reticulum. Although this recently discovered "microsomal" ATPase most likely represents an F_o-F_1 type, the lack of detailed structural information prevents definitive assignment to one or another type.

E_1-E_2

Proton pumps of the E_1-E_2 type are present in yeast and fungal plasma membranes (Scarborough & Addison 1984; Goffeau & Slayman 1981; Serrano 1984) and the gastric microsomal and plasma membranes (Sachs et al 1982). These ATPases contain one large transmembrane protein ($\sim$100 kDa) and occasionally another small subunit. The large subunit contains both ion-conducting and catalytic domains. The E_1-E_2 family of ATPases also includes the Na,K-ATPase and the Ca-ATPase. These ATPases catalyze cation exchange with a variety of stoichiometries. Some of these pumps are "electrogenic," i.e. they are able to generate a membrane potential in the absence of any ion gradients. The gastric enzyme is a neutral K^+-H^+ exchanger; the yeast and fungal enzyme is electrogenic. These enzymes are phosphorylated on an aspartyl residue during the catalytic cycle, and dephosphorylation is necessary for completion of the cycle.

These ATPases are inhibited by vanadate, a transition metal anion that probably substitutes for phosphate, resulting in stable vanadylation of the enzyme and consequent inhibition of the reaction. During the reaction cycle the enzyme enters various conformational states. In one of these, the "occluded state," the ions are buried deep in the protein structure and are not accessible from either side of the membrane; this has best been demonstrated in the Na,K-ATPase. The deduced amino acid sequences show remarkable structural similarity among these enzymes. The Na,K-ATPase (Shull et al 1985; Kawakami et al 1985), Ca-ATPase (MacLennan et al 1985), and yeast plasma membrane H^+-ATPase (Serrano & Fink 1986) are diverse enzymes whose structural similarity suggests common evolutionary ancestry. The most dramatic instance of conserved homology is in the sequence surrounding the aspartyl residue that is phosphorylated during the cycle. Proton ATPases that belong to this class of pumps have not been identified in prokaryotes.

F_o-F_1

(For recent reviews see Kagawa et al 1979; Senior & Wise 1983; Futai & Kanazawa 1983; Amzel & Pedersen 1983.) The F_o-F_1 pumps only catalyze electrogenic proton transport. This enzyme is composed of two portions: a transmembrane portion (F_o) that acts as a proton channel and a catalytic portion (F_1) that can be easily released from the membrane. The F_o domain is composed of as many as three proteins, with the following approximate stoichiometry (a, b, c_{10}). The c subunit (the proteolipid or DCCD-binding protein) is clearly a proton channel; the a and b subunits probably function to stabilize the channel structure and to allow binding of F_1 (Aris et al 1985; Schneider & Altendorff 1985). The conductance of F_o, purified from thermophilic bacteria and reconstituted into liposomes, increases as a function of the H^+ concentration, which suggests that the species of ion conducted is the proton rather than OH^- (Kagawa et al 1979). The large catalytic domain is composed of at least five proteins, with the following approximate stoichiometry (α_3, β_3, γ, δ, ε). The α and β subunits contain the nucleotide binding and hydrolyzing sites.

Electron microscopy of negatively stained specimens shows that the F_1 domain protrudes as a sphere above the bilayer surface, to which it is connected by a stalk. This "lollipop" appearance is characteristic of F_o-F_1 proton pumps. The F_o-F_1 enzymes are not phosphorylated during the catalytic cycle and no occluded state has been discovered. Only F_o-F_1 enzymes isolated from bacteria have been completely sequenced, and the sequences suggest that the nucleotide binding domain has extensive homology to other ATP-requiring or -utilizing proteins, such as adenylate kinase, tRNA synthetase, recA protein, and RNA polymerase (Futai & Kanazawa 1983). The genes for all the subunits of the ATPase are located near one of the origins of replication of the *Escherichia coli* genome in the *atp* or *unc* operon.

In eukaryotes, where F_o-F_1 proton ATPases are present in mitochondria and chloroplasts, the genes coding for most of the subunits are present in the nuclear DNA rather than the mitochondrial or chloroplast DNA. Although the gene for the proteolipid is in the mitochondrial DNA in one species (yeast), it is located in the nuclear DNA in *Neurospora*. The subunits of the ATPase are synthesized in practically all cases on free polysomes. Attached polysomes were found only in yeast in some special cases (Schatz & Butow 1983). The proteins are then posttranslationally inserted into the mitochondria (Schleyer & Neupert 1985). Despite the immense amount of information available on this ATPase, there is at present no knowledge of the physical mechanism by which ATP hydrolysis (or synthesis) is coupled to proton transport, although there are some interesting speculations (see, for instance, Mitchell 1985).

Microsomal Proton-Translocating ATPase

Although it had been known for some time that the contents of lysosomes are acidic, it was only recently that this organelle and many others were found to be acidified by proton-translocating ATPases. These proton pumps are present in endosomes, secretory granules, Golgi elements, endo-plasmic reticulum, and in the plasma membrane of many eukaryotic cells (appropriate references are given in Table 1). They are also present in yeast vacuoles and plant vacuolar membranes (Serrano 1984; Sze 1985). The inhibitor characteristics of these enzymes are shown in Table 1. The lack of sensitivity to vanadate suggests that they do not belong to the E_1-E_2 type of enzyme.

None of these enzymes have been purified to homogeneity; however, preliminary results suggest that they are multisubunit complexes composed of five or six proteins (Percy et al 1985; Gluck & Caldwell 1986; Xie & Stone 1986). In the case of the chromaffin granule ATPase the molecular masses were 140, 70, 57, 41, 33, and 16 kDa. The smallest subunit bound ^{14}C-DCCD, which suggests that it may be an analogue of the proteolipid of the F_o-F_1 ATPase (Percy et al 1985). The clathrin-coated vesicle ATPase was found to have proteins of molecular masses of 116, 70, 58, 40, 34, 33, and 15 kDa (Xie & Stone 1986). Experiments using kidney medulla microsomes suggest that the proton pump has at least five subunits with molecular masses of 70, 56, 44, 35, and 17 kDa (Gluck & Caldwell 1986). "Lollipop" type images were also found on negatively stained specimens in these studies. Similar images were also found in neurosecretory vesicles (Stadler et al 1984). The finding that this ATPase is a multisubunit protein, one of whose subunits binds DCCD, provides evidence, albeit indirect, that this enzyme belongs to the F_o-F_1 class of proton pumps.

However, it will be necessary to demonstrate that this complex contains a proton-conducting portion that is an integral membrane protein(s) and a catalytic domain that can be released from the bilayer without detergent solubilization. One would expect that the catalytic portion be able to bind to the proton-conducting domain and occlude it, thereby reducing proton permeability. One of the important questions in this area is whether the proton pumps in the various organelles are the same enzyme or are simply similar proteins. So far they all appear to be similar in their drug sensitivities. It has been difficult to demonstrate that the lysosomal ATPase is electrogenic, but that is probably due to the fact that lysosomal membranes are electrically "leaky" (Harikumar & Reeves 1983). It also appears that the lysosomal ATPase can use GTP as well as ATP for pumping protons (Ohkuma et al 1982); the chromaffin granule enzyme can also use ITP, though less well than ATP (Percy et al 1985). If this turns out to be true of the purified enzymes, it would imply that there are related types of the microsomal ATPase that can be separated on the basis of their nucleotide specificity.

BIOENERGETICS AND THE PROTON ATPASE

Ion-translocating ATPases are, in principle at least, reversible, i.e. they can function as ATP synthases. The driving force for reversal is a proton electrochemical gradient whose magnitude is greater than the available energy derived from ATP hydrolysis. When the imposed proton electro- chemical gradient is equal to the available energy from ATP hydrolysis, no net transport of protons occurs, nor will there be ATP hydrolysis. In other words, this energy converter does not spend any energy while "idling." The proton electrochemical gradient at this state of minimum energy expenditure is termed the reversal potential. It is this ability of the proton ATPase to transport protons in two directions that allows it to play such a central role in bioenergetics.

To synthesize ATP the membrane must be clamped to a "proton motive force" (pmf) that is greater than the reversal potential of the enzyme. Protons will then be transported towards the ATP side of the pump (in intracellular organelles, this would be the cytoplasmic surface). In mitochondria the electron transport chain generates such a proton motive force which is composed of a pH difference and a membrane potential. This is accomplished by pumping protons from the matrix to the cytoplasm. The F_1 portion of the ATPase, i.e. the ATP binding site, faces the matrix space. The pmf generated by the electron transport chain is larger than the reversal potential of the ATPase, which allows "backward" flow through the F_o-F_1 ATPase with consequent ATP synthesis. Addition of proton ionophores (uncouplers) reduces the proton electrochemical gradient to a level below the reversal potential, hence the F_o-F_1 proton pump will function as an ATPase, and protons will be transported from the ATP side to the other. In intracellular organelles, this is the usual mode of operation, i.e. ATP hydrolysis and proton transport to the intravesicular medium.

If the proton pump is electrogenic, a membrane potential develops in addition to the pH difference: The ATP binding surface (cytoplasmic side) becomes alkaline and negatively charged with respect to the other side. If we allow an ATPase to turn over from a state of zero electrochemical difference, the rate of transport will decrease as the gradient develops because the orientation of the gradient is not favorable for transport. If the proton and electrical permeability of the membrane is very low the ATPase can eventually generate a gradient sufficiently large to stop trans- port, i.e. it will reach the reversal potential. The implications of this behavior for the energy economy of the cell are discussed later.

The reversal potential of the proton ATPase is given by:

$$\text{pmf} = (\Delta\tilde{\mu}_H)_{J_H=0} = 2.303 \ RT\Delta pH + F\Delta\psi \qquad 1.$$

$$\text{pmf} = N\Delta G'_{ATP}, \qquad 2.$$

where $\Delta\tilde{\mu}_H$ is the electrochemical proton gradient, $\Delta\psi$ is the membrane potential, F is the Faraday constant, R is the universal gas constant, T is the absolute temperature, and N the H/ATP stoichiometry. $\Delta G'_{ATP}$ is the free energy of ATP hydrolysis. Under the actual conditions studied it will equal

$$\Delta G'_{ATP} = \Delta G_0 + RT \ln([ATP]/[ADP][P_i]), \qquad\qquad 3.$$

where ΔG_0 is the standard free energy of ATP hydrolysis and the bracketed values are the cytoplasmic free concentrations of ATP, ADP, and P_i. These relations apply to a reversible engine with "perfect" coupling, i.e. the efficiency of energy conversion by the ATPase is 100%. Surprisingly, where the efficiency was measured, it turned out to have this maximal value. The $\Delta G'_{ATP}$ of the cytosol is now known to be ~ 58 kJ/mol, which is enough free energy to generate a gradient of about 9 pH units. Since the H/ATP stoichiometry of many proton ATPases is 2 or 3, the maximal gradient is 3 or 4 pH units. (Unfortunately, the stoichiometry of many of the proton ATPases is not known with the certainty that one would like.) The higher the H/ATP stoichiometry the easier it becomes to reverse the flow of protons and to synthesize ATP. It is likely that this fact played an important role in the evolution of F_o-F_1 type ATPases whose stoichiometry is 2 or 3 H/ATP. The gastric H^+-ATPase is likely to have a stoichiometry of 1 H/ATP to allow it to build the massive pH gradients seen in the stomach. Recall, however, that if the proton ATPase is an exchange pump, as in the case of the gastric H,K-ATPase, one must include the K electrochemical gradient in these calculations.

Role of Membrane Conductance in the Generation of pH Differences

The electrochemical gradient generated by an electrogenic proton pump is composed of a pH difference and a membrane potential. The contribution of each of these components to the total electrochemical gradient depends on the electrical conductivity of the membrane in which the ATPase is located. If the conductance of the membrane is sufficiently high, no membrane potential can be generated, and all of the pmf can be expressed as a pH difference. Conversely, if the membrane conductance is low, a large membrane potential can develop rapidly enough to stop the ATPase from generating a pH difference. Transfer of as few as 600 ions can generate a membrane potential of 100 mV. Since the turnover of the ATPase is likely to be of the order of 100 ions/sec, it would not take more than a few seconds to generate a large membrane potential. Considering that most vesicles have some buffering power, their internal pH will not drop to any significant degree initially. Hence, within a short time the pmf, now composed solely of a membrane potential, will reach the limiting level and turn the ATPase off.

Many intracellular vesicles that contain a proton ATPase also contain a chloride channel. Opening the chloride conductance could change the state of a vesicle from one in which there is a large membrane potential (positive inside) and the pH of the vesicle is similar to that of the cytoplasm to a state in which the vesicle pH is quite acidic and the membrane potential is near zero. Such switching could be achieved by regulation of the chloride channel conductance rather than by regulation of the ATPase. Such regulation of the chloride channel protein(s) could entail its phosphorylation and/or binding to an ion.

Recent measurement of the pH inside a variety of intracellular vesicles has shown that many populations of vesicles have a pH near 6 or 6.5, while endosomes have a pH near 5 and the pH of lysosomes is even lower (Yamashiro et al 1984; Schwartz & Al-Awqati 1985). The most likely explanation for this finding is different states of regulation of chloride channels. Lysosomes are known to have both high K conductance and high chloride conductance, which ensures that their internal pH will be maximally acid (Harikumar & Reeves 1983). Whether regulation of chloride channel conductance in these vesicles is a physiologically significant mechanism is a question that needs further study.

Role of the pmf in Cellular Energy Metabolism

The energy expenditure of a cell is the sum of all the energy-consuming processes it engages in. Ion transport is the main such process: It accounts for an average of 40–50% of the total ATP utilized. In some cells that are "professional transporters," e.g. the kidney epithelial cell, ion transport can use as much as 90% of the ATP produced. The stoichiometry of all ion-translocating ATPases varies between one and three ions per ATP hydrolyzed. Since the passive permeability of most plasma membranes allows many ions to enter down their concentration gradients, it follows that the expenditure of energy needed to keep the intracellular environment constant and different from the outside milieu is large. The exorbitant energy price that cells must pay for osmotic regulation indicates the primacy of this process, without which no cell could have evolved. Upon initial examination, the presence of proton ATPases in endocytic vesicles appears to be a very wasteful system: The proton ATPases present in the membrane are internalized in endocytic vesicles and are carried to endosomes which they acidify. They are later reinserted in the plasma membrane in a process of continuous membrane recycling. However, on further reflection, this process can be seen to be elegantly thrifty in energy expenditure. When the proton ATPase is in the plasma membrane (the coated pit), it faces the extracellular medium, whose pH is generally greater than that of the cell. The electrical potential across the plasma membrane is around 60 mV (outside positive). The net proton electrochemical gradient imposed on the proton ATPase in the plasma membrane is low, much lower than its reversal potential, hence the proton pump will extrude protons out of the cell at a brisk rate and at much energy expenditure. Once the coated pit is internalized, the proton ATPase will rapidly acidify the vesicle contents and build a membrane potential that is positive on the inside. Because of the small volume (endocytic vesicles with a radius of 0.1 μm have a volume of $\sim 4 \times 10^{-15}$ liters) and reasonably high turnover number (~ 100 ions/sec), the proton ATPase could acidify a vesicle to a pH of 5 in less than one minute, even if one proton ATPase were present per vesicle and the buffering power of that vesicle were high. Hence, the pmf across the vesicle membrane will rapidly reach the reversal potential, thereby stopping the turnover of the ATPase and with it ATP hydrolysis. It is not surprising that endocytosis enjoys such ubiquity: for a process that depends on ion transport, it is very economical.

REGULATION OF THE PROTON ATPASE

Although ion transport is a heavily regulated cellular function, it has been difficult to demonstrate direct regulation of the ion-transporting molecules themselves. In large part this is due to the fact that at present we do not have biochemical tools that will allow us to demonstrate specific mechanisms of regulation. It has been suggested that a variety of ion-transporting molecules are phosphorylated. In a few convincing cases one can demonstrate the effects of protein kinases on the properties of single channels, e.g. serotonin-sensitive K channels in *Aplysia* (Shuster et al 1985). In others one can demonstrate phosphorylation of the protein, e.g. in the voltage-gated Na channel, but it is not clear what effect phosphorylation has on the channel behavior (Catterall 1984). Nor is the situation for ion-transporting ATPases any clearer. There is no convincing evidence that covalent modification of any ATPase can lead to a change in its kinetics. However, there exists a "mitochondrial inhibitory protein" that inhibits the F_o-F_1 ATPase and whose inhibitory effect is neutralized by calmodulin (Pedersen & Hullihen 1984). This protein is not associated with the prokaryotic enzyme, and its physiological role in oxidative phosphorylation is unknown. As discussed in detail above, the electrochemical gradient across a membrane can regulate the rate of proton pumping.

Proton transport in urinary epithelia is effected by proton-translocating ATPases that are located in a polarized manner either in the luminal (Dixon & Al-Awqati 1979; Gluck et al 1982a) or in the basolateral domain of the plasmalemma (Schwartz et al 1985). Transepithelial H^+ transport is stimulated by a variety of physiological factors, including the ambient

CO_2. This regulation may result from an increase in the turnover of individual proton pumps in the plasma membrane or from an increase in the number of these ATPases. One usually associates an increase in the number of pumps with an increase in the synthesis, or a reduction in the degradation of the pumps. However, we recently discovered that the number of proton ATPases in the plasma membrane of certain epithelial cells can be rapidly increased by exocytotic insertion of proton pumps and can be decreased by endocytic retrieval.

In the collecting tubule of the kidney and the urinary bladder of amphibia and reptiles, H^+ is transported into the urine by the intercalated (mitochondria-rich) cells. These cells are enriched in carbonic anhydrase and have a proton ATPase in the apical plasma membrane that has the same physiological and pharmacological characteristics as the microsomal proton ATPase (Gluck et al 1982b). We found that this proton pump is endocytosed into small vesicles located under the apical membrane (Gluck et al 1982a; Schwartz & Al-Awqati 1985) and that there is continuous endocytosis and exocytosis of these vesicles. When we increased the ambient CO_2 tension, the rate of H^+ transport rapidly increased due to rapid exocytosis of these vesicles and consequent insertion of proton pumps into the apical plasma membrane. CO_2 exerted its effect by initially acidifying the cell, which led to an increase in cell calcium. It was the increase in cell calcium that was the proximate cause of exocytosis: Buffering the increase in calcium prevented the exocytosis induced by CO_2 and other weak acids but had no effect on their ability to acidify the cytoplasm (Cannon et al 1985). How lowering cell pH leads to an increase in cell calcium is unknown at present, but one possible mechanism is that it depolarizes the membrane by closing pH-sensitive K channels.

Exocytotic insertion of proton pumps also leads to an increase in cell pH, since the number of pumps in the plasma membrane becomes greater than before the addition of the stimulus (van Adelsberg & Al-Awqati 1986). Hence, this process is a new mechanism for the regulation of cell pH. Exocytotic insertion of pumps is similar to other regulated secretory phenomena, except that it is the vesicle membrane rather than its contents that is the object of regulation. Since many secretory vesicles contain proton ATPases, regulated exocytosis could increase the number of proton pumps in the plasma membrane in the secretory cells. We therefore expect that stimulated secretion will be associated with an increase in cell pH. Indeed, this has been shown at least for the case of the β cells of pancreatic islets (Lindstrom & Sehlin 1984). Elevation of cell pH may have dramatic effects on protein and DNA synthesis, thus the insertion of proton ATPase into the plasmalemma by exocytosis may underlie the mechanisms by which cells respond to chronic stimulation with an increase in synthesis of hormones and membranes.

ROLE OF PROTON GRADIENTS IN VESICLE FUNCTION

The presence of a proton electrochemical gradient in a variety of intracellular vesicles raises a question concerning the role these gradients play in the function of the vesicles. A few such functions are clearly established; many others may be present, but more experimental data are needed to establish them.

Enzyme Function

The majority of lysosomal enzymes have acid pH optima, hence the role of the proton gradient here is twofold : The acid pH may denature ingested proteins, thereby allowing them to be digested by the lysosomal enzymes. The low pH also provides biochemical compartmentalization (in addition to structural compartmentalization). If the lysosome membrane ruptures, the alkaline cytoplasmic pH will protect the cell from damage by inactivating the acid hydrolases. Further, many pancreatic zymogens are inactive at acid pH, which protects the granule from autodigestion. The optimum pH for many Golgi complex enzymes is less than 7, which fits well with the recent discovery that some Golgi compartments, especially the trans cisternae, may be acidic (Glickman et al 1983 ; Anderson et al 1984). Note also that cells with mutations that impair their acidification function appear to have problems in protein glycosylation, which implies that the low Golgi pH may be needed for this function (Robbins et al 1984). (This observation by itself, however, need not imply that the reason for abnormal glycosylation is poor catalytic function of the enzymes.)

Ligand Binding, Dissociation, and Transport

Lysosomal enzymes are phosphorylated to contain a mannose-6-phosphate group, which allows the newly synthesized protein to bind to a mannose-6-phosphate receptor on the transport vesicle membrane. The binding of the enzyme to its receptor is pH sensitive: it decreases at low pH. Hence, when transport vesicles fuse with the lysosomes, they can deliver the enzymes to that compartment. As importantly, if these vesicles fuse with the plasma membrane, the enzyme will face the alkaline extracellular fluid and will remain bound to the receptor. This mechanism prevents dissociation and extracellular loss of the enzyme if the vesicle is fused with the plasma membrane. For a very large number of ligands binding to receptors is pH sensitive. Only two such ligands will be mentioned here ; a recent review should be consulted for a detailed description of this process (Goldstein et al 1985). For instance, low-density lipoprotein binds to its receptor at high pH (i.e. at the cell surface) and dissociates from its receptor at low pH (i.e. in the endosome). This allows delivery of the ligand to other compartments. The case of transferrin is a particularly interesting example: The iron-loaded protein can bind to its receptor at high pH; when internalized into an acid compartment it loses the iron but the desferroxy form remains tightly bound even at low pH. When it recycles back to the surface, the desferroxy ligand has a reverse pH optimum, i.e. its binding is weakest at high pH. Hence, it will be released to the extracellular space where it will bind iron. Iron-loaded transferrin can then bind to its receptor to complete the cycle. Transepithelial transport of IgG by neonatal intestinal epithelia requires acid vesicles to allow binding and to prevent shuttling of the immunoglobulin to the lysosome (Rodewald & Kraehenbuhl 1984).

Virus and Toxin Entry

Diphtheria toxin enters cells by binding to surface membrane receptors via its B-subunit. The toxin-receptor complex is then internalized into acid endosomes. Entry of the active toxin, the A-subunit, into the cytosol requires acid endosomes, since alkalinizing these organelles prevents cell intoxication. Endocytosis is required only to expose the toxin to an acid pH, since exposure of the cell to an acid pH in the absence of endocytosis promotes intoxication (Sandvig & Olsnes 1980). Cells with defective endosomal acidification are resistant to toxin (Robbins et al 1984). Although many bacterial and plant toxins require the acid endosome, this is by no means a general mechanism since it is clear that some toxins do not require exposure to an acidic environment to affect the cells (Sandvig & Olsnes 1982). Many enveloped viruses also enter the cytosol via acidic vesicles. Much as toxins do, the virus binds to the surface and is then internalized into acidic vesicles. The low pH induces a conformational change in the fusion protein of the virus, which then initiates viral penetration (reviewed by Kielian & Helenius 1985).

Solute Transport

The presence of proton ATPases that can generate a pH gradient and a membrane potential suggests that this electrochemical gradient could be used to accumulate or repel other ions or solutes. The uptake of amino acids and calcium into yeast vacuoles by a mechanism that utilizes the proton electrochemical gradient is well-documented (Ohsumi & Anraku 1981, 1983; Sato et al 1984). The best-studied system in organelles of higher eukaryotes is uptake of catecholamines and serotonin into secretory granules (Beers et al 1982). These amines are translocated by carriers located in the organellar membrane by a mechanism that depends on a proton electrochemical gradient.

Although they have not been specifically studied, nucleotide sugars may also be transported across microsomal membranes by coupling to this gradient. Recent studies suggest that a carrier exchanges the nucleotide sugar (or nucleotide sulfate) with the nucleoside phosphate with a 1:1 stoichiometry (Capasso & Hirschberg 1984a). The nucleotide-sulfate co-transporter is inhibited by atractyloside, a specific inhibitor of the mitochondrial ADP-ATP exchanger (Capasso & Hirschberg 1984b). Transport by this latter carrier is electrogenic (Klingenberg 1980), and if the microsomal carrier also mediates electrogenic transport, it could be driven by the positive membrane potential likely to be present inside Golgi elements.

At neutral pH both species to be transported, nucleoside phosphate and nucleotide sugar (or sulfate), are negatively charged, though with unequal valence. Note that the species that is transported out of the Golgi complex is usually less negatively charged. A low intravesicular pH would further reduce the charge on the nucleotide, the species destined for outward transport, allowing the membrane potential to drive it out of the vesicle. Capasso & Hirschberg found that the exchange of these solutes was not inhibited by ionophores; however, these experiments were performed in the absence of ATP where no pH gradients can be present. It would be interesting to compare the charge balance of all the pairs of coupled nucleotides and nucleotide sugars and test the hypothesis that some pairs could be transported by a membrane potential while others would require titration of the charge by low pH in the relevant Golgi compartments. Since recent studies have suggested that only the most distal stacks of the Golgi complex are acidic (Anderson et al 1984), this mechanism might provide an attractive way to segregate the uptake of nucleotide sugars. Whether similar exchange reactions are involved in dolichol-mediated transport is unclear at present.

Protein Transport

Transport of many proteins across the membranes of mitochondria and bacteria frequently requires the presence of a membrane potential (Schatz & Butow 1983). Whether the endoplasmic reticulum of higher eukaryotes also has this requirement is an intriguing possibility that has not been directly addressed. Cotranslational transport into dog pancreas microsomes can be demonstrated in vitro. I have found that these membranes are electrically very leaky and cannot maintain a membrane potential. Further, the addition of electrogenic proton and K ionophores has no effect. However, the fact that microsomes are active in transporting proteins but cannot maintain a voltage need not imply that such a gradient is not required in vivo. A membrane potential may be required to accelerate rather than initiate this process. It is interesting in this regard that the rate of in vitro protein translation is apparently much lower than the estimates for such rates in vivo, which suggests that some factor is absent in vitro. However, at present there is no compelling reason, other than personal bias, to invoke any role for a membrane potential in the cotranslational transport of proteins in the eukaryotic endoplasmic reticulum.

Exocytosis

Fusion of vesicles with the plasma membrane implies three fundamental steps: approach, adhesion, and fusion. The last step is likely to involve osmotic swelling of the adherent vesicle with consequent rupture. Studies aimed at measuring the size of the vesicle just before rupture suggest that osmotic swelling occurs (Zimmerberg & Whittaker 1985). The mechanism for generation of swelling must include an increase in the osmotic pressure inside the vesicle either due to pumping of solutes into the vesicles, breakdown of large molecular weight compounds to smaller molecules, release of bound material, or change in the physical state of the granular contents. The presence of a proton electrochemical gradient across the vesicle membrane raises the possibility that it may act as a driving force for solute uptake. Cell excitation may lead to opening of anion channels, which would allow uptake of chloride into these vesicles driven by the positive membrane potential.

In recent studies, the proton electrochemical gradient across the secretory vesicle was measured simultaneously with exocytosis in permeabilized chromaffin cells (Knight & Baker 1985; Holz 1986). Reduction of this gradient by ionophores did not result in any decrease in exocytosis, except when the gradient was nearly abolished. These results imply that if a proton electrochemical gradient is involved, even small gradients can produce the effect and that there is no quantitative relationship between exocytosis and the magnitude of the gradient.

EVOLUTION OF PROTON ATPASES

One of the advantages of writing a review is that it allows one to speculate on certain subjects, which is looked at askance when one publishes "original" results. The accumulating wealth of detailed structural information on a few proteins has recently made it possible to make some defensible statements regarding the origin and evolution of ion transporting pumps. These results help to convert these molecules into "living fossils" that allow us to make certain preliminary deductions. The origin of cells is intimately linked to the origin of ion-transporting molecules in general and to proton transport in particular. When the first few molecules that constituted a cell were enclosed by a semipermeable membrane, that original cell faced two fundamental problems. First, the presence of impermeant intracellular molecules created a colloid osmotic pressure that tended to pull water into the cell. Further, most of these molecules were (and are) anionic, hence they created a Donnan potential that attracted cations from the primeval seas. A second problem, which is less commonly discussed, is that hydrolysis of energy-rich molecules, e.g. ATP, is an acid-producing reaction, and hence the intracellular environment will rapidly become acidified.

Recently, T. H. Wilson suggested that the original proton ATPase that arose must have been a simple, one subunit protein, probably similar to the yeast and *Neurospora* type of plasma membrane proton ATPase (Wilson & Lin 1980). This enzyme would be the precursor of all cation ATPases of the E_1-E_2 type. In parallel with this enzyme a Na-H exchanger would have developed to extrude Na. This enzyme would then accomplish the task of cell volume regulation. However, I do not think that this is a satisfactory scenario because of the following: Most prokaryotes do not contain any E_1-E_2 proton-translocating enzyme. (A K-ATPase exists, however, in some aerobic organisms and has remarkable homology to the Na,K-ATPase and Ca-ATPase.) Further, a proton-translocating ATPase of the F_o-F_1 type is present in all ancient bacteria, such as the anaerobes. It is possible that each of the subunits in this complex enzyme has a different evolutionary past. I suggest that the proteolipid proton channel (F_o) was the first molecule to evolve. Of all subunits of the F_o-F_1 ATPases, it is the one whose sequence has been most highly conserved in a variety of bacteria, mitochondria, and chloroplasts. Further, there are stretches of cytochrome oxidase (another proton pump) that have much homology with the proteolipid. Since the F_1 portion of the ATPases has much homology with a variety of ATP-utilizing or -binding enzymes, e.g. adenylate kinase, it is very likely that the whole proton-pumping ATPase complex arose as a combination of the two independently derived portions, a proton channel and a catalytic subunit. It will be important to test whether the ATP-binding domain bears any homology to that of "ancient" enzymes, such as those involved in glycolysis. The role that such a proton pump might play in proton balance is quite obvious. Its role in osmotic balance of the cell would require, as Wilson suggested, the development of a cation-proton exchanger.

We await with great interest the sequencing of the microsomal proton ATPase, which as discussed above, is very likely to be of the F_o-F_1 type, so that we can compare its sequences to that of the prokaryotic ATPase. If it turns out to be homologous it will raise a number of interesting questions regarding the origin of intracellular organelles. Mitochondria and chloroplasts are likely to have developed by "endosymbiosis" as a consequence of invasion of an anaerobic prokaryote by an aerobic one. Recent studies suggest that many intracellular organelles contain an electron transport chain in parallel to the proton ATPase. In one instance, the chromaffin granule, this electron transport chain is deployed in such a way as to be able to generate a membrane potential by what must be proton transport (Harnadek et al 1985). (Recall that electron transport is a "solid-state" phenomenon; its equivalent in aqueous media is proton transport.)

Other organelles also contain electron transport chains, e.g. microsomes contain cytochrome P450, and a variety of secretory vesicles in neutrophils and macrophages use redox reactions to generate oxygen radicals. Whether these electron transport mechanisms are organized in a manner to allow them to generate membrane potentials by vectorial electron (or proton) transport is not known. Organelles that have proton ATPases in parallel to an electron transport chain, e.g. secretory granules and possibly endosomes and lysosomes, are similar to the aerobic prokaryotes. If that is their origin, they must have arisen by fusion with, rather than invasion of, the fermenting prokaryote. Fusion must also have been followed by invagination and endocytosis of the fused membrane. This is suggested by the fact that the ATP-binding site of the ATPase faces the cytoplasm in these organelles, while it faces away from the cytoplasm in mitochondria and chloroplasts. Further, none of the intracellular organelles (other than mitochondria and chloroplasts) contain their own DNA. The proposed fusion and invagination would have delivered the DNA of the aerobic prokaryote to the cytoplasm of the fermenting prokaryote, thereby allowing the merging of the two DNAs into a common genome.

Even in the case of the symbiont invaders such as mitochondria and chloroplasts, the cognate genome is only partially retained. Most of it is transferred to the genome of the invaded cell. One evidence for this transfer of mitochondrial DNA is the observation that the majority of the genes coding for the mitochondrial F_o-F_1 ATPase are indeed in the nuclear genome.

ACKNOWLEDGMENT

I am grateful to Marc Reitman for many helpful discussions and to George Young for help in preparation of Table 1. I wish to thank J. Barasch, D. Landry, M. Reitman, and G. Young for critical reading of the manuscript.

Literature Cited

Amzel, L. M., Pedersen, P. L. 1983. Proton ATPases: Structure and mechanism. *Ann. Rev. Biochem.* 52: 801–24

Anderson, R. G. W., Falck, J. R., Goldstein, J. L., Brown, M. S. 1984. Visualization of acidic organelles in intact cells by electron microscopy. *Proc. Natl. Acad. Sci. USA* 81: 4838–42

Aris, J. P., Klionsky, D. J., Simoni, R. D. 1985. The F_o subunits of the *Escherichia coli* F_1F_o-ATP synthase are sufficient to form a functional proton pore. *J. Biol. Chem.* 260: 11207–15

Beers, M. F., Cart, S. E., Johnson, R. G., Scarpa, A. 1982. H^+-ATPase and cate-choleamine transport in chromaffin granules. *Ann. NY Acad. Sci.* 402: 116–33

Bennett, A. B., Spanswick, R. M. 1983. Solubilization and reconstitution of an anion sensitive H^+-ATPase from corn roots. *J. Membr. Biol.* 75: 21–31

Bowman, E. J., Bowman, B. J. 1982. Identification and properties of an ATPase in vacuolar membranes of *Neurospora crassa*. *J. Bacteriol.* 151: 1326–37

Cannon, C., van Adelsberg, J., Kelly, S., Al-Awqati, Q. 1985. Carbon-dioxide-induced exocytotic insertion of H^+ pumps in turtle-bladder luminal membrane: Role of cell pH and calcium. *Nature* 314: 443–46

Capasso, J. M., Hirschberg, C. B. 1984a. Mechanisms of glycosylation and sulfation in the golgi apparatus: Evidence for nucleotide sugar/nucleoside monophosphate and nucleotide sulfate/nucleoside monophosphate antiports in the golgi apparatus membrane. *Proc. Natl. Acad. Sci. USA* 81: 7051–55

Capasso, J. M., Hirschberg, C. B. 1984b. Effects of atractyloside, palmitoyl coenzyme A and anion transport inhibitors on translocation of nucleotide sugars and nucleotide sulfate into golgi vesicles. *J. Biol. Chem.* 259: 4263–66

Catterall, W. A. 1984. Molecular basis of neuronal excitability. *Science* 223: 653–61

Dean, G. E., Fishkes, H., Nelson, P. J., Rudnick, G. 1984. The hydrogen ion pumping adenosine triphosphatase of platelet dense granule membranes. Differences from F_o-F_1 and phosphoenzyme type ATPases. *J. Biol. Chem.* 259: 13567–72

Dixon, T. E., Al Awqati, Q. 1979. Urinary acidification in turtle bladder is due to a reversible proton-translocating ATPase. *Proc. Natl. Acad. Sci. USA* 76: 3135–38

Flatmark, T., Gronberg, M., Husebye, E. Jr., Berg, S. V. 1982. Inhibition of N-ethylmaleimide of the MgATP-driven proton pump of the chromaffin granules. *FEBS Lett.* 149: 71–74

Forgac, M., Cantley, L., Wiedenmann, B., Altstiel, L., Branton, D. 1983. Clathrin-coated vesicles contain an ATP-dependent proton pump. *Proc. Natl. Acad. Sci. USA* 80: 1300–3

Futai, M., Kanazawa, H. 1983. Structure and function of proton-translocating adenosine triphosphatase (F_oF_1): Biochemical and molecular biological ap-proaches. *Microbiol. Rev.* 47: 285–312

Galloway, C. J., Dean, G. E., Rudnick, G., Mallman, I. 1983. Acidification of macrophage and fibroblast endocytic vesicles in vitro. *Proc. Natl. Acad. Sci. USA* 80: 3334–38

Glickman, J., Croen, K., Kelly, S., Al-Awqati, Q. 1983. Golgi membranes contain an electrogenic H^+ pump in parallel to a chloride conductance. *J. Cell Biol.* 97: 1303–8

Gluck, S., Al-Awqati, Q. 1984. An electrogenic proton-translocating adenosine triphosphatase from bovine kidney medulla. *J. Clin. Invest.* 73: 1704–10

Gluck, S., Caldwell, J. 1986. Isolation, characterization and reconstitution and bovine medulla H^+-ATPase. *Kidney Int.* 29: 366 (Abstr.)

Gluck, S., Cannon, C., Al-Awqati, Q. 1982a. Exocytosis regulates urinary acidification in turtle bladder by rapid insertion of H^+ ATPase into the luminal membrane. *Proc. Natl. Acad. Sci. USA* 79: 4327–31

Gluck, S., Kelly, S., Al-Awqati, Q. 1982b. The proton translocating ATPase responsible for urinary acidification. *J. Biol. Chem.* 257: 9230–33

Goffeau, A., Slayman, C. W. 1981. The proton-translocating ATPase of the fungal plasma membrane. *Biochim. Biophys. Acta* 639: 58–61

Goldstein, J. L., Brown, M. S., Anderson, R. G. W., Russell, D. W., Schneider, W. J. 1985. Receptor-mediated endocytosis. *Ann. Rev. Cell Biol.* 1: 1–40

Harikumar, P., Reeves, J. P. 1983. The lysosomal proton pump is electrogenic. *J. Biol. Chem.* 258: 10403–10

Harlos, P., Lee, D. A., Stadler, H. 1984. Characterization of a Mg^{2+}-ATPase and a proton pump in cholinergic synaptic vesicles from the electric organ of *Torpedo marmorata*. *Eur. J. Biochem.* 144: 441–46

Harnadek, G. J., Callahan, R. E., Barone, A. R., Njus, D. 1985. An electron transfer dependent membrane potential in chromaffin vesicle ghosts. *Biochemistry* 24: 384–89

Holtz, R. W. 1986. The role of osmotic forces in exocytosis from adrenal chromaffin cells. *Ann. Rev. Physiol.* 48: 175–90

Kagawa, Y., Sone, N., Hirata, H., Yoshida, M. 1979. Structure and function of H^+-ATPase. *J. Bioenerg. Biomembr.* 11: 39–78

Kaunitz, J. D., Gunther, R. D., Sachs, G. 1985. Characterization of an electrogenic ATP and chloride-dependent proton translocating pump from rat renal medulla. *J. Biol. Chem.* 260: 11567–73

Kawakami, K., Noguchi, S., Noda, M., Takahashi, H., Ohta, T., et al. 1985. Primary structure of the α-subunit of Torpedo californica Na⁺K⁺ ATPase deduced from cDNA sequence. Nature 316: 733–36

Kielian, M. C., Helenius, A. 1985. Entry of alpha viruses. In The Virus Series, ed. H. Fraenkel-Conrad, R. Wagner. New York: Plenum

Klingenberg, M. 1980. ADP-ATP translocation in mitochondria, a membrane potential controlled transport. J. Membr. Biol. 56: 97–105

Knight, D. E., Baker, P. F. 1985. The chromaffin granule proton pump and calcium-dependent exocytosis in bovine adrenal medullary cells. J. Membr. Biol. 83: 147–56

Lindstrom, P., Sehlin, J. 1984. Effect of glucose on the intracellular pH of pancreatic islet cells. Biochem. J. 218: 887–92

MacLennan, D. H., Brandl, C. J., Korczak, B., Green, N. M. 1985. Amino acid sequence of $Ca^{2+} + Mg^{2+}$-dependent ATPase from rabbit muscle sarcoplasmic reticulum deduced from its complementary DNA sequence. Nature 316: 696–700

Mitchell, P. 1985. Molecular mechanics of protonmotive F_oF_1 ATPase. Rolling well and turnstile hypothesis. FEBS Lett. 182: 1–7

Ohkuma, S., Moriyama, Y., Takanom, T. 1982. Identification and characterization of a proton pump on lysosomes by fluorescein-isothiocyanate-dextran fluorescence. Proc. Natl. Acad. Sci. USA 79: 2758–62

Ohsumi, Y., Anraku, Y. 1981. Active transport of basic amino acids driven by a proton motive force in vacuolar membrane vesicles of Saccharomyces cerevisiae. J. Biol. Chem. 256: 2079–82

Ohsumi, Y., Anraku, Y. 1983. Calcium transport driven by a proton motive force in vacuolar membrane vesicles of Saccharomyces cerevisiae. J. Biol. Chem. 258: 5614–17

Pedersen, P. L., Hullihen, J. 1984. Inhibitor peptide of mitochondrial proton adenosine triphosphatase. Neutralization of its inhibitory action by calmodulin. J. Biol. Chem. 259: 15148–53

Percy, J. M., Pryde, J. G., Apps, D. K. 1985. Isolation of ATPase I, the proton pump of chromaffin-granule membrane. Biochem. J. 231: 557–64

Perlin, D. S., Latchney, L. R., Senior, A. E. 1985. Inhibition of Escherichia coli H⁺-ATPase by venturicidin, oligomycin and ossamycin. Biochim. Biophys. Acta 807: 238–44

Rees-Jones, R., Al-Awqati, Q. 1984. Proton-translocating adenosinetriphosphatase in rough and smooth microsomes from rat liver. Biochemistry 23: 2236–40

Robbins, A. R., Oliver, C., Bateman, J. L., Krag, S. S., Galloway, C. J., Mellman, I. 1984. A single mutation in chinese hamster ovary cells impairs both golgi and endosomal functions. J. Cell Biol. 99: 1296–1308

Rodewald, R., Kraehenbuhl, J. P. 1984. Receptor-mediated transport of IgG. J. Cell Biol. 99: 159s–64s

Russell, J. T. 1984. Delta pH, H⁺ diffusion potentials, and Mg^{2+} ATPase in neurosecretory vesicles isolated from bovine neurohypophyses. J. Biol. Chem. 259: 9496–97

Sachs, G., Faller, L. D., Rabon, E. 1982. Proton/hydroxyl transport in gastric and intestinal epithelia. J. Membr. Biol. 64: 123–35

Sandvig, K., Olsnes, S. 1980. Diphtheria toxin entry into cells is facilitated by low pH. J. Cell Biol. 87: 828–32

Sandvig, K., Olsnes, S. 1982. Entry of the toxic proteins abrin, ricin, modeccin, and diphtheria toxin into cells. II. Effect of pH, metabolic inhibitors, and ionophores and evidence for toxin penetration from endocytic vesicles. J. Biol. Chem. 257: 7504–13

Sato, T., Ohsumi, Y., Anraku, Y. 1984. Substrate specificities of active transport system for amino acids in vacuolar-membrane vesicles of Saccharomyces cerevisiae. Evidence of seven independent proton/amino acid antiport systems. J. Biol. Chem. 259: 11505–8

Scarborough, G. A., Addison, R. 1984. On the subunit composition of the Neurospora plasma membrane H⁺-ATPase. J. Biol. Chem. 259: 9109–14

Schatz, G., Butow, R. A. 1983. How are proteins imported into mitochondria? Cell 32: 316–18

Schleyer, M., Neupert, W. 1985. Transport of proteins into mitochondria: Translocational intermediates spanning contact sites between outer and inner membranes. Cell 43: 339–50

Schneider, D. L. 1983. ATP-dependent acidification of membrane vesicles isolated from purified rat liver lysosomes. Acidification activity required phosphate. J. Biol. Chem. 258: 1833–38

Schneider, E., Altendorf, K. 1985. All three subunits are required for the reconstitution of an active proton channel (F_o) of Escherichia coli ATP synthase (F_1F_o). EMBO J. 4: 515–18

Schwartz, G. J., Al-Awqati, Q. 1985. Carbon dioxide causes exocytosis of vesicles containing H⁺ pumps in isolated perfused proximal and collecting tubules. J. Clin. Invest. 75: 1638–44

Schwartz, G. J., Barasch, J., Al-Awqati, Q. 1985. Plasticity of functional epithelial polarity. *Nature* 318: 368–71

Selman-Reimer, S., Duhe, R. J., Selman, B. R. 1985. *N*-Ethylmaleimide inhibition of the catalytic activities of the *Dunaliella salina* coupling factor 1 (CF1) and the restoration of the inhibition of the CF1 ATPase activity by *N*-ethylmaleimide. *Biochim. Biophys. Acta* 810: 325–31

Senior, A. E., Wise, J. G. 1983. The proton ATPase of bacteria and mitochondria. *J. Membr. Biol.* 73: 105–24

Serrano, R. 1984. Plasma membrane ATPase of fungi and plants as a novel type of proton pump. *Curr. Top. Cell. Regul.* 23: 87–126

Serrano, R., Kielland-Brandt, M. C., Fink, G. R. 1986. Yeast plasma membrane ATPase is essential for growth and has homology with Na⁺K, K, and Ca ATPases. *Nature* 319: 689–93

Shull, G. E., Schwartz, A., Lingrel, J. B. 1985. Amino acid sequence of the catalytic subunit of the Na⁺ + K⁺ ATPase deduced from a complementary DNA. *Nature* 316: 691–95

Shuster, M. J., Camardo, J. S., Siegelbaum, S. A., Kandel, E. R. 1985. Cyclic AMP-dependent protein kinase closes the serotonin-sensitive K⁺ channels of *Aplysia* sensory neurons in cell-free membrane patches. *Nature* 313: 392–95

Stadler, H., Tsukita, S. 1984. Synaptic vesicles contain an ATP-dependent proton pump and show "knob-like" protru-sions on their surface. *EMBO J.* 3: 3333–37

Stone, D. K., Xie, X.-S., Racker, E. 1983. An ATP-driven proton pump in clathrin-coated vesicles. *J. Biol. Chem.* 258: 4059–62

Sze, H. 1985. H⁺ translocating ATPases: Advances using membrane vesicles. *Ann. Rev. Plant Physiol.* 36: 175–208

van Adelsberg, J., Al-Awqati, Q. 1986. Regulation of cell pH by calcium-mediated exocytosis of H⁺ ATPases. *J. Cell Biol.* 102: 1638–45

Van Dyke, R. W., Hornick, C. A., Belcher, J., Scharschmidt, B. F., Havel, R. J. 1985. Identification and characterization of ATP-dependent proton transport by rat liver multivesicular bodies. *J. Biol. Chem.* 260: 11021–26

Wilson, T. H., Lin, E. C. 1980. Evolution of membrane bioenergetics. *J. Supramol. Struct.* 13: 421–46

Xie, X.-S., Stone, D. K. 1986. Isolation and reconstitution of the clathrin-coated vesicle proton translocating complex. *J. Biol. Chem.* 261: 2492–95

Yamashiro, D. J., Tycko, B., Fluss, S. R., Maxfield, F. R. 1984. Segregation of transferrin to a mildly acidic (pH 6.5) para-golgi compartment in the recycling pathway. *Cell* 37: 789–800

Zimmerberg, J., Whittaker, M. 1985. Irreversible swelling of secretory granules during exocytosis caused by calcium. *Nature* 315: 581–82

Ann. Rev. Cell Biol. 1986. 2 : 201–29

REGION-SPECIFIC CELL ACTIVITIES IN AMPHIBIAN GASTRULATION

John Gerhart and Ray Keller

Departments of Molecular Biology and Zoology, University of California, Berkeley, California 94720

CONTENTS

INTRODUCTION

Gastrulation transforms the unicellular organization of the egg into the embryonic organization of the multicellular organism. In vertebrates, it produces a distinctive body plan with three tissue layers, an extended anteroposterior axis, and an accumulation of cells on the dorsal midline from which the notochord, somites, and neural tube will form at neurulation. Here we review the various region-specific cellular activities that function in amphibian gastrulation and emphasize the role of convergent extension as the major driving force of this morphogenesis. Since the location and intensity of these cellular activities affect the process as a

201

0743–4634/86/1115–0201$02.00

whole, we also review the means by which the oocyte and egg set up the initial conditions of gastrulation, and we emphasize the role of inductive interactions between cells cleaved from egg regions of differing cytoplasmic content. Finally, we draw attention to gastrulation as a process of pattern formation as well as morphogenesis. We have chosen to cover *Xenopus laevis* because its gastrulation is simpler and better understood than that of other amphibians, and the stages before gastrulation have been better analyzed.

OUTLINE OF EARLY *XENOPUS* DEVELOPMENT

The *Xenopus* egg (1.2 mm in diameter) provides a clear example of the localization of cytoplasmic materials within a single cell. The brownish *animal hemisphere* contains common organelles and cytosolic materials, and a layer of melanin granules near the plasma membrane. The metaphase II spindle and chromosomes reside at its pole. The yellowish *vegetal hemisphere* has relatively smaller amounts of organelles and cytosol but contains the bulk of the nutrient reserves of the egg, particularly membrane-bounded yolk platelets. Germ plasm granules reside near its pole. This

---

Figure 1 Outline of early *Xenopus* development. The upper and middle rows show development from fertilization until neurulation. The egg is oriented with the dark animal hemisphere upward and the unpigmented vegetal hemisphere downward. The sperm is arbitrarily shown to enter on the right of the animal hemisphere. It leaves a mark termed the sperm entry point (SEP). During the first cell cycle after fertilization, the grey crescent (GrC) forms at an equatorial position most distant from the SEP. The dorsal midline (D) of the embryonic body axis is expected to arise on the egg meridian bisecting the crescent, i.e. on the left side of the figure. The ventral midline (V) will be centered on the meridian through the SEP, i.e. on the right side of the figure. In the bottom row a hatched tadpole at approximately 3 days of age is shown, with the dorsal (D) and ventral (V) midlines indicated in the orientation predicted from the fertilized egg shown in the top row. In the center of the bottom row is schematized a young oocyte, derived from four serial mitoses of the oogonium. It retains a division bridge (DB) to sister oocytes. A prominent mitochondrial cloud (MC) surrounds the centrosome. Golgi vesicles also abound in this vicinity. The nucleus (N) contains meiotic chromosomes oriented with their tips toward the centrosome. Notice the axis of organelles at this earliest stage. To the right a full grown oocyte is shown in the orientation supposedly derived from the young oocyte. The nucleus (germinal vesicle or GV) with a clear cytoplasmic region (C) at its base occupies the center of the animal hemisphere, and the materials of the mitochondrial cloud including germ plasm granules (G) occupy the vegetal hemisphere near the pole. Yolk platelets differ in size and amount in the different regions of the oocyte: In the animal hemisphere they are small (4 μm maximum length) and low in abundance; in the vegetal hemisphere they are large (up to 8 μm in length) or very large (up to 12 μm) and are almost close-packed. The largest platelets will later mark the boundary of the blastopore. A yolk-excluding cortex encircles the oocyte, beneath the plasma membrane. Pigment (P) granules are located in the cortex of the animal hemisphere.

hemispheric organization of the egg is referred to as its primary polarity, or animal-vegetal polarity.

The organization of the single-celled egg is topographically related to the tissue organization of the embryo. The animal half gives rise to the nervous system and skin of the tadpole (ectodermal organs), whereas the vegetal hemisphere gives rise to the digestive and respiratory systems (endodermal organs). From materials of the equatorial boundary of the two hemispheres will develop the skeletal, muscular, circulatory, and excretory systems (mesodermal organs). Only at fertilization, though, can one predict the dorsoventral fates of various regions of the egg (Figure 1). The single sperm enters anywhere in the animal hemisphere, leaving a darkly pigmented sperm entry point (SEP) (see Elinson 1980). A meridian from pole to pole through the SEP coincides with the eventual ventral midline of the embryo and is referred to as the prospective ventral midline.

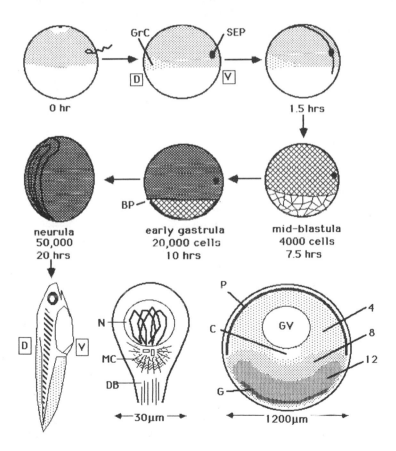

The opposite meridian of the egg is the prospective dorsal midline, on which the most distinctive structures of the vertebrate body will be centered. Since sperm entry is random, one could conclude that the singular selection of the embryo's future dorsoventral organization is determined by the sperm.

During the first cell cycle after fertilization, there is a major reorganization of the egg cytoplasm, manifested externally by the *grey crescent*, an equatorial zone of reduced pigmentation farthest from the SEP. Thereafter the egg has bilateral symmetry instead of its original cylindrical symmetry, and various regions differ in their ability to develop dorsoventral structures. If the egg is experimentally bisected to separate the grey crescent half from the SEP half, the former develops a more complete set of dorsal and anterior structures of the body axis than does the latter (Spemann 1938; Kageura & Yamana 1984; Cooke & Webber 1985).

Development to the 4000-cell midblastula stage proceeds entirely on maternal materials and cytoplasmic control circuits. After a first cell cycle of 100 min (19°C), the egg cleaves every 30 min for 11 cycles (Figure 1). Cleavages are asymmetric in the animal-vegetal direction, leaving much larger cells in the vegetal polar region than near the equator and in the animal hemisphere. A liquid-filled blastocoel forms inside the animal hemisphere due to the osmosis of water into ion-rich intercellular spaces that are isolated from the external medium by tight junctions of the epithelial cell layer at the embryo's surface (Kalt 1971). Then, 7–8 hr postfertilization, cell division becomes slower and asynchronous due to the exhaustion of maternal cytoplasmic materials needed for the formation of zygotic nuclei (Newport & Kirschner 1982). With the emergence of G1/G2 phases in the cell cycle, gene expression begins at a high level (Kimmelman et al 1986). During the mid and late blastula stages, a period of 2–3 hr, cells of the vegetal hemisphere induce cells of the animal hemisphere to take on the various morphogenetic activities which, when integrated, produce gastrulation (Nieuwkoop 1973, 1977).

Gastrulation begins at 10 hr with the appearance of the black blastopore pigment line at 50° of latitude from the vegetal pole, when the embryo contains about 20,000 cells. In discussing the organization of this stage, we distinguish five regions. Each is cylindrically symmetric about the animal-vegetal axis and is characterized by a particular type of motile activity of its constituent cells. In several of the regions, cells gain their characteristic activity earlier and with greater intensity in that part of the region closest to the prospective dorsal midline of the embryo. The placement, timing, and graded intensity of the five types of region-specific cellular activity are all essential conditions for successful gastrulation. The following summary refers to Figure 2 and derives from the analyses of

Nieuwkoop & Florshutz (1950), Keller (1975, 1976), and Keller & Schoen-wolf (1977):

1. The *animal cap* (AC) cells, derived from the pigmented animal hemi-sphere of the egg, engage in *epiboly*, an isotropic expansion of approxi-mately 1.7-fold, to cover 40% of the late gastrula surface.

2. In the vegetal hemisphere, 50° from the vegetal pole, a thin annulus of epithelial cells, the *bottle cells* (BC), contract their apices greatly on the side facing the external medium while elongating basally, thus acquiring a bottle shape. Apical contraction concentrates scattered pigment gran-ules into a small area, giving a black *blastoporal pigment line* (BPL). Contraction begins at the prospective dorsal midline (shown at the left side of each embryo, in the visual plane of the page in Figure 2) and progresses bidirectionally over a period of 2 hr to reach the ventral midline (shown at the embryo's right side in Figure 2).

3. The *marginal zone* of the early gastrula is a wide annulus of cells between the blastoporal pigment line and the animal cap. The cells are small and 4–5 layers deep. There are two subzones. The more vegetal half of the marginal zone will turn inside the gastrula (the process of involution) and is called the *involuting marginal zone* (IMZ). This portion of the annulus turns inside out. The epithelial layer comes to form the roof of the archenteron (AR), an elongated internal cavity that later constitutes the lumen of the gut. In contrast, the more equa-torial half of the marginal zone does not involute but spreads over the surface vacated by the IMZ. This *noninvoluting marginal zone* (NIMZ) reaches the edge of the blastopore by the end of gastrulation. Its border with the internalized IMZ is the *limit of involution* (LI). The IMZ and NIMZ share important properties: Their cell populations converge strongly toward the dorsal midline with concomitant extension in the animal-vegetal direction. This *convergent extension* begins midway in gastrulation, first and most vigorously at the prospective dorsal midline (embryo's left side in Figure 2) and then spreads laterally.

4. The *deep zone* (DZ) lies internal to the IMZ as a ring of cells apposed to the vegetal core. Deep cells engage in spreading migration along the blastocoel wall and roof. They lead involution and come to occupy anterior positions in the embryo. They adhere tightly to the yolky cells, dragging them forward as they migrate. Their migration occurs first at the prospective dorsal midline and spreads ventrally. The extent of displacement is greatest on the dorsal midline.

5. The above-mentioned *vegetal base* (VB) of cells is the core of large yolky cells of the vegetal hemisphere, which extend conically from the subblastoporal vegetal epithelial surface to the blastocoel (BLC) floor.

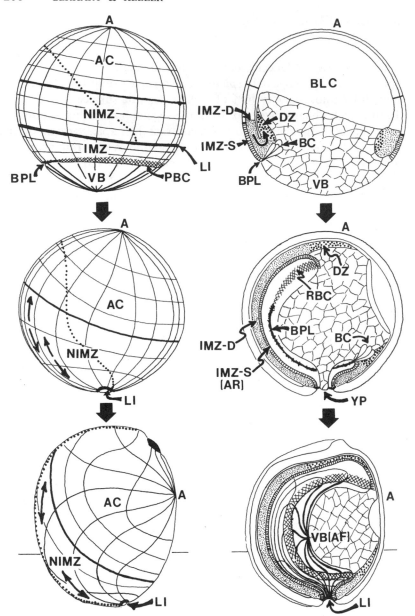

Figure 2 *Xenopus* development is shown at the early gastrula (*top panel*), late gastrula (*middle panel*), and late neurula (*bottom panel*) stages. Morphogenetic movements are shown from a surface viewpoint (*far left column*) and from a midsagital viewpoint (*left center*). Involution of the IMZ-D and DZ (*right center*) and IMZ-S (*far right*) are also shown. Embryos are arbitrarily oriented to have the prospective dorsal midline on the left (in the visual plane of the page) and the prospective ventral midline on the right. Arrows indicate

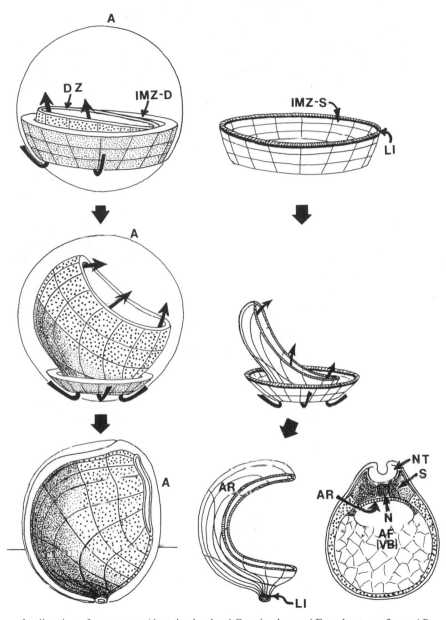

the direction of movement. (A, animal pole; AC, animal cap; AF, archenteron floor; AR, archenteron roof; BC, bottle cells; BLC, blastocoel; BPL, blastopore pigment line; DZ, deep zone; IMZ, involuting marginal zone; IMZ-D, IMZ deep layer; IMZ-S, IMZ superficial layer; LI, limit of involution; N, notochord; NIMZ, noninvoluting marginal zone; NT, neurula tube; PBC, prospective bottle cells; RBC, respread bottle cells; S, somitic mesoderm; VB, vegetal base; YP, yolk plug.)

During gastrulation this region is rolled inward, toward the prospective ventral side of the embryo, as the blastopore closes over its surface. The epithelial surface finally forms the floor of the archenteron (AF), meeting the archenteron walls composed of respread bottle cells (RBC). The archenteron roof is composed of epithelial cells of the converged, extended IMZ (IMZ-S).

Neurulation begins at 15 hr as a smooth continuation of gastrulation; the NIMZ and the underlying IMZ continue to converge toward the dorsal midline and to extend in the animal-vegetal direction, now the anteroposterior axis of the embryo. Simultaneously the dorsal NIMZ thickens and flattens into the posterior portion of the *neural plate*, which then rolls into a tube (NT) that is the rudiment of the central nervous system. The wide anterior portion of the plate forms from animal cap cells underlain by the dorsal-most deep zone (DZ) of spreading, migrating, adhering cells that lead involution. The head and trunk regions of the neural plate are induced by, and form from, functionally different cell populations. Also during neurulation the most dorsally converged IMZ coalesces into a rod, the notochord (N), alongside which thickening tissue forms somites (S). By 24 hr postfertilization neurulation is complete, and the still-external animal cap and NIMZ tissue expand to cover the entire surface of the embryo, eventually becoming the epidermis. Neurulation occurs only on the dorsal side; there is no related or attenuated event on the ventral side. Thus, with each stage of development, the prospective dorsal region of the embryo becomes progressively different from its ventral counterpart, although the two were identical in the cylindrically symmetric unfertilized egg.

CYTOPLASMIC LOCALIZATION IN OOGENESIS

In order to inquire into the locations, types of activity, and graded intensity of the five functional regions of the early gastrula, we must trace their origins to the earliest stage of amphibian development, the oocyte. This small single cell (30 μm in diameter) arises by mitosis from a self-perpetuating stem cell precursor, the oogonium. Regional cytoplasmic differences are evident even at this stage. At one end of the cell is the cytoplasmic bridge that remains from incomplete cytokinesis (Coggins 1973). Next comes a centrosome surrounded by numerous mitochondria, germ plasm precursors, and Golgi vesicles, and farthest from the cytoplasmic bridge is the nucleus. The chromosomes orient their tips toward the centrosome at an early meiotic ("bouquet") stage (Figure 1).

This organellar axis typifies many kinds of eukaryotic animal cells, at least transiently in the cell cycle. This general eukaryotic organization may be exploited by the oocyte to localize cytoplasmic materials as it grows in the ovary and differentiates the animal-vegetal axis. In the full grown oocyte the nucleus does in fact occupy the animal half, and the cluster of mitochondria and associated germ plasm granules occupy the vegetal pole (Wylie et al 1985). The cytoplasmic bridge disappears early in oogenesis, and its relation to the vegetal pole is unknown. Thus, there is circumstantial evidence that the axis of organelles of the earliest oocyte organizes the primary polarity of the egg, which in turn holds a regular topographic relationship to the eventual tissue organization of the embryo.

There are four obvious regions of the full-sized oocyte. First, there is the germinal vesicle, the 300-μm diameter nucleus which contains nuclear proteins and nuclear membrane in amounts equal to the sum of the contents of several thousand blastula nuclei (Gerhart 1980). When the nucleus breaks down at meiotic maturation, these proteins are distributed throughout the animal hemisphere and enter the periphery of the vegetal hemisphere (Hausen et al 1985).

Second, there is a mass of large, closely packed, membrane-bounded yolk platelets in the vegetal hemisphere. Yolk proteins constitute 80% of the oocyte's protein, 75% of which is in the vegetal half (Danilchik & Gerhart 1986). A yolk precursor is made in the liver as a soluble lipophosphoprotein, vitellogenin, which is released into the blood and internalized by the oocyte. This precursor is packaged in endocytotic intermediates, the platelets, in which it crystallizes after a specific proteolytic modification (Wallace 1983). Vitellogenin is taken up equally over the entire oocyte surface, and platelets formed from it in the animal half move steadily into the vegetal half and accumulate there (Danilchik & Gerhart 1986). Thus, a directional flow of materials within the cell accounts for the regional localization of these membrane-bounded organelles. The oocyte has radially directed intermediate filaments at early stages and various microtubule arrays (Wylie et al 1985; Palacek et al 1985), which may be involved in such flow. The polar subregion of the vegetal hemisphere (Figure 1) contains the largest platelets. These platelets differ from others in their buoyant density, which perhaps indicates a difference in composition (Radice et al 1981). The cytoplasmic boundary of these large platelets probably marks the future level of the blastopore in the gastrula, in which subblastoporal cells contain much larger platelets than neighboring marginal zone and deep zone cells (Harris 1964; Nakatsuji 1975).

The periphery of the oocyte is also a specialized region, distinct from the deeper cytoplasm. A thick actin mat underlies the plasma membrane (Franke et al 1976), and cytokeratins accumulate in the periphery during maturation (Franz et al 1983). The periphery is known as the *cortex* of the cell. Materials of the cortex are incorporated into the continuous and functionally distinct epithelial cell layer of the embryo. Cytokeratins are typical components of desmosome-linked epithelial cells. In the oocyte and egg, the cortex is probably a dynamic structure whose thickness and composition changes with the cell cycle and developmental stage. It consists of two or more layers at certain stages (Merriam et al 1983). The cortex of the two hemispheres of the oocyte differs, thus giving us a third and fourth major localization. The plasma membrane of the animal hemisphere is less fluid and protrudes in more and longer microvilli (Dictus et al 1984). Pigment granules occupy the submembrane region of this hemisphere in much greater number than in the vegetal. Cortical granules of the cortex of the two hemispheres also differ in number and size (Campanella & Andreucetti 1977).

Nieuwkoop (1977, 1985) has proposed that the oocyte consists of only two developmentally important regions: the animal and vegetal "moieties." As discussed later, this view is consistent with evidence that the further regional differences achieved by the time of gastrulation arise from the interactions of cells cleaved from these two parts. We would modify this proposal in two minor ways. One, the cortex should be distinguished from the deeper cytoplasm, since epithelial and nonepithelial cells of the embryo derive directly from each, respectively. And two, the animal and vegetal moieties are not spatially equivalent to the animal and vegetal hemispheres of the oocyte and egg as defined by pigment granules and the geometric equator. Perhaps the domain of the vegetal moiety is delineated by the large yolk platelets within the vegetal hemisphere, and other subregions of this hemisphere should be considered part of the animal moiety. The functional material differences of the animal and vegetal localizations are far from known. They may include the ratio of yolk platelets relative to other organelles and cytosolic materials, or region-specific determinative materials. A few RNA species, which are presumably translatable, are localized to one hemisphere or the other (see Rebagliati et al 1985), although most RNAs are rather uniformly distributed. Finally, the "two moieties" proposal may be too simple; in particular the equatorial zone may be a specialized third region (Gurdon et al 1985).

REGIONAL SPECIALIZATION IN THE FIRST CELL CYCLE

The cytoplasmic reorganization of the egg after fertilization was first studied in detail by Ancel & Vintemberger (1948) using eggs of the frog *Rana fusca*. They cauterized the egg surface at two or three points and watched the wounds move during the first cell cycle as the grey crescent formed. The egg surface rotated as a unit by 30° of arc in the animal-vegetal direction, relative to the underlying cytoplasm. They found that the grey crescent is a local manifestation of this global rotation. As the darkly pigmented cortex of the animal hemisphere rotates toward the animal pole on one side, it no longer covers the less pigmented equatorial cytoplasm, which appears grey. Elinson (1980) has also observed this rotation in *Rana* eggs by measuring the displacement of the sperm trail and maturation spot. Recently, these classic studies were repeated with *Xenopus* eggs stained with grid patterns of fluorescent dyes applied to the egg surface or subcortical core of cytoplasm (Vincent et al 1986). Between 45 and 90 min after fertilization the entire surface and core move as rigid units relative to one another, with a net displacement of 30° (or a 300-μm linear distance). Although *Rana* and *Xenopus* eggs engage in equivalent movements, the *Xenopus* egg does not form a distinct grey crescent, simply because the core contains all the pigment granules at the time of rotation.

The direction of rotation predicts the position of the future dorsal midline of the embryo more accurately than does the SEP. That meridian of the egg along which the subcortical cytoplasm moves farthest in a vegetal direction (and along which the cortex moves farthest in an animal direction) coincides within 20° with the centerline of the neural plate, which forms one day later in development (Vincent et al 1986). Rotation is an interesting geometric operation: the animal-vegetal polarity of the egg surface is offset 30° relative to the animal-vegetal polarity of the core, creating a bilaterally unique pattern of new appositions between the two rigid units. Thus, a secondary polarization of the egg is generated by the rotational displacement of two layers of the egg's primary polarity.

Ancel & Vintemberger (1948) found that artificially activated *Rana* eggs, which lack a sperm, rotate their contents correctly and form a grey crescent. With dye-marked *Xenopus* eggs normal rotation can also be seen to occur at approximately the right time and to the normal extent (Roberts et al 1986). Thus, the egg does not need sperm components for the rotation. However, in artificially activated eggs the rotation occurs in an unpredictable direction, with no relation to the point of the activating needle puncture, whereas it would normally align with the SEP. Thus, the sperm, as a determinant of the future dorsoventral organization of the embryo, seems only to provide a cue which is used but not needed by the egg to establish a direction of rotation. The sperm centrosome, which originates near the entry point and organizes a large aster, is perhaps the organelle that biases rotation in fertilized eggs (Manes & Barbieri 1977; Ubbels et al 1983). It is possible to bias the direction of rotation of the contents of an artificially activated *Rana* egg: Prior to the rotation period the egg is briefly tipped so that the animal-vegetal axis is horizontal rather than vertical. In this orientation the vegetal yolk mass is out of gravitational equilibrium and slips slightly relative to the cortex. The egg is then allowed to return to gravitational equilibrium. In the subsequent rotation period, the egg moves its internal contents by 30° in the direction of the prior gravitational displacement (Ancel & Vintemberger 1948).

The importance of rotation for the later region-specific cell activities of the early gastrula is shown by experiments in which it is prevented (Manes & Elinson 1980; Scharf & Gerhart 1980, 1983; Scharf·1986a). Inhibitory nonlethal treatments include brief cold shock (2°C, 4 min), high hydrostatic pressure (6000 psi, 6 min), low concentrations of nocodazole (5 μg/ml, 4 min), and ultraviolet irradiation of the vegetal surface. The inhibited egg develops into a cylindrically symmetric embryo that lacks a vertebrate body axis (Grant & Wacaster 1972; Malacinski et al 1977). The "invertebrate" embryo has no notochord, somites, or neural tube, the distinctive dorsal elements of the body axis. Even without rotation, the egg engages in many processes of early development. It cleaves normally and gastrulates from the usual latitude, 50° from the vegetal pole. Thus, rotation is not needed for the founding of the five functional regions of the gastrula. However, early intense dorsal activity is absent in the marginal zone, deep zone, and the bottle cells. While each of these cylindrically symmetric regions retains its characteristic type of cellular activity, the activity is expressed late and weakly, as it would be in the vicinity of the prospective ventral midline of the normal gastrula. The IMZ and NIMZ do not engage in convergent extension, so the body axis does not elongate. Neurulation does not occur. Later the cylindrically symmetric embryo develops a simple three-layered anatomy with a ciliated epidermis, red blood cells, mesenchyme, and a short gut. Since these tissue types characterize the ventral regions of normal embryos, the treated embryos are said to be "ventralized." Ventral development seems to be a default pathway or ground state generated directly from the cellular organization of the oocyte, whereas dorsal development (leading to characteristic vertebrate structures) depends entirely on the rotation of egg materials after fertilization.

Intermediate forms of embryos are obtained when rotation is partially inhibited or partially restored by gravity or centrifugal force (Scharf & Gerhart 1980, 1983; Black & Gerhart 1986). After a small displacement of the egg cortex relative to the core, the embryo develops only a tail and is otherwise ventralized. When rotation is intermediate, the embryo later develops a tail and trunk, but no head. And finally, at greater displacements, it develops a tail, trunk and head, with the hindbrain, midbrain, and forebrain, appearing in that order. Thus, rotation specifies not only the position at which the body axis will develop, but also the anteroposterior completeness of the dorsal structures of the axis. As discussed later, rotation controls quantitatively, though indirectly, the onset and intensity of those cellular activities in the marginal zone that are to exceed the basic ventral level of activity.

Natural and artificial rotations probably do not localize special axis-specifying substances to one region of the egg, but cue "local activation" of widespread but latent cytoplasmic agents needed for early, vigorous gastrulation by the marginal zone. This suggestion is supported by the finding that the entire egg circumference can be activated to develop with dorsal characteristics. For example, when D_2O ("heavy water") is applied to eggs early in the first cell cycle, cylindrically symmetric "dorsalized" embryos are formed (Scharf 1986b). These gastrulate early and vigorously around the entire marginal zone; they represent the developmental opposite of ventralized embryos. Presumably, then, rotation of the egg's cytoplasmic contents normally activates this dorsalizing capacity in only one limited region, or two in the case of forced twinning (Black & Gerhart 1986). In summary, the first cell cycle includes a stage when the egg, still a single cell, reorganizes the cortical and subcortical layers of its cytoplasm in a systematic way and creates further regional cytoplasmic differences, perhaps by the local activation of maternal materials. Cortical/cytoplasmic rotation is a major morphogenetic process.

PREPARATIONS FOR GASTRULATION

Morphogenetic processes of the embryo before the midblastula stage operate from maternal cytoplasmic regulatory circuits and materials, as shown by the early normal development of embryos in the presence of α-amanitin (Brachet et al 1972; Newport & Kirschner 1982). These processes include the rotation discussed above, asymmetric cleavage, the centripetal movement of germ plasm granules to the vegetal pole, ingression of cytoplasmic materials along the new cleavage membranes, epithelialization of the surface, and blastocoel formation. The transition to the midblastula stage is itself under maternal posttranslational control (Kimelman et al 1986).

In the 2–3 hr period from the midblastula stage to gastrulation, inductive interactions of animal and vegetal cells define the location and activity of the five functional regions of the gastrula. It is not known exactly when these interactions start under normal conditions of development, but under experimental conditions the 2–3 hr period prior to gastrulation is sufficient (Nieuwkoop 1973; Gimlich 1985). Nieuwkoop (1973, 1977) and his colleagues demonstrated that the vegetal base of cells of the midblastula is a source of inductive signals that cause adjacent animal hemisphere cells to engage in marginal zone activities and, if subsequent events are normal, to develop eventually into mesodermal and endodermal structures. This interaction is termed endo-mesoderm induction. As Nieuwkoop showed, when animal cap cells and vegetal base cells develop separately from one another from early cleavage stages, the animal cap develops into a ciliated epidermis, rather like belly skin, whereas the vegetal base cells cleave well but do not produce identifiable structures. Gastrulation does not occur in either place, and mesoderm is not formed. Nonetheless, when these two types of cell groups are placed in contact from the midblastula stage onward, the animal cap cells engage in gastrulation and neurulation and develop the full spectrum of dorsal structures of the vertebrate body axis. The inductions cause marginal zone cells to undergo convergent extension, spreading migration, and bottle cell formation, instead of epiboly. Furthermore, vegetal base cells induce the dorsoventral differences of intensity and time of onset of cellular activity in the marginal zone.

Gimlich & Gerhart (1984) demonstrated the dorsoventral component of these inductions in another way. First, they inhibited the cortical/cytoplasmic rotation of the first cell cycle so that eggs would develop as ventralized embryos. As described previously, such embryos do establish the five functional regions of the gastrula, but the marginal zone has no dorsal gradation of early intense activity. Second, at the 64-cell stage they replaced cells (two neighboring cells at a time) with their counterparts from a normal embryo of the same age. Vegetal cells from the prospective dorsal midline of the donor greatly improved the development of the recipient and sometimes made it normal. When the host was rescued by donor cells taken from the dorsal-most quadrant of the tier of cells closest to the vegetal pole, the body axis was formed entirely by progeny of host cells, not donor cells, as shown by fluorescent lineage tracers. The rescuing donor cells induced nearby marginal zone cells of the host to undertake early extensive gastrulation movements and subsequent differentiation of the entire vertebrate body axis.

In many transplants, rescue was only partial; the body axes were truncated at various anteroposterior levels, in exactly the same series as observed for embryos from eggs partially inhibited in their rotation in the first cell cycle. Weyer et al (1978) have suggested that induction by the vegetal base is quantitatively graded in a dorsoventral direction: At the mid and late blastula stages the highest levels of induction would promote dorsal endomesodermal development and the lowest levels, ventral development. Regarding gastrulation, the stronger the induction, the earlier and more vigorous will be the marginal zone activity. Gimlich & Gerhart (1984) adopted this notion of quantitative gradations of induction to suggest that the anteroposterior extent of dorsal development in their partially rescued embryos was directly related to the intensity of induction by the graft cells.

The role of cytoplasmic localizations and inductive interactions in establishing the five distinctive regions of the early gastrula can be summarized as follows:

1. The *animal cap* is derived from the original animal hemisphere of the oocyte, one of the two major regions specialized by cytoplasmic localization in oogenesis. The animal cap, although responsive to inductions, is not exposed to them during the blastula stages because the blastocoel separates it from the vegetal source. The cap proceeds autonomously along a cytoplasmically defined developmental path until it is affected by other tissues. Epiboly may be the activity of noninduced cells of the animal moiety during gastrulation.

2. The *vegetal base* is derived from the second major region of cytoplasmic localization of the oocyte. It can be identified by its large, closely packed yolk platelets. A subregion of the vegetal hemisphere is further specialized by the cortical/cytoplasmic rotation in the first cell cycle. The cellularized vegetal base as a whole induces nearby animal hemisphere cells to become marginal zone cells, and the specialized subregion is the strongest source.

3. *Bottle cells* form when three conditions are met. Bottle cells are epithelial cells of the animal moiety, located at the interface with the vegetal moiety (Nakatsuji 1975). These cells may represent the most strongly induced epithelial cells. Bottle cells at different positions around the blastopore differ only in the time at which their apices contract.

4. The spreading, migrating, *deep zone* cells are derived from nonepithelial cells of the "animal moiety." In their normal location, they are unique in having no neighboring epithelial cells and in having ongoing contact with vegetal base cells. They may be highly induced due to their proximity to the vegetal core. Around the circumference of this zone, deep zone cells appear to differ from one another only in the intensity and time of onset of movement, depending on their distance from the specialized vegetal subregion.

5. The convergent, extending *marginal zone* can be formed by any cells of the animal moiety if, by transplantation, they come into inductive contact with the vegetal base. Marginal zone cells closest to the most inductive subregion of the vegetal base begin their convergent extension earliest and most vigorously in comparison to marginal zone cells at greater distances from this subregion. However, under normal conditions in the cleaving egg, the marginal zone may develop its activities autonomously and may not require induction from the vegetal base (Gurdon et al 1985), perhaps because it is a region of overlapping animal and vegetal cytoplasmic localizations or because it contains unique localized material. In the IMZ, epithelial cells retain epithelial

function throughout gastrulation and come to line the archenteron and form endoderm. The underlying nonepithelial cells of the IMZ converge and extend during gastrulation and eventually form mesoderm. The earliest, most vigorous nonepithelial cells of the IMZ become the "Spemann organizer region" of the gastrula, the region that inductively organizes the neurula stage. Epithelial and nonepithelial members of this zone may differ in behavior and fates depending on the cortical or subcortical type of cytoplasm received from the oocyte. The NIMZ resembles the IMZ, except that its cells are farther from the vegetal base. At the late blastula stage they occupy the blastocoel wall and may receive inductive signals through the continuous cell layers they share with the IMZ, rather than directly from the vegetal base.

TIMING OF GASTRULATION

Two preconditions must be met before cells of the marginal zone initiate gastrulation activities. First, gene expression must have started, as indicated by the fact that α-amanitin-inhibited embryos do not gastrulate even though they develop to the late blastula stage (Brachet et al 1972; Newport & Kirschner 1982). R. Gimlich (personal communication) recently found that amanitin-inhibited vegetal base cells do not induce marginal zone cells to gastrulate. Thus, induction may require zygotic gene expression, which can only occur after the midblastula stage. Since most new transcripts are similar or identical to transcripts stored in the egg (Dawid et al 1985), zygotic gene expression may be needed primarily to replace maternal materials exhausted in the pregastrula stages of development. There are also a few new RNA species unique to the mid and late blastula periods (Krieg & Melton 1985), which may encode required gastrulation-specific materials. While gene expression is necessary, it is not sufficient to explain the time of onset of gastrulation. In normal development new transcripts can be detected at 7.5 hr after fertilization in all cells (Newport & Kirschner 1982), and yet the dorsal marginal zone begins gastrulation at 10 hr whereas its ventral counterpart does not begin until 12 hr, almost twice as long an interval of gene expression. Furthermore, even when the transition to the midblastula stage (and presumably to concomitant gene expression) is moved forward by one or two cell cycles, gastrulation does not start earlier (Kobayakawa & Kubota 1981). Thus the initiation of gastrulation must also depend upon the completion of maternal cytoplasmic events. Maternal mRNA's for histone H1 (Woodland et al 1979), lamin 3 (Stick & Hausen 1985), and fibronectin (Lee et al 1984) are increasingly recruited into polysomes starting at the midblastula stage. Fibronectin synthesis accelerates on time, even in artificially activated eggs, which suggests that the cytoplasmic schedule is independent of cleavage and the midblastula transition (Lee et al 1984). New maternal products and new zygotic products may be needed before gastrulation can start; neither alone is sufficient.

Dorsal marginal zone cells may fulfill the cytoplasmic requirements earlier than do ventral cells, and this difference may account for the graded times of onset of gastrulation. If this is true, the earlier completion by the dorsal marginal zone cells must be a consequence of the stronger induction they receive from vegetal cells of the specialized subregion of the vegetal base. The first sign of effective rescue of ventralized host embryos by grafts of normal vegetal cells from this subregion is seen at gastrulation when host cells involute early at blastopore positions closest to the graft (Gimlich & Gerhart 1984). The immediate effect of vegetal induction on marginal zone cells is to control the time and place of gastrulation. The relationship of the time of onset and the level of activity is unknown.

CELLULAR MECHANISMS OF GASTRULATION

We must now discern which morphogenetic changes at gastrulation are generated by *forces* within a particular one of the five regions and whether that force is attributable to changes in individual cell shape or to cell repacking. Also, we must discern whether the exact morphogenesis of a region is determined autonomously within that region or is context dependent.

Epiboly by the Animal Cap

Epithelial cells of the superficial layer of the cap expand isotropically in area, flatten, and divide in the superficial plane so that both daughter cells occupy the epithelial sheet (Keller 1978). Expansion begins 2 hr before the blastopore pigment line is visible. The superficial layer thins in proportion to its increased area (see Figure 3). Cells are linked by tight junctions and desmosomes, and seem never to leave the layer. Neither are underlying cells added to the layer in *Xenopus*, although they are in the axolotl (see Keller 1986).

In *Xenopus*, the several layers of cells under the epithelium intercalate with one another to form fewer layers of greater area during epiboly. These pleimorphic cells are interconnected by short filiform and lamelliform extensions (Keller 1980). On the side facing the blastocoel, the cells are flattened and their extensions lap across the surfaces of their neighbors. This blastocoelic face is covered by a thin extracellular lamina containing laminin (Nakatsuji et al 1985) and fibronectin (Lee et al 1984; Nakatsuji 1984; Boucaut et al 1985). The lamina is thought to provide a substratum on which postinvoluted deep cells migrate, particularly in urodeles (Nakatsuji & Johnson 1984; Boucaut et al 1985), and to provide a boundary across which cells do not mix as they pass each other in gastrulation.

Isotropic expansion has long been considered an autonomous activity of the animal cap since the isolated cap throws itself into folds (Spemann 1938; Keller 1986). However, expansion has never been measured directly

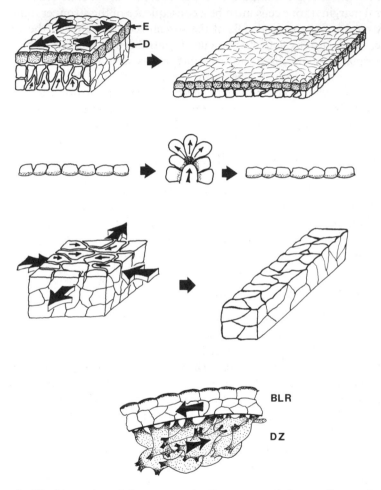

Figure 3 The four major cellular processes in *Xenopus* gastrulation are illustrated: Cell intercalation and cell spreading during epiboly of the animal cap (*top*); formation of bottle cells and their subsequent respreading (*top center*); cell intercalation during convergent extension of the involuting and noninvoluting marginal zones (*bottom center*); traction of the stream of involuting deep zone cells on the roof of the blastocoel (*bottom*). Arrows indicate direction of cell movements or tissue displacements. (BLR, blastocoel roof; E, superficial epithelial cells; D, deep nonepithelial cells; DZ, deep zone cells.)

in explants, and shrinkage within the epithelium might create the same appearance. Thus, we can only surmise that epiboly is due to intrinsic cell activities of the cap. Subepithelial cells of the cap may wedge between one another and spread on the epithlium, generating a force for epibolic expansion (Keller 1980). Epiboly may occur in vivo because the marginal zone pulls the animal cap downward, and as the cap expands over the incompressible blastocoelic fluid, subepithelial cells just interdigitate to fill the available space. In summary, little is known about the regional autonomy or cellular basis of epiboly.

Convergent Extension by the Marginal Zone

The autonomous activity of this region is well demonstrated. If dorsal sectors of the marginal zones of two early gastrulae are excised and "sandwiched" with their epithelial sides outward, they undergo substantial convergent extension, even though the IMZ cells are prevented from involution.

One may say without exaggeration that convergent extension is the major regional activity driving gastrulation of the embryo as a whole. It drives involution of the marginal zone, closure of the blastopore, elongation of the anteroposterior axis, and accumulation of cells on the dorsal midline (Keller et al 1985). The series of gastrulation events can be reconstructed as follows: Initially, the dorsal-most bottle cells and deep zone cells cause the dorsal IMZ to involute at its vegetal edge. At the midgastrula stage, the not yet involuted dorsal IMZ begins convergent extension in a vegetal direction and pushes cells over the lip. Since the IMZ converges strongly toward the dorsal midline and extends vegetally, the blastopore narrows in circumference and closes eccentrically near the ventral edge of the vegetal base. The dorsal NIMZ assists in closure, for as soon as its basal surface is contacted by involuted extending IMZ, it also begins convergent extension. The dorsal NIMZ maps to the center line of the mid and posterior neural plate, a region called the notoplate by Jacobson et al (1985), who first drew attention to its convergent extension, its close association with the underlying notochord, and its ultimate fate as the floor of the posterior neural tube.

Convergent extension by the marginal zone is probably driven by the subepithelial cells (IMZ-D in Figure 2), not those of the epithelial layer (IMZ-S) since the latter can be replaced by epithelial cells of the animal cap without adverse effect, whereas replacement of the subepithelial cells stops movement immediately (Schechtman 1942; Keller 1981, 1986). However, the epithelial layer is not altogether dispensable, since isolated subepithelial cells fail to converge and extend without this covering. Epithelial cells of the IMZ and NIMZ differ from those of the animal cap in being able to change planar neighbors readily when the sheet is distorted by the convergent extension of underlying nonepithelial cells; cells of the animal cap epithelium maintain rigid spatial relations to their neighbors.

What intrinsic cell activity drives convergent extension of the population? Waddington (1940) proposed that cells repack themselves into narrower, longer arrays, since there is no proportionate change of cell shape. Keller et al (1985) marked subpopulations of the NIMZ and IMZ with fluorescent lineage tracers and observed their extensive interdigitation with unmarked neighbors during mediolateral convergence in vivo. Recently, interdigitation has been observed in vitro in marginal zone explants cultured in a blastocoel-like medium (Keller et al 1985). The dorsal part of the zone converges and extends much more vigorously than does the ventral part. Dorsal cells engage in extensive blebbing and protrusive activity, and fractures open and close rapidly in the lengthening direction of the tissue fabric. Cells frequently move into the openings.

Although cell repacking certainly occurs, it is not clear what intrinsic cell activity drives directional population behavior. Holtfreter (1944) suggested that there is an initial vectorial organization of the tissue of the dorsal marginal zone. This idea was based on the finding that disaggregated-reaggregated marginal zone cells do not undergo extension in culture, even though they eventually differentiate as notochord cells. In summary, convergent extension is an autonomous activity of the marginal zone, but the cellular and intercellular basis for this important directional morphogenetic activity is far from understood. We know only that cell repacking occurs.

Spreading Migration by the Deep Zone

In the early gastrula, deep zone cells lie as an annulus between the IMZ and the vegetal core. During their spreading migration they lead the involuted IMZ along the walls and roof of the blastocoel (see Keller 1986). The cells are rotund and largely separate, though they are connected to one another and to the blastocoel wall by large filiform and lamelliform protrusions. In *Xenopus*, these cells constitute a cohesive population on the periphery of the vegetal core. As they migrate on the blastocoel wall, they seem to pull the vegetal core with them (Keller 1986), rather like the tugging effect ascribed by Holtfreter (1943) to bottle cells. This traction is not quantitatively important for gastrulation, for if the blastocoel wall and roof are removed from a *Xenopus* early gastrula, the residual piece still gastrulates. The IMZ continues involution and convergent extension, and the blastopore closes (Keller et al 1985). Nonetheless, deep zone cells may be important in the initial involution of the IMZ, without which convergent extension would occur entirely on the surface of the gastrula, yielding an "exogastrula."

Deep zone cells in the axolotl appear loosely organized and individually migratory, more so than in *Xenopus*. Nakatsuji & Johnson (1984) have suggested that these migrating cells, especially in urodeles, are guided by contact with extracellular matrix fibrils aligned in an animal-vegetal direction on the wall of the blastocoel. Kubota & Durston (1978) present an alternative idea. They filmed the leading edge of deep cells in opened axolotl gastrulae and found that this edge advances in waves across the blastocoel wall. Lone cells on the wall migrate little and randomly; forward migration seems to require contact stimulation from the cells behind those at the leading edge. Thus, they propose that directed migration of the population is an outcome of cell-cell interactions and available surface space. According to either proposal, the direction and extent of movement of the deep zone is strongly context dependent.

Apical Contraction by the Bottle Cells

Throughout gastrulation the bottle cells retain an apical connection to the continuous epithelial cell sheet. There are two classical ideas about the function of bottle cells: Firstly, their change of shape from columnar to bottle-like was thought to cause a bending of the epithelial sheet at the blastoporal groove; secondly, their active penetration between interior cells of the embryo was thought to allow them to pull the epithelial sheet inward to form the cavity of the archenteron (see Holtfreter & Hamburger 1955; Keller 1986). This tugging action was thought to be a major driving force for gastrulation. When bottle cells are removed from the *Xenopus* gastrula, however, morphogenesis is almost normal, except that the archenteron is truncated (Cooke 1975; Keller 1981). The lengthening of the archenteron seems to be driven mainly by convergent extension. The IMZ epithelial surface is in fact the roof of the archenteron (see Keller 1981). Bottle cells in the wall of the archenteron lose their bottle shape midway through gastrulation, long before the archenteron has fully extended.

Although bottle cells do not drive gastrulation, they probably do orient other movements in the way mentioned previously for deep zone cells. The initial invagination of the epithelial sheet at the blastoporal groove and the initial small involution of the IMZ are caused by the apical contraction of the bottle cells and by their active penetration between deeper cells of the embryo. Without this initial involution, exogastrulation would occur (see Keller 1986). After this initial effect, however, bottle cells are probably dispensable in *Xenopus*, except that they contribute additional surface area to the archenteron walls when they later respread as typical epithelial cells.

Each bottle cell contracts its apex 10- to 50-fold in an animal-vegetal direction as the blastoporal pigment line is formed. The contracting apex is directly derived from the original vegetal cortex of the egg cell. The integrity of cortical material during the contraction is remarkable: Scattered pigment granules are concentrated into a visible black line, and microvilli become clustered on the external surface (Baker 1965; Perry & Waddington 1966). In the blastoporal groove, the bottle cells extend inward and gain a greatly elongated shape. However, when bottle cells are cultured as small explants in vitro, they do not elongate but acquire only a spherical shape, even though they contract their apices to the full extent (J. Hardin & R. E. Keller, unpublished). The bottle shape is probably assumed by cells in vivo because of mechanical constraints imposed on them by the vegetal base on one side and IMZ cells on the other. Furthermore, although bottle cells contract their apices mostly in an animal-vegetal direction in vivo, they contract isotropically in vitro, which probably reflects the absence of mechanical constraints from the vegetal core. Thus, although apical contraction is an autonomous activity, the well-known shape of bottle cells is probably context dependent.

The Vegetal Base

The mass of large, yolky cells of the vegetal hemisphere undergoes less deformation than any other region of the gastrula. The vegetal base cells show no characteristic autonomous movement except for a gradual decrease of surface area throughout gastrulation and neurulation (see Keller 1978). The vegetal epithelial surface becomes covered by the blastopore lip and eventually comprises the floor of the archenteron. In the course of internalization, this region elongates in the anteroposterior direction, probably as a passive response to the elongating IMZ (the archenteron roof), to which it is connected through the bottle cells of the archenteron walls. If the bulk of the vegetal core is removed, gastrulation proceeds with little disruption (Keller 1978; Gimlich 1986).

Given the importance of cell migration, surface protrusion, and apical contraction in gastrulation, it should be no surprise that inhibitors such as cytochalasin B block the process (Nakatsuji 1979). It is perhaps more remarkable that antimicrotubule drugs such as colcemid (Nakatsuji 1979) and nocodazole (J. Newport, unpublished) have little effect. This may indicate that IMZ cells do not depend on individual cell polarity for repacking but depend on contact with other cells and the extracellular matrix. In this regard, it is revealing that inhibitors of matrix-cell interactions block convergent extension very effectively in vivo (Boucaut et al 1985; Gerhart et al 1984). Inhibitors of cell division, such as mitomycin or deoxynucleosides (or colcemid) have no effect on gastrulation as long as they are applied within an hour before gastrulation (Cooke 1975). IMZ cells divide infrequently during gastrulation anyway, and the cell cycle is blocked in G2 (Cooke 1979).

In summary of this section, the marginal zone, deep zone, and bottle cells are autonomous regions of specialized cell activity, but the details of cell shape and population deformation are highly context dependent and not encoded within the region. We have no explicit information about the intrinsic cell activities that generate the particular regional behaviors, except in the case of apical contraction by the bottle cells.

CELLULAR AND REGIONAL DETERMINATION IN THE EARLY GASTRULA

The fate map of the early gastrula is usually labelled in terms of the ultimate anatomical differentiation of each region, such as the notochord, somites, or neural tube. One might infer from such maps that the regions of the early gastrula are already determined for these fates, or even that individual cells of the early gastrula are determined for their final positions and cell types in the differentiated embryo. The word "fate" implies determination of the outcome in advance. If this were true, gastrulation would be a mere playing out of a predetermined pattern, with no addition of complexity to the pattern. Several recent experiments argue against this interpretation:

1. Single cells of the animal cap and marginal zone of the early *Xenopus* gastrula will take on the final fates of cells of any region if they are removed from a donor embryo, labelled with a fluorescent lineage tracer, and seeded singly into the blastocoel of an intact early gastrula which is then allowed to develop (see Heasman et al 1985). These results confirm and extend classic studies of small tissue fragments of the early gastrula transplanted to different ectopic positions (see Spemann 1938). The results also accord with Holtfreter's descriptions of the differentiation of explanted small pieces of the early gastrula. For example, the isolated dorsal IMZ eventually differentiates to notochord, somites, neural tube, and epidermis, although the normal fate of this region in situ is solely notochord (Holtfreter & Hamburger 1955). Thus, individual cells of the early gastrula can still follow a wide variety of developmental paths depending on their interactions with other cells. As mentioned before, animal hemisphere cells in isolation from early stages follow autonomously a path ending finally in differentiation as ciliated epidermis. As development proceeds these cells lose the ability to depart from this path.

2. When dorsal deep zone cells are explanted at the start of gastrulation, they can induce ectoderm of the same age to form precursors of trunk neural structures, and they differentiate as trunk mesoderm. However, the same cells explanted later in gastrulation induce head neural structures and themselves differentiate as head mesoderm (Kaneda & Hama 1979; Suzuki et al 1984). These cells seem to change fates and inductive properties during gastrulation as the result of transient interactions with cells of the IMZ and NIMZ.

3. When convergent extension is blocked by high doses of agents that interfere with the interactions of cells with each other and with components of the extracellular matrix (antifibronectin antibody, peptide analogs to fibronectin, Boucaut et al 1985; trypan blue, Gerhart et al 1984), the embryo fails to gastrulate or to differentiate dorsal mesodermal structures, such as notochord or muscle, despite apparently normal development up to the early gastrula stage. When gastrulation is blocked at successively later times in its course, the final body axis of the embryo is truncated at ever more anterior levels (Gerhart et al 1984). This seems paradoxical since the prospective anterior dorsal cells are the first to gastrulate. As shown by the results of Hama & Kaneda, deep zone cells and IMZ cells seem to acquire more dorsal and anterior fates in the course of gastrulation as they move and change their contacts with overlying cells. At the start of gastrulation a marginal zone cell may only have an identity with regard to its type of gastrulation activity. It may later gain an anteroposterior and mediolateral identity based on the outcome of gastrulation. Spreading migration by the deep cells and convergent extension by the IMZ do no less than create the anteroposterior and mediolateral dimensions of the body axis. Perhaps these axis-generating processes also serve to define cell type.

4. When a dorsal blastopore lip (i.e. dorsal IMZ and deep zone) is grafted into the marginal zone of an early gastrula, the embryo will often develop a partially doubled body axis. The extent of doubling depends on the original angle separating the host dorsal lip and the graft (Cooke 1972). When the lips are opposite, marginal zone cells of the host seem to go in equal numbers to the two convergence sites and form separate sets of axial dorsal structures. Ventral IMZ cells participate in forming the secondary notochord, whereas they would have formed only posterior somites and lateral plate mesoderm in the absence of the graft (Gimlich & Cooke 1983). When the two lips are originally separated by 90–150° in the marginal zone, the two final body axes unify at the posterior head or trunk level. When the two lips are still less separate, the final embryo has only a single axis with almost normal proportions except for minor anterior widening. Thus, convergent extension seems

to be a highly adaptive form-generating process. This behavior is contrary to that expected for populations of fully predetermined individual cells.

Although an early gastrula cell may not have a fixed commitment to a final cell fate, it of course has some level of commitment to one of the five region-specific gastrulation activities, its "dynamic determination" (Spemann 1938). Also, cells of three of the regions differ dorsoventrally in intensity and time of initiation, although these differences may be more quantitative than qualitative. As discussed before, a region is not determined for the final pattern of deformation, but for the deformation process. A final example is given here: Explants of the dorsal IMZ will always converge and extend in culture, but the direction is variable; it sometimes departs 90° from the expected animal-vegetal direction (Keller et al 1985). Thus, although IMZ cells are determined to engage in convergent extension, the actual direction and extent of morphogenesis are context dependent. They are not accurately encoded in the region, much less in the individual cells. Since the final position of a cell after gastrulation is subject to all these contingencies, it seems reasonable that ultimate cell fate is determined by the gastrulation processes that place the cell in its ultimate position. The early gastrula may be determined just enough to engage in the gastrulation processes that establish the next stage and its processes.

We can summarize the morphogenetic events before gastrulation in terms of the effect of each on gastrulation. In general, if any prior process fails, marginal zone cells will begin gastrulation later and less intensely than normal and will eventually differentiate as more posterior and ventral embryonic structures. Explicitly stated: (a) the more extensive the cortical cytoplasmic rotation, the more activated the vegetal cytoplasm will be in the first cell cycle; (b) the more activated the vegetal cytoplasm is, the more inductive the vegetal cells cleaved from it will be; (c) the more intense the induction, the earlier and more vigorously will nearby marginal zone cells undertake convergent extension, spreading migration, and apical contraction; (d) the more these marginal zone cells move and interact with each other, with new neighbors, and with the extracellular matrix during gastrulation, the more anterior and dorsal will be their neural plate–inducing activity and final mesodermal self-differentiation. Thus, the early embryo increases its regional differences, and each stage creates the context for the next.

ACKNOWLEDGMENTS

This research has been supported by research grants to JG (USPHS GM-19363) and to RK (NSF 81-10985). The authors thank M. Danilchik,

J.-P. Vincent, S. Scharf, R. Gimlich, J. Roberts, and B. Rowning for their suggestions on the manuscrupt and for their communication of unpublished results.

Literature Cited

Ancel, P., Vintemberger, P. 1948. Recherches sur le determinisme de la symetrie bilaterale dans l'oeuf de Amphibiens. *Bull. Biol. Fr. Belg.* 31 : 1–182 (Suppl.)

Baker, P. 1965. Fine structure and morphogenetic movements in the gastrula of the treefrog, *Hyla regilla. J. Cell Biol.* 24 : 95–116

Black, S., Gerhart, J. 1986. High-frequency twin embryos from centrifugation of *Xenopus laevis* eggs. *Dev. Biol.* 115 : In press

Boucaut, J. C., Darribere, T., Li, S. D., Boulekbache, H., Yamada, K. M., Theiry, J. P. 1985. Evidence for the role of fibronectin in amphibian gastrulation. *J. Embryol. Exp. Morphol.* 89 : 211–27 (Suppl.)

Brachet, J., Hubert, E., Lievens, A. 1972. The effects of α-amanitin and rifampicins on amphibian egg development. *Rev. Suisse Zool.* 79 : 47–63

Campanella, C., Andreucetti, P. 1977. Ultrastructural observations on cortical endoplasmic reticula and on residual cortical granules in the egg of *Xenopus laevis. Dev. Biol.* 56 : 1–10

Coggins, L. W. 1973. An ultrastructural and autoradiographic study of early oogenesis in the toad, *Xenopus laevis. J. Cell Sci.* 12 : 71–86

Cooke, J. 1972. Properties of the primary organization field in the embryo of *Xenopus laevis.* II. Positional information for axial organization in embryos with two head organizers. *J. Embryol. Exp. Morphol.* 28 : 27–46

Cooke, J. 1975. Local autonomy of gastrulation movements after dorsal lip removal in two anuran amphibians. *J. Embryol. Exp. Morphol.* 33 : 147–57

Cooke, J. 1979. Cell number in relation to primary pattern formation in the embryo of *Xenopus laevis.* II. Sequential cell recruitment and control of the cell cycle during mesoderm formation. *J. Embryol. Exp. Morphol.* 53 : 269–89

Cooke, J., Webber, J. A. 1985. Dynamics of the control of body pattern in *Xenopus laevis.* I. Timing and pattern in the development of single blastomeres (presumptive lateral halves) isolated at the 2-cell stage. *J. Embryol. Exp. Morphol.* 88 : 85–112

Danilchik, M., Gerhart, J. 1986. Differentiation of the animal-vegetal axis in oocytes of *Xenopus laevis. Dev. Biol.* In press

Dawid, I. B., Haynes, S. R., Jamrich, M., Jonas, E., Miyatani, S., et al. 1985. Gene expression in *Xenopus* embryogenesis. *J. Embryol. Exp. Morphol.* 89 : 113–24 (Suppl.)

Dictus, W. J. A. G., van Zoelen, E. J. J., Tetteroo, P. A. J., Tertoolen, L. G. J., DeLaat, S. W., Bluemink, J. G. 1984. Lateral mobility of plasma membrane lipids in *Xenopus* eggs; regional differences related to animal/vegetal polarity become extreme upon fertilization. *Dev. Biol.* 101 : 201–11

Elinson, R. 1980. The amphibian egg cortex in fertilization and early development. *Symp. Soc. Dev. Biol.* 38 : 217–34

Franke, W. W., Rathke, P. C., Sieb, E., Trendelenburg, M. F., Osborn, M., Weber, K. 1976. Distribution and mode of arrangement of microfilamentous structures and actin in the cortex of the amphibian oocyte. *Cytobiologie* 14 : 111–30

Franz, J. K., Gall, L., Williams, M., Picheral, B., Franke, W. 1983. Intermediate-size filaments in a germ cell: Expression of cytokeratins in oocytes and eggs of the frog *Xenopus. Proc. Natl. Acad. Sci.* 80 : 6254–58

Gerhart, J. 1980. Mechanisms regulating pattern formation in the amphibian egg and early embryo. In *Biological Regulation and Development,* ed. R. Goldberger, 2 : 133–316. New York : Plenum

Gerhart, J., Vincent, J.-P., Scharf, S. R., Black, S. D., Gimlich, R. L., Danilchik, M. 1984. Localization and induction in early development of *Xenopus. Philos. Trans. R. Soc. London Ser. B* 307 : 319–30

Gimlich, R. L. 1985. Cytoplasmic localization and chordamesoderm induction in the frog embryo. *J. Embryol. Exp. Morphol.* 89 : 89–111 (Suppl.)

Gimlich, R. L. 1986. Acquisition of developmental autonomy in the equatorial zone of the *Xenopus* embryo. *Dev. Biol.* 115 : 340–52

Gimlich, R. L., Cooke, J. 1983. Cell lineage and the induction of second nervous systems in amphibian development. *Nature* 306 : 471–73

Gimlich, R. L., Gerhart, J. 1984. Early cellular interactions promote embryonic axis formation in *Xenopus laevis. Dev. Biol.* 104 : 117–30

Grant, P., Wacaster, J. F. 1972. The amphibian grey crescent—a site of developmental information? *Dev. Biol.* 28: 454–71

Gurdon, J. B., Mohun, T. J., Brennan, S., Cascio, S. 1985. Actin genes in *Xenopus* and their developmental control. *J. Embryol. Exp. Morphol.* 89: 125–36 (Suppl.)

Harris, T. M. 1964. Pregastrular mechanisms in the morphogenesis of the salamander *Ambystoma maculatum. Dev. Biol.* 10: 247–68

Hausen, P., Wang, Y. H., Dreyer, C., Stick, R. 1985. Distribution of nuclear proteins during maturation of the *Xenopus* oocyte. *J. Embryol. Exp. Morphol.* 89: 17–34 (Suppl.)

Heasman, J., Snape, A., Smith, J. C., Wylie, C. C. 1985. Single cell analysis of commitment in early embryogenesis. *J. Embryol. Exp. Morphol.* 89: 297–316 (Suppl.)

Holtfreter, J. 1943. A study of the mechanics of gastrulation. Part I. *J. Exp. Zool.* 94: 261–318

Holtfreter, J. 1944. A study of the mechanics of gastrulation. Part II. *J. Exp. Zool.* 95: 171–212

Holtfreter, J., Hamburger, V. 1955. Embryogenesis, progressive differentiation in amphibians. In *Analysis of Development,* ed. B. H. Willier, P. Weiss, V. Hamburger, pp. 230–96. Philadelphia: Saunders

Jacobson, A., Odell, G., Oster, G. 1985. The cortical tractor model for epithelial folding: Application to the neural plate. In *Molecular Determinants of Animal Form,* ed. G. M. Edelman, *UCLA Symp. Mol. Cell. Biol.,* 31: 143–66. New York: Liss

Kageura, H., Yamana, K. 1984. Pattern regulation in defect embryos of *Xenopus laevis. Dev. Biol.* 101: 410–15

Kalt, M. 1971. The relationship between cleavage and blastocoel formation in *Xenopus laevis.* II. Electron microscopic observations. *J. Embryol. Exp. Morphol.* 26: 51

Kaneda, T., Hama, T. 1979. Studies on the formation and state of determination of the trunk organizer in the newt, *Cynops pyrrhogster. Wilhelm Roux Arch. Entwicklungsmech. Org.* 187: 25–34

Keller, R. E. 1975. Vital dye mapping of the gastrula and neurula of *Xenopus laevis.* I. Prospective areas and morphogenetic movements of the superficial layer. *Dev. Biol.* 42: 222–41

Keller, R. E. 1976. Vital dye mapping of the gastrula and neurula of *Xenopus laevis.* II. Prospective areas and morphogenetic movements in the deep region. *Dev. Biol.* 51: 118–37

Keller, R. E. 1978. Time-lapse cinemicrographic analysis of superficial cell behavior during and prior to gastrulation in *Xenopus laevis. J. Morphol.* 157: 223–48

Keller, R. E. 1980. The cellular basis of epiboly: An SEM study of deep cell rearrangement during gastrulation in *Xenopus laevis. J. Embryol. Exp. Morphol.* 60: 201–34

Keller, R. E. 1981. An experimental analysis of the role of bottle cells and the deep marginal zone in gastrulation of *Xenopus laevis. J. Exp. Zool.* 216: 81–101

Keller, R. E. 1986. The cellular basis of amphibian gastrulation. In *Developmental Biology: A Comprehensive Synthesis,* ed. L. Browder, 2: 241–327. New York: Plenum

Keller, R. E., Danilchik, M., Gimlich, R., Shih, J. 1985. The function and mechanism of convergent extension during gastrulation of *Xenopus laevis. J. Embryol. Exp. Morphol.* 89: 185–209 (Suppl.)

Keller, R. E., Schoenwolf, G. C. 1977. An SEM study of cellular morphology, contact, and arrangement, as related to gastrulation in *Xenopus laevis. Wilhelm Roux Arch. Entwicklungsmech. Org.* 182: 165–86

Kimmelman, D., Scherson, T., Kirschner, M. 1986. Precocious transcription in *Xenopus laevis* eggs blocked by protein synthesis inhibitors. In preparation

Kobayakawa, Y., Kubota, H. Y. 1981. Temporal pattern of cleavage and the onset of gastrulation in amphibian embryos developed from eggs with reduced cytoplasm. *J. Embryol. Exp. Morphol.* 62: 83–94

Kricg, P., Melton, D. 1985. Developmental regulation of a gastrula-specific gene injected into fertilized *Xenopus* eggs. *FMBO J.* 4. 3463–71

Kubota, H., Durston, A. J. 1978. Cinematographical study of cell migration in the opened gastrula of *Ambystoma mexicanum. J. Embryol. Exp. Morphol.* 44: 71–80

Lee, G., Hynes, R. O., Kirschner, M. 1984. Temporal and spatial regulation of fibronectin in early *Xenopus* development. *Cell* 36: 729–40

Malacinski, G. M., Brothers, A. J., Chung, H.-M. 1977. Destruction of components of the neural induction system of the amphibian egg with ultraviolet irradiation. *Dev. Biol.* 56: 24–39

Manes, M. E., Barbieri, F. D. 1977. On the possibility of sperm aster involvement in dorso-ventral polarization and pronuclear migration in the amphibian egg. *J. Embryol. Exp. Morphol.* 40: 187–97

Manes, M., Elinson, R. P. 1980. Ultraviolet light inhibits grey crescent formation in the frog egg. *Wilhelm Roux Arch. Dev. Biol.* 198 : 73–76

Merriam, R. W., Sauterer, R. A., Christensen, K. 1983. A subcortical, pigment-containing structure in *Xenopus* eggs with contractile properties. *Dev. Biol.* 95 : 439–46

Nakatsuji, N. 1975. Studies on the gastrulation of amphibian embryos : Cell movement during gastrulation in *Xenopus laevis* embryos. *Wilhelm Roux Arch. Dev. Biol.* 178 : 1–14

Nakatsuji, N. 1979. Effects of injected inhibitors of microfilament and microtubule function on the gastrulation movements in *Xenopus laevis*. *Dev. Biol.* 68 : 140–50

Nakatsuji, N. 1984. Cell locomotion and contact guidance in amphibian gastrulation. *Am. Zool.* 24 : 615–27

Nakatsuji, N., Hashimoto, K., Hayashi, M. 1985. Laminin fibrils in newt gastrulae, visualized by immunofluorescent staining. *Dev. Growth Differ.* 27 : 639–43

Nakatsuji, N., Johnson, K. 1984. Experimental manipulation of a contact guidance system in amphibian gastrulation by mechanical tension. *Nature* 307 : 453–55

Newport, J., Kirschner, M. 1982. A major developmental transition in early *Xenopus* embryos : I. Characterization and timing of cellular changes at the midblastula stage. *Cell* 30 : 675–86

Nieuwkoop, P. D. 1973. The "organisation center" of the amphibian embryo ; its origin, spatial organisation and morphogenetic action. *Adv. Morphog.* 10 : 2–39

Nieuwkoop, P. D. 1977. Origin and establishment of embryonic polar axes in amphibian development. *Curr. Top. Dev. Biol.* 11 : 115

Nieuwkoop, P. D. 1985. Inductive interactions in early amphibian development and their general nature. *J. Embryol. Exp. Morphol.* 89 : 333–47 (Suppl.)

Nieuwkoop, D. P., Florshutz, P. 1950. Quelques caracteres speciaux de la gastrulation et de la neurulation de l'oeuf de *Xenopus laevis*, Daud. et de quelques autres Anoures. Première partie. Etude descriptive. *Arch. Biol.* 61 : 113–50

Palacek, J., Habrova, V., Nedvidek, J., Romanovsky, A. 1985. Dynamics of tubilin structures in *Xenopus laevis* oogenesis. *J. Embryol. Exp. Morphol.* 87 : 75–86

Perry, M., Waddington, C. H. 1966. Ultrastructure of the blastoporal cells in the newt. *J. Embryol. Exp. Morphol.* 15 : 317–30

Radice, G. P., Neff, A. W., Malacinski, G.

M. 1981. Different classes of yolk platelets in *Xenopus laevis* eggs, separated by density equilibrium centrifugation. *Physiologist* 24 : 79–80

Rebagliati, M. R., Weeks, D. L., Harvey, R. P., Melton, D. A. 1985. Identification and cloning of localized maternal RNAs from *Xenopus* eggs. *Cell* 42 : 769–77

Roberts, J., Rowning, B., Vincent, J. P. 1986. The sperm is not required for dorso-ventral polarization of the *Xenopus* egg. In preparation

Scharf, S. 1986a. Ventralized embryos of *Xenopus laevis* from eggs treated with nocodazole. *Nature.* In press

Scharf, S. 1986b. Production of hyperdorsal embryos by treatment of *Xenopus* eggs with D_2O. *Dev. Biol.* In press

Scharf, S. R., Gerhart, J. 1980. Determination of the dorsal-ventral axis in eggs of *Xenopus laevis*. Rescue of UV-impaired eggs by oblique orientation. *Dev. Biol.* 79 : 181–98

Scharf, S., Gerhart, J. 1983. Axis determination in eggs of *Xenopus laevis*: A critical period before first cleavage, identified by the common effects of cold, pressure and ultraviolet irradiation. *Dev. Biol.* 99 : 75–87

Schechtman, A. J. 1942. The mechanism of amphibian gastrulation. I. Gastrulation-promoting interactions between various regions of an anuran egg (*Hyla regilla*). *Univ. Calif. Berkeley Publ. Zool.* 51 : 1–39

Spemann, H. 1938. *Embryonic Development and Induction.* New Haven : Yale Univ. Press ; Reprinted 1962, New York : Hafner

Stick, R., Hausen, P. 1985. Changes in the nuclear lamina composition during early development of *Xenopus laevis. Cell* 41 : 191–200

Suzuki, A. S., Mifune, Y., Kaneda, T. 1984. Germ layer interactions in pattern formation of amphibian mesoderm during primary embryonic induction. *Dev. Growth Differ* 26 : 81–94

Ubbels, G. A., Hara, K., Koster, C. H., Kirschner, M. W. 1983. Evidence for a functional role of the cytoskeleton in determination of the dorsoventral axis in *Xenopus laevis* eggs. *J. Embryol. Exp. Morphol.* 77 : 15–37

Vincent, J.-P., Oster, G., Gerhart, J. C. 1986. Kinematics of grey crescent formation in amphibian eggs : The displacement of subcortical cytoplasm relative to egg surface. *Dev. Biol.* 113 : 484–500

Waddington, C. H. 1940. *Organizers and Genes*, p. 109. Cambridge : Cambridge Univ. Press

Wallace, R. A. 1983. Interactions between somatic cells and the growing oocyte of *Xenopus laevis*. In *Current Problems in Germ Cell Differentiation*, ed. A. McLaren, C. C. Wylie, *Symp. Brit. Soc. Dev. Biol.*, pp. 285–306. Cambridge: Cambridge Univ. Press

Weyer, C. J., Nieuwkoop, P. S., Lindenmayer, A. 1978. A diffusion model for mesoderm induction in amphibian embryos. *Acta Biotheor.* 26: 164–80

Woodland, H. R., Flynn, J. M., Wyllie, A.

J. 1979. Utilization of stored mRNAs in *Xenopus* embryos and its replacement by newly synthesized transcripts: Histone H1 synthesis using interspecies hybrids. *Cell* 18: 165–71

Wylie, C. C., Brown, D., Godsave, S. F., Quarmby, J., Heasman, J. 1985. The cytoskeleton of *Xenopus* oocytes and its role in development. *J. Embryol. Exp. Morphol.* 89: 1–15 (Suppl.)

Ann. Rev. Cell Biol. 1986. 2: 231–53

T-CELL ACTIVATION

H. Robson MacDonald

Ludwig Institute for Cancer Research, Lausanne Branch,
1066 Epalinges, Switzerland

Markus Nabholz

Genetics Unit, Swiss Institute for Experimental Cancer Research,
1066 Epalinges, Switzerland

CONTENTS

INTRODUCTION[1]

The major clearly identified role of the immune system is the defense of
the organism against invading microorganisms and viruses. To carry out
this function it relies on a number of different highly mobile cell types that

[1] Literature references for this general introduction can be found in Nabholz & MacDonald 1983.

231

0743–4634/86/1115–0231$02.00

can reach and penetrate almost any part of the body and mount an immune response in the invaded organ. The cells that are responsible for the control of this system and its exquisite specificity are the lymphocytes. Every lymphocyte expresses on its surface an antigen receptor with a single type of specificity or "combining site." In physiological circumstances a lymphocyte will become "active," i.e. express its latent functional properties and undergo cell division, only when it encounters an antigenic determinant that it can "recognize." Its daughter cells will express antigen receptors with the same combining site.

Activated lymphocytes can carry out many different functions. To some extent this complexity is reflected in the heterogeneity among these cells: The function of one major class, the B lymphocytes (or B cells)—named for their site of differentiation in birds, the bursa of Fabricius—is the secretion of antibodies. Among T lymphocytes, which mature in the thymus, at least two subclasses can be distinguished. The so-called T helper cells (TH) can, when activated, secrete a number of different protein molecules with hormonelike properties. These lymphokines, which include interleukin-2 (IL-2), interferon-γ (γ-IFN), colony stimulating factor, and B-cell differentiation factors, are mediators by which TH control and coordinate the participation of other cells in an ongoing immune response. The other main, well-characterized subclass of T cells is the precursors (pCTL) of cytolytic T lymphocytes (CTL). When these pCTL are activated they can acquire the capacity to lyse target cells that carry the antigenic determinant recognized by the CTL. Most or all pCTL will also secrete γ-IFN in response to antigenic stimulation. While there are claims for the existence of other T lymphocyte subclasses, their existence as a separate subset is much less clearly established.

The hypothesis that lymphocytes respond to antigenic stimulation by proliferation and subsequent generation of clonal progeny with the same antigenic specificity (the clonal selection theory of Burnet) was based on in vivo experiments, but it led to attempts to demonstrate antigen-stimulated proliferation in culture. In time these resulted in the development of the mixed leukocyte culture (MLC) system, in which T lymphocytes of one individual are stimulated by coculture with leukocytes of another (antigenically different) one, and the discovery that T-cell proliferation could be induced with certain plant lectins such as concanavalin A (Con A) and phytohemagglutinin (PHA), so called polyclonal activators. Thus it became feasible, in principle, to analyze the minimal cellular requirements for lymphocyte activation. This proved to be a difficult task, and many of the controversies that originated in the early phases of this work are still not definitively resolved. The obstacles were of three kinds: (*a*) the extent of the heterogeneity of the cellular composition of the immune system was only gradually recognized; (*b*) this heterogeneity made it almost impossible to obtain really pure preparations of the different cell types; and (c) the cellular interactions that take place during an MLC- or lectin-induced T-cell response are indeed very complex. Another source of experimental difficulties was the ill-defined nature of the determinants recognized as antigen by T cells. For example, it is still not possible to stimulate pure T cells with antigenic molecules in solution, and for most experimental purposes T-cell antigen is an incompletely defined molecular entity on the surface of a living "stimulator" cell.

In the course of time it became increasingly clear that T-cell responses depended on, or were at least strongly influenced by, the presence (and presumably the specific functions) of other cell types and that different T-cell classes could influence each other. Several reports indicated that at least some of these interactions were mediated by soluble factors, but it was only after the discovery of a so-called T-cell growth factor, later renamed IL-2, that the essential role of such mediators was generally recognized.

Today it seems probable that the proliferation of all T cells depends on IL-2, which is itself a T-cell product, and it has recently been shown that this molecule can also stimulate the proliferation of B cells. The early observations that T cells can respond to IL-2 only after prior "activation" by a mitogenic lectin and that only such activated T cells are able to specifically bind IL-2 led to the formulation of a model of T-cell growth control. The basic features of this model are generally accepted today (although many of its molecular details have yet to be worked out): Resting, mature T cells, upon encountering antigen or a functional substitute, are activated to express IL-2 receptors (IL-2Rec) and to become IL-2 responsive. IL-2 then induces such cells to enter the cell cycle and undergo cell division.

Antigenic stimulation of resting T cells thus leads to three different kinds of responses: It induces the cells (a) to secrete soluble mediators, proteins with hormonelike properties; (b) to express IL-2Rec and become responsive to IL-2; and (c) it is required for the acquisition of cytolytic activity by pCTL.

The main purpose of this review is to discuss what we consider the relevant experimental evidence bearing on the precise requirements and the molecular basis of these induction phenomena.

CELLULAR AND MOLECULAR REQUIREMENTS FOR T-CELL ACTIVATION

The T-Cell Antigen Receptor Complex

To understand how an antigenic stimulus is delivered to a lymphocyte, it is necessary to identify and characterize the antigen-specific receptor on these cells. The main reason the T-cell receptor (TCR) resisted identification by classical biochemical techniques was probably related to the elusive nature of the antigenic determinants recognized by T cells. It is very difficult to demonstrate specific binding of soluble antigen molecules to T cells, although periodically there are claims to the contrary. In part this difficulty is due to the phenomenon usually referred to as associative recognition or major histocompatibility complex (MHC) restriction (Zinkernagel & Doherty 1974). T cells recognize antigens only when these are "associated" with cell surface molecules coded for by genes of the MHC. MHC restriction was for a long time defined purely by functional tests (e.g. the proliferative response of TH to an antigenic stimulus or the target specificity of CTL), and its molecular basis was the subject of a great deal of often highly esoteric discussion, which is still not resolved (for recent reviews see Norcross 1984; Schwartz 1985).

But the recent molecular characterization of the TCR should help solve this puzzle. TCR molecules were finally identified by two different approaches: the search for genes that undergo somatic rearrangements in T cells from a T-cell-specific cDNA library (Hedrick et al 1984a) and the selection of monoclonal antibodies (MAbs) that react specifically with individual T-cell clones (so-called clonotypic antibodies) (Allison et al 1982; Haskins et al 1983; Meuer et al 1983b; Kaye et al 1983; Samelson et al 1983). The work that followed has led to the discovery that TCR genes are homologous to the genes coding for antibody molecules (Yanagi et al 1984; Hedrick et al 1984b). There are three loci that are both rearranged and expressed specifically in T cells: The α and β genes code for the individual chains of a disulfide-linked, dimeric cell surface glycoprotein; no protein product of the third locus (γ locus) has so far been found. In T cells all three genes undergo rearrangements very similar to those of immunoglobulin genes in B cells (reviewed in Hood et al 1985).

There is no doubt that the α and β genes contribute to T-cell antigen specificity, and recent DNA transfection experiments (Dembic et al 1986) strongly suggest that the entire specificity of T cells is controlled by these two genes, i.e. that the α-β dimer accounts for both antigen specificity and MHC restriction. Despite this fact, it is also clear that the functional TCR complex includes additional polypeptide chains (Borst et al 1983; Kanellopoulos et al 1983; Meuer et al 1983a). Recent work from several laboratories has shown that in the plasma membrane of both human and mouse T cells, α-β heterodimers are physically linked with the peptides that together form the cell surface molecular complex known as T3 in man (Brenner et al 1985; Allison & Lanier 1985; Samelson et al 1985a). Antibodies against this antigenic complex have the same biological effects on T cells as anti-clonotypic reagents that react with the TCR (see below), and expression of the α-β heterodimer and T3 is interdependent (Meuer et al 1983b; Weiss & Stobo 1984). The respective roles of the different components of the TCR complex in the generation and transmission of an antigen-induced stimulus are not known. That the T3 molecule may play an important role in the transmission of the antigenic signal from the outside to the inside of the T cell is suggested by the fact that the α and β chains of the T-cell receptor have only a very short cytoplasmic tail, whereas the T3 subunits have much larger cytoplasmic domains (van den Elsen et al 1984).

T-Cell Activation by Antibodies and Lectins

Physiological activation or triggering of resting T cells is initiated via interaction of the TCR complex with antigen and MHC molecules on the surface of an antigen-presenting cell. This primary interaction has proved to be rather difficult to approach experimentally because (a) T cells with a given antigenic specificity are present only at a very low frequency in the nonimmune population; and (b) presentation of antigen is a complicated process that frequently involves intracellular processing and expression of antigenic determinants in association with MHC (reviewed in Unanue 1984).

This complexity has led to attempts to mimic T-cell triggering by exposing T cells to soluble substances that react with the TCR complex. Thus, MAbs directed against T3 are capable of inducing polyclonal proliferation of resting human T lymphocytes (van Wauwe et al 1980). Furthermore, certain MAbs directed against the α-β heterodimer of human leukemic T-cell lines react with a significant proportion (1–10%) of T cells in normal individuals and selectively induce proliferation of these cells (Acuto et al 1985; Bigler et al 1985; Boylston & Cosford 1985; Moretta ct al 1985a). Similarly, a MAb reactive with 20% of mouse T lymphocytes (and thereby defining a family of variable region genes of the β-chain of the TCR) stimulates the proliferation of T cells bearing this antigenic determinant (Crispe et al 1985).

Before the advent of MAbs, it was known that T-cell proliferation could be induced polyclonally by mitogenic lectins. In view of the different sugar specificities of these molecules and the wide variety of cell surface glycoproteins to which they could bind, we consider it beyond the scope of this review to attempt a detailed comparison of lectin- versus antigen- (or MAb-) induced T-cell activation. Of particular relevance, however, is the question of whether or not lectin binding to the TCR complex is sufficient to explain all of the observed mitogenic effects of lectins. Although no definitive answer to this question is currently available, it has been shown that PHA binds to the human TCR complex (Kanellopoulos et al 1985). More importantly, Weiss & Stobo (1984) have demonstrated that mutants of the human T-cell leukemia line JURKAT, which do not express the TCR complex, are not activated (as defined by IL-2 production) by exposure to a combination of PHA and phorbol myristate acetate (PMA). This defect does not reflect an intrinsic inability of the mutant to produce IL-2, since exposure to a combination of PMA and calcium ionophore results in normal IL-2 production. Moreover, reconstitution of TCR expression in these mutants via DNA transfection (Ohashi et al 1985) results in a restored capacity to respond to PHA plus PMA. Taken together, these data strongly suggest that the mitogenic effect of PHA (at least in this model system) is dependent upon its ability to bind directly to the TCR complex. But the participation of other glycoproteins in the PHA-mediated stimulatory effect cannot be formally excluded.

Whereas the ability of lectins or MAbs directed against the TCR complex to activate T cells can be readily explained on the basis of "mimicking" the physiological effects of antigen-MHC binding, it is also true that a number of MAbs directed against cell surface molecules not known to be associated with the TCR complex are capable of activating T cells. For example, rabbit antisera to mouse brain stimulate T-cell proliferation in the mouse (Norcross & Smith 1979; Larsson & Coutinho 1980; Jones & Janeway 1981). A number of lines of evidence suggest that the stimulatory determinant recognized by these polyclonal sera is in fact the cell surface glycoprotein Thy-1. Most convincingly, several investigators have recently raised rat MAbs to Thy-1 that stimulate resting T cells (Gunter et al 1984; MacDonald et al 1985; Pont et al 1985). Similarly, in the human several other cell surface glycoproteins have been identified as potential activators. Thus MAbs directed against the T11 (Meuer et al 1984) or T44 (Hara et al 1985; Moretta et al 1985b) molecules are able to induce polyclonal T-lymphocyte proliferation. The case of T11 is of particular interest because a combination of MAbs (recognizing independent epitopes) is required to induce proliferation of resting cells.

The ability of MAbs directed against three apparently unrelated, non-antigen-receptor-associated structures (Thy-1, T11, T44) to activate T lymphocytes raises questions about the physiological relevance of these activation pathways. It is important to emphasize that the physiological ligands recognized by these molecules (if they exist) have not been identified. Thus the possibility that they function in "alternate pathways" of T-cell activation that play a role in vivo remains open. Further comparison of the molecular consequences of activation of T cells by TCR-associated or other structures should help clarify this issue.

Other Cell Surface Structures (Accessory Molecules) Involved in Antigen Recognition and Activation

In addition to the TCR complex and other structures directly involved in activating T lymphocytes, there are a number of other cell surface molecules that appear to play an ancillary role in antigen recognition and subsequent activation. The best characterized of these molecules are T4 (L3T4 in mouse), T8 (Lyt-2/3 in mouse), and LFA-1.

T4 and T8 are of particular interest in that they divide mature T cells into two nonoverlapping subpopulations that recognize foreign antigens predominantly in the context of MHC class II or class I molecules, respectively (reviewed in Swain 1983; Reinherz et al 1983). Antibodies directed against T4 and T8 and their mouse homologues have been known for some time to inhibit T-cell effector functions as well as activation. The precise mechanism of action of these "accessory" molecules is unknown, but a widely held hypothesis (MacDonald et al 1982) is that they increase the avidity of the TCR for the antigen-MHC complex, perhaps by interacting directly with nonpolymorphic determinants on the MHC molecules themselves (Swain 1983). In any case, it is clear that the dependency of T-cell activation on these molecules is relative, since clones and hybridomas have been isolated that react with class I or class II antigens in the absence of any detectable requirement for T8 or T4 molecules (MacDonald et al 1981; Marrack et al 1983).

In contrast to T4 and T8, LFA-1 (Kurzinger et al 1981) is expressed on all T lymphocytes (as well as other hematopoietic cells) and appears to function primarily as a nonspecific adhesion promoting molecule (Mentzer et al 1985). Interestingly, recognition of hematopoietic cells by either T4$^+$ or T8$^+$ T cells appears to require participation of the LFA-1 molecule, whereas recognition of certain other cells does not (Golde et al 1985; Shimonkevitz et al 1985).

Minimal Requirements for T-Cell Activation

Most studies of T-cell activation (whether induced by MAbs or mitogenic lectins) have used heterogeneous populations of responding cells containing macrophages and other cell types in addition to T lymphocytes. There has been considerable controversy as to whether these contaminating cells (or soluble products released by them) might be involved in T cell activation. These controversies are still not completely resolved (Hünig et al 1983; Roosnek et al 1985a,b); however, recent progress in cell purification procedures, combined with the molecular cloning of several lymphokines, has allowed a reductionist approach to this question. Here we discuss our interpretation of the data bearing on the minimal cellular and molecular requirements for T-cell activation. In so doing, we try to distinguish between the different criteria used to measure activation (i.e. proliferation, IL-2 secretion, IL-2Rec expression, CTL activity). Furthermore, the possibility that the major T-cell subsets (TH, pCTL) exhibit distinct activation requirements (Gullberg et al 1983; Czitrom et al 1983; Larsson 1984) will be considered.

One of the most controversial issues surrounding T-cell activation is the role played by so-called accessory cells (i.e. macrophages, dendritic cells, and other Ia$^+$ and/or Fc receptor-positive cells). In human T-cell proliferation induced by antibodies that bind to the T-cell surface protein T3, an obligatory role for accessory cells was demonstrated by exploiting the observation that human macrophages were shown to have a polymorphism for binding of mouse IgG$_1$ via their Fc receptors (Tax et al 1983; 1984). Only T cells from those individuals whose macrophages were capable of binding appropriate IgG$_1$ anti-T3 antibodies could be stimulated to proliferate by these antibodies, and macrophages from positive individuals could restore the responsiveness of T cells from negative individuals. More recently, it has been shown that PMA, an activator of protein kinase C (Pk-C) (Kraft & Anderson 1983; Niedel et al 1983) can replace the accessory cell requirement in this system (Hara & Fu 1985). The implications of this latter observation are discussed in a subsequent section.

In apparent contrast to studies of human T cells activated with anti-T3 MAbs, it has been clearly established that highly purified mouse T cells can be induced to proliferate by lectins in the absence of macrophages or PMA. However, this proliferation is dependent upon the addition of exogenous IL-2 and is restricted to the Lyt-2$^+$ subset (Gullberg & Larsson 1983; Erard et al 1985a; Vohr & Hünig 1985). Proliferation of purified L3T4$^+$ cells, in contrast, requires (in addition to lectin) the presence of macrophages or PMA (Erard et al 1985c). Furthermore, the production of IL-2 by either subset of T cells is only detected when both lectin and PMA are present. Taken together with the human data, these findings are consistent with a model in which (a) IL-2Rec expression by the Lyt-2$^+$ (T8$^+$) subset of T cells is induced directly by triggering the TCR complex, whereas IL-2Rec expression by L3T4$^+$ (T4$^+$) cells requires macrophages in addition (or an artificial substitute such as PMA); (b) in all T cells IL-2 production requires both TCR triggering and the presence of macrophages (or PMA). According to such a model, any T-cell proliferation in the absence of exogenously added IL-2 should be macrophage- or PMA-dependent.

The experiments showing that the accessory cell function of macrophages in the anti-T3-induced T-cell response depends on the presence of Fc receptor on macrophages (Tax et al 1983; 1984) suggest that at least one function of macrophages is to present antigen (mimicked in the above experiments by operationally equivalent MAbs) in a way that allows extensive multipoint contacts with the responding T cells. Support for this idea comes from the demonstration in several systems that the requirement for macrophages can be overcome by coupling stimulatory MAbs to Sepharose beads (Meuer et al 1983c; Crispe et al 1985). But there is evidence that this "cross-linking" of antigen is not the sole function of antigen-presenting cells and that an additional requirement may be the production of interleukin-1 (IL-1), a macrophage-derived hormone-like protein (Mizel 1982). It has recently been shown that purified IL-1 is required for proliferation when highly purified human T lymphocytes are stimulated with Sepharose-coupled anti-T3 MAbs (Williams et al 1985; Palacios 1985; Scheurich et al 1985).

The role of IL-1 in T-cell activation has been further explored using cloned T-cell lines. In these studies, two distinct effects of IL-1 have been observed. The first is the ability of IL-1 to induce (or enhance) IL-2 production in the presence of other cofactors (such as lectins) that are not by themselves stimulatory. This effect has been clearly demonstrated with both murine (Smith et al 1980; Gillis & Mizel 1981) and human (Kasahara et al 1985; Manger et al 1985) T-cell leukemia lines. A second effect of IL-1 was first demonstrated by Kaye et al (1984) for a mouse T-cell clone derived from normal TH cells (D10G2). These authors showed that IL-1 was mainly acting to induce (or enhance) expression of IL-2Rec in D10G2 cells, whereas antibodies to the TCR mainly stimulated IL-2 production. Effects of IL-1 on the induction of IL-2Rec expression were also reported in other systems, including a rat × mouse T-cell hybrid (Erard et al 1984).

At present it is not known whether the above two effects of IL-1 during T-cell activation are mediated by the same or different induction pathways. Using an EL4 thymoma subline as a model system, Lowenthal et al (1986a) recently showed that high levels of both IL-2 secretion and IL-2Rec expression can be induced in an IL-1-dependent fashion. By varying the conditions utilized for stimulation, they were able to obtain IL-2 production in the absence of IL-2Rec expression, but the converse was never seen. Pending the outcome of similar experiments on highly purified populations of resting L3T4$^+$ cells, it may tentatively be concluded that a physiological role of IL-1 is to augment IL-2 production and/or IL2-Rec expression, depending upon the availability of other environmental cofactors.

It should be noted that all of the clearly defined effects of IL-1 to date involve cells of the TH (T4/L3T4) phenotype. That Lyt-2$^+$ (T8$^+$) cells may not respond to IL-1 is further suggested by the fact that they are able to express IL-2Rec in a macrophage-independent fashion (Malek et al 1985). It is thus tempting to speculate that among T cells responsiveness to IL-1 (and expression of receptors for IL-1) may be restricted to the T4 subset. With the recent development of a ligand binding assay for IL-1 (Dower et al 1985), direct testing of this hypothesis should be feasible.

Clearly the minimal activation requirements differ between the two major T cell subsets: Lyt-2$^+$ (T8$^+$) cells can be induced to respond to IL2 by triggering of the TCR complex in the absence of accessory cells, whereas activation of L3T4$^+$ (T4$^+$) cells is accessory cell dependent. The role of accessory cells in T-cell activation is not clear, but probably involves at least two operationally distinct functions: (a) "cross-linking" of antigen and (b) secretion of IL-1.

CONSEQUENCES OF T-CELL ACTIVATION

Expression of "Activation" Antigens

Upon stimulation by antigens or mitogenic lectins, T lymphocytes express a series of surface molecules that are not (or are only weakly) expressed on resting cells. In general, these "activation antigens" were defined initially by MAbs that reacted preferentially with activated T cells. Such antigens would be expected to fall into one of the following two categories: (a) molecules involved in differentiated function of effector T cells, or (b) molecules involved in the enhanced metabolism of activated (versus resting) lymphocytes.

To our knowledge, all activation antigens that have so far been defined in molecular terms belong to the second category. The IL-2Rec (Uchiyama et al 1981) and the transferrin receptor (Goding & Burns 1981) were originally defined as T-cell activation antigens. Moreover, the 4F2 antigen (Haynes et al 1981), which is one of the earliest cell surface markers expressed following activation of human T cells (Cotner et al 1983), now appears to be the Na$^+$-Ca^{2+} exchanger (Michalak et al 1986). It will be of interest to see whether other T-cell activation antigens will also turn out to be components of the metabolic machinery of proliferating cells. So far no activation antigens have been identified that are relevant to differentiated functions of T cells, but our current knowledge of the molecular basis of T-cell killing and T-cell help is insufficient to predict whether any unique cell surface components (other than the TCR complex) are required for these functions.

Mechanisms of Signal Transmission via the TCR Complex

Only little is known, as yet, about the pathway(s) by which the interaction between an antigen and its receptor elicits a response in T lymphocytes. It is clear, however, that the biological consequences of TCR stimulation occur very rapidly. For example, the lytic effect of CTL, which may be mediated by the triggering of the secretion of specific molecules (see Henkart 1985 for review), is irreversibly initiated within one minute of contact with the target cell (MacDonald 1975). Furthermore, in a CTL clone that requires antigenic stimulation and IL-2 to proliferate, a 30–60 min exposure to antigen is sufficient to induce IL-2 responsiveness (Lowenthal et al 1985b). In this model system the effects of a relatively short (5 hr) exposure to antigen are long lasting in that they allow the stimulated cells to proceed through 7–8 cell generations over a 5-day period. Ultimately, however, the number of IL-2Rec decreases (even in the presence of excess IL-2), and the cells stop proliferating unless they are restimulated with specific antigen or lectin (Lowenthal et al 1985b; see also Cantrell & Smith 1983; Depper et al 1984).

There are no published data that directly trace the fate of newly synthesized TCR complexes after they have reached the cell surface. In particular, it is not known whether these complexes are internalized and whether such internalization plays a role in antigenic stimulation. However, there is indirect evidence that this may be the case: A short treatment of T cells with anti-clonotypic or anti-T3 MAbs induces a rapid reduction in the number of TCR complexes on the cell surface (Meuer et al 1983b). This reduction persists for up to 24 hr. In other receptor systems a similar phenomenon is called down-regulation and has been shown to be due to a redistribution of receptors among the plasma membrane and internal membrane systems (Goldstein et al 1979; Gordon et al 1980). That the molecular basis for the modulation of TCR complexes may be similar is suggested by recent experiments concerning the effects of activators of Pk-C on T cells. Activation of Pk-C augments phosphorylation of two T3 polypeptides and induces modulation of the TCR (Cantrell et al 1985; Samelson et al 1985a). Phosphorylation of one T3 chain is also increased by antigenic stimulation, but it is not yet known whether this occurs on the same amino acid (Samelson et al 1985b). Treatment of fibroblasts with PMA has been shown to induce a Pk-C-dependent phosphorylation that correlates with an increased internalization of other surface receptors (Klausner et al 1984; Cochet et al 1984; see Hunter 1984). If TCR complexes are internalized in response to antigen, it will be interesting to consider the fate of the ligand, particularly as the antigen is itself (at least in some cases) believed to be an integral membrane protein.

There is very good evidence that antigenic stimulation induces a rise in the concentration of cytosolic free Ca^{2+} ($[Ca^{2+}]_i$) and that this constitutes part of the antigen-induced signal. Treatment of fresh T cells (Tsien et al 1982) or T-cell lines (Weiss et al 1984a) with mitogenic lectins or anti-T3 antibodies induces a rapid increase in $[Ca^{2+}]_i$, and several recent reports show that antigenic stimulation has the same effect (O'Flynn et al 1985; Shapiro et al 1985). Calcium ionophores can mimic certain effects of antigenic stimulation. Weiss and his collaborators (see 1986 for review) have analyzed the molecular requirements for the $[Ca^{2+}]_i$ signal using a human leukemia line with TH characteristics (JURKAT). These cells respond to stimulation with PHA by secretion of IL-2 and γ-IFN. The

Pk-C activator PMA has no effect on this secretion by itself but strongly increases the effect of PHA. MAbs against the TCR as well as anti-T3 antibodies in combination with PMA induced the cells to secrete IL-2. Mutants that no longer expressed TCR complexes on the cell surface did not respond to these stimuli and (as mentioned earlier) displayed a much-reduced responsiveness to PHA, but wild type and mutant cells responded equally well to stimulation with a combination of PMA and calcium ionophore. Measurements of $[Ca^{2+}]_i$ using Quin-2 showed that PHA and anti-TCR antibodies induced a rapid and transient increase in $[Ca^{2+}]_i$ in the wild type but not in the mutant cells. This increase was not dependent on the presence of external calcium and was accompanied by an increased synthesis of inositol triphosphate (IP_3) (Imboden & Stobo 1985). Treatment of permeabilized cells with IP_3 induced a similar rise in $[Ca^{2+}]_i$.

While these reports make a good case for the importance of a rapid mobilization of $[Ca^{2+}]_i$ from internal stores as one result of antigenic stimulation, this signal is almost certainly not the only one generated by the interaction of antigen with the TCR. Firstly, there are indications that treatment with anti-TCR antibodies also induces a longer lasting rise in $[Ca^{2+}]_i$ that is dependent on calcium influx (Oettgen et al 1985). Secondly, as discussed earlier, activation of TH cells including JURKAT by anti-TCR MAbs requires accessory cells that bind the antibodies via Fc receptors and produce IL-1. Calcium ionophores cannot eliminate either of these requirements, but PMA substitutes for both (Weiss et al 1984a ; Hara & Fu 1985). Induction of proliferation (Truneh et al 1985; Kaibuchi et al 1985) or IL-2 responsiveness (Erard et al 1985b) in TH also requires both a calcium ionophore and a Pk-C activator, whereas murine pCTL will respond to either type of mediator alone, with little synergy between the two (Erard et al 1985b). The experiments with JURKAT show that anti-TCR antibodies (and probably antigens) induce breakdown of phosphatidylinositol biphosphate. This should result in the production of the natural Pk-C activator diacylglycerol (Berridge & Irvine 1984; Nishizuka 1984) but apparently not in sufficient amounts to overcome the need for PMA.

In summary, the mechanism of signal transmission resulting from the interaction between antigen and the TCR complex is still poorly understood. Although mobilization of intercellular calcium is a clearly demonstrated consequence of this interaction, it has not been formally proven that calcium flux is an obligatory event in T-cell activation, nor has the precise cascade of events initiated by calcium mobilization been delineated. The relationship between Pk-C activation and T-cell triggering is likewise unclear. Whereas Pk-C activators such as PMA can clearly substitute for accessory cell function in certain systems of T-cell activation, there is as yet no direct evidence that (*a*) the TCR interaction (as opposed to some other accessory cell related event) triggers Pk-C activation or (*b*) that Pk-C activation is an obligatory event for T-cell stimulation.

The Mode of Control of Gene Expression During T-Cell Activation

Many of the lymphokine genes have been molecularly cloned, including the IL-2 receptor gene and a number of genes whose expression is known to be associated with activation in other cell types. This makes it possible to study at which level the expression of these genes is regulated during activation. The results published so far show that induction of secretion of IL-2 and γ-IFN and of the expression of IL-2 receptors is associated with the appearance of detectable mRNA one to two hours after exposure to an antigenic stimulus or its substitute (Efrat & Kaempfer 1984; Weiss et al 1984b; Wiskocil et al 1985; Leonard et al 1985b; Yamamoto et al 1985). Stimulation with mitogenic lectins also induces the rapid appearance of detectable levels of *c-myc* mRNA (Kelly et al 1983). Experiments in which the rate of transcription is measured (rather than the steady-state level of mRNA) indicate that stimulation of human T cells with PHA and PMA results in a rapid induction of transcription of the genes for γ-IFN, IL-2, and the IL-2Rec to levels 30- to 60-fold above the threshold of detection (Krönke et al 1985). This induction is not affected by the protein synthesis inhibitor cycloheximide. In contrast, the increase of transcription of the transferrin receptor gene, which occurs much later, requires protein synthesis. According to these authors, *c-myc* expression is also controlled at the level of transcription: It has, however, recently been reported that the dramatic increase in *c-myc* mRNA induced in serum-deprived density-arrested fibroblasts by stimulation with growth factors is not due to an increase in the rate of transcription but probably to regulation of mRNA half-life (Blanchard et al 1985).

The Mechanism of Action of IL-2

The best studied biological effect of IL-2 is its capacity to stimulate proliferation in IL-2-responsive lymphocytes (reviewed in Smith 1980). In IL-2-dependent cell lines, IL-2 deprivation results in an accumulation of the cells in the G1 phase of the cell cycle (Sekaly et al 1982). Reexposure to IL-2 induces the arrested cells to reenter the cell cycle after a lag time of 8–10 hr (Sekaly et al 1984). IL-2 has also been found to induce morphological changes (Sekaly et al 1982), to increase levels of IL-2Rec mRNA and IL-2Rec expression on the cell surface (Malek & Ashwell 1985), and to induce cytolytic activity in resting T cells previously stimulated in MLC (Lefrançois et al 1984). But in IL-2-dependent lymphocytes, IL-2 deprivation can also result in a dramatic decrease in amino acid incorporation into protein (M. Nabholz, unpublished results), and it is difficult to determine which of the other effects that IL-2 has on these cells are in fact secondary consequences of its influence on protein synthesis. However, there are certain hybrids between IL-2-dependent mouse CTL lines and a rat thymoma that will proliferate in the absence of IL-2, while responding to IL-2 (in combination with IL-1) by expressing IL-2Rec and cytolytic activity (Erard et al 1984). These cells will probably be useful tools for the elucidation of the molecular mechanisms of action of both interleukins.

Recent work from two laboratories has shown that the affinity of IL-2Rec expressed by homogeneous cell populations (including cloned lines) is heterogeneous (Robb et al 1984; Lowenthal et al 1985a,c). Mature, exponentially growing T cells can express about 10^5 IL-2Rec per cell. Of these approximately 10% are high affinity receptors (HAR) with a dissociation constant of about 5 pM and dissociation half-time of 60 min. The remaining 90% are low affinity receptors (LAR); their affinity is about 50 to 100 times lower, and their dissociation is correspondingly faster (Lowenthal et al 1985a). The growth-promoting effect of IL-2 seems to depend on binding to HAR only (Robb et al 1981). The structural basis for the difference between HAR and LAR has not been elucidated, but it is clear from studies with monoclonal antibodies and the characterization of IL-2Rec gene clones that both receptors contain an IL-2-binding surface glycoprotein coded for by the same gene (Leonard et al 1984; Nikaido et al 1984; Shimuzu et al 1985; Miller et al 1985; Ishida et al 1985; Leonard et al 1985a). So far, there is no evidence that this gene gives rise to functional translation products with different amino acid sequences.

IL-2Rec cDNA clones can be expressed in mouse fibroblasts. In these cells, they give rise to cell surface molecules detectable with monoclonal anti-IL-2Rec antibodies and capable of binding IL-2 but only with low affinity (Greene et al 1985). However, a recent report indicates that a subline of the mouse T lymphoma EL-4 transfected with a human IL-2Rec cDNA clone does express both high and low affinity receptors (Hatakayama et al 1985). These findings suggest that the formation of HAR for IL-2 depends on at least one additional gene that is expressed in lymphocytes but not in fibroblasts. It may be that the product of this gene associates with the IL-2Rec to form HAR and that it is involved in the transduction of the IL-2 signal. The sequence analyses of mouse and human IL-2Rec cDNA's indicate that the protein has a cytoplasmic domain of only 13 amino acids (containing no tyrosines but two conserved serines), which is probably too short to function, on its own, as a signal transmitter.

IL-2 is internalized by IL-2Rec-bearing cells and is eventually degraded

(Robb et al 1981). Recent studies indicate that such internalization is mediated by HAR only (M. Nahbolz & M. Combe, manuscript in preparation). IL-2-dependent cell lines that express about 10^4 HAR on their surface internalize about 500 IL-2 molecules/cell/min. It is not yet known what the fate of the internalized receptor molecules is or whether cell surface LAR can be converted into HAR. There is no evidence that IL-2 has an effect on the half-life of IL-2Rec or on their distribution between the surface of the cell and its interior.

It is unlikely that IL-2 exerts its effects via a rise in $[Ca^{2+}]_i$ or an activation of Pk-C. Direct measurements indicate that IL-2 has no short or long term effects on $[Ca^{2+}]_i$ in IL-2-responsive cell lines or normal T cells, but the published data do not rule out a very rapid transient change (Mills et al 1985). Attempts to find an increase in phosphatidylinositol phosphate metabolism during IL-2 stimulation of CTL lines have only given negative results (Kozumbo et al 1986). Finally, while the calcium ionophore ionomycin and/or PMA can induce IL-2Rec expression and IL-2 responsiveness, they cannot replace IL-2 as a signal for proliferation except in situations in which they contribute to the production of IL-2.

Acquisition of Cytolytic Activity

In discussing the factors involved in T-cell activation, we have used the induction of lymphokine secretion and IL-2 responsiveness as the parameters of T-cell activation. However, it should be noted that a number of investigators have addressed the question of whether or not different factors may be involved in the acquisition of other functions by T cells. The best-studied (although still highly controversial) system is the development of cytolytic activity by Lyt-2$^+$ cells stimulated by lectins or alloantigens. Using bulk cultures and lectin stimulation, several groups have reported a requirement for a soluble factor distinct from IL-2 in the induction of cytolytic activity (Raulet & Bevan 1982; Wagner et al 1982; Kanagawa 1983; Folk et al 1983). Other workers using purified Lyt-2$^+$ cells have been unable to confirm these observations: IL-2 was found to be sufficient for optimal generation of cytolytic activity in low density microcultures (Erard et al 1985a, 1985b; Vohr & Hünig 1985). The reasons for these apparent discrepancies remain obscure, but the possibility exists that the soluble factor acts by increasing IL-2 concentrations and that IL-2 was limiting in the original studies. Any differentiation factors that are produced in sufficient quantities by the responding Lyt-2$^+$ cells themselves would of course be missed in all of these studies. A case in point is the study of Simon et al (1986), which indicates that endogenous production of γ-IFN may be necessary for the induction of cytolytic activity in the Lyt-2$^+$ subset.

ANALOGIES WITH IMMATURE T CELLS

Although the major aim of this article is to review the processes involved in the activation of mature T lymphocytes, it is instructive to digress briefly to the situation of developmentally immature T cells. Mature T lymphocytes are derived by a poorly understood process from precursor or "stem" cells, which migrate from bone marrow to thymus (reviewed in Mathieson & Fowlkes 1984; Scollay et al 1984; Rothenberg & Lugo 1985). No precise phenotypic identification of such stem cells is available, but evidence from several sources suggests that they belong to a minor sub-population that does not express either TCR proteins or T4/T8. In the thymus, immature precursors acquire T-cell-specific recognition structures (the α and β chains of the heterodimeric TCR) and other cell surface molecules such as T4 and T8. Selection of T cells within the thymus is inferred from the high level of cell death in situ, but the underlying mechanism remains completely obscure. Cells putatively emerging from the selection process share phenotypic and functional properties with mature peripheral T cells.

In the context of this review, it is useful to address the question of the developmental regulation of those genes that control proliferation by activated T lymphocytes (i.e. IL-2 and IL-2Rec). In the so-called stem cell compartment of the thymus (i.e. $T4^-T8^-$ thymocytes), a significant proportion of cells ($\sim 50\%$) express the IL-2Rec in situ (Ceredig et al 1985; Raulet 1985; Lugo et al 1985). Although this finding suggests a physiological role for IL-2 even at this early stage of development, several observations cast doubt on this hypothesis. Binding studies using purified, radiolabeled IL-2 indicate that the affinity of IL-2Rec on isolated $T4^-T8^-$ thymocytes is 3- to 5-fold lower than on activated peripheral T cells (Ceredig et al 1985). Furthermore, this thymocyte subpopulation is unable to respond directly (via proliferation) to IL-2 (Palacios & von Boehmer 1986), nor is there any significant internalization of IL-2 by these cells (J. Lowenthal & H. R. MacDonald, unpublished data). Nevertheless, it cannot be excluded that the IL-2Rec on $T4^-T8^-$ cells are functional in situ, since the isolation procedure required to enrich for these $IL-2Rec^+$ cells in vitro may result in abolition of receptor function (for example, if IL-2Rec expression is maintained by cellular interactions or soluble factors unique to the thymic microenvironment).

An alternative possibility is that the IL-2:IL-2Rec interaction on $T4^-T8^-$ thymocytes triggers a differentiation event distinct from cell proliferation. Whatever the role of IL-2Rec on immature thymocytes may be, its expression cannot be induced via stimulation of the TCR since these cells do not yet express the TCR complex on the cell surface (see Rothenberg & Lugo 1985 for review). However, IL-2Rec expression may be the consequence of stimulation by another unidentified type of ligand. Obvious candidates for molecules that might function as receptors in such an interaction are antigens such as Thy-1, T11, and T44, since they are expressed on immature T cells and antibodies against them have been shown to trigger similar responses as anti-TCR MAbs in mature T cells (see above).

Attempts to demonstrate in situ IL-2 production by immature thymocytes have not been successful. However, at least some of these cells are already capable of secreting IL-2, since stimulation of thymic blast cells or isolated $T4^-T8^-$ thymocytes with PMA plus ionomycin results in significant IL-2 titers (Caplan & Rothenberg 1984; Rothenberg & Lugo 1985; R. Ceredig & H. R. MacDonald, unpublished data). IL-2 production by immature thymocytes under these conditions is quantitatively deficient in comparison with that of peripheral T cells. Furthermore, there is evidence that the $T4^-T8^-$ cells are heterogeneous with regard to their requirements for induction of IL-2 secretion. Whereas $IL-2Rec^-$ cells produce IL-2 in response to PMA plus ionophore, purified $IL-2Rec^+$ cells require IL-1 in addition (R. Howe & H. R. MacDonald, unpublished data). It is not clear which (if any) of these nonspecific activation regimens corresponds to a physiological stimulus for IL-2 production in vivo. However, the utilization of sensitive in situ hybridization techniques (using high specific activity RNA probes) may shed some light on this important question.

From the above discussion, it is apparent that IL-2 and IL-2Rec genes can be expressed at relatively early stages of T lymphocyte differentiation, even though the physiological relevance of this expression is unclear. More detailed information concerning the lineage relationships of developing precursors within the thymus is necessary to determine whether the inducibility of expression of these genes is acquired in an ordered sequence.

ACKNOWLEDGMENT

We gratefully acknowledge Paskale Brunet for excellent (and patient) assistance in preparing the manuscript.

Literature Cited

Acuto, O., Camper, T. J., Royer, H. D., Hussey, R. E., Poole, C. B., Reinherz, E. L. 1985. Molecular analysis of T cell receptor (Ti) variable region (V) gene expression. Evidence that a single Tiβ V gene family can be used in formation of V domains on phenotypically and functionally diverse T

cell populations. *J. Exp. Med.* 161: 1326

Allison, J., McIntyre, B., Bloch, D. 1982. Tumor-specific antigen of murine T-lymphoma defined with monoclonal antibody. *J. Immunol.* 129: 2293

Allison, J. P., Lanier, L. L. 1985. Identification of antigen receptor-associated structures on murine T cells. *Nature* 314: 107

Berridge, M. J., Irvine, R. F. 1984. Inositol trisphosphate, a novel second messenger in cellular signal transduction. *Nature* 312: 315

Bigler, R. D., Posnett, D. N., Chiorazzi, N. 1985. Stimulation of a subset of normal resting T lymphocytes by a monoclonal antibody to a crossreactive determinant of the human T cell antigen receptor. *J. Exp. Med.* 161: 1450

Blanchard, J.-M., Piechaczyk, M., Dani, C., Chambard, J.-C., Franchi, A., et al. 1985. c-myc gene is transcribed at high rate in G_0-arrested fibroblasts and is post-transcriptionally regulated in response to growth factors. *Nature* 317: 443–45

Borst, J., Alexander, S., Elder, J., Terhorst, C. 1983. The T3 complex on human T lymphocytes involves four structurally distinct glycoproteins. *J. Biol. Chem.* 258: 5135

Boylston, A. W., Cosford, P. 1985. Growth of normal human T lymphocytes induced by monoclonal antibody to the T cell antigen receptor. *Eur. J. Immunol.* 15: 738–42

Brenner, M. B., Trowbridge, I. S., Strominger, J. L. 1985. Cross-linking of human T cell receptor proteins: Association between the T cell idiotype β subunit and the T3 glycoprotein heavy subunit. *Cell* 40: 183

Cantrell, D. A., Smith, K. A. 1983. Transient expression of interleukin 2 receptors: Consequences for T cell growth. *J. Exp. Med.* 158: 1895

Cantrell, D. A., Davies, A. A., Crumpton, M. J. 1985. Activators of protein kinase C down-regulate and phosphorylate the T3/T-cell antigen receptor complex of human T lymphocytes. *Proc. Natl. Acad. Sci. USA* 82: 8158–62

Caplan, B., Rothenberg, E. 1984. High-level secretion of interleukin-2 by a subset of proliferating thymic lymphoblasts. *J. Immunol.* 2: 38–45

Ceredig, R., Lowenthal, J., Nabholz, M., MacDonald, H. R. 1985. Expression of interleukin-2 receptors as a differentiation marker on intrathymic stem cells. *Nature* 314: 98–100

Cochet, C., Gill, G. N., Meisenhelder, J., Cooper, J. A., Hunter, T. 1984. C-kinase phosphorylates the epidermal growth factor receptor and reduces its epidermal growth factor-stimulated tyrosine protein kinase activity. *J. Biol. Chem.* 259: 2553–58

Cotner, T., Williams, J., Christenson, L., Shapiro, H. M., Strom, T. B., Strominger, J. L. 1983. Simultaneous flow cytometric analysis of human T cell activation antigen expression and DNA content. *J. Exp. Med.* 157: 461

Crispe, I. N., Bevan, M. J., Staerz, U. D. 1985. Selective activation of Lyt-2$^+$ precursor cells by ligation of the antigen receptor. *Nature* 317: 627

Czitrom, A. A., Sunshine, G. H., Reme, T., Ceredig, R., Glasebrook, A. L., et al. 1983. Stimulator cell requirements for allospecific T cell subsets: Specialized accessory cells are required to activate helper but not cytolytic T lymphocyte precursors. *J. Immunol.* 130: 546

Dembic, Z., Haas, W., Weiss, S., McCubrey, J., Kiefer, H., et al. 1986. Transfer of specificity by murine α and β T-cell receptor genes. *Nature* 320: 232

Depper, J. M., Leonard, W. J., Krönke, M., Noguchi, P. D., Cunningham, R. E., et al. 1984. Regulation of interleukin 2 receptor expression: Effects of phorbol diester, phospholipase C and reexposure to lectin or antigen. *J. Immunol.* 133: 3054

Dower, S. K., Kronheim, S. R., March, C. J., Conlon, P. J., Hopp, T. P., et al. 1985. Detection and characterization of high affinity plasma membrane receptors for human interleukin 1. *J. Exp. Med.* 162: 501

Efrat, S., Kaempfer, R. 1984. Control of biologically active interleukin 2 messenger RNA formation in induced human lymphocytes. *Proc. Natl. Acad. Sci. USA* 81: 2601–5

Erard, F., Corthésy, P., Smith, K. A., Fiers, W., Conzelmann, A., Nabholz, M. 1984. Characterization of soluble factors that induce the cytolytic activity and the expression of T cell growth factor receptors of a T cell hybrid. *J. Exp. Med.* 160: 584

Erard, F., Corthésy, P., Nabholz, M., Lowenthal, J. W., Zaech, P., et al. 1985a. Interleukin 2 is both necessary and sufficient for the growth and differentiation of lectin-stimulated cytolytic T lymphocyte precursors. *J. Immunol.* 134: 1644

Erard, F., Nabholz, M., Dupuy-d'Angeac, A., MacDonald, H. R. 1985b. Differential requirements for the induction of interleukin 2 responsiveness in L3T4$^+$ and Lyt-2$^+$ T cell subsets. *J. Exp. Med.* 162: 1738

Erard, F., Nabholz, M., MacDonald, H. R. 1985c. Antigen stimulation of cytolytic T lymphocyte precursors: Minimal requirements for growth and acquisition of cytolytic activity. *Eur. J. Immunol.* 15: 798

Folk, W., Männel, D. N., Dröge, W. 1983. Activation of CTL requires at least two spleen cell-derived factors besides interleukin 2. *J. Immunol.* 130: 2214

Gillis, S., Mizel, S. B. 1981. T-cell lymphoma

model for the analysis of interleukin 1-mediated T-cell activation. *Proc. Natl. Acad. Sci. USA* 78: 1133

Goding, J. W., Burns, G. F. 1981. Monoclonal antibody OKT-9 recognizes the receptor for transferrin on human acute lymphocytic leukemia cells. *J. Immunol.* 127: 1256

Golde, W., Kappler, J., Greenstein, J., Malissen, B., Hood, L., Marrack, P. 1985. The MHC-restricted antigen receptor on T-cells. VIII. The role of the LFA-1 molecule. *J. Exp. Med.* 161: 635

Goldstein, J. L., Anderson, R. G. W., Brown, M. S. 1979. Coated pits, coated vesicles and receptor-mediated endocytosis. *Nature* 279: 679

Gordon, P., Carpentier, J.-L., Freychet, P., Orci, L. 1980. Internalization of polypeptide hormones. Mechanism, intracellular location and significance. *Diabetologica* 18: 263

Greene, W. C., Robb, R. J., Svetlik, P. B., Rusk, C. M., Depper, J. M., Leonard, W. 1985. Stable expression of cDNA encoding the human interleukin 2 receptor in eukaryotic cells. *J. Exp. Med.* 162: 363–68

Gullberg, M., Larsson, E. L. 1983. Con A-induced TCGF reactivity is selectively acquired by Lyt-2 positive T cell precursors. *J. Immunol.* 131: 19

Gullberg, M., Pobor, G., Bandeira, A., Larsson, E. L., Coutinho, A. 1983. Differential requirements for activation and growth of unprimed cytotoxic and helper T lymphocytes. *Eur. J. Immunol.* 13: 719

Gunter, K., Malek, T. R., Shevach, E. M. 1984. T cell activating properties of an anti-Thy-1 monoclonal antibody. Possible analogy to OKT3/LEU-4. *J. Exp. Med.* 159: 716

Hara, T., Fu, S. M. 1985. Human T cell activation. I. Monocyte-independent activation and proliferation induced by anti-T3 monoclonal antibodies in the presence of tumor promotor 12-O-tetradecanoyl phorbol-13-acetate. *J. Exp. Med.* 161: 641

Hara, T., Fu, S. M., Hansen, J. A. 1985. Human T cell activation. II. A new activation pathway used by a major T cell population via a disulphide-bonded dimer of a 44 kilodalton polypeptide (9.3 antigen). *J. Exp. Med.* 161: 1513

Haskins, K., Kubo, R., White, J., Pigeon, M., Kappler, J., Marrack, P. 1983. The major histocompatibility complex-restricted antigen receptor on T cells. *J. Exp. Med.* 157: 1149

Hatakayama, M., Minamoto, S., Uchiyama, T., Hardy, R. R., Yamada, G., Taniguchi,

T. 1985. Reconstitution of functional receptor for human interleukin-2 in mouse cells. *Nature* 318: 467–70

Haynes, B. F., Hemler, M. E., Mann, D. L., Eisenbarth, G. S., Shelhamer, J., et al. 1981. Characterization of a monoclonal antibody (4F2) that binds to human monocytes and to a subset of activated lymphocytes. *J. Immunol.* 126: 1409

Hedrick, S., Cohen, D., Nielsen, E., Davis, M. 1984a. Isolation of cDNA clones encoding T cell-specific membrane-associated proteins. *Nature* 308: 149

Hedrick, S., Nielsen, E., Kavaler, J., Cohen, D., Davis, M. 1984b. Sequence relationships between putative T-cell receptor polypeptides and immunoglobulins. *Nature* 308: 153

Henkart, P. A. 1985. Mechanism of lymphocyte-mediated cytotoxicity. *Ann. Rev. Immunol.* 3: 31–58

Hood, L., Kronenberg, M., Hunkapiller, T. 1985. T cell antigen receptors and the immunoglobulin supergene family. *Cell* 40: 225

Hünig, T., Loos, M., Schimpl, A. 1983. The role of accessory cells in polyclonal T cell activation. I. Both induction of interleukin 2 production and of interleukin 2 responsiveness by concanavalin A are accessory cell dependent. *Eur. J. Immunol.* 13: 1

Hunter, T. 1984. The epidermal growth factor receptor gene and its product. *Nature* 311: 414–16

Imboden, J., Stobo, J. D. 1985. Transmembrane signalling by the T cell antigen receptor. Perturbation of the T3-antigen receptor complex generates inositol phosphates and releases calcium ions from intracellular stores. *J. Exp. Med.* 161: 446–56

Ishida, N., Kanamori, H., Noma, T., Nikaido, T., Save, H., et al. 1985. Molecular cloning and structure of the human interleukin 2 receptor gene. *Nucl. Acids Res.* 13: 7579–89

Jones, B., Janeway, C. A. 1981. Functional activities of antibodies against brain-associated T cell antigen. I. Induction of T cell proliferation. *Eur. J. Immunol.* 11: 584

Kaibuchi, K., Takai, Y., Nishizuka, Y. 1985. Protein kinase C and calcium ion in mitogenic response of macrophage-depleted human peripheral lymphocytes. *J. Biol. Chem.* 260: 1366–69

Kanagawa, O. 1983. Three different signals are required for the induction of cytolytic T lymphocytes from resting precursors. *J. Immunol.* 131: 606

Kanellopoulos, J. M., Wigglesworth, N. M., Owen, M. J., Crumpton, M. J. 1983. Biosynthesis and molecular nature of the T3 antigen of human T lymphocytes. *EMBO J.* 2: 1807

Kanellopoulos, J. M., De Petris, S., Leco, G., Crumpton, M. 1985. The mitogenic lectin from *Phaseolus vulgaris* does not recognize the T3 antigen of human T lymphocytes. *Eur. J. Immunol.* 15: 479

Kasahara, T., Mukaida, N., Hatake, K., Motoyoshi, K., Kawai, T., Shiori-Nadano, K. 1985. Interleukin 1 (IL-1)-dependent lymphokine production by human leukemic T cell line HSB.2 subclones. *J. Immunol.* 134: 1682

Kaye, J., Porcelli, S., Tite, J., Jones, B., Janeway, C. Jr. 1983. Both a monoclonal antibody and antisera specific for determinants unique to individual cloned helper T cell lines can substitute for antigen and antigen-presenting cells in the activation of T cells. *J. Exp. Med.* 158: 836

Kaye, J., Gillis, S., Mizel, S. B., Shevach, E. M., Malek, T. R., et al. 1984. Growth of a cloned helper T cell line induced by a monoclonal antibody specific for the antigen receptor: Interleukin 1 is required for the expression of receptors for interleukin 2. *J. Immunol.* 133: 1339

Kelly, K., Cochran, B. H., Stiles, C. D., Leder, P. 1983. Cell-specific regulation of the *c-myc* gene by lymphocyte mitogens and platelet-derived growth factor. *Cell* 35: 603–10

Klausner, R. D., Harford, J., van Renswoude, J. 1984. Rapid internalization of the transferrin receptor in K562 cells is triggered by ligand binding or treatment with a phorbol ester. *Proc. Natl. Acad. Sci. USA* 81: 3005–9

Kozumbo, W. J., Harris, D. T., Gromkowski, S., Cerottini, J.-C., Cerutti, P. A. 1986. Molecular mechanisms involved in T-cell activation: II. The phosphatidylinositol signal transducing mechanism mediates antigen-induced lymphokine production but not interleukin-2-induced proliferation in cloned murine cytotoxic T lymphocytes. *J. Immunol.* In press

Kraft, A. S., Anderson, W. B. 1983. Phorbol esters increase the amount of Ca^{2+}, phospholipid-dependent protein kinase associated with plasma membrane. *Nature* 301: 621

Krönke, M., Leonard, W. J., Depper, J. M., Greene, W. C. 1985. Sequential expression of genes involved in human T lymphocyte growth and differentiation. *J. Exp. Med.* 161: 1593–98

Kurzinger, K., Reynolds, T., Germain, R., Davignon, D., Martz, E., Springer, T. 1981. A novel lymphocyte function-associated antigen (LFA-1): Cellular distribution, quantitative expression and structure. *J. Immunol.* 127: 596

Larsson, E. L. 1984. Activation and growth requirements for cytotoxic and noncytotoxic T lymphocytes. *Cell. Immunol.* 89: 223

Larsson, E. L., Coutinho, A. 1980. Mechanism of T cell activation. I. A screening of "step one" ligands. *Eur. J. Immunol.* 10: 93

Lefrançois, L., Klein, J. R., Paetkau, V., Bevan, M. J. 1984. Antigen-independent activation of memory cytotoxic T cells by interleukin 2. *J. Immunol.* 132: 1845

Leonard, J. W., Depper, J. M., Crabtree, G. R., Rudikoff, S., Pumphrey, J., et al. 1984. Molecular cloning and expression of cDNAs for the human interleukin-2 receptor. *Nature* 311: 626–31

Leonard, W. J., Depper, J. M., Kanehisa, M., Krönke, M., Peffer, N. J., et al. 1985a. Structure of the human interleukin-2 receptor gene. *Science* 230: 633–39

Leonard, W. J., Krönke, M., Peffer, N. J., Depper, J. M., Greene, W. C. 1985b. Interleukin 2 receptor gene expression in normal human T lymphocytes. *Proc. Natl. Acad. Sci. USA* 82: 6281–85

Lowenthal, J. W., Corthésy, P., Tougne, C., Lees, R. K., MacDonald, H. R., Nabholz, M. 1985a. High and low affinity IL2 receptors: Analysis by IL2 dissociation rate and reactivity with monoclonal anti-receptor antibody PC61. *J. Immunol.* 135: 3988–94

Lowenthal, J. W., Tougne, C., MacDonald, H. R., Smith, K. A., Nabholz, M. 1985b. Antigenic stimulation regulates the expression of IL-2 receptors in a cytolytic T lymphocyte clone. *J. Immunol.* 134: 931–39

Lowenthal, J. W., Zubler, R. H., Nabholz, M., MacDonald, H. R. 1985c. Similarities between interleukin 2 receptor number and affinity on activated B and T lymphocytes. *Nature* 315: 669

Lowenthal, J. W., Cerottini, J.-C., MacDonald, H. R. 1986. Interleukin 1-dependent induction of both interleukin 2 secretion and interleukin 2 receptor expression by thymoma cells. *J. Immunol.* In press

Lugo, J. P., Krishman, S. N., Sailor, R. D., Koen, P., Malek, T., Rothenberg, E. 1985. Proliferation of thymic stem cells with and without receptors for interleukin 2: Implications for intrathymic antigen recognition. *J. Exp. Med.* 161: 1048

MacDonald, H. R. 1975. Early detection of potentially lethal events in T cell-mediated cytolysis. *Eur. J. Immunol.* 5: 251

MacDonald, H. R., Thiernesse, N., Cerottini, J.-C. 1981. Inhibition of T cell-mediated cytolysis by monoclonal antibodies directed against Lyt-2: Heterogeneity of inhibition at the clonal level. *J. Immunol.* 126: 1671

MacDonald, H. R., Glasebrook, A. L., Bron, C., Kelso, A., Cerottini, J.-C. 1982. Clonal heterogeneity in the functional requirement for Lyt-2/3 molecules on cytolytic T lymphocytes (CTL): Possible implications for the affinity of CTL antigen receptors. *Immunol. Rev.* 68: 89

MacDonald, H. R., Bron, C., Rousseaux, M., Horvath, C., Cerottini, J.-C. 1985. Production and characterization of monoclonal anti-Thy-1 antibodies that stimulate lymphokine production by cytolytic T cell clones. *Eur. J. Immunol.* 15: 495

Malek, T. R., Ashwell, J. D. 1985. Interleukin 2 upregulates expression of its receptor on a T cell clone. *J. Exp. Med.* 161: 1575–80

Malek, T. R., Schmidt, J. A., Shevach, E. M. 1985. The murine IL 2 receptor. III. Cellular requirements for the induction of IL 2 receptor expression on T cell subpopulations. *J. Immunol.* 134: 2405

Manger, B., Weiss, A., Weyand, C., Goronzy, J., Stobo, J. D. 1985. T cell activation: Differences in the signals required for IL-2 production by nonactivated and activated T cells. *J. Immunol.* 135: 3669 73

Marrack, P., Endres, R., Shimonkevitz, R., Zlotnik, A., Dialynas, D., et al. 1983. The MHC-restricted antigen receptor on T cells. II. Role of the L3T4 product. *J. Exp. Med.* 158: 1077

Mathieson, B. J., Fowlkes, B. J. 1984. Cell surface antigen expression on thymocytes: Development and phenotypic differentiation of intrathymic subsets. *Immunol. Rev.* 82: 141–73

Mentzer, S., Gromkowski, S., Krensky, A., Burakoff, S., Martz, E. 1985. LFA-1 membrane molecule in the regulation of homotypic adhesions of human B lymphocytes. *J. Immunol.* 135: 9

Meuer, S. C., Acuto, O., Hussey, R. E., Hodgdon, J. C., Fitzgerald, K. A., et al. 1983a. Evidence for the T3-associated 90 KD heterodimer as the T cell antigen receptor. *Nature* 303: 808

Meuer, S. C., Fitzgerald, K., Hussey, R., Hodgdon, J., Schlossman, S., Reinherz, E. 1983b. Clonotypic structures involved in antigen-specific human T cell function. *J. Exp. Med.* 157: 705

Meuer, S. C., Hodgdon, J. C., Hussey, R. E., Protentis, J. P., Schlossman, S. F., Reinherz, E. L. 1983c. Antigen-like effects of monoclonal antibodies directed at receptors on human T cell clones. *J. Exp. Med.* 158: 988

Meuer, S. C., Hussey, R. E., Fabbi, M., Fox, D., Acuto, O., et al. 1984. An alternative pathway of T cell activation: A functional role for the 50 KD T11 sheep erythrocyte receptor protein. *Cell* 36: 897

Michalak, M., Quackenbush, E. J., Letarte, M. 1986. Inhibition of Na^+/Ca^{2+} exchanger activity in cardiac and skeletal muscle sarcolemmal vesicles by monoclonal antibody 44D7. *J. Biol. Chem.* 261: 92–95

Miller, J., Malek, T. R., Leonard, W. J., Greene, W. C., Shevach, E. M., Germain, R. N. 1985. Nucleotide sequence and expression of a mouse interleukin 2 receptor cDNA. 134: 4212–17

Mills, G. B., Cheung, R. K., Grinstein, S., Gelfand, E. W. 1985. Interleukin 2-induced lymphocyte proliferation is independent of increases in cytosolic-free calcium concentrations. *J. Immunol.* 134: 2431–35

Mizel, S. B. 1982. Interleukin 1 and T cell activation. *Immunol. Rev.* 63: 51

Moretta, A., Pantaleo, G., Lopez-Botet, M., Mingari, M. C., Moretta, L. 1985a. Anticlonotypic monoclonal antibodies induce proliferation of clonotype-positive T cells in peripheral blood human T lymphocytes: Evidence for a phenotypic (T4/T8) heterogeneity of the clonotype-positive proliferating cells. *J. Exp. Med.* 162: 1393–98

Moretta, A., Pantaleo, G., Lopez-Botet, M., Moretta, L. 1985b. Involvement of T44 molecules in an antigen-dependent pathway of T cell activation: Analysis of the correlation to the T cell antigen-receptor complex. *J. Exp. Med.* 162: 823–38

Nabholz, M., MacDonald, H. R. 1983. Cytolytic T lymphocytes. *Ann. Rev. Immunol.* 1: 273

Niedel, J. E., Kuhn, L. J., Vandenbark, G. R. 1983. Phorbol diester receptor copurifies with protein kinase C. *Proc. Natl. Acad. Sci. USA* 80: 36

Nikaido, T., Shimizu, A., Ishida, N., Sabe, H., Teshigwara, K., et al. 1984. Molecular cloning of cDNA encoding human interleukin-2 receptor. *Nature* 311: 631–34

Nishizuka, Y. 1984. Turnover of inositol phospholipids and signal transduction. *Science* 225: 1365

Norcross, M. A., Smith, R. T. 1979. Regulation of T cell mitogen activity of anti-lymphocyte serum by a B-helper cell. *J. Immunol.* 122: 1620

Norcross, M. A. 1984. A synaptic basis for T-lymphocyte activation. *Ann. Immunol. Inst. Pasteur* 1350: 113–33

Oettgen, J., Terhorst, C., Cantley, L., Rosoff, P. 1985. Stimulation of the T3-T cell receptor complex induces a membrane-potential-sensitive calcium influx. *Cell* 40 : 583

O'Flynn, K., Zanders, E., Lamb, J., Beverley, P., Wallace, D., et al. 1985. Investigation of early T cell activation: Analysis of the effect of specific antigen, interleukin 2 and monoclonal antibodies on intracellular free calcium concentration. *Eur. J. Immunol.* 15 : 7

Ohashi, P. S., Mak, T. W., Van den Elsen, P., Yanagi, Y., Yoshikai, Y., et al. 1985. Reconstitution of an active surface T3/T-cell antigen receptor by DNA transfer. *Nature* 316 : 606

Palacios, R. 1985. Mechanisms by which accessory cells contribute in growth of resting T lymphocytes initiated by OKT3 antibody. *Eur. J. Immunol.* 15 : 645–51

Palacios, R., von Boehmer, H. 1986. Requirements for growth of immature thymocytes from fetal and adult mice *in vitro. Eur. J. Immunol.* 16 : 12–19

Pont, S., Regnier-Vigouroux, A., Naquet, P., Blanc, D., Pierres, A., et al. 1985. Analysis of the Thy-1 pathway of T cell hybridoma activation using 17 rat monoclonal antibodies reactive with distinct Thy-1 epitopes. *Eur. J. Immunol.* 15 : 1222–28

Raulet, D. H., Bevan, M. J. 1982. A differentiation factor required for the expression of cytotoxic T-cell function. *Nature* 196 : 754

Raulet, D. H. 1985. Expression and function of interleukin 2 receptors on immature thymocytes. *Nature* 314 : 101

Reinherz, E. L., Meuer, S. C., Schlossman, S. F. 1983. The delineation of antigen receptors on human T lymphocytes. *Immunol. Today* 4 : 5

Robb, R. J., Munck, A., Smith, K. A. 1981. T cell growth factor receptors. *J. Exp. Med.* 154 : 1455–74

Robb, R. J., Greene, W. C., Rusk, C. M. 1984. Low and high affinity cellular receptors for interleukin 2. Implications for the level of Tac antigen. *J. Exp. Med.* 160 : 1126

Roosnek, E. E., Brouwer, M. C., Aarden, L. A. 1985a. T cell triggering by lectin. I. Requirements for interleukin 2 production; lectin concentration determines the accessory cell dependency. *Eur. J. Immunol.* 15 : 652–56

Roosnek, E. E., Brouwer, M. C., Aarden, L. A. 1985b. T cell triggering by lectin. II. Stimuli for induction of interleukin 2 responsiveness and interleukin 2 production differ only in quantitative aspects. *Eur. J. Immunol.* 15 : 657–61

Rothenberg, E., Lugo, J. P. 1985. Differentiation and cell division in the mammalian thymus. *Dev. Biology* 112 : 1

Samelson, L., Germain, R., Schwartz, R.

1983. Monoclonal antibodies against the antigen receptor on a cloned T-cell hybrid. *Proc. Natl. Acad. Sci. USA* 80 : 6972

Samelson, L. E., Harford, J. B., Klausner, R. D. 1985a. Identification of the components of the murine T cell antigen receptor complex. *Cell* 43 : 223–31

Samelson, L. E., Harford, J., Schwartz, R. H., Klausner, R. D. 1985b. A 20-kDa protein associated with the murine T-cell antigen receptor is phosphorylated in response to activation by antigen or concanavalin A. *Proc. Natl. Acad. Sci. USA* 82 : 1969–73

Scheurich, U., Ucer, U., Wrann, M., Pfizenmaier, K. 1985. Early events during primary activation of T cells : Antigen receptor cross-linking and interleukin 1 initiate proliferative response of human T cells. *Eur. J. Immunol.* 15 : 1091–95

Schwartz, R. H. 1985. T lymphocyte recognition of antigen in association with gene products of the major histocompatibility complex. *Ann. Rev. Immunol.* 3 : 239

Scollay, R., Bartlett, P., Shortman, K. 1984. T cell development in the adult murine thymus : Changes in the expression of the surface antigens Ly2, L3T4 and B2A2 during development from early precursor cells to emigrants. *Immunol. Rev.* 82 : 79–103

Sekaly, R. P., MacDonald, H. R., Zaech, P., Nabholz, M. 1982. Cell cycle regulation of cloned cytolytic T cells by T cell growth factor: Analysis by flow microfluorometry. *J. Immunol.* 129 : 1407–15

Sekaly, R. P., MacDonald, H. R., Nabholz, M., Smith, K. A., Cerottini, J.-C. 1984. Regulation of the rate of cell cycle progression in quiescent cytolytic T cells by T cell growth factor: Analysis by flow microfluorometry. *J. Cell Physiol.* 121 : 159–66

Shapiro, D. N., Adams, B. S., Niederhuber, J. E. 1985. Antigen-specific T cell activation results in an increase in cytoplasmic free calcium. *J. Immunol.* 135 : 2256–61

Shimonkevitz, R., Cerottini, J.-C., MacDonald, H. R. 1985. Variable requirement for murine lymphocyte function-associated-antigen-1 (LFA-1) in T cell-mediated lysis depending upon the tissue origin of the target cells. *J. Immunol.* 135 : 1555–57

Shimuzu, A., Shigeru, K., Shun-ichi, T., Yodoi, J., Ishida, N., et al. 1985. Nucleotide sequence of mouse IL-2 receptor cDNA and its comparison with the human IL-2 receptor sequence. *Nucl. Acids Res.* 13 : 1505–16

Simon, M. M., Landolfo, S., Diamantstein, T., Hochgeschwender, U. 1986. Antigen and lectin sensitized cytolytic T lymphocyte precursors require both interleukin 2 and endogenously produced immune (γ) interferon for their growth and differentiation into effector cells. *Curr. Top. Microbiol. Immunol.* 126 : 173

Smith, K. A. 1980. T cell growth factor. *Immunol. Rev.* 51 : 336

Smith, K. A., Gilbride, K. J., Favata, M. F. 1980. Lymphocyte activating factor promotes T-cell growth factor production by cloned murine lymphoma cells. *Nature* 287 : 853

Swain, S. L. 1983. T cell subsets and the recognition of MHC class. *Immunol. Rev.* 74 : 129

Tax, W., Willems, H., Reekers, P., Capel, P., Koene, R. 1983. Polymorphism in mitogenic effect of IgG1 monoclonal antibodies against T3 antigen on human T cells. *Nature* 304 : 445

Tax, W. J. M., Hermes, F. F. M., Willems, R. W., Capel, P. J. A., Koene, R. A. P. 1984. Fc receptors for mouse IgG1 on human monocytes: Polymorphism and role in antibody-induced T cell proliferation. *J. Immunol.* 133 : 1185

Truneh, A., Albert, F., Golstein, P., Schmitt-Verhulst, A. 1985. Early steps of lymphocyte activation bypassed by synergy between calcium ionophores and phorbol ester. *Nature* 313 : 318

Tsien, R. Y., Pozzan, T., Rink, T. J. 1982. T-cell mitogens cause early changes in cytoplasmic free Ca^{2+} and membrane potential in lymphocytes. *Nature* 295 : 68–71

Uchiyama, T., Broder, S., Waldmann, T. 1981. A monoclonal antibody (anti-Tac) reactive with activated and functionally mature human T cells. I. Production of anti-Tac monoclonal antibody and distribution of Tac^+ cells. *J. Immunol.* 126 : 1393

Unanue, E. R. 1984. Antigen-presenting function of the macrophage. *Ann. Rev. Immunol.* 2 : 395–428

van den Elsen, P., Shepley, B. A., Borst, J., Coligan, J., Markham, A., et al. 1984. Isolation of cDNA clones coding for the 20 kD T3 glycoprotein associated with antigen specific receptors on the surface of human T lymphocytes. *Nature* 312 : 413

Vohr, H. W., Hünig, T. 1985. Induction of proliferative and cytotoxic responses in resting $Lyt-2^+$ T cells with lectin and recombinant interleukin 2. *Eur. J. Immunol.* 15 : 332

Wagner, H., Hardt, C., Rouse, B. T., Rölling-hoff, M., Scheurich, P., Pfizenmaier, K. 1982. Dissection of the proliferative and differentiative signals controlling murine cytotoxic T lymphocyte responses. *J. Exp. Med.* 155 : 1876

van Wauwe, F. P., DeMay, J. R., Goossener, J. G. 1980. OKT3: A monoclonal anti-human T lymphocyte antibody with potent mitogenic properties. *J. Immunol.* 124 : 2708

Weiss, A., Stobo, J. D. 1984. Requirement for the coexpression of T3 and the T cell antigen receptor on a malignant human T cell line. *J. Exp. Med.* 160 : 1284

Weiss, A., Imboden, J., Shoback, D., Stobo, J. 1984a. Role of T3 surface molecules in human T cell activation: T3 dependent activation results in a rise in cytoplasmic free calcium. *Proc. Natl. Acad. Sci. USA* 81 : 4169

Weiss, A., Wiskocil, R., Stobo, J. 1984b. The role of T3 surface molecules in the activation of human T cells: A two stimulus requirement for IL-2 production reflects events occurring at a pre-translational level *J. Immunol.* 133 : 123

Weiss, A., Imboden, J., Hardy, K., Manger, B., Terhorst, C., Stobo, J. 1986. The role of the T3/antigen receptor complex in T cell activation. *Ann. Rev. Immunol.* 4 : 593–619

Williams, J. M., Deloria, D., Hansen, J. A., Dinarello, C. A., Loertscher, R., et al. 1985. The events of primary T cell activation can be staged by use of sepharose-bound anti-T3 (64.1) monoclonal antibody and purified interleukin 1. *J. Immunol.* 135 : 2249–55

Wiskocil, R., Weiss, A., Imboden, J., Kamin-Lewis, R., Stobo, J. 1985. Activation of a human T cell line: A two-stimulus requirement in the pretranslational events involved in the coordinate expression of interleukin 2 and γ-interferon genes. *J. Immunol.* 134 : 1599–1603

Yamamoto, Y., Ohmura, T., Fujimoto, K., Onone, K. 1985. Interleukin 2 mRNA induction in human lymphocytes: Analysis of the synergistic effect of a calcium ionophore A23187 and a phorbol ester. *Eur. J. Immunol.* 15 : 1704–8

Yanagi, Y., Yoshikai, Y., Leggett, K., Clark, S., Aleksander, I., Mak, T. 1984. A human T cell-specific cDNA clone encodes a protein having extensive homology to immunoglobulin chains. *Nature* 308 : 145

Zinkernagel, R. M., Doherty, P. C. 1974. Restriction of in vitro T cell mediated cytotoxicity in lymphocytic choriomeningitis within a syngeneic or semi-allogenic system. *Nature* 248 : 701

Ann. Rev. Cell Biol. 1986. 2 : 255–313

ANCHORING AND BIOSYNTHESIS OF STALKED BRUSH BORDER MEMBRANE PROTEINS: Glycosidases and Peptidases of Enterocytes and Renal Tubuli

Giorgio Semenza

Laboratorium für Biochemie der ETH, ETH-Zentrum,
CH-8092 Zurich, Switzerland

CONTENTS

0743–4634/86/1115–0255$02.00

INTRODUCTION

Our knowledge of the biosynthesis and the mode of anchoring of several intrinsic membrane proteins of the brush borders found on the epithelia of the small intestine and renal proximal convolution has made major progress during the past decade. This was rendered possible by technical advances, some specific to this area of research and some common to others, among which at least the following deserve mention:

(*a*) One important development is a quick and efficient procedure to prepare, with higher quality and yield, essentially closed brush border membrane vesicles (BBMVs) that retain their original orientation. Under appropriate conditions, Ca^{2+} or Mg^{2+} is added to the homogenate to precipitate intracellular and basolateral membranes. BBMV are then recovered from the supernatant by differential centrifugation (Schmitz et al 1973; Louvard et al 1973; Booth & Kenny 1974; Kessler et al 1978; for reviews see Kenny & Booth 1978; Murer & Kinne 1980). Subsequently (Danielsen et al 1981a) it was also found that early forms in the biosynthetic pathway of the brush border proteins can be obtained from Ca^{2+}-precipitated membranes, which made characterization of these proteins possible.

(*b*) Immunoelectrophoresis and immunochromatography using polyclonal (e.g. Skovbjerg et al 1978; Danielsen et al 1982c; Skovbjerg et al 1981) or monoclonal (Gee et al 1983) antibodies have proved to be extremely valuable techniques, both for analytical and preparative purposes. (Note that the presence of plentiful and varied proteases in the cell lysates makes fast separation procedures almost mandatory.)

(*c*) A number of important techniques for the characterization of hydrophobic or amphipathic proteins, in solution or when anchored in the membrane, were applied successfully to the study of the intrinsic proteins of the brush border membranes—indeed, some of these techniques were developed in connection with these proteins. They include the following: charge-shift electrophoresis (Helenius & Simons 1977); the formation of single bilayer proteoliposomes (Brunner et al 1978) (the optimal composition of lipids may be different for different brush border enzymes, see e.g. Tiruppathi et al 1985); a nitrene generator transported across the membrane to hydrophilically photolabel from the cytosolic side (Booth & Kenny 1980) [a similar technique developed for the same purpose by Louvard et al (1976) seems to have not met with much success, possibly because it may artifactually label the luminal side as well (Booth & Kenny 1980)]; and hydrophobic labeling. This last technique has proven extremely useful in identifying a membrane protein as being intrinsic, in establishing which of its portions interact with the hydrophobic membrane layer, in handling hydrophobic peptides, etc (for reviews see Khorana 1980; Brunner 1981, 1987; Bayley 1983; Bisson & Montecucco 1985). At present, Brunner's reagent, 3-(trifluoromethyl)-3-(m[125I]-iodophenyl) diazirine (TID) (Brunner & Semenza 1981), probably best meets the stringent requirements for a hydrophobic photolabeling reagent (Bayley 1983). Finally, cell biology techniques, such as tissue culture and cell-free translation systems, have found successful application in this area.

The proteins of the brush border cytoskeleton have been reviewed recently (Mooseker 1985; Bretscher 1983; Mooseker et al 1983; Coudrier et al 1983; Matsudaira 1983; Mooseker & Howe 1982). The brush borders of the small-intestine and kidney proximal convoluted tubuli are endowed with remarkably similar sets of intrinsic membrane proteins; some of these catalyze the transport of solutes, and others are involved in the last steps of digestion. (The biological significance of "digestive enzymes" in the renal tubuli brush border membrane is not clear, although the peptidases may contribute to the degradation and inactivation, when that is the case, of some circulating peptides.) The proteins responsible for transport across the brush border membranes are the least characterized and are not dealt with here. However, it suffices to say that very often the same transport systems occur in both intestinal and renal brush border membranes, and they are usually subject to the same genetic control mechanisms. There are some differences, however, both quantitative (for discussions see, for example, Murer & Kinne 1980; Semenza et al 1984; Semenza & Corcelli 1986) and qualitative. For example, the kidney tubuli brush borders are endowed with a Na^+-dependent transport system for ascorbate, which does not appear in the small intestine of most mammals, but they lack the transport system for D-fructose, which is present in most, if not all, epithelia of mammalian small intestines.

Of the digestive enzymes of small-intestinal and renal brush borders, this review covers only glycosidases and proteases-peptidases (including γ-glutamyltransferase) because only for these enzymes do we have adequate knowledge of their mode of anchoring in the membrane and their mode of biosynthesis. Little is known, for example, of these aspects for alkaline phosphatase (Colbeau & Maroux 1978) (see below) or of the enzymatic equipment of other brush border membranes, namely those of the colonic mucosa and of the placenta. [The latter has 5'-nucleotidase and alkaline phosphatase but is nearly devoid of peptidase or disaccharidase activities (Booth et al 1980).]

As discussed later, all the brush border enzymes whose mode of anchoring is known in some detail are "stalked" intrinsic membrane proteins. Most of their protein mass, including the catalytic site(s), stands out from the extracellular luminal side. Their anchoring segment is located at the

N-terminal region of the polypeptide chain and crosses the membrane bilayer once. This mode of anchoring is quite different from that of glycophorin, the first stalked plasma membrane protein to be characterized in detail. Thus, the anchoring of these enzymes does not conform to the original, fairly straightforward postulates on synthesis and insertion of intrinsic membrane proteins; hence other proposed models have attempted to accommodate the mode of insertion of brush border proteins.

The complement of intrinsic proteins of the brush border membrane differs somewhat along the intestine. The ratios of the disaccharidase activities vary somewhat in the various intestinal segments (Newcomer & McGill 1966). The glucoamylases (unlike the other glycosidases), dipeptidyl peptidase IV and aminopeptidase N, increase in the ileum as compared to jejunum (Norén et al 1980; Skovbjerg 1981b; Miura et al 1983; Triadou et al 1983); endopeptidase 24.11 has maximal specific activity in the jejunum and low activity in the duodenum and ileum (Gee et al 1985); and γ-glutamyltransferase is fairly equally distributed along the intestine. Of course, the transport systems for bile acids and for Vitamin B_{12} are present only in the ileum (from which the receptor for the intrinsic factor cobalamin has been isolated; Seetharam et al 1981, 1982). Also, renal BBMVs, as prepared by standard procedures, are somewhat heterogeneous and can be fractionated by free-flow electrophoresis and density gradient centrifugation (Kinne et al 1975; Mamelok et al 1980). These procedures bring about a partial separation of alkaline phosphatase and γ-glutamyltransferase (Mamelok et al 1980) (but see more below). This separation may reflect a different distribution of these enzymes in the renal brush border membrane of the outer cortex as compared to the outer medulla, which by independent procedures has been shown to have Na^+–D-glucose cotransporters of different properties (Turner & Moran 1982a,b).

The elaborate structure of the brush border greatly expands the surface area of the apical plasma membrane, making it better suited to carry out the final steps in digestion and absorb the resulting products (in the intestine) or recover filtered solutes (in the renal tubuli). The tight packing of microvilli, however, also results in very large unstirred layers (to which mucus also contributes in the intestine), which are probably reduced, at least in part, by the movements of the microvillar region (see Mooseker 1985). The presence of unstirred layers, while sizeably increasing the apparent K_m values for both the digestive enzymes and the transport systems, also ensures that enough Na^+ originating from the paracellular pathways is present at the luminal surface of brush border membrane. Obviously, a transmembrane gradient of $\tilde{\mu}_{Na^+}$ is needed for absorption and accumulation of a number of solutes.

In the last few years a number of reviews have been devoted to various aspects of the hydrolases of small-intestinal and renal brush borders, including their biosynthesis (Kenny & Booth, 1978; Semenza 1968; Sacktor 1977; Semenza 1979a,b, 1981a,b; Kenny & Maroux 1982; Hauser & Semenza 1983; Brunner et al 1983; Danielsen et al 1984a; Norén et al 1986b). The proceedings of a Ciba Foundation symposium on brush border membranes have appeared (Porter & Collins 1983).

I present in the following some general aspects of brush border hydrolases, including their gross positioning in the membrane, solubilization procedures, and the determination of their subunit structure. I then concentrate on the details of the positioning of glycosidases and of proteases-peptidases, including γ-glutamyltransferase. Table 1 lists these brush border hydrolases and some of their properties.

GROSS POSITIONING OF BRUSH BORDER HYDROLASES AND THEIR SOLUBILIZATION

Glycosidases, peptidases and proteases, γ-glutamyltransferase, and alkaline phosphatase are "intrinsic membrane proteins," as operationally defined by the fact that they cannot be extracted from the membrane by water solutions of low or high ionic strength or of moderately acidic or alkaline pH values. They are freely accessible to their substrates from the lumen [the functional evidence concerning intestinal glycosidases was reviewed some years ago (Semenza 1968)]. Negative-stained brush borders of the small intestine (Johnson 1967; Nishi et al 1968) and of the kidney (Kenny et al 1983) are studded with a great number of particles that have been identified with at least some of these hydrolases. Papain, as ascertained for small-intestinal BBMVs (Ca^{2+} precipitation procedure), acts solely from the outer, luminal surface and does not affect the membrane permeability (Tannenbaum et al 1977); it severs many of these particles and their associated hydrolase activities.

A particularly careful study was carried out by Kenny et al (1983) on BBMVs from pig kidney. Prior to papain treatment the microvilli were coated with particles of stalk lengths ranging from 2.5 to 9 nm. After treatment with the protease only the particles with stalks of 2–3 nm were still associated with the microvillus surface. At least some of these short-stalked particles must represent endopeptidase 24.11, an enzyme totally resistant to papain solubilization. The stalks of enzymes solubilized by papain, i.e. aminopeptidase N and sucrase-isomaltase (in the small intestine), were longer, 5 nm (Hussain et al 1981) and at least 4 nm (Brunner et al 1978; Nishi & Takesue 1978; Cowell et al 1986), respectively. Since the substrate specificity of papain is fairly broad (Glazer & Smith 1971), the authors concluded that a critical factor in determining whether a stalk is attacked by papain is whether or not it is long enough to accommodate the protease, which measures $5.0 \times 3.7 \times 3.7$ nm (Drenth et al 1971), in

Table 1 Molecular forms of microvillar enzymes during biosynthesis (from Danielsen et al 1984, modif.). [Most apparent M_r values listed here were estimated from SDS-PAGE. They thus occasionally differ somewhat from those calculated from hydrodynamic parameters (see e.g. Hauser & Semenza 1983).]

| Enzyme | Species | Tissue | $10^{-3} \times M_r$ of: | | Endo-glycosidase H sensitivity | Mature form[a] ($10^{-3} \times M_r$) | Transport time (min) | References |
			Primary translation product, non-glycosylated form	Transient form				
Amino-peptidase N (EC 3.4.11.2)	Pig	Intestine	115	140	+	166[c,d] (123^c+62)	60–90	Danielsen (1982); Danielsen et al (1982b, 1983b, 1984a); Sjöström et al (1978)
	Rat	Intestine	—	120[b]	Not detected	130	60–180	Ahnen et al (1982)
	Rat	Intestine	—	<130	+	130	>180	Ahnen et al (1983)
	Rabbit	Intestine	—	110	+	115–128	30–90	Feracci et al (1985)
	Pig	Kidney	99	106	+	125	—	Massey & Maroux (1985)
	Pig	Kidney	115	140	+	160[e]	90	Stewart & Kenny (1984a)
Amino-peptidase A (EC 3.4.11.7)	Pig	Intestine	—	140	+	170[e]	60–90	Danielsen et al (1983a)
	Pig	Kidney	—	145	+	170[e]	90	Stewart & Kenny (1984a)
Dipeptidyl peptidase IV (EC 3.4.14.5)	Pig	Intestine	—	115	+	137	60–90	Danielsen et al (1983a)
	Pig	Kidney	—	115	+	130[c,d]	90	Stewart & Kenny (1984a)
Endopeptidase 24.11 (EC 3.4.24.11)	Pig	Kidney	—	88	+	93	90	Stewart & Kenny (1984a)

Enzyme (EC)	Species	Tissue						References
γ-Glutamyl transferase (EC 2.3.2.2)	Rat	Kidney	63[e]	78	Not detected	(51+22)	~120	Nash & Tate (1982, 1984)
	Rat	Kidney	63[e]	78	Not detected	(50[c,a]+23)	—	Kuno et al (1983); Matsuda et al (1983b)
	Rat	Kidney	—	75	+, −	(49.5+29)	—	Capraro & Hughey (1983); Hughey & Capraro (1984)
Sucrase-isomaltase complex (EC 3.2.1.48-10)	Rabbit	Intestine	≃200[f]	≃230	+	275[c] (120+140[c,d])	—	Wacker et al (1981); Ghersa et al (1986); Hauri et al (1979); Brunner et al (1979)
	Pig	Intestine	225	240	+	265[c] (140+150[c])	60–90	Danielsen (1982); Danielsen & Cowell (1984)
Maltase-glucoamylase (Glucoamylase complex) (EC 3.2.1.20)	Pig	Intestine	200	225	+	245 (125+135[c])	60–90	Danielsen et al (1983a); Danielsen & Cowell (1984)
Lactase-phlorizin hydrolase (β-Glycosidase complex) (EC 3.2.1.23-62)	Pig	Intestine	210	225	+	245 (160)	60–90	Danielsen et al (1984b); Sjöström et al (1983)

[a] Values in parentheses indicate the M_r of proteolytically cleaved mature forms.

[b] This transient form was soluble rather than membrane-bound (see text).

[c] Strong indication for anchoring via the N-terminal region is available (i.e. change in the N-terminal amino acid residue during proteolytic solubilization and/or sizeable hydrophobic sequence in the N-terminal region), although not necessarily for the species indicated in the second column. (For references on anchoring and subunit composition of the mature forms and pro forms, see text.)

[d] Evidence for the nonparticipation of the C-terminal region in the anchoring is available (for references see text).

[e] The M_r of the unglycosylated one-chain precursor, as deduced from cDNA cloning and sequencing, is 61,800 (Laperche et al 1986).

[f] The M_r of the unglycosylated pro-sucrase-isomaltase, as deduced from cDNA cloning and sequencing, is 203,000 (Hunziker et al 1986).

the gap between the membrane surface and the "body" of the protein. Other proteolytic treatments used occasionally include digestion with elastase (which generally yields the same results as papain) or trypsin, and "autolysis."[1]

Alternative solubilization procedures include the use of detergents [Triton X-100, Emulphogen BC 720, deoxycholate, cholate, sodium dodecylsulfate (SDS)]; freezing and thawing; and the use of organic solvents (butanol, toluene). (For reviews see Semenza 1981a; Kenny & Maroux 1982.) Generally, intestinal or renal brush border membranes are (partially) purified before being used as the substrate for solubilization. When very little material is available, solubilization may be carried out using the total sediment from the whole homogenate, in which case some precautions must be taken (Schlegel-Haueter et al 1972).

Proteolytic solubilization implies that the stalk connecting the body of the enzyme to the membrane fabric is severed, the other treatments mentioned above can be presumed not to bring about covalent changes. Thus, the macromolecular properties of enzymes solubilized by proteolysis are usually different from those of the same enzymes solubilized by detergents. In general, the detergent-solubilized forms are amphipathic, as documented by charge-shift electrophoresis; hydrophobic chromatography; their behavior in "cloud point" forming detergents, such as Triton X-114 (Bordier 1981); their tendency to form aggregates in the absence of detergents; and their capacity to form proteoliposomes (e.g. Pattus et al 1976; Wacker et al 1976; Brunner et al 1978; Kenny et al 1983); etc. Conversely, enzymes solubilized by proteolysis usually behave like ordinary hydrophilic proteins. However, note that the hydrophobic "anchor" only accounts for 3% or less of the protein mass (see more

---

[1] An unusual procedure from the distant past may be worth mentioning. Thirty years ago a purification of cysteinyl glycinase, leucine aminopeptidase, and alkaline phosphatase from hog kidney was described. It involved prolonged digestion by autolysis followed by denaturation by chloroform treatment of most of the proteins in the extract (Binkley 1952; Binkley et al 1957). The preparations obtained were thought to be nucleic acids (Binkley 1952). However, this was disproved when the advent of protein chromatography clearly showed that the activity was associated with a protein and not with the nucleic acids present in the preparation (Semenza 1957). Recently, these cysteinyl glycinase and leucine aminopeptidase activities were identified with aminopeptidase M (also known as N) (Rankin et al 1980). Naturally, prolonged proteolysis, which undoubtedly partially modifies the very protein to be purified, is not an appealing step. This, plus the claim of a "nonproteic nature" asserted by Binkley (Binkley 1952; Binkley et al 1957), certainly helped this unusual procedure sink into oblivion. However, the resistance of these preparations at room temperature to further proteolytic degradation or the denaturing action of chloroform [almost certainly due to a protective action of nucleic acids (Semenza 1957) or polysaccharides] makes this unusual procedure worth a second look.

below); therefore, the amphipathic behavior may not always be easy to document in a clear-cut way.

SUBUNIT STRUCTURE AND MACROMOLECULAR PROPERTIES

A discussion of the subunit structure of these enzymes goes beyond the scope of the present review (the reader is referred to Sacktor 1977; Semenza 1981a; Kenny & Maroux 1982; Hauser & Semenza 1983; Sjöström et al 1983; Norén et al 1986b). We recommend some caution in applying information on the oligomeric structure of a given enzyme in detergent solution to its state in the original membrane. It is conceivable, for example, that the detergent dissociates oligomers existing in the membrane or that proteins occurring as monomers in the membrane aggregate to oligomers when forming mixed micelles with detergents. In principle, these problems can be solved by cross-linking the enzyme in the membrane prior to solubilization, with dimethyl-suberimidate for example, if the cross-linking reaction is fast compared to the lateral diffusion of the protein in the membrane. By this criterion, amino peptidase N was believed to occur as a dimer in the original membrane (Svensson 1979). Recent radiation inactivation analysis of renal aminopeptidases A and N and of dipeptidyl peptidase IV shows that the targets are monomeric, i.e. that dimerization is not needed for activity. In the case of γ-glutamyltransferase the target is the sum of both (unequal) subunits. (I. S. Fulcher & A. J. Kenny, unpublished results.)

The heterodimers that originate from a single-chain precursor (sucrase-isomaltase, glucoamylases, lactase-glycosylceramidase, γ-glutamyltransferase; see more below) must occur as such, or as higher oligomers, in the original membrane, particularly since in those heterodimers in which only one of the subunits is anchored to the membrane fabric, the other subunit has an extracellular location.

Endopeptidases that are present in the extract or that act already in vivo can be an important source of error because they can generate artifactual "subunits." (Thus the advantage of the fast and specific immunochromatographic methods.) Often, comparison between small-intestinal and renal enzymes is quite informative because only the former are exposed to pancreatic proteases.

In most cases, S-S bridges are apparently not involved in the interaction between subunits. This has been shown for sucrase-isomaltase (Brunner et al 1979), lactase-glycosylceramidase (Skovbjerg et al 1981), aminopeptidase N (Maroux et al 1973), and dipeptidyl peptidase IV (MacNair & Kenny 1979). No universally successful procedure is known to allow dissociation of the subunits with recovery of the enzymatic activity, although the subunits of aminopeptidase N recover activity after dissociation with urea (Wacker et al 1976; pig kidney enzyme), or with diiodosalicylate (Benajiba & Maroux 1981; pig intestinal enzyme) and those of rabbit small-intestinal sucrase-isomaltase do so after citraconylation/deacylation (Brunner et al 1979; Braun et al 1975; Takesue et al 1977). Some procedures dissociate the two subunits of sucrase-isomaltase with simultaneous, selective destruction of one (Cogoli et al 1973; Quaroni et al 1975). Little is known of the forces that hold sucrase and isomaltase together. However, they probably consist of hydrophobic and/ or electrostatic interactions.

As mentioned above, enzymes solubilized from the membrane (or from the proteoliposomes) by limited proteolysis behave as ordinary hydrophilic proteins, whereas those solubilized by detergent treatment are amphipathic. In the absence of detergent the latter proteins form fairly regular protein micelles of apparent M_r larger than one million (Hussain 1985; H. Sigrist, M. Müller, and G. Semenza, unpublished observation; M. A. Bowes & A. J. Kenny, unpublished observation). In the presence of detergent they bind an amount of detergent usually corresponding to a micelle (e.g. approximately 150 molecules of Triton X-100), which entails an increase in apparent M_r of about 95 kDa (Hughey & Curthoys 1976; Feracci & Maroux 1980; Wacker et al 1981). Thus, the difference in apparent M_r of detergent-solubilized forms as compared to protease-solubilized forms is made up by the anchor proper, the stalk, and the bound detergent.

Bound detergent, the tendency to aggregate, and the variable amounts of SDS bound in SDS-PAGE (all these proteins are glycosylated) may explain the contrasting M_r values reported in the literature for the same enzyme (comparison of values for sucrase-isomaltase can be found in the review by Hauser & Semenza 1983 and for aminopeptidase N in that of Kenny & Maroux 1982). Rosenbusch has introduced detergents with favorable characteristics for the study of the hydrodynamic properties of amphipathic proteins. They have been used in one such study of sucrase-isomaltase (Spiess et al 1981).

ANCHORING OF GLYCOSIDASES IN THE MEMBRANE

The Sucrase-Isomaltase Complex

The sucrase-isomaltase (SI) complex is one of the major proteins of the small-intestinal brush border; it accounts for approximately 8–10% of the latter's proteins (Kessler et al 1978). It does not occur in the kidney. The sucrase-isomaltase complex accounts for approximately 80% of the maltase activity, most of the isomaltase, and all of the sucrase activity in the small intestine. It is composed of two subunits (each a glycosylated, single polypeptide chain). One subunit [~120 kilodaltons (kDa)] splits mainly maltose and sucrose (the "sucrase" subunit); the other (~140 kDa) splits mainly maltose and the 1,6-α-glucopyranosidic bonds in a number of oligosaccharides (the "isomaltase" subunit). The minimum catalytic mechanism for both sucrase and isomaltase includes the protonation of the glycosidic oxygen (presumably by a γ-Glu-COOH), followed by the formation of an oxocarbonium ion, which is temporarily stabilized by a β-Asp-COO$^-$ (see Cogoli & Semenza 1975; Quaroni & Semenza 1976; see also Table 2, below).

Like all other glycosidases of the intestine and of other tissues, sucrase and isomaltase are inhibited competitively by the unprotonated form of tris(hydroxymethyl)-aminomethane (Tris) (Semenza & Balthazar 1974). Sucrase undergoes a slow hysteretic interaction with some high-affinity competitive inhibitors (Hanozet et al 1981). The macromolecular (Semenza 1981a; Hauser & Semenza 1983) and the catalytic properties (Semenza 1976) of sucrase-isomaltase have been reviewed elsewhere.

The sucrase-isomaltase complex is brought into solution by limited proteolysis (usually papain is used) or by detergent treatment (usually Triton X-100) (see Sigrist et al 1975 and literature quoted therein; Semenza 1981a; Hauser & Semenza 1983). The membrane-bound enzyme (the one solubilized by either papain or detergent) and its isolated subunits have identical kinetic properties (Cogoli et al 1973), including K_m and k_{cat} values, Na$^+$ activation, and Arrhenius activation energy (Semenza 1976).

The anchoring of the sucrase-isomaltase complex to the brush border membrane is drawn schematically in Figure 1a. Note that the sucrase subunit does not interact directly with the membrane fabric and has its C- and N-termini exposed to the outer, luminal water phase (and not to the cytosol). The "anchor" is confined to the N-terminal region of the isomaltase subunit and crosses the hydrophobic layer once. The N-terminal of isomaltase is located on the cytosolic side of the membrane, and the C-terminal is on the luminal side. The anchoring segment is very hydrophobic (see a partial sequence in Figure 2). The evidence for this positioning in the membrane is detailed below.

Sucrase-isomaltase (Frank et al 1978; Brunner et al 1979) and renal dipeptidyl peptidase IV (Kenny & Booth 1978; MacNair & Kenny 1979) were the first plasma membrane proteins shown to be anchored in the membrane *solely* via a segment located at the N-terminal region of their polypeptide chains.[2] The C-terminal regions of the two subunits of sucrase-isomaltase and the N-terminal region of the sucrase subunit are not involved in membrane anchoring. Comparison of the Triton-solubilized and papain-solubilized forms of sucrase-isomaltase shows that papain, which only acts from the outer, luminal side, and does not detectably alter the permeability properties of the brush border membrane (Tannenbaum et al 1977), does not produce any change in the C-terminal region of either subunit (the same amino acids are released by the action of carboxy-peptidase Y) nor in the N-terminal region of sucrase [its N-terminal, Ile, and its apparent M_r are the same for both forms (Brunner et al 1979)]. The peripheral positioning of sucrase is further demonstrated by (a) its preferential solubilization by citraconylation from native BBMVs (Brunner et al 1979); (b) the fact that it is not labeled by TID, a photolabel that is totally confined to, and exclusively reactive within, the hydrophobic core of the membrane (Spiess et al 1982); and (c) the electron microscope observation that anti-sucrase antibody conjugates are 4–5 nm more remote from the apparent "unit membrane" of the brush border than those formed by anti-isomaltase antibodies (Nishi & Takesue 1978).

The "anchor" is a highly hydrophobic segment of the isomaltase subunit located not far from the N-terminus. Papain solubilization changes the N-terminal amino acid of isomaltase from Ala to a microheterogeneous terminal of Gly, Glu, Tyr, and reduces the apparent M_r of isomaltase from 140 kDa to 120 kDa (Brunner et al 1979), with simultaneous disappearance of the amphipathic properties. Edman degradation of the isomaltase subunit prepared from the Triton-solubilized complex reveals a very hydrophobic sequence starting at position 12 (Frank et al 1978; Hauri et al 1982; Sjöström et al 1982) (Figure 2, see more below). Hydrophobic photolabeling with TID is confined to the isomaltase subunit prepared from the Triton-solubilized complex. No hydrophobic photolabel is found in papain-solubilized complexes (Spiess et al 1982). The complete amino

---

[2] The occurrence of a hydrophobic segment is an important indication of interaction with the hydrophobic layer of the membrane, but of course does not constitute final evidence for it, and less so that it is the *sole* anchoring segment. For example, as pointed out in previous reviews (Kenny & Maroux 1982; Semenza et al 1983), a different N-terminal amino acid in the detergent-solubilized enzyme as compared with the enzyme form solubilized by limited proteolysis and the occurrence of a hydrophobic domain in the N-terminal region are strong indications that the protein is anchored via the N-terminal region. Only in the cases of small-intestinal sucrase-isomaltase (Brunner et al 1979; Hunziker et al 1986), renal dipeptidyl peptidase IV (MacNair & Kenny 1979), renal γ-glutamyltransferase (Matsuda et al 1983a; Laperche et al 1986), and pig intestinal aminopeptidase N (Norén & Sjöström 1980; Sjöström & Norén 1982) has the possible additional participation of the C-terminal region been ruled out.

acid sequence of pro-sucrase-isomaltase, recently obtained from cDNA cloning and sequencing, clearly shows a *single* hydrophobic stretch, located between positions 12 and 31 (Figure 2) (Hunziker et al 1986). Thus, the evidence is consistent: pro-sucrase-isomaltase interacts with the hydrophobic layer of the brush border membrane solely via a single hydrophobic stretch located at the N-terminal region of the isomaltase subunit (and of the isomaltase region in the case of pro-sucrase-isomaltase; see below).

The hydrophobic anchor has a helical conformation, as shown by circular dichroism (CD) spectra in SDS or deoxycholate in water, liposomes, or 2-chloroethanol (Spiess et al 1982).

The anchor emerges into the cytosol, where it can be labeled by [^{125}I]-3,5-diiodo-4-azidobenzene sulfonate([^{125}I]DIABS) (A. G. Booth & A. J.

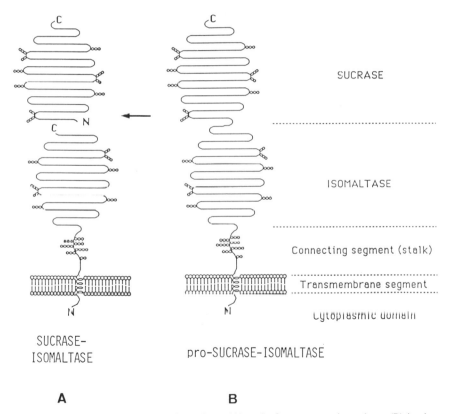

A   **B**

Figure 1 Positioning of sucrase-isomaltase (A) and of pro-sucrase-isomaltase (B) in the small-intestinal brush border membrane. The (unspecified) interactions within and between the sucrase and isomaltase domains (or subunits, respectively) are not indicated. ∞ = sugar chains. (Adapted from Hunziker et al 1986.)

```
                        1           5          10              15              20              25              30              35

HOG ProSI [c]     Ala-Arg-Lys-Ser/Thr-Phe-Ser-Gly-Leu-Glu-Ile—X—Leu-Ile-Val-Leu-Phe-Ala-Ile-Val-

HOG 1 [c]         Ala-Arg-Lys-Lys-Phe-Ser-Gly-Leu-Glu-Ile-X—Leu-Ile-Val-Leu-Phe-Ala-Ile-Val-Val-X-Ala-Ser-Lys—X—Pro-Ala-Val-

RAT ProSI [a]     Ala-Lys-Lys-Lys-Phe-Arg-Ala-Glu-Ile-X—Leu-Ile-Val-Leu-Phe-Ile-Ile-

RAT 1 [a]         Ala-Lys-Lys-Lys-Phe-Ser-Ala-Leu-Glu-Ile-X—Leu-Ile-Val-Leu-Phe-Ile-Ile-Ala-Ile-Ala-Leu-Val-Leu-Val-

RABBIT 1 [b]      Ala-Lys-Arg-Lys-Phe-Ser-Gly-Leu-Glu-Ile-Thr-Leu-Ile-Val-Leu-Phe-Val-Ile-Val-Phe-Ile-Ile-Val-Ile-Ala-Val-Leu-Ile-Ala-Val-Leu-Ala-Thr-Lys-Thr-Pro-Ala-Val-

                                                              40              45              50              55              60
                  Glu-Glu-Val-Asn-Pro-Ser-Ser-Ser-Thr-Pro-Thr-Thr-Ser-Pro-Thr-Thr-Thr-Ser-Gly-Ser-Val-Ser-

HOG S [f]         Ile-Lys-Leu-Pro-Ser-Asp-Pro-Thr-Leu-Arg-Val-Lys-Tyr-His-Lys-Thr-Asp-Tyr-Met-Leu-Glu-Phe-X—Arg-Tyr-Asp-Pro-Glu-Arg-
                                              Val-Glu/Met/Thr

RAT S [a]         Ile-Lys-Leu-Pro-Ser-Asn-Pro-Ile-Arg-Val-Ala-Glu-Val-Val-Thr—X—Thr-Asn-Arg-Val-Leu-Gln-Phe-Arg-Ile-Tyr-Arg-Ala-Glu—X—X—Gly-
                                                                                        Glu

RABBIT S [f]      Ile-Thr-Leu-Pro-Ser-Glu-Pro-Ile-Thr-Asn-Leu-Arg-Val-Lys-Lys-Asn-Asp-Met-Gln-Phe-Lys-Ile-

                        1           5          10              15              20              25              30              35

HOG GLUCO-
AMYLASE [i]       —X—Arg-Lys-Lys-Leu-Lys-Phe-Thr-Asp-Ser-Glu-Leu-Met-Leu-Ser-Val-Leu-Leu-Val-Val-Phe-Ile-Val-

RAT KIDNEY γ-GLUTAMYL TRANSFERASE [g]

HUMAN GLYCOPHORIN [c]    —His-His-Phe-Ser-Glu-Pro-Glu-Ile-Thr-Leu-Ile-Ile-Phe-Gly-Val-Met-Ala-Gly-Val-Ile-
                           66    70         75            80            85

HOG GLYCOPHORIN [d]      -Gln-Asp-Phe-Ser-His-Ala-Glu-Ile-Thr-Ile-Ile-Phe-Ala-Val-Met-Ala-Gly-Leu-Leu-
                           56              80         85            90            95

HORSE GLYCOPHORIN [e]   —His-Asp-Phe-Ser-Gln-Pro-Val-Ile-Thr-Val-Ile-Ile-Thr-Val-Met-Ala-Gly-Ile-Ile
                           44              50            55            60

RAT KIDNEY γ-GLUTAMYL TRANSFERASE [g]

Heavy subn. detrg. solub.:  Met-Lys-Asn-Arg-Phe-Leu-Val-Leu-Gly-Leu-Ala-Val-Leu-Val-Leu-Val-Phe-Ile-Ile-Gly-Leu-Cys-Ile-Trp-Leu-Pro-Thr-Ser-Gly-Lys-Pro-Asp-
                                                   5              10              15              20              25              30
           papain solub.:   Gly-Pro-Pro-Leu-

Light subn. detrg. solub.:  Thr-Ala—X—Leu-Ser-Met-Val-
           papain solub.:   Thr-Ala—X—Leu-

PIG KIDNEY DIPEPTIDYL PEPTIDASE [h]

Detergent solubilized:   Leu-Gly-Phe-Ala-Leu-Ala-Phe-Ile-
Solubil. by autolysis:   Ser-Thr-Ser-Thr-Tyr-Thr-Leu-Thr—X—Tyr-Leu-Lys—X—X—Phe-Leu-
                                        5              10              15

RABBIT INTESTINAL
AMINOPEPTID. N [j]       Tyr-Ile-Ser-Lys-Ala-Leu-Gly-Ile-Leu-Gly-Phe—X—Leu-Gly-
```

Kenny, unpublished observations; quoted in Kenny & Maroux 1982), a nitrene-generator that enters the brush border vesicles via a Na^+-dependent transport system (Booth & Kenny 1980). The N-terminus of the anchoring segment (i.e. of the isomaltase subunit) and the following ten or so amino acid residues are located on the cytosolic side of the membrane. This location was inferred by the amino acid sequence deduced from cDNA (Hunziker et al 1986), and from very recent chemical labeling studies using an impermeant reagent and intact small intestine rather than BBMVs (J. Brunner, T. Zürcher, and G. Semenza, unpublished results). A different localization of the N-terminus of isomaltase was reported previously (Bürgi et al 1983). [Thr-11, Figure 2, contrary to our own previous reports (Frank et al 1978) is not glycosylated (H. Wacker et al 1986, unpublished results).]

On the *luminal* side of the hydrophobic anchor one finds a series of fairly hydrophilic amino acid residues. A conspicuous segment of 23 residues (positions 43–64) contains not less than 17 serine or threonine residues, at least some of which are glycosylated (Hunziker et al 1986). Indeed, the major reason for placing this segment on the extracellular side of the membrane is that it is glycosylated. (For the topological implications of the glycosylation of membrane proteins see Kornfeld & Kornfeld 1985; Hanover & Lennarz 1981; Weiser & Wilson 1981.) We propose that the polypeptide segment between positions ~ 32 and ~ 64 forms a connecting piece (stalk) between the membrane anchor and the bulk of the protein. This is in excellent agreement with the length of the stalk segment as determined by electron microscopy (3–3.5 nm, which corresponds to

---

Figure 2 Partial N-terminal sequences of some stalked membrane proteins of intestinal pro-sucrase-isomaltases, isomaltases, and sucrases; of intestinal glucoamylase; of three glycophorins; of renal γ-glutamyltransferase; of renal dipeptidyl peptidase IV; and of intestinal aminopeptidase N. The species is indicated. Unless indicated otherwise, the original enzyme was solubilized by a detergent. Most sequences were determined by Edman degradation; two were confirmed and extended to completion by cDNA cloning and sequencing. If the identification of a residue is uncertain, a question mark appears above it. Hydrophobic stretches are indicated by ⌃⌃⌃⌃. Residues in the glycophorins that are identical to residues in rabbit isomaltase are underlined. [[a] From Hauri et al 1982. [b] From Frank et al 1978; Sjöström et al 1982; Hunziker et al (1986) provided the whole sequence of the 1827 amino acid residues of pro-sucrase-isomaltase by cDNA cloning and sequencing. The initiation Met (not indicated in the figure) is at position -1. [c] From Furthmayr et al 1978. [d] From Honma et al 1980. [e] From Murayama et al 1982. [f] From Sjöström et al 1982; the sequence of rabbit sucrase was confirmed and extended by Hunziker et al 1986. [g] From Matsuda et al 1983a. Laperche et al (1986) provided the whole sequence of the 568 amino acid residues of pro-γ-glutamyltransferase by cDNA cloning and sequencing. The initiation Met is identical with the Met-1 residue identified by Matsuda et al (1983a). [h] From Macnair & Kenny 1979. [i] From Norén et al 1986a. [j] From Feracci et al 1982.]

17–37 residues; Cowell et al 1986). The isomaltase sequence from position 1 to approximately 70 (which encompasses the cytosolic segment, the hydrophobic anchor, and the "stalk") has no homology with any other segment of pro-sucrase-isomaltase (see below for more on the homology between the sucrase and the isomaltase domains).

In the brush border membrane the sucrase-isomaltase complex occurs as a dimer via interaction of at least the hydrophilic luminal "lollipops" of its two monomers (Cowell et al 1986). This also agrees with the observation that the papain-solubilized sucrase-isomaltase complexes form dimers under appropriate conditions [as deduced from their hydrodynamic behavior (Mosimann et al 1973) and their appearance in electron micrographs (Nishi et al 1968)].

The Glucoamylase Complex

The glucoamylase complex accounts for approximately 20% of the maltase, a few percent of the isomaltase, and all of the glucoamylase (γ-amylase) activity in the small intestine.

The apparent contradictions in number, size, and subunit composition of intestinal glucoamylase(s) (for reviews see Semenza 1968, 1981a; Kenny & Maroux 1982) now seem to have been reconciled. The glucoamylase complex has an apparent M_r of 240–260 kDa and is composed of two subunits of apparent M_r's of 125 and 135 kDa (Sørensen et al 1982; C. Hu, M. Spiess, and G. Semenza, in preparation). The complex has two active sites, which differ in heat stability (Sjöström et al 1983). [These enzymes were originally classified as "heat-stable" maltases because they denatured at temperatures of 55°C or higher (Dahlqvist 1962; Auricchio et al 1965).] The glucoamylase complex seems to be a heterodimer, like sucrase-isomaltase, if it is assumed that one active site occurs in each of the subunits. It has been purified by detergent and proteolytic solubilization from the small intestines of man (Kelly & Alpers 1973b; De Burlet et al 1979), rat (Schlegel-Haueter et al 1972; Lee et al 1980; Tsuboi et al 1979), pig (Sørensen et al 1982; Taravel et al 1983), rabbit (Sivakami & Radhakrishnan 1973), pigeon (the glucoamylase from pigeon intestine consists of a single polypeptide chain having an apparent M_r of ~200 kDa, Prakash et al 1983), and chicken (C. Hu, M. Spiess, and G. Semenza, in preparation).

The mode of anchoring of the glucoamylase complex to the brush border membrane is known in less detail than that of sucrase-isomaltase, but it seems to be strongly reminiscent of it. As one N-terminus of the enzyme is changed by papain solubilization, the anchoring portion is most likely confined to the N-terminal region of one of the subunits (Maroux & Louvard 1976). Indeed, TID labeling of the chicken enzyme shows that the hydrophobic anchor occurs in the larger subunit only (C. Hu, M. Spiess, and G. Semenza, in preparation), and the most recent sequence data on the N-terminal region of the pig enzyme show a short hydrophilic, positively charged (presumably cytosolic) segment followed by a long hydrophobic segment, which presumably represents the "membrane anchor" (Norén et al 1986a) (Figure 2).

Renal neutral maltase has been purified after proteolytic solubilization from human (De Burlet & Sudaka 1976), horse (Giudicelli et al 1980), and rat (Reiss & Sacktor 1981) kidneys, and after detergent solubilization from human (De Burlet et al 1979) and horse kidneys (Giudicelli et al 1985). As judged from SDS-PAGE results, the enzyme is composed of a single polypeptide chain of ~ 335 kDa. This chain is even larger (possibly due to the larger size of the sugar moiety) than the polypeptide chain of intestinal glucoamylase in normal pigeons (Prakash et al 1983), or in pigs whose pancreata have been disconnected from their duodena (Sørensen et al 1982). It is also longer than the largest polypeptide commonly found as a minor or major component in preparations of small-intestinal gluco-amylase of normal pigs (Sørensen et al 1982) and chickens (C. Hu, M. Spiess, and G. Semenza, in preparation). Rat renal maltase reportedly does not split starch (Reiss & Sacktor 1981), although the human (De Burlet & Sudaka 1977) and horse enzymes do (Giudicelli et al 1980, 1985).

Trehalase

This enzyme is responsible for all the trehalase activity present in the small-intestinal and renal brush borders. Rat intestinal trehalase has been isolated (Sasajima et al 1975; Nakano et al 1977), with conflicting results as to its molecular weight and subunit composition. Nakano (1980, 1982) has also purified trehalase from rabbit kidney, using a pretreatment with papain (which does not solubilize this enzyme) followed by solubilization with a mixture of deoxycholate and Triton X-100. At least four forms of the enzyme have been reported, two of them glycosylated (Nakano & Sacktor 1985). However, Galand (1984) and Yokota et al (1986) recently purified rabbit small-intestinal and renal trehalase to homogeneity. These enzymes, which were obtained by detergent solubilization of brush border membranes, have (by gel filtration) an apparent M_r of 85 kDa (Galand 1984, renal and small-intestinal enzymes, no detergent present) or 330 kDa (Yokota et al 1986, renal enzyme, in Triton X-100), or of 75 kDa when denatured by SDS-PAGE (both papers). When in detergent solution, trehalase easily loses its hydrophobic anchor (Yokota et al 1986), which probably explains the different M_r values reported for the native enzymes by the two groups and the fact that trehalase was prepared without any explicit solubilization step (Labat et al 1974) from hog and human kidneys.

From the apparent M_r values of the native hydrophobic and hydrophilic forms and from the micelle weight of Triton X-100 it can be deduced that trehalase forms dimers (in Triton) and that its hydrophobic anchor must be very small (less than 5 kDa) (Yokota et al 1986). Nothing further is known about their possible oligomeric state in the membrane or their mode of anchoring. (See note added in proof.)

The trehalases, like all other brush border enzymes investigated thus far, are glycosylated. They are minor components, accounting for only 0.1 or 0.3% of the total protein.

Both renal and intestinal trehalases are inhibited competitively by Tris, sucrose, and phlorizin (see older literature in previous reviews, e.g. Semenza 1976, 1981a). Kinetic analysis of competition among inhibitors has allowed partial mapping of the subsites in renal trehalase (Nakano & Sacktor 1984).

The β-Glycosidase Complex

The β-glycosidase complex, lactase–phlorizin hydrolase (i.e. lactase-glyco-sylceramidase), only occurs in the small-intestinal brush border, where it accounts for all the lactase and nearly all the phlorizin hydrolase activity. It carries two active sites, which can be differentiated by heat inactivation and kinetic criteria (Schlegel-Haueter et al 1972; Birkenmeier & Alpers 1974; Colombo et al 1973; Skovbjerg et al 1981, 1982). One site splits lactose, cellobiose, cellotriose, and cellulose (the latter only slowly) with little activity against aryl-β-glycosides; the other splits phlorizin and other aryl-β-glycosides. The "natural substrates" of phlorizin hydrolase have been identified with glycosylceramides (Leese & Semenza 1973).

The M_r of human and pig native, detergent-solubilized, amphiphilic β-glycosidase complex is 320 kDa; in SDS-PAGE it yields a single band with an apparent M_r of ~160 kDa. It seems, therefore, that the two subunits of this complex have identical or very similar apparent M_rs. It has not yet been established whether both active sites occur on the same or on different polypeptide chains.

Papain treatment of the native amphipathic β-glycosidase complex yields a hydrophilic form with an apparent M_r of ~280 kDa, i.e. some 40 kDa smaller than the amphiphilic form (Skovbjerg et al 1981, 1982). This finding indicates that one or both subunits carry hydrophobic detergent-binding anchoring segment(s). This anchor spans the membrane, as documented by labeling from the cytosolic side with [^{125}I]DIABS (A. G. Kenny, unpublished results quoted in Kenny & Maroux 1982). It is not known which subunit is anchored to the membrane or at which point in the polypeptide chain the anchoring segment is located.

However, one can speculate on the phylogeny of the β-glycosidase complex along the same lines as for sucrase-isomaltase (see below). Thus, one would expect the β-glycosidase complex to be anchored to the brush border membrane via its glycosylceramidase subunit, perhaps at the N-terminal region. Phlorizin hydrolase, i.e. glycosylceramidase, activity is present in the small intestine of all vertebrates examined (reviewed by Semenza 1968, 1981a), much as isomaltase activity is, whereas lactase is confined to mammals (Semenza 1968), as sucrase is to most, but not all, terrestrial vertebrates (Semenza 1968) (see below).

BIOSYNTHESIS OF GLYCOSIDASES

Pro-Sucrase-Isomaltase

This section does not discuss the general aspects of the biosynthesis and insertion of intrinsic membrane proteins because the subject has been adequately covered by a number of recent important reviews. Rather, I indicate some specific features in the biosynthesis and membrane insertion of the brush border hydrolases (see also Danielsen et al 1984a). Any proposed biosynthetic mechanism for sucrase-isomaltase production must explain: (a) the peripheral position of sucrase; (b) the hydrophobic "anchor" at the N-terminal region of the polypeptide chain with the N-terminus at the cytosolic side of the membrane; (c) the fact that the anchor spans the membrane only once; (d) the finding that the N- and C-termini of sucrase face the luminal side of the cell; (e) the absence of any significant role of the C-terminal regions in anchoring to the membrane fabric; (f) the coordination of the biosynthesis and assembly of the two subunits (as shown by their simultaneous appearance in development or in response to dietary stimuli, and by their related genetic control; for reviews see Semenza 1981a,b; Koldovský 1981); (g) their very similar catalytic properties (listed in Semenza 1976, 1981a; see also Table 2); and (h) their distribution in the small intestine of vertebrates: Whereas isomaltase activity has been found in all species investigated, sucrase is only present in (most) terrestrial mammals, birds, and reptiles, and apparently is absent in (the few investigated) amphibians (reviewed in Semenza 1968, 1981a).

In 1977–1978 I presented the "one-chain precursor hypothesis" (Semenza 1978, 1979a,b; Frank et al 1978; Brunner et al 1979), which can be formulated as follows: (a) An ancestral gene coded for a one–polypeptide chain, one–active site enzyme that split both maltose and isomaltose (a "simple isomaltase"). (b) Partial gene duplication led to a gene coding for a long, single polypeptide chain with two identical active sites, which each split maltose and isomaltose (a "double isomaltase"). (c) Point mutation(s) and/or deletion(s) changed one of these active sites from an isomaltase-maltase into the sucrase-isomaltase of the present times. Thus a long polypeptide chain was formed, carrying two similar but not identical active sites. (For this "precursor" the name "pro-sucrase-

Table 2 Comparison of some properties of the subunits in rabbit small-intestinal sucrase-isomaltase

Property	Comparison	References
Number of active sites	One in sucrase subunit, one in isomaltase subunit	Cogoli et al 1973; Quaroni et al 1974
Anomeric form of glucose liberated during hydrolysis	α-Glucopyranose, by both subunits	Semenza et al 1969
Bond split	That between C_1 of glucose and glycosyl oxygen	Stefani et al 1975
Kinetic mechanism	Ping-pong BiBi or Iso-Ping-pong BiBi; rapid equilibrium	Semenza & Balthazar 1974; Janett 1974
Transglucosidase activity	Present in both sucrase and isomaltase	Semenza & Balthazar 1974
Na^+ activation	Increase of k_{cat} by approximately 30% with slight decrease in K_m	Takesue 1969; Semenza 1968; Semenza & Balthazar 1974
Substrate specificity	Both subunits split maltose, maltotriose, maltitol, and a number of aryl α-glucopyranosides. In addition, sucrase splits sucrose; isomaltase splits isomaltose, isomaltulose, and oligo 1,6-α-dextrins	Kolinská & Semenza 1967; Cogoli & Semenza 1975
Competitive inhibitors	Tris, glucono-1, 5-lactone, α-phenylthioglucopyranoside Dextran is either a pseudosubstrate or a competitive inhibitor, mainly for the isomaltase moiety Sephadex G-200 retards isomaltase considerably more than sucrase (affinity chromatography)	Janett 1974; Cogoli & Semenza 1975 Kolinská & Semenza 1967 Takesue 1969; Cogoli et al 1973; Kolinská & Semenza 1967
Irreversible competitive inhibitor (affinity label)	Conduritol β-epoxide binds irreversibly at the active site of sucrase and isomaltase, forming an ester bond with the β-carboxylate of an aspartate residue. The isomaltase site reacts approximately eight times faster than the sucrase site	Quaroni et al 1974

Partial sequence around the COO^- at the active sites (the residue bound to the label is italicized)	For both sucrase and isomaltase: Asp-Gly-Leu-Trp-Ile-*Asp*-Met-Asn-Glu	Quaroni & Semenza 1976; Hunziker et al 1986
A deprotonated group participates in the catalysis	Its pK'_a (from double logarithmic Dixon plots of both sucrase and isomaltase) corresponds to that of the carboxylate reacting with conduritol β-epoxide in each of the two active sites	Quaroni et al 1974; P. Vanni, P. L. Luisi & G. Semenza, unpublished
Additional groups identified at the active sites	One more COO^- (or COOH), as identified by reaction with $R—N=C=N—R'$ No positive evidence for other residues	Braun et al 1977
Secondary deuterium effect at C_1 of glucopyranosyl residue	The k_D/k_H for p-chlorophenyl-α-glucopyranoside ranges between 1.14 and 1.20 for both activities	Cogoli & Semenza 1975
Hammett-Hansch equations	Sucrase: $\log k_{cat} = 0.253\pi + 0.093\sigma + 2.22$ Isomaltase: $\log k_{cat} = -0.009\pi - 0.394\sigma + 2.52$	Cogoli & Semenza 1975
A proposed minimum mechanism	Protonation of the glycosyl oxygen by an as yet unidentified acid (perhaps a COOH); splitting of the bond between C_1 of glucose and glycosyl oxygen, with formation of an oxocarbonium ion (which is temporarily stabilized by the β-COO^- of an Asp) and liberation of the aglycone; final stabilization of the oxocarbonium by a nucleophile from the solvent (OH^- in the case of hydrolysis), with reformation of the original α-configuration at C_1 of glucose	Cogoli & Semenza 1975

isomaltase" was suggested, without any implication as to whether or not it is enzymatically active.) (*d*) This single-chain pro-sucrase-isomaltase is synthesized and inserted in the membrane of the endoplasmic reticulum and is finally transferred to the brush border membrane (during these processes glycosylation occurs). (*e*) Posttranslational modification of this single, long polypeptide chain (by intracellular or extracellular protease(s), e.g. from the pancreas), leads to the two subunits that make up the "final" SI complex. The subunits remain associated with one another via interactions formed during the folding of the original single-chain pro-sucrase-isomaltase (Figure 1b).

This hypothetical chain of events implies that an extensive homology exists between the two subunits of the sucrase-isomaltase complex. The large number of functional similarities between sucrase and isomaltase (Table 2) made extensive homology between them a very real possibility. Recent cDNA cloning and sequencing (Hunziker et al 1986) have elucidated the total sequence of the 1827 amino acid residues that form the single polypeptide chain of pro-sucrase-isomaltase. A significant homology was found between positions 70–860 (the isomaltase region) and 950–1750 (the sucrase region). Optimal alignment, obtained by introducing some gaps, showed that 41% of the aligned residues are identical, and 40% are nonidentical residues representing conservative changes, of which 28% result from single base mutations. The homology at the cDNA level is 52% when the same alignment is used as for the amino acids. Interestingly, the cytosolic segment, the hydrophobic anchor, and the (Thr/Ser-rich) "stalk" piece do not have homologous counterparts anywhere in the 1827-residue sequence. There can be little doubt, therefore, that a *partial* genetic duplication has occurred in the phylogenetic history of sucrase-isomaltase.

The existence of a single polypeptide chain (pro-sucrase-isomaltase of $M_r \approx 260$ kDa) during biosynthesis, insertion, and transfer of the "final" sucrase-isomaltase complex is now generally accepted. First, Hauri et al (1979) found that in vivo labeling with ^{3}H-fucose led to the appearance of a fast-labeled, high–molecular weight polypeptide in the Golgi membranes, which after Triton solubilization precipitated with antibodies directed against the final complex. This large polypeptide was split by pancreatic elastase into two bands of apparent molecular weights close to those of the subunits of final sucrase-isomaltase. The kinetics of transfer of the label from the Golgi membranes into brush border membranes, with simultaneous splitting into two bands, was also compatible with a single-chain pro-sucrase-isomaltase appearing in the Golgi membranes, being transferred to the brush border membrane with concomitant or subsequent splitting.

Direct evidence for the single-chain pro-sucrase-isomaltase was provided by observations made on hogs whose pancreas had been totally disconnected from the duodenum 3–4 days prior to sacrifice. The sucrase and isomaltase activities in the small-intestinal brush border membranes of these animals were associated with a single large (~ 260 kDa) polypeptide chain (which we call "pro-sucrase-isomaltase," although it is enzymatically fully active). Treatment with minute amounts of pancreatic proteases splits this chain into two subunits of apparent molecular weights similar, but not identical, to those of the subunits of the final sucrase-isomaltase complex (Sjöström et al 1980). Pro-sucrase-isomaltase has the same catalytic and immunological properties as final sucrase-isomaltase, which suggests that the pro form and the final complex have very similar or identical secondary and tertiary structures (Sjöström et al 1980; Hauri et al 1982). Pro-sucrase-isomaltase was also found in rats in which the pancreatic ducts had been bypassed with a catheter placed in the common bile duct (Riby & Kretchmer 1985).

Hauri et al (1980, 1982) and Montgomery et al (1981) transplanted the small intestine of fetal rats under the skin of adult animals and obtained a fast, fucose-labeled, large polypeptide that cross-reacted with anti-sucrase-isomaltase antibodies and could be converted by elastase treatment to bands similar to those of the final complex. This pro-sucrase-isomaltase has been purified from rat transplants (Hauri et al 1982). Nude mice inoculated with human malignant colonic cells develop a single-chain unsplit sucrase-isomaltase (Zweibaum et al 1983).

Human pro-sucrase-isomaltase has been demonstrated in fetal intestine obtained prior to development of significant pancreatic proteolytic activity (Skovbjerg 1982; Triadou & Zweibaum 1985) and in fetal colon (Triadou & Zweibaum 1985).

The sucrase-isomaltase developed in in vitro tissue cultures (Danielsen 1982; Danielsen et al 1982a) and in CaCo-2 cells (Hauri et al 1985b) and HT-29 cells (Zweibaum et al 1985) (CaCo-2 and HT-29 cells are established human colon carcinoma lines) is also a pro-sucrase-isomaltase with an apparent M_r in the 260 kDa range.

Preparations of detergent-solubilized sucrase-isomaltase from most species contain variable amounts of unsplit pro-sucrase-isomaltase. Unsplit pro-sucrase-isomaltase is enriched in the "intracellular and basolateral" membrane faction that precipitates with Ca^{2+} in the first step of the preparation of BBMVs (Danielsen et al 1981a,b).

Finally, cell-free translation of pro-sucrase-isomaltase has been achieved in the reticulocyte system using total RNA with or without microsomal membranes. The translation product was precipitated with antibodies raised against final sucrase-isomaltase (Wacker et al 1981; Helms et al 1984; Alpers et al 1986). (This translation product was the largest identified membrane polypeptide chain successfully translated in a cell-free system from total RNA.) In the absence of microsomal membranes pro-sucrase-isomaltase is synthesized as a polypeptide chain of apparent M_r of approximately 200 kDa, which is totally degraded by added proteinase K. In the presence of membranes, the synthesized product has an apparent M_r of approximately 220 kDa, is associated with the membranes, and is inaccessible to proteolytic degradation. The 220 kDa peptide is glycosylated (Ghersa et al 1986).

There can be no doubt that whenever sucrase-isomaltase is synthesized by a cell or cell extracts without being exposed to pancreatic or other proteases, it is always in the form of a single polypeptide chain of 200 kDa or larger. The splitting of this chain into the sucrase and isomaltase subunits is clearly a posttranslational, extracellular event.

Pro-sucrase-isomaltase begins with the isomaltase portion, as shown by the identity (not mere homology) between its N-terminal sequence and that of the final isomaltase subunit in the same species (rat, pig; Figure 2) (Hauri et al 1982; Sjöström et al 1982). This identity also strongly indicates that the mode of anchoring of pro-sucrase-isomaltase in the brush border membrane is essentially the same as that of the final sucrase-isomaltase. This is also borne out by the complete primary structure of pro-sucrase-isomaltase, which shows a single hydrophobic sequence in the N-terminal region (Hunziker et al 1986) that is identical with that already established by Edman degradation of pro-sucrase-isomaltase and the isomaltase subunit (Figure 2). In particular, no hydrophobic sequence is found in the segment joining the sucrase and isomaltase portions in pro-sucrase isomaltase or in the C-terminal region.

In a polypeptide of the size of pro-sucrase-isomaltase it is difficult to determine whether a pre-piece is synthesized prior to the N-terminus of the mature protein. The length of an average cleaved signal would be less than 1% of that of the whole pro-sucrase-isomaltase. However, the cDNA sequence of pro-sucrase-isomaltase has recently resolved this issue. In the cDNA-deduced sequence, the N-terminus of mature pro-sucrase-isomaltase (Ala-1) is preceded by the initiation codon Met, and is only 9 base pairs (bp) downstream from an in-frame stop codon (Hunziker et al 1986). Clearly, the initiation Met only is cleaved during biosynthesis and insertion of pro-sucrase-isomaltase. The only reasonable candidate to target this protein to the endoplasmic reticulum membrane during biosynthesis and insertion is the hydrophobic segment Leu-12 through Ala-31 (Figure 2), which eventually spans the brush border membrane (Figure 1). That no cleavable signal (other than the initiation Met) ever exists in pro-sucrase-isomaltase biosynthesis is also shown by the observation that [³H]Leu is incorporated into the cell-free in vitro translation product at positions 8, 12, 15, 26, and 30, i.e. in exactly the same positions as Leu is found in the final, mature, pro-sucrase-isomaltase (Ghersa et al 1986).

Figure 3 shows a minimum model for this kind of insertion that is compatible with the present concepts of the biosynthesis and insertion of membrane proteins (Austen 1979; Di Rienzo et al 1978; Inouye & Hale-goua 1980; Engelman & Steitz 1981; Wickner 1979, 1980; Blobel 1980; Sabatini et al 1982; Walter et al 1984; Gilmore & Blobel 1985; Wickner & Lodish 1985; von Heijne & Blomberg 1979; von Heijne 1985). The figure indicates the dual role of the hydrophobic sequence as "uncleaved signal" and eventual anchor.

This kind of dual role for the hydrophobic sequence was suggested long ago (Semenza 1978; 1979b), at the same time as Kenny & Booth (1978) and MacNair & Kenny (1979) suggested a similar model for renal dipeptidyl peptidase IV. During the last few years, however, we favored the hypothesis of a cleavable signal due to the position of the N-terminus of isomaltase, which was thought to be extracellular. As pointed out in a previous section, we now know that the cytosolic positioning of this terminus indicated in Figure 1 is correct.

A computer-assisted search for homologies of the sequences shown in Figure 2 with other published sequences was carried out. The only significant homology found is between segment 68–80 of human glycophorin and segment 5–17 of the isomaltases: They have 61% identity. If one restricts the comparison to segment 68–76 of glycophorin and segment 5–13 of rabbit isomaltase the identity becomes as high as 78%. The analogy between the isomaltase and glycophorin sequences extends much further to encompass His-66 and His-67 in glycophorin at positions corresponding to Arg-3 and Lys-4 in rabbit isomaltase, and the long hydrophobic sequences starting at position 76 in glycophorin and position 12 in iso-maltase. Note that the homology and analogy are based on the parts of glycophorin and isomaltase that are highly conserved in the species considered. The significance (if any) of this homology is not obvious, particularly since glycophorins and pro-sucrase-isomaltases are inserted in the bilayer in opposite directions.

Some days prior to the expression of sucrase-isomaltase, the intestinal brush border membrane in baby rabbits (Dubs et al 1975) and baby rats (Kolínská et al 1984 and work quoted therein) contains an enzymatically inactive protein that cross-reacts with final sucrase-isomaltase. The same or similar cross-reacting material was found by some workers to be associated with the crypt cells (Dubs et al 1975; Silverblatt et al 1974; Henning et al 1975); however, others have failed to detect such a protein (Hauri et

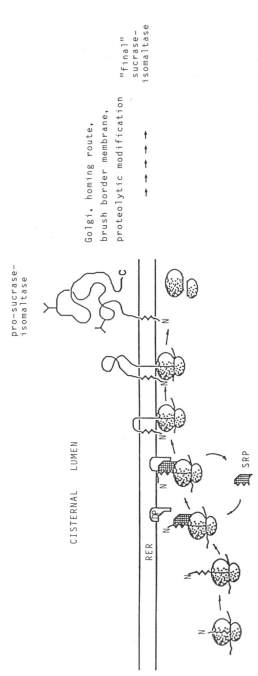

Figure 3 Suggested minimum mechanism for the synthesis and membrane assembly of pro-sucrase-isomaltase. The highly hydrophobic stretch between positions 12 and 31 (see Figure 2) is suggested to play a dual role of (uncleaved) signal during biosynthesis and of membrane anchor in the final pro-sucrase-isomaltase. SRP = signal recognition particle. DP = docking protein. The sugar chains in pro-sucrase-isomaltase are shown as branches (see Hunziker et al 1986 and Ghersa et al 1986).

al 1980; Skovbjerg 1981a). At any rate, it seems unlikely that these immunologically cross-reactive, enzymatically inactive protein(s) are biosynthetic intermediate(s) of pro-sucrase-isomaltase, because they have never been transformed into final sucrase or isomaltase and biosynthesis of sucrase-isomaltase cannot be detected in crypts (Danielsen 1984).

Pro Forms of Other Glycosidase Complexes

As indicated previously, there is no conclusive evidence that the two active sites in the glucoamylase or β-glycosidase complex are located on one subunit each, as they are in sucrase-isomaltase. However, this is probably so, because in each of the three complexes the two active sites can be clearly differentiated by their sensitivity to heat, and often to other denaturing agents as well (see, for example, Lee & Forstner 1985). The analogy among the three intestinal glycosidase complexes is also evident in their mode of biosynthesis: Like sucrase-isomaltase, the glucoamylase and β-glycosidase (lactase-glycosylceramidase) complexes are synthesized as single, gigantic polypeptide chains, which are then split into the final subunits. Pro-glucoamylase (apparent M_r approximately 260 kDa) is the sole glucoamylase present in the small intestine of normal pigeons (Prakash et al 1983) and in hogs whose pancreas has been disconnected from the duodenum (Sørensen et al 1982). It occurs in variable amounts as an additional large M_r component in preparations of detergent-solubilized glucoamylases of both pig (Sørensen et al 1982) and chicken (C. Hu, M. Spiess, and G. Semenza, in preparation). The pro form of glucoamylase appears in intracellular membranes during biosynthesis (Danielsen et al 1982a). Pro-glucoamylase is split into the two final subunits by pancreatic proteases, in particular, by elastase. [A recent report on the cell-free in vitro translation of glucoamylase as a 145-kDa peptide is still difficult to rationalize (Alpers et al 1986).] In the kidney, as mentioned in a previous section, the single-chain glucoamylase is not split into subunits.

The case of pro-lactase–phlorizin hydrolase is somewhat different. In organ cultures from human (Skovbjerg et al 1984) and pig small intestine (Danielsen et al 1984b) lactase–phlorizin hydrolase appears first as a high-M_r band (~ 245 kDa or larger), which is split intracellularly into its final subunits (each of $M_r \approx 160$ kDa) after trimming and complex glycosylation. Leupeptin inhibits the proteolytic splitting but does not prevent the homing of the enzyme into the brush border membrane. In the presence of leupeptin the lactase–phlorizin hydrolase that appears in the brush border membrane has an apparent M_r of ~ 265 kDa. It is not known whether or not pro-lactase–phlorizin hydrolase is enzymatically active.

No information is available on the biosynthetic mechanism of either renal or intestinal trehalase.

SOME PHYLOGENETIC CONSIDERATIONS

Clearly, intestinal (and renal) glycosidases are closely related to one another. Although the catalytic mechanism has only been identified in some detail for sucrase and isomaltase (Semenza 1976; Cogoli & Semenza 1975; Stefani et al 1975; Quaroni et al 1974; Quaroni & Semenza 1976; Braun et al 1977), whatever information is available on the others points to a strong similarity with sucrase and isomaltase. These similarities include the same pH dependence; the same type of inhibition by the nonprotonated form of Tris [fully competitive, also for phlorizin hydrolase if tested at the proper pH (Colombo et al 1973)]; affinity labeling by conduritol-B-expoxide (for glucoamylase, O. Norén and H. Sjöström, unpublished; for lactase-glucosylceramidase, Sjöström et al 1983); and restoration of the same configuration at the carbonyl carbon in the product as in the substrate (Semenza et al 1969). Furthermore, as mentioned above, the biosynthetic mechanism of all glycosidases investigated proceeds via the formation of a "double" enzyme, i.e. a polypeptide chain carrying two similar but not identical active sites, which eventually becomes a heterodimer.

The distribution of small-intestinal glycosidase activities among vertebrates is fairly informative, although detailed information is available almost exclusively for mammals (reviewed by Semenza 1968, 1981a). Activities of the α-glucosidases maltase and isomaltase and the β-glycosidase phlorizin hydrolase have been found in all species examined. Sucrase is apparently confined to terrestrial mammals, birds, and reptiles; lactase is present in the first weeks of life of nearly all mammalian species.

The species distribution of sucrase activity deserves special comment. A few mammals, particularly some species belonging to Pinnipedia, have isomaltase but not sucrase activity (Sunshine & Kretchmer 1964). This fact and the chain of events postulated in the "one-chain pro-sucrase-isomaltase hypothesis" (see above) prompted a study of the isomaltase occurring in the small intestine of one of these Pinnipedia, the California sea lion (*Zalophus californianus*) (Wacker et al 1984). The study found that this isomaltase is composed of a long single polypeptide chain carrying two identical active sites that each split both isomaltose and maltose. Because of a number of properties, this isomaltase clearly belongs to the family of sucrase-isomaltase. However, the position of sea lions in the pedigree of mammals does not allow placement of its isomaltase in the evolutionary line leading to the more common (pro-)sucrase-isomaltases. Rather, sea lion isomaltase probably originated by way of a back mutation that restored a protein with characteristics identical to, or reminiscent of, those of the actual double isomaltase ancestor.

The same phylogenetic considerations inherent in the pro-sucrase-isomaltase hypothesis also prompted the study of an avian sucrase-isomaltase. The chicken enzyme (and also the enzyme from pigeon, Prakash et al 1983, and from Beijing duck, Hu et al 1985) had been prepared by papain solubilization earlier (Siddons 1970; Mizuno et al 1982; Maisushita 1983), but information on the detergent-solubilized form, its anchoring and possible mode of biosynthesis, was obtained only recently (C. Hu, M. Spiess, and G. Semenza, in preparation). This work can be summarized as follows: Like the mammalian enzyme, chicken sucrase-isomaltase is composed of two subunits, the anchor is confined to isomaltase, and a long single polypeptide chain (pro-sucrase-isomaltase) is also present in the brush border membrane.

In the avian enzyme, sucrase is larger than isomaltase; however, this is the only difference detected in comparison with the mammalian enzymes. (Incidentally, this should be taken as a warning that SDS-PAGE bands should not be interpreted as sucrase or isomaltase merely on the basis of their molecular size without going through the chore of collecting direct criteria of identification.) Thus, the only difference detected thus far between the avian and the mammalian enzymes seems to be the site or extent of glycosylation and/or the site of splitting between the two domains in pro-sucrase-isomaltase, which perhaps is due to a difference in amino acid composition. The similarity of the mammalian and avian enzymes indicates that the sequence of events that led to the development of pro-sucrase-isomaltase from the hypothetical "one-chain isomaltase" (see above) took place prior to the evolutionary separation of these groups of vertebrates, i.e. more than 300 million years ago, and perhaps after the separation of amphibia.

Much too little information is available as to the other families of intestinal glycosidases to speculate on their possible evolutionary pedigrees. However, the species distributions of lactase and of phlorizin hydrolase (glycosylceramidase) activities mentioned above suggest, by analogy with the sucrase-isomaltase family, a similar chain of events. Possibly, a one-chain, one-active site phlorizin hydrolase partially doubled at some point to a one-chain, two–active site enzyme, and still later (in mammals only) one of the active sites mutated so as to split lactose.

BIOLOGICAL REGULATION OF SMALL-INTESTINAL GLYCOSIDASE ACTIVITIES

An extensive discussion of the genetic, hormonal, and dietary regulation of these enzymes is beyond the scope of the present review. A number of autosomic recessive conditions are known in man in which these activities are missing or strongly reduced. They include sucrose-isomaltose malabsorption, which is probably genetically heterogeneous (reviewed by

Semenza 1981a,b; Rey & Frézal 1967; Auricchio 1986); trehalose malabsorption, which is rare (reviewed by Semenza 1981a); lactose malabsorption of the adult, which is extremely frequent; very much reduced lactase and phlorizin hydrolase activity (Lorenz-Meyer et al 1972), which is due, as shown by SDS-PAGE and by quantitative immunological examination, to a low amount of β-glycosidase complex in the brush border membrane, not to a modified, inactive enzyme (Crane et al 1976; Skovbjerg et al 1980a; reviewed by Semenza 1981a; Hämmerli et al 1965; Kretchmer 1971; Asp & Dahlqvist 1974; Paige & Bayless 1981); and congenital lactose malabsorption, which is very rare (Holzel et al 1959; Asp et al 1973). (See also Dahlqvist & Semenza 1985.)

In most mammals lactase activity is highest around birth and decreases at the time of weaning. Caucasians are an exception to this rule: Their activity levels for these enzymes remain fairly high into adulthood. α-Glucosidase activities are usually absent at birth and develop at the time of weaning (around the second week of extrauterine life in rats and rabbits). Humans develop α-glucosidase activities in fetal life and are thus equipped with the full complement of maltase, sucrase, isomaltase, and trehalase activities, plus lactase and β-glucosidase activities, at birth. (For reviews see Semenza 1968; Koldovský 1969, 1981; Henning 1985.)

The appearance of α-glucosidases at the time of weaning is subject to hormonal control: The precocious appearance of sucrase (Doell et al 1964), isomaltase (G. Galand, unpublished), glucoamylase (Malo & Ménard 1980), trehalase (Malo & Ménard 1979a), and γ-glutamyltransferase (Ménard et al 1981a) can be triggered by administering cortisone-like steroids or ACTH some days prior to the time of their normal development. The decrease in lactase activity after weaning is accelerated by administering thyroid hormones (Raul et al 1983; Malo & Ménard 1979b). The interplay of these and other hormones (e.g. insulin, EGF) is very complex: The reader is referred to two representative papers (Ménard et al 1981b; Malo & Ménard 1982) and to some important recent reviews (Koldovský 1981; Henning 1985; Kedinger et al 1986).

Equally complex and little understood is the dietary regulation of both α- and β-glycosidases. Removal of sugars from the diet leads to a decrease in sucrase and maltase activities, which are restored to normal in a few days upon readministration of the sugars (see Riby & Kretchmer 1984, for example). Much effort has been devoted to trying to reinduce intestinal lactase after its decline in adulthood or to prevent or delay its decline at the time of weaning. The administration of lactose has only a very limited effect on this decline, if any (for reviews see Semenza 1978; Koldovský 1981). The similarities and differences in the responses of sucrase and lactase to sucrose and lactose have been critically investigated recently by Goda et al (1985).

Intestinal glycosidase activity shows a circadian rhythm (Stevenson et al 1975; reviewed by Koldovský 1981). In the rat, at 6 PM the activity may be twice that present at 6 AM. These circadian changes are primarily related to changes in the enzyme's degradation rate (Kaufman et al 1980).

The rates of turnover of intestinal disaccharidases are fast as a result of a number of factors (see Semenza 1981a; Alpers & Kinzie 1973). First, the enterocytes themselves, to which the disaccharidases are confined, take approximately 4–5 days to reach the top of the villus and be shed. When labeled leucine or glucosamine is administered intravenously, the half-life of the whole homogenate is approximately 31 hr, that of the brush border fraction approximately 18 hr, and that of the individual disaccharidase 11.5 hr (James et al 1971). In general, the large proteins of the brush border membrane, which include both the glycosidases and the proteases/peptidases, have faster turnover rates than the smaller ones (Alpers & Tedesco 1975; Forstner & Galand 1976). Important reviews on the relationship between biosynthesis of intestinal hydrolases and the ontogenesis on villus cells appeared recently (Haffen et al 1986; Kedinger et al 1986a; Lipkin 1985).

Subtotal pancreatectomy or pancreatic bypass in rats causes 50% elevation of maltase, sucrase, and lactase activity and a sizeable decrease in turnover rate (Alpers & Tedesco 1975; Riby & Kretchmer 1985). It is not clear which of the two subunits of sucrase-isomaltase is degraded first by the luminal content. In rats some researchers (Goda & Koldovský 1985) have reported that sucrase is degraded first, whereas others (Abe et al 1985) found more isomaltase-free sucrase in the ileal content than in the jejunal fluid.

ANCHORING OF PEPTIDASES-PROTEASES IN THE MEMBRANE

Unlike the glycosidases, the peptidases-proteases of the brush border membranes are a fairly heterogeneous group of enzymes. Of course, γ-glutamyltransferase has a much higher transferase than hydrolase activity. The peptidases-proteases can be differentiated by judicious choice of substrate (see Kenny & Maroux 1982) and by the use of specific inhibitors or affinity labels.

Aminopeptidase N, the most abundant enzyme of this group, is a Zn^{2+} protein that contains one metal atom per (undegraded) subunit (Lehky et al 1973). It has a rather broad substrate specificity; it prefers amino acid residues with an uncharged side chain. Aminopeptidase A is a Ca^{2+} enzyme; it splits Asp-X and Glu-X peptide bonds (Benajiba & Maroux 1980; Danielsen et al 1980). Dipeptidyl peptidase IV is a serine enzyme that is irreversibly inactivated by diisopropyl-fluorophosphate (Kenny et al 1976); it splits Gly-Pro-X peptide bonds. Endopeptidase 24.11 (that is EC 3.4.24.11) splits nonterminal peptide bonds that involve the amino groups of hydrophobic amino acid residues. It is a Zn^{2+} enzyme and is specifically inhibited by phosphoramidon (Kenny 1977; Kenny et al 1981; Kenny & Fulcher 1983; Fulcher et al 1982) and thiorphan (Fulcher et al 1982) at concentrations as low as 0.1–0.01 μM. A second endopeptidase, "meprin," which is insensitive to phosphoramidon, occurs in rat and mouse kidney brush borders (Kenny et al 1981; Beynon et al 1981; Kenny & Fulcher 1983; Beynon & Bond 1983), but not in those of rabbit, pig, or man. Carboxypeptidase P, a metalloenzyme of the renal microvillus membrane, has been purified recently after both detergent and proteolytic solubilization (Hedeager-Sørensen & Kenny 1986). Recently, a brush border antigen revealed by a monoclonal antibody has been identified as a novel aminopeptidase (named aminopeptidase W). It preferentially splits short substrates at peptide bonds involving the amino group of Trp residues (Gee & Kenny 1985). γ-Glutamyltransferase transfers the γ-glutamyl residue to an acceptor amino group. It is inactivated by the affinity label 6-diazo-5-oxo-L-norleucine, an analogue of glutamine (Inoué et al 1977; Tate & Meister 1977; Inoué et al 1979), and by phenylmethyl-sulfonyl-fluoride (Relton et al 1983).

All these enzymes occur in the brush border membranes of both kidney and small intestine (with the possible exception of the endopeptidase "meprin," for which data are available for the kidney only). Indeed, the distribution of endopeptidase 24.11 is quite vast and interesting (Kenny & Fulcher 1983). It has been isolated from the brain (Relton et al 1983) and from a number of other tissues. It has been identified with "enkephalinase" (Fulcher et al 1982; Relton et al 1983) and can split a number of neuro-peptides (Matsas et al 1984; Kenny et al 1985; Kenny 1986; Gee et al 1985). It is particularly abundant in lymph nodes (Gee et al 1985; Bowes & Kenny 1986).

γ-Glutamyltransferase may occur in a different population of renal brush borders than alkaline phosphatase (Mamelok et al 1980; however, see Curthoys 1983; Marathe et al 1979). For an excellent review of the pre-1976 literature on this enzyme see Meister & Tate (1976).

Aminopeptidases N and A and Dipeptidyl Peptidase IV

These enzymes have similar subunit structures and modes of anchoring in the membrane and have often been investigated simultaneously. Therefore, I discuss them as a group. Determination of the size and number of subunits has often been complicated by partial proteolysis, which occurs both in vitro during isolation and in vivo (insofar as an in vivo cleavage can be regarded as "artifactual"). Aminopeptidase N in all species investigated (man, rabbit, rat, and pig) is composed of a single type of subunit (M_r of approximately 160 kDa, detergent form), which occurs as a monomer in rabbits and as a homodimer in other species (Feracci & Maroux 1980; Kim & Brophy 1976; Hiwada et al 1980). This has been substantiated for the pig intestinal enzyme by covalent cross-linking (Svensson 1979) and holds true for both the renal and the small intestinal enzymes. However, proteolysis, which in the pig small intestine occurs in vivo unless the pancreas is disconnected from the duodenum (Sjöström et al 1978), produces a different subunit composition (Maroux et al 1973; Sjöström et al 1978); Vannier et al 1976) than that of the renal enzymes (Wacker et al 1976).

In most cases examined aminopeptidase A also occurs as a homodimer, with a subunit M_r of approximately 170 kDa in pig intestine and kidney and in rabbit intestine (see Table 1; Gorvel et al 1980; Stewart & Kenny 1984a). For aminopeptidase A, different subunit compositions and sizes have been reported, most probably arising from partial proteolytic degradation, which apparently also occurs in vivo in the kidney (Danielsen et al 1980). For a critical and detailed discussion on the subunit composition of aminopeptidase N and A the reader is referred to the Kenny & Maroux review (1982). Important information is also provided by the M_r of the mature forms obtained in the pulse-chase experiments (see Table 1 and below).

These proteolytic alterations in the subunit structure of aminopeptidases N and A (and of those occurring in γ-glutamyltransferase, see below), even those occurring in vivo, should not be confused with the proteolytic processing of the proglycosidases discussed above, which leads to heterodimers in which each subunit has enzymatic activity. The partial proteolytic attacks of aminopeptidases N and A and of γ-glutamyltransferase are akin, rather, to those leading to the three chains in α-chymotrypsin, for example.

Dipeptidyl peptidase IV from pig kidney is a homodimer with a subunit M_r of approximately 130 kDa (MacNair & Kenny 1979). The enzyme from pig small intestine is also a homodimer (Svensson et al 1978).

Aminopeptidase N and A and dipeptidyl peptidase IV can all be solubilized from both renal and small-intestinal brush borders by partial proteolysis or detergent treatment. The protease-solubilized forms do not have amphiphatic characteristics, so comparison of protease-solubilized and detergent-solubilized forms yields important clues as to the mode of anchoring in the original membrane. For all three enzymes, whether solubilized from kidney or small-intestinal brush border, and independent of species, it has been consistently found that proteolytic solubilization changes the N-terminal amino acid(s). At least the first 8 residues (but probably more) in the detergent form of pig kidney dipeptidyl peptidase IV are hydrophobic (MacNair & Kenny 1979) (Figure 2), as are residues 5 to at least 11 or higher in the detergent form of pig intestinal aminopeptidase N (Feracci et al 1982). These observations strongly suggest that these proteins are anchored via a hydrophobic segment located in the N-terminal region of the polypeptide chain.

In the case of dipeptidyl peptidase IV (MacNair & Kenny 1979) this segment is certainly the only anchor because the C-terminal regions of its detergent- and protease-solubilized forms do not differ and are thus not involved in anchoring to the membrane. For pig intestinal aminopeptidase N, evidence is also available that the C-terminal regions of the subunits are not involved in anchoring to the membrane fabric. Labeling in the hydrophobic layer using [^{125}I]-iodonaphthylazide does not introduce any label into subunit C (Norén & Sjöström 1980), which follows subunit B at its C-terminus. Since the subunit composition $A + B + C$ is derived from the splitting of the subunits of an original homodimer of the form $A + A$ (Sjöström & Norén 1982), this observation can only indicate that the C-terminal region of the original A subunits (i.e. the "artifactual" C subunit) is not embedded in the hydrophobic layer, where it would have been accessible to iodonaphthylazide.

Aminopeptidases N and A from the kidney (Booth & Kenny 1980) and small intestine (A. G. Booth and A. J. Kenny, unpublished results quoted by Kenny & Maroux 1982) and dipeptidyl peptidase IV from the kidney (Booth & Kenny 1980) are transmembrane proteins. This was elegantly shown by labeling the anchor, i.e. the part of the polypeptide chain that remains in the membrane during proteolytic solubilization, using [^{125}I]DIABS.

How long are the hydrophobic anchors of these enzymes? A rough upper estimate can be obtained by comparing the decrease in apparent M_r of the polypeptide chain brought about by different methods of proteolytic solubilization, i.e. comparing the sizes of the subunits solubilized by detergent or protease. This difference, after correction for the bound detergent, corresponds to the size of the "stalk," or part of it, plus the anchor proper (even in the absence of additional cuts elsewhere in the polypeptide chain). The anchor, like most hydrophobic peptides, is usually difficult to handle, and determinations of its M_r may be ambiguous. Using an isotope dilution method, the M_r of the hydrophobic anchoring peptide of rabbit intestinal aminopeptidase N was estimated to be 3.8 kDa (Feracci & Maroux 1980), that of the pig intestinal enzyme 3.5 kDa (Benajiba & Maroux 1981), and that of pig intestinal aminopeptidase A 4.5 kDa (Benajiba & Maroux 1980). These M_rs are approximately half of those previously estimated by the same laboratory based on more classical procedures. However, the isotope dilution method is also subject to sources of error because it depends on accurate determination of the amino acid composition, which may be difficult to obtain in the case of very long, very hydrophobic peptides since they are often fairly resistant to acid hydrolysis (M. Spiess,

J. Brunner, and G. Semenza, unpublished results). The M_r of the hydrophobic anchor of pig kidney dipeptidyl peptidase IV, as estimated by SDS-PAGE, is approximately 3.5 kDa (MacNair & Kenny 1979).

Hydrophobic segments are likely to form α- or 3_{10}-helical structures in the hydrophobic bilayer. In principle, a segment 35–40 amino acid residues long could cross the hydrophobic layer twice, particularly if it were in a 3_{10}-helical configuration. However, the suggestion that the anchors of dipeptidyl peptidase IV (Figure 4A) and aminopeptidases N and A span the membrane only once is more likely to be accurate if the hydrophobic stretch is short. We must await the determination of the sidedness of the N-terminal residue by independent means or the determination of sequences longer than those mentioned above. In the meantime, the suggested limited homology among the anchors of glycosidases and peptidases (Figure 2) implies that they have similar modes of anchoring and biosynthetic insertion into the membrane (see also Sjöström et al 1982; Norén et al 1986b).

Some information is also available on the length of the connecting piece between the anchor and the body of the protein (the stalk), which protrudes towards the luminal side like the stick of a lollipop. In reconstituted proteoliposomes of pig intestinal aminopeptidase N, the stalk is about 5 nm long (Hussain et al 1981; Hussain 1985). Papain can solubilize aminopeptidases N and A and dipeptidyl peptidase IV. It has been clearly shown (Kenny et al 1983) that the stalks of these proteins in the original renal brush border membrane and in reconstituted proteoliposomes are 4 nm or longer. Thus, in the gap between the body of the protein and the membrane surface, papain can have access to the stalk. This is not so for endopeptidase 24.11 (see next section).

As mentioned above, these three kinds of enzymes occur in the membrane most probably as homodimers. As far as we can judge from the chemical evidence available and the known biosynthetic mechanisms (see below), each monomer in the homodimers is anchored to the membrane by its own anchoring segment.

Endopeptidase 24.11

This enzyme, which has been purified from and identified in various tissues and species (e.g. Kerr & Kenny 1974a,b; Kenny 1977; Kenny & Fulcher 1983), occurs as a monomer in rabbits and as a dimer in other species. The monomer has an apparent M_r of approximately 87–93 kDa. The conditions required for solubilization differ from those of other brush border hydrolases. It can be solubilized by detergents such as Triton X-100, but it is totally impervious to proteolytic solubilization unless the brush border membrane has been pretreated with toluene. Only then can solubilization be achieved with trypsin. The toluene-trypsin-solubilized endopeptidase has hydrophilic properties, but that obtained by trypsin treatment of the detergent-solubilized enzyme retains its amphipathic properties. This experimental limitation makes it more difficult to infer the mode of anchoring for this endopeptidase than for enzymes that are readily released by trypsin (see also Kenny & Fulcher 1983; Fulcher & Kenny 1983). However, the simplest model is shown in Figure 4B. One established difference is

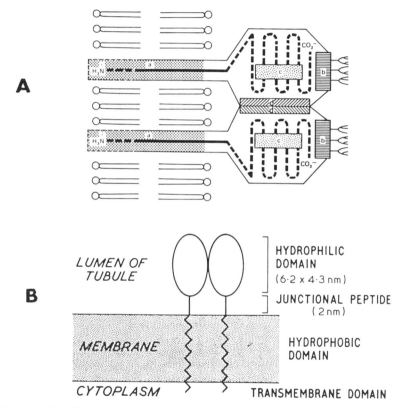

Figure 4 Models of the membrane topology of kidney dipeptidyl peptidase IV and of endopeptidase 24.11. A: *Dipeptidyl peptidase IV*. The model assumes that the enzyme has a dimeric structure in the membrane. The uncertainty concerning the extent of penetration of the anchor across the membrane is indicated by the broken lines near the internal surface of the membrane. Four domains on each subunit are represented schematically: (a) a hydrophobic domain, at or near the N-terminus, anchoring the protein to the lipid bilayer. The model is not to scale; the anchor comprises no more than 2–3% of the polypeptide chain; (b) a glycosylated region; (c) a catalytic site containing the reactive serine residue; and (d) a surface participating in association of the two monomers. (From Macnair & Kenny 1979). B: *Endopeptidase 24.11*. Note that the 2-nm gap between the hydrophilic domain and the membrane surface is too narrow to accommodate papain, which probably explains this protease's failure to solubilize endopeptidase 24.11. (From Kenny 1985.)

the length of the stalk : In the original brush border membrane and in reconstituted proteoliposomes it is as short as 2 nm (Kenny et al 1983). Thus, the gap between the body of the enzyme and the membrane surface may be too small to accommodate papain. This is a perfectly adequate explanation for the fact that this enzyme, unlike most other hydrolases, is not solubilized by papain.

Endopeptidase 24.11 also is a transmembrane protein, as shown by labeling from the cytosol with [^{125}I]DIABS (Booth & Kenny 1980). At the moment, it is not known in which region of the polypeptide chain the anchoring segment is located. One difficulty is the fact that the N-terminus is blocked (I. S. Fulcher & A. J. Kenny, unpublished results).

γ-Glutamyltransferase

This enzyme can be solubilized from the brush border membranes of kidney and small intestine by detergent or proteolytic treatment (brome-lain, papain) (Hughey & Curthoys 1976; Tate & Meister 1975). Both the detergent- and protease-solubilized forms are heterodimers. The apparent M_r of the smaller subunit is the same in both forms (~ 27 kDa in rat and 21 kDa in humans), whereas the larger subunit is ~ 54 kDa in rat (62 kDa in humans) in the detergent form and ~ 51 kDa in the protease form (Hughey & Curthoys 1976; Tate & Meister 1976; Tate & Ross 1977). Neither the N- nor the C-terminal region of the smaller subunit nor the C-terminal region of the larger subunit is changed by the proteolytic treatment leading to solubilization (Tsuji et al 1980; Matsuda et al 1983a); however, the N-terminal region of the larger subunit is changed (Figure 2). The difference in the apparent M_r value of the larger subunit in the detergent as compared to the protease form is 3–6.5 kDa (Hughey & Curthoys 1976; Tate & Meister 1976; Tsuji et al 1980), which is in fair agreement with the estimated size of the hydrophobic fragment released by proteolytic treatment of the proteoliposomes (Frielle et al 1982; Frielle & Curthoys 1983a,b). This segment encompasses the anchor proper plus a part of the stalk.

The N-terminal region of the larger subunit of the detergent form is the sole portion labeled by the hydrophobic photolabel [^{125}I]-TID in the original membrane or in proteoliposomes prepared from the detergent form (the papain-solubilized form does not form proteoliposomes) (Frielle et al 1982; Frielle & Curthoys 1983a,b). Its N-terminal sequence, as determined by Edman degradation, strongly suggests that the N-terminus is located on the cytosolic side of the membrane (three positive changes at positions 1–4) and that the hydrophobic sequence between positions 5 and approximately 21 is the membrane anchor (Matsuda et al 1983a).

These conclusions have now been fully substantiated by cDNA cloning and sequencing of the single-chain pro-γ-glutamyltransferase (Laperche et al 1986): The sequences established by Edman degradation of the large and small subunits by Matsuda's group were fully confirmed. A single hydrophobic segment longer than 10 amino acid residues was found. It is 18 (or 20) residues long, and it is located between positions 5 and 22 (or 24) of the larger subunit in "mature" γ-glutamyltransferase (Figure 2). Thus, both the proform and the "mature" enzyme are anchored to the membrane fabric via a single hydrophobic stretch located near the N-terminus of the proform (i.e. of the larger subunit). (See note added in proof.)

Despite some similarities between γ-glutamyltransferase and the sucrase-isomaltase complex (Figure 1), there are also differences. First, the subunits of γ-glutamyltransferase are much smaller. Second, the anchoring subunit, which in both cases is the larger of the two, is enzymatically active in the case of sucrase-isomaltase, but is inactive in γ-glutamyltransferase in the native state (although it may be endowed with latent activity, Horiuchi et al 1980), as shown by the fact that only the smaller subunit is labeled by affinity labels (Tate & Meister 1977; Gardell & Tate 1980). The hydrophobic anchoring segment is located at an N-terminal region both in isomaltase and in the larger subunit of γ-glutamyltransferase. This segment crosses the membrane once in each case, and the N-terminus is located on the cytosolic side of the brush border membrane.

BIOSYNTHESIS OF PROTEASES-PEPTIDASES

As discussed above, aminopeptidases N and A, dipeptidyl peptidase IV, and endopeptidase 24.11 occur in the brush border membranes mainly as homodimers (save for occasional secondary proteolytic splitting of a "normal" subunit into two or more fragments), with each subunit having its own anchoring segment. Dipeptidyl peptidase IV (Figure 4A) and the aminopeptidases are anchored to the membrane via a fairly short anchoring piece at the N-terminus, which is exposed to the cytosolic phase.

The approximate sizes of the pro forms of these enzymes have been established using various experimental approaches, including in vivo labeling followed by intestinal organelle fractionation (for aminopeptidase N, Ahnen et al 1983); cell-free translation (for intestinal aminopeptidase N, Danielsen et al 1982b; for kidney endopeptidase 24.11, I. S. Fulcher and A. J. Kenny, unpublished results); use of intestinal brush border membranes prepared from hogs whose pancreas had been disconnected from the duodenum some days prior to sacrifice (Sjöström et al 1978); investigation of the intracellular membranes precipitated with Ca^{2+} from small-intestinal homogenate from normal animals (Danielsen et al 1981a,b); use of tissue cultures of both small intestine (Danielsen 1982; Danielsen et al 1982a,b, 1983a–d; Danielsen & Cowell 1984) and kidney

(Stewart & Kenny 1984a); and use of CaCo-2 in culture (Hauri et al 1985b). The results are strikingly consistent: The subunits of these enzymes are all synthesized as forms whose M_r values are close to those of the final subunits that occur in the brush border membranes of untreated animals. (Data on small-intestinal endopeptidase 24.11 are still lacking, but it would be most surprising if they were at variance.) In no case has the synthesis of a "double enzyme" comparable to the "double glycosidase" pro form discussed in a previous section been observed. All the evidence is that each subunit is independently synthesized, inserted into the rough endoplasmic reticulum membrane, and glycosylated (see below), and that at some point it may form the homodimer eventually found for several of these proteases/peptidases.

γ-Glutamyltransferase, which is a heterodimer anchored to the membrane via the larger, non-enzymatically active subunit (Figure 2) (see above), is somewhat different. Its two subunits are synthesized as a single precursor. This was shown consistently by in vivo labeling (Capraro & Hughey 1983; Matsuda et al 1983b; Kuno et al 1983), in vitro labeling using kidney slices (Nash & Tate 1982, 1984), and in vitro translation in a cell-free system (Nash & Tate 1984; Finidori et al 1984; Barouki et al 1984; this last paper deals with the hepatoma enzyme). In the single-chain precursor, the domain corresponding to the final larger subunit is synthesized first, and thus corresponds to its N-terminal portion (Nash & Tate 1984). It is not clear at what stage the proteolytic cleavage of the single-chain precursor into the final subunits takes place: in the rough endoplasmic reticulum (Nash & Tate 1984), before (Capraro & Hughey 1983) or after (Nash & Tate 1982) the Golgi complex, or in the brush border membrane (Kuno et al 1983, 1984).

This pro form has only approximately 2% of the activity of the "mature" form (Tate 1986) (which contrasts with pro-sucrase-isomaltase, which is fully active, see above). Possibly, chain cleavage and a conformational change are needed to powerfully increase the activity of the catalytic site, as is true of pancreatic zymogens. In "mature" γ-glutamyltransferase the active site, although mainly located in the smaller subunit, extends into the intersubunit region (Gardell & Tate 1980); the enzyme activity may depend critically on the relative position of the subunits (or domains). These conclusions rest on the assumption that the proglutamyltransferase studied by Tate (1986) was undenatured. As the antibodies used for its isolation had been raised against a denatured antigen, this point is not quite secured.

Using a synchronized, cell-free translation system, the "signal" sequence (pre-piece) has been localized in the first 25% of aminopeptidase N (Danielsen et al 1983c). In the cases of dipeptidyl peptidase IV (Kenny & Booth 1978; Macnair & Kenny 1979), amino peptidase N and A (Feracci et al 1982), and γ-glutamyltransferase (Matsuda et al 1983a), it has been suggested that the hydrophobic anchor is an unsplit signal. This interesting possibility rests on the suggested mode of anchoring of these enzymes. Indeed, in the case of γ-glutamyltransferase, decisive additional information has been obtained recently; cDNA cloning and sequencing of pro-γ-glutamyltransferase mRNA unequivocally show (Laperche et al 1986) that the initiation methionine (and the following segment, including the hydrophobic sequence) are identical with Met-1 and the following amino acid residues, which Matsuda's group had sequenced at the N-terminal region of the "mature," larger subunit (Matsuda et al 1983a). No indication of a hydrophobic, potentially cleavable signal is found upstream. Also, the hydrophobic anchor of pro-γ-glutamyltransferase is not split by *Escherichia coli* leader peptidase (Finidori et al 1984), an enzyme known to cleave several eukaryotic preproteins (Watts et al 1983).

It seems, therefore, that a pattern of fundamental unity in the mode of anchoring and membrane insertion of brush border glycosidases and peptidases is emerging. As discussed in detail above, they are anchored to the membrane via a hydrophobic sequence located at the N-terminal region of the polypeptide chain. The N-terminus is on the cytosolic side of the membrane, as inferred from cDNA cloning and sequencing in the case of pro-γ-glutamyltransferase and pro-sucrase-isomaltase and from chemical labeling experiments (for the latter) and as suggested by quite reasonable indirect evidence for most peptidases. Pro-γ-glutamyltransferase and pro-sucrase-isomaltase (and in all likelihood the other enzymes as well) are synthesized and inserted without a cleavable signal. Thus a dual role for the hydrophobic sequence in the N-terminal region can be suggested: that of a "signal" during biosynthesis and insertion and that of the "anchor" in the final, "mature" enzyme (see, e.g., Figures 1 and 3). No hydrophobic anchor has yet been found in the C-terminal region or in the middle of the polypeptide chain of any of these enzymes. In some cases it has been positively rule out (see footnote 2).

Recently other proteins have been reported to be synthesized without a cleavable signal, but with a permanent hydrophobic sequence in the N-terminal region. These proteins are influenza virus neuraminidase (Bos et al 1984), the asialoglycoprotein receptor (Spiess & Lodish 1986), and the β-subunit of Na,K-ATPase (Noguchi et al 1986; Ovchinnikov et al 1986). For the first two proteins this dual role of the hydrophobic stretch located at the N-terminal region has been explicitly suggested.

Within the framework of this fundamental unity, differences are also emerging. The three glycosidase complexes are as a rule subjected to a posttranslational proteolytic processing (with the exception of the gluco-amylase of brush border kidney), which leads to heterodimers whose two subunits (probably) have a high degree of homology, with one active site each, but only one carrying the anchoring segment. The posttranslational proteolytic processing of γ-glutamyltransferase is similar to that of the glycosidase complexes; the major differences are that γ-glutamyltransferase is a much smaller molecule and that only one of the subunits in the final heterodimer is endowed with a catalytically active site. Finally, the peptidases are often homodimers, with each subunit carrying its anchoring segment. In some peptidase homodimers one subunit is split into two fragments.

It is still too early to say whether *all* brush border membrane proteins are synthesized and anchored to the membrane in the same ways glycosidases and proteases are. A cleavable signal appears during biosynthesis and insertion of placental alkaline phosphatase (Ito & Chou 1983). (This enzyme is, however, immunologically different from that of the brush border.) Also, the peculiar membrane anchoring of renal alkaline phosphatase, via phosphatidylinositol (Low & Finean 1978; Low & Zilversmit 1980), implies some pecularities, or at least additional events, in its mode of biosynthesis and primary membrane insertion. Recently it was shown that small intestinal alkaline phosphatase is not solubilized by either phospholipase C or D (Seetharam 1985), whereas the enzyme from renal brush borders is solubilized by phosphatidylinositol-specific phospholipase C (Yusufi et al 1983). (See note added in proof.)

GLYCOSYLATION AND HOMING ROUTE OF GLYCOSIDASES AND PEPTIDASES

The glycosidases, peptidases, proteases, and γ-glutamyltransferase of brush border membranes are all glycoproteins; the sugar can represent quite a sizeable percentage of the total molecular weight. The sugar chains are often responsible for the blood-group specificity of these enzymes (Kelly & Alpers 1973a; Feracci et al 1985). In most glycoproteins investigated both N- and O-glycosidic linkages have been found (see Norén et al 1986; Massey & Maroux 1985). Some of these proteins have terminal sialic acids, some do not.

Isozymes of these hydrolases having different M_r values in SDS-PAGE have been found along the small intestine (e.g. for lactase, Cousineau & Green 1980), in different tissues and strains (e.g. for endopeptidase 24.11, Fulcher et al 1983; Relton et al 1983; Stewart et al 1984), or in the same tissue at different stages of development (e.g. intestinal α-glucosidases, dipeptidyl peptidase IV, γ-glutamyltransferase, Kraml et al 1983, 1984; see also Mahmood & Torres-Pinedo 1983; Torres-Pinedo & Mahmood 1984). In the extreme case of γ-glutamyltransferase, twelve isozymes have been counted for the rat kidney enzyme (Tate & Meister 1976; Matsuda et al 1980). Indeed, the oligosaccharide chains of γ-glutamyltransferase are exceptionally heterogeneous (Yamashita et al 1983a), although they are bound at only three sites in the polypeptide chain (all three are N-glycosidic). These isozymes are apparently identical polypeptide chains with different complements of associated sugars. Thus, treatment with neuraminidase leads to disappearance of the differences in the apparent M_r values for intestinal lactases (Cousineau & Green 1980) and renal glucoamylase (De Burlet et al 1979) and eliminates the differences in apparant M_rs of intestinal enzymes during development (but see more in Auricchio 1983). The different mobilities of rat kidney γ-glutamyltransferases correlate with their sialic acid content (Yamashita et al 1983b). Deglycosylation of endopeptidase 24.11 from different tissues using trifluoromethane sulfonic acid leads in all cases to polypeptide chains of 77 kDa, although the untreated forms have apparent M_r values ranging from 89–95 kDa (Stewart et al 1984). These observations suggest that many of the differences and discrepancies in the apparent sizes of subunits of the same enzyme in various organs may be due to differences in the complement of sugars more often than previously thought, rather than to different sizes of the polypeptide chains.

In cell-free translation systems using dog pancreas microsomal membranes, translocation and glycosylation of the translation product have been shown for pro-sucrase-isomaltase (Ghersa et al 1986), intestinal aminopeptidase N (Danielsen et al 1983c), and γ-glutamyltransferase (Nash & Tate 1984; Finidori et al 1984).

A few minutes after a pulse of radioactive precursor, newly synthesized brush border membrane proteins are found already solely associated with intracellular membranes. This has been shown in vivo for small-intestinal sucrase-isomaltase (Hauri et al 1979), and for aminopeptidase N (Feracci et al 1985); in small-intestinal organ or cell cultures for aminopeptidases N and A, dipeptidyl peptidase IV, pro-sucrase-isomaltase, pro-glucoamylase, and pro-β-glycosidase complex (Danielsen 1982; Danielsen et al 1982a, 1983a; Sjöström et al 1983; Hauri et al 1985b), and in kidney slices for aminopeptidases N and A, dipeptidyl peptidase IV, endopeptidase 24.11 (Stewart & Kenny 1984a), and γ-glutamyltransferase (Nash & Tate 1982). In all fairness, at least some of the authors (Ahnen et al 1982) who suggested the existence of non-membrane-associated, cytosolic, transient precursors indicated that the apparent precursors could have arisen artifactually by being proteolytically dislodged from intracellular membranes during the processing of the tissue—a possibility the present reviewer indicated earlier (Semenza 1981a).

In intestinal tissue cultures [and in vivo (Ahnen et al 1983)] and kidney slices, the earliest products detected are already N-glycosylated to high-mannose forms (presumably cotranslationally), as evidenced by their endo-H sensitivity. Pulse chase experiments (Danielsen et al 1983b) (Figure

5A) showed that these high-mannose forms prevail for 30–60 min, a time that presumably reflects their movement from the rough endoplasmic reticulum to the Golgi complex (Hubbard & Ivatt 1981). In the latter, the last "trimming" steps, complex glycosylation, and O-glycosylation take place (Danielsen et al 1983d; Danielsen & Cowell 1984), which lead to the final forms (or in some cases pro forms) of the brush border proteins. These glycoproteins are no longer sensitive to endo-H. Thus, the apparent M_rs of the translation products change en route from the rough endoplasmic reticulum to the brush border membranes (see Table 1). If we take the example of the small-intestinal aminopeptidase N (Danielsen et al 1982b), the M_r of its primary translation product is 115 kDa, that of the high mannose form is 140 kDa, and that of the final complex, glycosylated form is 160 kDa.

The kidney slices system has been equally fruitful. It has allowed identification of the routes of renal brush border aminopeptidases N and A, dipeptidyl peptidase IV, and endopeptidase 24.11, which are strikingly similar to those established for the small-intestinal enzymes (Stewart & Kenny 1984a).

The conclusions quoted above, which are based on pulse-chase experiments carried out mainly in tissue cultures, agree very well with observations made with known inhibitors of the individual steps in the glycosylation assembly line. In pig intestine tissue cultures, tunicamycin, although severely reducing their expression, leads to the appearance of

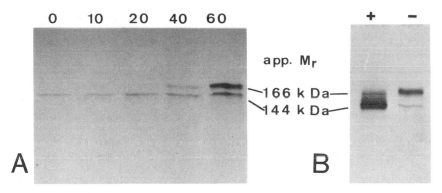

Figure 5 Biosynthesis of aminopeptidase N in pig small-intestinal organ cultures. A: *Pulse-chase labeling*, showing conversion from transient to mature form. The pulse ([^{35}S]methionine) lasted 10 min, and the chase was for the time indicated (in min) at the top of the lane. The total membranes of the explants were solubilized by Triton X-100 and precipitated with anti-aminopeptidase N IgG + protein-A-sepharose. Separation was in SDS-PAGE after denaturation in hot SDS plus mercaptoethanol (from Danielsen 1982). B: *Effect of monensin* on conversion from transient to mature form. In the presence of monensin (+) a blur of polypeptides is seen extending below and above the transient form. Conversion to the mature form is not completed (from Danielsen et al 1983b).

polypeptides of apparent M_r values of 115 kDa for aminopeptidase N, 225 kDa for pro-sucrase-isomaltase, 200 kDa for pro-lactase-glycosyl-ceramidase, and 200 kDa for glucoamylase, all of which correspond to unglycosylated primary translation product values (Danielsen & Cowell 1984). Swainsonine, an inhibitor of mannosidase II, interferes with the trimming process and leads to hybrid forms that are still sensitive to Endo-H but are complex glycosylated (Danielsen et al 1983d). Monensin, which causes gross ultrastructural changes in the Golgi apparatus, stops the processing of high-mannose forms to the complex, glycosylated forms (Figure 5B) (Danielsen et al 1983b). In kidney slices (Stewart & Kenny 1984b), monensin and swainsonine produce alterations similar to those they produce in the small intestine. The glycosylation of brush border glycosidases and peptidases thus follows the pattern generally accepted for integral membrane proteins (for reviews see Hanover & Lennarz 1981; Weiser & Wilson 1981; Kornfeld & Kornfeld 1985).

As to the enzymatic activity of the transient forms, Sjöström et al (1985) isolated from the Ca^{2+} precipitate the "high-mannose" forms of intestinal aminopeptidase N, pro-glucoamylase, and pro-sucrase-isomaltase. No difference in enzyme activities was found between the high-mannose and complex glycosylated forms of aminopeptidases N (see also Feracci et al 1985) or pro-glucoamylases, but high-mannose pro-sucrase-isomaltase is about three times less active than the complex glycosylated form. This difference may reflect a stabilizing effect of complex glycosylation (plus O-glycosylation).

How the fully glycosylated brush border membrane proteins reach their final destination is unclear. Colchicine does not detectably affect the synthesis and posttranslational modification of intestinal aminopeptidase N, pro-sucrase-isomaltase, or glucoamylase, but it prevents their final homing to the brush border membrane, thus microtubuli may play a role in this process (Danielsen et al 1983b; for experiments in vivo using colchicine see Quaroni et al 1979b). However, at smaller concentrations and shorter exposure times colchicine does not have this effect in the kidney slices system (Stewart & Kenny 1984b).

Peptidases travel considerably faster than glycosidases from the rough endoplasmic reticulum (RER) to the Golgi membranes; but the two groups of enzymes take about the same time to move from the Golgi complex to the brush border membranes (Danielsen & Cowell 1985a; Hauri et al 1985b). Which route do these enzymes follow between the Golgi apparatus and the brush border membranes? The possibility that they may first be incorporated into the basolateral membrane (Hauri et al 1979; Quaroni et al 1979a) has now been effectively ruled out by Danielsen & Cowell (1985b; see also Danielsen et al 1986). These authors used a highly sensitive detection procedure (biosynthetic labeling with [^{35}S]-

methionine) to show that the extracellular face of right-side-out vesicles of basolateral membranes is devoid of aminopeptidase N.

Brush border proteins must, therefore, reach their final destination from the Golgi membranes by crossing the terminal web, a route for which highly suggestive electron micrographs have been produced (Mooseker et al 1983). The cytoskeleton of the brush border, rather than representing an obstacle, may help the final homing of brush border membrane proteins, perhaps by directing the movement of relevant smooth-surfaced vesicles, or perhaps originating from non–clathrin coated vesicles (Danielsen et al 1986). Whatever route is followed by the brush border membrane proteins from the Golgi complex, it does not involve either CURL or lysosomes, since chloroquine does not affect the surface expression of aminopeptidase N in small-intestinal tissue cultures (Danielsen et al 1985). A discussion of this very interesting topic, which is obviously related to the generation and maintenance of cell polarity in epithelia, is premature at the present state of knowledge. The "address" responsible for the sorting out (and perhaps for the correct homing) of the brush border membrane proteins is still unknown. (For a discussion see Simons & Fuller 1985; Sjöström et al 1986.) Local microtubules may be a part of the machinery of the post-Golgi transport (Quaroni et al 1979b; Blok et al 1981; Danielsen et al 1983b).

A case of sucrose-isomaltose malabsorption in which pro-sucrase-iso-maltase is interpreted not to proceed beyond the Golgi apparatus has been recently reported (Hauri et al 1985a). Although the published electron micrographs are difficult to interpret, and although it is premature to suggest testable hypotheses as to the nature of the block involved, sucrose-isomaltose malabsorption may again prove an important tool in the investigation of the cell biology of the enterocyte. (For a review of the pre-1980 literature see Semenza 1981b.)

ACKNOWLEDGMENTS

Thanks are due to Drs. A. J. Kenny (Leeds), H. Sjöström and O. Norén (Copenhagen), J. Brunner, M. Spiess, and H. Wacker (Zürich) for critically reading the manuscript, and to the SNSF (Berne) and Nestlé Alimentana (Vevey) for having supported in part the reviewer's work quoted.

NOTE ADDED IN PROOF   Coloma & Pitot (1986) have published the sequence of a cDNA clone of pro-γ-glutamyltransferase (not extending, however, upstream from the initiation Met), which confirms that found by Laperche et al (1986). Recently, Takesue et al (1986) reported that trehalase and alkaline phosphatase are solubilized from small-intestinal and renal brush border membranes by phosphatidylinositol-specific phospholipase C. Sucrase, maltase, endopeptidase, and aminopeptidase N are not. (See also

Yasufi et al 1983 : This phospholipase C does not solubilize renal leucine aminopeptidase, γ-glutamyltransferase, or maltase.) Very recently, Lee & Forstner (1986) used the hydrophobic photolabel [^{125}I] TID to label the anchoring segment of rat small-intestinal maltase-glucoamylase. In this species the anchor is found in the larger of the two subunits.

Literature Cited

Abe, M., Yamada, K., Hosoya, N., Moriuchi, S. 1985. Some properties of luminal sucrase and sucrase-isomaltase complex in rat small intestine. *J. Nutr. Sci. Vitaminol.* 31 : 243–52

Ahnen, D. J., Mircheff, A. K., Santiago, N. A., Yoshioka, Ch., Gray, G. M. 1983. Intestinal surface aminooligopeptidase. Distinct molecular forms during assembly on intracellular membranes in vivo. *J. Biol. Chem.* 258 : 5960–66

Ahnen, D. J., Santiago, N. A., Cezard, J.-P., Gray, G. M. 1982. Intestinal aminooligopeptidase. In vivo synthesis on intracellular membranes of rat jejunum. *J. Biol. Chem.* 257 : 12129–35

Alpers, D. H., Helms, D., Seetharam, S., May, V. L., Strauss, A. W. 1986. In vitro translation of intestinal sucrase-isomaltase and glucoamylase. *Biochem. Biophys. Res. Commun.* 134 : 37–43

Alpers, D. H., Kinzie, J. L. 1973. Regulation of small-intestinal protein metabolism. *Gastroenterology* 64 : 471–96

Alpers, D. H., Tedesco, F. J. 1975. The possible role of pancreatic proteases in the turnover of intestinal brush border proteins. *Biochim. Biophys. Acta* 401 : 28–40

Asp, N. G., Dahlqvist, A. 1974. Intestinal β-galactosidases in adult low lactase activity and in congenital lactase deficiency. *Enzyme* 18 : 84–102

Asp, N. G., Dahlqvist, A., Kuitunen, P., Launiala, K., Visakorpi, J. K. 1973. Complete deficiency of brush border lactase in congenital lactose malabsorption. *Lancet* 2 : 329–30

Auricchio, S. 1983. "Fetal" forms of brush border enzymes in the intestine and meconium. *J. Pediatr. Gastroent. Nutr.* 3 : 28–36

Auricchio, S. 1986. Brush border enzymes. In *Paediatric Gastroenterology*, ed. Ch. Anderson, V. Burke, M. Gracey. Carlton, Victoria, Australia : Blackwell Sci. 2nd Ed. In press

Auricchio, S., Semenza, G., Rubino, A. 1965. Multiplicity of human intestinal disaccharidases. II. Characterization of the individual maltases. *Biochim. Biophys. Acta* 96 : 498–507

Austen, B. M. 1979. Predicted secondary structures of amino-terminal extension sequences of secreted proteins. *FEBS Lett.* 103 : 308–13

Barouki, R., Finidori, J., Chobert, M. N., Aggerbeck, M., Laperche, Y., Hanoune, J. 1984. Biosynthesis and processing of γ-glutamyl transpeptidase in hepatoma tissue culture cells. *J. Biol. Chem.* 259 : 7970–74

Bayley, H. 1983. *Photogenerated Reagents in Biochemistry and Molecular Biology*. Amsterdam : Elsevier. 187 pp.

Benajiba, A., Maroux, S. 1980. Purification and characterization of an aminopeptidase A from hog intestinal brush border-membrane. *Eur. J. Biochem.* 107 : 381–88

Benajiba, A., Maroux, S. 1981. Subunit structure of hog intestinal brush border aminopeptidase N. *Biochem. J.* 197 : 573–80

Beynon, R. J., Bond, J. S. 1983. Deficiency of a kidney metalloproteinase activity in inbred mouse strains. *Science* 219 : 1351–53

Beynon, R. J., Shannon, J. D., Bond, J. S. 1981. Purification and characterization of a metallo-endoproteinase from mouse kidney. *Biochem. J.* 199 : 591–98

Binkley, F. 1952. Evidence for the polynucleotide nature of cysteinyl-glycinase. *Exp. Cell Res. Suppl.* 2 : 145–57

Binkley, F., Alexander, V., Bell, F. E., Lea, Ch. 1957. Peptidases and alkaline phosphatase of swine kidney. *J. Biol. Chem.* 228 : 559–67

Birkenmeier, E., Alpers, D. H. 1974. Enzymatic properties of rat lactase-phlorizin hydrolase. *Biochim. Biophys. Acta* 350 : 100–12

Bisson, E., Montecucco, C. 1985. Use of photoreactive phospholipids for the study of lipid-protein interactions. In *Progress in Protein-Lipid Interactions*, ed. A. Watts, J. J. H. H. M. de Pont, pp. 259–87. Amsterdam : Elsevier Biomedical

Blobel, G. 1980. Intracellular protein topogenesis. *Proc. Natl. Acad. Sci. USA* 77 : 1496–1500

Blok, J., Mulder-Stapel, A. A., Ginsel, L. A., Daems, W. T. 1981. Endocytosis in absorptive cells of cultured human small-intestinal tissue. Horseradish peroxidase and ferritin as markers. *Cell Tissue Res.* 216: 1–13

Booth, A. G., Kenny, A. J. 1974. A rapid method for the preparation of microvilli from rabbit kidney. *Biochem. J.* 142: 575–81

Booth, A. G., Kenny, A. J. 1980. Proteins of the kidney microvillar membrane. Asymmetric labelling of the membrane by lactoperoxidase-catalyzed radioiodination and by photolysis of 3,5-di[^{125}I]iodo-4-azidobenzenesulphonate. *Biochem. J.* 187: 31–44

Booth, A. G., Olaniyan, R. O., Vanderpuye, O. A. 1980. An improved method for the preparation of human placental syncytiotrophoblast microvilli. *Placenta* 1: 327–67

Bordier, C. 1981. Phase separation of integral membrane proteins in TX-114 solution. *J. Biol. Chem.* 256: 1604–7

Bowes, M. A., Kenny, A. J. 1986. Endopeptidase-24.11 in pig lymph nodes: Purification and immunochemical localization in reticular cells. *Biochem. J.* 236: 801–10

Braun, H., Legler, G., Deshusses, J., Semenza, G. 1977. Stereospecific ring opening of conduritol-B-epoxide by an active site aspartate residue of sucrase-isomaltase. *Biochim. Biophys. Acta* 483: 135–40

Bretscher, A. 1983. In *Molecular Architecture of the Microvillus Cytoskeleton*, ed. R. Porter, G. Collins, pp. 164–79

Brunner, J. 1981. Labelling the hydrophobic core of membranes. *Trends Biochem. Sci.* 6: 44–46

Brunner, J. 1987. Photochemical labeling of the apolar phase of membranes. *Methods Enzymol.* In press

Brunner, J., Hauser, H., Braun, H., Wilson, K. J., Wacker, H., O'Neill, B., Semenza, G. 1979. The mode of association of the enzyme complex sucrase-isomaltase with the intestinal brush-border membrane. *J. Biol. Chem.* 254: 1821–28

Brunner, J., Hauser, H., Semenza, G. 1978. Single bilayer lipid-protein vesicles formed from phosphatidylcholine and small intestinal sucrase-isomaltase. *J. Biol. Chem.* 253: 7538–48

Brunner, J., Semenza, G. 1981. Selective labeling of the hydrophobic core of membranes with 3-trifluoromethyl-3-(m-[^{125}I]iodophenyl)diazirine, a carbene-generating reagent. *Biochemistry* 20: 7174–82

Brunner, J., Wacker, H., Semenza, G. 1983. Sucrase-isomaltase of the small-intestinal brush border membrane: Assembly and biosynthesis. *Methods Enzymol.* 96: 386–406

Bürgi, R., Brunner, J., Semenza, G. 1983.

A chemical procedure for determining the sidedness of the NH$_2$ terminus in a membrane protein. The small-intestinal sucrase-isomaltase. *J. Biol. Chem.* 258: 15114–19

Capraro, M. A., Hughey, R. 1983. Processing of the propeptide form of rat renal γ-glutamyltranspeptidase. *FEBS Lett.* 157: 139–43

Cogoli, A., Eberle, A., Sigrist, H., Joss, Ch., Robinson, E., et al. 1973. Subunits of the small-intestinal sucrase-isomaltase complex and separation of its enzymatically active isomaltase moiety. *Eur. J. Biochem.* 33: 40–48

Cogoli, A., Semenza, G. 1975. A probable oxocarbonium ion in the reaction mechanism of small intestinal sucrase and isomaltase. *J. Biol. Chem.* 250: 7802–9

Colbeau, A., Maroux, S. 1978. Integration of alkaline phosphatase in the intestinal brush-border membrane. *Biochem. Biophys. Acta* 511: 39–51

Colombo, V., Lorenz-Meyer, H., Semenza, G. 1973. Small-intestinal phlorizin hydrolase: The "β-glycosidase complex." *Biochim. Biophys. Acta* 327: 412–24

Coudrier, E., Reggio, H., Louvard, D. 1983. Characterization of membrane glycoproteins involved in attachment of microfilaments to the microvillar membrane. See Porter & Collins 1983, pp. 216–30

Cousineau, J., Green, J. R. 1980. Isolation and characterization of the proximal and distal forms of lactase-phlorizin hydrolase from the small intestine of the suckling rat. *Biochim. Biophys. Acta* 615: 147–57

Cowell, G. M., Sjöström, H., Norén, O., Tranum-Jensen, J. 1986. Topology and quarternary structure of pro-sucrase-isomaltase and final form sucrase-isomaltase. *Biochem. J.* 237: In press

Crane, R. K., Menard, D., Preiser, H., Cerda, J. 1976. The molecular basis of brush border membrane disease. In *Membranes and Disease*, ed. L. Bolis, J. F. Hoffman, A. Leaf, pp. 229–41. New York: Raven

Curthoys, N. P. 1983. See Porter & Collins 1983, p. 29

Dahlqvist, A. 1962. Specificity of the human intestinal disaccharidases and implications for hereditary disaccharide intolerance. *J. Clin. Invest.* 41: 463–70

Dahlqvist, A., Semenza, G. 1985. Disaccharidases of small-intestinal mucosa (Editorial). *J. Pediatr. Gastrent. Nutr.* 4: 857–56

Danielsen, E. M. 1982. Biosynthesis of intestinal microvillar proteins. Pulse-chase labeling studies on aminopeptidase N and sucrase-isomaltase. *Biochem. J.* 204: 639–45

Danielsen, E. M. 1984. Biosynthesis of intestinal microvillar proteins. Expression of aminopeptidase N along the crypt-villus axis. *Eur. J. Biochem.* 145: 653–58

Danielsen, E. M., Cowell, G. M. 1984. Biosynthesis of intestinal microvillar proteins. Further characterization of the intracellular processing and transport. *FEBS Lett.* 166: 28–32

Danielsen, E. M., Cowell, G. M. 1985a. Biosynthesis of intestinal microvillar proteins. The intracellular transport of aminopeptidase N and sucrase-isomaltase occurs at different rates *pre*-Golgi but at the same rate *post*-Golgi. *FEBS Lett.* 190: 69–72

Danielsen, E. M., Cowell, G. M. 1985b. Biosynthesis of intestinal microvillar proteins. Evidence for an intracellular sorting taking place in, or shortly after, exit from the Golgi complex. *Eur. J. Biochem.* 152: 493–99

Danielsen, E. M., Cowell, G., Hansen, J. H., Jorr, S. U., Sjöström, H., Norén, O. 1986. Role of the Golgi complex and characteristics of post-Golgi transport in the biosynthesis of intestinal microvillar proteins. *Biochem. Soc. Trans.* 14: 165–70

Danielsen, E. M., Cowell, G. M., Norén, O., Sjöström, H. 1984a. Biosynthesis of microvillar proteins. *Biochem. J.* 221: 1–14

Danielsen, E. M., Cowell, G. M., Norén, O., Sjöström, H., Dorling, P. R. 1983d. Biosynthesis of intestinal microvillar proteins. The effect of swainsonine on posttranslational processing of aminopeptidase N. *Biochem. J.* 216: 325–31

Danielsen, E. M., Cowell, G. M., Poulsen, S. S. 1983b. Biosynthesis of intestinal microvillar proteins. Role of the Golgi complex and microtubules. *Biochem. J.* 216: 37–42

Danielsen, E. M., Norén, O., Sjöström, H. 1982b. Biosynthesis of intestinal microvillar proteins. Translational evidence *in vitro* that aminopeptidase N is synthesized as a M_r 115,000 polypeptide. *Biochem. J.* 204: 323–27

Danielsen, E. M., Norén, O., Sjöström, H. 1983c. Biosynthesis of intestinal microvillar proteins. Processing of aminopeptidase N by microsomal membranes. *Biochem. J.* 212: 161–65

Danielsen, E. M., Norén, O., Sjöström, H., Ingram, J., Kenny, A. J. 1980. Aspartate aminopeptidase: Purification by immunoadsorbant chromatography and properties of the detergent- and proteinase-solubilized forms. *Biochem. J.* 189: 591–603

Danielsen, E. M., Sjöström, H., Norén, O. 1981a. Biosynthesis of intestinal microvillar proteins. Putative precursor of microvillus aminopeptidase and sucrase-isomaltase isolated from Ca^{2+}-precipi-

tated enterocyte membranes. *FEBS Lett.* 127: 129–32

Danielsen, E. M., Sjöström, H., Norén, O. 1982c. Hypotonic elution, a new desorption principle in immunoadsorbent chromatography. *J. Immunol. Methods* 52: 223–31

Danielsen, E. M., Sjöström, H., Norén, O. 1983a. Biosynthesis of microvillar proteins. Pulse-chase labeling studies on maltase-glucoamylase, aminopeptidase A and dipeptidyl peptidase IV. *Biochem. J.* 210: 389–93

Danielsen, E. M., Sjöström, H., Norén, O. 1985. Biosynthesis of intestinal microvillar proteins. Surface expression of aminopeptidase N is not affected by chloroquine. *FEBS Lett.* 179: 160–64

Danielsen, E. M., Sjöström, H., Norén, O., Bro, B., Dabbelsteen, E. 1982a. Biosynthesis of intestinal microvillar proteins. Characterization of intestinal explants in organ culture and evidence for the existence of pro-forms of microvillar proteins. *Biochem. J.* 202: 647–54

Danielsen, E. M., Skovbjerg, H., Norén, Sjöström, H. 1981b. Biosynthesis of intestinal microvillar proteins. Nature of precursor forms of microvillar enzymes from Ca^{2+}-precipitated enterocyte membranes. *FEBS Lett.* 132: 197–200

Danielsen, E. M., Skovbjerg, H., Norén, O., Sjöström, H. 1984b. Biosynthesis of intestinal microvillar proteins. Intracellular processing of lactase-phlorizin hydrolase. *Biochem. Biophys. Res. Commun.* 122: 82–90

De Burlet, G., Vannier, C., Giudicelli, J., Sudaka, P. 1979. Neutral α-glucosidase from human kidney. *Biochimie* 61: 1177–83

De Burlet, G., Sudaka, P. 1977. Propriétés catalytiques de l'α-glucosidase neutre du rein humain. *Biochimie* 59: 7–14

De Burlet, G., Sudaka, P. 1976. Préparation de la maltase neutre rénale humaine. *Biochimie* 58: 621–23

Di Rienzo, J. M., Nakamura, K., Inouye, M. 1978. The outer membrane proteins of gram-negative bacteria: Biosynthesis, assembly, and functions. *Ann. Rev. Biochem.* 47: 481–532

Doell, R. G., Kretchmer, N. 1964. Intestinal invertase: Precocious development of activity after injection of hypercortisone. *Science* 143: 42–44

Drenth, J., Jansonius, J. N., Koekoek, R., Wolthers, B. G. 1971. Papain, X-ray structure. *Enzymes* 3: 485–99

Dubs, R., Gitzelmann, R., Steinmann, B., Lindenmann, J. 1975. Catalytically inactive sucrase antigen of rabbit small intestine: The enzyme precursor. *Helv. Paediat. Acta* 30: 89–102

Engelman, D. M., Steitz, J. 1981. The spontaneous insertion of proteins into and across membranes: The helical hairpin hypothesis. *Cell* 23: 411–22

Feracci, H., Maroux, S. 1980. Rabbit intestinal aminopeptidase N. Purification and molecular properties. *Biochim. Biophys. Acta* 599: 448–63

Feracci, H., Maroux, S., Bonicel, J., Desnuelle, P. 1982. The amino acid sequence of the hydrophobic anchor of rabbit intestinal brush border aminopeptidase N. *Biochim. Biophys. Acta* 684: 133–36

Feracci, H., Rigal, A., Maroux, S. 1985. Biosynthesis and intracellular pool of aminopeptidase N in rabbit enterocytes. *J. Membr. Biol.* 83: 139–46

Finidori, J., Laperche, Y., Haguenauer-Tsapis, R., Barouki, R., Guellaën, G., Hanoune, J. 1984. In vitro biosynthesis and membrane insertion of γ-glutamyl transferase transpeptidase. *J. Biol. Chem.* 259: 4687–90

Forstner, G., Galand, G. 1976. The influence of hydrocortisone on the synthesis and turnover of microvillus membrane glycoproteins in suckling rat intestine. *Can. J. Biochem.* 54: 224–32

Frank, G., Brunner, J., Hauser, H., Wacker, H., Semenza, G., Zuber, H. 1978. The hydrophobic anchor of small-intestinal sucrase-isomaltase. N-terminal sequence of the isomaltase subunit. *FEBS Lett.* 96: 183–88

Frielle, T., Brunner, J., Curthoys, N. P. 1982. Isolation of the hydrophobic membrane binding domain of rat renal γ-glutamyl transpeptidase selectively labeled with 3-trifluoromethyl - 3 - ($m[^{125}I]$iodophenyl) diazirine. *J. Biol. Chem.* 257: 14979–92

Frielle, T., Curthoys, N. P. 1983a. Specific-labeling of the hydrophobic domain of rat renal γ-glutamyl-transferase. See Porter & Collins 1983, pp. 73–83

Frielle, T., Curthoys, N. P. 1983b. Characterization of the membrane binding domain of γ-glutamyltranspeptidase by specific labeling techniques. *Biochemistry* 22: 5709–14

Fulcher, I. S., Chaplin, M. F., Kenny, A. J. 1983. Endopeptidase-24.11 purified from pig intestine is differently glycosylated from that in kidney. *Biochem. J.* 215: 317–23

Fulcher, I. S., Kenny, A. J. 1983. Proteins of the kidney microvillar membrane. The amphipathic forms of endopeptidases purified from pig kidney. *Biochem. J.* 211: 743–53

Fulcher, I. S., Matsas, R., Turner, A. J., Kenny, A. J. 1982. Kidney neutral endopeptidase and the hydrolysis of enkephalin by synaptic membranes show similar

sensitivity to inhibitors. *Biochem. J.* 203: 519–22

Furthmayr, H., Galardy, R. E., Tomita, M., Marchesi, V. T. 1978. The intramembranous segment of human erythrocyte glycophorin A. *Arch. Biochem. Biophys.* 185: 21–29

Galand, G. 1984. Purification and characterization of kidney and intestinal brush border membrane trehalases from the rabbit. *Biochim. Biophys. Acta* 789: 10–19

Gardell, S. J., Tate, S. S. 1980. Affinity labeling of γ-glutamyl-transpeptidase by glutamine antagonists. *FEBS Lett.* 122: 171–74

Gee, N. S., Bowes, M. A., Buck, P., Kenny, A. J. 1985. An immunoradiometric assay for endopeptidase-24.11 shows it to be a widely distributed enzyme in pig tissues. *Biochem. J.* 228: 119–26

Gee, N. S., Kenny, A. J. 1985. The 130 kDa protein in pig kidney, recognized by monoclonal antibody GK5Cl, is an ectoenzyme with aminopeptidase activity. *Biochem. J.* 230: 753–64

Gee, N. S., Matsas, R., Kenny, A. J. 1983. A monoclonal antibody to kidney endopeptidase-24.11. Its application in immunoadsorbent purification of the enzyme and immunofluorescent microscopy of kidney and intestine. *Biochem. J.* 214: 377–86

Ghersa, P., Huber, P., Semenza, G., Wacker, H. 1986. Cell-free synthesis, membrane integration and glycosylation of pro-sucrase-isomaltase. *J. Biol. Chem.* 261: 7969–74

Gilmore, R., Blobel, G. 1985. Translocation of secretory proteins across the microsomal membrane occurs through an environment accessible to aqueous perturbants. *Cell* 42: 497–505

Giudicelli, J., Boudouard, M., Delque, P., Vannier, Ch., Sudaka, P. 1985. Comparison of amphipathic properties of purified horse kidney neutral α-glucosidase after solubilization by Triton X-100 or by proteases. *Biochim. Biophys. Acta.* In press

Giudicelli, J., Emiliozzi, R., Vannier, C., De Burlet, G., Sudaka, P. 1980. Purification by affinity chromatography and characterization of a neutral α-glucosidase from horse kidney. *Biochim. Biophys. Acta* 612: 85–96

Glazer, A. N., Smith, E. L. 1971. Papain and other plant sulfhydryl proteolytic enzymes. *Enzymes* 3: 119–64

Goda, T., Bustamante, S., Koldovský, O. 1985. Dietary regulation of intestinal lactase and sucrase in adult rats: Quantitative comparison of effect of lactose and sucrose. *J. Pediatr. Gastroent. Nutr.* 4: 998–1008

Goda, T., Koldovský, O. 1985. Evidence of degradation process of sucrase-isomaltase in jejunum of adult rats. *Biochem. J.* 229: 751–58

Gorvel, J. P., Benajiba, A., Maroux, S. 1980. Purification and characterization of the rabbit intestinal brush border aminopeptidase A. *Biochim. Biophys. Acta* 615: 271–74

Haemmerli, U. P., Kistler, H., Ammann, R., Marthaler, T., Semenza, G., et al. 1965. Acquired milk intolerance in the adult caused by lactose malabsorption due to a selective deficiency of intestinal lactase activity. *Am. J. Med.* 38: 7–30

Haffen, K., Kedinger, M., Lacroix, B. 1986. Cytodifferentiation of the intestinal villus epithelium. In *Molecular and Cellular Basis of Digestion*, ed. P. Desnuelle, H. Sjöström, O. Norén, pp. 303–22. Amsterdam: Elsevier

Hanover, J. A., Lennarz, W. J. 1981. Transmembrane assembly of membrane and secretory glycoproteins. *Arch. Biochem. Biophys.* 211: 1–19

Hanozet, H., Pircher, H. P., Vanni, P., Oesch, B., Semenza, G. 1981. An example of enzyme hysteresis. The slow and tight interaction of some fully competitive inhibitors with small intestinal sucrase. *J. Biol. Chem.* 256: 3703–11

Hauri, H.-P. 1983. Biosynthesis and transport and plasma membrane glycoproteins in the rat intestinal epithelial cell: Studies with sucrase-isomaltase. See Porter & Collins 1983, pp. 132–47

Hauri, H.-P., Quaroni, A., Isselbacher, K. J. 1979. Biogenesis of intestinal plasma membrane: Posttranslational route and cleavage of sucrase-isomaltase. *Proc. Natl. Acad. Sci. USA* 76: 5183–86

Hauri, H.-P., Quaroni, A., Isselbacher, K. J. 1980. Monoclonal antibodies to sucrase-isomaltase: Probes for the study of postnatal development and biogenesis of the intestinal microvillus membrane. *Proc. Natl. Acad. Sci. USA* 77: 6629–33

Hauri, H.-P., Wacker, H., Rickli, E. E., Bigler-Meier, B., Quaroni, A., Semenza, G. 1982. Biosynthesis of sucrase-isomaltase. Purification and NH$_2$-terminal amino acid sequence of the rat sucrase-isomaltase precursor (pro-sucrase-isomaltase) from fetal intestinal transplants. *J. Biol. Chem.* 257: 4522–28

Hauri, H.-P., Roth, J., Sterchi, E. E., Lentze, M. J. 1985a. Transport to cell surface of intestinal sucrase-isomaltase is blocked in the Golgi apparatus in a patient with congenital sucrase-isomaltase deficiency. *Proc. Natl. Acad. Sci. USA* 82: 4423–27

Hauri, H.-P., Sterchi, E. E., Bienz, D., Fransen, J. A. M., Marxer, A. 1985b. Expression and intracellular transport of microvillous membrane hydrolases in human epithelial cells. *J. Cell Biol.* 101: 838–51

Hauser, H., Semenza, G. 1983. Sucrase-isomaltase: A stalked intrinsic protein of the brush border membrane. *CRC Crit. Rev. Biochem.* 14: 319–45

Hedeager-Sørensen, S., Kenny, A. J. 1986. Proteins of the kidney microvillar membrane. Purification and properties of carboxypeptidase P from pig kidney. *Biochem. J.* 229: 251–57

Helenius, A., Simons, K. 1977. Charge shift electrophoresis: Simple method for distinguishing between amphiphilic and hydrophobic proteins in detergent solution. *Proc. Natl. Acad. Sci. USA* 74: 529–32

Helms, D. B., Alpers, D. H., Seetharam, S., May, V. L., Strauss, A. W. 1984. *Fed. Proc.* 43: 2024 (Abstr.)

Henning, S. J. 1985. Ontogeny of enzymes in the small intestine. *Ann. Rev. Physiol.* 47: 231–45

Henning, S. J., Helman, T. A., Kretchmer, N. 1975. Studies on normal and precocious appearance of jejunal sucrase in suckling rats. *Biol. Neonate* 26: 249–62

Hiwada, K., Ito, T., Yokoyama, M. 1980. Isolation and characterization of membrane bound arylamidases from human placenta and kidney. *Eur. J. Biochem.* 104: 155–65

Holzel, A., Schwarz, V., Sutcliffe, K. W. 1959. Defective lactose absorption causing malnutrition in infancy. *Lancet* 1: 1126–28

Honma, K., Tomita, M., Hamada, A. 1980. Amino acid sequence and attachment sites of oligosaccharide units of porcine erythrocyte glycophorin. *J. Biochem.* 88: 1679–91

Horiuchi, S., Inoué, M., Morino, Y. 1980. Latent active site in rat kidney γ-glutamyl transpeptidase. *Eur. J. Biochem.* 105: 93–102

Hu, C., Shi, Q., Wei, Y., Lu, Z. 1985. The preparation and circular dicroism studies of sucrase-isomaltase complex from duck intestinal brush border. *Acta Biochim. Biophys. Sinica* 17: 16–22 (In Chinese)

Hubbard, S. C., Ivatt, R. J. 1981. Synthesis and processing of aspargine-linked oligosaccharides. *Ann. Rev. Biochem.* 50: 555–83

Hughey, R. P., Capraro, M. A. 1984. *Fed. Proc.* 43: 2024 (Abstr.)

Hughey, R. P., Curthoys, N. P. 1976. Comparison of the size and physical properties of γ-glutamyl transpeptidase purified from rat kidney following solubilization with papain or with Triton X-100. *J. Biol. Chem.* 251: 7863–69

Hunziker, W., Spiess, M., Semenza, G., Lodish, H. 1986. The sucrase-isomaltase complex: Primary structure, membrane-orientation and evolution of a stalked intrinsic brush border protein. *Cell*. In press

Hussain, M. M. 1985. Reconstitution of purified dipeptidyl peptidase IV. A comparison with aminopeptidase N with respect to morphology and influence of anchoring peptide on function. *Biochim. Biophys. Acta* 815: 306–12

Hussain, M. M., Tranum-Jensen, J., Norén, O., Sjöström, H., Christiansen, K. 1981. Reconstitution of purified amphiphilic pig-intestinal microvillus aminopeptidase. Mode of membrane insertion and morphology. *Biochem. J.* 199: 179–86

Inoué, M., Horiuchi, S., Morino, Y. 1977. Affinity labeling of rat kidney γ-glutamyl transpeptidase. *Eur. J. Biochem.* 73: 335–42

Inoué, M., Horiuchi, S., Morino, Y. 1979. Affinity labeling of rat kidney γ-glutamyl transpeptidase by 6-diazo-5-oxo-d-norleucine. *Eur. J. Biochem.* 99: 169–77

Inouye, M., Halegoua, S. 1980. Secretion and membrane localization of proteins in *Escherichia coli*. *CRC Crit. Rev. Biochem.* 10: 339–71

Ito, F., Chou, J. Y. 1983. Biosynthesis and processing of placental alkaline phosphatase. *Biochem. Biophys. Res. Commun.* 111: 611–18

James, W. P. T., Alpers, D. H., Gerber, J. E., Isselbacher, K. J. 1971. The turnover of disaccharidases and brush border proteins in rat intestine. *Biochim. Biophys. Acta* 230: 194–203

Janett, M. 1974. *Identifikation der durch Saccharase und Isomaltase gespaltenen Bindung im Substrat. Steady state Kinetik der Isomaltase*. Thesis. ETH, Zurich. 45 pp.

Johnson, C. F. 1967. Disaccharidase: Localization in hamster intestine brush borders. *Science* 155: 1670

Kaufman, M. A., Korsmo, H. A., Olsen, W. A. 1980. Circadian rhythm of intestinal sucrase activity in rats. *J. Clin. Invest.* 65: 1174–81

Kedinger, M., Haffen, K., Simon Assmann, P. 1986a. Control mechanisms in the ontogenesis of villus cells. In *Molecular and Cellular Basis of Digestion*, ed. P. Desnuelle, H. Sjöström, O. Norén, pp. 323–34. Amsterdam: Elsevier

Kelly, J. J., Alpers, D. H. 1973a. Blood group antigenicity of purified human intestinal disaccharidases. *J. Biol. Chem.* 248: 8216–21

Kelly, J. J., Alpers, D. H. 1973b. Properties of human intestinal glucoamylase. *Biochim. Biophys. Acta* 315: 113–20

Kenny, A. J. 1977. Endopeptidases in the brush border of the kidney proximal tubule. In *Peptide Transport and Hydro-*

lysis. Ciba Found. Symp., ed. K. Elliott, M. O'Connor, 50: 209–19. Amsterdam: Elsevier/Excerpta Medica/North-Holland

Kenny, A. J. 1986. Endopeptidase-24.11: An ectoenzyme capable of hydrolyzing regulatory peptides at the surface of many different cell types. In *Cellular Biology of Ectoenzymes*, ed. G. W. Kreutzberg, M. Reddington, H. Zimmermann, pp. 257–71. Heidelberg/New York/Tokyo: Springer-Verlag

Kenny, A. J., Booth, A. G. 1978. Microvilli: Their ultrastructure enzymology and molecular organization. *Essays Biochem.* 14: 1–44

Kenny, A. J., Booth, A. G., George, S. G., Ingram, J., Kershaw, D., et al. 1976. Dipeptidyl peptidase IV, a kidney brush border serine peptidase. *Biochem. J.* 157: 169–82

Kenny, A. J., Bowes, M. A., Gee, N. S., Matsas, R. 1985. Endopeptidase-24.11: A cell-surface enzyme for metabolizing regulatory peptides. *Biochem. Soc. Trans.* 13: 293–95

Kenny, A. J., Fulcher, I. S. 1983. Microvillar endopeptidases, an enzyme with special topological features and a wide distribution. See Porter & Collins 1983, pp. 12–25

Kenny, A. J., Fulcher, I. S., McGill, K., Kershaw, D. 1983. Proteins of the kidney microvillar membrane. Reconstitution of endopeptidase in liposomes shows that it is a short-stalked protein. *Biochem. J.* 211: 755–62

Kenny, A. J., Fulcher, I. S., Ridgwell, K., Ingram, J. 1981. Microvillar membrane neutral endopeptidases. *Acta Biol. Med. Germ.* 40: 1465–71

Kenny, J. A., Maroux, S. 1982. Topology of microvillar membrane hydrolases of kidney and intestine. *Physiol. Rev.* 62: 91–128

Kerr, M. A., Kenny, A. J. 1974a. The purification and specificity of neutral endopeptidase from rabbit kidney brush border. *Biochem. J.* 137: 477–88

Kerr, M. A., Kenny, A. J. 1974b. The molecular weight and properties of a neutral metallo-endopeptidase from rabbit kidney brush border. *Biochem. J.* 137: 489–95

Kessler, M., Acuto, O., Storelli, C., Murer, H., Müller, M., Semenza, G. 1978. A modified procedure of the rapid preparation of efficiently transporting vesicles from small intestinal brush border membranes. *Biochim. Biophys. Acta* 506: 136–54

Khorana, H. G. 1980. Chemical studies of biological membranes. *Bioorganic Chem.* 9: 363–405

Kim, Y. S., Brophy, E. J. 1976. Rat intestinal brush border membrane peptidases. I.

Solubilization, purification and physicochemical properties of five different forms of enzyme. *J. Biol. Chem.* 251: 3199–3205

Kinne, R., Murer, H., Kinne-Saffran, E., Thees, M., Sachs, G. 1975. Sugar transport by renal plasma membrane vesicles. *J. Membr. Biol.* 21: 375–95

Koldovský, O. 1969. *Development of the Functions of the Small Intestine in Mammals and Man.* New York/Basel: Karger. 204 pp.

Koldovský, O. 1981. Developmental, dietary and hormonal control of intestinal disaccharidases in mammals (including man). In *Carbohydrate Metabolism and its Disorders,* ed. J. P. Randle, D. F. Steiner, W. J. Whelan, 3: 418–522. London: Academic

Kolínská, J., Kraml, J., Zákostelecká, M., Lojda, Z. 1984. Low molecular weight antigens of sucrase-isomaltase in the intestinal mucosa of suckling rats. *Mol. Physiol.* 5: 133–48

Kolínská, J., Semenza, G. 1967. Studies on intestinal sucrase and on intestinal sugar transport. V. Isolation and properties of sucrase-isomaltase from rabbit small intestine. *Biochim. Biophys. Acta* 146: 181–95

Kornfeld, R., Kornfeld, S. 1985. Assembly of aspargine-linked oligosaccharides. *Ann. Rev. Biochem.* 54: 631–64

Kraml, J., Kolínská, J., Kadlecová, L., Zákostelecká, M., Lojda, Z. 1983. Analytical isoelectric focusing of rat intestinal brush border enzymes: Postnatal changes and effect of neuraminidase in vitro. *FEBS Lett.* 151: 193–96

Kraml, J., Kolínská, J., Kadlecová, L., Zákostelecká, M., Lojda, Z. 1984. Effect of hydrocortisone on the desialylation of intestinal brush-border enzymes of the rat during postnatal development. *FEBS Lett.* 172: 25–28

Kretchmer, N. 1971. Memorial lecture: Lactose and lactase—A historical perspective. *Gastroenterology* 61: 805–13

Kuno, T., Matsuda, Y., Katunuma, N. 1983. The conversion of the precursor form of γ-glutamyltranspeptidase to its subunit form takes place in brush border membranes. *Biochem. Biophys. Res. Commun.* 114: 889–95

Kuno, T., Matsuda, Y., Katunuma, N. 1984. Characterization of a processing protease that converts the precursor form of γ-glutamyltranspeptidase to its subunits. *Biochem. Int.* 8: 581–88

Labat, J., Baumann, F., Courtois, J. F. 1974. La tréhalase rénale du porc et de l'homme: Purification et quelques propriétés. *Biochimie* 56: 805–13

Laperche, Y., Bulle, F., Aissani, T., Chobert, M. N., Aggerbeck, M., et al. 1986. Molecular cloning and nucleotide sequence of rat kidney γ-glutamyl transpeptidase cDNA. *Proc. Natl. Acad. Sci. USA* 83: 937–41

Lee, L., Forstner, G. 1985. *Can. J. Biochem. Cell Biol.* 63: 257–62

Lee, L. M. Y., Salvatore, A. K., Flanagan, P. R., Forstner, G. G. 1980. Isolation of a detergent-solubilized maltase/glucoamylase from rat intestine and its comparison with a maltase-glucoamylase solubilized by papain. *Biochem. J.* 187: 437–46

Leese, H. J., Semenza, G. 1973. On the identity between the small-intestinal enzymes phlorizin-hydrolase and glycosylceramidase. *J. Biol. Chem.* 248: 8170–73

Lehky, P., Lisowski, J., Wolf, D. P., Wacker, H., Stein, E. A. 1973. Pig kidney particulate aminopeptidase: A zinc metalloenzyme. *Biochim. Biophys. Acta* 321: 274–81

Lorenz-Meyer, H., Blum, A. L., Haemmerli, H. P., Semenza, G. 1972. A second enzyme defect in acquired lactase deficiency. Lack of small-intestinal phlorizin-hydrolase. *Eur. J. Clin. Invest.* 2: 326–31

Louvard, D., Maroux, S., Baratti, J., Desnuelle, P., Mutaftschiev, S. 1973. On the preparation and some properties of closed membrane vesicles from hog duodenal and jejunal brush border. *Biochim. Biophys. Acta* 291: 747–63

Louvard, D., Sémériva, M., Maroux, S. 1976. The brush-border intestinal aminopeptidase, a transmembrane protein as probed by macromolecular photolabelling. *J. Mol. Biol.* 106: 1023–35

Low, M. G., Finean, J. B. 1978. Specific release of plasma membrane enzymes by a phosphatidylinositol-specific phospholipase C. *Biochim. Biophys. Acta* 508: 565–70

Low, G. L., Zilversmit, D. B. 1980. Role of phosphatidylinositol in attachment of alkaline phosphatase to membranes. *Biochemistry* 19: 3913–18

Macnair, R. D., Kenny, A. J. 1979. Proteins of the kidney microvillar membrane. The amphipathic form of dipeptidyl peptidase IV. *Biochem. J.* 179: 379–95

Mahmood, A., Torres-Pinedo, R. 1983. Postnatal changes in lectin binding to microvillus membranes from rat intestine. *Biochem. Biophys. Res. Commun.* 113: 400–6

Maisushita, S. 1983. Purification and partial characterization of chick intestinal sucrase. *Comp. Biochem. Physiol.* 76B: 465–70

Malo, C., Ménard, D. 1979a. Effects of cortisone and thyroxine on intestinal trehalase activity in infant mouse. *Experientia* 35: 874–75

Malo, C., Ménard, D. 1979b. Opposite effects of one and three injections of cortisone or thyroxine on intestinal lactase activity in suckling mice. *Experientia* 35: 493–94

Malo, C., Ménard, D. 1980. Hormonal control of intestinal glucoamylase activity in suckling and adult mice. *Comp. Biochem. Physiol.* 65B: 169–72

Mamelok, R. D., Groth, D. F., Prusiner, St. B. 1980. Separation of membrane-bound γ-glutamyl transpeptidase from brush border transport and enzyme activities. *Biochemistry* 19: 2367–73

Marathe, G. V., Nash, B., Hashemeyer, R. H., Tate, S. S. 1979. Ultrastructural localization of γ-glutamyltranspeptidase in rat kidney and jejunum. *FEBS Lett.* 107: 436–40

Maroux, S., Louvard, D. 1976. On the hydrophobic part of aminopeptidase and maltases which bind the enzyme to the intestinal brush border membrane. *Biochim. Biophys. Acta* 419: 189–95

Maroux, S., Louvard, D., Baratti, J. 1973. The aminopeptidase from hog intestinal brush border. *Biochim. Biophys. Acta* 321: 282–95

Massey, D., Maroux, S. 1985. The carbohydrate moiety of aminopeptidase N of rabbit intestinal brush border membrane. *FEBS Lett.* 181: 207–10

Matsas, R., Kenny, A. J., Turner, A. J. 1984. The metabolism of neuropeptides. The hydrolysis of peptides, including enkephalins, tachykinins and their analogues by endopeptidase-24.11. *Biochem. J.* 223: 433–40

Matsuda, Y., Tsuji, A., Katunuma, N. 1983a. Studies on the structure of γ-glutamyltranspeptidase. III. Evidence that the amino terminus of the heavy subunit is the membrane binding segment. *J. Biochem.* 93: 1427–33

Matsuda, Y., Tsuji, A., Kuno, T., Katunuma, N. 1983b. Biosynthesis and degradation of γ-glutamyltranspeptidase of rat kidney. *J. Biochem.* 94: 755–65

Matsudaira, P. T. 1983. Structural and functional relationship between the membrane and the cytoskeleton in brush border microvilli. See Porter & Collins 1983, pp. 233–42

Meister, A., Tate, S. S. 1976. Glutathione and related γ-glutamyl compounds: Biosynthesis and utilization. *Ann. Rev. Biochem.* 45: 559–604

Ménard, D., Malo, C., Calvert, R. 1981a. Development of γ-glutamyltranspeptidase activity in mouse small intestine: Influence of cortisone and thyroxine. *Biol. Neonate* 40: 70–77

Miura, S., Song, I. S., Morita, A., Erickson, R. H., Kim, Y. S. 1983. Distribution and biosynthesis of aminopeptidase N and dipeptidyl aminopeptidase IV in rat small intestine. *Biochim. Biophys. Acta* 761: 66–75

Mizuno, K., Moriuchi, S., Hosoya, N. 1982. Demonstration of sucrase-isomaltase complex in chick intestine. *J. Nutr. Sci. Vitaminol.* 28: 599–608

Montgomery, R. K., Sybicki, A. A., Forcier, A. G., Grand, R. J. 1981. Rat intestinal microvillus membrane sucrase-isomaltase is a single high molecular weight protein and fully active enzyme in the absence of luminal factors. *Biochim. Biophys. Acta* 661: 346–49

Mooseker, M. S. 1985. Organization, chemistry, and assembly of the cytoskeletal apparatus of the intestinal brush border. *Ann. Rev. Cell Biol.* 1: 209–41

Mooseker, M. S., Howe, C. L. 1982. The brush border of intestinal epithelium: A model system for analysis of cell-surface architecture and mobility. *Methods Cell Biol.* 25: 144–75

Mooseker, M. S., Keller, T. C. S. III, Hirokawa, N. 1983. Regulation of cytoskeletal structure and contractility in the brush border. See Porter & Collins 1983, pp. 195–210

Mosimann, H., Semenza, G., Sund, H. 1973. Hydrodynamic properties of the sucrase-isomaltase complex from rabbit small intestine. *Eur. J. Biochem.* 36: 489–94

Murayama, J.-I., Tomita, M., Hamada, A. 1982. Primary structure of horse erythrocyte glycophorin HA. Its amino acid sequence has a unique homology with those of human and porcine erythrocyte glycophorins. *J. Membr. Biol.* 64: 205–15

Murer, H., Kinne, R. 1980. The use of isolated membrane vesicles to study epithelial transport processes. *J. Membr. Biol.* 55: 81–95

Nakano, M. 1980. Purification and properties of rabbit renal brush border membrane trehalase. *Fed. Proc.* 39: 1649

Nakano, M. 1982. Effect of inorganic anions on the inhibition of trehalase activity by mercuric chloride. *Biochim. Biophys. Acta* 707: 115–20

Nakano, M., Sacktor, B. 1984. Renal trehalase: Two subsites at the substrate-binding site. *Biochim. Biophys. Acta* 791: 45–49

Nakano, N., Sacktor, B. 1985. Isolation and characterisation of four forms of trehalase from rabbit kidney cortex. *J. Biochem.* 97: 1329–35

Nakano, M., Sumi, Y., Miyakawa, M. 1977. Purification and properties of the trehalase from rat intestinal mucosa cells. *J. Biochem.* 81: 1041–49

Nash, B., Tate, S. S. 1982. Biosynthesis of rat renal γ-glutamyl transpeptidase. Evidence

308 SEMENZA

for a common precursor of the two subunits. *J. Biol. Chem.* 257: 585–88

Nash, B., Tate, S. S. 1984. In vitro translation and processing of rat kidney γ-glutamyl transpeptidase. *J. Biol. Chem.* 259: 678–85

Newcomer, A. D., McGill, D. B. 1966. Distribution of disaccharidase activity in the small bowel of normal and lactase-deficient subjects. *Gastroenterology* 51: 481–88

Nishi, Y., Takesue, Y. 1978. Localization of intestinal sucrase-isomaltase complex on the microvillus membrane by electron microscopy using nonlabeled antibodies. *J. Cell Biol.* 79: 516–25

Nishi, Y., Yoshida, T. O., Takesue, Y. 1968. Electron microscope studies on the structure of rabbit intestinal sucrase. *J. Mol. Biol.* 37: 441–44

Noguchi, S., Noda, M., Takahashi, H., Kawakami, K., Ohta, T., Nagano, K., et al. 1986. Primary structure of the β-subunit of *Torpedo californica* (Na⁺,K⁺)-ATPase deduced from the cDNA sequence. *FEBS Lett.* 196: 315–20

Norén, O., Sjöström, H. 1980. The insertion of pig microvillus aminopeptidase into the membrane as probed by [¹²⁵I]iodonaphthylazide. *Eur. J. Biochem.* 104: 25–31

Norén, O., Sjöström, H., Cowell, G., Tranum-Jensen, J., Hansen, O. C., Welinder, K. G. 1986a. Pig intestinal microvillar maltase-glucoamylase. Structure and membrane insertion. *J. Biol. Chem.* In press

Norén, O., Sjöström, H., Danielsen, E. M., Cowell, G., Skovbjerg, H. 1986b. The enzymes of the enterocyte plasma membrane. In *Molecular and Cellular Biology of Digestion*, ed. P. Desnuelle, H. Sjöström, O. Norén, pp. 335–65. Amsterdam: Elsevier

Norén, O., Sjöström, H., Gudmand-Høyer, E., Skovbjerg, H., Andersen, B. 1980. Peptidase activities in the functioning jejunum and ileum before and after jejunoileal bypass in morbid obesity. *Scand. J. Gastroenter.* 15: 825–32

Ovchinnikov, Y. A., Modyanov, N. N., Broude, N. E., Petrukhin, K. E., Grishin, A. V., et al. 1986. Pig kidney Na⁺,K⁺-ATPase. Primary structure and spatial organization. *FEBS Lett.* 201: 237–45

Paige, D. M., Bayless, T. M., eds. 1981. Lactose digestion: Clinical and nutritional consequences. Baltimore: Johns Hopkins Univ. Press

Pattus, F., Verger, R., Desnuelle, P. 1976. Comparative study of the interactions of the trypsin and detergent form of the intestinal aminopeptidase with liposomes. *Biochem. Biophys. Res. Commun.* 69: 718–23

Porter, R., Collins, G., eds. 1983. Brush border membranes. *Ciba Found. Symp.* 95. London: Pitman

Prakash, K., Patil, S. D., Hegde, S. N. 1983. Studies on the intestinal disaccharidases of the pigeon. III. Separation, purification and properties of sucrase-isomaltase and maltase-glucoamylase. *Arch. Int. Physiol.* 91: 379–90

Quaroni, A., Gershon, E., Semenza, G. 1974. Affinity labeling of the active sites in the sucrase-isomaltase complex from small intestine. *J. Biol. Chem.* 249: 6424–33

Quaroni, A., Gershon-Quaroni, E., Semenza, G. 1975. Tryptic digestion of native small-intestinal sucrase-isomaltase complex: Isolation of the sucrase subunit. *Eur. J. Biochem.* 52: 481–86

Quaroni, A., Kirsch, K., Weiser, M. M. 1979a. Synthesis of membrane glycoproteins in rat small intestinal villus cells. Redistribution of l-(1,5,6-³H)fucose-labeled membrane glycoproteins among Golgi lateral basal and microvillus membranes in vivo. *Biochem. J.* 182: 203–12

Quaroni, A., Kirsch, K., Weiser, M. M. 1979b. Synthesis of membrane glycoproteins in rat small intestinal villus cells. Effect of colchicine on the redistribution of l-(1,5,6-³H)fucose-labeled membrane glycoproteins among Golgi, lateral basal and microvillus membranes. *Biochem. J.* 182: 213–21

Quaroni, A., Semenza, G. 1976. Partial amino acid sequences around the essential carboxylate in the active sites of the intestinal sucrase-isomaltase complex. *J. Biol. Chem.* 251: 3250–53

Rankin, B., McIntyre, T. M., Curthoys, N. P. 1980. Brush border membrane hydrolysis of S-benzyl-cysteine-p-nitroanilide, an activity of aminopeptidase M. *Biochem. Biophys. Res. Commun.* 96: 991–96

Raul, F., Noriega, R., Nsi-Emvo, E., Doffoel, M., Grenier, J. F. 1983. Lactase activity is under hormonal control in the intestine of adult rat. *Gut* 24: 648–52

Reiss, U., Sacktor, B. 1981. Kidney brush border membrane maltase: Purification and properties. *Arch. Biochem. Biophys.* 209: 342–48

Relton, J. M., Gee, S. N., Matsas, R., Turner, A. J., Kenny, A. J. 1983. Purification of endopeptidase-24.11 ("enkephalinase") from pig brain by immunoadsorbent chromatography. *Biochem. J.* 215: 519–23

Rey, J., Frézal, J. 1967. Les anomalies des disaccharidases. *Arch. Fr. Pediatr.* 24: 65–101

Riby, J. E., Kretchmer, N. 1984. The effect of dietary sucrose on synthesis and degradation of intestinal sucrase. *Am. J. Physiol.* 246: G757–64

Riby, J. E., Kretchmer, N. 1985. Participation of pancreatic enzymes in the degradation of intestinal sucrase-isomaltase. *J. Pediat. Gastroenterol. Nutr.* 4: 971–79

Sabatini, D. D., Kreibich, G., Takashi, M., Adesnik, M. 1982. Mechanisms for the incorporation of proteins in membranes and organelles. *J. Cell Biol.* 92: 1–22

Sacktor, B. 1977. The brush border of the renal proximal tubule and the intestinal mucosa. In *Mammalian Cell Membranes*, ed. G. A. Jamieson, D. E. Robinson, 4: 221–54. London: Butterworths

Sasajima, K., Kawachi, T., Sato, S., Sugimura, T. 1975. Purification and properties of α,α-trehalase from the mucosa of the rat small intestine. *Biochim. Biophys. Acta* 403: 139–46

Schlegel-Haueter, S., Hore, P., Kerry, K. R., Semenza, G. 1972. The preparation of lactase and glucoamylase of rat small intestine. *Biochim. Biophys. Acta* 258: 506–19

Schmitz, J., Preiser, H., Maestracci, D., Ghosh, B. K., Cerda, J. J., Crane, R. K. 1973. Purification of the human intestinal brush border membrane. *Biochim. Biophys. Acta* 323: 98–112

Seetharam, B., Alpers, D. H., Allen, R. H. 1981. Isolation and characterization of the ileal receptor for intrinsic factor–cobalamin. *J. Biol. Chem.* 256: 3785–90

Seetharam, B., Bagur, S. S., Alpers, D. H. 1982. Isolation and characterization of proteolytically derived ileal receptor for intrinsic factor–cobalamin. *J. Biol. Chem.* 257: 183 89

Seetharam, B., Tiruppathi, C., Alpers, D. H. 1985. Membrane interactions of rat intestinal alkaline phosphatase: Role of polar head groups. *Biochemistry* 24: 6603–8

Semenza, G. 1957. Chromatographic purification of cysteinyl-glycinase. *Biochim. Biophys. Acta* 24: 401–13

Semenza, G. 1976. Small intestinal disaccharidases: Their properties and role as sugar translocators across natural and artificial membranes. In *The Enzymes of Biological Membranes*, ed. A. Martonosi, 3: 349–82. New York: Plenum

Semenza, G. 1968. Intestinal oligosaccharidases and disaccharidases. In *Handbook of Physiology*, Sect. 6, Vol. V, ed. C. F. Code, pp. 2543–66. Washington, DC: Am. Physiol. Soc.

Semenza, G. 1978. The sucrase-isomaltase complex, a large dimeric amphipathic protein from the small intestinal brush border membrane: Emerging structure-function relationships. In *Structure and Dynamics of Chemistry*, ed. P. Ahlberg, L.-O. Sundelöf. Presented *Symp. 500th Jubilee, Univ. Uppsala, Sweden, 1977.* pp. 226–40

Semenza, G. 1979a. The mode of anchoring of sucrase-isomaltase to the small-intestinal brush border membrane and its biosynthetic implications. In *Proc. 12th FEBS Meet. Dresden 1978*, ed. S. Rapoport, T. Schewe, 53: 21–28. Oxford/New York: Pergamon

Semenza, G. 1979b. Mode of insertion of the sucrase-isomaltase complex in the intestinal brush border membrane: Implications for the biosynthesis of this stalked intrinsic membrane protein. In *Development of Mammalian Absorptive Processes. Ciba Found. Symp.*, ed. K. Elliott, J. Whelan, 70: 133–44. Amsterdam: Excerpta Medica

Semenza, G. 1981a. Intestinal oligo- and disaccharidases. In *Carbohydrate Metabolism and its Disorders*, ed. P. J. Randle, D. F. Steiner, W. J. Whelan, 3: 425–79. London: Academic

Semenza, G. 1981b. Molecular pathophysiology of small-intestinal sucrase-isomaltase. *Clin. Gastroenterol.* 10: 691–706

Semenza, G. 1986. A unifying concept in the phylogenesis, biosynthesis, membrane insertion and physiopathology of sucrase-isomaltase and other brush border proteins. In *Ion Gradient-coupled Transport*, ed. F. Alvarado, C. H. van Os, pp. 41–48. Amsterdam: Elsevier

Semenza, G., Balthazar, A.-K. 1974. Steady state kinetics of rabbit intestinal sucrase: Kinetic mechanism, Na$^+$-activation, inhibition by Tris (hydroxylmethyl)-aminomethane at the glucose subsite, with an appendix on interactions between enzyme inhibitors: A kinetic test for simple cases. *Eur. J. Biochem.* 41: 149 62

Semenza, G., Brunner, J., Wacker, H. 1983. Biosynthesis and assembly of the largest and major intrinsic polypeptide of the small intestinal brush border. See Porter & Collins 1983, pp. 92–107

Semenza, G., Corcelli, A. 1986. The absorption of sugars and amino acids across the small intestine. In *Molecular and Cellular Basis of Digestion*, ed. P. Desnuelle, H. Sjöström, O. Norén, pp. 381–412. Amsterdam: Elsevier

Semenza, G., Curtius, H.-Ch., Raunhardt, O., Hore, P., Müller, M. 1969. The configuration at the anomeric carbon of the reaction products of some digestive carbohydrases. *Carbohydr. Res.* 10: 417–28

Semenza, G., Kessler, M., Hosang, M., Weber, J., Schmidt, U. 1984. Biochemistry of the Na$^+$, D-glucose cotransporter of the small-intestinal brush border membrane.

The state of art in 1984. *Biochim. Biophys. Acta* 779 : 343–79

Siddons, R. C. 1970. Heat inactivation and sephadex chromatography of the small-intestine. Disaccharidases of the chick. *Biochem. J.* 116 : 71–78

Sigrist, H., Ronner, P., Semenza, G. 1975. A hydrophobic form of the small-intestinal sucrase-isomaltase complex. *Biochim. Biophys. Acta* 406 : 433–46

Silverblatt, E. R., Conklin, K., Gray, G. M. 1974. Sucrase precursor in human jejunal crypts. *J. Clin. Invest.* 53 : 76a

Simons, K., Fuller, S. D. 1985. Cell surface polarity in epithelia. *Ann. Rev. Cell Biol.* 1 : 243–88

Sivakami, S., Radhakrishan, A. M. 1973. Purification of rabbit intestinal glucoamylase by affinity chromatography on Sephadex G-200. *Ind. J. Biochem. Biophys.* 10 : 283–84

Sjöström, H., Norén, O. 1982. Changes in the quaternary structure of microvillus aminopeptidase in the membrane. *Eur. J. Biochem.* 122 : 245–50

Sjöström, H., Norén, O., Christiansen, L., Wacker, H., Semenza, G. 1980. A fully active, two-active site, single-chain-sucrase-isomaltase from pig small intestine. *J. Biol. Chem.* 255 : 11332–38

Sjöström, H., Norén, O., Christiansen, L. A., Wacker, H., Spiess, M., et al. 1982. Membrane anchoring and partial structure of small-intestinal pro-sucrase-isomaltase. A possible biosynthetic mechanism. *FEBS Lett.* 148 : 321–25

Sjöström, H., Norén, O., Danielsen, E. M. 1985. The enzymatic activity of "high-mannose" glycosylated forms on intestinal microvillar hydrolases. *J. Pediat. Gastroent. Nutr.* 4 : 980–83

Sjöström, H., Norén, O., Danielsen, E. M., Gorr, S.-U. 1986. Intracellular transport of newly synthesized secretory and membrane proteins. In *Molecular and Cellular Biology of Digestion*, ed. P. Desnuelle, H. Sjöström, O. Norén, pp. 61–77. Amsterdam : Elsevier

Sjöström, H., Norén, O., Danielsen, E. M., Skovbjerg, H. 1983. See Porter & Collins 1983, pp. 50–69

Sjöström, H., Norén, O., Jeppesen, L., Staun, M., Svensson, B., Christiansen, L. 1978. Purification of different amphiphilic forms of a microvillus aminopeptidase from pig small intestine using immunoadsorbent chromatography. *Eur. J. Biochem.* 88 : 503–11

Skovbjerg, H. 1981a. Immunoelectrophoretic studies on human small-intestinal brush border proteins. Relation between enzyme activity and immunoreactive enzyme along the villus-crypt axis. *Biochem. J.* 193 : 887–90

Skovbjerg, H. 1981b. Immunoelectrophoretic studies on human small intestinal brush border proteins. The longitudinal distribution of peptidases and disaccharidases. *Clin. Chim. Acta* 112 : 205–12

Skovbjerg, H. 1982. High-molecular weight pro-sucrase-isomaltase in human fetal intestine. *Pediatr. Res.* 16 : 948–49

Skovbjerg, H., Danielsen, E. M., Norén, O., Sjöström, H. 1984. Evidence for biosynthesis of lactase-phlorizin hydrolase as a single-chain high-molecular weight precursor. *Biochim. Biophys. Acta* 798 : 247–51

Skovbjerg, H., Gudmand-Høyer, E., Fenger, H. J. 1980a. Immunoelectrophoretic studies on human small intestinal brush border proteins. The amount of lactase protein in adult-free hypolactasia. *Gut* 21 : 360–64

Skovbjerg, H., Norén, O., Sjöström, H. 1978. Immunoelectrophoretic studies on human small intestinal brush border proteins. *Scand. J. Clin. Invest.* 38 : 723–29

Skovbjerg, H., Norén, O., Sjöström, H., Danielsen, E. M., Enevoldsen, B. S. 1982. Further characterization of intestinal lactase/phlorizin hydrolase. *Biochim. Biophys. Acta* 707 : 89–97

Skovbjerg, H., Sjöström, H., Norén, O. 1980b. Immunoelectrophoretic studies on human small intestinal brush border proteins. Cellular alterations in the level of brush border enzymes after jejunoileal bypass operation. *Gut* 21 : 662–68

Skovbjerg, H., Sjöström, H., Norén, O. 1981. Purification and characterization of amphiphilic lactase/phlorizin hydrolase from human small intestine. *Eur. J. Biochem.* 114 : 653–61

Sørensen, S. H., Norén, O., Sjöström, H., Danielsen, E. M. 1982. Amphiphilic pig intestinal microvillus maltase/glucoamylase. Structure and specificity. *Eur. J. Biochem.* 126 : 559–68

Spiess, M., Lodish, H. F. 1986. An internal signal sequence : The asialoglycoprotein receptor membrane anchor. *Cell* 44 : 177–85

Spiess, M., Brunner, J., Semenza, G. 1982. Hydrophobic labeling, isolation and partial characterization of the NH_2-terminal membranous segment of sucrase-isomaltase complex. *J. Biol. Chem.* 257 : 2370–77

Spiess, M., Hauser, H., Rosenbusch, J. P., Semenza, G. 1981. Hydrodynamic properties of phospholipid vesicles and of sucrase-isomaltase-phospholipid vesicles. *J. Biol. Chem.* 256 : 8977

Stefani, A., Janett, H., Semenza, G. 1975. Small-intestinal sucrase and isomaltase split the bond between glucosyl-C_1 and the glycosyl oxygen. *J. Biol. Chem.* 250 : 7800–13

Stevenson, N. R., Ferrigni, F., Parnicky, K., Day, S., Fierstein, J. S. 1975. Effect of changes in feeding schedule on the diurnal rhythms and daily activity levels of intestinal brush border enzymes and transport systems. *Biochim. Biophys. Acta* 406 : 131–45

Stewart, J. R., Chaplin, M. F., Kenny, A. J. 1984. Deglycosylation by trifluoromethanesulphonic acid of endopeptidase-24.11 purified from pig kidney and intestine. *Biochem. J.* 221 : 919–22

Stewart, J. R., Kenny, A. J. 1984a. Proteins of the kidney microvillar membrane. Biosynthesis of endopeptidase-24.11, dipeptidylpeptidase IV and aminopeptidases N and A in pig kidney slices. *Biochem. J.* 224 : 549–58

Stewart, J. R., Kenny, A. J. 1984b. Proteins of the kidney microvillar membrane. Effect of monensin, vinblastine, swainsonine and glucosamone on the processing and assembly of endopeptidase-24.11 and dipeptidylpeptidase IV in pig kidney slices. *Biochem. J.* 224 : 559–58

Sunshine, P., Kretchmer, N. 1964. Intestinal disaccharidases : Absence in two species of sea lions. *Science* 144 : 805–51

Svensson, B. 1979. Covalent cross-linking of porcine small intestine microvillar aminopeptidase. Subunit structure of the membrane-bound and the solubilized enzyme. *Carlsberg Res. Commun.* 44 : 417–30

Svensson, B., Danielsen, M., Staun, M., Jeppesen, L., Norén, O., Sjöström, H. 1978. An amphiphilic form of dipeptidyl peptidase IV from pig small intestinal brush-border membrane. *Eur. J. Biochem.* 90 : 489–98

Takesue, Y. 1969. Purification and properties of rabbit intestinal sucrase. *J. Biochem.* 65 : 545 52

Takesue, Y., Tamura, R., Nishi, Y. 1977. Immunochemical studies on the subunits of rabbit intestinal sucrase-isomaltase complex. *Biochim. Biophys. Acta* 483 : 375–85

Tannenbaum, C., Toggenburger, G., Kessler, M., Rothstein, A., Semenza, G. 1977. High-affinity phlorizin binding to brush border membranes from small intestine : Identity with (a part of) the glucose transport system, dependence on the Na^+-gradient, partial purification. *J. Supramol. Struct.* 6 : 519–33

Taravel, F. R., Datema, R., Woloszczuk, W., Marshall, J. J., Whelan, W. J. 1983. Purification and characterization of a pig intestinal α-limit dextrinase. *Eur. J. Biochem.* 130 : 147–53

Tate, S. S. 1986. Single-chain precursor of renal γ-glutamyl transpeptidase. *FEBS Lett.* 194 : 33–38

Tate, S. S., Meister, A. 1975. Identity of maleate-stimulated glutaminase with γ-glutamyl transpeptidase in rat kidney. *J. Biol. Chem.* 250 : 4619–27

Tate, S. S., Meister, A. 1976. Subunit structure and isozymic forms of γ-glutamyl transpeptidase. *Proc. Natl. Acad. Sci. USA* 73 : 2599–2603

Tate, S. S., Meister, A. 1977. Affinity labeling of γ-glutamyl transpeptidase and location of the γ-glutamyl binding site on the light subunit. *Proc. Natl. Acad. Sci. USA* 74 : 931–35

Tate, S. S., Ross, E. M. 1977. Human kidney γ-glutamyl transpeptidase : Catalytic properties, subunit structure, and localization of the γ-glutamyl binding site on the light subunit. *J. Biol. Chem.* 252 : 6042–45

Tiruppathi, C., Alpers, D. H., Seetharam, B. 1985. Interaction of intestinal disaccharidases with phospholipids : Effect of cholesterol. *J. Pediatr. Gastroent. Nutr.* 4 : 965–70

Torres-Pinedo, R., Mahmood, A. 1984. Postnatal changes in biosynthesis of microvillus membrane glycans of rat small intestine : I. Evidence of a developmental shift from terminal sialylation to fucosylation. *Biochem. Biophys. Res. Commun.* 125 : 546–53

Triadou, N., Bataille, J., Schmitz, J. 1983. Longitudinal study of the human intestinal brush border membrane proteins. Distribution of the main disaccharidases and peptidases. *Gastroenterology* 85 : 1326–32

Triadou, N., Zweibaum, A. 1985. Maturation of sucrase isomaltase complex in human fetal small and large intestine during gestation. *Pediatr. Res.* 19 : 136–38

Tsuboi, K. K., Schwartz, S. M., Burrill, P. H., Kwong, L. S., Sunshine, P. 1979. Sugar hydrolases of the infant rat intestine and their arrangement on the brushborder membrane. *Biochim. Biophys. Acta* 554 : 234–48

Tsuji, A., Matsuda, Y., Katanuma, N. 1980. Studies on the structure of γ-glutamyltranspeptidase. Location of the segment anchoring γ-glutamyl transpeptidase to the membrane. *J. Biochem.* 87 : 1567–71

Turner, R. J., Moran, A. 1982a. Heterogeneity of sodium-dependent D-glucose transport sites along the proximal tubule: Evidence from vesicle studies. *Am. J. Physiol.* 242: F406–14

Turner, R. J., Moran, A. 1982b. Further studies of proximal tubular brush border membrane D-glucose transport heterogeneity. *J. Membr. Biol.* 70: 37–45

Vannier, C., Louvard, D., Maroux, S., Desnuelle, P. 1976. Structural and topological homology between porcine and renal brush border aminopeptidase. *Biochim. Biophys. Acta* 455: 185–99

von Heijne, G. 1985. Structural and thermodynamic aspects of the transfer of proteins into and across membranes. *Current Top. Membr. Transp.* 24: 151–79

von Heijne, G., Blomberg, G. 1979. Transmembrane translocation of proteins. *Eur. J. Biochem.* 97: 175–81

Wacker, H., Aggeler, R., Kretchmer, N., O'Neill, B., Takesue, Y., Semenza, G. 1984. A two-active site one-polypeptide enzyme: The isomaltase from sea lion small intestinal brush border membrane. *J. Biol. Chem.* 259: 4878–84

Wacker, H., Jaussi, R., Sonderegger, P., Dokow, M., Ghersa, P., et al. 1981. Cell-free synthesis of the one-chain precursor of a major intrinsic protein complex of the small-intestinal brush border membrane (pro-sucrase-isomaltase). *FEBS Lett.* 136: 329–32

Wacker, H., Lehky, P., Vanderjaege, F., Stein, E. A. 1976. On the subunit structure of particulate aminopeptidase from pig kidney. *Biochim. Biophys. Acta* 429: 546–54

Wacker, H., Müller, F., Semenza, G. 1976. Incorporation of hydrophobic aminopeptidase from hog kidney into egg lecithin liposomes: Number and orientation of aminopeptidase molecules in the lecithin vesicles. *FEBS Lett.* 68: 145–52

Walter, P., Gilmore, R., Blobel, G. 1984. Protein translocation across the endoplasmic reticulum. *Cell* 38: 5–8

Watts, C., Wickner, W., Zimmermann, R. 1983. M13 procoat and a pre-immunoglobulin share processing specificity but use different membrane receptor mechanisms. *Proc. Natl. Acad. Sci. USA* 80: 2809–13

Weiser, M. M., Wilson, J. R. 1981. Serum levels of glycosyltransferases and related glycoproteins as indicators of cancer: Biological and clinical implications. *CRC Crit. Rev. Clin. Lab. Sci.* 14: 189–239

Wickner, W. 1979. The assembly of proteins into biological membranes: The membrane trigger hypothesis. *Ann. Rev. Biochem.* 48: 23–45

Wickner, W. 1980. Assembly of proteins into

membranes. *Science* 210: 861–68

Wickner, W. T., Lodish, H. F. 1985. Multiple mechanisms of protein insertion into and across membranes. *Science* 230: 400–7

Yamashita, K., Hitoi, A., Matsuda, Y., Tsuji, A., Katunuma, N., Kubata, A. 1983a. Structural studies of the carbohydrate moieties of rat kidney-glutamyl-transpeptidase. *J. Biol. Chem.* 258: 1089–1107

Yamashita, K., Tachibana, Y., Hitoi, A., Matsuda, Y., Tsuji, A., et al. 1983b. Difference in the sugar chains of two subunits and of isozymic forms of rat kidney γ-glutamyltranspeptidase. *Arch. Biochem. Biophys.* 227: 225–32

Yokota, K., Nishi, Y., Takesue, Y. 1986. Purification and characterization of amphiphilic trehalase from rabbit small intestine. *Biochim. Biophys. Acta* 881: 405–14

Zweibaum, A., Pinto, M., Chevalier, G., Dussaulx, E., Triadou, N., et al. 1985. Enterocytic differentiation of a subpopulation of the human colon tumor cell line HT-29 selected for growth in sugar-free medium and its inhibition by glucose. *J. Cell. Physiol.* 122: 21–29

Zweibaum, A., Triadou, N., Kediger, M., Augeron, Ch., Robine-Léon, S., et al. 1983. Sucrase-isomaltase: A marker of foetal and malignant epithelial cells of the human colon. *Int. J. Cancer* 32: 407–12

References added in proof

Bos, T. J., Davis, A. R., Nayak, D. P. 1984. NH₂-terminal hydrophobic region of influenza virus neuraminidase provides the signal function in translocation. *Proc. Natl. Acad. Sci. USA* 81: 2327–31

Braun, H., Cogoli, A., Semenza, G. 1975. Dissociation of small-intestinal sucrase-isomaltase complex into enzymatically active subunits. *Eur. J. Biochem.* 52: 475–80

Coloma, J., Pitot, H. C. 1986. Characterization and sequence of a cDNA clone of γ-glutamyltranspeptidase. *Nucl. Acids Res.* 14: 1393–1403

Lipkin, M. 1985. Growth and development of gastrointestinal cells. *Ann. Rev. Physiol.* 47: 175–97

Malo, C., Ménard, D. 1982. Influence of epidermal growth factor on the development of suckling mouse intestinal mucosa. *Gastroenterology* 82: 28–35

Matsuda, Y., Tsuji, A., Katanuma, K. 1980. Studies on the structure of γ-glutamyltranspeptidase. I. Correlation between sialylation and isozymic forms. *J. Biochem.* 87: 1243–48

Ménard, D., Malo, C., Calvert, R. 1981b. Insulin accelerates the development of intestinal brush border hydrolytic activities of suckling mice. *Dev. Biol.* 85: 150–55

Takesue, Y., Yokota, K., Nishi, Y., Taguchi, R., Ikesawa, H. 1986. Solubilization of trehalase from rabbit renal and intestinal brush-border membranes by a phos- phoinositol-specific phospholipase C. *FEBS Lett.* 201 : 5–8

Yusufi, A. N. K., Low, M. G., Turner, S. T., Dousa, T. P. 1983. Selective removal of alkaline phosphatase from renal brush-border membrane and sodium-dependent brush-border membrane transport. *J. Biol. Chem.* 258 : 5695–5701

Ann. Rev. Cell Biol. 1986. 2 : 315–36

COTRANSLATIONAL AND POSTTRANSLATIONAL PROTEIN TRANSLOCATION IN PROKARYOTIC SYSTEMS

Catherine Lee and Jon Beckwith

Department of Microbiology and Molecular Genetics, Harvard Medical School, 25 Shattuck Street, Boston, Massachusetts 02115

CONTENTS

INTRODUCTION

An era of intense study into the details of the mechanism of protein secretion began with the formulation of the signal hypothesis by Blobel & Dobberstein in 1975 (Blobel & Dobberstein 1975a). According to their model, eukaryotic secretory proteins are transferred across the membrane

315

0743–4634/86/1115–0315$02.00

of the rough endoplasmic reticulum (RER) concomitant with their synthesis. Several lines of evidence support this model. First, secretory proteins are synthesized on polysomes that are bound to the RER membrane (Blobel & Dobberstein 1975a). Further evidence came from the analysis of an in vitro secretion system (Blobel & Dobberstein 1975b). In this system, secretory proteins could be transported into the interior of microsomal vesicles during their synthesis. When such vesicles were added to already completed polypeptide chains, no transfer of protein into the vesicles was observed, which suggested that posttranslational secretion was not possible. Yet, it could be argued that the presence of vesicles was only necessary to allow the nascent chains to associate with the membrane, and that the actual traversal of the membrane occurred after translation was completed. The most direct experiments that verified the cotranslational hypothesis were those showing cotranslational glycosylation of secretory proteins (Katz et al 1977; Glabe et al 1980). Since glycosylation occurs on the luminal side of the RER, any portion of the polypeptide chain that is glycosylated a priori must have traversed the membrane. Thus, cotranslational glycosylation means cotranslational secretion.

One of the early questions concerning the cotranslational mechanism of protein secretion in eukaryotes was how the ribosomes are brought to the RER membrane. This problem was apparently solved with the discovery of the signal recognition particle (SRP), a complex of six proteins and a 7S RNA (Walter & Blobel 1983). The SRP is thought to recognize signal sequences as they emerge from translating ribosomes and, via its affinity for an RER membrane protein, promote the binding of ribosomes to the RER (Gilmore & Blobel 1983). In bacteria, genetic studies have identified several genes whose products appear to be essential for protein export. Mutations in the *secA*, *secB*, *secD*, and *prlA* (also called *secY*) genes cause pleiotropic defects in protein secretion (Emr et al 1981; Oliver & Beckwith 1981; Kumamoto & Beckwith 1983; Shiba et al 1984; C. Gardel, in preparation). In addition, an *Escherichia coli* in vitro system developed by Muller & Blobel requires a 12S particle for transfer of secretory proteins into membrane vesicles (Muller & Blobel 1984a). While the protein secretion machinery in bacteria could function to facilitate a cotranslational process analogous to that in the RER, to date such a hypothesis is by no means proven.

It was not long after the presentation of evidence for the signal hypothesis in eukaryotes that a controversy emerged over whether cotranslational secretion occurs in prokaryotes. Wickner and coworkers, using the bacteriophage M13 coat protein, concluded that this protein was transferred posttranslationally in *E. coli* (Date & Wickner 1981; Ito et al 1979, 1980; Goodman et al 1981). Studies that questioned these conclusions were presented by Model and coworkers (Chang et al 1978, 1979; Russel & Model 1981). Since that time, a large number of new studies have accumulated that speak to the question of whether bacteria export their proteins co- or posttranslationally. In this review, we summarize such studies. We conclude that while posttranslational secretion can be demonstrated to occur in *E. coli*, usually under abnormal conditions, it is not yet clear that the normal mode of protein export involves posttranslational transfer.

Three Models of Protein Secretion

STRICTLY COTRANSLATIONAL   It has been proposed that a ribosome that is translating the nascent chain of a secretory protein associates with the cytoplasmic face of the membrane bilayer. The elongation of the nascent chain then results in the extrusion of the amino acid sequences through the membrane in a linear fashion (Blobel & Dobberstein 1975a; Inouye & Halegoua 1980; Davis & Tai 1980; Meyer 1982). This strictly cotranslational secretion model has several attractive features: (*a*) It specifies that a minimal amount of hydrophilic polypeptide will reside in the hydrophobic membrane at any one time, rather than demanding the translocation of longer lengths of a protein that might fold into extensive tertiary structures; (*b*) it avoids the accumulation in the cytoplasm of noncytoplasmic proteins, which might be deleterious; and (*c*) it states that the energy of translational elongation is used for extrusion of the secretory protein. According to this model, ribosomes synthesizing secretory proteins are attached to the membrane and cause the passage of the nascent polypeptide, amino acid by amino acid, through the membrane bilayer.

STRICTLY POSTTRANSLATIONAL   In contrast, it has been proposed that the synthesis of precursors to secretory proteins is completed in the cytoplasm. One hypothesis has been proposed to explain how posttranslational secretion might occur. According to the membrane-trigger hypothesis, secretory proteins initially fold into a soluble cytoplasmic form. Interaction with the membrane then triggers the refolding of the secretory protein into a membrane-compatible form, which would then promote the translocation of the precursor across the membrane (Wickner 1979, 1980). Recent studies of nuclear-encoded mitochondrial proteins have shown that these proteins are imported into mitochondria in a strictly posttranslational manner, and support the proposal that large protein structures can cross biological membranes after their synthesis has been completed (Reid 1985).

INTERMEDIATE MODEL   Another model combines features of the strictly cotranslational and posttranslational models for secretion. As in the strictly cotranslational model, it has been proposed that the ribosome and nascent secretory chain associate with the cytoplasmic face of the membrane at an early step in secretion. However, according to this model, stretches of the polypeptide are transferred across the membrane post-translationally (Randall & Hardy 1984a, b), not amino acid by amino acid as in the strictly cotranslational model. Thus, in this partly cotranslational model, actual translocation of certain portions of a protein occurs post-translationally.

Studies in Escherichia coli

Protein secretion in *E. coli* has been studied using a variety of experimental systems. For example, the phage M13 coat protein has been inserted into vesicles and liposomes (Geller & Wickner 1985; Goodman et al 1981), and more recently, secretion of periplasmic and outer membrane proteins into inverted membrane vesicles has been reconstituted in vitro (Muller & Blobel 1984a,b; Chen et al 1985; Chen & Tai 1985). In vivo protein secretion using whole cells, spheroplasts, and mutant cells has been characterized. As a result of the comparison of properties of mutant and wild-type cells, important features of the signal sequence of secretory proteins and the cellular components that are required for secretion have been identified (Bankaitis et al 1985).

Several experimental techniques have been used to measure protein secretion in *E. coli*. Secretory proteins contain an amino-terminal signal peptide that is essential for translocation and is subsequently removed by a specific peptidase outside of the cytoplasmic membrane (Zimmerman et al 1982; Ohno-Iwashita & Wickner 1983). Thus, the processing of a precursor secretory protein to its mature form is evidence that the protein has traversed the membrane. The translocation of protein across the inner membrane can also be measured by its accessibility or inaccessibility to externally added reagents or proteases that cannot cross the membrane bilayer. For example, in *E. coli* spheroplasts, proteins are secreted outward and become accessible to external reagents (Smith et al 1977; Randall 1983). Conversely, in reconstituted *E. coli* protein secretion systems, proteins are secreted into inverted membrane vesicles and become inaccessible to external reagents (Muller & Blobel 1984a; Chen et al 1985). Biochemical fractionation procedures have also been used to determine the location of secretory proteins in *E. coli*. For example, soluble secretory proteins within the periplasmic space can be released by osmotic shock or by removing the outer membrane after treatment with lysozyme and EDTA (Copeland et al 1982; Neu & Heppel 1965).

However, cell fractionation studies in gram-negative bacteria have many possible artifacts. Thus, the claim that a precursor intermediate is *cyto-*

plasmic must be critically examined. Koshland & Botstein (1982) have shown that by using fractionation procedures alone it is possible to mistakenly conclude that a protein is in the cytoplasm when it is in fact loosely adhering to the extracytoplasmic surface of the cytoplasmic membrane. The most careful studies have checked the sensitivity of proteins to externally added proteases to verify their location as intra- or extracytoplasmic. In some cases, the interpretation of these studies also can be misleading due to the protease-resistance of many periplasmic proteins. Furthermore, it seems possible that precursor molecules that appear to be in the cytoplasm, as judged by fractionation and by protease resistance, could actually be in the membrane or even on the outside of the membrane in some protected state. Recently, several laboratories have used agents such as gold-labelled antibodies to observe the cellular location of certain proteins with the electron microscope (Pages et al 1984; Tommassen et al 1985). In some cases, the results conflict with those obtained by standard fractionation techniques. While these EM studies appear to be more definitive, they are certainly not free of artifacts either.

To study the kinetics of export and processing, *E. coli* cells can be pulse labelled with radioactive amino acids and then incubated without labelling for various times. Individual secretory proteins are then analyzed during this chase period by immunoprecipitation. Using this experimental approach in wild-type cells, secretory proteins are observed to be processed very rapidly. For most of the secretory proteins studied, a 30-sec pulse of label, followed by antibody precipitation of cell extracts and autoradiography after SDS-PAGE, reveals essentially no precursor form. In contrast, posttranslational export can be observed in a variety of cases in which the normal rapid pathway of secretion is inhibited either by mutation or by special treatment of cells. In these cases, such pulse labelling usually shows secretory proteins entirely or almost entirely in the form of precursor. A slow accumulation of mature exported protein is eventually observed with longer chase times. The kinetics of this posttranslational export process exhibits half-times of 0.5–10 min or longer.

Some cautions about the interpretation of such experiments should be mentioned. First, the chase experiments are usually not conducted in the presence of chloramphenicol or any other agent that prevents protein synthesis. Therefore it is conceivable that in some cases the appearance of the label in mature protein is due to completion of labelled nascent chains, not due to processing of already completed precursor molecules. However, whenever chase experiments were conducted in the presence of chloramphenicol, posttranslational appearance of mature protein was still observed.

STUDIES OF POSTTRANSLATIONAL EXPORT

Phage M13 Coat Protein

The M13 coat protein is a small transmembrane peptide of 50 amino acids that is initially synthesized with a 23-amino acid signal sequence. The precursor form of the coat protein is only transiently observed when cells infected with wild-type M13 phage are pulse labelled at 30 or 37°C (Ito et al 1979, 1980; Russel & Model 1981). Unfortunately, the small amount of labelled precursor and its rapid disappearance made it difficult to determine whether these proteins were truly precursors to mature coat protein in the membrane. However, later studies by Date & Wickner (1981) showed that at 42°C M13 procoat is exported posttranslationally with a $t_{1/2}$ of approximately 30 sec and that certain M13 phages processed their coat proteins much more slowly, with a $t_{1/2}$ of 5 min. The mutant phages contain amber mutations in gene 1, 5 or 7 that disrupt phage assembly and extrusion and eventually cause cell death. By pulse labelling cells 1 hr postinfection, the precursor coat protein of these mutant M13 phages was readily observed and demonstrated to be soluble in the cytoplasm. Furthermore, the coat protein eventually appeared in the membrane fraction in its processed form. Consequently, Wickner and his coworkers concluded that the M13 coat protein can be inserted into the membrane in a strictly posttranslational fashion.

Later studies by Russel & Model (1981) suggest that the cells infected with the mutant M13 phages were severely abnormal at 1 hr postinfection and challenge the significance of previous studies. These investigators argued that the cells were not viable and that the membranes were stressed by excess coat protein that had not been removed by the normal phage-extrusion process. In order to test this hypothesis, they analyzed the coat protein of the same mutant M13 phages at earlier times postinfection, when cell viability had not decreased. They found that the slowed post-translational secretion kinetics did not occur until later in the infection, at which time they presume cell physiology is abnormal.

Russel & Model (1981) studied another mutant M13 phage that produces a mutant coat protein with a changed amino acid at position 2 of the mature protein. This mutant coat protein, R6, was found to mature slowly, possibly due to a decrease in its recognition by the signal peptidase (however, Wickner and his coworkers do not agree with this hypothesis). Unlike the mutant phage studied by Ito et al (1979, 1980) this coat protein mutant displayed slow maturation kinetics very soon after infection and apparently did not decrease cell viability for at least 1 hr postinfection. In contrast to the findings of Ito et al (1979, 1980), the precursor form of the mutant coat protein was found associated with the membrane at the earliest time point analyzed. Although Russel & Model (1981) demonstrated that insertion can occur independent of maturation, their result does not necessarily refute the hypothesis that export of the coat protein is strictly posttranslational. It is likely that, as with the wild-type M13 coat protein, the insertion of the mutant coat protein is far too rapid to distinguish whether or not the precursor coat protein transiently exists in soluble form in the cytoplasm.

It is clear from the studies of Wickner and coworkers that insertion and maturation of the M13 coat protein can occur in a strictly posttranslational manner. They have also demonstrated in vitro that purified coat protein precursor can be inserted posttranslationally into membrane vesicles and purified liposomes (Geller & Wickner 1985; Goodman et al 1981). However, as others have suggested, the posttranslational process observed in some of these experiments might not be relevant to the events in wild-type cells and phage. For example, posttranslational export may occur only when the normal mechanism is disrupted by abnormal cellular physiology or by incomplete reconstitution in vitro. However, the observation that the M13 procoat is exported posttranslationally at 42°C indicates that this process can occur normally (Date & Wickner 1981).

Elucidation of the mechanism of coat protein export may reveal a unique process that is not applicable to the export of periplasmic and outer membrane proteins. In fact, the export of coat protein does not appear to be affected by the *secA* or the *secY* mutations that block the export of all periplasmic and outer membrane proteins examined thus far (Wolfe et al 1985).

Signal Sequence Mutants

Nearly all of the signal sequence mutants that have been obtained are only partially defective. Except when the signal sequence is subject to extreme disruption (e.g. by its complete deletion or by deletion of its hydrophobic core), such mutants always export some fraction of the mutant secretory protein. In several cases, the mode by which this fraction is exported has been studied by conducting pulse-chase experiments. Such analysis shows that the export of most signal sequence mutant proteins is strictly posttranslational.

Koshland et al (1982) have generated a number of alterations in the signal sequence of the TEM β-lactamase using in vitro mutagenesis with bisulfite. Most of these mutations alter the region near the carboxyl-terminal processing site of the signal peptide, and the slower kinetics of conversion of precursor to mature β-lactamase may be due mainly to slower processing and not slower export. This is suggested by protease accessibility studies that showed that the mutant precursors were, even at early times in pulse-chase experiments, protease-sensitive in spheroplasts, i.e. they had been secreted through the membrane. However, one mutant [*fs(7,9)*] is altered toward the amino terminus of the hydrophobic region and contains the charged amino acid arginine in what is normally a hydrophobic sequence. In this mutant, no mature β-lactamase was seen after a 30-sec pulse of incorporation of radiolabel. Analysis of the kinetics of conversion of *fs(7,9)* precursor to mature β-lactamase showed a half-time of processing of approximately 6 min, with 80% conversion by 20 min. Furthermore, in contrast to mutants that affect processing of the signal peptide, the precursor seen in mutant *fs(7,9)* was insensitive to protease added to spheroplasts, which indicates that the precursor was intra-cytoplasmic and was secreted posttranslationally.

A series of mutants of the major lipoprotein of *E. coli* have been obtained by Inouye and coworkers using oligonucleotide-directed mutagenesis. Some of these alter the amino-terminal charged region of the signal sequence (Inouye et al 1982; Vlasuk et al 1983). In most cases these mutations lead to altered kinetics of export and processing of lipoprotein, although the changes are not as dramatic as those caused by mutations that introduce charged amino acids into the hydrophobic region of certain other secretory proteins (see below). For instance, lipoprotein mutants I-4 and I-7 have half-times of export of ~ 0.5 and 2 min, respectively. Most of the precursor is converted to mature lipoprotein by 2 min in the case of I-4, and by 8 min in the case of I-7.

Signal sequence mutants that reduce the export of alkaline phosphatase to the *E. coli* periplasm have also been examined for their kinetics of export (Michaelis et al 1986). In the five mutants studied, the steady state amounts of alkaline phosphatase in the periplasm range from 1.5 to 30% of wild-type levels. In each case, the protein was transported posttranslationally. The kinetics of appearance of the mature alkaline phosphatase was found to vary according to the extent of the defect in the signal sequence. However, it was difficult to precisely determine half-times of export, since internalized alkaline phosphatase precursor was degraded with a half-life of about 5 min. Thus, the failure to see increasing amounts of mature alkaline phosphatase after longer chase periods may merely have been due to the absence of any precursor to be chased rather than the cell's inability to secrete alkaline phosphatase at long chase times. At a minimum, the half-times of processing were estimated to be 5 min.

The most extensive analysis of the kinetics of export in signal sequence mutants has been conducted by Bassford and coworkers, who have examined mutants of the periplasmic maltose binding protein (Ryan & Bassford 1985). These mutants were selected in vivo for severe defects in export. In addition, Bassford and coworkers have examined revertants of certain of these signal sequence mutants (Bankaitis et al 1984) and the behavior of signal sequence mutants when combined with an extragenic suppressor of such mutants, *prlA* (Ryan & Bassford 1985). As was found in studies of alkaline phosphatase, the signal sequence mutants of maltose binding protein all exhibited posttranslational export with variable kinetics that depended upon the extent of the defect in the signal sequence. For instance, the least altered mutant, 16–1, which has a single amino acid substitution in the hydrophobic core of the signal sequence, shows a small amount of mature protein even at the earliest time point (1 min). The half-time for appearance of the mature maltose binding protein in the 16–1 mutant was approximately 5 min. However, only a minority of the total maltose binding protein is found in the mature form even after 60 min. The amount of mature protein appears to reach a plateau about 15 min after initiation of the pulse-chase and only increases slowly after that time (Ryan & Bassford 1985; P. Bassford, personal communication). The other maltose binding protein signal sequence mutants exhibit much longer half-times for export (some as high as 30 min or more), and, with some, precursor molecules appeared to continue to be chased into mature form after 60 min, the longest chase time used in the experiments.

The kinetics of export of mutant maltose binding protein was also examined in strains that contain the *prlA* mutation, which suppresses the secretion defects of signal sequence mutations. Interestingly, while this suppression was very efficient in terms of increasing the amount of maltose binding protein exported, the export was still posttranslational. The half-times of export of the mutant secretory protein were much shorter in the *prlA* strains, ranging from 2.5 to 10 min.

Ryan & Bassford propose an explanation for the kinetics of conversion of precursor to mature form observed in signal sequence mutants. They suggest that in such mutants there are two pools of precursor in the cytoplasm. One of these is available for export, and the other is destined to remain in the cytoplasm indefinitely or until it is degraded. They suggest that in signal sequence mutants only a fraction of the nascent polypeptide chains associate with the cell's secretion machinery and are, therefore, competent for export. The other, export-incompetent fraction may have assumed a conformation that prevents their interaction with the secretion machinery, and therefore they remain in the cytoplasm. Ryan & Bassford also propose that even the precursor molecules that have associated with the secretion machinery are not exported via the normal, rapid process. The mutant precursor protein is thought to be less effective at passing through the membrane, so slow posttranslational export kinetics is observed. Thus, in this model, the signal sequence functions at two steps in the secretion process, first in the recognition of the secretion machinery and then in passage through the membrane.

The Ryan-Bassford model predicts that each signal sequence mutant will exhibit a finite half-life and fraction of conversion of precursor to mature secretory protein that depend on how much precursor is in a state that will allow export. However, from studies of maltose binding protein and other systems, there may be only one case where such a prediction is clearly fulfilled, that of the mutant 16–1 of maltose binding protein described above. In the other cases, even at the longest time point of the chase (as long as 60 min), it is possible that export and processing are still taking place. For example, if the chase were to be continued for hours, most or all of the precursors might be converted to the mature form. Thus, the evidence for two fractions of precursors is limited.

In the Ryan-Bassford model, it seems possible that export-competent precursor would be loosely associated with the membrane, which might keep it in a conformation or position that would facilitate its ultimate passage through the membrane. Thus far there is no biochemical evidence for such an export-competent precursor that is membrane bound. When fractionation studies have been conducted on signal sequence mutants, the precursors are usually found in the cytoplasmic fraction. However, in certain cases in which secretion is partially blocked, precursor has been found in both membrane and cytoplasmic fractions. In fact, Koshland & Botstein (1982) have proposed that there is a membrane-bound form of β-lactamase precursor that is an intermediate in the export process. Thus, it is possible that fractionation artifacts have led to improper identification of other precursors as soluble in the cytoplasm. A recent analysis of *E. coli* spheroplasts revealed two forms of maltose binding protein precursor, a protease-resistant form and a protease-sensitive form (L. L. Randall & S. J. S. Hardy, submitted for publication). An intriguing correlation was found between the amount of maltose binding protein precursor that was in the protease-resistant form and the inability to export maltose binding protein. Randall & Hardy suggest that only the protease-sensitive form of precursor can be exported. This conformation of the maltose binding protein precursor may correspond to the export-competent state, as proposed by Ryan & Bassford. The location of the protease-sensitive precursors has not yet been determined. The model predicts that they are located inside of the spheroplasts, possibly at the membrane.

The Ryan-Bassford model assumes that the fraction of precursors exported in signal sequence mutants follow the normal secretion pathway. An alternative explanation for these data is that another, distinct post-translational pathway is used when the normal pathway is disrupted. If this were the case, the fact that *prlA* suppressor mutants speed up the *posttranslational* pathway would suggest that the *prlA* gene product, which appears to be required for normal secretion, is also required for the alternative pathway. The slow posttranslational kinetics of the *prlA*-dependent alternative process may be a result of the inability of signal sequence mutant precursors to utilize other gene products that function in the normal, rapid secretion pathway. This alternative model is also based on the assumption that the signal sequence functions at two steps in the normal export process.

Pleiotropic Mutants Defective in Protein Secretion

Many of the mutants altered in the *sec* and *prl* genes cause substantial accumulation of precursors of exported proteins in *E. coli* (Oliver & Beckwith 1981; Kumamoto & Beckwith 1983; Shiba et al 1984; Gardel et al, in preparation). In certain cases it has been shown that these precursors are exported and processed by a posttranslational pathway. In a *secA(Ts)* mutant, early studies indicated that the precursor of maltose binding protein found in the cytoplasm when the mutant was incubated at a high temperature could not be chased into mature maltose binding protein (Oliver & Beckwith 1981). However, subsequent experiments with the same mutant showed posttranslational export of both maltose binding protein and ribose binding protein with a half-time of approximately 5 min (Garwin & Beckwith 1982). An explanation for these different results may be found in the time of incubation at the high temperature and the resultant physiological state of the cells. After long times of incubation at high temperatures, the *secA(Ts)* mutation eventually has severe effects on cellular physiology. These effects may include depletion of the proton-motive force and serious membrane perturbations. In this state, the cell may be incompetent for either co- or posttranslational export. Whether such a state is reached may well depend on the conditions of the experiment.

In a mutant with an insertion of the *Tn5* transposable element in the *secB* gene, the precursor maltose binding protein that accumulates in the cytoplasm can also be chased efficiently into mature protein by a posttranslational mechanism (Kumamoto & Beckwith 1985). Again, the half-time of processing is approximately 5 min. The *secA(Ts)* mutation (located in the *prlA* gene) also exhibits slow export and processing of the maltose binding protein precursor, with a half-time of approximately one to two min (Ito et al 1983; Shiba et al 1984).

In all of these cases, the kinetics of export can be explained by either of the two hypotheses devised to explain the posttranslational export observed in signal sequence mutants. For example, if the *sec* mutants have a partially defective normal pathway, the secretory proteins may utilize this normal pathway but with slower kinetics due to the defective secretion machinery. Alternatively, the secretory proteins in *sec* mutants may be exported via a separate posttranslational pathway. It might appear that these two hypotheses can be easily distinguished. For example, if the latter hypothesis were correct, a null mutation in one of the *sec* genes might have no effect on the alternative posttranslational pathway. In contrast, if the normal pathway were utilized in the *sec* mutants, a null mutation would eliminate all export. Yet, in the case of *secA*, analysis of a null mutation has not shown whether the SecA protein is required for posttranslational export in vivo. The problem is that it is difficult to study *secA* null mutations because the *secA* gene is essential, and defects in protein secretion have many secondary effects on cellular physiology (Strauch et al 1986).

β-Lactamase

Koshland & Botstein (1980, 1982) showed that the wild-type *bla* gene codes for a β-lactamase that is exported posttranslationally. After a 30-sec pulse of radioactivity, chase experiments showed that β-lactamase was posttranslationally processed with a half-time of approximately 1 min. However, even after 5.5 min of chase (the last time point analyzed), there still appeared to be a small amount of precursor remaining in the cytoplasm. These properties of β-lactamase are similar to those of some of the signal sequence mutants of lipoprotein studied by Inouye and coworkers. It could be that the β-lactamase signal sequence is naturally partially defective, causing slower posttranslational export. Koshland & Botstein point out that this enzyme did not evolve in *E. coli* or *Salmonella* in which it has been studied and, therefore, may not be well-fitted to the secretion machineries of these organisms.

The possibility that the β-lactamase signal sequence is partially defective could be studied by examining the export of other proteins fused to the β-lactamase signal sequence and of hybrids in which the mature portion of β-lactamase is fused to the signal sequence of proteins that are processed more rapidly. Certain hybrids between lipoprotein and β-lactamase have been studied, although not for this purpose, and the results of those experiments do not reveal the exact component of β-lactamase that is responsible for its slow export (Ghrayeb & Inouye 1984; Coleman et al 1985).

Overproduction of Secretory Proteins

At least two cases have been reported in which overproduction of a periplasmic protein coded on a multicopy plasmid causes a block in export.

In the case of a plasmid carrying the *phoA* gene, when cells were starved for phosphate, which causes derepression of the *phoA* gene, only the precursor of the gene product, alkaline phosphatase, was observed (Inouye et al 1981). Similarly, when a strain carrying the *phoS* gene on a plasmid was starved for phosphate, significant amounts of PhoS precursor were observed (Pages et al 1984). The PhoS precursor was found in both the cytoplasmic and membrane fractions. The authors concluded that the membrane-bound precursor is on the cytoplasmic face of the membrane and can be exported and processed posttranslationally. However, the evidence for the location of this precursor is quite indirect. The data are also consistent with the possibility that the PhoS precursor is on the periplasmic face of the membrane and is processed only slowly due to saturation of the signal peptidase enzyme. The cytoplasmic precursor appears to be partially degraded and not exported.

Disruption of the Proton-Motive Force

Several studies have demonstrated that disruption of the transmembrane proton-motive force in whole cells or in minicells prevents protein secretion. By treatments with a variety of reagents that dissipate the proton-motive force, Randall and her colleagues found that the periplasmic proteins—maltose binding protein, arabinose binding protein, and β-lactamase—and the outer membrane proteins, OmpA and OmpF, are accumulated in their precursor forms (Bakker & Randall 1984; Enequist et al 1981). Daniels et al (1981) similarly demonstrated that the maturation of the precursor of the leucine-specific binding protein was inhibited by disruption of the membrane potential. Fractionation and protease accessibility analyses have been interpreted to indicate that the precursor molecules that accumulate in these studies are in the cytoplasm or at the cytoplasmic face of the inner membrane (Daniels et al 1981; Randall 1983). After removal of the inhibitor, these precursors were found to slowly chase to their mature forms in a strictly posttranslational manner.

The processing of the M13 coat protein was also found to be inhibited by uncouplers of the membrane potential (Zimmerman et al 1982). Further analysis of the mutant coat protein R6, which is not affected by uncouplers, may reveal what step in export is blocked by disruption of the membrane potential.

In Vitro Reconstitution

As mentioned above, the in vitro insertion of M13 coat protein into membrane vesicles and purified liposomes can occur in a strictly post-translational manner. Secretion of periplasmic and outer membrane proteins of *E. coli* has also been demonstrated in vitro by combining the soluble components for transcription and translation of secretory proteins with inverted membrane vesicles. By allowing protein translation and translocation to occur simultaneously in vitro, Muller & Blobel (1984a) found that 25% of two periplasmic proteins (alkaline phosphatase and a truncated β-lactamase) and an outer membrane protein (lambda receptor protein) were secreted into inverted vesicles and correctly processed. Chen et al (1985) also reported in vitro translocation and processing of alkaline phosphatase and the outer membrane protein OmpA. Both translocation systems were able to secrete protein in a strictly posttranslational manner. Muller & Blobel (1984a) partially purified the precursor secretory proteins in a postribosomal supernatant and again found that 25% of the proteins were translocated and processed when inverted membrane vesicles were added posttranslationally. Chen et al (1985) temporally separated protein translation and posttranslational translocation by allowing synthesis of the precursor secretory proteins in the absence of vesicles and then adding chloramphenicol before or during the subsequent addition of the membrane vesicles.

Unfortunately, since posttranslational secretion can occur in vitro, it is difficult to test whether cotranslational secretion is also occurring in the reaction mixtures in which translation and translocation occur simultaneously. Evidence for cotranslational secretion in vitro might be obtained if posttranslational secretion could be differentially blocked or if the cotranslational secretion process occurred more rapidly or more efficiently than the posttranslational secretion process. Chen et al (1985) attempted to distinguish cotranslational and posttranslational secretion in vitro by comparing the extent of translocation during a short period of simultaneous protein translation and translocation with that in the same translational mixture incubated with vesicles posttranslationally after treatment with chloramphenicol. Under a particular set of conditions, they reported that the cotranslational secretion process was rapid, whereas the posttranslational process showed a lag phase. Thus, they were able to estimate that the efficiency of in vitro translocation of alkaline phosphatase was 15% posttranslationally and 23% cotranslationally, while that of OmpA protein was 22% cotranslationally and 33% posttranslationally. These studies show that a cotranslational secretion process can be masked in an in vitro system in which posttranslational secretion also occurs.

STUDIES OF COTRANSLATIONAL EXPORT

Isolation of Membrane-Bound Polysomes

It has been established that almost all exported proteins in E. coli are synthesized on polysomes that are bound to the membrane (Randall &

Hardy 1984a, b). There has been much speculation on the nature and function of this membrane association. With regard to the mechanism of export, proteins that are synthesized at the membrane might conceivably cross the membrane posttranslationally, strictly cotranslationally, or partly cotranslationally. The following two sections describe experiments that attempted to determine the transmembrane disposition of the nascent chains on these polysomes.

External Labelling of Nascent Chains

To identify nascent secretory proteins that are extruded cotranslationally across the membrane bilayer, Smith et al (1977) radiolabelled amino groups exposed on the surface of *E. coli* spheroplasts with a membrane-impermeable reagent. When the membrane-associated polysomes were purified, 2% of the total label remained. This radioactive fraction appeared to represent a heterogeneous population of nascent polypeptides that had been labelled from the outside of the spheroplasts. The labelled nascent chains displayed a broad molecular weight distribution, and 20% of the completed translation products reacted with antibody directed against the periplasmic protein alkaline phosphatase. Thus, Smith et al (1977) concluded that alkaline phosphatase and other secretory proteins were translocated across the membrane in a strictly cotranslational manner.

These experiments demonstrate that nascent polypeptide chains are outside the inner membrane of *E. coli* spheroplasts and that a fraction of these represent a secretory protein, alkaline phosphatase. However, these results are not exclusively consistent with a strictly cotranslational mode of protein secretion for alkaline phosphatase. For example, since a gradient of lengths of external nascent chains of alkaline phosphatase was not identified, the secretion of alkaline phosphatase via the intermediate mode cannot be eliminated. Furthermore, these experiments cannot exclude the possibility that some alkaline phosphatase precursors are secreted posttranslationally since the analysis was limited to only those molecules that could be externally labelled.

Protease Accessibility of Nascent Chains

Studies by Randall (1983), unlike those of Smith et al (1977), were designed to detect both internal nascent secretory proteins and those translocated across the inner membrane. Randall (1983) similarly prepared *E. coli* spheroplasts, but from cells that had been pulse labelled and then immediately frozen. The location of labelled nascent chains of the periplasmic protein maltose binding protein was determined by testing their accessibility to externally added proteinase K. All nascent chains of maltose binding protein that carried the amino-terminal signal sequence were proteinase K resistant, i.e. were inside the spheroplasts, whereas those nascent chains that had been processed were outside and were accessible to the externally added protease. In previous studies, Josefsson & Randall (1981a, b) demonstrated that 80–100% of the full length of maltose binding protein is synthesized before the signal sequence is removed. Thus, Randall (1983) concluded that large nascent polypeptides of maltose binding protein (at least 80% of the full length) that carry the signal sequence first exist inside the spheroplasts and then are translocated to the outer face of the membrane where the signal sequence is removed even before the synthesis of the maltose binding protein is completed. These results are not consistent with a strictly cotranslational nor a strictly posttranslational mode of protein secretion, so Randall (1983) proposes an intermediate model in which large polypeptide domains of nascent secretory proteins are translocated across the membrane bilayer. In addition, Randall and her colleagues suggest that the synthesis of 80% of maltose binding protein is required before translocation and processing can occur and postulate that a translocation-competent state is normally achieved by synthesis of a "required" polypeptide domain. However, it is clear that a special polypeptide domain is not absolutely required for translocation since Josefsson & Randall (1981b) found that even the shortest nascent chains of precursor maltose binding protein detected were processed after a one minute chase in the presence of chloramphenicol, which prevents the completion of further synthesis during the chase.

Conclusions from Studies of Cotranslational Export

The studies by Smith et al (1977) and Randall (1983) attempted to analyze the in vivo translocation process and thereby circumvent the problems of interpreting the physiological significance of studies in abnormal cells or in vitro systems. The results of Smith et al (1977) are consistent with the strictly cotranslational or the intermediate mode of secretion of the periplasmic protein alkaline phosphatase. Their studies cannot address whether some or even most of the alkaline phosphatase precursors can be secreted posttranslationally, since the nascent chains analyzed were limited to those labelled externally.

The results of Randall (1983) led her to propose that the periplasmic maltose binding protein is secreted both posttranslationally and via the intermediate mode of secretion. The accuracy of her conclusions depends upon the validity of her interpretation of proteinase K accessibility. For example, one could argue that maltose binding protein is secreted in a strictly cotranslationally manner, if the signal sequence at the amino terminus of a cotranslationally secreted nascent chain causes the external polypeptide to associate closely with the outer face of the membrane so as to hinder digestion of the protein by proteinase K. The demonstration that the polypeptide becomes proteinase K–sensitive after treatment with detergent or sonication may not be sufficient proof that the polypeptide is inside the spheroplasts, since these treatments may also disrupt the association of the signal sequence with the membrane. Furthermore, it is possible that removal of the signal sequence is the limiting step, such that in rapidly growing cells 80% of the full length of maltose binding protein, on average, is synthesized and translocated before it encounters the signal peptidase. In this scenario, once the signal sequence is removed by the peptidase, the amino terminus of the external nascent chain is freed, and the polypeptide becomes accessible to digestion by the externally added proteinase K.

This alternative hypothesis to explain the results of Randall (1983) is also consistent with the finding that, after a brief incubation in the absence of translation, processing and proteinase K accessibility of short nascent chains are observed, since the incubation increases the time during which the signal peptidase may be encountered. Continued analysis of the association of the nascent chains at the spheroplast membrane may reveal whether the large precursor domains are at the inner face as proposed by Randall (1983) or at the outer face as we suggest is possible here. For example, a technique using chemical probes could be devised to determine the location of precursor nascent chains to test the validity of the results obtained using proteinase K accessibility. However, these types of experiments are especially difficult in *E. coli* due to the presence of the outer membrane. In addition, the appropriate membrane-impermeable reagents currently available are limited and difficult to apply to these particular analyses, which require immunoprecipitation and limited proteolysis to examine amino-terminal fragments of nascent chains.

In the studies by Smith et al (1977) and Randall (1983), it was assumed that the in vivo process was precisely halted during the in vitro procedures required to form spheroplasts and treat the spheroplasts with labelling reagent or proteinase K. Is it possible that in vivo secretion occurs in a strictly posttranslational manner and that the internalized nascent chains are artifactually translocated across the membrane during the time required for the in vitro manipulations? This is unlikely since Josefsson & Randall (1981a,b) harvested cells in trichloroacetic acid and still found that many nascent chains of secretory proteins had been cotranslationally processed by the signal peptidase. This result argues that in vivo secretion of periplasmic proteins occurs by either a strictly cotranslational or an intermediate mode. However, other aspects of the conclusions may be biased by the possibility that other artifacts are introduced during the experimental procedures. For example, besides the assumptions that the translocation process is halted during spheroplast formation and that all exported polypeptides are sensitive to externally added protease, these studies rely on the quantitative recovery and non-alteration of nascent chains during manipulation of the spheroplasts and during immuno-precipitation. If a certain group of external nascent chains were not recovered during one of the procedures, the results of these experiments would be unrepresentative.

CONCLUSIONS

We have summarized the evidence for co- and posttranslational export of proteins in gram-negative bacteria. Two questions have been studied: (*a*) What mechanisms are possible in these organisms, and (*b*) what is the normal process by which proteins are exported?

There is now overwhelming evidence from both in vivo and in vitro studies that posttranslational passage of proteins through membranes can occur in gram-negative bacteria. Except in the cases of export of β-lactamase and M13, this mode of export has only been observed under abnormal conditions or in in vitro systems. However, the fact that post-translational export can occur does not indicate whether or not it is the normal export pathway. The posttranslational pathway may only come into play under conditions in which the normal cotranslational pathway is blocked by mutation, chemical treatment, or some nonphysiological aspect of an in vitro system. Conversely, the normal mechanism may be posttranslational, and the experimental systems may simply slow the pro-cess enough to allow its observation. For example, it is striking that signal sequence mutants that export their proteins posttranslationally do so with a spectrum of half-lives that approaches at its low end that of protein export in normal cells. (See for instance the lipoprotein mutant I-4 described above.) Furthermore, the signal sequence, which is required for normal secretion, appears to be recognized by the posttranslational pathway since the efficiency and kinetics of posttranslational export vary according to the extent of the defect in the signal sequence.

With regard to the second question, we feel that there is not enough evidence at this point to determine the normal mode of protein export in these systems. If β-lactamase is a normally exported protein in *Salmonella*, then export is posttranslational. However, because the export of β-lac-tamase appears somewhat inefficient, it is probably not a good system to generalize from. It has been shown that the export of the phage M13 coat protein can occur in a strictly posttranslational manner. Although it has been argued that the posttranslational process is not the normal pathway for this protein, even if it were, the unusually small size of the coat protein might allow a unique export mechanism not applicable to large secretory proteins. Randall's studies of normal periplasmic proteins suggest that many are exported by a partly cotranslational mechanism. These are the most sophisticated studies of normal secretion to date; but we await further experiments using other approaches to verify the conclusions.

A major question in studying the problem of the secretory mechanism is whether the cell contains a secretion machinery. Analysis of such machinery, if it exists, should eventually reveal the mechanism of secretion. For example, the properties of the components in the eukaryotic in vitro system suggest that secretion into the endoplasmic reticulum normally occurs cotranslationally. We describe in the Introduction the genetic and in vitro evidence that a secretion machinery exists in prokaryotes. However, the components of this machinery have not been as well defined as some of the components in the eukaryotic in vitro system. It remains to be determined whether there is a soluble complex analogous to the SRP (signal recognition particle) that interacts with nascent signal sequences to facilitate the export process. Both co- and posttranslational mechanisms may require these components. The Ryan-Bassford model proposes that both modes of export require the presence of the same secretion machinery. The in vitro results of Muller & Blobel indicate that this may be the case. But so far, there has been no correlation made between phenomena observed in the in vitro system and those observed in vivo. Such a correlation will presumably be possible when the components of the in vitro system are purified and their relation to the *sec* and *prl* gene products is tested.

A major advantage of eukaryotic systems for studies of the mechanism of protein secretion is the existence of protein glycosylation. This allows the ready identification of those portions of the protein that have passed through the RER membrane. Glycosylation does not occur in the bacteria that have been studied. (If it did, we would know today the answer to many of the questions raised in this review.) However, there are some modifications of cell envelope proteins that may be useful for analogous studies. For example, the lipoproteins of bacteria are modified on a cysteine at their amino terminus by the attachment of a glyceride that contains two ester-linked fatty acid residues and an amide-linked fatty acid (Halegoua & Inouye 1979). In addition, at least two periplasmic proteins contain disulfide bonds (Schlesinger & Barrett 1965; Pollitt & Zalkin 1983), and some evidence suggests that these bonds can form only in the periplasm (Pollitt & Zalkin 1983). If so, the analysis of disulfide linkages close to the amino terminus of an exported protein may provide another tool to study the kinetics of traversal of the membrane. Alternatively, some imaginative genetic engineer may be able to construct bacterial strains that contain novel protein-modifying enzymes in the periplasmic space.

Clearly, studies of *E. coli* would benefit from the development of additional analytical techniques. For example, several studies have raised the possibility that precursors of secreted proteins exist in two or more forms and in different intracellular or intramembrane locations (Koshland & Botstein 1982; Pages et al 1984; Ryan & Bassford 1985; L. L. Randall & S. J. S. Hardy, submitted for publication). We anticipate the development of more sophisticated approaches for examination of the precise cellular location and structural features of these precursors of secretory proteins.

Many of our suggestions argue for a more intensive cell biology of bacteria such as *E. coli*, including cytological studies, improvements in cellular fractionation techniques, and better methods for differentiation of intracellular compartments. Clearly, the genetics and biochemistry of *E. coli* are much more sophisticated than those of higher organisms. But in the case of cell biology, the reverse is true.

ACKNOWLEDGMENT

This research has been supported by the National Science Foundation and the American Cancer Society. C.L. was a Fellow of The Jane Coffin Childs Memorial Fund for Medical Research.

Literature Cited

Bakker, E. P., Randall, L. L. 1984. The requirement for energy during export of β-lactamase in *Escherichia coli*. *EMBO J.* 3:895–900

Bankaitis, V. A., Rasmussen, B. A., Bassford, P. J. Jr. 1984. Intragenic suppressor mutations that restore export of maltose binding protein with a truncated signal peptide. *Cell* 37:243–52

Bankaitis, V. A., Ryan, J. P., Rasmussen, B. A., Bassford, P. J. Jr. 1985. The use of genetic techniques to analyze protein export in *Escherichia coli*. *Curr. Top. Membr. Transp.* 24:105–50

Blobel, G., Dobberstein, B. 1975a. Transfer of proteins across membranes. I. presence of proteolytically processed and nonprocessed nascent immunoglobulin light chains on membrane-bound ribosomes of murine myeloma. *J. Cell Biol.* 67:835–51

Blobel, G., Dobberstein, B. 1975b. Transfer of proteins across membranes. II. reconstitution of functional rough microsomes from heterologous components. *J. Cell Biol.* 67:852–62

Chang, N. C., Blobel, G., Model, P. 1978. Detection of prokaryotic signal peptidase in an *Escherichia coli* membrane fraction: Endoproteolytic cleavage of nascent f1 pre-coat protein. *Proc. Natl. Acad. Sci. USA* 75:361–65

Chang, N. C., Model, P., Blobel, G. 1979. Membrane biogenesis: Cotranslational integration of the bacteriophage f1 coat protein into an *Escherichia coli* membrane fraction. *Proc. Natl. Acad. Sci. USA* 76:1251–55

Chen, L., Rhoads, D., Tai, P. C. 1985. Alkaline phosphatase and OmpA protein can be translocated posttranslationally into membrane vesicles of *Escherichia coli*. *J. Bacteriol.* 161:973–80

Chen, L., Tai, P. C. 1985. ATP is essential for protein translocation into *Escherichia coli* membrane vesicles. *Proc. Natl. Acad. Sci. USA* 82:4384–88

Coleman, J., Inukai, M., Inouye, M. 1985. Dual functions of the signal peptide in protein transfer across the membrane. *Cell* 43:351–60

Copeland, B. R., Richter, R. J., Furlong, C. E. 1982. Renaturation and identification of periplasmic proteins in two-dimensional gels of *Escherichia coli*. *J. Biol. Chem.* 257:15065–71

Daniels, C. J., Bole, D. G., Quay, S. C., Oxender, D. L. 1981. Role for membrane potential in the secretion of protein into the periplasm of *Escherichia coli*. *Proc. Natl. Acad. Sci. USA* 78 : 5396–5400

Date, T., Wickner, W. T. 1981. Procoat, the precursor of M13 coat protein, inserts posttranslationally into the membrane of cells infected by wild-type virus. *J. Virol.* 37 : 1087–89

Davis, B. D., Tai, P.-C. 1980. The mechanism of protein secretion across membranes. *Nature* 283 : 433–38

Emr, S. D., Hanley-Way, S., Silhavy, T. J. 1981. Suppressor mutations that restore export of a protein with a defective signal sequence. *Cell* 23 : 79–88

Enequist, H. G., Hirst, T. R., Harayama, S., Hardy, S. J. S., Randall, L. L. 1981. Energy is required for maturation of exported proteins in *Escherichia coli*. *Eur. J. Biochem.* 116 : 227–33

Garwin, J. L., Beckwith, J. 1982. Secretion and processing of ribose-binding protein in *Escherichia coli*. *J. Bacteriol.* 149 : 789–92

Geller, B. L., Wickner, W. 1985. M13 procoat inserts into liposomes in the absence of other membrane proteins. *J. Biol. Chem.* 260 : 13281–85

Ghrayeb, J., Inouye, M. 1984. Nine amino acid residues at the NH$_2$-terminal of lipoprotein are sufficient for its modification, processing, and localization in the outer membrane of *Escherichia coli*. *J. Biol. Chem.* 259 : 463–67

Gilmore, R., Blobel, G. 1983. Transient involvement of signal recognition particle and its receptor in the microsomal membrane prior to protein translocation. *Cell* 35 : 677–85

Glabe, C. G., Hanover, J. A., Lennarz, W. J. 1980. Glycosylation of ovalbumin nascent chains: The spatial relationship between translation and glycosylation. *J. Biol. Chem.* 255 : 9236–42

Goodman, J. M., Watts, C., Wickner, W. 1981. Membrane assembly: Posttranslational insertion of M13 procoat into *E. coli* membranes and its proteolytic conversion to coat protein in vitro. *Cell* 24 : 437–41

Halegoua, S., Inouye, M. 1979. Biosynthesis and assembly of outer membrane proteins. In *Bacterial Outer Membranes: Biogenesis and Functions*, ed. M. Inouye, pp. 67–113. New York : Wiley

Inouye, H., Michaelis, S., Wright, A., Beckwith, J. 1981. Cloning and restriction mapping of the alkaline phosphatase structural gene (*phoA*) of *Escherichia coli* and generation of deletion mutants in vitro. *J. Bacteriol.* 146 : 668–75

Inouye, M., Halegoua, S. 1980. Secretion and membrane localization of proteins in *Escherichia coli*. *CRC Crit. Rev. Biochem.* 7 : 339–71

Inouye, S., Soberon, X., Franceschini, T., Nakamura, K., Itakura, K., Inouye, M. 1982. Role of positive charge on the amino-terminal region of the signal peptide in protein secretion across the membrane. *Proc. Natl. Acad. Sci. USA* 79 : 3438–41

Ito, K., Date, T., Wickner, W. 1980. Synthesis, assembly into the cytoplasmic membrane, and proteolytic processing of the precursor of coliphage M13 coat protein. *J. Biol. Chem.* 255 : 2123–30.

Ito, K., Mandel, G., Wickner, W. 1979. Soluble precursor of an integral membrane protein: Synthesis of procoat protein in *Escherichia coli* infected with bacteriophage M13. *Proc. Natl. Acad. Sci. USA* 76 : 1199–1203

Ito, K., Wittekind, M., Nomura, M., Shiba, K., Yura, T., et al. 1983. A temperature-sensitive mutant of *E. coli* exhibiting slow processing of exported proteins. *Cell* 32 : 789–97

Josefsson, L.-G., Randall, L. L. 1981a. Processing in vivo of precursor maltose-binding protein in *Escherichia coli* occurs posttranslationally as well as co-translationally. *J. Biol. Chem.* 256 : 2504–7

Josefsson, L.-G., Randall, L. L. 1981b. Differential exported proteins in *E. coli* show differences in the temporal mode of processing in vivo. *Cell* 25 : 151–57

Katz, F. N., Rothman, J. E., Lingappa, V. R., Blobel, G., Lodish, H. F. 1977. Membrane assembly in vitro: Synthesis, glycosylation and asymmetric insertion of a transmembrane protein. *Proc. Natl. Acad. Sci. USA* 74 : 3278–82

Koshland, D., Botstein, D. 1980. Secretion of beta-lactamase requires the carboxy end of the protein. *Cell* 20 : 749–60

Koshland, D., Botstein, D. 1982. Evidence for posttranslational translocation of β-lactamase across the bacterial inner membrane. *Cell* 30 : 893–902

Koshland, D., Sauer, R. T., Botstein, D., 1982. Diverse effects of mutations in the signal sequence on the secretion of β-lactamase in *Salmonella typhimurium*. *Cell* 30 : 903–14

Kumamoto, C. A., Beckwith, J. 1983. Mutations in a new gene, *secB*, cause defective protein localization in *Escherichia coli*. *J. Bacteriol.* 154 : 253–60.

Kumamoto, C. A., Beckwith, J. 1985. Evidence for specificity at an early step in protein export in *Escherichia coli*. *J. Bacteriol.* 163 : 267–74

Meyer, D. I. 1982. The signal hypothesis—a working model. *TIBS* 7 : 320–21

Michaelis, S., Hunt, J. F., Beckwith, J. 1986. Effects of signal sequence mutations on the kinetics of alkaline phosphatase export to the periplasm in *Escherichia coli*. *J. Bacteriol.* 167: 160–67

Muller, M., Blobel, G. 1984a. In vitro translocation of bacterial proteins across the plasma membrane of *Escherichia coli*. *Proc. Natl. Acad. Sci. USA* 81: 7421–25

Muller, M., Blobel, G. 1984b. Protein export in *Escherichia coli* requires a soluble activity. *Proc. Natl. Acad. Sci. USA* 81: 7737–41

Neu, H. C., Heppel, L. A. 1965. The release of enzymes from *Escherichia coli* by osmotic shock and during formation of spheroplasts. *J. Biol. Chem.* 240: 3685–92

Ohno-Iwashita, Y., Wickner, W. 1983. Reconstitution of rapid and asymmetric assembly of M13 coat protein into liposomes which have bacterial leader peptidase. *J. Biol. Chem.* 258: 1895–1900

Oliver, D. B., Beckwith, J. 1981. *E. coli* mutant pleiotropically defective in the export of secreted proteins. *Cell* 25: 2765–72

Pages, J.-M., Anba, J., Berndac, A., Shinagawa, H., Nakata, A., Lazdunski, C. 1984. Normal precursors of periplasmic proteins accumulated in the cytoplasm are not exported post-translationally in *Escherichia coli*. *Eur. J. Biochem.* 143: 499–505

Pollitt, S., Zalkin, H. 1983. Role of primary structure and disulfide bond formation in β-lactamase secretion. *J. Bacteriol.* 153: 27–32

Randall, L. L. 1983. Translocation of domains of nascent periplasmic proteins across the cytoplasmic membrane is independent of elongation. *Cell* 33: 231–40

Randall, L. L., Hardy, S. J. S. 1984a. Export of protein in bacteria. *Microbiol. Rev.* 48: 290-98

Randall, L. L., Hardy, S. J. S. 1984b. Export of protein in bacteria: Dogma and data. *Modern Cell Biology* 3: 1–20

Reid, G. A. 1985. Transport of proteins into mitochondria. *Curr. Top. Memb. Transp.* 24: 295–336

Russel, M., Model, P. 1981. A mutation downstream from the signal peptidase cleavage site affects cleavage but not membrane insertion of phage coat protein. *Proc. Natl. Acad. Sci. USA* 78: 1717–21

Ryan, J. P., Bassford, P. J. Jr. 1985. Post-translational export of maltose-binding protein in *Escherichia coli* strains harboring *malE* signal sequence mutations

and either *prl*⁺ or *prl* suppressor alleles. *J. Biol. Chem.* 260: 14832–37

Schlesinger, M. J., Barrett, K. 1965. The reversible dissociation of the alkaline phosphatase of *Escherichia coli*. I. Formation and reactivation of subunits. *J. Biol. Chem.* 240: 4284–92

Shiba, K., Ito, K., Yura, T., Cerreti, D. P. 1984. A defined mutation in the protein export gene within the *spc* ribosomal protein operon of *Escherichia coli*: Isolation and characterization of a new temperature-sensitive *secY* mutant. *EMBO J.* 3: 631–36

Smith, W. P., Tai, P.-C., Thompson, R. C., Davis, B. D. 1977. Extracellular labeling of nascent polypeptides traversing the membrane of *Escherichia coli*. *Proc. Natl. Acad. Sci. USA* 74: 2830–34

Strauch, K. L., Kumamoto, C. A., Beckwith, J. 1986. Does *secA* mediate coupling between secretion and translation in *Escherichia coli*? *J. Bacteriol.* 166: 505–12

Tommassen, J., Leunissen, J., van Damme-Jongsten, M., Overduin, P. 1985. Failure of *E. coli* K-12 to transport PhoE-LacZ hybrid proteins out of the cytoplasm. *EMBO J.* 4: 1041–47

Vlasuk, G. P., Inouye, S., Ito, H., Itakura, K., Inouye, M. 1983. Effects of the complete removal of basic amino acid residues from the signal peptide on secretion of lipoprotein in *Escherichia coli*. *J. Biol. Chem.* 258: 7141–48

Walter, P., Blobel, G. 1983. Disassembly and reconstitution of signal recognition particle. *Cell* 34: 525–33

Walter, P., Gilmore, R., Blobel, G. 1984. Protein translocation across the endoplasmic reticulum. *Cell* 38: 5–8

Wickner, W. 1979. The assembly of proteins into biological membranes: The membrane trigger hypothesis. *Ann. Rev. Biochem.* 48: 23–45

Wickner, W. 1980. Assembly of proteins into membranes. *Science* 210: 861–68

Wolfe, P. B., Rice, M., Wickner, W. 1985. Effects of two *sec* genes on protein assembly into the plasma membrane of *Escherichia coli*. *J. Biol. Chem.* 260: 1836–41

Zimmerman, R., Watts, C., Wickner, W. 1982. The biosynthesis of membrane-bound M13 coat protein. *J. Biol. Chem.* 257: 6529–36

Ann. Rev. Cell Biol. 1986. 2 : 337–65
Copyright © 1986 by Annual Reviews Inc. All rights reserved

THE DIRECTED MIGRATION
OF EUKARYOTIC CELLS

S. J. Singer and Abraham Kupfer

Department of Biology, University of California at San Diego,
La Jolla, California 92093

CONTENTS

INTRODUCTION

The in vivo migration of eukaryotic cells over surfaces is important to a
wide range of physiological processes, such as the development of the
embryo, defense against infections, wound healing, the homing of lym
phocytes to lymphoid organs, and tumor metastases (cf Trinkaus 1984).
Despite much study, the directed migration of cells is not yet well understood, no doubt because it is indeed a complex problem. Part of the
complexity is due to the great variety of locomotory phenotypes that are
expressed by different types of motile cells. Many different cells can migrate
under appropriate circumstances: In this article we refer to the directed
migration of fibroblasts, endothelial cells, epithelial cells, macrophages,
granulocytes, fish epidermal keratocytes, and amebae. Many of these cell
types show distinct locomotory characteristics. However, most investigators agree that common mechanisms must operate in all cell migration.

337

The problem then is to elucidate such common features. For this purpose, it is essential to recognize as uncommon those mechanisms that are either superimposed upon, or circumvent, the common features. For example, the stress-fiber arcs (Heath 1983) that are swept backwards under the fibroblast cell surface as the cell's leading edge moves forward are most probably an uncommon feature of cell locomotion since most motile cells other than fibroblasts do not exhibit stress fibers. Another example is fish epidermal keratocytes (Euteneuer & Schliwa 1984), which appear to be permanently polarized and therefore may circumvent the need to achieve the polarization that is required by an amorphous cell in order for it to migrate. As a practical matter, this task of culling out the common features of cell migration is not made any the easier by the fact that many individual investigators work with only one type of migratory cell which sometimes, ipso facto, thereby assumes the mantle of generality.

Besides the diversity of migratory phenotypes, another complexity is that cell migration is almost certainly a multifactorial process. Even if we restrict our attention to those mechanistic features that are common to a wide variety of migratory phenotypes, it is still likely that several different factors must operate simultaneously and coordinately for cells to migrate. If any one of these factors is inoperative or is inhibited, the cell will not show directional migration, according to this view. The thesis of this review, not a particularly original one, is that for amorphous cells (i.e. those not exhibiting permanent intrinsic polarity) that show persistent and long-term migration (rather than a short-term migratory burst), there are at least three distinct categories of mechanisms that must generally operate simultaneously for cell locomotion to occur: (a) intracellular force-generation mechanisms involving the cytoskeleton; (b) polarity-determining mechanisms involving the insertion of new membrane mass into the leading edge; and (c) adhesion mechanisms involving interactions of the cell surface with the substratum. The exploration of this thesis at the molecular level is a primary objective of this review. One important result of this analysis is that the extension of axons by neurones can be viewed mechanistically as only a minor variation of cell migration. This is discussed towards the end of the article.

INTRACELLULAR FORCE GENERATION

Many types of migrating cells in culture, including fibroblasts, macrophages, granulocytes, and epithelial cells, send out a flattened, irregular projection called a lamellipodium in the direction of migration (see Figure 1). The forward movement of the lamellipodium appears to generate tension at the rear of the cell, ultimately causing it to be pulled forward.

Thus the extension and propulsion of the leading edge in the anterior portion of the cell is generally thought to be of primary importance in cell migration. The cell posterior, in this view, is not the site of primary force generation, but rather responds secondarily to the extension of the leading edge (Chen 1981).

It is generally accepted that the leading edge is internally propelled by force-generating systems that are components of the cytoskeleton. The experimental support for this, in part, is that drugs that affect the cytoskeleton inhibit cell migration (see below). In the last decade much has been learned about the cytoskeletal meshwork that exists in nonmuscle cells and of the molecular interactions that regulate the dynamics of that meshwork. Of the three major filamentous structures that have so far been discovered to constitute the meshwork—actin microfilaments, microtubules, and intermediate filaments—it is clear that actin is involved in cell migration, there is some uncertainty about the role of microtubules, and intermediate filaments are not significantly implicated.

Actin Involvement

It has been observed in many types of migratory cells that cytochalasin B inhibits not only directional migration, but all protrusive activity (blebbing). It is currently thought (MacLean-Fletcher & Pollard 1980) that the primary effect of this drug is to attach to ("cap") the barbed ends of F-actin filaments. Such binding can also lead to the severing of actin filaments (Hartwig & Stossel 1979). Therefore, a role for actin in cell migration is strongly implied by these findings.

The actin system includes not only actin monomers (G-actin) and microfilaments (F-actin), but a host of other proteins that interact with actin in many different ways (for reviews see Weeds 1982; Stossel et al 1985). Our current information about the system is undoubtedly incomplete; new proteins that are associated with actin and its functions are continually being discovered. Limitations of space preclude any detailed analysis of the entire system in this article. Suffice it to say that it is already clear that there are several actin-related mechanisms that could be involved in force-generation for cell migration. They are listed below.

1. The reversible polymerization of G- to F-actin could participate in force generation. The prototype for this mechanism of generating thrust is the remarkable elongated process that rapidly develops from the acrosome of thyone sperm when it is activated (Tilney 1978). In this system, the G- to F-actin conversion may be accompanied by bundling of the F-actin in order to achieve stiffening and elongation (mechanism 2, below), but that has not been directly demonstrated. That G- to F-actin interconversions may also be involved in directed cell migration is consistent

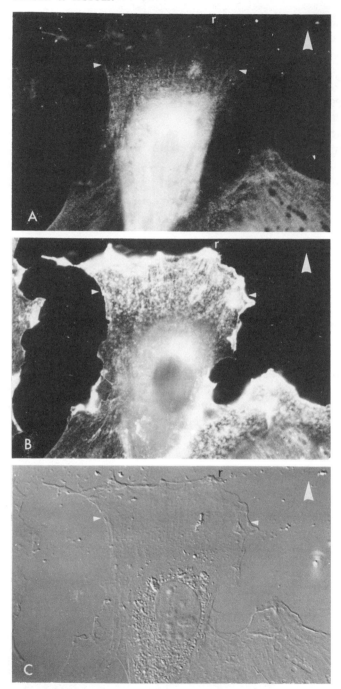

with the finding in *Dictyostelium discoideum* amebae that actin becomes associated with the Triton-insoluble cytoskeleton within a few seconds after stimulation with the chemoattractant c-AMP (McRobbie & Newell 1984). The extreme rapidity with which these effects follow stimulation and their transient nature make it unclear, however, exactly how they are related to persistent locomotion. Similar results have been obtained with granulocytes stimulated with the oligopeptide chemoattractant FMLP (Fechheimer & Zigmond 1983; Jesaitis et al 1984). The demonstration that microinjected actin subunits become incorporated (presumably into F-actin) at the leading edge of fibroblasts (Wang 1985) and subsequently move towards the cell interior may be related to these observations.

2. The bundling of a number (~ 20–100) of individual loosely coiled filaments of F-actin into a stiff fiber (stress fiber) could, in principle, itself impart a forward thrust, if one end of the bundle was fixed posterior to the leading edge of the cell. However, the highly motile ruffles at the leading edge of fibroblasts appear to contain F-actin in a largely unorganized state (Abercrombie et al 1971). Bundles of F-actin could be indirectly involved by providing binding sites for myosin molecules (Figure 1) if an actomyosin sliding-filament mechanism functioned in cell propulsion (see below). Another possible role for bundles of F-actin is associated with cell adhesions. F-actin bundles appear to be attached to the cell membrane just posterior to the ruffle, where strong adhesions (focal contacts) are made by the ventral cell surface to the substratum (Abercrombie et al 1971). These adhesions are discussed later; they could serve as sites where tension is exerted at the leading edge on more posterior regions of the motile cell.

3. The reversible gelation and solation of three-dimensional actin gels seem very likely to be involved in cell migration. Actin-binding proteins have been discovered (Stossel et al 1979) that appear to mediate end-to-side interactions of actin filaments to form a gel network. Other proteins are known (Weeds 1982; Stossel et al 1985) that can cause actin filaments to be severed in a Ca^{2+}-dependent reaction, thus helping to solate the gel. These and other proteins can therefore mediate gel-sol transformations of actin filaments in response to transient ion fluxes. The gel phase should be

---

Figure 1   Double fluorescence labeling for myosin (A) and actin (B) of a rat A10 smooth muscle cell (shown in Nomarski optics in C) migrating out of the edge of an experimental wound ($\times$ 750). The large white arrowhead at the top right of each panel denotes the direction of migration. The small white arrowheads in each panel demarcate the region of the lamella ahead of which (for a distance of about 10 μm) the myosin (A) is severely depleted or absent and the actin (B) is concentrated. The symbol r in the top center of each panel indicates the ruffling edge of the lamella. (From Gotlieb et al 1979.)

relatively stiff and resistant to compression and/or shear; the formation of the sol may produce osmotic swelling (Oster 1984), and the sol should be more deformable than the gel (Taylor et al 1979; Stossel 1982). However, two of the advocates for the participation of actin gel-sol transformations in force generation at the leading edge of the motile cell have opposite views as to the primary event: one (Stossel 1982) believes that the sol → gel conversion is involved, the other (Taylor et al 1979) that the gel → sol conversion is primary.

4. An actomyosin contractile system is probably involved (cf Berlot et al 1985), but exactly how is not entirely clear. An early suggestion (Huxley 1973) was that actin filaments might be anchored to the cell membrane and by the sliding-filament mechanism act to propel soluble myosin molecules and cytoplasm into the leading edge. However, it appears that myosin is absent or severely depleted from a large part of the leading edge of motile cells where actin is highly concentrated (Gotlieb et al 1979; Jockusch et al 1983) (Figure 1); these results are not consistent with such a scheme. More recently it has been suggested (Taylor et al 1979; Oster 1984) that actomyosin contraction is coupled to the gel-sol transformation of actin gels (mechanism 3 above) (the solation-contraction coupling hypothesis), as mediated by the separate effects of Ca^{2+} on the actomyosin and gel-sol systems.

While there is a general consensus that actin, and probably the actomyosin system, is involved in directed cell migration, there is no consensus as to the mechanism. New kinds of experiments with intact cells are needed, such as observations of the effects of the microinjection of Fab fragments of monospecific antibodies to one or another putative participant in force generation. If a component is an essential participant in cell motility, the in vivo binding of its Fab might result in an inhibition of motility. The integration of these actin-related mechanisms will also require that attention be paid to temporal and spatial factors: certain mechanisms may precede others (McRobbie & Newell 1984), and various participating molecules may be physically segregated (Figure 1) in the locomoting cell.

Microtubule Involvement

The directed migration of many types of cells, but not all, is inhibited by microtubule-depolymerizing drugs such as colchicine and vinblastine (Vasiliev et al 1970; Goldman 1971; Bandmann et al 1974; Cheung et al 1978). (The few exceptions are discussed later.) These drugs do not inhibit blebbing, however, as does cytochalasin B. While such results may appear to suggest a role for microtubules in force generation for directed migration, most investigators do not believe this to be the case, mainly because microtubules are generally depleted from the cell periphery at the leading edge of motile cells (see, however, Geiger et al 1984). In a later section, we suggest that microtubule-depolymerizing drugs inhibit directed cell migration because of their concomitant effects on the Golgi apparatus.

PERSISTENCE PHENOMENA

There are a number of reasons to infer that intracellular force generation is not the only factor involved in directed cell migration. Several lines of evidence indicate that locomoting cells can retain directional "memory" for times as long as many minutes. We refer to these effects as *persistence*. If actin-related force generation were the sole cause of cell migration, one would not expect such persistence. Processes such as polymerization-depolymerization of actin (Pollard & Mooseker 1981) and gel-sol trans-formations of actin should occur in a few seconds.

One kind of persistence phenomenon involves chemotactic cells. With cells as different as granulocytes (Zigmond et al 1981) and *D. discoideum* amebae (Swanson & Taylor 1982), a rapid change in the direction of the chemotactic gradient often does not lead to the rapid generation of a new leading edge on the cell. When the cAMP gradient was reversed 180° for example, some amebae reversed their direction but only after some delay, whereas other cells made a complete U-turn instead. If a redistribution of c-AMP receptors in the fluid membrane, along with changes in the actomyosin cytoskeleton, were the only things necessary for the amebae to reverse direction, it seems likely that all the cells would have put out a new leading edge and reversed direction in less than a minute. Very similar U-turns were obtained upon reversal of a chemotactic gradient on granulocytes (Zigmond et al 1981).

Another type of persistence phenomenon involves the migration of the two daughter cells newly arisen from the cell division of a fibroblast (Albrecht-Buehler 1977). A significant fraction of the daughter cells were observed to move apart in nearly opposite directions for some considerable distance. It was suggested that this might be due to the fact that during mitosis the two pairs of centrioles of the developing daughter cells are aligned on the spindle pole axis, imparting an exactly opposite polarity to the two fully separated daughter cells. The pair of centrioles is usually located within the microtubule-organizing center (MTOC) in such cells, and these results therefore suggested that the MTOC might somehow play a role in directed cell migration and in persistence phenomena.

To gain further insight into the phenomenon of persistence, we first consider the membrane at the leading edge of a locomoting cell.

MEMBRANE INSERTION AND RECYCLING

In a cell being propelled forward by an intracellular force acting on its leading edge, one might expect that the plasma membrane would extend passively in the direction of migration, i.e. flow forward from the rear of the cell to the leading edge. However, over the years a variety of experiments have suggested the contrary: As a cell moves forward, the membrane appears to flow backward from the leading edge toward the rear of the cell, on both its dorsal and ventral surfaces. For example, small inert particles in the path of a motile cell are picked up on the cell surface and are transported towards the rear at velocities exceeding the rate of forward extension of the cell (Shaffer 1963; Abercrombie et al 1970). Lectins such as concanavalin A, when bound to the surface of a locomoting cell, are cleared from the leading edge of the cell (Abercrombie et al 1972; Vasiliev et al 1976). To account for such results it was suggested that as interphase cells locomote they insert new membrane mass into the leading edge of a cell. Since the cell's surface area must be maintained nearly constant, old membrane mass would be retrieved into the cell interior (endocytosis) somewhere towards the rear (Abercrombie et al 1970). Such physically separated insertion and retrieval events would therefore result in a rearward flow of membrane mass from the leading edge, carrying particles and bound lectins backward over the cell surface.

Most of these results have been obtained with fibroblasts. It has more recently been convincingly demonstrated, however, that the rearward flow of particles on motile fibroblasts is mediated not by membrane flow but by an actin-mediated rearward sweeping of the cell surface (Heath 1983). Discrete bundles of actin filaments ("arcs") continuously form more or less parallel to the leading edge immediately under the dorsal surface of a motile fibroblast, and move towards the rear of the cell. When a particle becomes attached to the cell surface, it probably induces the formation of a small cluster of integral proteins in the membrane, much as lectins are known to do. Such induced clusters have been shown to become attached through the membrane to bundles of actin filaments underlying the fibroblast membrane (Ash & Singer 1976; Ash et al 1977; Singer et al 1978). Hence particles (and lectins) can become attached across the fibroblast membrane to the actin "arcs" and be swept backward over the cell surface with them (Heath 1983).

These results with fibroblasts, by providing an alternative explanation for particle displacements, have tended to discredit the proposal that membrane mass insertion occurs at the leading edge of a motile cell. However, rearward particle displacements have also been observed with other types of cells such as macrophages (Dembo & Harris 1981), which do not exhibit well-developed stress fibers or "arcs." It is interesting that the rate of rearward movement of particles on fibroblasts is about five times faster than on macrophages, although fibroblasts migrate about five times more slowly. While the rapid rearward particle movement on fibroblasts most likely reflects the mediation of the actin "arcs," the slower rearward particle movement on migrating macrophage surfaces may therefore indeed reflect rearward flow of membrane mass. Furthermore, with cells of *Fundulus* that exhibit blebbing (Tickle & Trinkaus 1977), particles on the blebs were seen to move *forward* on the bleb rather than rearward, which suggests that the extension of a leading edge on a migrating cell and the protrusion of a bleb involve quite different flow characteristics in the surface membrane. However, a different type of experiment, discussed below, has demonstrated directly that new membrane mass is indeed inserted at the leading edge of a motile fibroblast.

Codistribution of the Golgi Apparatus (GA) and the Microtubule-Organizing Center (MTOC) in Interphase Cells

Much attention has been paid in recent years to the MTOC in interphase cells, a structure defined by experimental observation as that region of the cell from which all the cytoplasmic microtubules grow out (for a review see Brinkley 1985). This region of the cell, examined by transmission electron microscopy, is rather amorphous in appearance and usually includes the pair of centrioles. It is a structure that is often located in a relatively compact mass to one side of the nucleus (see below; Figure 2). In recent years it has been largely overlooked, however, although it is implicit in some very old observations (cf Wilson 1925), that the MTOC is always closely juxtaposed to the GA in interphase cells. The ubiquitous nature of this GA-MTOC codistribution can be clearly visualized at the light microscope level of resolution by double immunofluorescence labeling for the MTOC and the GA inside many different types of cells (Figure 2). The codistribution probably reflects some molecular interactions between components associated with the two organelles (Rogalski & Singer 1984; Sandoval et al 1984). One significant feature of the GA-MTOC codistribution is that various effects on cell motility of alterations in microtubules or in the MTOC in many cases are likely to be primarily due to associated changes in the GA, as discussed below.

Coordinate and Rapid Reorientation of the GA and MTOC in Cells Stimulated to Migrate

Malech et al (1977) observed by electron microscopy that upon imposing a gradient of a chemotactic agent on granulocytes, the pair of centrioles became rapidly reoriented (within minutes) in the direction of the gradient.

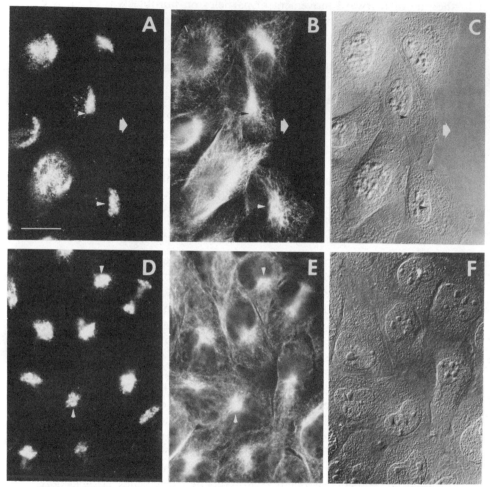

Figure 2   NRK cells at the edge of an experimental wound (A–C) 5.5 hr after wounding and within the confluent portion of the same monolayer (D–F). The same cells were examined in Nomarski optics (C and F) and by double indirect immunofluorescence with antibodies that labeled the Golgi apparatus (A and D) and the microtubules and MTOC (B and E). The blunt arrows in A–C give the direction of cell migration perpendicular to the wound. The small white and black arrowheads denote the same positions in A–C and in D–F. Note that in D–F the Golgi apparatus and the MTOC (arrowheads point to examples) are colocalized in a perinuclear position in every cell, but are oriented randomly in the plane of the figure. In A–C, in cells at the edge of the wound the GA and MTOC are also colocalized but now face the direction of migration. (Bar = 20 μm.) (From Kupfer et al 1982.)

Gotlieb et al (1981) and Kupfer et al (1982) observed cultured endothelial and fibroblastic cells, respectively, at the edge of an experimentally induced wound and found that the perinuclear MTOC was oriented in the direction of migration. Similar results were obtained with locomotory amebae (Rubino et al 1984). Kupfer et al (1982), however, demonstrated several additional points: (*a*) that the orientation of the MTOC was due to an intracellular reorientation of the organelle and not to the rotation of the entire cell; (*b*) that the reorientation of the MTOC was accompanied by the coordinate reorientation of the GA (Figure 2A–C); and (*c*) that the reorientation of the two organelles was very rapid (with a half-time of several minutes, but with a broad dispersion) upon stimulation and occurred long before a leading edge was formed. These observations led them to suggest that the purpose served by the GA-MTOC reorientation was to direct the insertion of new membrane mass, via vesicles derived from the GA, to the region of the cell surface that was to become the leading edge.

The Insertion of New Membrane Mass at the Leading Edge

This was subsequently directly demonstrated to occur (Bergmann et al 1983). By means of a model system composed of motile fibroblasts infected with a mutant strain of vesicular stomatitis virus (VSV) that was temperature sensitive for the cell surface expression of the G membrane glycoprotein of VSV, the insertion of new membrane mass could be both synchronized and detected. Immunofluorescence observations showed that the first appearance of the G protein at the cell surface in these infected motile cells was at the leading edge (Figure 3A), opposite the reoriented perinuclear GA. As the cell surface G protein is derived from the GA via fairly large transfer vesicles (Bergmann & Singer 1983), which presumably contain in their membranes other components (phospholipids and other integral proteins besides the G protein), bound for fusion with the plasma membrane of the cell, these results demonstrate that new membrane mass is indeed inserted at the leading edge of locomoting cells. There have been a number of other observations of cell surface components that are entirely consistent with this conclusion (Marcus 1962; Harris et al 1969; Walter et al 1980; Weinbaum et al 1980; Wilkinson et al 1980; Bretscher 1983; Jacobson et al 1984; Hopkins 1985). An experiment in which no membrane insertion was detected at the leading edge (Middleton 1979) was very likely too insensitive. In the latter experiment the attempt was made to find an increase in a protein already present in the cell surface, in contrast to the experiments of Bergmann et al (1983) in which the protein inserted into the leading edge was not previously present in the cell surface. Furthermore, in recent experiments on the migration of macrophages in a chemotactic gradient (Nemere et al 1985) it was shown that monensin, the cationic ionophore that inhibits membrane insertion and recycling (Tartakoff & Vassalli 1977), inhibits chemotaxis, although it does not inhibit the GA-MTOC reorientation in the chemotactic gradient. These results therefore suggest that membrane insertion is *required* for directed migration to occur.

The insertion of new membrane mass into the leading edge implies an equivalent rate of retrieval of old membrane mass towards the rear of the

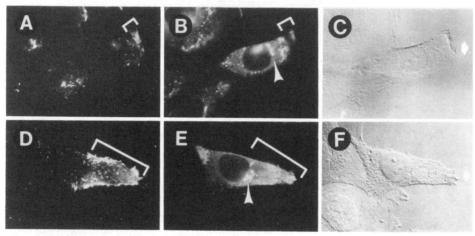

Figure 3 NRK cells at the edge of an experimental wound infected with the 0–45 temperature-sensitive mutant of vesicular stomatitis virus (VSV). The cells were first maintained for 3.5 hr at the nonpermissive temperature, 39.7°C, during which time the G integral membrane protein of VSV was synthesized but was unable to be transported out of the endoplasmic reticulum and never reached the cell surface (not shown). The temperature was then reduced to 32°C, and 22 min after the temperature shift the cells were fixed and labeled by double indirect immunofluorescence for the G protein, both inside the cell (B and E) and on the cell surface (A and D), separately. Two different cells at the wound edge are shown in A–C and in D–F (depicted in Nomarski optics in C and F). (The blunt arrows indicate the direction of cell migration.) Twenty-two min after the temperature shift was the earliest time that a substantial fraction of the infected cells showed any G protein on their surfaces, but the surface appearance was not precisely synchronized from one cell to another in the population. In A, the G protein at the surface was confined to the leading edge (*bracket*) facing the Golgi apparatus inside the cell (*B, arrowhead*). In D, the G protein at the surface was confined to the front of the cell (*bracket*) in front of the nucleus, where the Golgi apparatus was located (*E, arrowhead*). The results are consistent with the cell in A–C representing a slightly earlier stage of surface expression of the G protein than the cell in D–F. The G protein thus first appeared on the cell surface at the leading edge (A–C) and then flowed backwards to a "sink" over the Golgi apparatus (D–F). At later times (35 min after the temperature shift), the infected cells at the edge of the wound mostly showed a uniform distribution of the G protein over the entire cell surface (not shown). (From Bergmann et al 1983, which should be consulted for further details.)

motile cell, so that the total surface area remains the same. Retrieval may occur mainly near the region of the cell surface lying over the GA, since the Golgi is likely to be involved (Farquhar 1985). (This could be the significance of the half-cell surface distribution of the G protein seen in Figure 3D, which suggests that there is a sink for the membrane over the GA.)

All cells, motile or not, exhibit rapid and extensive recycling of their plasma membrane (Steinman et al 1983). Therefore, the proposal that membrane mass is inserted into the leading edge and retrieved towards the rear of a motile cell is not just an ad hoc one, unrelated to normal cell physiology. It is only necessary to stipulate that ordinary membrane recycling be polarized, i.e. that at least some of the processes of membrane insertion and retrieval (recycling) that normally occur *uniformly* over the surface of a nonmotile cell, be *polarized* so that insertion always occurs at the leading edge and retrieval towards the rear. This is what is presumably achieved by the GA-MTOC reorientation.

These observations raise the next question. Can the insertion of new membrane mass into the leading edge entirely account for all of the membrane extension of the leading edge or only a part of that extension? (In the latter case, forward flow of the surface membrane would then have to make up the rest.) All that has been firmly demonstrated by the results in Figure 3 (Bergmann et al 1983) is that membrane mass is indeed inserted at the leading edge, but how much is inserted during a given time interval can not be determined from these experiments.

The amount of membrane inserted is not simple to quantitate. For one thing, not all of the membrane mass inserted into the leading edge would be expected to lead to extension, because much of that mass might fold backward and flow into the dorsal surface of the cell (Harris 1973a). Secondly, membrane insertion and recycling may occur by several distinct mechanisms (cf Huet et al 1980 ; Robbins et al 1983), and not all of these need be involved in extension of the leading edge. Coated-pit mechanisms (Anderson et al 1982), for example, are exceedingly rapid ; individual cycles of internalization and surface reappearance of membrane must therefore occur in close physical proximity to one another. To extend a leading edge, however, insertion and retrieval events must be well separated from each other on the cell surface.

A qualitative analysis of the problem may nevertheless be useful. Direct stereologic determination of pinocytosis in fibroblastic L cells (Steinman et al 1976) showed that these cells internalize the equivalent of their entire cell surface area about every 2 hr. The locomotion of L cells varies somewhat with the culture conditions, but a reasonable value for their average rate is 25 μm/hr (Sugimoto & Hagiwara 1979). A cell of 50 μm in length would therefore move a cell length in 2 hr. An L cell could therefore recycle its entire membrane in moving an entire cell length. For our purposes, this correspondence between the rates of membrane recycling and locomotion indicates simply that it is not unreasonable to suggest that the polarization of normal membrane recycling could continually introduce sufficient membrane mass into the leading edge to account for the membrane extension that occurs during the locomotion of the L cell. In this connection, it is interesting that chemotacting macrophages move about five times faster than fibroblasts, while by stereological analysis, resting macrophages recycle their surface membrane about four times faster than fibroblasts (Steinman et al 1976).

A requirement for polarized membrane recycling in cell locomotion can also provide a new explanation for the findings with many types of migrating cells that drugs that depolymerize their microtubules inhibit directed migration. Such drugs, by disrupting the MTOC, simultaneously cause a dispersion of the Golgi complex from its normal compact perinuclear configuration into the entire periphery of the cell (Robbins & Gonatas 1964; Rogalski & Singer 1984). Without the MTOC, elements of the GA are apparently no longer confined to a perinuclear site. This dispersion of elements of the GA does not, however, inhibit new membrane mass from being inserted into the plasma membrane, but it *destroys the polarity of that insertion* (Rogalski et al 1984). New membrane mass is now introduced more or less uniformly over the entire cell surface, and therefore no leading edge can form and continue to be extended.

The persistence phenomena in cell migration that are cited in a previous section can also be explained. The orientation of the GA (along with the MTOC), by designating the region of the cell surface that by membrane insertion will become the leading edge, provides a steering mechanism for the motile cell. In the experiments of Kupfer et al (1982) with fibroblasts, the half-time for the reorientation of the GA-MTOC after receiving the polarized signal to move was an average of several minutes, with a fairly broad dispersion about the average. After reversal of the chemotactic gradient acting on amebae (Swanson & Taylor 1982) or on granulocytes (Zigmond et al 1981) the GA-MTOC may require several minutes to reorient 180° to initiate cell migration in the opposite direction. Many such cells might then make a U-turn in the reoriented gradient before their GA-MTOC could undergo intracellular reorientation. Regarding the experiments of Albrecht-Buehler (1977) on the persistent migratory patterns of daughter cells recently formed from a fibroblast mitotic division, not only are the MTOCs in the two daughter cells aligned on the spindle axis, as expected, but so are their newly redistributed GAs (Figure 4). The consequent mirror-image insertion of new membrane mass into the two daughter cells would therefore be expected to produce oppositely oriented leading edges, and hence for as long as the organelles remain so oriented, the two daughter cells will move in directions 180° apart.

We further contend that cells that cannot polarize their membrane recycling are generally unable to locomote. In this connection, we recently observed that isolated chick embryonic cardiac myocytes in culture, although they actively produce protuberances all over their surfaces, do not exhibit net directed migration (P. J. Kronebusch & S. J. Singer, unpublished). Interestingly, these myocytes exhibit a very rare property: their MTOC and GA are both spread more or less uniformly over the entire nuclear envelope (P. J. Kronebusch & S. J. Singer, unpublished) instead of being confined to the more typical compact perinuclear configuration. The absence of cell migration but not of the kind of actin-mediated protrusive activity seen in migratory cells is correlated with the absence of a compact and polarizable GA and hence the incapacity to polarize membrane recycling in these cells.

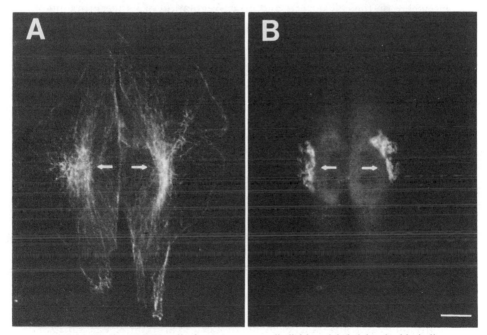

Figure 4 Daughter NRK cells from a recent cell division, labeled by double indirect immunofluorescence, as in Figure 2, for microtubules and the MTOC (A) and the Golgi apparatus (GA) (B). Before the daughter cells move apart, the Golgi apparatus in each cell is already fully formed and colocalized with its MTOC (*arrows*), with the organelles in the daughter cells aligned in mirror-image orientation along the mitotic spindle axis.

All migratory cells when they enter mitosis become immotile. While there are many possible explanations for this fact, it is interesting that during mitosis cells can no longer insert new membrane mass into their surfaces (Warren et al 1983).

Possible Functions of Membrane Mass Insertion at the Leading Edge

If membrane mass insertion per se, and not just the secretion that accompanies it (see below), is critical to directed cell migration, what functions other than simply directing lamellepodial extension might it serve? Several interesting possibilities suggest themselves, as discussed in the following paragraphs.

The binding of a chemotactic ligand to its receptor, as in many other ligand-receptor systems, has been shown to lead to the clustering of that receptor in the cell membrane followed by its removal from the surface by endocytosis (down-regulation) (Zigmond 1981). In a chemotactic gradient, the concentration of ligand would be higher at the leading edge than elsewhere on the cell surface, and down-regulation of its receptors might therefore occur more rapidly at the leading edge than elsewhere. This could create problems, however, because a higher concentration of the receeptor may be required at the leading edge in order to sense the gradient (Zigmond 1981; Sullivan et al 1984). This paradoxical situation could be resolved if new membrane mass, containing recycled but undegraded receptor that had earlier been endocytosed, and/or newly synthesized receptor, were continuously being inserted into the leading edge of the cell. This mechanism could maintain a higher steady-state concentration of the receptor at the leading edge than elsewhere on the cell surface.

Ion fluxes at the leading edge, particularly of H^+ and/or Ca^{2+}, have often been implicated in the directional triggering of actin-mediated force-generating mechanisms. It is possible that new membrane, in the form of vesicles derived from the trans face of the GA, contains relatively high concentrations of a proton pump (Reggio et al 1984; Robbins et al 1984; Anderson & Pathak 1985; Tougard et al 1985). Vesicular fusion that was confined to the leading edge could then maintain a higher steady-state concentration of the pump there than elsewhere on the cell surface. The presence of a proton pump in GA-derived vesicles could explain why monensin, a cationic ionophore, appears to affect rather specifically the morphology of trans-Golgi elements in a cell (Tartakoff & Vassalli 1977) at the same time that it inhibits membrane recycling: It could dissipate a proton gradient that was important to the function of the trans-Golgi elements.

Adhesion mechanisms (discussed below) probably involve specific integral membrane proteins in the cell surface binding to extracellular adhesion molecules that are attached to the substratum. However, adhesions have to be continually made and broken in order for the leading edge of a cell to move forward. The insertion of new membrane mass at the leading edge could therefore continuously insert fresh adhesive integral proteins where they are needed.

These speculations about the functions that might be served by membrane mass insertion at the leading edge are clearly more stimulating than compelling at this time.

The Mechanism of GA and MTOC Reorientation

The reorientation of the GA and MTOC that occurs within minutes after polar stimulation of a cell (Malech et al 1977; Kupfer et al 1982) involves a massive reorientation within the cytoplasm that probably includes the nucleus as well. These effects are probably the same as those observed in other types of experiments by Soranno & Bell (1982). There is hardly any information, however, about the mechanism involved. In experiments with macrophages in a chemotactic gradient (Nemere et al 1985), extracellular Ca^{2+} was required for the reorientation to occur. A local influx of Ca^{2+} at the leading edge may have a variety of effects on the cytoskeleton, including a rapid depolymerization of microtubules in the vicinity (Mitchison & Kirschner 1984) and solation of the actin gel (Pollard 1984). Such alteration in the local cytoskeletal structure may result in a torque being exerted via microtubules attached to the MTOC, causing the organelles to rotate.

The reorientation of the GA-MTOC in response to a polar signal is not unique to cells stimulated to undergo directed migration. It also occurs inside natural killer cells and cytotoxic lymphocytes in response to a polarized signal delivered by a bound target cell that will subsequently be lysed (Kupfer et al 1983, 1985). We speculate that there may be a subset of membrane receptors, including chemotactic receptors on motile cells and T-cell receptors on cytotoxic lymphocytes, whose polarized engagement results in the reorientation of the GA-MTOC inside these cells.

THE MOTILE BEHAVIOR OF CELL FRAGMENTS

Investigators have sought to gain insight into the mechanisms of cell motility by studying fragments of different types of motile cells made in a variety of ways. The results have to some extent been contradictory and confusing. The preceding results and discussion, however, may help to resolve the confusion. We have stressed the importance of a polarized Golgi apparatus and consequent polarized membrane recycling to directed cell migration; we contend that a cell fragment must contain at least a part of the GA in a functionally active state to retain for a considerable time both the protrusive activity and the migratory behavior of the parent cell. In most cell fragmentation studies, however, the presence or functional state of the GA in the fragments was not determined. Nevertheless, most of the results that have been obtained appear to be consistent with our expectations. For example, the surgical bisection of a polarized ameba (Swanson & Taylor 1982) resulted in a nucleated fragment that retained chemotactic activity and an anucleate fragment that retained protrusive activity but showed no net migration in the gradient. We suggest that the perinuclear GA-MTOC remained in the nucleated fragment, which was therefore still able to migrate in the chemotactic gradient but that the anucleate fragment lacking these two organelles lost the capacity to migrate, although it retained actin-mediated protrusive activity. We similarly interpret the inability of small fragments of fibroblasts to locomote (Albrecht-Buehler 1980), although they retained the ability to produce filopodia and exhibited ruffling and blebbing.

However, Keller & Bessis (1975) produced anucleate fragments of granulocytes that retained for some time the chemotactic activity of the parent cell. Such fragments, which have been termed cytokineplasts (Malawista & de Boisfleury Chevance 1982; Dyett et al 1985), are produced by first polarizing the cells, which are anchored to glass, in a chemotactic gradient and then heating them to 45°C for about 10 min. The leading lamellipodium then breaks off, leaving the nucleus behind. It is not clear, however, what happens to the GA upon such heat treatment. It was observed by electron microscopy (Malawista & de Boisfleury Chevance 1982) that heating appears to affect the centrosomal region (associated with the MTOC) of the granulocyte, and this may release the GA from its normal perinuclear association with the MTOC. The cytokineplast may therefore contain at least part of the GA in a functional state. It is also worth noting that anucleate fragments of granulocytes produced by use of cytochalasin B retained chemotactic receptors in their membranes, as well as several actin-mediated functions, but lost chemotactic activity (Roos et al 1983; Dyett et al 1985). We suggest that the cytochalasin B method of fragmenting granulocytes either excludes the GA from the cytoplasts or includes it in a nonfunctional state.

An intriguing set of experiments has been carried out with fish epidermal keratocytes and their anterior, anuclear cytoplasmic fragments, which retain the directed migration of the parent cells (Euteneuer & Schliwa 1984). However, these may be rather atypical motile cells, since their directed migration is unaffected by microtubule depolymerizing drugs, which inhibit the directed migration of most motile cells. It is conceivable that in special cases a motile cell might become stably polarized (i.e. with differently structured membrane domains at the leading edge as compared to the lateral and posterior surfaces of the cell) in a manner that persists for relatively long times. This is not the case with more typical motile cells, which are amorphous and capable of blebbing anywhere on their surfaces. The migratory fish epidermal keratocytes, however, show an unusual, "canoe-like" morphology (Euteneuer & Schliwa 1984), which suggests that there are specific constraints on different regions of the cell surface. If such membrane constraints exist, the protrusive activity of the intracellular (actin-mediated) force-generation mechanism may be channeled to act only on the leading edge. The polarized membrane insertion and retrieval characteristic of most motile cells may thus be circumvented: The surface membrane may simply flow forward passively in response to the directed intracellular thrust. We suggest that this exceptional situation is the case with fish epidermal keratocytes and is retained in their cytoplasts. It would be interesting to use these cells in particle-displacement experiments of the type conducted by Abercrombie et al (1970) with fibroblasts and by Dembo & Harris (1981) with macrophages. In the light of the preceding discussion, one might find that particles placed on the dorsal surface move *forward* over the migrating keratocyte instead of *backward* as they do on fibroblasts and macrophages. In any event, the motile phenotype of fish epidermal keratocytes appears to be quite unusual and may therefore be of only limited relevance to more typical motile phenotypes.

ADHESION TO THE SUBSTRATUM

It has long been recognized that adhesions formed between the cell surface and its substratum play important roles in cell migration. As the leading edge of a cell is extended, the surface of the extension must be attached to the substratum, otherwise the extension might be transient and fold back into the cell body. However, if such adhesions were irreversible, they could inhibit continued cell extension (Kolega et al 1982; Couchman et al 1985). Thus, cycles of adhesion and de-adhesion are likely required for effective locomotion. Another function of adhesion could be to exert traction to pull up the rear of the motile cell after its leading edge has been extended. That fibroblast adhesions do exert traction was quite dramatically demonstrated by Harris et al (1980).

There is a voluminous literature (cf Carter 1965; Harris 1973b; Davis 1980) showing that manipulation of the adhesive properties of the substratum can have a marked influence on the rate and direction of cell migration in culture. Recently, however, physiologically more relevant studies of the role of adhesion in cell motility have focused on specific cell components that have been identified as adhesion molecules. There has been great interest in components of the extracellular matrix (ECM), such as fibronectin (for review see Yamada 1983), laminin (Timpl et al 1983), collagen, proteoglycans, and hyaluronic acid (Hay 1981). Fibronectin, for example, has been directly implicated in several types of morphogenetic cell movements during embryogenesis (Thiery et al 1985). Monoclonal anti-fibronectin antibodies and peptide fragments that function as competitive antagonists of fibronectin binding specifically interfere with the migration of mesodermal cells during amphibian gastrulation in vivo. A monoclonal antibody directed against a complex of 140-kDa integral membrane adhesion proteins (Greve & Gottlieb 1982; Neff et al 1982; Chen et al 1985; Knudsen et al 1985) that is most likely the cell surface receptor for fibronectin (see below) had marked effects on neural crest cell migration in chick embryos in vivo (Bronner-Fraser 1985). Many experiments in vitro have also implicated fibronectin and laminin in cell migration (cf Yamada 1983; Donaldson et al 1985).

The Molecular Nature of Adhesions

There is increasing evidence that adhesions between the cell surface and its substratum are often mediated by transmembrane interactions involving, on the outside of the cell, one or more components of the ECM and, on the inside of the cell, actin microfilaments. Such transmembrane interactions are observed, for example, with fibronectin, which was found to be lined up on the outside surface of fibroblasts directly over actin filaments situated under the membrane (Figure 5) (Heggeness et al 1978; Hynes & Destree 1978).

It was suggested some time ago (Singer et al 1978) that such interactions across the membrane are probably mediated by specific transmembrane integral proteins in the membrane, whose extracellular domains are linked, directly or indirectly, to ECM components, and whose cytoplasmic domains are associated directly or indirectly with actin filaments. In accordance with this idea, a complex of 140-kDa integral membrane glycoproteins has recently been implicated as the cell surface receptor for fibronectin (Pytela et al 1985), and a similar, if not identical, protein has been shown to be a transmembrane molecule associated with actin filaments (Rogalski & Singer 1985). By virtue of such transmembrane interactions, adhesions between the cell surface and the substratum involving the ECM component fibronectin could be directly linked mechanically to actin-mediated force-generating mechanisms inside the cell. This could be a way to exert tension from adhesions near the leading edge that would act on the rear of the cell (Chen 1981).

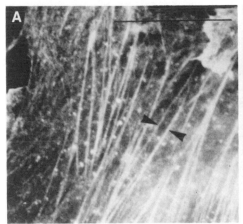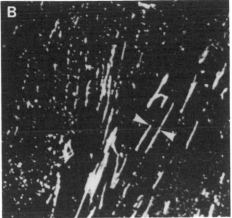

Figure 5 WI38 fibroblasts 24 hr after plating, double fluorescently labeled for intracellular actin (A) and for surface fibronectin (B). The rod-like structures containing surface fibronectin were generally lined up directly over intracellular bundles of actin filaments (examples marked by corresponding arrowheads). (Bar in A represents 25 μm.) (From Heggeness et al 1978.)

The molecular biology of cell adhesions, however, is still in an early stage of development. With any one cell, several adhesion mechanisms may coexist (Izzard & Lochner 1980). These different adhesions may well involve distinct sets of components engaged in transmembrane linkages (Chen & Singer 1982). Furthermore, different cell types undoubtedly have their own adhesion systems, each of which may play a role in the migratory behavior of a particular cell. For example, the Mac-1/LFA-1 family of integral membrane glycoproteins, which are characteristic of leucocytes and are distinct from the fibronectin receptors on other cells, have been convincingly demonstrated to be involved in both the adhesion and directed migration of leucocytes (Springer & Anderson 1985).

Cell Surface Expression of Adhesive Components

The evidence briefly discussed above indicates that both ECM components and the integral membrane proteins to which they bind function in cell adhesion. In vivo the ECM components may be part of a matrix laid down by other cells, or they may be secreted by the migrating cell itself as it moves. In either case, the integral membrane protein to which an ECM component binds to generate an adhesion must be expressed on the surface of the migrating cell. The continuous insertion of new membrane mass into the leading edge of such a cell, by introducing the adhesive integral protein(s) into the leading edge, could be a mechanism for continuously producing and concentrating the new adhesions required to attach the leading edge to the substratum. Furthermore, in response to successive signals along its migratory path, a locomoting cell could rapidly insert successively different specific integral adhesive proteins into its leading edge, which could provide a mechanism for "steering" the cell along a chemically complex, tortuous path. (This could also be of great importance in the axonal extension of neurones, see below.)

If the locomoting cell made its own adhesive ECM components, these would also very likely be introduced at the leading edge. The insertion of new membrane mass would be expected to produce simultaneous secretion, since it has been shown (Strous et al 1983) that the same Golgi-derived vesicles whose membranes include proteins bound for incorporation into the plasma membrane also contain soluble secretory proteins. The fusion of the vesicle with the plasma membrane that inserts new mass into the membrane therefore must also release the vesicle's contents to the cell exterior. These secreted contents could include ECM components, as well as proteases (cf Chen et al 1984), that might in an appropriately timed sequence destroy previously formed adhesions (de-adhesion).

Recent studies (Turley & Torrance 1985) on the distributions of hyaluronic acid (HA) and its putative cell surface receptor (HABP) on chick heart fibroblasts are consistent with these suggestions. HA has been implicated as an ECM adhesive component (Toole 1982), and HABP may be an integral protein of the plasma membrane to which HA binds. By immunofluorescence observations, both proteins appear to be concentrated in the leading edge of motile fibroblasts but are uniformly distributed on the surfaces of stationary cells. These results are to be expected if secretion of HA is polarized and new membrane mass containing HABP is simultaneously inserted at the leading edge of the motile cell.

AXONAL EXTENSION OF NEURONES

The resemblance between the phenomena of cell migration and neurite extension has been pointed out many times (cf Wessells et al 1973); neurite extension is pictured as similar to the advance of a migratory cell, except that the cell body of the neurone remains stationary. The growth cone at the growing tip of the neurite (cf Landis 1983) has been likened to the lamellipodium at the leading edge of a locomoting fibroblast, although these structures appear quite different morphologically. Whether these resemblances are more superficial than real has, however, been unclear.

Recent evidence suggests that they may indeed be real and may reflect the operation of similar mechanisms in neurite extension and cell migration. For example, microtubule depolymerizing drugs inhibit both processes (Solomon 1980). There are also striking similarities between certain persistence phenomena (see earlier section on persistence phenomena) arising in neurite extensions and fibroblast migration: Daughter cells arising from a recent division of a neuronal cell often send out neurite extensions that are mirror-images of one another (Solomon 1979). These results bear a striking resemblance to the persistence of the migratory paths that are taken by the two daughter cells after a recent fibroblast division (Albrecht-Buehler 1977). It also appears that new membrane mass for the growing neurite is added in the growth cone (Feldman et al 1981; Small & Pfenninger 1984). This corresponds to the conclusion that new membrane mass is inserted into the leading edge of locomotory cells, as discussed above.

All of this evidence is consistent with the proposal that the GA-MTOC in the cell body of the neurone, upon receipt of a polar signal to extend a neurite, becomes oriented to face the region of neurite extension, much as these organelles become oriented in locomotory cells to face the leading edge. If such polarization occurred in neurones it would have a result similar to that in cell locomotion: the insertion of new membrane mass (and secretion) at the growth cone and the leading edge, respectively. In this view, the principal mechanistic differences between neurite extension and cell locomotion would then be reduced simply to differences in membrane recycling. As was suggested earlier, locomotory cells must retrieve (endocytose) the equivalent of the membrane mass that they insert into the leading edge from regions of the cell surface well to the rear of that edge. This leaves the overall surface area constant and will ultimately result in the rear of the motile cell catching up with the forward-moving front. The entire cell therefore locomotes. Conversely, if neurones that are sending out neurites by inserting new membrane mass into the growth cone do not retrieve membrane mass from the neurites at sites far removed from the growth cone (i.e. if any membrane recycling that does go on is localized to the growth cone itself), then the total surface area of the neurites would continually increase, and the cell body would remain stationary, while the neurite continually extended. In the absence of endocytosis towards the rear of the neurite, no backward flow of membrane mass would be expected to occur over the surface of the neurite. The finding that small particles placed on a neurite remain stationary as the neurite extends (Bray 1970), in contrast to the case in migratory cells (Abercrombie et al 1970; Dembo & Harris 1981), is consistent with this scenario. If this picture is correct in its essentials, neurones in the process of initiating neurite extension would have to suspend the regulatory mechanisms that normally function to couple quantitatively the amount of surface membrane mass retrieved (by pinocytosis?) to the amount inserted. (Any membrane recycling processes whose insertion and retrieval stages were both confined to the growth cone, however, could continue.)

Conceptually, the processes of neurite extension and cell locomotion may indeed be closely related, as many investigators have thought, but perhaps they are even more so at the level of membrane mechanisms than has hitherto been suspected.

CONCLUDING REMARKS

In this necessarily brief review, we attempted to collect and integrate some of the major currents in the study of directed cell migration, viewed insofar as possible from the vantage of molecular biology. The major theme has been that intracellular force generation, insertion of membrane mass and secretion at the leading edge, and adhesion mechanisms probably all operate simultaneously in extending the front end of a locomotory cell. We particularly stressed polarized membrane insertion as a critical process in locomotion, because it has been a controversial matter and needed emphasis. As an indication of the obligatory cooperation of these several processes, either cytochalasin B or monensin can inhibit the directed migration of macrophages (Nemere et al 1985). The former disrupts actin-mediated force generation without affecting membrane insertion (Rogalski et al 1984), whereas monensin interferes with membrane recycling without apparently affecting the cytoskeleton. Such a holistic view of cell migration may lead to fruitful experiments that a narrower view may cause to be overlooked. Finally, many other subjects that are of great importance in the molecular biology of cell migration, such as the structure and effector functions of chemotactic receptors and the enzyme biochemistry of chemotactic signal transduction, have been consciously omitted from this review in order to focus on the main theme.

ACKNOWLEDGMENT

We would like to acknowledge several helpful discussions and correspondence with Drs. George F. Oster and Albert K. Harris. The work from our laboratory cited in this review was supported by US Public Health Service grants AI-06659 and GM-15971.

SJS wishes to acknowledge the fruitful tenure of the Newton-Abraham Chair at Oxford University during the academic year 1984–1985, during which most of the preparation for this article was carried out.

Literature Cited

Abercrombie, M., Heaysman, J. E. M., Pegrum, S. M. 1970. The locomotion of fibroblasts in culture. III. Movements of particles on the dorsal surface of the leading lamella. *Exp. Cell Res.* 62 : 389–98

Abercrombie, M., Heaysman, J. E. M., Pegrum, S. M. 1971. The locomotion of fibroblasts in culture. IV. Electron microscopy of the leading lamella. *Exp. Cell Res.* 67 : 359–67

Abercrombie, M., Heaysman, J. E. M., Pegrum, S. M. 1972. The locomotion of fibroblasts in culture. V. Surface marking with concanavalin A. *Exp. Cell Res.* 73 : 536–39

Albrecht-Buehler, G. 1977. Daughter 3T3 cells. Are they mirror images of each other? *J. Cell Biol.* 72 : 595–603

Albrecht-Buehler, G. 1980. Autonomous movements of cytoplasmic fragments. *Proc. Natl. Acad. Sci. USA* 77 : 6639–43

Anderson, R. G. W., Brown, M. S., Beisiegel, V., Goldstein, J. L. 1982. Surface distribution and recycling of the low density lipoprotein receptor as visualized with antireceptor antibodies. *J. Cell Biol.* 93 : 523–31

Anderson, R. G. W., Pathak, R. K. 1985. Vesicles and cisternae in the *trans* Golgi apparatus of human fibroblasts are acidic compartments. *Cell* 40 : 635–43

Ash, J. F., Louvard, D., Singer, S. J. 1977. Antibody-induced linkages of plasma membrane proteins to intracellular actomyosin-containing filaments in cultured fibroblasts. *Proc. Natl. Acad. Sci. USA* 74 : 5584–88

Ash, J. F., Singer, S. J. 1976. Concanavalin A–induced transmembrane linkage of concanavalin A surface receptors to intracellular myosin-containing filaments. *Proc. Natl. Acad. Sci. USA* 73 : 4575–79

Bandmann, U., Rydgren, L., Norberg, B. 1974. The difference between random movement and chemotaxis. Effects of antitubulins on neutrophil granulocyte locomotion. *Exp. Cell Res.* 88 : 63–73

Bergmann, J. E., Kupfer, A., Singer, S. J. 1983. Membrane insertion at the leading edge of motile fibroblasts. *Proc. Natl. Acad. Sci. USA* 80 : 1367–71

Bergmann, J. E., Singer, S. J. 1983. Immunoelectron microscopic studies of the intracellular transport of the membrane glycoprotein (G) of vesicular stomatitis virus in infected Chinese hamster ovary cells. *J. Cell Biol.* 97 : 1777–87

Berlot, C. H., Spudich, J. A., Devreotes, P. N. 1985. Chemoattractant-elicited increases in myosin phosphorylation in *Dictyostelium*. *Cell* 43 : 307–14

Bray, D. 1970. Surface movements during the growth of single explanted neurons. *Proc. Natl. Acad. Sci. USA* 65 : 905–10

Bretscher, M. S. 1983. Distribution of receptors for transferrin and low density lipoprotein on the surface of giant HeLa cells. *Proc. Natl. Acad. Sci. USA* 80 : 454–58

Brinkley, B. R. 1985. Microtubule organizing centers. *Ann. Rev. Cell Biol.* 1 : 197–24

Bronner-Fraser, M. 1985. Alterations in neural crest migration by a monoclonal antibody that affects cell adhesion. *J. Cell Biol.* 101 : 610–17

Carter, S. B. 1965. Principles of cell motility : The direction of cell movement and cancer invasion. *Nature* 208 : 1183–87

Chen, W.-T. 1981. Mechanism of the retraction of the trailing edge during fibroblast movement. *J. Cell Biol.* 90 : 187–200

Chen, W.-T., Greve, J. M., Gottlieb, D. I., Singer, S. J. 1985. Immunocytochemical localization of a 140 kd cell adhesion molecule in cultured chicken fibroblasts, and in chicken smooth muscle and intestinal epithelial tissues. *J. Histochem. Cytochem.* 33 : 576–86

Chen, W.-T., Olden, K., Bernard, B. A., Chu, F.-F. 1984. Expression of transformation-associated protease(s) that degrade fibronectin at cell contact sites. *J. Cell Biol.* 98 : 1546–55

Chen, W.-T., Singer, S. J. 1982. Immunoelectron microscopic studies of the sites of cell-substratum and cell-cell contacts in cultured fibroblasts. *J. Cell Biol.* 95 : 205–22

Cheung, H. T., Contarow, W. D., Sundharadas, G. 1978. Colchicine and cytochalasin B effects on random movement, spreading and adhesion of mouse macrophages. *Exp. Cell Res.* 111 : 95–103

Couchman, J. R., Lenn, M. R., Rees, D. A. 1985. Coupling of cytoskeleton functions for fibroblast locomotion. *Eur. J. Cell Biol.* 36 : 182–94

Davis, E. M. 1980. Translocation of neural crest cells within a hydrated collagen lattice. *J. Embryol. Exp. Morphol.* 55 : 17–31

Dembo, M., Harris, A. K. 1981. Motion of particles adhering to the leading lamella of crawling cells. *J. Cell Biol.* 91 : 528–36

Donaldson, D. J., Mahan, J. T., Hasty, D. L., McCarthy, J. B., Furcht, L. T. 1985. Location of a fibronectin domain involved in newt epidermal cell migration. *J. Cell Biol.* 101 : 73–78

Dyett, D. E., Malawista, S. E., van Blaricom, G., Melnick, D. A., Malech, H. L. 1985. Functional integrity of cytokineplasts :

Specific chemotactic and capping responses. *J. Immunol.* 135:2090–94

Euteneuer, U., Schliwa, M. 1984. Persistent, directional motility of cells and cytoplasmic fragments in the absence of microtubules. *Nature* 310:58–61

Farquhar, M. G. 1985. Progress in unraveling pathways of Golgi traffic. *Ann. Rev. Cell Biol.* 1:447–88

Fechheimer, M., Zigmond, S. H. 1983. Changes in cytoskeletal proteins of polymorphonuclear leukocytes induced by chemotactic peptides. *Cell Motil.* 3:349–61

Feldman, E. L., Axelrod, D., Schwartz, M., Heacock, A. M., Agranoff, B. W. 1981. Studies on the localization of newly added membrane in growing neurites. *J. Neurobiol.* 12:591–98

Geiger, B., Avnur, Z., Rinnerthaler, G., Hinssen, H., Small, V. J. 1984. Microfilament-organizing centers in areas of cell contact: Cytoskeletal interactions during cell attachment and locomotion. *J. Cell Biol.* 99:83s–91s

Goldman, R. D. 1971. The role of three cytoplasmic fibers in BHK-21 cell motility. I. Microtubules and the effects of colchicine. *J. Cell Biol.* 51:752–62

Gotlieb, A. I., Heggeness, M. H., Ash, J. F., Singer, S. J. 1979. Mechanochemical proteins, cell motility, and cell-cell contacts: the localization of mechanochemical proteins inside cultured cells at the edge of an in vitro "wound." *J. Cell. Physiol.* 100:563–78

Gotlieb, A. I., May, L. M., Subrahmanyan, L., Kalnins, V. I. 1981. Distribution of microtubule organizing centers in migrating sheets of endothelial cells. *J. Cell Biol.* 91:589–94

Greve, J. M., Gottlieb, D. I. 1982. Monoclonal antibodies which alter the morphology of cultured chick myogenic cells. *J. Cell Biochem.* 25:109–21

Harris, A. K. 1973a. Cell surface movements related to cell locomotion. In *Locomotion of Tissue Cells* 14:3–26. New York: Ciba Found./Elsevier

Harris, A. K. 1973b. Behavior of cultured cells on substrata of variable adhesiveness. *Exp. Cell Res.* 77:285–97

Harris, A. K., Wild, P., Stopak, D. 1980. Silicone rubber substrata: A new wrinkle in the study of cell locomotion. *Science* 208:177–79

Harris, H., Sidebottom, E., Grace, D. M., Bramwell, M. E. 1969. The expression of genetic information: A study with hybrid animal cells. *J. Cell Sci.* 4:499–525.

Hartwig, J. H., Stossel, T. P. 1979. Cytochalasin B and the structure of actin gels. *J. Mol. Biol.* 134:539–54

Hay, E. D. 1981. *Cell Biology of Extracellular Matrix.* New York: Plenum

Heath, J. P. 1983. Behavior and structure of the leading lamella in moving fibroblasts. I. Occurrence and centripetal movement of arc-shaped microfilament bundles beneath the dorsal cell surface. *J. Cell Sci.* 60:331–54

Heggeness, M. H., Ash, J. F., Singer, S. J. 1978. Transmembrane linkage of fibronectin to intracellular actin-containing filaments in cultured human fibroblasts. *Ann. NY Acad. Sci.* 312:414–17

Hopkins, C. R. 1985. The appearance and internalization of transferrin receptors at the margins of spreading human tumor cells. *Cell* 40:199–208

Huet, C., Ash, J. F., Singer, S. J. 1980. The antibody-induced clustering and endocytosis of HLA antigens on cultured human fibroblasts. *Cell* 21:429–38

Huxley, H. E. 1973. Muscular contraction and cell motility. *Nature* 243:445–49

Hynes, R. O., Destree, A. T. 1978. Relationships between fibronectin (LETS protein) and actin. *Cell* 15:875–86

Izzard, C. S., Lochner, L. R. 1980. Formation of cell-to-substrate contacts during fibroblast motility: An interference reflection study. *J. Cell Sci.* 42:81–116

Jacobson, K., O'Dell, D., Holifield, B., Murphy, T. L., August, J. T. 1984. Re-distribution of a major cell surface glycoprotein during cell movement. *J. Cell Biol.* 99:1613–23

Jesaitis, A. J., Naemura, J. R., Sklar, L. A., Cochrane, C. G., Painter, R. G. 1984. Rapid modulation of N-formyl chemotactic peptide receptors on the surface of human granulocytes: Formation of high-affinity ligand-receptor complexes in transient association with the cytoskeleton. *J. Cell Biol.* 98:1378–87

Jockusch, B. M., Haemmerli, G., Albon, A. 1983. Cytoskeletal organization in locomoting cells of the V2 rabbit carcinoma. *Exp. Cell Res.* 144:251–63

Keller, H. V., Bessis, M. 1975. Migration and chemotaxis of anucleate cytoplasmic leukocyte fragments. *Nature* 258:73–74

Knudsen, K., Horwitz, A., Buck, C. 1985. A monoclonal antibody identifies a glycoprotein complex involved in cell substratum adhesion. *Exp. Cell Res.* 157:218–26

Kolega, J., Shure, M. S., Chen, W.-T., Young, N. D. 1982. Rapid cellular translocation is related to close contacts formed between various cultured cells and their substrata. *J. Cell Sci.* 54:23–34

Kupfer, A., Dennert, G., Singer, S. J. 1983. Polarization of the Golgi apparatus and the microtubule-organizing center within cloned natural killer cells bound to their targets. *Proc. Natl. Acad. Sci. USA* 80: 7224–28

Kupfer, A., Dennert, G., Singer, S. J. 1985. The reorientation of the Golgi apparatus and the microtubule-organizing center in the cytotoxic effector cell is a prerequisite in the lysis of bound target cells. *J. Mol. Cell. Immunol.* 2:37–49

Kupfer, A., Louvard, D., Singer, S. J. 1982. Polarization of the Golgi apparatus and the microtubule-organizing center in cultured fibroblasts at the edge of an experimental wound. *Prroc. Natl. Acad. Sci. USA* 79:2603–7

Landis, S. C. 1983. Neuronal growth cones. *Ann. Rev. Physiol.* 45:567–80

MacLean-Fletcher, A., Pollard, T. D. 1980. Mechanism of action of cytochalasin B on actin. *Cell* 20:329–41

Malawista, S. E., de Boisfleury Chevance, A. 1982. The cytokineplast: Purified, stable, and functional motile machinery from human blood polymorphonuclear leukocytes. *J. Cell Biol.* 95:960–73

Malech, H. L., Root, R. K., Gallin, J. I. 1977. Structural analysis of human neutrophil migration: Centriole, microtubule, and microfilament orientation and function during chemotaxis. *J. Cell Biol.* 75:666–93

Marcus, P. I. 1962. Dynamics of surface modification in myxovirus-infected cells. *Cold Spring Harbor Symp. Quant. Biol.* 27:351–65

McRobbie, S. J., Newell, P. C. 1984. Chemoattractant-mediated changes in cytoskeletal actin of cellular slime molds. *J. Cell Sci.* 68:139–51

Middleton, C. A. 1979. Cell surface labelling reveals no evidence for membrane assembly and disassembly during fibroblast locomotion. *Nature* 282:203–5

Mitchison, T., Kirschner, M. 1984. Dynamic instability of microtubule growth. *Nature* 312.237–42

Neff, N. T., Lowrey, C., Decker, C., Tovar, A., Damsky, C., et al. 1982. A monoclonal antibody detaches embryonic skeletal muscle from extracellular matrices. *J. Cell Biol.* 95:654–66

Nemere, I., Kupfer, A., Singer, S. J. 1985. Reorientation of the Golgi apparatus and the microtubule-organizing center inside macrophages subjected to a chemotactic gradient. *Cell Motil.* 5:17–29

Oster, G. F. 1984. On the crawling of cells. *J. Embryol. Exp. Morph.* 83:329–64

Pollard, T. D. 1984. Polymerization of ADP-actin. *J. Cell Biol.* 99:769–77

Pollard, T. D., Mooseker, M. S. 1981. Direct measurement of actin polymerization rate constants by electron-microscopy of actin filaments nucleated by isolated microvillus cores. *J. Cell Biol.* 88:654–59

Pytela, R., Pierschbacher, M., Ruoslahti, E. 1985. Identification and isolation of a 140 kd cell surface glycoprotein with properties expected of a fibronectin receptor. *Cell* 40:191–98

Reggio, H., Bainton, D., Harms, E., Coudrier, E., Louvard, D. 1984. Antibodies against lysosomal membranes reveal a 100,000-mol-wt protein that cross-reacts with purified H^+, K^+ ATPase from gastric mucosa. *J. Cell Biol.* 99:1511–26

Robbins, A. R., Oliver, C., Bateman, J. L., Krag, S. S., Galloway, C. J., et al. 1984. A single mutation in Chinese hamster ovary cells impairs both Golgi and endosomal functions. *J. Cell Biol.* 99:1296–1308

Robbins, A. R., Peng, S. S., Marshall, J. L. 1983. Mutant Chinese hamster ovary cells pleiotropically defective in receptor-mediated endocytosis. *J. Cell Biol.* 96:1064–71

Robbins, E., Gonatas, N. K. 1964. Histochemical and ultrastructural studies on HeLa cell cultures exposed to spindle inhibitors with special reference to the interphase cell. *J. Histochem. Cytochem.* 12:704–11

Rogalski, A. A., Bergmann, J. E., Singer, S. J. 1984. Effect of microtubule assembly status on the intracellular processing and surface expression of an integral protein of the plasma membrane. *J. Cell Biol.* 99:1101–9

Rogalski, A. A., Singer, S. J. 1984. Associations of elements of the Golgi apparatus with microtubules. *J. Cell Biol.* 99:1092–1100

Rogalski, A. A., Singer, S. J. 1985. An integral glycoprotein associated with the membrane attachment sites of actin microfilaments. *J. Cell Biol.* 101:785–801

Roos, D., Voetman, A. A., Meerhof, L. J. 1983. Functional activity of enucleated human polymorphonuclear leukocytes. *J. Cell Biol.* 97:368–77

Rubino, S., Fighetti, M., Unger, E., Cappuccinelli, P. 1984. Location of actin, myosin, and microtubular structures during directed locomotion of *Dictyostelium* amebae. *J. Cell Biol.* 98:382–90

Sandoval, I. V., Bonafacino, J. S., Klausner, R. D., Henkart, M., Wehland, J. 1984. Role of microtubules in the organization and localization of the Golgi apparatus. *J. Cell Biol.* 99:1135–85

Shaffer, B. M. 1963. Behavior of particles adhering to amoebae of the slime mold *Polysphondylium violaceum* and the fate of the cell surface during locomotion. *Exp. Cell Res.* 32:603–6

Singer, S. J., Ash, J. F., Bourguignon, L.

W., Heggeness, M. H., Louvard, D. 1978. Transmembrane interactions and the mechanisms of transport of proteins across membranes. *J. Supramol. Struct.* 9:378–89

Small, R. K., Pfenninger, K. H. 1984. Components of the plasma membrane of growing axons. I. Size and distribution of intramembrane particles. *J. Cell Biol.* 98:1422–33

Solomon, F. 1979. Detailed neurite morphologies of sister neuroblastoma cells are related. *Cell* 16:165–69

Solomon, F. 1980. Neuroblastoma cells re-capitulate their detailed neurite morphologies after reversible microtubule disassembly. *Cell* 21:333–38

Soranno, T., Bell, E. 1982. Cytostructural dynamics of spreading and translocating cells. *J. Cell Biol.* 95:127–36

Springer, T. A., Anderson, D. A. 1985. The Mac-1, LFA-1 family of leucocyte adherence. In *Biochemistry of Macrophages,* ed. M. O'Connor, pp. 102–26. London: Ciba Found.

Steinman, R. M., Brodie, S. E., Cohn, Z. A. 1976. Membrane flow during pinocytosis. A stereological analysis. *J. Cell Biol.* 68:665–87

Steinman, R. M., Mellman, I. S., Muller, W. A., Cohn, Z. A. 1983. Endocytosis and the recycling of plasma membrane. *J. Cell Biol.* 96:1–27

Stossel, T. P. 1982. The spatial organization of cortical cytoplasm in macrophages. In *Spatial Organization of Eukaryotic Cells,* ed. J. R. McIntosh, pp. 203–23. New York: Liss

Stossel, T. P., Chaponnier, C., Ezzell, R. M., Hartwig, J. H., Janmey, P. A., et al 1985. Nonmuscle actin-binding proteins. *Ann. Rev. Cell Biol.* 1:353–402

Stossel, T. P., Hartwig, J. H., Yin, H. L., Davies, W. A. 1979. Actin-binding protein. In *Cell Motility: Molecules and Organization,* ed. S. Hatano, H. Ishikawa, H. Sato, pp. 189–209. Tokyo: Univ. Tokyo Press

Strous, G. J. A. M., Willemsen, R., van Kerkhof, P., Slot, J. W., Geuze, H. J., et al. 1983. Vesicular stomatitis virus glycoprotein, albumin, and transferrin are transported to the cell surface via the same Golgi vesicles. *J. Cell Biol.* 97:1815–22

Sugimoto, Y., Hagiwara, A. 1979. Cell locomotion on differently charged substrates. *Exp. Cell Res.* 120:245–52

Sullivan, S. J., Daukas, G., Zigmond, S. H. 1984. Asymmetric distribution of the chemotactic peptide receptor on polymorphonuclear leukocytes. *J. Cell Biol.* 99:1461–67

Swanson, J. A., Taylor, D. L. 1982. Local and spatially coordinated movements in *Dictyostelium discoideum* amoebae during chemotaxis. *Cell* 28:225–32

Tartakoff, A. M., Vassalli, P. 1977. Plasma cell immunoglobulin secretion. Arrest is accompanied by alterations of the Golgi complex. *J. Exp. Med.* 146:1332–45

Taylor, D. L., Hellewell, S. B., Virgin, H. W., Heiple, J. M. 1979. The solation-contraction coupling hypothesis of cell movements. In *Cell Motility: Molecules and Organization,* ed. S. Hatano, H. Ishikawa, H. Sato, pp. 363–77. Tokyo: Univ. Tokyo Press

Thiery, J. P., Duband, J. L., Tucker, G. C. 1985. Cell migration in the vertebrate embryo: Role of cell adhesion and tissue environment in pattern formation. *Ann. Rev. Cell Biol.* 1:91–113

Tickle, C., Trinkaus, J. P. 1977. Some clues as to the formation of protrusions by *Fundulus* deep cells. *J. Cell. Sci.* 26:139–50

Tilney, L. G. 1978. Polymerization of actin. V. A new organelle, the actomere, that initiates the assembly of actin filaments in *Thyone* sperm. *J. Cell Biol.* 77:551–64

Timpl, R., Engel, J., Martin, G. R. 1983. Laminin—a multifunctional protein of basement membranes. *Trends Biochem. Sci.* 8:207–9

Toole, B. P. 1982. Developmental role of hyaluronate. *Connect. Tissue Res.* 10:93–100

Tougard, C., Louvard, D., Picart, R., Tixier-Vidal, A. 1985. Antibodies against a lysosomal membrane antigen recognize a prelysosomal compartment involved in the endocytic pathway in cultured prolactin cells. *J. Cell Biol.* 100:786–93

Trinkaus, J. P. 1984. *Cells into Organs: The Forces that Shape the Embryo.* Englewood Cliffs, NJ: Prentice-Hall, 2nd ed.

Turley, E. A., Torrance, J. 1985. Localization of hyaluronate and hyaluronate-binding protein on motile and non-motile fibroblasts. *Exp. Cell Res.* 161:17–28

Vasiliev, J. M., Gelfand, I. M., Domnina, L. V., Dorfman, N. A., Pletyushkina, O. Y. 1976. Active cell edge and movements of concanavalin A receptors of the surface of epithelial and fibroblastic cells. *Proc. Natl. Acad. Sci. USA* 73:4085–89

Vasiliev, J. M., Gelfand, I. M., Domnina, L. V., Ivanova, O. Y., Komm, S. G., et al. 1970. Effect of colcemid on the locomotory behaviour of fibroblasts. *J. Embryol. Exp. Morph.* 24:625–40

Walter, R. J., Berlin, R. D., Oliver, J. M. 1980. Asymmetric Fc receptor distribution on human PMN oriented in a chemotactic gradient. *Nature* 286:724–25

Wang, Y.-L. 1985. Exchange of actin subunits at the leading edge of living fibroblasts: Possible role of treadmilling. *J. Cell Biol.* 101:597–602

Warren, G., Featherstone, C., Griffiths, G., Burke, B. 1983. Newly synthesized G protein of vesicular stomatitis virus is not transported to the cell surface during mitosis. *J. Cell Biol.* 97:1623–28

Weeds, A. 1982. Actin-binding proteins—regulators of cell architecture and motility. *Nature* 296:811–16

Weinbaum, D. L., Sullivan, J. A., Mandell, G. L. 1980. Receptors for concanavalin A cluster at the front of polarized neutrophils. *Nature* 286:725–27

Wessells, N. K., Spooner, B. S., Luduena, M. A. 1973. Surface movements, microfilaments and cell locomotion. In *Locomotion of Tissue Cells*, ed. R. Porter, D. W. Fitzsimons, pp. 53–82. New York:

Elsevier

Wilkinson, P. C., Michl, J., Silverstein, S. C. 1980. Receptor distribution in locomoting neutrophils. *Cell Biol. Int. Rep.* 4:736

Wilson, E. B. 1925. *The Cell in Development and Heredity*, p. 50. New York: Macmillan, 3rd ed.

Yamada, K. M. 1983. Cell surface interactions with extracellular materials. *Ann. Rev. Biochem.* 52:761–99

Zigmond, S. 1981. Consequences of chemotactic peptide receptor modulation for leukocyte orientation. *J. Cell Biol.* 88:644–47

Zigmond, S. H., Levitsky, H. I., Kreel, B. J. 1981. Cell polarity: An examination of its behavioral expression and its consequences for polymorphonuclear leukocyte chemotaxis. *J. Cell Biol.* 89:585–92

Ann. Rev. Cell Biol. 1986. 2 : 367–90

PROTEIN IMPORT INTO THE CELL NUCLEUS

Colin Dingwall and Ronald A. Laskey

CRC Molecular Embryology Research Group, Department of Zoology, University of Cambridge, Downing Street, Cambridge CB2 3EJ, England

CONTENTS

INTRODUCTION

It has been known for some time that the uptake of proteins by the nucleus is extremely selective (Gurdon 1970; Bonner 1975b) and that mature nuclear proteins are able to accumulate in the nucleus (Gurdon 1970; Bonner 1975b; De Robertis et al 1978; Dabauvalle & Franke 1982). Nuclear proteins must therefore contain within their final structure a signal that specifies selective accumulation in the nucleus (De Robertis et al 1978).

The study of the mechanisms by which proteins are selectively accumulated in the cell nucleus has lagged behind the study of other transport processes, such as those that direct proteins to the cell exterior or to cytoplasmic organelles. The late development of this field may be due to the widespread acceptance of a model for nuclear localization in which

367

0743–4623/86/1115–0367$02.00

any protein could diffuse freely into the nucleus, but only nuclear proteins would be retained by specific binding to nondiffusible nuclear components.

In the first part of this review we emphasize the limitations of the old model, and we examine the growing evidence that the accumulation of many nuclear proteins is mediated by their selective entry through the nuclear envelope rather than by selective retention within the nucleus after entry by diffusion. In the second part of the review we discuss investigations of the identity of the sequences in nuclear proteins that specify their accumulation within the nucleus. Finally, we discuss experiments that demonstrate mechanisms that allow the entry of proteins into the nucleus to be controlled during development and in response to intracellular signals.

PERMEABILITY PROPERTIES OF THE NUCLEAR ENVELOPE

The permeability of the nuclear envelope has been studied extensively by microinjecting various molecular probes into large cells, such as oocytes, and observing which probes enter the nucleus. These experiments have been reviewed thoroughly by Bonner (1978), so we only summarize the major findings below.

Early electrical measurements showed a resistance across the nuclear envelope, which suggested that the nuclear pores were not open channels. However, these measurements can now be interpreted as indicating that the resistance of individual nuclear pores is the same as that of the cytoplasm. Thus there appears to be no barrier to the passage of ions through nuclear pores (reviewed by Bonner 1978). This conclusion has been confirmed by injecting small radiolabelled molecules into the cytoplasm of frog oocytes and showing that potassium ions (Century et al 1970), aminoisobutyric acid (Frank & Horowitz 1975), sucrose, and inulin (Horowitz 1972) all diffuse into nuclei as fast as they pass through cytoplasm.

The nuclear envelope apparently has the properties of a sieve. Molecules larger than a certain size are excluded from the nucleus while smaller molecules are able to enter (see Table 1 and Bonner 1978). Hence there is a major qualitative difference in the permeability properties of the nuclear envelope compared with other membrane systems. The rate of accumulation of probes that are smaller than this size limit is inversely related to their size. For example, small dextrans (1.2-nm radius) enter the nucleus faster than larger dextrans (2.3- or 3.5-nm radius) (see Figure 1 and Paine et al 1975). The maximum size limit for entry into the nuclear compartment is consistent with a channel radius of 4.5 nm, which corresponds to a globular protein of approximately 60,000 molecular weight (see Table 1).

Table 1 Protein entry into nuclei after microinjection into cell cytoplasm[a]

Protein	Approximate mol. wt.	% in nucleus at 24 hr
Cytochrome C	13,000	12
Lysozyme	14,000	12
Myoglobin	18,000	12
Ovalbumin	45,000	5 (12 at 3 days)
Bovine serum albumin	68,000	< 1
γ-Globulin	160,000	< 1

[a] References: Bonner 1975a; Paine & Feldherr 1972; Paine 1975

There is reasonable agreement on this value among studies of nuclei from diverse organisms, including *Xenopus*, *Periplaneta*, and *Amoeba* (reviewed by Bonner 1978), and more recently from studies of rat hepatocytes using fluorescent microphotolysis (Peters 1984). There is also reasonable agreement among studies using probes of different types, ranging from cytoplasmic or secreted proteins to labelled dextrans or colloidal gold particles (reviewed by Bonner 1978).

Taken together these findings indicate that the nuclear envelope is not a barrier to small proteins but restricts access of large proteins. We believe that this distinction must be considered carefully when comparing studies of nuclear proteins with widely different molecular weights. Thus the next section considers a model that can satisfactorily account for nuclear accumulation of small proteins and that has attracted widespread support; however, we believe that it is inadequate to account for the accumulation of large nuclear proteins.

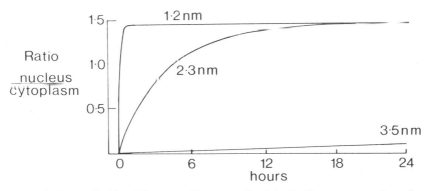

Figure 1 Entry of tritiated dextrans of known radius into the *Rana* oocyte nucleus after microinjection into the cytoplasm, as a function of time. (Redrawn from Paine et al 1975).

THE DIFFUSION AND BINDING MODEL FOR PROTEIN UPTAKE INTO THE NUCLEUS

Small non-nuclear proteins can enter the nucleus at rates that suggest that only their size, not their sequence, is limiting (see Table 1). Although such small proteins can enter the nucleus and equilibrate between the nucleus and cytoplasm, only nuclear proteins accumulate in the nucleus against a concentration gradient (Gurdon 1970; Bonner 1975b).

These observations are consistent with the possibility that selective nuclear accumulation is mediated by free diffusion of any small protein into the nucleus, with selective retention of nuclear proteins by binding to some component of the nucleus that is not diffusible.

Additional support for this model of protein accumulation in nuclei comes from a number of investigations.

Merriam & Hill (1976) showed that after sequential salt extraction of *Xenopus* oocyte nuclei a salt-resistant protein matrix or gel remained. They suggested that the soluble nucleoplasmic proteins are retained in the nucleus by binding to this nondiffusible material.

The nuclear envelope can be damaged in vivo without affecting the uptake of endogenous nuclear proteins in the cell. This was demonstrated by Feldherr & Pomerantz (1978) and by Feldherr & Ogburn (1980) and probably provides the best supporting evidence for this model. These authors punctured the *Xenopus* oocyte nucleus in situ with glass needles and showed that these damaged nuclei accumulated radiolabelled nuclear proteins from the cytoplasm. Hence the accumulation of nuclear proteins appears not to be controlled by the nuclear envelope but rather by selective binding within the nucleus.

EVIDENCE FOR A SELECTIVE ENTRY MECHANISM

The view that selective accumulation of proteins in the nucleus can be attributed to selective retention after entry by diffusion is difficult to reconcile with the behavior of large nuclear proteins. Many of the proteins present in the nucleus are much larger than the 60-kDa mass limit determined using exogenous macromolecular probes. The large nuclear proteins N1 and N2 ($M_r = 149,000$) accumulate in *Rana* oocyte nuclei much more rapidly than can be accounted for by simple diffusion through nuclear pores and binding within the nucleus (Feldherr et al 1983). This conclusion holds even when molecular dimensions are assumed that permit maximum diffusibility through the nuclear pore complex.

However, it has been suggested that large nuclear proteins could diffuse into the nucleus rapidly if they had large axial ratios and could enter "end on" (discussed in Bonner 1978) or if they could change shape during transit. In view of the wide range of nuclear proteins, we consider both suggestions unlikely. Furthermore, this explanation is excluded for experiments using colloidal gold coated with nucleoplasmin (described below).

Nucleoplasmin, which appears as a disc in electron micrographs with a radius of approximately 3.7 nm (Earnshaw et al 1980), accumulates in the *Xenopus* oocyte nucleus extremely rapidly (Dingwall et al 1982; Laskey et al 1985)—much more rapidly than a dextran that is of similar dimensions (3.5 nm, shown in Figure 1) and enters by diffusion. This finding argues directly against diffusion as an adequate explanation for the uptake of nucleoplasmin and is not subject to the assumptions that underlie the theoretical calculations used by Feldherr et al (1983) to analyze the uptake of N1 and N2 (see above).

For nucleoplasmin there is direct evidence of selective entry through the nuclear envelope. Nucleoplasmin is a large ($M_r = 165,000$), pentameric, thermostable protein present in high concentrations in the soluble phase of the *Xenopus* oocyte nucleus (Mills et al 1980). When microinjected into the *Xenopus* oocyte cytoplasm it rapidly accumulates in the nucleus (Dingwall et al 1982). Proteolysis of nucleoplasmin reveals that each subunit has two structural domains, a relatively protease-sensitive "tail" region and a relatively protease-resistant "core" region. A pentamer that has been proteolyzed sufficiently to remove all the tail domains remains pentameric and is unable to enter the nucleus (Figure 2b). All partially cleaved pentamers, even those bearing only a single intact subunit, retain the ability to accumulate in the nucleus (Figure 2a). The isolated tail domain also accumulates in the nucleus after microinjection into the cytoplasm. These results demonstrate that the "tail" domain bears the signal for accumulation in the nucleus and that the signal is a discrete polypeptide domain of the protein.

The pentameric core molecule was microinjected directly into the oocyte nucleus to determine whether the tail domain specifies selective entry or selective retention within the nucleus. If the tail domain specifies binding in the nucleus, we would expect the core molecules lacking this binding domain to diffuse back into the cytoplasm. However, the core molecules remained in the nucleus (Figure 3). This demonstrates that the tail region cannot specify binding but must specify entry into the nucleus.

THE NUCLEAR PORE AS A ROUTE OF PROTEIN ENTRY

The nuclear envelope consists of two layers of membrane, which are lined on the inside with lamins and perforated by nuclear pores (reviewed by Franke et al 1981). Each pore complex consists of a ring of eight subunits surrounding the central channel, which may contain a central plug (Unwin & Milligan 1982). The pore complexes are obvious candidates for the route of protein import, and recently direct evidence has been provided that the nuclear pore is the route of entry of nucleoplasmin into the nucleus of *Xenopus* oocytes (Feldherr et al 1984). Feldherr and his colleagues coated colloidal gold particles with nucleoplasmin and injected the suspension into *Xenopus* oocyte cytoplasm. They sectioned oocytes at various intervals after injection and examined the location of the gold particles by electron microscopy. At early times there was a remarkable clustering of colloidal gold particles at the nuclear pores (Figure 4); at later times the gold particles accumulated in the nucleus. Both gold particles coated with polyvinyl pyrolidone instead of nucleoplasmin and those coated with the pentameric protease-resistant core of nucleoplasmin remained in the cytoplasm and did not cluster at the nuclear pores or accumulate within the nucleus (Figure 4).

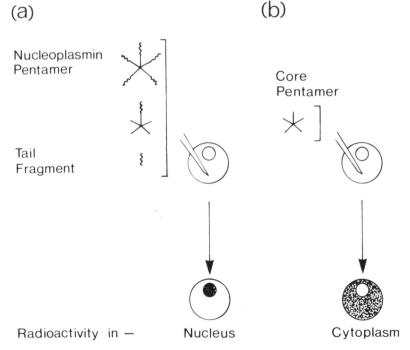

Figure 2 Diagram illustrating the transport of nucleoplasmin molecules into the *Xenopus* oocyte nucleus. (*a*) The intact nucleoplasmin pentamer, a pentamer with a single intact subunit, and the isolated "tail" fragment can all accumulate in the nucleus after microinjection into the cytoplasm. (*b*) The nucleoplasmin "core" molecule cannot enter the nucleus after microinjection into the cytoplasm.

These experiments provide direct evidence for passage of nucleoplasmin through nuclear pores. They also demonstrate that structural deformation of large nuclear proteins need not be invoked to explain their transit into the nucleus, since the pore complex can transport rigid gold particles that are much larger than the apparent pore orifice. The experiments of Feldherr et al raise some interesting questions. First, are all pores functionally equivalent, or are there separate classes for import and export or for different groups of transported molecules? All the pore complexes seen in the small regions of the nuclear envelope shown in the electron micrographs (Figure 4) have gold particles associated with them and are thus capable of transporting nucleoplasmin. This suggests that if there are different functional classes of pore they are clustered. The alternative is that all pores are functionally equivalent.

A second question raised by these experiments concerns the clustering and alignment of gold particles through the nuclear pore. What causes multiple nucleoplasmin-coated gold particles to be aligned through the

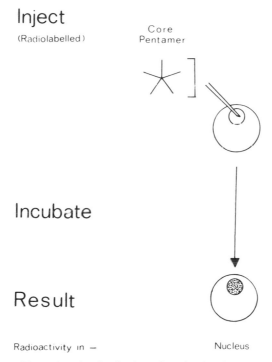

Figure 3 Diagram illustrating the distribution of nucleoplasmin "core" molecules after microinjection into the *Xenopus* oocyte nucleus.

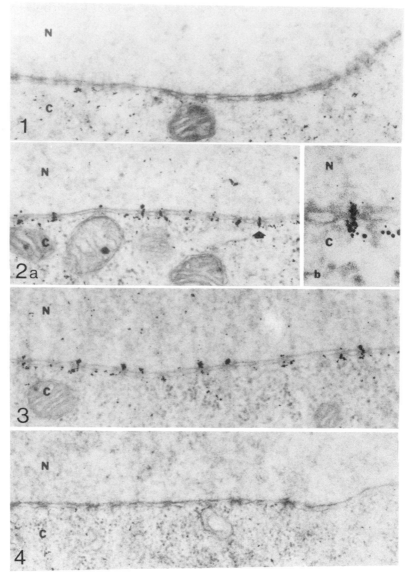

Figure 4 Colloidal gold particles coated with nucleoplasmin microinjected into the cyto-plasm of *Xenopus* oocytes N = nucleus, C = cytoplasm. *Panel 1*, oocytes fixed 10 sec after microinjection. *Panel 2a*, oocytes fixed 15 min after injection. *Panel 2b*, enlargement of nuclear pore complex indicated by arrow in panel 2a. *Panel 3*, oocytes fixed 1 hr after injection. *Panel 4*, oocytes injected with colloidal gold coated with the trypsin-resistant nucleoplasmin core and fixed 1 hr after injection.

nuclear pore? The alignments seen in the figures of Feldherr et al (1984) and Figure 4 would not be expected if there were only one binding site per pore or even a ring of binding sites for each pore complex. Instead, the patterns suggest that additional binding sites exist on the cytoplasmic surface of the pore complex. These sites may hold molecules in a queue awaiting transport through the pore. If this is the case, it will be interesting to determine the molecules and mechanisms involved.

If, as Feldherr's micrographs suggest, molecules destined for transport through the pores are collected near the cytoplasmic surface of the pore complex, then a further question arises of whether nuclear proteins diffuse through the cytosol to the nucleus or are transported there through the cytoplasm. Calculations show that a change in the diffusion process from a three-dimensional random walk in the cytoplasm to a two-dimensional random walk would only bring about a twofold increase in the rate of accumulation (Dingwall et al 1986). Hence there is no need to postulate directed transport (through the cytoplasm to the nuclear envelope) to account for the observed rate of movement of nucleoplasmin. Furthermore, nucleoplasmin appears to spread uniformly from the site of injection in the oocyte cytoplasm and not to migrate preferentially towards the nucleus (A. D. Mills, personal communication). The high degree of structural conservation of nuclear pore complexes (Franke et al 1981) contrasts with the apparent lack of conservation in sequences specifying entry into the nucleus (Table 2) and suggests that the relationship between pore complexes and such sequences will not be a simple one (Rine & Barnes 1985).

SEQUENCES THAT SPECIFY SELECTIVE ACCUMULATION IN THE NUCLEUS

Partial proteolysis has established that the sequence of nucleoplasmin responsible for its localization in the nucleus resides in the third of the polypeptide chain near the carboxy terminus (Dingwall et al 1982). Recombinant DNA techniques allow us to determine much more precisely which protein sequences are responsible for nuclear localization. Recently genes encoding several nuclear proteins have been subjected to deletion or base substitution mutagenesis in vitro to identify potential signal sequences. To date the most informative studies of a protein sequence responsible for nuclear localization are those using SV40 large T antigen (a tetramer with subunits of $M_r = 90,000$). The sequence demonstrated to function as a nuclear location signal in this protein can be regarded as a prototype for other such signals (Smith et al 1985).

Table 2 Amino acid sequences that act as signals for translocation to the cell nucleus

Source	Nuclear protein	Location of signal in parent protein	Transported nonnuclear protein	Deduced signal sequence	Reference
Yeast	Mat α2	amino terminus (residues 1–13)	E. coli β-galactosidase	lys^3-ile-pro-ile-lys	Hall et al 1984
SV40	Large T antigen	amino terminus (residues 126–132)	Chicken muscle pyruvate kinase	pro-lys-lys^{128}-lys-arg-lys-val	Kalderon et al 1984b
Yeast	GAL 4	amino terminus 74 residues	E. coli β-galactosidase	—	Silver et al 1984
Yeast	Ribosomal protein L3	amino terminus 21 residues	E. coli β-galactosidase	pro^{18}-arg-lys-arg	Moreland et al 1985
Influenza virus (expressed in Xenopus oocytes)	Nucleoprotein (NP)	carboxy terminus (residues 336–345)	Chimpanzee α-globin	ala^{336}-ala-phe-glu-asp-leu-arg-val-leu-ser	Davey et al 1985
Polyoma virus	Large T antigen	carboxy terminus residues around lysine 192 and lysine 282	Chicken muscle pyruvate kinase	val-ser-arg^{192}-arg-pro-arg-pro pro-pro-lys-lys^{282}-ala-arg-glu-asp	Richardson et al 1986

SV40 Large T Antigen

A SINGLE AMINO ACID SUBSTITUTION ABOLISHES TRANSPORT INTO THE NUCLEUS   A mutant SV40-adenovirus hybrid (para cT) virus has been described (Rapp et al 1969; Butel et al 1969) that blocks the accumulation of the large T antigen in the nucleus of the infected cell. DNA sequence analysis of para cT (Lanford & Butel 1984) revealed that the lysine residue at position 128 in the wild-type large T antigen sequence was substituted by an asparagine residue in cT.

Lanford & Butel (1980) showed by cellular fractionation that the mutant large T antigen (cT) is soluble in the cytoplasm and retains the ability to bind DNA: Hence its failure to accumulate in the nucleus is not due to binding to cytoplasmic components or an inability to bind DNA: Additionally, these authors showed that the cT antigen, like the wild-type protein, is associated with the cell surface (Lanford & Butel 1982).

The finding that the smallest carboxy-terminal fragment capable of localization in the nucleus contains 162 amino acids and extends from the carboxy terminus to amino acid 47 suggests that the sequence near lysine 128 is the only determinant of nuclear localization in the large T antigen (Lewis & Rowe 1971).

Kalderon et al (1984a) independently generated a series of point mutations in SV40 large T antigen using mixed oligonucleotide mutagenesis and showed that mutation of lysine 128 to threonine abolished transport into the nucleus. This mutant protein binds DNA and is present in a soluble form in the cytoplasm (Lanford & Butel 1980; Paucha et al 1985). Therefore the abolition of transport into the nucleus is not due to either failure to bind DNA in the nucleus or to irreversible binding to cytoplasmic components.

It is remarkable that such different approaches both identified the same amino acid as crucial for nuclear localization. Mutation of other amino acids in the region of lysine 128 influenced the nuclear localization of the large T antigen: Cells expressing the mutant protein showed different fluorescent staining patterns when probed with a monoclonal antibody to large T antigen. Some cells showed predominantly nuclear staining, some exclusively cytoplasmic, some exclusively nuclear, and some showed equal staining in the nucleus and cytoplasm. This result is thought to indicate that mutations in the region of lysine residue 128 influence the rate of accumulation of the mutant proteins in the nucleus.

Kalderon et al (1984a,b) identified a minimum protein sequence from SV40 large T antigen that will specify entry into the nucleus, and they have examined its properties in detail.

THE NUCLEAR LOCATION SEQUENCE IS EXTREMELY SHORT   The precise sequence requirements for nuclear location of SV40 large T antigen have been determined. Using recombinant DNA techniques it has been demonstrated that:

1. The shortest amino acid sequence derived from SV40 large T antigen that confers nuclear localization on a nonnuclear protein (chicken muscle pyruvate kinase) when linked to it at the amino terminus is pro-lys-lys[128]-lys-arg-lys-val.
2. This sequence confers nuclear localization on a protein that also bears a "mutant" sequence, i.e. a sequence differing only in the substitution of the crucial lysine residue by a threonine residue. The ability of the wild-type sequence to function when a mutant sequence is present in the same molecule indicates that the failure of the mutant signal to bring about nuclear localization is not due to nonspecific binding to cytoplasmic components, which confirms the earlier finding of Lanford & Butel (1980) (described above).
3. This sequence can confer nuclear localization on a mutated large T antigen protein when the sequence is located at the protein's amino terminus or when it is linked to the amino terminus of pyruvate kinase.

This protein sequence also has the remarkable property of conferring nuclear localization on bovine serum albumin to which it has been chemically cross-linked (Goldfarb et al 1986). Goldfarb et al (1986) examined the nuclear transport properties of two synthetic polypeptides twelve amino acids long, one of which contained the SV40 nuclear localization signal described above and the other contained the "mutant" SV40 sequence in which the crucial lysine (lysine 128) residue was replaced by threonine. These peptides were iodinated and chemically cross-linked to bovine serum albumin through a carboxy-terminal cysteine residue. The radiolabelled peptide-protein cross-linked product was microinjected into the cytoplasm of *Xenopus* oocytes. Only the protein cross-linked to the peptide having the wild-type SV40 sequence accumulated in the *Xenopus* oocyte nucleus. This accumulation showed saturation kinetics. The rate of accumulation of this peptide-linked protein was markedly reduced by coinjection with the free "wild-type" peptide, while the mutant peptide when coinjected had a much less dramatic effect on this rate. These results argue strongly for the existence of a carrier-mediated transport process.

A remarkable feature of the nuclear localization sequence of SV40 T antigen is its ability to function in quite different protein environments. It can function when it is structurally constrained within a protein sequence (e.g. in the normal position in SV40 large T antigen) and when it is not constrained (e.g. when it is chemically cross-linked through its carboxy terminus to the surface of another protein or as an amino terminal extension of another protein). This suggests that in order for the sequence to act it is not necessary for it to adopt a fixed secondary structure, but perhaps it adopts a particular configuration on binding to its receptor. If so, it would be analogous to the binding of peptide hormones to their receptors (Rose et al 1985). Many different backbone conformations of these peptides are consistent with biological activity, and it is the spatial disposition of the amino acid side chains that appears to determine activity. In addition, many of these peptides contain amino acids that promote bending of the peptide backbone. It is interesting to note that the SV40 sequence contains proline, which can be substituted by a glycine residue: Both of these residues promote bending.

HOMOLOGOUS SEQUENCES IN OTHER NUCLEAR PROTEINS   Smith et al (1985) conducted a computer search of a data base of 2511 protein sequences for homologies to the SV40 prototype sequence. The majority of proteins selected as containing a homologous sequence were viral proteins, including adenovirus E1A and SV40 VP1, VP2, and VP3 (see below). Interestingly, the search found five possible candidate sequences in polyoma large T.

A sequence in the amino-terminal region of the *Drosophila* 70-kDa heat-shock protein was also selected as a possible homologous region. However, it has been shown that both the amino-terminal and carboxy-terminal domains of this protein molecule have the ability to accumulate in the nucleus (Munro & Pelham 1984). Hence a nuclear location signal that is not related to the SV40 signal must be located in the carboxy-terminal portion of the molecule. We have examined the amino acid sequence of the nucleoplasmin tail region and have identified two regions that show very close homology to the SV40 prototype sequence and two regions that show very close homology to the yeast MAT α2 sequence (Dingwall et al 1986). It is not yet clear which of these, if any, confer nuclear location.

Nuclear Location Signal Sequences in other Nuclear Proteins

The type of experimental approach described above for SV40 large T antigen has been used in a number of studies to link sequences derived from nuclear proteins to nonnuclear proteins. In this way nuclear location signal sequences have been identified in several nuclear proteins (Table 2). The disappointing but not unexpected result of these investigations is that there is no signal sequence common to all the nuclear proteins studied. The significance of this is discussed later (see conclusions and prospects section). We do not describe all of these experiments in detail but discuss a number of examples that either raise interesting questions about the process of nuclear transport or present potentially valuable systems for further investigation.

YEAST MAT α2   The first report of an amino acid sequence that could target nonnuclear proteins to the nucleus was from Hall et al (1984), who linked amino-terminal fragments of the yeast MAT α2 protein to *Escherichia coli* β-galactosidase. The shortest fragment of MAT α2 that would translocate the fusion protein to the yeast nucleus corresponds to the thirteen amino-terminal residues. A consensus sequence was deduced from a comparison of this sequence with other nuclear protein sequences (Table 2). Subsequently, M. Hall (personal communication) has shown that there is a second nuclear localization determinant, which lies in the carboxy-terminal portion of the protein.

Certain MAT-α2/β-galactosidase hybrids exhibit a lethal phenotype when overproduced in the cell. This may result from the association of fusion proteins with the nuclear periphery, thus excluding other proteins from entering the nucleus. Mutant cells that do not exhibit the lethal phenotype could be used to identify those components in the nuclear envelope involved in the transport process.

YEAST RIBOSOMAL PROTEIN L3   Moreland et al (1985) showed that the first 21 amino acids of the ribosomal protein L3 are sufficient to translocate *E. coli* β-galactosidase to the yeast nucleus. β-Galactosidase-L3 fusion proteins lacking only the 14 amino acids from the carboxy terminus of L3 were localized to the nucleus. They were then assembled into 60S ribosomal subunits that were subsequently exported to the cytoplasm. However, this was only observed when a "bridge" peptide containing proline and glycine was inserted between the β-galactosidase and L3 moieties, which presumably allowed each to fold correctly and independently, thus exposing the nuclear location signal.

This system offers a potential means of detecting and characterizing mutants defective in the nuclear transport process, since cells exhibiting an imbalance in the synthesis of ribosomal components and hence protein synthesis could have mutations in the sequences responsible for transport processes.

INFLUENZA VIRUS NUCLEAR PROTEIN (NP)   By linking portions of the influenza nucleoprotein virus (NP) DNA to chimpanzee α-globin, Davey et al (1985) identified a sequence in NP that specifies accumulation in the nucleus. These authors demonstrated that all of the fusion proteins they produced had the ability to enter the nucleus, but only those bearing a particular sequence from the carboxy-terminal end of the NP accumulated significantly in the *Xenopus* oocyte nucleus (see Table 2). These experiments indicate that in NP the functions of entry into and accumulation in the nucleus are encoded by separate polypeptide domains. The conclusion drawn from these results (Davey et al 1985) is that the nuclear accumulation of this protein may require two signals, one that allows proteins to enter the nucleus more rapidly than would be predicted on the basis of size and another that allows its intranuclear accumulation.

POLYOMA VIRUS LARGE T ANTIGEN   The studies of the nuclear transport of polyoma virus large T antigen revealed that, like the yeast MAT α2 protein (see above), it contains two nuclear location signals (Richardson et al 1986). They are located near residues 192 and 282. The latter position is equivalent to that of the nuclear location sequence in SV40 large T antigen. Deletion of either sequence does not abolish nuclear transport, but deletion of both does. A short stretch of amino acids encoding the sequence in the region of residue 192 can transport chicken muscle pyruvate kinase to the cell nucleus. However, a short sequence of amino acids from the region centered around amino acid 282 cannot. This putative signal sequence differs from the SV40 prototype sequence at three amino acid positions (see Table 2). These amino acids have been mutated in SV40 large T antigen (Smith et al 1985). These mutations did not abolish transport into the nucleus, but some cells showed cytoplasmic as well as nuclear fluorescence. This indicates that, although a particular residue can be of crucial importance, neighboring amino acids also play an important role (see above). The investigation of the polyoma large T antigen nuclear location signal sequences supports this idea. A polyoma large T antigen from which the candidate sequence at residue 192 has been deleted but the sequence at residue 282 has been left intact is located in the nucleus. However, a small change in the amino acid sequence upstream of this site (282) can cause polyoma large T to be located both in the nucleus and cytoplasm.

Although the SV40 nuclear location signal sequence is a particularly well characterized example, studies of other proteins illustrate that we are not yet able to formulate general rules about the requirements for a sequence specifying nuclear localization.

ADENOVIRUS E1A   The adenovirus *E1A* gene enclodes a 289–amino acid protein and a 243–amino acid protein. The smaller protein differs from the larger protein by the internal deletion of 46 amino acids (residues 140–185). A sequence showing homology to the SV40 nuclear location sequence is located at residue 230 (Smith et al 1985) in the protein domain encoded by the second exon of the gene. Deletion of the carboxy-terminal region of the protein shows that this region is essential for rapid accumulation of the protein in the nucleus of monkey kidney cells (Krippl et al 1985). This finding is consistent with the idea that this sequence is a nuclear localization signal.

However, each major mRNA (13S and 12S) from the *E1A* gene produces two major polypeptides, which can be resolved on SDS polyacrylamide gels. A posttranslational modification is involved, because it has been shown (Richter et al 1985) that when the proteins are expressed in *E. coli*, they have the same electrophoretic mobilities as the lower molecular weight species seen in infected HeLa cells. When the proteins expressed in *E. coli* are injected into *Xenopus* oocytes, they undergo a secondary modification that causes them to electrophorese with the higher molecular weight species. Only this species is able to translocate to the oocyte nucleus, and even its rate of accumulation is slow. The secondary modification that causes this apparent increase in molecular weight of 2 to 4 kDa occurs in the cytoplasm of the oocyte and in in vitro translation reaction mixtures. However, the domain of the protein that undergoes this modification is the domain encoded by the first exon, not the domain that contains the sequence homologous to the SV40 nuclear location sequence.

These results suggest that the carboxy-terminal (second exon) domain encodes a critical determinant for rapid nuclear localization in somatic (mammalian) cells. The slow accumulation in *Xenopus* oocytes has been attributed to diffusion and retention by binding because the modification maps to the region of E1A shown to be important for interaction of the protein with nuclear factors. Thus the mechanism of nuclear accumulation of this protein may differ in oocytes and somatic cells, perhaps because of differences in the availability of specific factors that function in alternate pathways of nuclear localization.

SV40 STRUCTURAL PROTEINS   The SV40 viral capsid contains three major polypeptides, VP1, VP2, and VP3. It has been shown that the proportions of these three proteins in the nucleus in infected cells is the same as their proportion in the virus capsid. The ratio of the three proteins is constant in the nucleus despite fluctuating ratios in the cytoskeleton and the cytosol. Therefore mechanisms must exist that transport the correct proportions of these proteins from the cytoplasm to the nucleus for virus assembly (Lin et al 1984).

VP1 contains a nuclear localization signal in the 94 amino-terminal amino acids, and it does not require VP2 or VP3 for efficient nuclear transport (Wychowski et al 1985; J. Takahashi & H. Kasamatsu, personal communication). VP2 and VP3 can be transported into the nucleus in the absence of VP1 but less efficiently. The signals for their nuclear transport lie at the carboxy-terminal end of the protein.

Sequences homologous to the nuclear location sequence of SV40 large T antigen occur in the amino-terminal end of VP1 and at the carboxy-terminal end of VP2 and VP3 (Smith et al 1985). This is consistent with the experimental findings summarized above, but it leaves unsolved the interesting problem of how the relative proportions of these proteins in the nucleus are controlled when they all appear to contain related putative nuclear location sequences.

PROGRAMMED CHANGES IN THE SUBCELLULAR LOCATION OF NUCLEAR PROTEINS

Proteins that are normally associated with snRNA molecules to form snRNP particles are stockpiled in the cytoplasm of *Xenopus* oocytes (Zeller et al 1983). When snRNA is injected into the oocyte cytoplasm these proteins associate with the snRNA and are transferred to the nucleus. Thus the mature snRNP particle accumulates in the nucleus, but its constituent parts do not. This presumably involves a conformational change in, or secondary modification of, the snRNP that is characteristic of its mature (nuclear) form.

A similar mechanism may underlie the precisely timed accumulation of other proteins in the nuclei of developing *Xenopus* embryos. In early amphibian development the oocyte nucleus (germinal vesicle) accumulates a variety of proteins that are shed into the cytoplasm when the oocyte matures into an egg. As the total volume of all nuclei present in the embryo during cleavage and early blastula is smaller than the volume of the original oocyte nucleus, these nuclei cannot contain all of the proteins originally present in the oocyte nucleus.

Monoclonal antibodies against germinal vesicle proteins have been used to follow the fate of these oocyte nuclear proteins during *Xenopus* development. Dreyer et al (1981, 1982, 1983) found that there are two classes of nuclear antigens. The "early migrating antigens," which include nucleoplasmin, accumulate in nuclei upon pronucleus formation. The "late migrating antigens" appear to be excluded from the nucleus during cleavage, but they move into the nucleus later, each at a charcteristic developmental stage of the embryo. Comparable observations have been made in *Drosophila* embryos (Dequin et al 1984). In *Drosophila* the nucleus of the embryo undergoes thirteen synchronous divisions to produce a syncytial blastoderm. In later embryonic stages, DNA replication slows down and transcription begins.

In the experiments of Dequin et al the distribution of four antigens was followed during early developmental stages using monoclonal antibodies. The antigens were localized to the cytoplasm in early stages and showed distinct granular staining. Shortly before transcription was detected (although it may have been occurring earlier at a low level), a dramatic shift of these antigens into the nuclei of the syncytium was observed. The possible link between the onset of transcription and the movement of these antigens into the nucleus led Dequin et al to suggest that the antigens are stored in the cytoplasm for their later use in transcription-related processes.

These studies may indicate that the movement of many proteins into the nucleus is analogous to the migration of snRNP particles. However, other explanations for the behavior of these antigens have been proposed. Dreyer et al (1986) suggested that the explanation for the behavior of *Xenopus* oocyte nuclear antigens may be partly kinetic. They posited that the early-migrating class of antigens possesses more potent (or multiple) signal sequences and that the late-migrating class has less efficient signal sequences that cause only a slow nuclear accumulation. Consistent with this hypothesis is the finding that the early-migrating antigens are more concentrated in the oocyte nucleus than the late-migrating antigens. However, extension of the cell cycle length at the early blastula stage with cycloheximide and aphidicolin apparently does not allow the late-migrating antigens to accumulate further in the blastula cell nuclei.

Goldstein (1974) has suggested that changes in the selectivity of the nuclear envelope may occur during the cell cycle or during development that would account for differential uptake of proteins. Dreyer et al (1986) point out that several gradual changes in the selectivity of the nuclear envelope would be needed to account for their observations. They suggest that a causal link between selectivity and nuclear lamina composition is unlikely as all the nucleoplasmic proteins they studied are accumulated by the germinal vesicle, which lacks the lamins characteristic of somatic cell nuclei.

The sequential shift of specific proteins to the nuclei of cells in the embryo could also be brought about by covalent addition of a signal to the protein at defined times. However, in contrast to the adenovirus E1A protein example discussed above, two-dimensional gel analysis indicates that no structural change occurs in the late-migrating antigens (Dreyer & Hausen 1983). This finding agrees with that of Dabauvalle & Franke (1982), who showed that proteins translated in vitro can migrate into nuclei.

Another example of regulated nuclear localization is provided by the catalytic subunit of the cAMP-dependent protein kinase. This is a tetramer of two regulatory subunits and two catalytic subunits. Binding of cAMP to the regulatory subunits causes dissociation of the tetramer, thus activating the catalytic subunits, which then phosphorylate cellular proteins. Using indirect immunofluorescence, Nigg et al (1985a) recently demonstrated that the regulatory subunit of the type II protein kinase is associated with the Golgi complex (and with the centromeres in mitosis). When intracellular cAMP levels are raised by administration of forskolin, the distribution of the regulatory subunit is unchanged, but the catalytic subunit accumulates in the cell nucleus (Nigg et al 1985b). This effect is reversible, since the catalytic subunits rapidly reassociate with the regulatory subunits when the drug is removed.

There has been evidence for some time that such a nuclear translocation occurs, but the evidence was provided by cell fractionation studies, which have the inherent danger of artifacts due to redistribution (reviewed by Lohmann & Walter 1984). However, the evidence discussed above appears convincing.

This result raises the question of the topology of the enzyme with respect to the Golgi membrane. The regulatory subunit is a membrane-associated protein whose sequence is known (Takio et al 1982). The enzyme could be situated in the membrane such that the catalytic subunits face the cytosol. Their accumulation in the nucleus after dissociation from the regulatory subunits could therefore proceed via the cytosol. The amino acid sequence of the catalytic subunit, investigated by Shoji et al (1981), shows no obvious homology to the sequences known to translocate proteins to the cell nucleus (Table 2). However, the catalytic subunit is small enough to diffuse passively through nuclear pores ($M_r = 40,580$), so perhaps its accumulation can be accounted for by passive diffusion and selective binding.

Alternatively, if the catalytic subunits face the lumen of the Golgi, then their accumulation in the nucleus must involve their translocation across the inner nuclear membrane and would indicate the existence of an alternative (nonnuclear pore) route into the nucleus.

Interestingly, early responses to elevated cAMP levels in many tissues include stimulation of secretory processes, while later responses often include changes in nuclear activity.

IN VITRO PROTEIN TRANSPORT

An in vitro system in which isolated nuclei take up proteins from the surrounding medium is a prerequisite for detailed biochemical analysis of the transport process. There are a number of problems to be overcome in the establishment of such a system. Firstly, proteins are rapidly lost from isolated nuclei, even in buffer systems that closely match the ionic conditions of the cell (Paine et al 1983). Secondly, as pointed out by Lang & Peters (1984), the loss of a single pore complex from a nucleus would create a large hole in the nuclear envelope, which should cause a twofold increase in the flux of molecules into the nucleus.

An alternative approach that circumvents these problems is the use of cell-free systems derived from amphibian eggs (*Rana* or *Xenopus*). Egg homogenates can promote extensive nuclear swelling (Barry & Merriam 1972; Lohka & Masui 1983, 1984), apparently by causing protein accumulation. Similar systems are able to reconstitute nucleus-like structures from purified DNA (Forbes et al 1983). Newmeyer (1986) studied nucleoplasmin transport in this system and showed that ATP is required for transport. Dreyer et al (1986) demonstrated by immunofluorescent techniques that in this system the nuclei take up proteins of the "rapidly migrating" class (see above) from the egg homogenate but that those of the "late migrating" class are not accumulated.

A recent report (Burke & Gerace 1986) describes a cell-free system from CHO cells that assembles telophase-like nuclear envelopes around mitotic chromosomes. This system lacks many of the components that are stockpiled in large quantity in the amphibian egg for use in early development. Hence it is less complex biochemically and may be ideally suited to analysis and fractionation.

CONCLUSIONS AND PROSPECTS

A range of studies have clearly shown that the information that specifies accumulation of large proteins in the cell nucleus is encoded in a small region of each protein. The clearest example of this is provided by SV40 large T antigen, in which the sequence pro-lys-lys-lys-arg-lys-val is sufficient to specify nuclear accumulation. Studies of other viral proteins that contain homologous sequences and of other nonviral proteins have affirmed the significance of this sequence, but we cannot yet formulate reliable general rules about sequence requirements for nuclear localization. However, the SV40 large T antigen sequence currently provides the best model for identifying candidate sequences.

It is unclear why a single consensus sequence has not emerged, but there are several possible explanations. It appears unlikely that it can be explained by species specificity of the transport process, since there have been many examples of correct transport into nuclei of foreign species. However, there may be several different pathways within a cell. This view is supported by the many observations that small proteins can diffuse into nuclei, while large proteins cannot. Thus for small proteins (less than 60 kDa), selective accumulation may not require signal sequences for nuclear entry but may only require binding to another nuclear component. The binding site within the nucleus need not be insoluble (like the lamina or matrix)—it could be a large soluble protein. For example, the pentameric core fragments of nucleoplasmin are soluble and diffuse freely through either the nucleus or the cytoplasm, yet they are unable to cross the nuclear envelope in either direction. Therefore small diffusible molecules could be retained in the nucleus by interacting with nucleoplasmin or similar proteins. Such binding could also account for histone accumulation in oocyte nuclei or for accumulation of other nuclear proteins small enough to enter the nucleus by diffusion.

However, entry by diffusion is inadequate to account for the import of large nuclear proteins (greater than 60 kDa), some of which enter the nucleus much too rapidly to be accounted for by diffusion. This suggests that accumulation of large proteins may occur by selective entry. A discrete polypeptide domain ("tail") of nucleoplasmin has been shown to allow this protein to be selected for entry into the nucleus. Nucleoplasmin entry occurs through the center of the nuclear pore complex, as shown by studies using colloidal gold coated with the protein. These same experiments exclude models that rely on protein unfolding to explain the transport of large proteins through the nuclear membrane, and they establish that particles as large as 16 nm in diameter can enter the nucleus through pores that have resting channels with an effective diameter of only 9 nm.

It remains to be seen how the nuclear pore can increase the size of its channel to allow a selected group of large proteins to enter. One possibility is that it functions as a sphincter that opens in response to contact with an appropriate signal sequence. It will be important to determine whether there is also an active translocation mechanism that propels selected molecules through the pore.

We do not know if all large nuclear proteins enter the nucleus by the same mechanism or if there are different mechanisms for different groups of proteins. Nor do we know if recognition by the nuclear pore complex is direct or is mediated by binding to an intermediate "adaptor" molecule. Such an adaptor might recognize nuclear proteins in the cytoplasm and bind to them and serve as a vehicle for entry.

It is encouraging that approaches to each of these questions have been developed in the last five years; hopefully the next five years of research will provide the crucial answers.

ACKNOWLEDGMENTS
We are grateful to several colleagues for the communication of results prior to publication, to Carl Feldherr and the Rockefeller Press for permission to use Figure 4, and to Julian Blow and Stephanie Cascio for comments on the manuscript. We thank the Cancer Research Campaign for their generous support.

388 DINGWALL & LASKEY

Literature Cited

Barry, J. M., Merriam, R. W. 1972. Swelling of hen erythrocyte nuclei in cytoplasm from *Xenopus* eggs. *Exp. Cell Res.* 71:90–96

Bonner, W. M. 1975a. Protein migration into nuclei. I. Frog oocyte nuclei *in vivo* accumulate microinjected histones, allow entry to small proteins, and exclude large proteins. *J. Cell Biol.* 64:421–30

Bonner, W. M. 1975b. Protein migration into nuclei. II. Frog oocyte nuclei accumulate a class of microinjected oocyte nuclear proteins and exclude a class of microinjected oocyte cytoplasmic proteins. *J. Cell Biol.* 64:431–37

Bonner, W. M. 1978. Protein migration and accumulation in nuclei. In *The Cell Nucleus*, ed. H. Busch, Vol. 6, Part C, pp. 97–148. New York: Academic

Burke, B., Gerace, L. 1986. A cell free system to study reassembly of the nuclear envelope at the end of mitosis. *Cell* 44:639–52

Butel, J. S., Guentzel, M. J., Rapp, F. 1969. Variants of defective simian papovavirus 40 (PARA) characterised by cytoplasmic localisation of simian papovavirus 40 tumor antigen. *J. Virol.* 4:632–41

Century, T. J., Fenichel, I. R., Horowitz, S. B. 1970. The concentrations of water, sodium and potassium in the nucleus and cytoplasm of amphibian oocytes. *J. Cell Sci.* 7:5–13

Dabauvalle, M. C., Franke, W. W. 1982. Karyophilic proteins. Polypeptides synthesised *in vitro* accumulate in the nucleus on microinjection into the cytoplasm of amphibian oocytes. *Proc. Natl. Acad. Sci. USA* 79:5302–6

Davey, J., Dimmock, N. J., Colman, A. 1985. Identification of the sequence responsible for the nuclear accumulation of the influenza virus nucleoprotein in *Xenopus* oocytes. *Cell* 40:667–75

Dequin, R., Saumweber, H., Sedat, J. W. 1984. Proteins shifting from the cytoplasm into the nuclei during early embryogenesis of *Drosophila melanogaster*. *Dev. Biol.* 104:37–48

De Robertis, E. M., Longthorne, R., Gurdon, J. B. 1978. Intracellular migration of nuclear proteins in *Xenopus* oocytes. *Nature* 272:254–56

Dingwall, C., Burglin, T. R., Kearsey, S. E., Dilworth, S., Laskey, R. A. 1986. Sequence features of the nucleoplasmin tail region and evidence for a selective entry mechanism for transport into the cell nucleus. *Proc. Workshop on Nucleo-cytoplasmic Transport, Heidelberg, 1986*, ed. R. Peters, M. F. Trendelenburg. In press

Dingwall, C., Sharnick, S. V., Laskey, R. A. 1982. A polypeptide domain that specifies migration of nucleoplasmin into the nucleus. *Cell* 30:449–58

Dreyer, C., Hausen, P. 1983. Two dimensional gel analysis of the fate of oocyte nuclear proteins in the development of *Xenopus laevis*. *Dev. Biol.* 100:412–25

Dreyer, C., Scholz, E., Hausen, P. 1982. The fate of oocyte nuclear proteins during early development of *Xenopus laevis*. *Wilhelm Roux Arch. Entwicklungsmech. Org.* 191:228–33

Dreyer, C., Singar, H., Hausen, P. 1981. Tissue specific nuclear antigens in the germinal vesicle of *Xenopus laevis* oocytes. *Wilhelm Roux Arch. Entwicklungsmech. Org.* 190:197–207

Dreyer, C., Stick, R., Hausen, P. 1986. Uptake of oocyte nuclear proteins by nuclei of *Xenopus* embryos. See Dingwall et al 1986

Earnshaw, W. C., Honda, B. M., Laskey, R. A., Thomas, J. O. 1980. Assembly of nucleosomes: The reaction involving *X. laevis* nucleoplasmin. *Cell* 21:373–83

Feldherr, C. M., Cohen, R. J., Ogburn, J. A. 1983. Evidence for mediated protein uptake by amphibian oocyte nuclei. *J. Cell Biol.* 96:1486–90

Feldherr, C. M., Kallenbach, E., Schultz, N. 1984. Movement of a karyophilic protein through the nuclear pores of oocytes. *J. Cell Biol.* 99:2216–22

Feldherr, C. M., Ogburn, J. A. 1980. Mechanism for the selection of nuclear polypeptides in *Xenopus* oocytes. II. Two dimensional gel analysis. *J. Cell Biol.* 87:589–93

Feldherr, C. M., Pomerantz, J. 1978. Mechanism for the selection of nuclear polypeptides in *Xenopus* oocytes. *J. Cell Biol.* 78:168–75

Forbes, D. J., Kirschner, M. W., Newport, J. W. 1983. Spontaneous formation of nucleus like structures around bacteriophage DNA microinjected into *Xenopus* eggs. *Cell* 34:13–23

Frank, M., Horowitz, S. B. 1975. Nucleocytoplasmic transport and distribution of an amino acid, *in situ*. *J. Cell Sci.* 19:127–39

Franke, W. W., Scheer, U., Khrone, G., Jarasch, E. 1981. The nuclear envelope and the architecture of the nuclear periphery. *J. Cell Biol.* 91:39s–50s

Goldfarb, D. S., Gariepy, J., Schoolnik, G., Kornberg, R. D. 1986. Synthetic peptides as nuclear localisation signals. *Nature* In press

Goldstein, L. 1974. Movement of molecules

between nucleus and cytoplasm. In *The Cell Nucleus*, ed. H. Busch, 1 : 387–438. New York : Academic

Gurdon, J. B. 1970. Nuclear transplantation and the control of gene activity in animal development. *Proc. R. Soc. London Ser. B* 176 : 303–14

Hall, M. N., Hereford, L., Herskowitz, I. 1984. Targeting of *E. coli* β-galactosidase to the nucleus in yeast. *Cell* 36 : 1057–65

Horowitz, S. B. 1972. The permeability of the amphibian oocyte nucleus, *in situ*. *J. Cell Biol.* 56 : 609–25

Kalderon, D., Richardson, W. D., Markham, A. T., Smith, A. E. 1984a. Sequence requirements for nuclear location of simian virus 40 large T antigen. *Nature* 311 : 33–38

Kalderon, D., Roberts, B. L., Richardson, W. D., Smith, A. E. 1984b. A short amino acid sequence able to specify nuclear location. *Cell* 39 : 499–509

Krippl, B., Ferguson, B., Jones, N., Rosenberg, M., Wesphal, H. 1985. Mapping of functional domains in adenovirus E1A proteins. *Proc. Natl. Acad. Sci. USA* 82 : 7480–84

Lanford, R. E., Butel, J. S. 1980. Biochemical characterization of nuclear and cytoplasmic forms of SV40 tumor antigens encoded by parental and transport-defective mutant SV40-adenovirus T hybrid viruses. *Virology* 105 : 314–27

Lanford, R. E., Butel, J. S. 1982. Intracellular transport of SV40 large tumor antigen : A mutation which abolishes migration to the nucleus does not prevent association with the cell surface. *Virology* 119 : 169–84

Lanford, R. E., Butel, J. S. 1984. Construction and characterisation of an SV40 mutant defective in nuclear transport of T antigen. *Cell* 37 : 801–13

Lang, I., Peters, R. 1984. Nuclear envelope permeability : A sensitive indicator of pore complex integrity. In *Information and Energy Transduction in Biological Membranes*, pp. 377–86. New York : Liss

Laskey, R. A., Dingwall, C., Mills, A. D., Dilworth, S. M. 1985. Transport and assembly of nuclear proteins. In *Nuclear Envelope Structure and RNA Maturation, UCLA Symp. on Mol. Cell. Biol. New Ser. 26*, ed. E. A. Smuckler, G. A. Clawson. New York : Liss

Lewis, A. M. Jr., Rowe, W. P. 1971. Studies on nondefective adenovirus-simian virus 40 hybrid viruses. I. A newly characterised simian virus 40 antigen induced by the Ad2+ND virus. *J. Virol.* 7 : 189–97

Lin, W., Hata, T., Kasamatsu, H. 1984. Subcellular distribution of viral structural proteins during simian virus 40 infection.

J. Virol. 50 : 363–71

Lohka, M. J., Masui, Y. 1983. Formation *in vitro* of sperm pronuclei and mitotic chromosomes induced by amphibian ooplasmic components. *Science* 220 : 719–21

Lohka, M. J., Masui, Y. 1984. Role of cytosol and cytoplasmic particles in nuclear envelope assembly and sperm pronuclear formation in cell free preparations from amphibian eggs. *J. Cell Biol.* 98 : 1220–30

Lohman, S. M., Walter, U. 1984. Regulation of the cellular and subcellular concentrations and distribution of cyclic nucleotide-dependent protein kinases. In *Advances in Cyclic Nucleotide and Protein Phosphorylation Research*, ed. P. Greengard, G. A. Robinson, 18 : 63–117. New York : Raven

Merriam, R. W., Hill, R. J. 1976. The germinal vesicle nucleus of *Xenopus laevis* oocytes as a selective storage receptacle for proteins. *J. Cell Biol.* 69 : 659–68

Mills, A. D., Laskey, R. A., Black, P., De Robertis, E. M. 1980. An acidic protein which assembles nucleosomes *in vitro* is the most abundant protein in *Xenopus* oocyte nuclei. *J. Mol. Biol.* 139 : 561–68

Moreland, R. B., Nam, H. G., Hereford, L. M., Fried, H. M. 1985. Identification of a nuclear localisation signal of a yeast ribosomal protein. *Proc. Natl. Acad. Sci. USA* 82 : 6561–65

Munro, S., Pelham, H. R. B. 1984. Use of peptide tagging to detect proteins expressed from cloned genes : Deletion mapping functional domains of *Drosophila* hsp 70. *EMBO J.* 3 : 3087–93

Newmeyer, D. D., Lucocq, J. M., Bürglin, T. R., De Robertis, E. M. 1986. Assembly in vitro of nuclei active in nuclear protein transport : ATP is required for nucleoplasmin accumulation. *EMBO J.* 5 : 501–10

Nigg, E. A., Hilz, H., Eppenberger, H. M., Dutley, F. 1985b. Rapid and reversible translocation of the catalytic subunit of cAMP-dependent protein kinase type II from the Golgi complex to the nucleus. *EMBO J.* 4 : 2801–6

Nigg, E. A., Schafer, G., Hilz, H., Eppenberger, H. M. 1985a. Cyclic AMP-dependent protein kinase type II is associated with the Golgi complex and with centrosomes. *Cell* 41 : 1039–51

Paine, P. L. 1975. Nucleocytoplasmic movement of fluorescent tracers microinjected into living salivary gland cells. *J. Cell Biol.* 66 : 652–57

Paine, P. L., Austerberry, C. F., Desjarlais, L. J., Horowitz, S. B. 1983. Protein loss during nuclear isolation. *J. Cell Biol.* 97 : 1240–42

Paine, P. L., Feldherr, C. M. 1972. Nucleo-cytoplasmic exchange of macromolecules. *Exp. Cell Res.* 74:81–98

Paine, P. L., Moore, L. C., Horowitz, S. B. 1975. Nuclear envelope permeability. *Nature* 254:109–14

Paucha, E., Kalderon, D., Richardson, W. D., Harvey, R. W., Smith, A. E. 1985. The abnormal location of cytoplasmic large T is not caused by failure to bind DNA or to p53. *EMBO J.* 4:3235–40

Peters, R. 1984. Nucleo-cytoplasmic flux and intracellular mobility in single hepatocytes measured by fluorescence microphotolysis. *EMBO J.* 3:1831–36

Rapp, F., Paulizzi, S., Butel, J. S. 1969. Variation in properties of plaque progeny of PARA (defective simian papovavirus 40)—adenovirus 7. *J. Virol.* 4:626–31

Richardson, W. D., Roberts, B. L., Smith, A. E. 1986. Nuclear location signals in polyoma virus large T. *Cell* 44:77–85

Richter, J. D., Young, P., Jones, N. C., Krippl, B., Rosenberg, M., Ferguson, B. 1985. A first exon-encoded domain of E1A sufficient for post-translational modification, nuclear localisation, and induction of adenovirus E3 promoter expression in *Xenopus* oocytes. *Proc. Natl. Acad. Sci. USA* 82:8436–38

Rine, J., Barnes, E. 1985. Localization of proteins to the nucleus. *Bioessays* 2:158–61

Rose, G. D., Gierasch, L. M., Smith, J. A. 1985. Turns in peptides and proteins. *Adv. Protein Chem.* 37:1–109

Shoji, S., Parnalee, D. C., Wade, R. D., Kumar, S., Ericsson, L. H., et al. 1981. Complete amino acid sequence of the catalytic subunit of bovine cardiac muscle cyclic AMP-dependent protein kinase. *Proc. Natl. Acad. Sci. USA* 78:848–51

Silver, P. A., Keegan, L. P., Ptashne, M. 1984. Amino terminus of the yeast *gal4* gene product is sufficient for nuclear localisation. *Proc. Natl. Acad. Sci. USA* 81:5951–55

Smith, A. E., Kalderon, D., Roberts, B. L., Colledge, W. H., Edge, M., et al. 1985. The nuclear location signal. *Proc. R. Soc. London Ser. B* 226:43–58

Takio, K., Smith, S. B., Krebs, E. G., Walsh, K. A., Titani, K. 1982. Primary structure of the regulatory subunit of type II cAMP dependent protein kinase from bovine cardiac muscle. *Proc. Natl. Acad. Sci. USA* 79:2544–48

Unwin, P. N. T., Milligan, R. 1982. A large particle associated with the perimeter of the nuclear pore complex. *J. Cell Biol.* 93:63–75

Wychowski, C., van der Werf, S., Girard, M. 1985. Nuclear localisation of poliovirus capsid polypeptide VP1 expressed as a fusion protein with SV40–VP1. *Gene* 37:63–71

Zeller, R., Nyffenegger, T., De Robertis, E. M. 1983. Nucleocytoplasmic distribution of snRNPs and stockpiled snRNA binding proteins during oogenesis and early development in *Xenopus laevis*. *Cell* 32:425–34

Ann. Rev. Cell Biol. 1986. 2 : 391–419

G PROTEINS: A FAMILY OF SIGNAL TRANSDUCERS

Lubert Stryer

Department of Cell Biology, Stanford University School of Medicine, Stanford, California 94305

Henry R. Bourne

Department of Pharmacology, University of California, San Francisco, California 94143

CONTENTS

OVERVIEW

The G proteins are a family of signal-coupling proteins that play key roles in many hormonal and sensory transduction processes in eukaryotes.

391

0743–4634/86/1115–0391$02.00

G proteins carry signals from activated membrane receptors to effector enzymes and channels (Table 1). The hormone-regulated adenylate cyclase system is the original source of our knowledge of this family of proteins. Adenylate cyclase is reciprocally controlled by a stimulatory G protein (G_s) and an inhibitory G protein (G_i). For example, the binding of epinephrine to the β-adrenergic receptor leads to the activation of G_s and the consequent stimulation of adenylate cyclase. When G_i is activated by the binding of a hormone such as somatostatin to its specific receptor, it inhibits adenylate cyclase. The phosphoinositide cascade, which controls calcium-sensitive processes such as secretion and chemotaxis, is also regulated by G proteins. The likely target is phospholipase C, which hydrolyzes phosphatidylinositol 4,5-bisphosphate to inositol trisphosphate (which mobilizes calcium ion) and diacylglycerol (which activates protein kinase C). Channels, as well as enzymes, appear to be directly controlled by G proteins. Potassium channels in cardiac pacemaker cells are opened by a G protein that is activated by the muscarinic acetylcholine receptor, which slows the cells' rate of firing.

G proteins also play major roles in sensory processes. Transducin, a member of the G protein family, mediates visual transduction in the rod and cone cells of vertebrate retinas. Photoexcitation of rhodopsin triggers the activation of transducin, which then stimulates a cGMP phosphodiesterase. The highly amplified hydrolysis of cGMP results in the closure of cation-specific channels in the plasma membrane and the generation of a hyperpolarizing nerve impulse. Transducin also plays a key role in invertebrate vision. Olfactory sensory neurones are rich in an odorant-sensitive adenylate cyclase that is controlled by a protein akin to G_s, which suggests that olfaction too is mediated by a G protein.

G proteins have a common design. They consist of three polypeptides: a guanyl-nucleotide binding α chain (39–52 kDa), a β chain (35–36 kDa), and a γ chain (8 kDa). G proteins cycle between an inactive GDP state and an active GTP state (Figure 1). When GDP is bound, α associates with β and γ to form a $G_{\alpha\beta\gamma}$ complex (denoted by G-GDP) that is membrane-bound. When GTP is bound, the α chain (G_α-GTP) dissociates from the β and γ chains ($G_{\beta\gamma}$). An essential feature of G proteins is that their conversion from the GDP to the GTP state is slow in the absence of the excited receptor (e.g. a hormone-receptor complex or photoexcited rhodopsin). The role of the excited receptor is to catalyze the activation of the G protein by markedly accelerating the rate of exchange of GTP for bound GDP. G_α-GTP released from $G_{\beta\gamma}$ then alters the activity of the target enzyme or channel. The α chain also has a built-in hydrolytic activity that converts bound GTP to GDP to terminate activation. The role of $G_{\beta\gamma}$ is to present G_α-GDP to the activated receptor. All G proteins are activated

Table 1 Examples of physiological processes mediated by G proteins

Stimulus	Receptor	G protein[a]	Effector	Physiological response	Reference
Epinephrine	β-Adrenergic recep or	G_s	Adenylate cyclase	Glycogen breakdown	Ross & Gilman 1980
Serotonin	Seroton n receptor	G_s	Adenylate cyclase	Behavioral sensitization and learning in *Aplysia*	Siegelbaum et al 1982
Light	Rhodopsin	Transducin	cGMP phospho-diesterase	Visual excitation	Stryer 1986
IgE-antigen complexes	Mast cell IgE receptor	G_{PLC}	Phospholipase C	Secretion	Smith et al 1985; Nakamura & Ui 1985
f-Met peptide	Chemotactic receptor	G_{PLC}	Phospholipase C	Chemotaxis	Krause et al 1985
Acetylcholine	Muscarinic receptor	G_K	Potassium channel	Slowing of pacemaker activity	Pfaffinger et al 1985; Breitwieser & Szabo 1985

[a] G_{PLC} and G_K refer to as yet unidentified G proteins in these cascades.

by receptors that are integral membrane glycoproteins. These receptors convert an external signal into a conformational change that triggers G-protein activation on the cytosolic face of the plasma membrane.

The G-protein family exhibits many additional structural and functional similarities. Their β chains are nearly identical, whereas their γ chains show some differences. The $\beta\gamma$ subunits of G_s, G_i, G_o, and transducin are functionally interchangeable. Their respective α chains contain common regions (e.g. the site for GTP binding and hydrolysis) and distinctive regions (e.g. binding sites for receptor and effector proteins). Furthermore, the α chains of G proteins can be specifically ADP-ribosylated by bacterial toxins. G_s and transducin are targets for cholera toxin, whereas G_i, G_o, and transducin are targets for pertussis toxin. It seems likely that the G-protein family evolved by duplication and divergence of a common ancestral gene. Sequence homologies among the prokaryotic elongation factor Tu, the RAS proteins of yeast, and the mammalian G proteins suggest that the controlled binding and release of proteins coupled to GTP-GDP exchange and hydrolysis was perfected early in evolution and elaborated over several billion years. This elegant molecular device now serves a wide

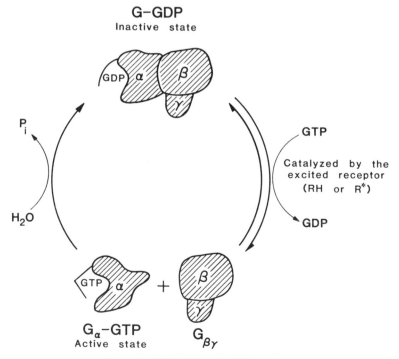

Figure 1   GTP-GDP cycle of G proteins.

variety of cellular functions, including translocation of macromolecules in protein synthesis, regulation of cell proliferation, and transduction of hormonal signals and sensory stimuli.

DISCOVERY OF THE G-PROTEIN FAMILY

The interplay of hormone and vision research led to the discovery of the G-protein family and the elucidation of a fundamental recurring mechanism of signal transduction. Four key experimental findings led to formulation of the concept of a GTP-GDP regulatory cycle (Figure 1): (1) GTP is required for hormonal stimulation of adenylate cyclase; (2) hydrolysis-resistant analogs of GTP produce persistent activation; (3) GTPase activity is stimulated in parallel with adenylate cyclase activity; and (4) the binding of agonists to hormone receptors stimulates GTP-GDP exchange (for reviews see Rodbell 1980; Ross & Gilman 1980; Schramm & Selinger 1984). The role of the activated receptor (the hormone-receptor complex) is to induce opening of the guanyl-nucleotide binding site of the regulatory component (G_s) to allow GDP release and GTP entry (Cassel & Selinger 1978). The subsequent hydrolysis of bound GTP terminates the activation of adenylate cyclase. Support for this postulated mechanism came from a study of the action of cholera toxin, which ADP-ribosylates G_s. This ADP-ribosylation slows down the hydrolysis of bound GTP and puts more G_s into the active GTP state, thus markedly enhancing adenylate cyclase activity (Cassel & Selinger 1977). Cholera toxin was also used to radiolabel G_s, which led to the identification of the G_s α chain (Cassel & Pfeuffer 1978; Johnson et al 1978). G_s has been purified by exploiting cultured S49 lymphosarcoma cells bearing cyc^-, a somatic mutation (Bourne et al 1975, 1982). Because cyc^- cells lack functional G_s, their membranes provide a bioassay for this G protein (Ross & Gilman 1977). G_s solubilized in detergent from membrane fractions of rabbit liver has been purified and characterized in detail (Sternweis et al 1981; Northup et al 1983a,b). Moreover, incorporation of purified β-adrenergic receptor and G_s into phospholipid vesicles reconstitutes hormone-stimulated GTP-GDP exchange and GTPase activities (Asano et al 1984). Incorporation of the pure catalytic unit (Smigel 1986) of adenylate cyclase (termed C) with the receptor and G_s reconstitutes hormone-sensitive synthesis of cAMP (May et al 1985); thus G_s alone suffices to transduce the hormonal signal from the receptor to the effector.

The G-protein story in photoreceptors began with the finding that retinal rod outer segments contain a light-activated cGMP phosphodiesterase (Bitensky et al 1975). The requirement of the phosphodiesterase for GTP and photoexcited rhodopsin (Yee & Liebman 1977), the presence of a light-activated GTPase activity (Wheeler & Bitensky 1977), and the occurrence of membrane-dependent GTP-GDP exchange in a soluble component (Godchaux & Zimmerman 1979) pointed to a G protein as a likely link in the activation sequence. It was then found that photoexcited rhodopsin (R*) catalyzes a highly amplified GTP-GDP exchange (Fung & Stryer 1980). A single R* triggers the binding of 500 molecules of GppNHp, a hydrolysis-resistant GTP analog. Transducin, the G protein of retinal rods, was purified and reconstituted with rhodopsin in phosphatidylcholine vesicles. This reconstituted system, like the native one, exhibited amplified GTP-GDP exchange (Fung et al 1981). Furthermore, the phosphodiesterase bound to dark membranes was activated by the addition of purified T_α-GppNHp. These experiments established that activated transducin is the first amplified intermediate in visual excitation and that T_α-GTP is the activator of the phosphodiesterase (Stryer et al 1981). The striking similarities between the structure and mechanism of action of transducin and G_s revealed that they are members of a family of signal-coupling proteins (Bitensky et al 1984; Stryer et al 1981; Gilman 1984) (Table 2).

Two more G proteins, G_i and G_o, were recently discovered. It was

Table 2   Functions and properties of purified G proteins

Functional and structural parameters	G_s	G_i	G_o	Transducin
Signal detector	β-Adrenergic receptor, glucagon receptor, and many others	Muscarinic receptor, opiate receptor, and many others	Unknown	Rhodopsin
Effector protein	Adenylate cyclase	Adenylate cyclase	Unknown	cGMP phospho-diesterase
Function	Stimulation	Inhibition	—	Stimulation
Subunit masses (kDa)				
α	45 and 52	41	39	39
β	35 and 36	35 and 36	35 and 36	36
γ	8	8	8	8
Toxin susceptibility	Cholera	Pertussis	Pertussis	Pertussis and cholera
Location	Nearly all cells	Nearly all cells	Brain	Retinal rod outer segments

known for some time that certain hormones, such as opiates, muscarinic agonists, and α_2-adrenergic amines, lower cAMP accumulation in intact cells and reduce adenylate cyclase activity in membrane preparations. These inhibitory agonists, like stimulatory agonists that act through G_s, also stimulated GTP hydrolysis in parallel with their inhibition of adenylate cyclase. Although a G protein was presumed to mediate inhibition by hormones, its identification as a protein distinct from G_s came from studies initially directed at a quite different goal, that of understanding the mechanism of action of the exotoxin of *Bordetella pertussis*, the bacterium that causes whooping cough (Ui et al 1984). Pertussis toxin (also called islet-activating protein) attenuates the inhibition of insulin release caused by the binding of epinephrine to α_2-adrenoreceptors. This toxin also blocks inhibition of adenylate cyclase by α_2-agonists and other inhibitory hormones by catalyzing the ADP-ribosylation of a 41-kDa polypeptide found in the plasma membrane (Katada & Ui 1982). This 41-kDa pertussis toxin substrate proved to be the α subunit of G_i, the inhibitory G protein, which is present in nearly all cells. The availability of a specific and convenient toxin labeling assay led rapidly to the purification and characterization of G_i from liver and erythrocyte plasma membranes (Codina et al 1983; Bokoch et al 1983, 1984).

Membrane fractions from bovine brain were subsequently found to contain a 39-kDa polypeptide substrate for pertussis toxin in addition to the 41-kDa substrate. The G proteins containing these α subunits were purified (Sternweis & Robishaw 1984; Neer et al 1984). The protein containing the 41-kDa α chain is likely to be the same as G_i, and the protein containing the 39-kDa α subunit has been noncommittally named G_o for G-other. Both G_i and G_o are abundant in brain; indeed, G_o comprises about 1% of the membrane protein. The roles of G_i, G_o, and other G proteins are discussed below.

HORMONAL CONTROL OF ADENYLATE CYCLASE

Stimulatory G Protein

G_s is a heterotrimer of α (45 or 52 kDa), β (35 or 36 kDa), and γ (~ 8 kDa) subunits. The 45 and 52 kDa forms of the α chain probably arise from alternative RNA splicing of a primary transcript encoded by a single gene (A. G. Gilman, personal communication). G_s from particulate extracts of rabbit liver has been solubilized in detergent and purified (Smigel et al 1984). Though detergent is needed to keep G_s in solution, less is bound to G_s than to solubilized integral membrane proteins. The α subunit of G_s is soluble in the absence of detergent (Sternweis 1986), which indicates that the $\beta\gamma$ subunit serves as the membrane anchor. Activation of detergent-solubilized G_s by GTP analogs or aluminofluoride ion (AlF_4^-) causes reversible dissociation of the $\beta\gamma$ subunit from the activated α subunit (Northup et al 1983a,b); addition of excess $\beta\gamma$ slows or reverses activation of the α subunit. The activated α subunit alone can stimulate cAMP synthesis by the catalytic unit (C) of adenylate cyclase (May et al 1985; Cerione et al 1984). Activation of adenylate cyclase by hormone-receptor complexes results from receptor-promoted binding of GTP to the α subunit of G_s, which is presumably followed by dissociation of the α and $\beta\gamma$ subunits (Codina et al 1984).

Inhibitory G Protein

Like G_s, G_i is a heterotrimer whose α subunit (41 kDa) dissociates from its $\beta\gamma$ subunit upon activation with GTP analogs or AIF_4^-. Purified G_i has been used to reconstitute GTP-dependent inhibition of adenylate cyclase activity by α_2-agonists in platelet membranes in which endogenous G_i has been inactivated by pertussis toxin (Katada et al 1984a,b,c). The results of addition of resolved α and $\beta\gamma$ subunits of G_i to platelet membranes led to the conclusion that inhibition of adenylate cyclase by G_i is mediated by the $\beta\gamma$ subunit. It was suggested that the dissociated $\beta\gamma$ subunit of G_i inactivates $G_{s\alpha}$-GTP by complexing it (Katada et al 1984a,b). However, this mechanism of inhibition does not account for the finding that GTP analogs and somatostatin plus GTP inhibit basal and forskolin-stimulated adenylate cyclase activity in cyc$^-$ membranes (Aktories et al 1983; Hildebrandt et al 1982, 1983). It is hard to imagine how the $\beta\gamma$ subunit of G_i could reduce adenylate cyclase activity by binding to $G_{s\alpha}$ in cyc$^-$, a mutant devoid of detectable $G_{s\alpha}$ mRNA (Harris et al 1985). In fact, recent studies of a reconstituted system show that $G_{i\alpha}$ bound to a hydrolysis-resistant GTP analog inhibits activation of resolved adenylate cyclase (Katada et al 1986). Moreover, the $\beta\gamma$ subunit of G_i appears to directly inhibit adenylate cyclase. Thus, the inhibitory action of G_i-GTP is mediated by both its α and $\beta\gamma$ subunits.

Diseases Produced by Altered G Proteins

A dominantly inherited mutation affecting G_s produces pseudohypoparathyroidism type Ia (PHP-Ia) (for reviews see Van Dop & Bourne 1983; Spiegel et al 1985). G_s activity, measured by in vitro complementation of the cyc$^-$ defect, is reduced by about 50% in erythrocytes, skin fibroblasts, lymphoblasts, and renal cells of PHP-Ia patients. These findings, in accord with the autosomal pattern of inheritance, suggest that PHP-Ia patients inherit one normal and one defective allele for a polypeptide component of G_s. Because the cyc$^-$ complementation assay measures the activity of $G_{s\alpha}$, the mutation is most likely in the $G_{s\alpha}$ gene. The endocrine defect in these patients is general: it causes reduced responsiveness to parathyroid hormone and many other hormones that act by stimulating adenylate cyclase. Hence, it seems likely that the α subunit of G_s is encoded by the same gene in most, if not all, endocrine target cells.

Altered G proteins are involved in the pathogenesis of at least two infectious diseases. The sometimes lethal secretion of salt and water into the gut of cholera patients results from cholera toxin–catalyzed ADP-ribosylation of G_s and the resulting increase in cAMP in intestinal mucosal cells (Fishman 1980). A heat-labile enterotoxin that also catalyzes ADP-ribosylation accounts for many cases of "traveler's diarrhea." It is evident that pertussis toxin plays a major role in the pathogenesis of whooping cough (Steinman et al 1985). However, neither the target cells in respiratory epithelium nor the critical molecular substrate (which could be G_i, G_o, or another G protein) has been identified.

VISUAL EXCITATION

Transducin Cycle in Vertebrate Photoreceptors

Retinal rod cells are single-photon detectors. The cis-trans isomerization of retinal in a single molecule of rhodopsin leads to the closure of hundreds of cation-specific channels in the plasma membrane. The influx of more than 10^6 Na^+ is blocked by a single photon, which results in a hyperpolarization that is conveyed to the synapse. A number of recent experiments have established that cGMP, not calcium ion, is the transmitter that carries the excitation signal from the disks to the plasma membrane (for reviews see Stryer 1986; Yau et al 1986). The flow of information in visual excitation is now known. Photoexcited rhodopsin (R*), formed by photoisomerization of rhodopsin (R), catalyzes GTP-GDP exchange in transducin. T_α-GTP, the activated form of transducin, then stimulates a cGMP phosphodiesterase (PDE*), which rapidly hydrolyzes cGMP to close sodium channels in the plasma membrane. Rhodopsin, a 40-kDa transmembrane glycoprotein (Dratz & Hargrave 1983), is located in the disks and plasma membrane of rod outer segments (ROS). Transducin and PDE are peripheral membrane proteins located on the cytosolic faces of disks. Transducin consists of α (39 kDa), β (36 kDa), and γ (8 kDa) chains (Kühn 1980; Fung et al 1981). PDE is made up of α (88 kDa), β (85 kDa), and γ (11 kDa) subunits (Baehr et al 1979). The rod outer segment is extremely rich in these proteins: The concentration of rhodopsin in ROS is 3 mM, which corresponds to a surface density of 27,000 μm^{-2}. The molar ratio of rhodopsin to transducin to PDE is 100:10:1.

The rhodopsin content of ROS is orders of magnitude higher than that of receptors for G proteins in other cells, and transducin is present at a much higher concentration than are the other G proteins. Transducin can be eluted from discs without using detergent, unlike other G proteins. Transducin can readily be prepared by taking advantage of its tight binding to R* in bleached ROS in the absence of GTP and its release from disc membranes by addition of GTP (Kühn 1980). These features make transducin an especially attractive target for biochemical studies.

The light-triggered amplification cycle involving rhodopsin, transducin, and the phosphodiesterase (Fung et al 1981; Stryer 1986) is shown in Figure 2. The interactions and mechanisms of this cycle are likely to be common to the action of most G proteins. The major features of this cycle are:

1. The cycle begins with the photoisomerization of the 11-*cis* retinal chromophore of rhodopsin to the all-*trans* form.

2. Photoexcited rhodopsin, in contrast with rhodopsin, has affinity for transducin. R* encounters T-GDP by lateral diffusion in the plane of the disk membrane and forms an R*-T-GDP complex.

3. GTP is then exchanged for bound GDP. This exchange proceeds through an R*-T intermediate devoid of a bound nucleotide. The role of R* is to open the guanyl-nucleotide binding site of transducin so that GDP can leave and GTP can enter.

4. The exchange of GTP for GDP markedly lowers the affinity of R* for transducin. Moreover, the entry of GTP greatly diminishes the affinity of the α and $\beta\gamma$ subunits of transducin for each other. These changes in binding affinities are essential for the efficient and rapid operation of the cascade. The released R* is free to interact with another T-GDP, which enables R* to act catalytically. T_α-GTP is liberated so that it can carry the excitation signal to PDE. About 500 molecules of T_α-GTP are formed per R*.

5. T_α-GTP then activates PDE, which very rapidly hydrolyzes cGMP. Phosphodiesterase activity is very low in the dark because the catalytic activity of its $\alpha\beta$ subunit is inhibited by its γ subunit. T_α-GTP activates PDE more than 100-fold by overcoming this inhibitory constraint. The kinetics of activation indicate that T_α-GTP binds rapidly to the holoenzyme, but it is uncertain whether T_α-GTP-PDE is the active enzyme or whether T_α-GTP displaces the inhibitory subunit of the PDE or carries it away from the $\alpha\beta$ subunits of the PDE to activate the enzyme.

6. The GTPase activity inherent in the α subunit of transducin converts

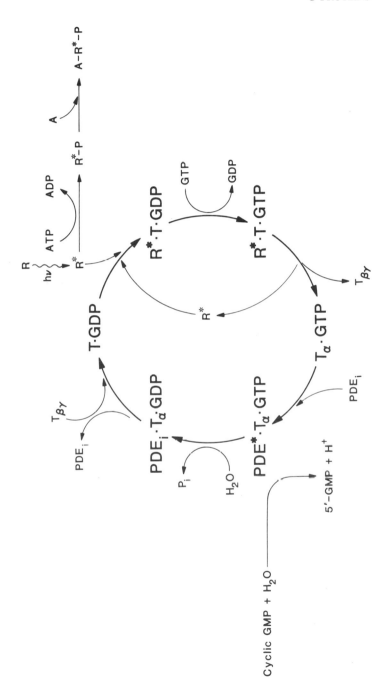

Figure 2 Light-triggered transducin cycle of vertebrate photoreceptors. (Based on Stryer et al 1981.) Abbreviation: A, arrestin; PDE$_i$ and PDE*, inhibited and activated forms of cGMP phosphodiesterase; R and R*, unexcited and photoexcited rhodopsin; T, transducin.

T_α-GTP to T_α-GDP, which rejoins $T_{\beta\gamma}$ to form T-GDP and inactive PDE. Hydrolysis, which takes less than a minute, is necessary for deactivation of the cascade. Reconstitution experiments using the purified subunits of transducin have shown that T_α-GDP alone does not bind to photolyzed or unphotolyzed disk membranes, nor does it undergo GTP-GDP exchange in the presence of R* (Fung 1983). Likewise, $T_{\beta\gamma}$ alone does not interact with R*. Hence, the role of $T_{\beta\gamma}$ is to present T_α-GDP to R*.

7. The return to the dark state also requires that R* be deactivated so that it does not continue to catalyze the activation of transducin. ATP is essential for rapid quenching of light-activated PDE (Liebman & Pugh 1980; Sitaramayya & Liebman 1983). Deactivation is accomplished by the sequential action of two cytosolic proteins, rhodopsin kinase (68 kDa) and arrestin (48 kDa) (Kühn 1980; Stryer 1986). Rhodopsin kinase binds to the cytosolic face of R* (but not to R) and catalyzes the phosphorylation of some nine serine and threonine residues in the carboxy-terminal tail. Arrestin (also known as the 48K protein) binds to phosphorylated R* but not to unphosphorylated R* nor to phosphorylated rhodopsin. The key point is that arrestin competes with transducin for binding to R*. Thus, arrestin acts as an inhibitory cap on phosphorylated R*, blocking its capacity to catalyze the activation of transducin (Wilden et al 1986).

Nucleotide exchange experiments have shown that the transducin cycle flows unidirectionally once triggered by R* (Yamanaka & Stryer 1987). A molecule of GTP that becomes bound to transducin is not released until it is hydrolyzed to GDP. The equilibrium for the reaction

$$\text{T-GDP} + \text{GTP} \rightleftharpoons T_\alpha\text{-GTP} + T_{\beta\gamma} + \text{GDP}$$

lies far to the right. Likewise, the hydrolysis of T_α-GTP is essentially irreversible. The ratio of T-GDP to T_α-GTP in the dark is greater than 1000. Thus, the phosphoryl potential of GTP is nicely partitioned to make both the formation and breakdown of T_α-GTP irreversible so that the level of this activated intermediate can be controlled kinetically. In the absence of R*, T-GDP undergoes GTP-GDP exchange very slowly, in times of hours, because there is a large activation barrier to be overcome in going to the state in which the nucleotide binding site is open for GDP release and GTP entry. R* catalyzes GTP-GDP exchange by stabilizing this transition state by some 10 kcal/mol. R* has high affinity for T devoid of a bound nucleotide but not for T-GDP or T-GTP. In other words, R*, like other effective catalysts, is complementary to the transition state but not to the substrate or product of the reaction. It is evident that the free energy profile of the transducin cycle has been optimized for (*a*) the rapid formation of the active T-GTP form, (*b*) a high degree of amplification, (*c*) effective operation over a very wide range of stimulus intensity (R*), and (*d*) maintenance of a very low level of T-GTP in the dark so that PDE is highly inhibited.

Transducin cDNA Sequences

The cloning of cDNAs for all three chains of transducin has shown that they are encoded by separate mRNAs. They are devoid of signal sequences and long stretches of hydrophobic amino acids, in accord with the fact that transducin is a peripheral membrane protein that cycles between membrane-bound and soluble forms. The analysis of cDNAs for the α chain led to the unexpected finding of two different but related sequences, now called $T_{\alpha 1}$ (Medynski et al 1985; Tanabe et al 1985; Yatsunami & Khorana 1985) and $T_{\alpha 2}$ (Lochrie et al 1985). These α chains contain 350 and 354 amino acids, respectively, are 80% identical, and contain extensive stretches of completely conserved residues. Staining by antibodies specific for each of these transducins has revealed that $T_{\alpha 1}$ is present in rod and $T_{\alpha 2}$ in cone outer segments (J. Hurley, personal communication). The phosphodiesterase of cones is similar but not identical to that of rods, as judged by immunocytochemical criteria (Hurwitz et al 1985). Likewise, cDNA cloning studies have shown that the transducin binding region of cone visual pigments is highly homologous to that of rhodopsin (Nathans et al 1986). Hence, it seems likely that the cGMP cascade of cones closely resembles that of rods.

cDNAs for the β subunit of transducin have also been cloned and sequenced (Sugimoto et al 1985). This 340-residue 37-kDa polypeptide is highly acidic and moderately hydrophilic. The amino acid sequence of bovine T_γ has been determined from cyanogen bromide fragments of the protein (Ovchinnikov et al 1985) and from the analysis of cDNAs (Van Dop et al 1984a; Hurley et al 1984; Yatsunami et al 1985). This 73-residue 8.4-kDa protein is very hydrophilic and acidic, with 19 acidic and 11 basic residues. An interesting feature of the sequence is the presence of a 44% homology between nucleotides 107–225 of T_γ cDNA and exon 4 of human *c-Ha-ras1* proto-oncogene (Hurley et al 1984). T_γ, like T_α and RAS, contains a cysteine four residues from the carboxy terminus.

G Proteins in Invertebrate Vision

The photoreceptor cells of vertebrates, molluscs, and arthropods are anatomically different. However, it is now evident that the photoreceptor proteins of these phyla are similar and that a transducin-like protein plays a key role in phototransduction in molluscs and arthropods too. *Drosophila* rhodopsin (O'Tousa et al 1985; Zucker et al 1985) has the same architectural plan as mammalian rhodopsin (Dratz & Hargrave 1983). Photoexcited rhodopsin from invertebrates can trigger the activation of vertebrate transducin (Ebrey et al 1980; Saibil & Michel-Villaz 1984). Furthermore, cholera and pertussis toxins ADP-ribosylate transducin-like proteins in invertebrates (Vandenberg & Montal 1984; Malbon et al 1984). Antibody specific for the β subunit of bovine transducin cross-reacts with a protein of the same size in squid photoreceptors (Tsuda et al 1986). The introduction of a hydrolysis-resistant GTP analog into the eyes of arthropods mimics the action of light (Fein & Corson 1981; Bolsover & Brown 1982). The kinetics of light-activated GTPase activity of fly eye membranes parallels that of the prolonged depolarizing afterpotential of photoreceptor cells, which indicates that a G protein directly participates in phototransduction (Blumenfeld et al 1985). In invertebrate photoreceptors, G proteins control phospholipase C rather than a cGMP phosphodiesterase. Phosphoinositides play a major role in phototransduction in arthropods (Fein et al 1984; Brown et al 1984).

RECENTLY DISCOVERED ACTIONS OF G PROTEINS

Criteria for G Protein Action in Transduction

G proteins transduce a wide array of extracellular signals in addition to those that utilize cAMP and cGMP as intracellular messenger molecules. Several criteria can be employed to determine whether a G protein participates in a transduction process: (*a*) GTP and hydrolysis-resistant analogs of GTP weaken the binding of hormones to receptors coupled to G proteins by promoting release of the G protein from the hormone-receptor complex (Lefkowitz et al 1983). (*b*) Binding of GTP analogs and GTP hydrolysis are stimulated by the hormone (or other extracellular signal, such as light or a chemoattractant). (*c*) The activity of the putative effector is altered on addition of hormone and GTP. (*d*) The effect of the hormone is altered by cholera toxin or pertussis toxin. (*e*) The effects of the hormone are mimicked by the introduction of GTPγS into cytosol. (*f*) Functional reconstitution provides the most direct evidence for the participation of a G protein in a transduction process. This stringent criterion, which has been met for adenylate cyclase and the cGMP phosphodiesterase, has not yet been fulfilled for the transduction processes described in this section.

Control of the Phosphoinositide Cascade

The hydrolysis of phosphatidylinositol 4,5-bisphosphate (PIP$_2$) by a specific membrane-bound phosphodiesterase (called phospholipase C or polyphosphoinositide phosphodiesterase) generates two intracellular signal molecules, diacylglycerol and inositol 1,4,5-trisphosphate (IP_3) (Nishizuka 1984; Berridge & Irvine 1984). Diacylglycerol activates protein kinase C, whereas IP_3 triggers the release of intracellular calcium ion into the cytosol. Recent evidence that G proteins carry the excitation signal from stimulated cell-surface receptors to phospholipase C in neutrophils, mast cells, and platelets is summarized in Table 3. In neutrophils and mast cells, pertussis toxin–catalyzed ADP-ribosylation of a 41-kDa membrane protein attenuates the ligand-stimulated mobilization of Ca^{2+} and the secretory response. This 41-kDa polypeptide could be the α subunit of G_i or a related G protein. Signals that stimulate secretion in these cell types do not appear to act by inhibiting adenylate cyclase. In platelet membranes, pertussis toxin modification of a 41-kDa protein attenuates inhibition of adenylate cyclase by epinephrine and thrombin (Jakobs et al 1985). The failure of pertussis toxin to ADP-ribosylate the polypeptide in intact platelets, however, has prevented testing of the hypothesis that Ca^{2+} mobilization and secretion are mediated by the pertussis toxin substrate. A feedback inhibition of G_i by protein kinase C is suggested by two observations: (1) phorbol esters inhibit epinephrine- and thrombin-induced inhibition of platelet adenylate cyclase (Jakobs et al 1985), and (2) protein kinase C can phosphorylate $G_{i\alpha}$ (Katada et al 1985).

Gating of Potassium Channels

The binding of agonists to cardiac muscarinic acetylcholine receptors (mAchR) leads to an increase in K^+ permeability, which in turn slows pacemaker activity. Table 3 summarizes evidence that this signal is transduced by a pertussis toxin–sensitive G protein. The change in K^+ permeability caused by muscarinic agonists is mediated neither by changes in cAMP or cGMP, nor by activation of phospholipase C. Muscarinic agonists do activate phospholipase C in cardiac cells, but by a pertussis toxin–insensitive mechanism (Masters et al 1985). Furthermore, patch-clamp studies have shown that the increase in K^+ permeability does not depend on diffusible cytosolic messengers. These results suggest that G_i or G_o (or both) carries the signal from the activated muscarinic receptor to the potassium channel without the intervention of cyclic nucleotides or soluble kinases (Pfaffinger et al 1985; Breitwieser & Szabo 1985). Voltage-dependent calcium channels are also regulated by G proteins (Holz et al 1986). These studies raise the possibility that channels are directly gated by G proteins.

Roles of G_o, G_i, and G_x

The signal-transducing role of G_o has not yet been elucidated, nor is the role of G_i in brain known. G_i is more rapidly ADP-ribosylated by pertussis

Table 3 Recently discovered action of G proteins[a]

Cell type	Response	Agonist	Effector	Criteria for G protein[b]	PT substrate for response (kDa)	Reference
Neutrophil	Secretion, chemotaxis	Chemoattractants	PLC	a, b, c, d	41	Krause et al 1985; Goldman et al 1986; Ohta et al 1985; Lad et al 1985
Mast cell	Secretion	48/80, antigen	PLC	c, d, e	41	Nakamura & Ui 1984, 1985; Fernandez et al 1984
Platelet	Secretion	Thrombin, epinephrine, ADP	PLC	a, c	41	Haslam & Davidson 1984
Liver	Glycogenolysis	Vasopressin	PLC	a, b, c	none	Pobiner et al 1985
Astrocytoma	cAMP hydrolysis	Muscarinic	PLC	a, c	none	Evans et al 1985
Pituitary	Prolactin	TRH	PLC	a, c	none	Schlegel et al 1985; Martin et al 1986; Yajima et al 1986
Heart	K⁺ permeability	Muscarinic	K⁺ channel	a, c, d, e	41, 39	Pfaffinger et al 1985; Breitwieser & Szabo 1985

[a] Abbreviations used: PLC, phospholipase C; PT, pertussis toxin; TRH, thyrotropin releasing hormone.
[b] See text for criteria of G protein action.

toxin than is G_o, and G_o hydrolyzes GTP more rapidly than does G_i (Neer et al 1984). Cloning of the cDNAs for the α chains of these proteins established that they are different, though highly homologous (Itoh et al 1986). Both G_i and G_o from bovine brain interact with muscarinic receptors to reconstitute high-affinity agonist binding that is sensitive to GTP (Florio & Sternweis 1985). This experiment suggests that G_i and G_o may interact with muscarinic receptors in vivo. The distribution of G_o in brain tissue (as measured by an anti-α_{39} antiserum) corresponds closely to that of protein kinase C (as assayed by the binding of phorbol ester) but differs from that of adenylate cyclase (as detected by the binding of forskolin) (Worley et al 1986). This result suggests that G_o may be a link between receptors and phospholipase C in the phosphoinositide cascade.

In at least three well-documented cases (Table 3), cell-surface receptors stimulate phospholipase C through a G protein, here called G_x, that does not stimulate adenylate cyclase and is insensitive to pertussis toxin. The possibility that G_x is actually G_i or G_o that is somehow impervious to the action of pertussis toxin cannot be excluded at this time. Recent studies of a pituitary cell line (Schlegel et al 1985; Martin et al 1986; Yajima et al 1986), however, make this appear unlikely. Thyrotropin releasing hormone (TRH) acts through a presumed G_x in these cells to stimulate phospholipase C, Ca^{2+} mobilization, and secretion of prolactin. In these cells, somatostatin and muscarinic agonists act on specific receptors to *inhibit* TRH-stimulated prolactin secretion; this inhibition is attenuated by pertussis toxin, which acts on a 41-kDa substrate. The simplest interpretation of these data is that somatostatin and muscarinic agonists act through a G_i-like protein whose activation inhibits phospholipase C or a subsequent component of its cascade; the α subunit of this protein may be distinct from that of the G_i protein that inhibits adenylate cyclase.

Olfaction

Sensory cilia from frog olfactory neuroepithelium contain 15 times more adenylate cyclase activity than do membranes from brain or whole olfactory epithelium (Pace et al 1985). GTP analogs or a mixture of odorant molecules plus GTP enhanced adenylate cyclase activity in cilial membranes. Cholera toxin labeled a 42-kDa protein that was highly enriched in the cilia compared with membranes derived from deciliated epithelium. These findings implicate G_s or a G_s-like protein in olfactory signal transduction. An abundant 95-kDa transmembrane glycoprotein specifically found in olfactory ciliary membranes may be the odorant receptor that interacts with the G protein to trigger a cascade resembling that of visual transduction (Chen et al 1986).

RECURRING MOTIFS IN G PROTEINS

Structural Similarities

The α chains of G proteins exhibit a high degree of homology, which indicates that they are variations on a common theme. Recent sequencing of cDNAs for the α chains of rod and cone transducins (Medynski et al 1985; Tanabe et al 1985; Yatsunami & Khorana, 1985; Lochrie et al 1985) and of G_s, G_i, and G_o (Robishaw et al 1986; Nukada et al 1986; Itoh et al 1986) provided a wealth of information that can be integrated with the results of previous chemical, immunological, and functional studies. The degree of sequence identity between various pairs of α chains is plotted as a function of residue number in Figure 3. It is evident that some regions are highly conserved, whereas others are divergent. The degree of homology, if one allows for conservative substitutions, is about 65% among transducin, G_o, and G_i, compared with about 45% between G_s and one of the other G-proteins (Itoh et al 1986). G_s, which contains several extra sequences, is the most divergent member of this group.

Functions can be ascribed to several regions of the α chains. The regions implicated in GTP-GDP binding and in interactions with the βγ subunit, the effector protein, and the excited receptor are marked in Figure 3, as are the sites modified by cholera toxin and pertussis toxin. The amino-terminal region is required for the binding of βγ because this interaction is lost following proteolytic removal of the first 18 residues (Fung 1983; Fung & Nash 1983). An 18-residue sequence beginning at position 43 is identical in all five chains. This conserved block contains the GXGXXGK sequence that is generally found in GTP-binding proteins (Halliday 1984). In the bacterial elongation factor Tu (EF-Tu), this lysine interacts with a phosphate group of GDP (Jurnak 1985) and in RAS, substitution of the first glycine results in decreased GTPase activity and oncogenesis (Gibbs et al 1984; McGrath et al 1984). GTP-binding proteins also contain the sequence NKXD, which begins at residue 290 in G_s. In EF-Tu, asparagine is located over the guanine ring of GDP, and aspartate interacts with the amino group of guanine. The other two GTP contact regions marked in Figure 3 are assigned by homology with EF-Tu and RAS.

The region between the first and second GTP contact sites in the G proteins corresponds to the aminoacyl tRNA binding region in EF-Tu (Laursen et al 1981). Because GTP regulates the affinity of EF-Tu for aminoacyl tRNA (Kaziro 1978), it is plausible that the analogous segment of the G-protein α chains interacts with both the effector and the βγ subunit in a GTP-dependent manner. Specifically, βγ may bind to the highly conserved region (residues 153–204 of $G_{s\alpha}$) preceding the second GTP-

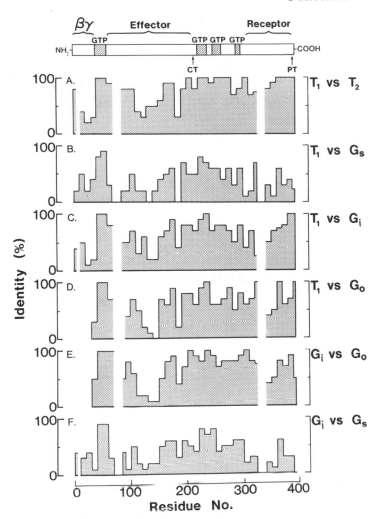

Figure 3 Comparison of the amino acid sequences of the α subunits of G proteins. The percent sequence identity averaged over sequential blocks of ten residues is plotted versus residue number for various pairs of G proteins. The x-axis refers to residue number in G_s, which is longer than the others, hence the gaps in the plots. Positions in the α chain (*top*) ADP-ribosylated by cholera and pertussis toxins are indicated by CT and PT, respectively.

binding region; this speculation is not indicated in Figure 3. In contrast, the highly variable sequence (residues 71–152 of $G_{s\alpha}$) in the stretch between the first and second GTP-binding sites is a good candidate for the effector binding site (e.g. the site for binding of adenylate cyclase or phospho-diesterase).

A region near the carboxy terminus probably participates in the binding of the excited receptor. ADP-ribosylation by pertussis toxin of the cysteine residue near the carboxyl terminus blocks the binding of photoexcited rhodopsin or of the hormone-receptor complex (Van Dop et al 1984b; Bokoch et al 1983). Furthermore, the carboxy-terminal region of transducin is homologous to that of arrestin, the 48-kDa protein that binds phosphorylated, photoexcited rhodopsin (Wistow et al 1986). A common feature of transducin and arrestin is their recognition of photoexcited rhodopsin. The reactivities of G_s, G_i, and transducin with the β-adrenergic receptor and photoexcited rhodopsin correlate with their carboxy-terminal sequences. G_s interacts much more effectively with the activated β-adren-ergic receptor than with photoexcited rhodopsin, whereas the reverse is true for transducin and G_i (Kanaho et al 1984; Cerione et al 1985). Correspondingly, of the 25 carboxy-terminal residues, 88% are identical in the α chains of G_i and transducin, compared with only 30% for G_s and rod transducin.

Antibodies prepared against purified subunits of G proteins and synthetic peptides corresponding to defined regions of these proteins have been used to probe homologies among members of this family (Pines et al 1985; Roof et al 1985; Mumby et al 1986). The 36-kDa β chains of G_s, G_i, and G_o, and the β chain of transducin are immunologically indistinguishable, which indicates that these subunits are very similar or identical. The 35-kDa β chains of G_s, G_i, and G_o, however, do not cross-react with antibody specific for the 36 kDa subunit. A second finding is that antibody specific for the γ subunit of transducin is not recognized by the other γ subunits. Furthermore, bovine retinal T_γ cDNA detects transcripts in retina but not in heart, liver, or brain (Van Dop et al 1984a). These findings suggest that the γ subunits of G_s, G_i, and G_o differ from that of transducin, despite their similarity in molecular weight. The extent of cross-reactivity of the α subunits depends on the antibody used. An antibody prepared against a synthetic peptide sequence common to the α subunits of G_o, G_i, transducin, and (with one amino acid substitution) to G_s reacts with all four proteins. In contrast, antibodies specific for particular α chains can be prepared using synthetic peptides that correspond to distinctive sequences. By immunological criteria, transducin appears to be more similar to G_i and G_o than to G_s, which agrees with their degree of sequence homology. The fact that G_i interacts with photoexcited rhodopsin (Tsai et al 1984; Cerione et al 1985) adds to the similarity of transducin and G_i and suggests that the contact sites of rhodopsin and the inhibitory hormone receptor for G proteins are structurally similar.

Mechanistic Similarities

We have discussed in detail the motifs of receptor-triggered GTP-GDP exchange, altered affinities of G-protein subunits for each other and for their receptors and effectors, and GTPase activity to restore the unstimulated state. The effects of cholera toxin and pertussis toxin provide further evidence for the kinship of the G-protein family. Cholera toxin ADP-ribosylates the α subunit of G_s (Cassel & Pfeuffer 1978; Johnson et al 1978) and transducin (Abood et al 1982; Navon & Fung 1984). Light and GppNHp are needed for modification of transducin by cholera toxin, just as hormone and GppNHp favor the labeling of G_s. ADP-ribosylation of these proteins by cholera toxin markedly inhibits the hydrolysis of bound GTP to GDP, keeping these signal-coupling proteins in their activated state. Pertussis toxin has the opposite effect on its target proteins. Pertussis toxin ADP-ribosylates the α subunit of G_i (Codina et al 1983; Bokoch et al 1983; Murayama & Ui 1983), G_o (Sternweis & Robishaw 1984; Neer et al 1984), and transducin (Van Dop et al 1984b; Watkins et al 1985). The inactive state of the G protein (G-GDP) is the substrate for pertussis toxin. Hormone or light blocks this type of ADP ribosylation, which indicates that G protein bound to the activated receptor is not modified by pertussis toxin. Also, the α subunit alone, whether in the GDP or GTP form, is not a substrate. G proteins ADP-ribosylated by pertussis toxin are permanently trapped in the G-GDP state because they are unable to bind to the excited receptor.

The activation of adenylate cyclase and retinal cGMP phosphodiesterase by fluoride is another expression of the kinship of the G-protein family. The α subunits of G_s and transducin are the targets of F^- (Stein et al 1985; Kanaho et al 1985), which acts by combining with trace amounts of aluminum to form AlF_4^-, the active species (Sternweis & Gilman 1982). AlF_4^- stoichiometrically activates transducin, releasing the α subunit, provided that GDP is bound (Bigay et al 1985). AlF_4^- has nearly the same shape as does PO_4^{-3}, and so it seems likely that AlF_4^- activates G proteins by binding next to GDP, where it mimics the γ-phosphate group of GTP (Bigay et al 1985). The activation of G_s, G_i, G_o, and transducin by aluminofluoride implies that their binding sites for GTP are very similar. Additional evidence for homology comes from the finding that the order of effectiveness in activating G_s and transducin is the same for GTP analogs substituted in the γ position: $GTP\gamma S > GTP\gamma F > GTP\gamma Me > GTP\gamma Ph$ (Pfeuffer & Eckstein 1976; Yamanaka et al 1985).

CHALLENGES AND PROSPECTS

The versatility of G proteins as transducers of external signals is now evident, a decade after the discovery of G_s. Four members of the rapidly growing family of G proteins—G_s, G_i, G_o, and rod transducin—have been purified and characterized. Specific signal-coupling roles can be assigned to G_s and transducin, and analogous but distinct functions can be inferred for G_i and G_o. cDNAs encoding subunits of all four proteins display common motifs and distinctive variations in their primary structure. By combining the tools of molecular genetics and immunology with those of biochemistry, investigators are now ready to tackle three complementary arrays of questions: the first addresses the common molecular mechanisms of G proteins; the second concerns the remarkable diversity of signals transduced in different cells, and the third considers the developmental facets of this family.

The biochemical studies reviewed here have delineated the flow of information in G-protein-mediated communication between receptors and effectors. The challenge now is to define the crucial structural elements of G proteins that enable them to interact with their triggers and targets. Activated receptors bind G proteins and induce opening of their guanyl-nucleotide binding site to allow GTP to be exchanged for GDP. How is the receptor site conformationally coupled to the nucleotide site? How does occupancy by GTP lead to mobilization of the activated α subunit? How does G_α-GTP modify the activity of the effector enzyme or channel? Precise answers to these questions await high-resolution structural analyses. The time is now ripe for x-ray crystallographic studies of the G-protein family. Site-specific mutagenesis will undoubtedly be a fruitful complementary approach in delineating the interactions of G proteins with their reaction partners in signal transduction cascades.

The second set of questions addresses the varied roles of G proteins in transmitting and regulating signal traffic within cells of metazoan organisms. How many different G proteins are there? Which ones regulate recently discovered effectors, such as phospholipase C and cardiac potassium channels? Why is the brain so rich in G_o? A related challenge is to delineate the interplay between signaling systems. Do the common $\beta\gamma$ subunits of G proteins form a link between G-protein-mediated processes? What are the interactions between tyrosine kinase cascades and G-protein cascades? Immunological and molecular genetic probes can now be used to pinpoint the precise cellular localization of different members of the G-protein family. Do they reside only in the plasma membrane? Are they confined to specific membrane domains, perhaps in complexes with receptors and effectors?

The developmental questions are equally intriguing. How are the amounts and functions of G proteins regulated in response to hormones during development, during cell differentiation and in the fully differentiated cell? What is the contribution of alternative RNA splicing in producing the repertoire of distinctive α chains? Some controls are undoubtedly exerted at the level of transcription of specific genes, as in setting the potentially critical stoichiometry of α and βγ subunits. How do individual subunits find their way to their distinctive locations? What are the roles of covalent modifications such as fatty acylation, phosphorylation, and endogenous ADP-ribosylation in regulating G protein localization and function?

A final challenge is to elucidate the origins of G proteins. Prokaryotic elongation factor Tu is likely to be one of the roots of this family. The use of a GTP-GDP cycle to drive the sequential binding of biological macromolecules to a series of reaction partners is indeed ancient. How did the earliest G protein make its way to the plasma membrane and for what purpose? How did G proteins evolve in tandem with receptors and effectors to create precisely controlled, highly amplifying cascades? By tracing the evolutionary history of G proteins, we will emerge with a deeper understanding of their design and function as central molecules in signal-processing arrays.

ACKNOWLEDGMENT

Research carried out in the authors' laboratories was supported by grants from the National Institute of General Medical Sciences and the National Eye Institute. We wish to thank Drs. Alfred Gilman, James Hurley, Toshiaki Katada, and Yoshito Kaziro for providing us with preprints of their manuscripts.

Literature Cited

Abood, M. E., Hurley, J. B., Pappone, M.-C., Bourne, H. R., Stryer, L. 1982. Functional homology between signal-coupling proteins: Cholera toxin inactivates the GTPase activity of transducin. *J. Biol. Chem.* 257 : 10540–43

Aktories, K., Schultz, G., Jakobs, K. H. 1983. Adenylate cyclase inhibition by somatostatin in S49 lymphoma *cyc⁻* variants are prevented by islet-activating protein. *FEBS Lett.* 158 : 169–73

Asano, T., Pedersen, S. E., Scott, C. W., Ross, E. M. 1984. Reconstitution of catecholamine-stimulated binding of guanosine 5′-0-(3-thiotriphosphate) to the stimulatory GTP-binding protein of adenylate cyclase. *Biochemistry* 23 : 5460–67

Baehr, W., Devlin, M. J., Applebury, M. L. 1979. Isolation and characterization of the cGMP phosphodiesterase from bovine rod outer segments. *J. Biol. Chem.* 254 : 11669–77

Berridge, M. J., Irvine, R. F. 1984. Inositol trisphosphate, a novel second messenger in cellular signal transduction. *Nature* 312 : 315–21

Bigay, J., Deterre, P., Pfister, C., Chabre, M. 1985. Fluoroaluminates activate transducin-GDP by mimicking the γ-phosphate of GTP in its binding site. *FEBS Lett.* 191 : 181–85

Bitensky, M. W., Miki, N., Keirns, J. J., Baraban, J. A., Wheeler, M. A., et al. 1975. Activation of photoreceptor disc membrane phosphodiesterase by light and ATP : Cyclic GMP as a modulator of sense receptor function. *Adv. Cyclic Nucleotide Res.* 12 : 227–38

Bitensky, M. W., Yamazaki, A., Wheeler, M. A., George, J. S., Rasenick, M. M. 1984. The mechanism of action of light-activated phosphodiesterase and evidence for homology with hormone-activated adenylate cyclase. *Adv. Cyclic Nucleotide Res.* 17:227–37

Blumenfeld, A., Erusalimsky, J., Heichal, O., Selinger, Z., Minke, B. 1985. Light-activated guanosinetriphosphatase in *Musca* eye membranes resembles the prolonged depolarizing afterpotential in photoreceptor cells. *Proc. Natl. Acad. Sci. USA* 82:7116–20

Bokoch, G. M., Katada, T., Northup, J. K., Hewlett, E. L., Gilman, A. G. 1983. Identification of the predominant substrate for ADP-ribosylation by islet activating protein. *J. Biol. Chem.* 258:2072–75

Bokoch, G. M., Katada, T., Northup, J. K., Ui, M., Gilman, A. G. 1984. Purification and properties of the inhibitory guanine nucleotide-binding regulatory component of adenylate cyclase. *J. Biol. Chem.* 259:3560–67

Bolsover, S. R., Brown, J. E. 1982. Injection of guanosine and adenosine nucleotides into *Limulus* ventral photoreceptor cells. *J. Physiol.* 332:325–42

Bourne, H. R., Beiderman, B., Steinberg, F., Brothers, V. M. 1982. Three adenylate cyclase phenotypes in S49 lymphoma cells produced by mutations of one gene. *Mol. Pharmacol.* 22:204–10

Bourne, H. R., Coffino, P., Tomkins, G. M. 1975. Selection of a variant lymphoma cell deficient in adenylate cyclase. *Science* 187:750–52

Breitwieser, G. E., Szabo, G. 1985. Uncoupling of cardiac muscarinic and β-adrenergic receptors from ion channels by a guanine nucleotide analogue. *Nature* 317:538–40

Brown, J. E., Rubin, L. J., Ghalayini, A. J., Tarver, A. P., Irvine, R. F., et al. 1984. Myo-inositol polyphosphate may be a messenger for visual excitation in *Limulus* photoreceptors. *Nature* 311:160–63

Cassel, D., Pfeuffer, T. 1978. Mechanism of cholera toxin action: Covalent modification of the guanyl nucleotide-binding protein of the adenylate cyclase system. *Proc. Natl. Acad. Sci. USA* 75:2669–73

Cassel, D., Selinger, Z. 1977. Mechanism of adenylate cyclase activation by cholera toxin: Inhibition of GTP hydrolysis at the regulatory site. *Proc. Natl. Acad. Sci. USA* 74:3307–11

Cassel, D., Selinger, Z. 1978. Mechanism of adenylate cyclase activation through the β-adrenergic receptor: Catecholamine-induced displacement of bound GDP by GTP. *Proc. Natl. Acad. Sci. USA* 75:4155–59

Cerione, R. A., Sibley, D. R., Codina, J., Benovic, J. L., Winslow, J., et al. 1984. Reconstitution of a hormone-sensitive adenylate cyclase system. *J. Biol. Chem.* 259:9979–82

Cerione, R. A., Staniszewski, C., Benovic, J. L., Lefkowitz, R. J., Caron, M. G., et al. 1985. Specificity of the functional interactions of the β-adrenergic receptor and rhodopsin with guanine nucleotide regulatory proteins reconstituted in phospholipid vesicles. *J. Biol. Chem.* 260:1493–1500

Chen, Z., Pace, U., Ronen, D., Lancet, D. 1986. Polypeptide gp95. A unique glycoprotein of olfactory cilia with transmembrane receptor properties. *J. Biol. Chem.* 261:1299–1305

Codina, J., Hildebrandt, J., Iyengar, R., Birnbaumer, L., Sekura, R. D., Manclark, C. R. 1983. Pertussis toxin substrate, the putative N_i component of adenylyl cyclases, is an α/β heterodimer regulated by guanine nucleotide and magnesium. *Proc. Natl. Acad. Sci. USA* 80:4276–80

Codina, J., Hildebrandt, J., Sunyer, T., Sekura, R. D., Manclark, C. R., et al. 1984. Mechanisms in the vectorial receptor-adenylate cyclase signal transduction. *Adv. Cyclic Nucleotide Res.* 17:111–25

Dratz, E. A., Hargrave, P. A. 1983. The structure of rhodopsin and the rod outer segment disk membrane. *Trends Biochem. Sci.* 8:128–32

Ebrey, T. G., Kilbride, P., Hurley, J. B., Calhoun, R., Tsuda, M. 1980. Light control of cyclic nucleotide concentrations in the retina. *Curr. Top. Membr. Transp.* 15:133–56

Evans, T., Martin, M. W., Hughes, A. R., Harden, T. K. 1985. Guanine nucleotide–sensitive, high affinity binding of carbachol to muscarinic cholinergic receptors of 1321N1 astrocytoma cells is insensitive to pertussis toxin. *Mol. Pharmacol.* 27:32–37

Fein, A., Corson, D. W. 1981. Excitation of *Limulus* photoreceptors by vanadate and by a hydrolysis-resistant analog of guanosine triphosphate. *Science* 212:555–57

Fein, A., Payne, R., Corson, D. W., Berridge, M. J., Irvine, R. F. 1984. Photoreceptor excitation and adaptation by inositol 1,4,5-triphosphate. *Nature* 311:157–60

Fernandez, J. M., Neher, E., Gomperts, B. D. 1984. Capacitance measurements reveal stepwise fusion events in degranulating mast cells. *Nature* 312:453–55

Fishman, P. H. 1980. Mechanism of action of cholera toxin: Events on the cell surface. In *Secretory Diarrhea*, ed. J. S. Fordtran, S. G. Schultz, pp. 85–106. Bethesda: Am. Physiol. Soc.

Florio, V. A., Sternweis, P. C. 1985.

Reconstitution of resolved muscarinic cholinergic receptors with purified GTP-binding proteins. *J. Biol. Chem.* 260: 3477–83

Fung, B. K.-K. 1983. Characterization of transducin from bovine retinal rod outer segments. I. Separation and reconstitution of the subunits. *J. Biol. Chem.* 258: 10495–10502

Fung, B. K.-K., Hurley, J. B., Stryer, L. 1981. Flow of information in the light-triggered cyclic nucleotide cascade of vision. *Proc. Natl. Acad. Sci. USA* 78: 152–56

Fung, B. K.-K., Nash, C. R. 1983. Characterization of transducin from bovine retinal rod outer segments. II. Evidence for distinct binding sites and conformational changes revealed by limited proteolysis with trypsin. *J. Biol. Chem.* 258: 10503–10

Fung, B. K.-K., Stryer, L. 1980. Photolyzed rhodopsin catalyzes the exchange of GTP for bound GDP in retinal rod outer segments. *Proc. Natl. Acad. Sci. USA* 77: 2500–4

Gibbs, J. B., Sigal, I. S., Poe, M., Scolnick, E. M. 1984. Intrinsic GTPase activity distinguishes normal and oncogenic *ras* p21 molecules. *Proc. Natl. Acad. Sci. USA* 81: 5704–8

Gilman, A. G. 1984. G proteins and dual control of adenylate cyclase. *Cell* 36: 577–79

Godchaux, W., III, Zimmerman, W. F. 1979. Membrane-dependent guanine nucleotide binding and GTPase activities of soluble protein from bovine rod cell outer segments. *J. Biol. Chem.* 254: 7874–84

Goldman, D. W., Chang, F. H., Gifford, L. A., Goetzl, E. J., Bourne, H. R. 1985. Pertussis toxin inhibition of chemotactic factor-induced calcium mobilization and function in human polymorphonuclear leukocytes. *J. Exp. Med.* 162: 145–56

Halliday, K. 1984. Regional homology in GTP-binding proto-oncogene products and elongation factors. *J. Cyclic Nucleotide Res.* 9: 435–48

Harris, B. A., Robishaw, J. D., Mumby, S. M., Gilman, A. G. 1985. Molecular cloning of complementary DNA for the alpha subunit of the G protein that stimulates adenylate cyclase. *Science* 229: 1274–77

Haslam, R. J., Davidson, M. L. M. 1984. Guanine nucleotides decrease the free $[Ca^{2+}]$ required for secretion of serotonin from permeabilized platelets. Evidence of a role for a GTP-binding protein in platelet activation. *FEBS Lett.* 174: 90–95

Hildebrandt, J. D., Hanoune, J., Birnbaumer, L. 1982. Guanine nucleotide inhibition of *cyc⁻* S49 mouse lymphoma cell membrane adenylyl cyclase. *J. Biol. Chem.* 257: 14723–25

Hildebrandt, J. D., Sekura, R. D., Codina, J., Iyengar, R., Manclark, C. R., Birnbaumer, L. 1983. Stimulation and inhibition and adenylyl cyclases mediated by distinct regulatory proteins. *Nature* 302: 706–9

Holz, G. G., Rane, S. G., Dunlap, K. 1986. GTP-binding proteins mediate transmitter inhibition of voltage-dependent calcium channels. *Nature* 319: 670–72

Hurley, J. B., Fong, H. K. W., Teplow, D. B., Dreyer, W. J., Simon, M. I. 1984. Isolation and characterization of a cDNA clone for the γ subunit of bovine retinal transducin. *Proc. Natl. Acad. Sci. USA* 81: 6948–52

Hurwitz, R. L., Bunt-Milam, A. H., Chang, M. L., Beavo, J. A. 1985. cGMP phosphodiesterase in rod and cone outer segments of the retina. *J. Biol. Chem.* 260: 568–73

Itoh, H., Kozasa, T., Nagata, S., Nakamura, S., Katada, T., et al. 1986. Molecular cloning and sequence determination of cDNAs coding for α subunits of G_s, G_i, and G_o proteins from rat brain. *Proc. Natl. Acad. Sci. USA* 83: 3776–80

Jakobs, K. H., Bauer, S., Watanabe, Y. 1985. Modulation of adenylate cyclase of human platelets by phorbolester. Impairment of the hormone-sensitive inhibitory pathway. *Eur. J. Biochem.* 151: 425–30

Johnson, G. L., Kaslow, H. R., Bourne, H. R. 1978. Genetic evidence that cholera toxin substrates are regulatory components of adenylate cyclase. *J. Biol. Chem.* 253: 7120–23

Jurnak, F. 1985. Structure of the GDP domain of EF-Tu and location of the amino acids homologous to *ras* oncogene proteins. *Science* 230: 32–36

Kanaho, Y., Moss, J., Vaughan, M. 1985. Mechanism of inhibition of transducin GTPase activity by fluoride and aluminum. *J. Biol. Chem.* 260: 11493–97

Kanaho, Y., Tsai, S. C., Adamik, R., Hewlett, E. L., Moss, J., Vaughan, M. 1984. Rhodopsin-enhanced GTPase activity of the inhibitory GTP-binding protein of adenylate cyclase. *J. Biol. Chem.* 259: 7378–81

Katada, T., Bokoch, G. M., Northup, J. K., Ui, M., Gilman, A. G. 1984a. The inhibitory guanine nucleotide-binding regulatory component of adenylate cyclase. Properties and function of the purified protein. *J. Biol. Chem.* 259: 3568–77

Katada, T., Bokoch, G. M., Smigel, M. D., Ui, M., Gilman, A. G. 1984c. The inhibitory guanine nucleotide-binding regulatory component of adenylate cyclase.

416   STRYER & BOURNE

Subunit dissociation and the inhibition of adenylate cyclase in S49 lymphoma cyc^- and wild-type membranes. *J. Biol. Chem.* 259:3586–95

Katada, T., Gilman, G., Watanabe, Y., Bauer, S., Jakobs, K. H. 1985. Protein kinase C phosphorylates the inhibitory guanine nucleotide-binding regulatory component and apparently suppresses its function in hormonal inhibition of adenylate cyclase. *Eur. J. Biochem.* 151:431–37

Katada, T., Northup, J. K., Bokoch, G. M., Ui, M., Gilman, A. G. 1984b. The inhibitory guanine nucleotide-binding regulatory component of adenylate cyclase. Subunit dissociation and guanine nucleotide-dependent hormonal inhibition. *J. Biol. Chem.* 259:3578–85

Katada, T., Oinuma, M., Ui, M. 1986. Mechanism for inhibition of the catalytic activity of adenylate cyclase by the guanine nucleotide binding proteins serving as the substrate of islet-activating protein, pertussis toxin. *J. Biol. Chem.* 261:5215–21

Katada, T., Ui, M. 1982. Direct modification of the membrane adenylate cyclase system by islet-activating protein due to ADP-ribosylation of a membrane protein. *Proc. Natl. Acad. Sci. USA* 79:3129–33

Kaziro, Y. 1978. The role of guanosine 5′-triphosphate in polypeptide chain elongation. *Biochim. Biophys. Acta* 505:95–127

Krause, K. H., Schlegel, W., Wollheim, C. B., Andersson, T., Waldvogel, F. A., Lew, P. D. 1985. Chemotactic peptide activation of human neutrophils and HL-60 cells. Pertussis toxin reveals correlation between inositol trisphosphate generation, calcium ion transients, and cellular activation. *J. Clin. Invest.* 76:1348–54

Kühn, H. 1980. Light- and GTP-regulated interaction of GTPase and other proteins with bovine photoreceptor membranes. *Nature* 283:587–89

Lad, P. M., Olson, C. V., Smiley, P. A. 1985. Association of the *N*-formyl-Met-Leu-Phe receptor in human neutrophils with a GTP-binding protein sensitive to pertussis toxin. *Proc. Natl. Acad. Sci. USA* 82:869–73

Laursen, R. A., L'Italien, J. J., Nagarkatti, S., Miller, D. L. 1981. The amino acid sequence of elongation factor Tu of *Escherichia coli*. The complete sequence. *J. Biol. Chem.* 256:8102–9

Lefkowitz, R. J., Stadel, J. M., Caron, M. G. 1983. Adenylate cyclase-coupled β-adrenergic receptors: Structure and mechanisms of activation and desensitization. *Ann. Rev. Biochem.* 52:159–86

Liebman, P. A., Pugh, E. N. Jr. 1980. ATP mediates rapid reversal of cyclic GMP phosphodiesterase activation in visual receptor membranes. *Nature* 287:734–36

Lochrie, M. A., Hurley, J. B., Simon, M. I. 1985. Sequence of the α subunit of photoreceptor G protein: Homologies between transducin, *ras*, and elongation factors. *Science* 228:96–99

Malbon, C. C., Kaupp, U. B., Brown, J. E. 1984. *Limulus* ventral photoreceptors contain a homologue of the α subunit of mammalian N$_s$. *FEBS Lett.* 172:91–94

Martin, T. F. J., Lucas, D. O., Bajjalieh, S. M., Kowalchyk, J. A. 1986. Thyrotropin-releasing hormone activates a Ca^{2+} dependent polyphosphoinositide phosphodiesterase in permeable GH$_3$ cells. GTPγS potentiation by a cholera and pertussis toxin-insensitive mechanism. *J. Biol. Chem.* 261:2918–27

Masters, S. B., Martin, M. W., Harden, T. K., Brown, J. H. 1985. Pertussis toxin does not inhibit muscarinic-receptor-mediated phosphoinositide hydrolysis or calcium mobilization. *Biochem. J.* 227:933–37

May, D. C., Ross, E. M., Gilman, A. G., Smigel, M. D. 1985. Reconstitution of catecholamine-stimulated adenylate cyclase activity using three purified proteins. *J. Biol. Chem.* 260:15829–33

McGrath, J. P., Capon, D. J., Goeddel, D. V., Levinson, A. D. 1984. Comparative biochemical properties of normal and activated human *ras* p21 protein. *Nature* 310:644–49

Medynski, D. C., Sullivan, K., Smith, D., Van Dop, C., Chang, F. H., et al. 1985. Amino acid sequence of the alpha subunit of transducin deduced from the cDNA sequence. *Proc. Natl. Acad. Sci. USA* 82:4311–15

Mumby, S. M., Kahn, R. A., Manning, D. R., Gilman, A. G. 1986. Antisera of designed specificity for subunits of guanine nucleotide-binding regulatory proteins. *Proc. Natl. Acad. Sci. USA* 83:265–69

Murayama, T., Ui, M. 1983. Loss of the inhibitory function of the guanine nucleotide regulatory component of adenylate cyclase due to its ADP ribosylation by islet-activating protein, pertussis toxin, in adipocyte membranes. *J. Biol. Chem.* 258:3319–26

Nakamura, T., Ui, M. 1984. Islet-activating protein, pertussis toxin, inhibits Ca^{2+}-induced and guanine nucleotide-dependent releases of histamine and arachidonic acid from rat mast cells. *FEBS Lett.* 173:414–18

Nakamura, T., Ui, M. 1985. Simultaneous inhibitions of inositol phospholipid breakdown, arachidonic acid release, and histamine secretion in mast cells by islet-activating protein, pertussis toxin. *J. Biol. Chem.* 260 : 3584–93

Nathans, J., Thomas, D., Hogness, D. S. 1986. Molecular genetics of color vision : The genes encoding blue, green, and red pigments. *Science* 232 : 193–202

Navon, S. E., Fung, B. K.-K. 1984. Characterization of transducin from bovine retinal rod outer segments. *J. Biol. Chem.* 259 : 6686–93

Neer, E. J., Lok, J. M., Wolf, L. G. 1984. Purification and properties of the inhibitory guanine nucleotide regulatory unit of brain adenylate cyclase. *J. Biol. Chem.* 259 : 14222–29

Nishizuka, Y. 1984. Turnover of inositol phospholipids and signal transduction. *Science* 225 : 1365–70

Northup, J. K., Smigel, M. D., Sternweis, P. C., Gilman, A. G. 1983a. The subunits of the stimulatory regulatory component of adenylate cyclase. Resolution of the 45,000-dalton alpha subunit. *J. Biol. Chem.* 258 : 11369–76

Northup, J. K., Sternweis, P. C., Gilman, A. G. 1983b. The subunits of the stimulatory regulatory component of adenylate cyclase. Resolution, activity, and properties of the 35,000-dalton β subunit. *J. Biol. Chem.* 258 : 11361–68

Nukuda, T., Tanabe, T., Takahashi, H., Noda, M., Hirose, T., et al. 1986. Primary structure of the α-subunit of bovine adenylate cyclase-inhibiting G-protein deduced from the cDNA sequence. *FEBS Lett.* 1971 : 305–10

Ohta, H., Okajima, F., Ui, M. 1985. Inhibition by islet-activating protein of a chemotactic peptide-induced early breakdown of inositol phospholipids and Ca^{2+} mobilization in guinea pig neutrophils. *J. Biol. Chem.* 260 : 15771–80

O'Touon, J., Baehr, W., Martin, R., Hirsch, J., Pak, W. L., Applebury, M. L. 1985. The *Drosophila ninaE* gene encodes an opsin. *Cell* 40 : 839–50

Ovchinnikov, Y. A., Lipkin, V. M., Shuvaeva, T. M., Bogachuk, A. P., Shemyakin, V. V. 1985. Complete amino acid sequence of γ-subunit of the GTP-binding protein from cattle retina. *FEBS Lett.* 179 : 107–10

Pace, U., Hanski, E., Salomon, Y., Lancet, D. 1985. Odorant-sensitive adenylate cyclase may mediate olfactory reception. *Nature* 316 : 255–58

Pfaffinger, P. J., Martin, J. M., Hunter, D. D., Nathanson, N. M., Hille, B. 1985. GTP-binding proteins couple cardiac muscarinic receptors to a K channel. *Nature* 317 : 536–38

Pfeuffer, T., Eckstein, F. 1976. Topology of the GTP-binding site of adenylyl cyclase

from pigeon erythrocytes. *FEBS Lett.* 67 : 354–58

Pines, M., Gierschik, P., Milligan, G., Klee, W., Spiegel, A. 1985. Antibodies against the carboxyl-terminal 5-kDa peptide of the α subunit of transducin crossreact with the 40-kDa but not the 39-kDa guanine nucleotide binding protein from brain. *Proc. Natl. Acad. Sci. USA* 82 : 4095–99

Pobiner, B. F., Hewlett, E. L., Garrison, J. C. 1985. Role of N_i in coupling angiotensin receptors to inhibition of adenylate cyclase in hepatocytes. *J. Biol. Chem.* 260 : 16200–9

Robishaw, J. D., Russell, D. W., Harris, B. A., Smigel, M. D., Gilman, A. G. 1986. Deduced primary structure of the α subunit of the GTP-binding stimulatory protein of adenylate cyclase. *Proc. Natl. Acad. Sci. USA* 83 : 1251–55

Rodbell, M. 1980. The role of hormone receptors and GTP-regulatory proteins in membrane transduction. *Nature* 284 : 17–22

Roof, D. J., Applebury, M. L., Sternweis, P. C. 1985. Relationships within the family of GTP-binding proteins isolated from bovine central nervous system. *J. Biol. Chem.* 260 : 16242–49

Ross, E. M., Gilman, A. G. 1977. Reconstitution of catecholamine-sensitive adenylate cyclase activity : Interaction of solubilized components with receptor-replete membranes. *Proc. Natl. Acad. Sci. USA* 74 : 3715–19

Ross, E. M., Gilman, A. G. 1980. Biochemical properties of hormone-sensitive adenylate cyclase. *Ann. Rev. Biochem.* 49 : 533–64

Saibil, H. R., Michel-Villaz, M. 1984. Squid rhodopsin and vertebrate rhodopsin cross-react with vertebrate photoreceptor enzymes. *Proc. Natl. Acad. Sci. USA* 81 : 5111–15

Schlegel, W., Wuarin, F., Zbaren, C., Wollheim, C. B., Zahnd, G. R. 1985. Pertussis toxin selectively abolishes hormone induced lowering of cytosolic calcium in GH_3 cells. *FEBS Lett.* 189 : 27–32

Schramm, M., Selinger, Z. 1984. Message transmission : Receptor controlled adenylate cyclase system. *Science* 225 : 1350–56

Siegelbaum, S. A., Camardo, J. S., Kandel, E. R. 1982. Serotonin and cyclic AMP close single K^+ channels in *Aplysia* sensory neurones. *Nature* 299 : 413–17

Sitaramayya, A., Liebman, P. A. 1983. Mechanism of ATP quench of phosphodiesterase activation in rod disc membranes. *J. Biol. Chem.* 258 : 1205–12

Smigel, M. D. 1986. Purification of the catalyst of adenylate cyclase. *J. Biol. Chem.* 261 : 1976–82

Smigel, M. D., Katada, T., Northup, J. K., Bokoch, G. M., Ui, M., Gilman, A. G. 1984. Mechanisms of guanine nucleotide-mediated regulation of adenylate cyclase activity. *Adv. Cyclic Nucleotide Protein Phosphorylation Res.* 17 : 1–18

Smith, C. D., Lane, B. C., Kusaka, I., Verghese, M. W., Snyderman, R. 1985. Chemoattractant receptor-induced hydrolysis of phosphatidylinositol 4,5-bisphosphate in human polymorphonuclear leukocyte membranes. *J. Biol. Chem.* 260 : 5875–78

Spiegel, A. M., Gierschik, P., Levine, M. A., Downs, R. W. Jr. 1985. Clinical implications of guanine nucleotide-binding proteins as receptor-effector couplers. *N. Engl. J. Med.* 312 : 26–33

Stein, P. J., Halliday, K. R., Rasenick, M. M. 1985. Photoreceptor GTP binding protein mediates fluoride activation of phosphodiesterase. *J. Biol. Chem.* 260 : 9081–84

Steinman, L., Weiss, A., Adelman, N., Lim, M., Zuniga, R., et al. 1985. Pertussis toxin is required for pertussis vaccine encephalopathy. *Proc. Natl. Acad. Sci. USA* 82 : 8733–36

Sternweis, P. C. 1986. The purified α-subunits of G_{do} and G_{di} from bovine brain require $\beta\gamma$ for association with phospholipid vesicles. *J. Biol. Chem.* 261 : 631–37

Sternweis, P. C., Gilman, A. G. 1982. Aluminum : A requirement for activation of the regulatory component of adenylate cyclase by fluoride. *Proc. Natl. Acad. Sci. USA* 79 : 4888–91

Sternweis, P. C., Northup, J. K., Smigel, M. D., Gilman, A. G. 1981. The regulatory component of adenylate cyclase. Purification and properties. *J. Biol. Chem.* 256 : 11517–26

Sternweis, P. C., Robishaw, J. D. 1984. Isolation of two proteins with high affinity for guanine nucleotides from membranes of bovine brain. *J. Biol. Chem.* 259 : 13806–13

Stryer, L. 1986. Cyclic GMP cascade of vision. *Ann. Rev. Neurosci.* 9 : 87–119

Stryer, L., Hurley, J. B., Fung, B. K.-K. 1981. Transducin : An amplifier protein in vision. *Trends Biochem. Sci.* 6 : 245–47

Sugimoto, K., Nukada, T., Tanabe, T., Takahashi, H., Noda, M., et al. 1985. Primary structure of the β-subunit of bovine transducin deduced from the cDNA sequence. *FEBS Lett.* 191 : 235–40

Tanabe, T., Nukada, T., Nishikawa, Y., Sugimoto, K., Suzuki, H., et al. 1985. Primary structure of the α subunit of transducin and its relationship to *ras* proteins. *Nature* 315 : 242–45

Tsai, S.-C., Adamik, R., Kanaho, Y., Hewlett, E. L., Moss, J. 1984. Effects of guanyl nucleotides and rhodopsin on ADP-ribosylation of the inhibitory GTP-binding component of adenylate cyclase by pertussis toxin. *J. Biol. Chem.* 259 : 15320–23

Tsuda, M., Tsuda, T., Terayama, Y., Fukada, Y., Akino, T., et al. 1986. Kinship of cephalopod photoreceptor G-protein with vertebrate transducin. *FEBS Lett.* 198 : 5–10

Ui, M., Katada, T., Murayama, T., Kurose, H., Yajima, M., et al. 1984. Islet-activating protein, pertussis toxin : A specific uncoupler of receptor-mediated inhibition of adenylate cyclase. *Adv. Cyclic Nucleotide Res.* 17 : 145–51

Vandenberg, C. A., Montal, M. 1984. Light-regulated biochemical events in invertebrate photoreceptors. *Biochemistry* 23 : 2339–52

Van Dop, C., Bourne, H. R. 1983. Pseudohypoparathyroidism. *Ann. Rev. Med.* 34 : 259–66

Van Dop, C., Medynski, D., Sullivan, K., Wu, A. M., Fung, B. K.-K., Bourne, H. R. 1984a. Partial cDNA sequence of the γ subunit of transducin. *Biochem. Biophys. Res. Commun.* 124 : 250–55

Van Dop, C., Yamanaka, G., Steinberg, F., Sekura, R. D., Manclark, C. R., et al. 1984b. ADP-ribosylation of transducin by pertussis toxin blocks the light-stimulated hydrolysis of GTP and cGMP in retinal photoreceptors. *J. Biol. Chem.* 259 : 23–26

Watkins, P. A., Burns, D. L., Kanaho, Y., Liu, T.-Y., Hewlett, E. L., Moss, J. 1985. ADP-ribosylation of transducin by pertussis toxin. *J. Biol. Chem.* 260 : 13478–82

Wheeler, G. L., Bitensky, M. W. 1977. A light-activated GTPase in vertebrate photoreceptors : Regulation of light-activated cyclic GMP phosphodiesterase. *Proc. Natl. Acad. Sci. USA* 74 : 4238–42

Wilden, U., Hall, S. W., Kuhn, H. 1986. Phosphodiesterase activation by photo-excited rhodopsin is quenched when rhodopsin is phosphorylated and binds 48 kDa-protein. *Proc. Natl. Acad. Sci. USA* 83 : 1174–78

Wistow, G. J., Katial, A., Craft, C., Shinohara, T. 1986. Sequence analysis of bovine retinal S-antigen. Relationships with α transducin and G proteins. *FEBS Lett.* 196 : 23–28

Worley, P. F., Baraban, J. M., Van Dop, C., Neer, E. J., Snyder, S. H. 1986. G_o, a guanine nucleotide-binding protein : Immunohistochemical localization in rat brain resembles distribution of second messengers. *Proc. Natl. Acad. Sci. USA* 83 : 4561–65

Yajima, Y., Akita, Y., Saito, T. 1986. Pertussis toxin blocks the inhibitory effects of somatostain on cAMP-dependent vasoactive intestinal peptide and cAMP-independent thyrotropin releasing hormone-stimulated prolactin secretion of GH_3 cells. *J. Biol. Chem.* 261 : 2684–89

Yamanaka, G., Eckstein, F., Stryer, L. 1985. Stereochemistry of the guanyl nucleotide binding site of transducin probed by phosphorothioate analogues of GTP and GDP. *Biochemistry* 24 : 8094–8101

Yamanaka, G., Stryer, L. 1987. Energetics of the light-activated transducin cycle. *Proc. Natl. Acad. Sci. USA* In press

Yatsunami, K., Khorana, G. 1985. GTPase of bovine rod outer segments : The amino acid sequence of the α subunit as derived from the cDNA sequence. *Proc. Natl. Acad. Sci. USA* 82 : 4316–20

Yatsunami, K., Pandya, B. V., Oprian, D. D., Khorana, H. G. 1985. cDNA-derived amino acid sequence of the γ subunit of GTPase from bovine rod outer segments. *Proc. Natl. Acad. Sci. USA* 82 : 1936–40

Yau, K.-W., Haynes, L. W., Nakatani, K. 1986. Rods of calcium and cyclic GMP in visual transduction. In *Membrane Control of Cellular Activity, Progress in Zoology*, Vol. 33, ed. H. Ch. Lüttgan. Stuttgart : Gustav Fischer Verlag; Massachusetts : Sinauer Assoc.

Yee, R., Liebman, P. A. 1978. Light-activated phosphodiesterase of the rod outer segment. Kinetics and parameters of activation and deactivation. *J. Biol. Chem.* 253 : 8902–9

Zucker, C. S., Cowman, A. F., Rubin, G. M. 1985. Isolation and nucleotide sequence of a rhodopsin gene from *Drosophila melanogaster. Cell* 40 : 851–58

Ann. Rev. Cell Biol. 1986. 2 : 421–57

MICROTUBULE-ASSOCIATED PROTEINS

J. B. Olmsted

Department of Biology, University of Rochester, Rochester, New York 14627

CONTENTS

INTRODUCTION

The work of the past two decades has demonstrated that the cytoplasm of cells is filled with an elaborate array of cytoskeletal elements. Much has been learned about the major constituents of microtubules, microfilaments, and intermediate filaments, and information on the less abundant components of these structures has been expanding rapidly. This review concentrates on microtubule-associated proteins (MAPs), a collection of varied molecules that have been defined on the basis of their binding and/or putative interaction with microtubules. Some of these proteins are well characterized, particularly those isolated by the in vitro polymerization of tubulin from brain extracts. Another class of proteins, as yet less well known in composition, can be more nebulously defined as those that interact, perhaps transiently, with microtubules. Proteins in this category include those associated with well-known microtubule-organizing sites, such as centrosomes or kinetochores (see review by Brinkley 1985), and those identified by immunostaining as being associated with micro-

421

0743–4634/86/1115–0421$02.00

tubule arrays but not found by in vitro assays. There are also proteins that have been characterized as interacting with, or modifying, the major MAPs. The purpose of this review is to describe the isolation and characterization of a variety of MAPs, primarily those of mammalian origin, and to discuss approaches to understanding how these molecules may modulate microtubule assembly and function in vivo (Table 1). Several reviews describe high-molecular-weight brain MAPs (Vallee 1984; Vallee & Bloom 1984; Vallee et al 1984a; Sloboda & Rosenbaum 1982; Wiche 1985) and the role of MAPs in the in vitro assembly of microtubules (Correia & Williams 1983; Kirschner 1978; McKeithan & Rosenbaum 1984; Purich & Kristofferson 1984).

ISOLATION AND IDENTIFICATION OF MAPs

Early Studies

The first attempts to identify proteins associated with microtubules paralleled the earliest investigations of the biochemical constituents of well-defined microtubule arrays. The first detailed study of one of these was performed by Gibbons (1965), who demonstrated that a protein termed dynein could be selectively released and rebound to axonemes from *Tetrahymena* (see Johnson et al 1984; Lefebvre & Rosenbaum 1986 for reviews of dynein chemistry and flagellar structure). The specific reassociation of proteins with the microtubule lattice is still one of the major criteria for identifying putative MAPs. Other studies (reviewed in Olmsted & Borisy 1973; Kuriyama et al 1984) focussed on isolating the mitotic apparatus, but these were less successful because of the lability of the structure. However, the hexylene glycol technique used for the stabilization of the mitotic spindle was subsequently employed by Kirkpatrick et al (1970) to purify intact microtubules from brain. In addition to tubulin, the purest fractions contained three major protein bands; two of these bands probably correspond to the proteins since identified as MAP 1 and MAP 2.

Cycled Assembly Purification of Microtubule Protein and MAPs

The discovery by Weisenberg (1972) that microtubules can be assembled in vitro from extracts of brain tissue laid the foundation for the identification of protein components that associated with the microtubule lattice. By enriching for microtubule fractions through repetitive cycles of temperature-dependent assembly and disassembly (Shelanski et al 1973; Borisy et al 1975), proteins were identified that copurify in constant stoi-

Table 1 Major classes of microtubule-associated proteins

Protein	Subspecies	Subunit mass (kDa)[a]	Primary source	Properties
MAP 1	1A, 1B, 1C	350	brain	thermolabile; projection on microtubule
Light chains		28, 30	brain	
MAP 2	2A, 2B	270	brain	thermostable; projection on microtubule; separable into projection (235 kDa) and binding (35 kDa) domains; phosphorylated; binds calmodulin; associated with MAP 2 projection domain
Type II cAMP-dependent protein kinase		53, 39		
Tau	3–5	55–62	brain	thermostable; number of polypeptides depends on age and species; phosphorylated; binds calmodulin
MAP 3	—	180	brain	
MAP 4; 210-kDa HeLa MAP; 205-kDa *Drosophila* MAP	3–4	200–240 depending on species	cultured mammalian cells; mouse tissues (MAP 4); *Drosophila* (205-kDa)	thermostable
125-kDa MAP	—	125	cultured mammalian cells	
Chartins	—	69, 72, 80	cultured mammalian cells; primary neurons	thermolabile; phosphorylated subspecies
STOPS	—	140, 72, 56	brain	associated with cold-stable microtubules
Sea urchin MAPS	—	37, 78, 30, 150, 200, 235	sea urchin eggs; sea urchin spindles	spindle localization
Kinesin	—	110 134	squid axoplasm sea urchin eggs	moves particles on microtubules

[a] Denatured mass of major polypeptides in each class.

chiometry with tubulin and whose association with microtubules depends upon the formation of polymer. Fractions enriched in copurifying proteins of high molecular mass (270–350 kDa) promoted the assembly of microtubules at much lower concentrations than did tubulin alone and stabilized existing polymers (Murphy & Borisy 1975; Murphy et al 1977; Sloboda et al 1976; Sloboda & Rosenbaum 1979). High-molecular-weight MAPs were shown to form projections on the surface of microtubules (Dentler et al 1975; Murphy & Borisy 1975; Sloboda & Rosenbaum 1979; Kim et al 1979; Herzog & Weber 1978) and were resolved into two major components, MAP 1 (300–350 kDa) and MAP 2 (270 kDa) (Sloboda et al 1975). Another copurifying protein factor, tau, was composed of several polypeptides with molecular masses between 55 and 68 kDa (Weingarten et al 1975; Cleveland et al 1977). Tau also promoted tubulin assembly in vitro. A number of reviews detail the early studies on these proteins (Kirschner 1978; Vallee 1984; Purich & Kristofferson 1984; Correia & Williams 1983) and provide further information on their chemistry. Recently, Matus and colleagues (Matus et al 1983; Huber et al 1985) identified a new brain MAP, MAP 3 (180 kDa), that copurifies with the other brain MAPs through multiple cycles of assembly.

Another class of proteins remains associated with cold-stable microtubules in brain microtubule preparations (Margolis & Rauch 1981); these microtubules depolymerize in the presence of calmodulin and low calcium concentrations (Job et al 1982). Associated proteins have molecular masses of 135, 70–82, and 56 kDa, and they are all retained on a calmodulin column (Job et al 1982); these proteins are now called STOPs (stable tubule only peptides). Recent studies suggest that the STOP proteins can slide on microtubules (Pabion et al 1984), and it has been postulated that these proteins, in concert with MAP, are important in regulating the stability of the microtubule polymer (Job et al 1985). A calmodulin-independent phosphorylation of the STOP proteins can occur (Job et al 1983). Recently, a calmodulin-dependent kinase has been identified in cold-stable microtubule preparations (Larson et al 1985). Whether these two enzymatic activities are coordinately regulated in generating stabilized microtubules is currently unknown.

Attempts at purifying MAPs from tissues other than brain using cycled assembly were largely unsuccessful, presumably due in part to the low concentration of tubulin in nonneuronal cell types. Microtubule protein preparations were ultimately obtained from HeLa cell extracts, and proteins of 125 and 210 kDa that fulfilled the criteria for MAPs were purified (Bulinski & Borisy 1979; Weatherbee et al 1980). Another protein, 215-kDa MAP, was found in microtubule protein preparations from differentiated mouse neuroblastoma cells (Olmsted & Lyon 1981). This protein has been identified as having a molecular mass closer to 220–240 kDa and is called MAP 4. It is probably the species-specific homolog of the 210-kDa HeLa MAP (Parysek et al 1984a ; Olmsted et al 1986).

Taxol-Driven Assembly of Microtubules

The discovery by Horwitz and collaborators (for overview see Horwitz et al 1982) that the drug taxol promotes the assembly of microtubules in vivo and in vitro circumvented the question of how to drive the self-assembly of microtubules in extracts with low tubulin concentrations. Vallee (1982) combined taxol with elements of the cycled assembly purification scheme to devise a taxol-based isolation method for microtubules. In this procedure proteins that sediment with microtubules are released from taxol-stabilized polymers by a buffer of moderate ionic strength. The salt-stripped proteins can then be used to drive the assembly of purified tubulin or to bind to taxol-stabilized microtubules to assess which are specific microtubule-associated components. While not all of the proteins that have been identified by this protocol have been demonstrated to promote tubulin assembly in vitro, the affinity for the tubulin lattice has led to their identification as MAPs. The following is a list of some of the proteins characterized using this procedure :

1. Proteins with molecular masses between 200 and 240 kDa have been found. This group includes the 210-kDa MAP from HeLa cells (Vallee 1982) and CHO cells (Brady & Cabral 1985), MAP 4 from neuroblastoma cells and a variety of mouse tissues (Parysek et al 1984a), a 205-kDa protein from a pituitary cell line (Bloom et al 1985a), and a 205-kDa MAP from *Drosophila* (Goldstein et al 1986). The majority of these proteins remain soluble after boiling (i.e. are thermostable), and all are distinct from the major brain MAPs.
2. Proteins from nonneural tissues or cell lines related to MAP 1 and MAP 2 have been identified. Taxol-polymerized microtubules from 3T6, CHO, C6, neuroblastoma, pituitary, heart, testis, and liver cells contained proteins that reacted with a polyclonal MAP 1 antibody. MAP 2 antibody reacted only with preparations obtained from skeletal muscle, neuroblastoma, and pituitary cells (Wiche et al 1984 ; Koska et al 1985). MAPs corresponding to MAP 1A, MAP 1B, and MAP 2 have recently been found in pituitary tissue and cultured pituitary cell lines (GH3) (Bloom et al 1985a).
3. Proteins from sea urchin egg extracts that coassemble with taxol-polymerized microtubules include those of 77, 100, and 120 kDa and a collection of proteins in the high-molecular-mass range (205–300 kDa) (Vallee & Bloom 1983). Monoclonal antibodies prepared against these

microtubule fractions react with four proteins of 235, 200, 150, and 37 kDa; all of these antibodies stained the spindles of sea urchin eggs. Proteins of 78 and 80 kDa (Scholey et al 1984, 1985a) have also been identified, and antibodies to these proteins react with the spindle. ATPase activities are also associated with taxol-polymerized microtubule preparations from sea urchin eggs (Scholey et al 1984; Hollenbeck et al 1984).

4. Motility-related MAPs have recently been discovered using new systems in which particle translocation on microtubules can be studied in vitro [e.g. see Allen et al 1985; Vale et al 1985a; Schroer & Kelly 1985 (review)]. Lasek & Brady (1985) discovered that some such movements are inhibited by the nonhydrolyzable ATP analog AMPPNP, and this observation has been used to identify proteins that interact with microtubules. Brady (1985) demonstrated that taxol-stabilized microtubules prepared from brain in the presence of AMPPNP contained a 130-kDa protein that could be released from the microtubules with ATP. Using a similar strategy, a 110-kDa MAP, now called kinesin, has been prepared from squid axoplasm and demonstrated to facilitate particle movements in a reconstituted system (Vale et al 1985a). Scholey et al (1985b) have identified a 134-kDa protein in sea urchin eggs that is antigenically related to kinesin; immunostaining demonstrates that this protein is localized to the mitotic spindle. These studies will be considered more fully in a future review in this series.

Selective Solubilization

While the above methods for identifying MAPs rely on the ability of the proteins to bind microtubules in vitro, another approach has been pioneered by Solomon and coworkers (Solomon et al 1979; Duerr et al 1981). They reasoned that proteins present in small amounts might be lost during in vitro assembly, so they devised a selective extraction scheme in which a sequence of solubilization treatments is used to fractionate cells into soluble, microtubule-associated, and insoluble fractions. By comparing the patterns of control cells with those of cells treated with agents that disrupt microtubule assembly, the proteins specific to the microtubule fraction could be deduced. In most cases, the capacity of these proteins to interact with microtubules was confirmed by coassembly with microtubules in vitro. The use of this technique led to the first identification of a number of MAPs in a variety of cultured cell lines (Duerr et al 1981). Proteins of 69 and 220 kDa are common to a number of rodent cell lines, and 125- and 200-kDa proteins, which are presumably identical to the HeLa MAPs, are present in human lines. Proteins of 55 and 80 kDa are found only in cells of neural origin. Pallas & Solomon (1982) demonstrated that proteins of 69, 72, and 80 kDa were actually a family of related polypeptides. The more highly phosphorylated forms of these proteins associated preferentially with microtubules. These proteins have recently been named chartins, and antibodies have been produced to facilitate further characterization (Magendantz & Solomon 1985).

The selective solubilization technique has also been employed to try to identify novel MAPs in cell types with well-defined functions. Zieve & Solomon (1982) found a protein of 150 kDa that is present only in microtubule fractions from mitotic cells and is also present in taxol-stabilized isolated spindles. Pillus & Solomon (1986) have just adapted this technique to the study of yeast spindles. Using two-dimensional gel analysis, Black & Kurdyla (1983) identified a number of microtubule-associated proteins (26) in cultured sympathetic neurons. Thermolabile proteins, ranging in molecular mass from 60 to 76 kDa, were a major part of the microtubule-associated components. These proteins were classified into four related groups, and the heterogeneity within each group was shown to be due to the degree of phosphorylation. At least some of these proteins are probably comparable to those described by Zieve & Solomon (1984) in studies on isolated neuronal cytoskeletons, and all may be chartins.

Identification of MAPs in Isolated Spindles

Attempts at identifying MAPs associated with well-defined, microtubule-containing structures other than flagella have largely been confined to analyses of the mitotic spindle. Keller & Rebhun (1982) found an 80-kDa protein from sea urchin egg spindle preparations that copurified by cycled assembly with the spindle tubulin, but this protein had no effect on the critical concentration for tubulin polymerization. Hirokawa et al (1985b) recently examined isolated sea urchin spindles and demonstrated that major proteins of 75, 245, and 250 kDa could be extracted from the lattice. Depolymerization and repolymerization of microtubules from the spindle yielded preparations enriched in these three proteins, and the structure of the reassembled microtubules, as analyzed by the rapid-freeze/deep-etch electron microscopy technique, resembled that seen in situ. Taxol-stabilized spindles have also been isolated from CHO cells. These preparations contain a 210-kDa thermostable protein that is antigenically similar to the 210-kDa HeLa MAP (Kuriyama et al 1984; Brady & Cabral 1985).

Putative MAPs Defined by Other Criteria

A number of proteins have been implicated as associated with microtubules using approaches different than those described above. Immunofluorescent staining studies demonstrated that antibodies to a number of proteins react with microtubule arrays in situ. Among the proteins defined as MAPs by this criterion are: calmodulin (Andersen et al 1978; Welsh et al 1979; Deery et al 1984; Vandert et al 1985), the regulatory subunit of cAMP kinase II (Browne et al 1982; Miller et al 1982), NUMA (Pettijohn et al 1984) and ankyrin (Bennett & Davis 1981). Purified proteins from several sources have also been identified as interacting with microtubules in vitro. These include synapsin I (Baines & Bennett 1986), fodrin (Fach et al 1985), ankyrin (Bennett & Davis 1981; Davis & Bennett 1984), and neuro-filament-associated proteins (Pytela & Wiche 1980; Runge et al 1979).

BIOCHEMICAL PROPERTIES OF MAPs

Of the MAPs identified to date, some have been thoroughly characterized, and others have not yet been purified to homogeneity. The following discussion is a synopsis of the major characteristics of each MAP that has been studied in detail.

MAP 1

Although MAP 1 has been distinguished from MAP 2 for over a decade, the routine purification of this protein has only recently been achieved (Kuznetsov et al 1981; Vallee & Davis 1983) MAP 1 alone stimulates microtubule formation; light chains of 28 and 30 kDa are associated with MAP 1 but are not required for binding (Vallee & Davis 1983). Recent studies demonstrated that MAP 1 is actually comprised of three poly-peptides, termed MAP 1A, 1B, and 1C in order of decreasing molecular weight on denaturing gels (Bloom et al 1984b; Herrmann et al 1985). These polypeptides have distinctive properties, as demonstrated by their sensitivity to proteolysis [MAP 1C is more resistant than 1A or 1B (Bloom et al 1984b)] and the results of one-dimensional peptide mapping (Bloom et al 1985b; Herrmann et al 1985). Epitopes exist that are unique to MAP 1A (Bloom et al 1984b) and MAP 1B (Bloom et al 1985b). Further, MAP 1B polymerizes much less efficiently with microtubules in extracts than does MAP 1A or MAP 2 (Bloom et al 1985b).

Several proteins have been identified that are related to MAP 1 by immunological criteria. Bonifacino et al (1985) have found that an anti-body to a nuclear protein cross-reacts with MAP 1 and decorates the mitotic apparatus. Conversely, monoclonal antibodies to MAP 1 react with a nuclear protein (Sato et al 1985). Other antibodies to MAP 1 react with a 205-kDa sulfoglycoprotein of the extracellular matrix (Briones & Wiche 1985) and with stress fibers (Asai et al 1985). Since it is possible for proteins with unrelated functions to share antigenic determinants, how MAP 1 is related to the proteins that are not associated with microtubules requires further investigation.

MAP 2

MAP 2, a thermostable brain MAP, has been extensively studied (see reviews by Vallee 1984; Vallee & Bloom 1984). The protein is an elongated molecule estimated to be between 90 ± 30 nm (Gottlieb & Murphy 1985) and 185 nm long (Voter & Erickson 1982). MAP 2 can be cleaved by gentle proteolysis into microtubule binding (35–40 kDa) and projection (240 kDa) domains (Vallee & Borisy 1977; Vallee 1980). Using antibodies specific to the "arm" or "stub" regions of MAP 2, Gottlieb & Murphy (1985) recently demonstrated that the microtubule-binding region of the molecule may occupy as much as one third of the total length. cAMP-dependent protein kinase is associated with the projection domain of MAP 2 (Vallee et al 1981), and calmodulin binds to the microtubule-binding region of the molecule (Lee & Wolff 1984a,b).

MAP 2 has been resolved into a pair of polypeptides (Kim et al 1979); these are now called MAP 2A and MAP 2B, in order of increasing mobility on gels. Peptide mapping (Schulman 1984a; Burgoyne & Cumming 1984; Herrmann ct al 1985) and reactivity with monoclonal antibodies (Binder et al 1984; Vallee & Bloom 1984) suggest that these two subspecies are essentially identical.

The phosphorylation of MAP 2 was first demonstrated by Sloboda et al (1975), and the sites at which this occurs have been identified (Burns & Islam 1984; Goldenring et al 1985; Schulman et al 1984b; Herrmann et al 1985). The phosphorylation state of MAP 2 affects its ability to stimulate microtubule assembly (Jameson & Caplow 1981; Burns et al 1984); extensive phosphorylation of MAP 2 inhibits its binding to formed microtubules (Murthy & Flavin 1983). Recently, Murthy et al (1985) investigated the question of whether there is a correlation between the phosphorylation sites defined in vitro and those that are modified in vivo. The majority of phosphates added in vitro in the presence of cAMP-dependent kinase are in the binding domain. A population of endogenous phosphate sites are localized on the projection domain; while these turn over in vivo, they are resistant to phosphatases that remove the majority of the phosphate groups added in vitro.

The enzymes that modulate the phosphorylation of MAP 2 have also been examined. Vallee and coworkers (Vallee et al 1981; Theurkauf & Vallee 1982) demonstrated that the projection domain of MAP 2 binds the regulatory subunit of cAMP-dependent kinase. The high affinity of this subunit for MAP 2 has been shown by cAMP affinity chromatography and gel overlay techniques (Lohman et al 1984), and the nucleotide specificity of the enzyme reaction has been determined (Richter-Landsberg & Jastroff 1985).

In addition to the cAMP-dependent kinase, another enzyme that affects MAP 2 phosphorylation has been defined. Theurkauf & Vallee (1983) showed that phosphorylation of MAP 2 could occur via both cAMP-dependent and cAMP-independent routes. A calmodulin-dependent kinase associated with microtubule preparations from brain has now been identified (Goldenring et al 1983; Schulman 1984a). This enzyme appears to have broad substrate specificity (Schulman et al 1985) and has been suggested to be more active than the cAMP-dependent kinase in the phosphorylation of MAP 2 (Yamamoto et al 1985). The calcium-calmodulin-dependent kinase and the cAMP-dependent kinase appear to phosphorylate MAP 2 at different sites in vitro (Vallano et al 1985; Yamamoto et al 1985; Schulman 1984b).

The possible identity of MAP 2 with other proteins is, as for MAP 1, based primarily on immunological criteria. Sheterline (1980) raised antibodies to electrophoretically purified MAP 2 and demonstrated that they failed to react with MAP 1 or tau. They did, however, react with a single protein in brain from a number of vertebrate species, which suggests that MAP 2 was conserved. Some monoclonal antibodies raised to MAP 2 show no reaction with MAP 1 (Binder et al 1984), and monoclonals raised to MAP 1 show no reaction with MAP 2 (Huber & Matus 1984a; Bloom et al 1984a, 1985b). Some groups (Herrmann et al 1984) have suggested that some peptide fragments of MAP 1 and MAP 2 are similar; but other researchers (Bloom et al 1985b; Kuznetsov et al 1981) have found little homology between MAP 1 and MAP 2 using peptide mapping. Once the clones coding for these proteins (Lewis et al 1986a,b) have been sequenced, these conflicting data may be resolved. Spectrin, a major structural protein of the erythrocyte, has also been thought to be homologous to MAP 2 on the basis of immunological cross-reaction (Davis & Bennett 1982). As discussed below, MAP 2 may interact with actin filaments, and it has been postulated that the homology between spectrin and MAP 2 may exist because of a common actin-binding site.

Tau

Tau was the first MAP identified that showed great heterogeneity: it consists of a complex of polypeptides. The number of tau polypeptides varies from species to species, ranging between three and six. The tau proteins are heat stable and are all related, as judged by peptide mapping (Cleveland et al 1977, 1979) and by shared epitopes recognized by monoclonal antibodies (Binder et al 1985). Tau proteins can be phosphorylated (Lindwall & Cole 1984a,b), and have calmodulin binding sites (Lee & Wolff 1984a).

Other MAPs

As outlined previously, a class of MAPs identified in a number of cells and tissues have molecular masses of 200–240 kDa and are thermostable. Bulinski & Borisy (1980a) demonstrated that the HeLa 210-kDa MAP, like MAP 2, is a highly asymmetric molecule. This protein can be resolved on SDS gels into three polypeptides of 220, 208, and 200 kDa. MAP 4 is also isolated as a complex of three polypeptides with molecular masses of 240, 235, and 220 kDa (Parysek et al 1984a), designated MAP 4A, B, and C, respectively. One-dimensional peptide mapping of the MAP 4 species has shown some homology among the three proteins and some differences (Olmsted et al 1986). Monoclonal and polyclonal antibodies react with all three bands, which suggests that conserved epitopes exist among the three proteins. The *Drosophila* 205-kDa MAP is a complex of four immunologically related polypeptides (Goldstein et al 1986). Little is yet known about how any of the polypeptide components in each of these thermostable MAPs are generated and whether any or all of them are post-translationally modified. Based on antibody cross-reactivity, it has been suggested that ankyrin may be structurally related to these proteins (Bennett & Davis 1981).

Chartins are the only other MAP polypeptides that have been studied in any detail (Magendantz & Solomon 1985; Pallas & Solomon 1982). They appear to be a complex set of proteins that are related in isoelectric point and extent of phosphorylation. The 80-kDa proteins and 69-kDa proteins share all the tryptic peptides except three (Pallas & Solomon 1982), but they have been shown to be immunologically distinct (Magendantz & Solomon 1985). In contrast to tau, chartins are thermolabile.

LOCATION OF MAPs

To understand how MAPs function, analyses have been carried out to determine how individual MAPs interact with the tubulin lattice in vitro and to characterize the distribution of these proteins in vivo.

In Vitro Studies on MAP-Tubulin Interaction

The interaction of brain MAPs with tubulin has been examined in detail. Binding experiments have shown that MAP 2 and tau compete with each other (Sandoval & Vandekerckhove 1981), as do MAP 1 and MAP 2 (Kuznetsov et al 1984). Although some of these data suggest that each of the major MAP molecules may occupy a similar site, results contradictory to this also exist. Kinetic analyses (Carlier et al 1984; Diez et al 1985) suggest that MAP 1 and MAP 2 are important in the formation of oligomers early in assembly, whereas tau, as shown previously (Cleveland et al 1977), appears to be preferentially bound to already-formed polymers. Connolly & Kalnins (1980b) postulated that high-molecular-mass MAPs and tau interact differently with the tubule lattice. They noted that proteins reacting with tau antibodies were not as readily removed from permeabilized cells as those reacting with antibodies to high-molecular-mass MAPs.

With the exception of work on MAP 2, little information exists on the physical arrangement of MAPs on the microtubule lattice. Using thin-section analyses, Kim et al (1979) demonstrated MAP 2 has an axial spacing of 32 ± 8 nm along the microtubule lattice, and Voter & Erickson (1982) found a period of 36 nm for the most clearly spaced molecules seen in shadowed preparations. In the presence of the light-chain components, MAP 1 forms projections on the microtubule surface that have a length of ~ 23 nm and a spacing of 21 ± 5 nm (Vallee & Davis 1983). Microtubules decorated with tau have no obvious projections (Zingsheim et al 1979). Diffraction studies (Amos 1977; Mandelkow et al 1985) indicate that MAPs may form a superlattice on the tubulin protofilaments, but the exact nature of the interactions among all of the MAP species remains to be resolved.

The site of interaction of MAPs with the tubulin subunit is being examined through the use of specific tubulin peptides (Ginzburg & Littauer 1984). These data suggest that tau and MAP 2 bind to a similar region of the carboxyl terminus of beta tubulin. Other sites that bind only tau have also been identified. Serrano et al (1984) have shown that removal of the carboxyl terminus of alpha tubulin destroys the site for MAP 2 interaction and that the 20-kDa polypeptide released from this region of the tubulin subunit will bind to MAP 2. Kumar & Flavin (1985) have recently determined that microtubules made with detyrosinated tubulin bind a smaller percentage of MAP 2 than microtubules made with tyrosinated tubulin. One way in which the in vivo interactions of MAPs with microtubules could be modulated would be by tubulin modification, and classes of tyrosinated and non-tyrosinated microtubules have now been identified in vivo (Gundersen et al 1984).

Analysis of MAP Distribution In Vivo

Early immunocytological analyses (Connolly et al 1977, 1978; Connolly & Kalnins 1980a,b; Sherline & Schiavone 1977; Sheterline 1978) were important in demonstrating that the major brain MAPs were associated with microtubules in vivo, as well as in vitro. The immunological approach has continued to be extremely powerful in the examination of MAP distribution in a number of cells and tissues. In the studies described in the following sections, a variety of polyclonal and monoclonal antibodies have been used, and various groups have obtained conflicting results. Clearly, a detailed understanding of the epitopes recognized by different antibodies will be necessary to unravel what various immunostaining patterns indicate about MAP function.

MAJOR BRAIN MAPs IN ADULT NERVOUS TISSUE   Examination of sections from adult brain stained with various MAP antibodies has revealed a remarkable compartmentation of MAPs within specific cell types. Biochemical studies suggested that MAP 1 was enriched in white matter relative to grey matter (Vallee 1982), and immunostaining analyses have universally confirmed that MAP 1 is present in neurons. However, reports on the intensity of staining within various regions and types of neurons and the presence of MAP 1 in glial cells vary. Bloom et al (1984b) showed that a monoclonal antibody to MAP 1A reacts with cell bodies and dendrites of neurons; the intensity of axonal staining varied with neuronal cell type. While only oligodendrocytes stained with an antibody against MAP 1A in situ, astrocytes were also found to react with the antibody when cultured brain cells were examined. Monoclonal antibodies to MAP 1B showed a distribution similar to that seen with MAP 1A antibody, except axonal staining was generally stronger (Bloom et al 1985b). Other monoclonal (Matus et al 1983; Huber & Matus 1984a,b) and polyclonal (Wiche et al 1983) antibodies to MAP 1 have also been demonstrated to label cell bodies, axons, and dendrites of neurons. However, Matus and coworkers (Matus et al 1983; Huber & Matus 1984a,b) found MAP 1 monoclonal antibodies did not react with glial cells. Whether or not only a specific subclass of MAP 1 is localized in glial cells remains to be determined.

The distribution of MAP 2 in adult brain tissue has also been thoroughly documented. Matus et al (1981) first showed that a polyclonal serum to high-molecular-weight brain MAP reacted with dendrites; this serum has since been found to react with MAP 2 (Matus et al 1983). A number of other studies demonstrated MAP 2 in dendrites (Huber & Matus 1984a,b; Bernhardt & Matus 1982, 1984; Wiche et al 1983; Miller et al 1982; De Camilli et al 1984; Caceres et al 1983) and in post-synaptic densities (Matus et al 1981; Wiche et al 1983; Caceres et al 1984b). These immunocytological data are consistent with the observation that the concentration of MAP 1 is approximately fivefold that of MAP 2 in white matter (Vallee 1982).

There are indications that MAP 2 may exist outside of dendritic processes and perikarya. MAP 2 has been found in the axons of spinal motor neurons but not in dorsal root axons or spinal cord white matter

(Papasozomenos et al 1985). Papasozomenos & Binder (1986) demon-strated that astrocytes in the periphery of the optic nerve labeled with MAP 2 antibodies, whereas astrocytes in the optic tract did not. These data suggest that neural regions outside the brain may possess different distributions of MAP than have been observed in the cerebellum and cerebrum.

Tau appears to have a distribution complementary to MAP 2; it is found almost exclusively in axons, and only weak labeling is seen in cell bodies and dendrites (Binder et al 1985). No glia are stained with tau antibodies. Quantitative immunoassays support these observations: tau is more abundant in white matter than in gray matter.

Immunocytological analyses of MAP 3 and MAP 4 in neural tissues have also been carried out. Antibodies to MAP 3 react strongly with glia and neurofilament-rich axons. MAP 3 is not in dendrites or in most nonmyelinated axons (Matus et al 1983; Huber et al 1985). MAP 4, a very minor component of cycled microtubule preparations compared to MAP 1 and MAP 2 (Parysek et al 1984a), is localized exclusively in glial cells in the CNS (Parysek et al 1984b, 1985). Both astrocytes and oligodendrocytes are labelled in situ, which suggests that MAP 4 may be a glial-specific marker in the adult brain.

Several studies have examined the distribution of MAPs in cells cultured from neural tissue. Immunocytological analyses with monoclonal anti-bodies to MAP 2 showed cell-type specificity of labeling in primary cultures of brain that was similar to that seen in situ. However, MAP 2 staining did not remain in a restricted cell domain once cells had been cultured (Izant & McIntosh 1980; Bloom & Vallee 1983; Bloom et al 1984a). In contrast, cultured hippocampal cells (Caceres et al 1984a) and granule cells (Cumming et al 1984; Alaimo-Beuret & Matus 1985) have labeling patterns that mimic those seen in vivo.

Peng et al (1986) have used a novel approach to examine MAP dis-tribution in neuronal cells. By growing sympathetic neurons in culture under conditions that cause extensive neurites to form, the focus of cell bodies and dendrites can be microdissected free from the axonlike neurites, and the components of these two compartments can be compared. Several proteins were equally distributed, including MAP 1A, MAP 1B, MAP 3, and a 60-kDa MAP. As predicted from immunological studies, tau was primarily in axons, whereas MAP 2 was in the cell body and in dendrites.

OCCURRENCE OF BRAIN MAPs OUTSIDE OF NEURONAL TISSUE   Although some studies have attempted to identify MAPs in nonneuronal cells on the basis of gel mobility, immunological approaches have been the most powerful in demonstrating whether various MAPs occur in a variety of cell types and tissues.

MAP 1 appears to be very widely distributed. Using affinity-purified polyclonal antisera and immunoblotting or immunostaining, Wiche and colleagues (Wiche et al 1984; Koska et al 1985; Zernig & Wiche 1985) found MAP 1 antibody–reactive material in a variety of cell and tissue types. Immunostaining was largely associated with typical microtubule arrays. Bloom et al (1984a) found similar breadth in the reaction of monoclonal antibodies to MAP 1A in cell lines, both with respect to species (rodent, human) and tissue (lung, heart, muscle, kidney) distribution. MAP 1A, as well as MAP 1B, has been localized to anterior pituitary (Bloom et al 1985a).

In addition to staining microtubule arrays, MAP 1 immunoreactive species have been identified in nonmicrotubular structures, including stress fibers (Asai et al 1985), microtubule organizing sites such as spindle poles and kinetochores (Mascardo et al 1982; De Mey et al 1984), nuclear flecks (Sato et al 1985), and Z lines of myocytes (Zernig & Wiche 1985). How these staining patterns may relate to microtubule function is still unknown.

MAP 2 distribution has also been analyzed in detail, but reports as to the presence of this protein outside of nervous tissue are disparate. Izant & McIntosh (1980) demonstrated by immunoassay that monoclonal antibodies to MAP 2 react strongly with mouse brain but not with other mouse tissues (kidney, liver) nor with a number of cultured cell lines (HeLa, 3T3, CHO, PtK1). Using a polyclonal antibody, Valdivia et al (1982) quantified the amount of MAP 2 in a variety of porcine tissues by immunoassay. They found very high levels in brain (~ 10 μg/mg protein) but much smaller amounts (100–1000-fold lower) in kidney, lung, spleen, and liver. Using both immunological and biochemical approaches, Weatherbee et al (1982) detected MAP 2 in HeLa cell extracts or cycled microtubules. However, they noted it was a minor component as compared to the heat stable 200-kDa MAP in HeLa cells. A number of other studies, primarily immunocytological, have indicated that MAP 2 is localized in: the marginal band of erythrocytes from chicken, toad, and newts (Sloboda & Dickersin 1980); secretory cells in anterior pituitary (Bloom et al 1985a); and chromaffin cells from the adrenal medulla (Burgoyne & Norman 1985). However, Northern blot analyses with cloned cDNA probes to MAP 2 indicate few transcripts are in tissues other than brain (Lewis et al 1986b).

The localization of tau in a number of cell types was demonstrated in early immunostaining papers (Connolly et al 1977; Connolly & Kalnins 1980a). Biochemical analyses (Cleveland et al 1979), immunological cross-reactivity of tau antibodies with proteins of many species (Wood et al 1986; Binder et al 1985), and Southern blot analyses using cloned probes (Drubin et al 1984a) demonstrate that the tau proteins are highly conserved.

LOCATION OF OTHER MAPs   The distribution of the 210-kDa HeLa MAP and MAP 4 has been extensively examined. Bulinski & Borisy (1980c) first demonstrated that the 210-kDa HeLa MAP is present in many cell types, including cell lines and primary cultures derived from brain, kidney, and various epithelia and endothelia. Both interphase and mitotic microtubule arrays are labeled (Bulinski & Borisy 1980b; De Brabander et al 1981), although only cells of primate and marsupial origin are immunoreactive. Izant et al (1982) found that one monoclonal antibody raised to the HeLa 210-kDa MAP stains only the mitotic apparatus and nuclear spots, whereas other monoclonal antibodies show staining on both mitotic and interphase arrays (Izant et al 1983). It is conceivable that each of these monoclonal antibodies recognizes epitopes unique to a subspecies of the 210-kDa HeLa MAP; however, these studies did not examine this possibility.

The species-specific homolog of the 210-kDa HeLa MAP, MAP 4, is also widely distributed, although antibodies to this mouse protein react only with mouse tissues. Based on evidence from both immunoblotting and immunocytology (Parysek et al 1984a; Olmsted et al 1986), MAP 4 is present in a variety of mouse cell lines and in a number of tissue types, including heart, liver, lung, and brain. It appears to be totally absent, however, from mature sperm and from all types of blood cells. Immunostaining of 0.5-μm tissue sections in which microtubules are visible shows that MAP 4 is restricted to microtubule arrays within a subset of cells in each tissue type (Parysek et al 1984b, 1985).

Antibodies to a 205-kDa thermostable MAP in *Drosphila* label microtubule arrays in interphase and mitotic Schneider cells (Goldstein et al 1986). The antibodies to this protein do not react with the human 210-kDa protein. This finding is consistent with other studies that suggest that this class of MAPs is highly species specific.

The location of chartins has just started to be analyzed. The biochemical data suggest that these MAPs are widely distributed across cell and tissue types (Duerr et al 1981; Pallas & Solomon 1982), and the low antigenicity of these proteins (Magendantz & Solomon 1985) suggests that they are highly conserved. Antibodies to the 69-kDa protein react with both spindle and interphase microtubule arrays.

Progress has recently been made in examining MAP distribution in sea urchin eggs. Vallee & Bloom (1983) found monoclonal antibodies that react with MAPs of 235, 205, 150, and 37 kDa that were all localized to the mitotic apparatus. Scholey et al (1985a) found two sea urchin MAPs that had slightly different distributions in situ. Antibodies to the 78-kDa MAP localized primarily to the spindle but appeared diffuse, whereas those to the 80-kDa MAP were more uniformly distributed, with astral regions staining more intensely than the main spindle body. The sea urchin homolog of kinesin, a molecule thought to be active in particle movements in axoplasm (Vale et al 1985a), has recently been localized in the spindle (Scholey et al 1985b). Dynein antibodies have also been shown to decorate microtubules in the spindle of sea urchins (Hirokawa et al 1985b). Whether any or all of these proteins are involved in generating the forces for mitotic movements is still unknown.

GENES FOR MAPs

Investigation of the genes that code for MAPs was begun only very recently. Drubin et al (1984a) enriched for tau message by immunoprecipitation of polysomes from mouse brain and prepared a cDNA clone that coded for tau in an in vitro translation system. Although the composition of tau is complex, there appears to be only one 6-kb gene that codes for all tau peptides. Southern blot analyses indicate that one fragment of the gene reacts strongly with mouse, human, chicken, and frog DNA, whereas another clone reacts weakly with human and chicken DNA. Both clones react little with either *Drosophila* or sea urchin DNA, which suggests that although regions of the gene are highly conserved, others may be divergent.

Using an expression library prepared from mouse brain cDNA, Lewis et al (1986a,b) recently obtained clones coding for both MAP 1 and MAP 2. There appears to be only one gene for each of these proteins, and Southern blot analyses show that homologous genes for MAP 1 or MAP 2 may exist in rat, human, and chicken; no reaction is seen with frog, *Drosophila*, or sea urchin DNA. Transcripts that hybridize with the MAP 2 clone are present only in brain and not in liver, kidney, spleen, stomach, or thymus. These data conflict with some of immunological and biochemical results described earlier on the presence of MAP 2 outside of the brain. In contrast, MAP 1 transcripts are found in all of these tissues, although the level in brain is 500-fold higher than in the other tissue types.

The sequence coding for the 205-kDa thermostable MAP from *Drosophila* has been isolated from a genomic expression library (Goldstein et al 1986). This MAP is coded for by one gene that has been localized to the polytene region 100-EF by in situ hybridization.

The four MAPs that have been cloned to date are all coded for by single genes. However, biochemical analyses show each of these MAPs is a complex of related polypeptides. As discussed below, several messages appear to be transcribed from the 6-kb gene for tau. However, for the other MAPs (MAP 1, MAP 2, and *Drosophila* 205-kDa MAP), the coding regions for the genes appear to be similar in size to those needed to code for a single protein. Some of the complexity in each MAP may arise from posttranslational modification; however, it remains to be determined whether some of the heterogeneity in the large MAPs is generated by alternative splicing of transcripts.

APPROACHES TO EXAMINING MAP FUNCTION

In this section, a variety of experiments are described that focus on how MAPs function in vivo. Although these studies are still largely descriptive, the changes in the location and heterogeneity of MAPs may yield important clues as to the function of the molecules.

Interaction of MAPs with Cytoplasmic Structures Other Than Microtubules

There have been a number of studies that have investigated whether MAPs serve as linkers between the other two major cytoskeletal systems in the cell, intermediate filaments and microfilaments, in addition to having a role in particle movement. Evidence for such associations is based on biochemical and morphological studies.

INTERMEDIATE FILAMENT–MICROTUBULE INTERACTIONS   There is a large body of morphological work demonstrating that microtubules and neurofilaments are in close proximity and that interconnecting links may bridge these two cytoskeletal structures (e.g. see Hirokawa et al 1985a and Runge et al 1981 for discussion of some of these studies). A number of biochemical analyses have been carried out to examine whether such an association could be reproduced in vitro. Runge et al (1979) demonstrated that a major copurifying component of cycled microtubule protein preparations from brain was the 68-kDa protein of neurofilaments. Subsequently, these workers (Runge et al 1981) showed that an ATP-dependent complex between microtubules and neurofilaments could form. Leterrier and coworkers (1982, 1984) have explored this interaction further using purified components. When microtubule protein preparations are mixed with purified neurofilaments, the assembly of microtubules is depressed; in contrast, glial filament protein has no effect. Analysis of the pellet components in these mixtures demonstrated that the neurofilament proteins appear to bind both MAP 1 and MAP 2. When ^{32}P-labeled MAP fractions are incubated with electrophoretically separated neurofilament proteins bound to nitrocellulose, the major binding site for MAP 2 is the 70-kDa polypeptide of the neurofilament complex (Heimann et al 1985). Microtubules do not compete with the 70-kDa proteins for binding of the

MAPs, suggesting that the sites for these interactions are different. Aamodt & Williams (1984) have shown that neurofilaments from brain contain proteins resembling MAPs that bind to microtubules. In contrast, neurofilaments from spinal cord do not possess these components, and no in vitro association is observed when spinal cord neurofilaments are mixed with microtubules. However, if the proteins from brain neurofilaments are added to spinal cord neurofilaments, microtubules bind. Since these proteins are not part of the neurofilament triplet, these data suggest that neurofilaments in various regions of the nervous system may have associated proteins that foster distinct types of interactions with microtubules.

In vivo data support the biochemical studies on the interaction between microtubules and neurofilaments or intermediate filaments. As reviewed by Lasek et al (1984), tubulin and neurofilament proteins migrate at the same rate during axoplasmic transport, which suggests that these elements are physically linked to one another. Evidence from a number of studies suggests that MAPs may provide a link between microtubules and intermediate filaments of various types. Early morphological work showed that the disruption of microtubules by various antimitotic drugs was paralleled by the reorganization of intermediate filaments (Goldman & Knipe 1972). Injection of antibodies to tubulin disrupts the microtubule arrays and causes the collapse of intermediate filaments into a perinuclear cap (Blose et al 1984). Several studies suggest that MAP 2 may modulate at least some of the interactions between filaments and microtubules. Using cultured brain cells, Bloom & Vallee (1983) demonstrated that MAP 2 staining distributes with both vimentin and microtubule arrays. Following vinblastine treatment, MAP 2 localizes with the juxtanuclear cap of intermediate filaments. In contrast, under similar conditions MAP 1A and 1B become dispersed and do not associate with the intermediate filaments (Vallee et al 1984a). The distribution of MAP has also been examined in β,β-iminodipropionitrile (IDPN)-treated axons. Following this treatment, the normally interdigitated array of neurofilaments and microtubules in the axon becomes stratified, with the neurofilaments at the perimeter and the microtubules in the center of the axon. Papasozomenos et al (1985) found that two different monoclonal antibodies to MAP 2 react with the microtubules in normal spinal motor axons. However, following IDPN treatment, one antibody reacts with microtubule arrays, and the other binds to neurofilaments. Although MAP 2 may bind with highest affinity to microtubules, the studies cited above demonstrate that there may be binding sites on this protein or its subspecies for neurofilaments. No such associations have yet been shown for any of the other MAPs, although recent studies have suggested that tau (Brion et al 1985; Wood et al 1986), as well as MAP 2 (Kosik et al 1984), may be associated with the neurofibrillary tangles found in nervous tissue from patients with Alzheimer's disease.

ACTIN-MICROTUBULE INTERACTIONS   Another set of experiments dealt with whether or not actin and microtubules interact. Studies by Griffith & Pollard (1978) first suggested that an actin-microtubule complex could form in vitro; this association appeared to be inhibited by ATP and dependent on MAPs. Subsequent analyses (Sattilaro et al 1981; Selden & Pollard 1983; Pollard et al 1984) demonstrated that both MAP 2 and tau could modulate the formation of high viscosity gels of microtubules and actin filaments. The extent of phosphorylation of the MAP apparently affected this cross-linking: at higher concentrations of MAP-bound phosphate there was less gelation (Nishida et al 1981; Selden & Pollard 1983; Pollard et al 1984). Recent studies suggest that calcium-calmodulin may affect the MAP interactions with actin filaments (Kotani et al 1985; Sobue et al 1985). The affinity of MAP for the F-actin filaments is much weaker than that for microtubules.

MAPs AND MICROTUBULE-ASSOCIATED MOVEMENTS   Microtubules have been implicated as structures important in the intracellular transport of organelles, and in vitro studies have provided evidence that suggests interactions between secretory granules and microtubules may be mediated by MAPs (Sherline et al 1977; Suprenant & Dentler 1982). Some of the MAPs in the pituitary have recently been identified (Bloom et al 1985), but how interactions between microtubules and granules are mediated in vivo is still unknown.

A role for MAPs in axonal transport has been postulated for many years, in part because axonal microtubules have prominent projections that could interact with other components. Lasek and colleagues (see review in Lasek et al 1984) have demonstrated that microtubules, neurofilaments, and tau all migrate at a similar rate in the axon, and they have suggested that these components are structurally linked to one another. Tytell et al (1984) showed that only two of the four tau proteins synthesized in the retina are transported; it is not yet known whether retinal ganglion cells synthesize only a subset of proteins, or whether these two tau proteins have specific functions in the axon.

Several studies investigated how microtubule-vesicle associations in axons might be made. Miller & Lasek (1985) recently demonstrated a physical association between microtubules and vesicles being transported in both directions in the axon. As described previously, a number of groups devised ingenious assays to try to characterize the nature of these crossbridges [see Allen et al 1985; Brady 1985; Vale et al 1985a,b; Schroer & Kelly 1985 (review)].

Role of MAPs in Microtubule Assembly and Organization

In vitro studies have shown that the assembly of tubulin is potentiated by the presence of MAPs and that MAPs are important in stabilizing microtubules (see Purich & Kristofferson 1984 for review). Whether or not MAPs have similar roles in vivo, either in regulating microtubule polymer formation or in organizing sites where microtubule initiation may occur, has been the subject of a number of investigations.

MAPs AND MICROTUBULE ASSEMBLY IN VIVO : STUDIES IN CULTURED CELLS   A number of studies have attempted to identify factors that might be important for microtubule assembly during neurite elongation. Seeds & Maccioni (1978) first showed that extracts from differentiated neuroblastoma cells promoted assembly of microtubules in vitro, whereas extracts of undifferentiated cells did not. Izant & McIntosh (1980) subsequently demonstrated that MAP 2 levels increase upon the cAMP-induced differentiation of neurites in a rat neuroblastoma cell line. Using cultured mouse neuroblastoma cells, Olmsted & Lyon (1981) showed that microtubules polymerized in differentiated cell extracts but not in those from undifferentiated cells. These two cell types contain the same total tubulin concentration, but the differentiated cells have four to five times more polymer than do the rounded cells (Olmsted 1981). Two-dimensional gel analyses and double label quantification did not detect MAP 4 in the extracts of suspension cultures; if MAP 4 is present, its concentration relative to tubulin is at least tenfold lower than in the neurite-bearing cells (Olmsted & Lyon 1981). MAP 4 synthesis increases dramatically after undifferentiated cells are plated onto a substrate (Olmsted et al 1984), which suggests that the synthesis of this protein is coordinated with the microtubule assembly that occurs during neurite formation. The synthesis of MAP 1 and tau also increases during neurite formation (Drubin et al 1984b).

Similar studies have examined MAP and tubulin distribution during NGF- (nerve growth factor) or cAMP-induced neurite differentiation in PC12 pheochromocytoma cells. Greene et al (1983) first demonstrated that MAP 1 is synthesized and phosphorylated during neurite outgrowth. Drubin et al (1984b, 1985) quantified the expression of MAP 1 and tau relative to the distribution of total tubulin in polymer form. Up to three days after NGF induction there is little neurite formation or change in microtubule mass. Thereafter, up to day 10, there is a coordinate increase in both microtubule mass and neurite length. The total pool of tubulin changes by less than 2.5-fold over this time course, whereas MAP 1 and tau increase 20-fold. However, the amounts of MAP 1 and tau increase in parallel with microtubule *mass*. The synthesis of the MAPs starts to increase at day 3 and continues to rise exponentially. If cells growing in suspension are treated with NGF, MAP 1 and tau proteins are synthesized, and upon plating these primed cells, neurite elongation occurs without a lag. All of these data strongly suggest that the appearance of the MAPs promotes the assembly of tubulin from the preexisting pool.

Whether the appearance of MAPs during neurite formation is transcriptionally or translationally controlled has also been examined. Neither tau (Drubin et al 1984b) nor MAP 1 (Lewis et al 1986a) mRNA levels change during PC12 induction. However, in contrast to the results discussed above, Lewis et al (1986a) did not find an increase in MAP 1 in differentiated PC12 cells. Although the origin of this disparity is not yet known, it is conceivable that two different MAP 1 subspecies are being detected in these studies and that the expression of one is translationally regulated and the other is not.

MAPs AND MICROTUBULE ASSEMBLY: STUDIES IN DEVELOPING ORGAN- ISMS   The distribution and activity of MAPs at various stages of brain development have now been studied for almost a decade. Early studies (Francon et al 1978; Lennon et al 1980) demonstrated that material from younger animals was generally less effective in polymerization assays than that from adults. Based upon the apparent differences in tau composition in the brains of 3-day-olds and adults (Mareck et al 1980), it was postulated that changes in tau polypeptides are related to the variation in the polymerization activity of the extracts. Subsequently, Francon et al (1982) showed that the peptide maps of the four tau species in adult rat brain are very similar, whereas the two tau species in young rats are dissimilar to each other and to the adult types. Using in vitro translation assays, Ginzburg et al (1982) found that mRNA from young animals synthesized five tau proteins; however, only two of these proteins were specific for the immature brain, whereas the other three were characteristic of adult brain. When mRNA from adult animals was analyzed, only the adult forms (four proteins) were synthesized. These data suggest that regulation occurs at both transcriptional and translational levels.

Drubin et al (1984a) recently examined the expression of tau using cloned cDNA probes. They also found changes in the complexity of the tau proteins during development but saw essentially no change in the abundance of the mRNA from day 5 to adulthood. Why the expression of various tau subspecies is important during brain development, and how it relates to tubulin expression and assembly, are currently being investigated.

Changes in the amount and/or polypeptide composition of the other major brain MAPs during development have also been documented. Binder et al (1984) and Burgoyne & Cumming (1984) showed that there is a single MAP 2 peptide (MAP 2B) in young animals, while there are two MAP 2 polypeptides in adults. Binder et al (1984) suggested that increased phosphorylation of MAP 2 subspecies may also occur during brain maturation. Recent analyses (Calvert & Anderton 1985) show that a specific form of MAP 1, designated MAP 1X, is present in neonatal brain but decreases fivefold through adulthood. Lewis et al (1986a) noted a substantial decrease in mRNA coding for MAP 1 during postnatal development. The relation of the neonatal MAP 1 to the more widely distributed MAP 1 species discussed earlier is not yet known.

Riederer & Matus (1985) compared three MAPs during brain development. As shown previously (Burgoyne & Cumming 1984; Binder et al 1984), MAP 2B is present from birth in relatively constant amount, whereas MAP 2A appears between days 10 and 15. MAP 1 increases steadily from birth to day 20 (11-fold) but decreases slightly in the adult. MAP 3A decreases from birth to adulthood ($\sim$ 30-fold), whereas MAP 3B peaks at day 10 and then decreases abruptly. MAP 4 also changes in complexity during brain development (Olmsted et al 1986). At fetal day 10, only MAP 4A is present, MAP 4B appears at day 13, and MAP 4C is present from day 21 (newborn) to adulthood.

To understand how these biochemical changes correlate with neural development, analyses of MAP distribution during brain maturation have been carried out. With one exception, however, all such studies employed antibodies that do not discriminate among the complex polypeptides. In immunocytological analyses of neonatal rat brain (Bernhardt & Matus 1982), the appearance of MAP 2 staining in dendrites was noted to antedate the appearance of microtubules in the process, and it was proposed that this MAP may function in organizing the microtubule array. Similar patterns were observed by Burgoyne & Cumming (1984), who postulated that since MAP 2B is the only form of MAP 2 present in early neonates, the efficient formation of MAP-microtubule interactions may not occur until the second form of MAP 2 (MAP 2A) appears later in development.

Comparative studies (Riederer & Matus 1985, Bernhardt et al 1985) on the distribution of three brain MAPs have shown that MAP 1 staining is weak at day 5 and increases steadily to adulthood in the neurons of the cerebrum, whereas MAP 2 is always present in dendrites. In contrast, the distribution of MAP 3 varies radically with age. Early in development (through day 10), the distribution of MAP 3 is primarily neuronal and is concentrated in the distal region of the dendrites. However, this pattern gradually changes to a glial-specific pattern by day 20. In cerebellum slightly different patterns are seen for MAPs 1 and 3. MAP 1 is initially present in axons and then becomes more concentrated in dendrites. MAP 3 is also initially found in axons, including parallel fiber and granule cells;

however, it is not present in these cell types in adult brain. How these changes correlate with changes in MAP subspecies and microtubule function in these cell types at various developmental stages remains to be elucidated.

Studies have recently been initiated to examine MAPs in early mouse embryogenesis (Bates & Kidder 1984; Olmsted et al 1986). Immunoblotting analyses of MAP 4 have demonstrated that MAP 4A is present in constant amount per embryo from zygote through early blastocyst, but MAP 4B and 4C are absent (Olmsted et al 1986). MAP 4 is not found in mature sperm (Parysek et al 1984a,b), but it has been localized in developing oocytes in the ovary (L. M. Parysek & J. B. Olmsted, unpublished observations). These data suggest that at least MAP 4A is a maternal gene product. At what stages of early development the other components of MAP 4 appear is a subject of current investigation.

MAPs AND MICROTUBULE STABILITY   Based upon in vitro evidence, it has been postulated that MAPs may stabilize microtubules by inhibiting disassembly. In lysed cell models, MAP-coated microtubules were more stable in the presence of added calcium than microtubules alone (Schliwa et al 1981). In vivo data also suggest that MAPs affect microtubule stability. Black & Greene (1982) noted that microtubules in PC12 cells became more resistant to microtubule depolymerizing agents after several days of neurite growth. Based upon the recent studies that show MAPs are assembled with microtubules during this time, it is conceivable that this increased stability reflects MAP interaction with formed microtubules. Brady et al (1984) showed that there is a population of cold-stable microtubules in axoplasm. Although the MAP composition of these microtubules was not determined, it has been suggested that the differential distribution of MAPs and STOPs could be important in generating microtubule arrays of varying stability (Job et al 1985). As already demonstrated (Pallas & Solomon 1982), phosphorylation may be one way in which the association of MAPs with microtubules is modulated in vivo.

MAPs AND ORGANIZING SITES   MAPs may foster either the initiation of microtubule assembly or the insertion of preformed microtubules at defined cellular organizing sites. Stearns & Brown (1979) isolated the microtubule-organizing center (MTOC) of the alga *Polytomella* and demonstrated that a set of four polypeptides with molecular masses of 190–210-kDa promoted the in vitro assembly of tubulin. A number of immunofluorescence studies suggest that well-defined MAPs may be localized in organizing sites. Material reactive with MAP 1 antibody has been found in centrosomes (De Mey et al 1984; Vallee et al 1984a; Mascardo et al 1982) and in kinetochores (De Mey et al 1984), and material that reacts with MAP 2 antibody has been localized in centrosomes (Mascardo et al 1982). Immunofluorescence analyses also suggest that MAP 4 may localize at discrete sites from which microtubules emanate following taxol treatment of neuroblastoma cells (Olmsted et al 1986). Maro et al (1985) noted that pericentriolar material stained with a human autoimmune serum localizes with random microtubule foci following treatment of the mouse oocytes with taxol. Nigg et al (1985) reported that cAMP-dependent kinase, the enzyme that binds to the projection arm of MAP 2, is associated with centrosomes. This later observation is particularly interesting in view of the finding that there are cell-cycle-specific changes in the phosphoproteins localized to centrosomes and kinetochores (Vandre et al 1984). In recent studies, Olmsted et al (1985) demonstrated that microinjected fluoresccinated MAP 2 associates with MTOCs following disruption of the microtubules by lowering the temperature. All of these data demonstrate that MAPs exist at microtubule organizing sites, but their function at these sites is still a mystery.

Investigations have been undertaken to determine whether templates other than well-defined MTOCs exist that dictate microtubule growth. In studies on the highly organized marginal band of nucleated erythrocytes, Miller & Solomon (1984) showed that the reassembly of microtubules into the marginal band appears to proceed from only one or two assembly points. Further, using cells extracted of endogenous tubulin, Swan & Solomon (1984) demonstrated that the pattern of reassembly into a marginal band is maintained when brain tubulin is added to extracted ghosts. These data suggest that there are intrinsic components associated with the periphery of the membrane that guarantee the proper assembly and alignment of the microtubules into bundles. The nature of these template molecules is not yet known, but they can be envisaged as linkers associated with the membrane that determine microtubule distribution. An intriguing, but as yet untested, concept is that MAPs may act as templates for the oriented growth of microtubules in other tissues.

MAP Distribution and Function Assayed in Living Cells

To understand how MAPs function, experiments that permit perturbation or analysis of living cells are necessary. The following section briefly reviews some of these approaches, which have only recently begun to be used.

MUTANT ANALYSES   The generation of mutations that affect particular proteins has been a powerful technique in the analysis of many cellular functions. Selection of mutants resistant to microtubule-active drugs has generated a number of tubulin mutants; most of these are affected in mitotic events. [For an excellent summary of this kind of work, see Section 2 of the volume edited by Borisy et al (1984).] In addition, mitotic mutants have been described that have no apparent tubulin defects (Cabral et al 1983). However, whether these mutations correspond to functional alterations in MAPs is still unknown.

ANTIBODY INJECTION   Antibodies can be used to neutralize protein activities, and one study has appeared in which the injection of monospecific antibodies to a MAP interfered with a microtubule-based phenomenon. Izant et al (1983) injected a monoclonal antibody to the 210-kDa HeLa MAP into PtK$_1$ cells. As assessed following fixation and secondary antibody labeling, the antibody binds to spindles and interphase arrays, but there is no evidence that any microtubule dissolution occurs. However, if this antibody is injected just prior to the onset of anaphase, chromosome movement is inhibited. This result was not found with another monoclonal antibody to the 210-kDa MAP. While these monoclonal antibodies were not characterized further with respect to binding sites, these data do suggest that either a subspecies or a modified form of HeLa 210-kDa MAP may be important in mitotic events.

FLUORESCENCE ANALOG CYTOCHEMISTRY   The use of fluorescence analog cytochemistry, in which the distribution of fluorophore-coupled molecules microinjected into cells is followed, has recently been applied to the study of MAPs. Scherson et al (1984) demonstrated that fluorescently labeled MAP 2 decorates interphase microtubules. Fluorescence redistribution following photobleaching (FRAP) experiments suggested that the half-time for redistribution of MAP 2 in interphase cells was approximately 5 min. In more recent studies (Olmsted et al 1985 and unpublished results), both MAP 2 and MAP 4 were analyzed. Both MAPs are rapidly distributed (within 1–2 min) into microtubule-like arrays in interphase or metaphase cells. Following cold treatment, microtubules disappear, and fluorescent MAPs become localized to centrosomes or spindle poles. Using FRAP analysis, the dynamics of MAP 2 and MAP 4 relative to each other and to tubulin were determined. These data show that MAP 2 and MAP 4 have similar rates of redistribution at a given stage of the cell cycle, but the half-time recovery in metaphase cells is approximately 3–4 times faster ($t_{1/2}$ = 20–30 sec) than in interphase cells ($t_{1/2}$ = 70–90 sec). In addition, MAPs in interphase cells turn over approximately 2–3 times more rapidly than tubulin (Saxton et al 1984). From these data we can infer that microtubules in interphase cells are constantly changing with respect to MAP composition. Analyses of MAP (Olmsted et al 1985) and tubulin

(Saxton et al 1984) dynamics in metaphase cells at 37°C show that both are extremely fast ($t_{1/2} \approx$ 11–20 sec); these data suggest that MAPs turn over rapidly because microtubules are constantly being depolymerized and repolymerized. However, preliminary experiments indicate that tubulin and MAP dynamics in metaphase cells at 26°C are separable. While some tubulin exchanges rapidly at 26°C, a population of microtubules exists that is much less dynamic. MAPs appear to interact rapidly with both populations. These data suggest that the cell modulates the rates of association and dissociation between microtubules and MAPs as a function of time in the cell cycle, and this modulation may be one of the ways in which the cell controls the polymerization state of tubulin.

SUMMARY AND PERSPECTIVES

The study of MAPs up to this point has largely been directed at understanding the detailed biochemistry of the isolated molecules and documenting where these proteins occur in cells. These studies have generated many unanswered questions. Why do so many types and subtypes of MAPs exist? All MAPs bind to microtubules in vitro; what do the differences in the cellular distribution of these molecules reflect in terms of function? Do some MAP subspecies, or modified states of MAPs, organize microtubule growth and others stabilize formed polymers or act as templates for subunit elongation? Since microtubules can be formed from tubulins of several different isotypes or modified states, do highly specific types of MAP interact with highly specific types of tubulin lattices to generate microtubule subclasses of as yet undefined function? Is there a causal or just a correlative relationship between the synthesis of MAPs and microtubule assembly during processes such as neurite extension in vitro or neuronal development in vivo, and why is more than one type of MAP expressed? Clearly, much work remains to be done before an understanding of how MAPs function in events as diverse as particle movement, microtubule growth, and mitosis is achieved.

ACKNOWLEDGMENTS

My thanks to colleagues who discussed unpublished results and sent preprints of their work to be included in this chapter. I am grateful to Mrs. Cheryl Cicero for her unflagging help and cheerfulness during the preparation of the manuscript. The work described from the author's lab was supported by NIH research grant GM22214. This review covers work published through January of 1986.

Literature Cited

Aamodt, E. J., Williams, R. C. Jr. 1984. Microtubule-associated proteins connect microtubules and neurofilaments in vitro. *Biochemistry* 23 : 6023–31

Alaimo-Beuret, D., Matus, A. 1985. Changes in the cytoplasmic distribution of microtubule-associated protein 2 during the differentiation of cultured cerebellar granule cells. *Neuroscience* 14 : 1103–16

Allen, R. D., Weiss, D. G., Hayden, J. H., Brown, D. T., Fujiwake, H., Simpson, M. 1985. Gliding movement of and bidirectional transport along single native microtubules from squid loligo-pealei axoplasm : Evidence for an active role of microtubules in cytoplasmic transport. *J. Cell Biol.* 100 : 1736–52

Amos, L. A. 1977. Arrangement of high molecular weight associated proteins on purified mammalian brain microtubules. *J. Cell Biol.* 72 : 642–54

Andersen, B., Osborn, M., Weber, K. 1978. Specific visualization of the calcium dependent regulatory protein of cyclic nucleotide phosphodiesterase (modulator protein) in tissue culture cells by immunofluorescence microscopy : Mitosis and intercellular bridge. *Eur. J. Cell Biol.* 17 : 354–64

Asai, D. J., Thompson, W. C., Wilson, L., Dresden, C. F., Schulman, H., Purich, D. L. 1985. Microtubule-associated proteins : A monoclonal antibody to MAP-1 decorates microtubules in vitro but stains stress fibers and not microtubules in vivo. *Proc. Natl. Acad. Sci. USA* 82 : 1434–36

Baines, A. J., Bennett, V. 1986. Synapsin I is a microtubule bundling protein. *Nature* 319 : 145–47

Bates, W. R., Kidder, G. M. 1984. Synthesis of putative microtubule-associated proteins by mouse blastocysts during early outgrowth in vitro. *Can. J. Biochem. Cell Biol.* 62 : 885–93

Bennett, V., Davis, J. 1981. Erythrocyte ankyrin : Immunoreactive analogues are associated with mitotic structures in cultured cells and with microtubules in brain. *Proc. Natl. Acad. Sci. USA* 78 : 7550–54

Bernhardt, R., Huber, G., Matus, A. 1985. Differences in the developmental patterns of three microtubule-associated proteins in the rat cerebellum. *J. Neurosci.* 5 : 977–91

Bernhardt, R., Matus, A. 1982. Initial phase of dendrite growth : Evidence for the involvement of high molecular weight microtubule-associated proteins (HMWP) before the appearance of tubulin. *J. Cell Biol.* 92 : 589–93

Bernhardt, R., Matus, A. 1984. Light and electron microscopic studies of the distribution of microtubule-associated protein 2 in rat brain : A difference between dendritic and axonal cytoskeletons. *J. Comp. Neurol.* 226 : 203–21

Binder, L. I., Frankfurter, A., Rebhun, L. I. 1985. The distribution of tau polypeptides in the mammalian central nervous system. *J. Cell Biol.* 101 : 1371–78

Binder, L. I., Frankfurter, A., Kim, H., Caceres, A., Payne, M. R., Rebhun, L. I. 1984. Heterogeneity of microtubule-associated protein 2 during rat brain development. *Proc. Natl. Acad. Sci. USA* 81 : 5613–17

Black, M., Kurdyla, J. T. 1983. Microtubule-associated proteins of neurons. *J. Cell Biol.* 97 : 1020–28

Black, M., Greene, L. A. 1982. Changes in the colchicine susceptibility of microtubules associated with neurite outgrowth : Studies with nerve growth factor-responsive PC12 pheochromocytoma cells. *J. Cell Biol.* 95 : 379–86

Bloom, G. S., Luca, F. C., Vallee, R. B. 1984a. Widespread cellular distribution of MAP 1A (microtubule-associated protein 1A) in the mitotic spindle and on interphase microtubules. *J. Cell Biol.* 98 : 331–40

Bloom, G. S., Luca, F. C., Vallee, R. B. 1985a. Identification of high molecular weight microtubule-associated proteins in anterior pituitary tissue and cells using taxol-dependent purification combined with microtubule-associated protein specific antibodies. *Biochemistry* 24 : 4185–91

Bloom, G. S., Luca, F. C., Vallee, R. B. 1985b. Microtubule-associated protein 1B : Identification of a major component of the neuronal cytoskeleton. *Proc. Natl. Acad. Sci. USA* 82 : 5404–8

Bloom, G. S., Schoenfeld, T. A., Vallee, R. B. 1984b. Widespread distribution of the major polypeptide component of MAP 1 (microtubule-associated protein 1) in the nervous system. *J. Cell Biol.* 98 : 320–30

Bloom, G. S., Vallee, R. B. 1983. Association of microtubule-associated protein 2 (MAP 2) with microtubules and intermediate filaments in cultured brain cells. *J. Cell Biol.* 96 : 1523–31

Blose, S. H., Meltzer, D., Feramisco, J. 1984. 10 nm filaments are induced to collapse in living cells microinjected with monoclonal and polyclonal antibodies against tubulin. *J. Cell Biol.* 98 : 847–58

Bonifacino, J. S., Klausner, R. D., Sandoval, I. V. 1985. A widely distributed nuclear protein immunologically related to the microtubule-associated protein MAP 1 is associated with the mitotic spindle. *Proc. Natl. Acad. Sci. USA* 82: 1146–50

Borisy, G. G., Cleveland, D., Murphy, D. B., eds. 1984. *Molecular Biology of the Cytoskeleton*. Cold Spring Harbor, New York: Cold Spring Harbor Lab. 512 pp.

Borisy, G. G., Marcum, J. M., Olmsted, J. B., Murphy, D. B., Johnson, K. A. 1975. Purification of tubulin and of associated high molecular weight proteins from porcine brain and characterization of microtubule assembly in vitro. *Ann. NY Acad. Sci.* 253: 107–32

Brady, R. C., Cabral, F. R. 1985. Identification of a 210-kDa microtubule-associated protein in chinese hamster ovary cells. *J. Cell Biol.* 101: 30a

Brady, S. T. 1985. A novel brain ATPase with properties expected for a fast axonal transport motor. *Nature* 317: 73–75

Brady, S., Tytell, M., Lasek, R. 1984. Axonal tubulin and axonal microtubules: Biochemical evidence for cold stability. *J. Cell Biol.* 99: 1716–25

Brinkley, B. R. 1985. Microtubule organizing centers. *Ann. Rev. Cell Biol.* 1: 145–72

Brion, J. P., Passareiro, H., Nunez, J., Filament-Durand, J. 1985. Immunological detection of tau protein in neurofibrillary tangles of Alzheimer's disease. *Arch. Biol.* 96: 229–35

Briones, E., Wiche, G. 1985. M_r 205,000 sulfoglycoprotein in extracellular matrix of mouse fibroblast cells is immunologically related to high molecular weight microtubule-associated proteins. *Proc. Natl. Acad. Sci. USA* 82: 5776–80

Browne, C., Lockwood, A., Steiner, A. 1982. Localization of the regulatory subunit of type II cyclic AMP-dependent protein kinase on the cytoplasmic microtubule network of cultured cells. *Cell Biol. Int. Rep.* 6: 19–28

Bulinski, J., Borisy, G. G. 1979. Self-assembly of microtubules in extracts of cultured HeLa cells and the identification of HeLa microtubule-associated proteins. *Proc. Natl. Acad. Sci. USA* 76: 293–97

Bulinski, J., Borisy, G. G. 1980a. Microtubule-associated proteins from cultured HeLa cells. Analysis of molecular properties and effects on microtubule polymerization. *J. Biol. Chem.* 255: 11570–76

Bulinski, J., Borisy, G. G. 1980b. Immunofluorescence localization of HeLa cell microtubule-associated proteins on microtubules in vitro and in vivo. *J. Cell Biol.* 87: 792–801

Bulinski, J., Borisy, G. G. 1980c. Widespread distribution of a 210,000 mol wt

microtubule associated protein in cells and tissues of primates. *J. Cell Biol.* 87: 802–8

Burgoyne, R. D., Cumming, R. 1984. Ontogeny of microtubule-associated protein 2 in rat cerebellum: Differential expression of the doublet polypeptides. *Neuroscience* 11: 156–67

Burgoyne, R. D., Norman, K. M. 1985. Presence of microtubule-associated protein 2 in chromaffin cells. *Neuroscience* 14: 955–62

Burns, R. G., Islam, K. 1984. Stoichiometry of microtubule-associated protein (MAP2): Tubulin and the localization of the phosphorylation and cysteine residues along the MAP2 primary sequence. *Eur. J. Biochem.* 141: 599–608

Burns, R. G., Islam, K., Chapman, R. 1984. The multiple phosphorylation of the microtubule-associated protein MAP2 controls the MAP2 tubulin interaction. *Eur. J. Biochem.* 141: 609–15

Cabral, F., Wible, L., Brenner, S., Brinkley, B. 1983. Taxol-requiring mutant of chinese hamster ovary cells with impaired mitotic spindle assembly. *J. Cell Biol.* 97: 30–39

Caceres, A., Banker, G., Steward, O., Binder, L., Payne, M. 1984a. MAP 2 is localized to the dendrites of hippocampal neurons which develop in culture. *Dev. Brain Res.* 13: 314–18

Caceres, A., Binder, L. I., Payne, M. R., Bender, P., Rebhun, L., Steward, O. 1984b. Differential subcellular localization of tubulin and the microtubule-associated protein MAP 2 in brain tissue as revealed by immunocytochemistry with monoclonal hybridoma antibodies. *J. Neurosci.* 4: 394–410

Caceres, A., Payne, M. R., Binder, L. I., Steward, O. 1983. Immunocytochemical localization of actin and microtubule associated protein MAP 2 in dendritic spines. *Proc. Natl. Acad. Sci. USA* 80: 1738–42

Calvert, R., Anderton, B. H. 1985. A microtubule-associated protein MAP 1 which is expressed at elevated levels during development of rat cerebellum. *EMBO J.* 4: 1171–76

Carlier, M. F., Simon, C., Pantaloni, D. 1984. Polymorphism of tubulin oligomers in the presence of microtubule-associated proteins. Implications in microtubule assembly. *Biochemistry* 23: 1582–90

Cleveland, D. W., Hwo, S.-Y., Kirschner, M. W. 1977. Purification of tau, a microtubule-associated protein that induces assembly of microtubules from purified tubulin. *J. Mol. Biol.* 116: 207–25

Cleveland, D. W., Spiegelman, B. M., Kirschner, M. W. 1979. Conservation of microtubule-associated proteins: Isolation and characterization of tau and HMW from chicken brain and from mouse fibroblasts and comparison to the corresponding mammalian brain proteins. *J. Biol. Chem.* 254: 12670–78

Connolly, J., Kalnins, V., Cleveland, D., Kirschner, M. 1977. Immunofluorescent staining of cytoplasmic and spindle microtubules in mouse fibroblasts with antibody to tau protein. *Proc. Natl. Acad. Sci. USA* 74: 2437–40

Connolly, J. A., Kalnins, V. I., Cleveland, D. W., Kirschner, M. W. 1978. Intracellular localization of the high molecular weight microtubule accessory protein by indirect immunofluorescence. *J. Cell Biol.* 76: 781–86

Connolly, J. A., Kalnins, V. I. 1980a. The distribution of tau and HMW microtubule-associated proteins in different cell types. *Exp. Cell Res.* 127: 341–50

Connolly, J., Kalnins, V. 1980b. Tau and HMW microtubule-associated proteins have different microtubule binding sites in vivo. *Eur. J. Cell Biol.* 21: 296–300

Correia, J. J., Williams, R. C. Jr. 1983. Mechanisms of assembly and disassembly of microtubules. *Ann. Rev. Biophys. Bioeng.* 12: 211–35

Cumming, R., Burgoyne, R. D., Lytton, N. A. 1984. Immunofluorescence distribution of alpha tubulin, beta tubulin and microtubule associated protein 2 during in vitro maturation of cerebellar granule cell neurones. *Neuroscience* 12: 775–82

Davis, J., Bennett, V. 1982. Microtubule-associated protein 2, a microtubule-associated protein from brain, is immunologically related to the alpha subunit of erythrocyte spectrin. *J. Biol. Chem.* 257: 5816–20

Davis, J., Bennett, V. 1984. Brain ankyrin. A membrane-associated protein with binding sites for spectrin, tubulin, and the cytoplasmic domain of the erythrocyte anion channel. *J. Biol. Chem.* 259: 13550–59

De Brabander, M., Bulinski, J. C., Geuens, G., De Mey, J., Borisy, G. G. 1981. Immunoelectron microscopic localization of the 210,000-mol wt microtubule-associated protein in cultured cells of primates. *J. Cell Biol.* 91: 438–45

De Camilli, P., Miller, P. E., Navone, F., Theurkauf, W. E., Vallee, R. B. 1984. Distribution of microtubule-associated protein 2 in the nervous system of the rat studied by immunofluorescence. *Neuroscience* 11: 817–46

Deery, W. J., Means, A. R., Brinkley, B. R. 1984. Calmodulin-microtubule association in cultured mammalian cells. *J. Cell Biol.* 98: 904–10

De Mey, J., Aerts, F., Moeremans, M., Geuens, G., Daneels, G., De Brabander, M. 1984. Anti-MAP 1 (microtubule associated protein 1) reacts with the centrosomes, kinetochores, midbody and spindle of mitotic PTK 2 cells. *J. Cell Biol.* 99: 447a

Dentler, W. L., Grannett, S., Rosenbaum, J. L. 1975. Ultrastructural localization of the high molecular weight proteins with in vitro assembled brain microtubules. *J. Cell Biol.* 65: 237–41

Diez, J. C., de la Torre, J., Avila, J. 1985. Differential association of the different brain microtubule proteins in different in vitro assembly conditions. *Biochim. Biophys. Acta* 838: 32–38

Drubin, D. G., Caput, D., Kirschner, M. 1984a. Studies on the expression of the microtubule-associated protein, tau, during mouse brain development, with newly isolated complementary DNA probes. *J. Cell Biol.* 98: 1090–97

Drubin, D., Kirschner, M., Feinstein, S. 1984b. Microtubule-associated tau protein induction by nerve growth factor during neurite outgrowth in PC12 cells. See Borisy et al 1984, pp. 343–56

Drubin, D., Feinstein, S., Shooter, E., Kirschner, M. 1985. Nerve growth induced outgrowth in PC12 cells involves the coordinate induction of microtubule assembly and assembly promoting factors. *J. Cell Biol.* 101: 1799–1807

Duerr, A., Pallas, D., Solomon, F. 1981. Molecular analysis of cytoplasmic microtubules in situ: Identification of both widespread and specific proteins. *Cell* 24: 203–11

Fach, B. L., Graham, S. F., Keates, R. A. 1985. Association of fodrin with brain microtubules. *Can. J. Biochem. Cell. Biol.* 63: 372–81

Francon, J., Fellous, A., Lennon, A., Nunez, J. 1978. Requirement for "factor(s)" for tubulin assembly during brain development. *Eur. J. Biochem.* 85: 43–53

Francon, J., Lennon, A. M., Fellous, A., Mareck, A., Pierre, M., Nunez, J. 1982. Heterogeneity of microtubule-associated proteins and brain development. *Eur. J. Biochem.* 129: 465–72

Gibbons, I. R. 1965. Chemical dissection of cilia. *Arch. Biol.* 76: 317–52

Ginzburg, I., Scherson, T., Giveon, D., Behar, L., Littauer, U. Z. 1982. Modulation of mRNA for microtubule-associated proteins during brain development. *Proc. Natl. Acad. Sci. USA* 79: 4892–96

Ginzburg, I., Littauer, U. 1984. Experimental cellular regulation of microtubule proteins. See Borisy et al 1984, pp. 357–66

Goldenring, J. R., Gonzalez, B., McGuire, J. S. Jr., DeLorenzo, R. J. 1983. Purification and characterization of a calmodulin-dependent kinase from rat brain cytosol able to phosphorylate tubulin and microtubule-associated proteins. *J. Biol. Chem.* 258 : 12632–40

Goldenring, J. R., Vallano, M. L., DeLorenzo, R. J. 1985. Phosphorylation of microtubule-associated protein 2 at distinct sites by calmodulin-dependent and cyclic-AMP-dependent kinases. *J. Neurochem.* 45 : 900–5

Goldman, R. D., Knipe, D. 1972. Function of cytoplasmic fibers in non-muscle cell motility. *Cold Spring Harbor Symp. Quant. Biol.* 37 : 523–34

Goldstein, L. S. B., Laymon, R. A., McIntosh, J. R. 1986. A microtubule-associated protein from *Drosophila melanogaster*: Identification, characterization, and isolation of coding sequences. *J. Cell Biol.* 102 : 2076–87

Gottlieb, R. A., Murphy, D. B. 1985. Analysis of the microtubule binding domain of microtubule-associated protein 2. *J. Cell Biol.* 101 : 1782–89

Greene, L. A., Liem, R. K., Shelanski, M. L. 1983. Regulation of a high molecular weight microtubule-associated protein in PC 12 cells by nerve growth factor. *J. Cell Biol.* 96 : 76–8;3

Griffith, L. M., Pollard, T. D. 1978. Evidence for actin filament microtubule interaction mediated by microtubule-associated proteins. *J. Cell Biol.* 78 : 958–65

Gundersen, G., Kalnoski, M., Bulinski, J. 1984. Distinct populations of microtubules: Tyrosinated and nontyrosinated alpha tubulin are distributed differently. *Cell* 38 : 779–89

Heimann, R., Shelanski, M., Liem, R. D. 1985. Microtubule-associated proteins bind specifically to the 70-kilodalton neurofilament protein. *J. Biol. Chem.* 260 : 12160 66

Herrmann, H., Dalton, J. M., Wiche, G. 1985. Microheterogeneity of microtubule-associated proteins, MAP 1 and MAP 2, and differential phosphorylation of individual subcomponents. *J. Biol. Chem.* 260 : 5797–5803

Herrmann, H., Pytella, R., Dalton, J., Wiche, G. 1984. Structural homology of microtubule-associated proteins 1 and 2 demonstrated by peptide mapping and immunoreactivity. *J. Biol. Chem.* 259 : 612–17

Herzog, W., Weber, K. 1978. Fractionation of brain microtubule-associated proteins. Isolation of two different proteins which stimulate tubulin polymerization in vitro. *Eur. J. Biochem.* 92 : 1–8

Hirokawa, N., Bloom, G. S., Vallee, R. B. 1985a. Cytoskeletal architecture and immunocytochemical localization of microtubule-associated proteins in regions of axons associated with rapid axonal transport: The beta beta-iminodipropionitrile-intoxicated axon as a model system. *J. Cell Biol.* 101 : 227–39

Hirokawa, N., Takemura, R., Hisanaga, S. 1985b. Cytoskeletal architecture of isolated mitotic apparatus with special reference to microtubule-associated proteins and dynein. *J. Cell Biol.* 101 : 1858–70

Hollenbeck, P. J., Suprynowicz, F., Cande, W. Z. 1984. Cytoplasmic dynein-like ATPase cross-links microtubules in an ATP-sensitive manner. *J. Cell Biol.* 99 : 1251–58

Horwitz, S. B., Parness, J., Schiff, P. B., Manfredi, J. J. 1982. Taxol: A new probe for studying the structure and function of microtubules. *Cold Spring Harbor Symp. Quant. Biol.* 46 : 219–26

Huber, G., Alaimo-Beuret, D., Matus, A. 1985. MAP 3 : characterization of a novel microtubule-associated protein. *J. Cell Biol.* 100 : 496–507

Huber, G., Matus, A. 1984a. Immunocytochemical localization of microtubule-associated protein 1 in rat cerebellum using monoclonal antibodies. *J. Cell Biol.* 98 : 777–81

Huber, G., Matus, A. 1984b. Differences in the cellular distributions of two microtubule-associated proteins, MAP 1 and MAP 2, in rat brain. *J. Neurosci.* 4 : 151–60

Izant, J. G., McIntosh, J. R. 1980. Microtubule-associated proteins: A monoclonal antibody to brain microtubule associated protein 2 binds to differentiated neurons. *Proc. Natl. Acad. Sci. USA* 77 : 4741–45

Izant, J., Weatherbee, J., McIntosh, J. R. 1983. A microtubule-associated protein antigen unique to mitotic spindle microtubules in PtK1 cells. *J. Cell Biol.* 96 : 424–34

Izant, J. G., Weatherbee, J. A., McIntosh, J. R. 1982. A microtubule-associated protein in the mitotic spindle and the interphase nucleus. *Nature* 295 : 248–50

Jameson, L., Caplow, M. 1981. Modification of microtubule steady-state dynamics by phosphorylation of the microtubule-associated proteins. *Proc. Natl. Acad. Sci. USA* 78 : 3413–17

Job, D., Rauch, C. T., Fischer, E. H., Margolis, R. L. 1982. Recycling of cold-stable microtubules: Evidence that cold stability is due to substoichiometric polymer blocks. *Biochemistry* 21 : 509–15

Job, D., Rauch, C., Fischer, E., Margolis, R. 1983. Regulation of microtubule cold stability by calmodulin dependent and calmodulin independent phosphorylation. *Proc. Natl. Acad. Sci.* 80 : 3894–98

Job, D., Pabion, M., Margolis, R. 1985. Generation of microtubule stability subclasses by microtubule-associated proteins: Implications for the microtubule "dynamic" instability model. *J. Cell Biol.* 101 : 1680–89

Johnson, K. A., Porter, M. E., Shimizu, T. 1984. Mechanism of force production for microtubule-dependent movements. *J. Cell Biol.* 99 : 132s–36s

Keller, T., Rebhun, L. I. 1982. *Strongylocentrotus purpuratus* spindle tubulin. I. Characterization of polymerization and depolymerization in vitro. *J. Cell Biol.* 93 : 788–96

Kim, H., Binder, L. I., Rosenbaum, J. L. 1979. The periodic association of microtubule-associated proteins with brain microtubules in vitro. *J. Cell Biol.* 80 : 266–76

Kirkpatrick, J. B., Hyams, L., Thomas, V. L., Howley, P. M. 1970. Purification of intact microtubules from brain. *J. Cell Biol.* 47 : 384–90

Kirschner, M. 1978. Microtubule assembly and nucleation. *Int. Rev. Cytol.* 54 : 1–71

Kosik, K. S., Duffy, L. K., Dowling, M. M., Abraham, C., McCluskey, A., Selkoe, D. J. 1984. Microtubule-associated protein 2: Monoclonal antibodies demonstrate the selective incorporation of certain epitopes into Alzheimer neurofibrillary tangles. *Proc. Natl. Acad. Sci. USA* 81 : 7941–45

Koska, C., Leichtfried, F. E., Wiche, G. 1985. Identification and spatial arrangement of high molecular weight proteins (Mr 300,000–330,000) co-assembling with microtubules from a cultured cell line (rat glioma C6). *Eur. J. Cell Biol.* 38 : 149–56

Kotani, S., Nishida, E., Kumagai, H., Sakai, H. 1985. Calmodulin inhibits interaction of actin with MAP2 and Tau, two major microtubule-associated proteins. *J. Biol. Chem.* 260 : 10779–83

Kumar, N., Flavin, M. 1985. Modulation of some parameters of assembly of microtubules in vitro by tyrosinolation of tubulin. *Eur. J. Biochem.* 128 : 215–22

Kuriyama, R., Keryer, G., Borisy, G. G. 1984. The mitotic spindle of Chinese hamster ovary cells isolated in taxol containing medium. *J. Cell Sci.* 66 : 265–76

Kuznetsov, S., Rodinov, V., Gelfand, V. 1981. *FEBS Lett.* 135 : 237–40

Kuznetsov, S., Rodinov, V., Gelfand, V., Rosenblat, V. 1984. Microtubule associated protein 2 competes with microtubule associated protein 1 for binding to microtubules. *Biochem. Biophys. Res. Commun.* 119 : 173–78

Larson, R., Goldenring, J., Vallano, M., DeLorenzo, R. 1985. Identification of endogenous calmodulin-dependent kinase and calmodulin binding proteins in cold-stable microtubule preparations from rat brain. *J. Neurochem.* 44 : 1566–74

Lasek, R., Brady, S. 1985. AMPPNP facilitates attachment of transported vesicles to microtubules in axoplasm. *Nature* 316 : 645–47

Lasek, R., Garner, J., Brady, S. 1984. Axonal transport of the cytoplasmic matrix. *J. Cell Biol.* 99 : 212–21s

Lee, Y. C., Wolff, J. 1984a. Calmodulin binds to both MAP 2 and tau proteins. *J. Biol. Chem.* 259 : 1226–30

Lee, Y. C., Wolff, J. 1984b. The calmodulin binding domain on microtubule associated protein 2. *J. Biol. Chem.* 259 : 8041–44

Lefebvre, P., Rosenbaum, J. L. 1986. *Ann. Rev. Cell Biol.* 2 : 517–46

Lennon, A., Francon, J., Fellous, A., Nunez, J. 1980. Rat, mouse and guinea pig brain development and microtubule assembly. *J. Neurochem.* 35 : 804–13

Leterrier, J. F., Liem, R., Shelanski, M. 1982. Interactions between neurofilaments and microtubule-associated proteins: A possible mechanism for intra-organellar bridging. *J. Cell Biol.* 95 : 982–86

Leterrier, J. F., Wong, J., Liem, R. K., Shelanski, M. L. 1984. Promotion of microtubule assembly by neurofilament-associated microtubule-associated proteins. *J. Neurochem.* 43 : 1385–91

Lewis, S. A., Sherline, P., Cowan, N. 1986a. A cloned cDNA encoding MAP 1 detects a single copy gene in mouse, and a brain abundant RNA whose level decreases during development. *J. Cell Biol.* 102 : 2106–14

Lewis, S. A., Villasante, A., Sherline, P., Cowan, N. 1986b. Brain specific expression of MAP 2 detected using a cloned cDNA probe. *J. Cell Biol.* 102 : 2098–2105

Lindwall, G., Cole, R. D. 1984a. Phosphorylation affects the ability of tau protein to promote microtubule assembly. *J. Biol. Chem.* 259 : 5301–5

Lindwall, G., Cole, R. D. 1984b. The purification of tau protein and the occurrence of two phosphorylation states of tau in brain. *J. Biol. Chem.* 259 : 12241–45

Lohman, S. M., De Camilli, P., Einig, I., Walter, U. 1984. High affinity binding of the regulatory subunit (RII) of cAMP-dependent protein kinase to microtubule-associated and other cellular proteins. *Proc. Natl. Acad. Sci. USA* 81 : 6723–27

Magendantz, M., Solomon, F. 1985. Analyzing the components of microtubules : Antibodies against chartins, associated proteins from cultured cells. *Proc. Natl. Acad. Sci.* 82 : 6581–85

Mandelkow, E. M., Herrmann, M., Ruehl, U. 1985. Tubulin domains probed by limited proteolysis and subunit-specific antibodies. *J. Mol. Biol.* 185 : 311–28

Mareck, A., Fellous, A., Francon, J., Nunez, J. 1980. Changes in composition and activity of microtubule-associated proteins during brain development. *Nature* 284 : 353–55

Margolis, R. L., Rauch, C. T. 1981. Characterization of rat brain crude extract microtubule assembly : Correlation of cold stability with the phosphorylation state of a microtubule-associated 64-kDa protein. *Biochemistry* 20 : 4451–58

Maro, B., Howlett, K., Webb, M. 1985. Non-spindle microtubule organizing centers in metaphase II–arrested mouse oocytes. *J. Cell Biol.* 101 : 1665–72

Mascardo, R. M., Sherline, P., Weatherbee, J. 1982. Localization of high molecular weight microtubule-associated proteins MAP 1 and MAP 2 in a HeLa microtubule organizing center. *Cytobios* 35 : 113–27

Matus, A., Bernhardt, R., Hugh-Jones, T. 1981. High molecular weight microtubule-associated proteins are preferentially associated with dendritic microtubules in brain. *Proc. Natl. Acad. Sci. USA* 78 : 3010–14

Matus, A., Huber, G., Bernhardt, R. 1983. Neuronal microdifferentiation. *Cold Spring Harbor Symp. Quant. Biol.* 48 : 775–82

McKeithan, T. W., Rosenbaum, J. L. 1984. The biochemistry of microtubules. A review. *Cell Muscle Motil.* 5 : 255–88

Miller, M., Solomon, F. 1984. Kinetics and intermediates of marginal band reformation : Evidence for peripheral determinants of microtubule organization. *J. Cell Biol.* 99 : 70s–78s

Miller, P., Walter, U., Theurkauf, W. E., Vallee, R. B., De Camilli, P. 1982. Frozen tissue sections as an experimental system to reveal specific binding sites for the regulatory subunit of type II cyclic AMP dependent protein kinase in neurons. *Proc. Natl. Acad. Sci. USA* 79 : 5562–66

Miller, R., Lasek, R. 1985. Cross-bridges mediate anterograde and retrograde vesicle transport along microtubules in squid axoplasm. *J. Cell Biol.* 101 : 2181–93

Murphy, D. B., Borisy, G. G. 1975. Association of high molecular weight proteins with microtubules and their role in microtubule assembly in vitro. *Proc. Natl. Acad. Sci. USA* 72 : 2696–2700

Murphy, D. B., Vallee, R., Borisy, G. 1977. Identity and polymerization-stimulatory activity of the nontubulin proteins associated with microtubules. *Biochemistry* 16 : 2598–2605

Murthy, A., Flavin, M. 1983. Microtubule assembly using the microtubule associated protein MAP 2 prepared in defined states of phosphorylation with protein kinase and phosphatase. *Eur. J. Biochem.* 137 : 37–46

Murthy, A. S., Bramblett, G. T., Flavin, M. 1985. The sites at which brain microtubule-associated protein 2 is phosphorylated in vivo differ from those accessible to cAMP-dependent kinase in vitro. *J. Biol. Chem.* 260 : 4364–70

Nigg, E., Schafer, G., Hilz, H., Eppenberger, H. 1985. Cyclic-AMP dependent protein kinase type II is associated with the Golgi complex and with centrosomes. *Cell* 41 : 1039–51

Nishida, E., Kuwaki, T., Sakai, H. 1981. Phosphorylation of microtubule associated proteins (MAPs) and pH of the medium control interaction between MAPs and actin filaments. *J. Biochem.* 90 : 575–78

Olmsted, J. B. 1981. Tubulin pools in differentiating neuroblastoma cells. *J. Cell Biol.* 89 : 418–23

Olmsted, J. B., Asnes, C. F., Parysek, L. M., Lyon, H. D., Kidder, G. M. 1986. Distribution of MAP 4 in cells and in adult and developing mouse tissues. *Ann. NY Acad. Sci.* 466 : 292–305

Olmsted, J. B., Borisy, G. G. 1973. Microtubules. *Ann. Rev. Biochem.* 42 : 507–40

Olmsted, J. B., Cox, J. V., Asnes, C. F., Parysek, L. M., Lyon, H. D. 1984. Cellular regulation of microtubule organization. *J. Cell Biol.* 99 : 28s–32s

Olmsted, J. B., Lyon, H. D. 1981. A microtubule-associated protein specific to differentiated neuroblastoma cells. *J. Biol. Chem.* 256 : 3507–11

Olmsted, J. B., Neighbors, B. W., Stemple, D. L., McIntosh, J. R. 1985. Derivatization and dynamics of thermostable microtubule-associated proteins. *J. Cell Biol.* 101 : 31a

Pabion, M., Job, D., Margolis, R. L. 1984. Sliding of STOP proteins on microtubules. *Biochemistry* 23 : 6642–48

Pallas, D., Solomon, F. 1982. Cytoplasmic microtubule-associated proteins : Phosphorylation at novel sites is correlated with their incorporation into assembled microtubules. *Cell* 30 : 407–14

Papasozomenos, S., Binder, L. 1986. Microtubule-associated protein 2 (MAP 2) is present in astrocytes of the optic nerve but is absent from astrocytes of the optic tract. *J. Neurosci.* In press

Papasozomenos, S. C., Binder, L. I., Bender, P. K., Payne, M. R. 1985. Microtubule-associated protein 2 within axons of spinal motor neurons: Associations with microtubules and neurofilaments in normal and beta beta iminodipropionitrile-treated axons. *J. Cell Biol.* 100: 74–85

Parysek, L. M., Asnes, C. F., Olmsted, J. B. 1984a. MAP 4: Occurrence in mouse tissues. *J. Cell Biol.* 99: 1309–15

Parysek, L. M., Wolosewick, J. J., Olmsted, J. B. 1984b. MAP 4: A microtubule-associated protein specific for a subset of tissue microtubules. *J. Cell Biol.* 99: 2287–96

Parysek, L. M., DelCerro, M., Olmsted, J. B. 1985. Microtubule-associated protein 4 antibody: A new marker for astroglia and oligodendroglia. *Neuroscience* 15: 869–76

Peng, I., Binder, L., Black, M. 1986. Biochemical and immunological analyses of cytoskeletal domains of neurons. *J. Cell Biol.* 102: 252–62

Pettijohn, D. E., Henzl, M., Price, C. 1984. Nuclear proteins that become part of the mitotic apparatus—a role in nuclear assembly. *J. Cell Sci.* 1: 187–201

Pillus, L., Solomon, F. 1986. Components of microtubular structures in *Saccharomyces cerevisiae*. *Proc. Natl. Acad. Sci. USA* 83: 2468–72

Pollard, T. D., Selden, S. C., Maupin, P. 1984. Interaction of actin filaments with microtubules. *J. Cell Biol.* 99: 33s–37s

Purich, D. L., Kristofferson, D. 1984. Microtubule assembly: A review of progress, principles and perspectives. *Adv. Protein Chem.* 36: 133–212

Pytela, R., Wiche, G. 1980. High molecular weight polypeptides from cultured cells are related to hog brain microtubule-associated proteins but copurify with intermediate filaments. *Proc. Natl. Acad. Sci. USA* 77: 4808–12

Richter-Landsberg, C., Jastorff, B. 1985. In vitro phosphorylation of microtubule-associated protein 2: Differential effects of cyclic AMP analogues. *J. Neurochem.* 45: 1218–22

Riederer, B., Matus, A. 1985. Differential expression of distinct microtubule-associated proteins during brain development. *Proc. Natl. Acad. Sci. USA* 82: 6006–9

Runge, M., Detrich, H. W., Williams, R. C. 1979. Identification of the major 68,000 dalton protein of microtubule preparations as a 10nm filament protein and its effects on microtubule assembly in vitro. *Biochemistry* 18: 1689–98

Runge, M., Laue, T., Yphantis, D., Lifsics,

M., Saito, A., et al. 1981. ATP-induced formation of an associated complex between microtubules and neurofilaments. *Biochemistry* 78: 1431–35

Sandoval, I. V., Vandekerckhove, J. S. 1981. A comparative study of the in vitro polymerization of tubulin in the presence of the microtubule-associated proteins MAP 2 and tau. *J. Biol. Chem.* 256: 8795–800

Sato, C., Tanabe, K., Nishizawa, K., Nakayama, T., Kobayashi, T., Nakamura, H. 1985. Localization of 350 kilodalton molecular weight and related proteins in both the cytoskeleton and nuclear flecks that increase during G-1 phase. *Exp. Cell Res.* 160: 206–20

Sattilaro, R. F., Dentler, W. L., LeCluyse, E. L. 1981. Microtubule-associated proteins (MAPs) and the organization of actin filaments in vitro. *J. Cell Biol.* 90: 467–73

Saxton, W., Stemple, D., Leslie, R., Salmon, E., Zavortink, M., McIntosh, J. R. 1984. Tubulin dynamics in cultured mammalian cells. *J. Cell Biol.* 99: 2175–86

Scherson, T., Kreis, T. E., Schlessinger, J., Littauer, U. Z., Borisy, G. G., Geiger, B. 1984. Dynamic interactions of fluorescently labeled microtubule-associated proteins in living cells. *J. Cell Biol.* 99: 425–34

Schliwa, M., Euteneuer, U., Bulinski, J. C., Izant, J. G. 1981. Calcium lability of cytoplasmic microtubules and its modulation by microtubule-associated proteins. *Proc. Natl. Acad. Sci. USA* 78: 1037–41

Scholey, J. M., Neighbors, B., McIntosh, J. R., Salmon, E. 1984. Isolation of microtubules and a dynein-like MgATPase from unfertilized sea urchin eggs. *J. Biol. Chem.* 259: 6516–25

Scholey, J. M., Neighbors, B., Grissom, P., McIntosh, J. R. 1985a. Localization of 2 spindle polypeptides which exhibit different distributions in dividing sea urchin eggs. *J. Cell Biol.* 101: 31a

Scholey, J. M., Porter, M. S., Grissom, P. M., McIntosh, J. R. 1985b. Identification of kinesin in sea urchin eggs, and evidence for its localization in the mitotic spindle. *Nature* 318: 483–86

Schroer, T., Kelly, R. 1985. In vitro translocation of organelles along microtubules. *Cell* 40: 729–30

Schulman, H. 1984a. Phosphorylation of microtubule associated proteins by calcium-calmodulin dependent protein kinase. *J. Cell Biol.* 99: 11–19

Schulman, H. 1984b. Differential phosphorylation of MAP 2 stimulated by calcium-calmodulin and cyclic AMP. *Mol. Cell Biol.* 4: 1175–78

Schulman, H., Kuret, J., Jefferson, A., Nose,

P., Spitzer, K. 1985. Calcium-calmodulin dependent microtubule associated protein 2 kinase: Broad substrate specificity and multifunctional potential in diverse tissues. *Biochemistry* 24: 5320–27

Seeds, N., Maccioni, R. 1978. Proteins from morphologically differentiated neuroblastoma cells promote tubulin polymerization. *J. Cell Biol.* 76: 547–55

Selden, S. C., Pollard, T. D. 1983. Phosphorylation of microtubule-associated proteins regulates their interaction with actin filaments. *J. Biol. Chem.* 258: 7064–71

Serrano, L., Avila, J., Maccioni, R. B. 1984. Controlled proteolysis of tubulin by subtilisin: Localization of the site for MAP 2 interaction. *Biochemistry* 23: 4675–81

Shelanski, M. L., Gaskin, F., Cantor, C. R. 1973. Assembly of microtubules in the absence of added nucleotide. *Proc. Natl. Acad. Sci. USA* 70: 765–68

Sherline, P., Schiavone, K. 1977. Immunofluorescence localization of proteins of high molecular weight along intracellular microtubules. *Science* 198: 1038–40

Sherline, P., Lee, Y., Jacobs, L. 1977. Binding of microtubules to pituitary secretory granules and granule membranes. *J. Cell Biol.* 72: 380–89

Sheterline, P. 1978. Localization of the major high-molecular weight protein on microtubules in vitro and in cultured cells. *Exp. Cell Res.* 115: 460–64

Sheterline, P. 1980. Immunological characterization of the microtubule-associated protein MAP 2. *FEBS Lett.* 111: 167–70

Sloboda, R. D., Dickersin, K. 1980. Structure and composition of the cytoskeleton of nucleated erythrocytes I. The presence of microtubule-associated protein 2 in the marginal band. *J. Cell Biol.* 87: 170–79

Sloboda, R. D., Rosenbaum, J. L. 1982. Purification and assay of microtubule-associated proteins (MAPs). *Meth. Enzymol.* 85: 409–16

Sloboda, R. D., Rudolph, S. A., Rosenbaum, J. L., Greengard, P. 1975. Cyclic AMP-dependent endogenous phosphorylation of a microtubule-associated protein. *Proc. Natl. Acad. Sci. USA* 72: 177–81

Sloboda, R. D., Dentler, W. L., Rosenbaum, J. L. 1976. Microtubule-associated proteins and the stimulation of tubulin assembly in vitro. *Biochemistry* 15: 4497–4505

Sloboda, R. D., Rosenbaum, J. L. 1979. Decoration and stabilization of intact, smooth-walled microtubules with microtubule-associated proteins. *Biochemistry* 18: 48–55

Sobue, K., Tanaka, T., Ashino, N., Kakiuchi, S. 1985. Ca^{2+} and calmodulin regulate microtubule-associated protein actin filament interaction in a flip-flop switch. *Biochim. Biophys. Acta.* 845: 366–72

Solomon, F., Magendantz, M., Salzman, A. 1979. Identification with cellular microtubules of one of the co-assembling microtubule-associated proteins. *Cell* 18: 431–38

Stearns, M. E., Brown, D. 1979. Purification of cytoplasmic tubulin and microtubule organizing center proteins functioning in microtubule initiation from the alga *Polytomella. Proc. Natl. Acad. Sci. USA* 76: 5745–49

Suprenant, K. A., Dentler, W. L. 1982. Association between endocrine pancreatic secretory granules and in vitro assembled microtubules is dependent upon microtubule-associated proteins. *J. Cell Biol.* 93: 164–74

Swan, J., Solomon, F. 1984. Reformation of the marginal band of avian erthrocytes in vitro using calf-brain tubulin: Peripheral determinants of microtubule form. *J. Cell Biol.* 99: 2108–13

Theurkauf, W. E., Vallee, R. B. 1983. Extensive cAMP dependent and cAMP-independent phosphorylation of microtubule-associated protein 2. *J. Biol. Chem.* 258: 7883–86

Theurkauf, W. E., Vallee, R. B. 1982. Molecular characterization of the cAMP-dependent protein kinase bound to microtubule-associated protein 2. *J. Biol. Chem.* 257: 3284–90

Tytell, M., Brady, S. T., Lasek, R. J. 1984. Axonal transport of a subclass of tau proteins: Evidence for the regional differentiation of microtubules in neurons. *Proc. Natl. Acad. Sci. USA* 81: 1570–74

Valdivia, M. M., Avila, J., Coll, J., Colaco, C., Sandoval, I. V. 1982. Quantitation and characterization of the microtubule-associated MAP 2 in porcine tissues and its isolation from porcine (PK 15) and human (HeLa) cell lines. *Biochem. Biophys. Res. Commun.* 105: 1241–49

Vale, R., Reese, T., Sheetz, M. 1985a. Identification of a novel force-generating protein, kinesin, involved in microtubule based motility. *Cell* 42: 39–50

Vale, R., Schnapp, B., Mitchison, T., Steuer, E., Reese, T., Sheetz, M. 1985b. Different axoplasmic proteins generate movement in opposite directions along microtubules in vitro. *Cell* 43: 623–32

Vallano, M. L., Goldenring, J. R., Buckholz, T. M., Larson, R. E. 1985. Separation of endogenous calmodulin and cAMP-dependent kinases from microtubule preparations. *Proc. Natl. Acad. Sci. USA* 82: 3202–6

Vallee, R. 1980. Structure and phosphorylation of microtubule-associated protein 2 (MAP2). *Proc. Natl. Acad. Sci. USA* 77: 3206–10

Vallee, R. 1982. A taxol-dependent procedure for the isolation of microtubules and microtubule-associated proteins (MAPs) *J. Cell Biol.* 92: 435–42

Vallee, R. B. 1984. MAP2 (microtubule-associated protein 2). *Cell Muscle Motil.* 5: 289–311

Vallee, R. B., Bloom, G. S. 1983. Isolation of sea urchin egg microtubules with taxol and identification of mitotic spindle microtubule-associated proteins with monoclonal antibodies. *Proc. Natl. Acad. Sci. USA* 80: 6259–63

Vallee, R. B., Bloom, G. S. 1984. High molecular weight microtubule-associated proteins. *Mod. Cell Biol.* 3: 21–76

Vallee, R. B., Bloom, G. S., Luca, F. C. 1984a. Differential cellular and subcellular distribution of microtubule-associated proteins. See Borisy et al 1984, pp. 111–30

Vallee, R. B., Bloom, G. S., Theurkauf, W. E. 1984b. Microtubule-associated proteins: Subunits of the cytomatrix. *J. Cell Biol.* 99: 38s–44s

Vallee, R. B., Borisy, G. G. 1977. Removal of the projections from cytoplasmic microtubules in vitro by digestion with trypsin. *J. Biol. Chem.* 252: 377–82

Vallee, R. B., Davis, S. D. 1983. Low molecular weight microtubule-associated proteins are light chains of microtubule-associated protein 1 (MAP 1). *Proc. Natl. Acad. Sci. USA* 80: 1342–46

Vallee, R. B., DiBartolomeis, M. J., Theurkauf, W. E. 1981. A protein kinase bound to the projection portion of MAP 2 (microtubule-associated protein 2). *J. Cell Biol.* 90: 568–76

Vandert, M., Lambert, A. M., De Mey, J., Picquot, P., Van Eldik, L. 1985. Characterization and immunocytochemical distribution of calmodulin in higher plant endosperm cells: Localization in the mitotic apparatus. *J. Cell Biol.* 101: 488–99

Vandre, D. D., David, F. M., Rao, P. N., Borisy, G. G. 1984. Phosphoproteins are components of mitotic microtubule organizing centers. *Proc. Natl. Acad. Sci. USA* 81: 4439–44

Voter, W. A., Erickson, H. P. 1982. Electron microscopy of MAP 2 (microtubule-associated protein 2). *J. Ultrastruct. Res.* 80: 374–82

Weatherbee, J. A., Sherline, P., Mascardo, R. N., Izant, J. G., Luftig, R. B., Weihing, R. R. 1982. Microtubule-associated proteins of HeLa cells: Heat stability of the 200,000 mol wt HeLa MAPs and detection of the presence of MAP 2 in HeLa cell extracts and cycled microtubules. *J. Cell Biol.* 92: 155–63

Weatherbee, J. A., Luftig, R. B., Weihing, R. R. 1980. Purification and reconstitution of HeLa cell microtubules. *Biochemistry* 19: 4116–23

Weingarten, M., Lockwood, A., Hwo, S., Kirschner, M. 1975. A protein factor essential for microtubule assembly. *Proc. Natl. Acad. Sci. USA* 72: 1858–62

Weisenberg, R. 1972. Microtubule formation in vitro in solutions containing low calcium concentrations. *Science* 177: 1104–5

Welsh, M., Dedman, J., Brinkley, B., Means, A. 1979. Tubulin and calmodulin. Effect of microtubule and microfilament inhibitors on localization in the mitotic apparatus. *J. Cell Biol.* 81: 624–34

Wiche, G. 1985. High molecular weight microtubule-associated proteins (MAPS): A ubiquitous family of cytoskeletal connecting links. *Trends Biochem. Sci.* 10: 67–70

Wiche, G., Briones, E., Hirt, H., Krepler, R., Artlieb, U., Denk, H. 1983. Differential distribution of microtubule-associated proteins MAP 1 and MAP 2 in neurons of rat brain and association of MAP 1 with microtubules of neuroblastoma cells (clone N2A). *EMBO J.* 2: 1915–20

Wiche, G., Briones, E., Koszka, C., Artlieb, U., Krepler, R. 1984. Widespread occurrence of polypeptides related to neurotubule-associated proteins (MAP 1 and MAP 2) in nonneuronal cells and tissues. *EMBO J.* 3: 991–98

Wood, J. G., Mirra, S., Pollock, N. J., Binder, L. I. 1986. Neurofibrillary tangles of Alzheimer's disease share antigenic determinants with the axonal microtubule-associated protein tau. *Proc. Natl. Acad. Sci. USA* 83: 4040–43

Yamamoto, H., Fukunaga, K., Goto, S., Tanaka, E., Miyamoto, E. 1985. Ca^{2+}, calmodulin-dependent regulation of microtubule formation via phosphorylation of microtubule-associated protein 2, tau factor, and tubulin, and comparison with the cyclic AMP-dependent phosphorylation. *J. Neurochem.* 44: 759–68

Zernig, G., Wiche, G. 1985. Morphological integrity of single adult cardiac myocytes isolated by collagenase treatment: Immunolocalization of tubulin, microtubule-associated proteins 1 and 2, plectin, vimentin and vinculin. *Eur. J. Cell Biol.* 38:113–22

Zieve, G., Solomon, F. 1982. Proteins specifically associated with the microtubules of the mammalian mitotic spindle. *Cell* 28:233–42

Zieve, G., Solomon, F. 1984. Direct isolation of neuronal microtubule skeletons. *Mol.* *Cell Biol.* 4:371–74

Zingsheim, H. P., Herzog, W., Weber, K. 1979. Differences in surface morphology of microtubules reconstituted from pure brain tubulin using two different microtubule-associated proteins: The high molecular weight MAP 2 proteins and tau proteins. *Eur. J. Cell Biol.* 19:175–83.

Ann. Rev. Cell Biol. 1986. 2 : 459–98

STRUCTURE AND FUNCTION OF NUCLEAR AND CYTOPLASMIC RIBONUCLEOPROTEIN PARTICLES

Gideon Dreyfuss

Department of Biochemistry, Molecular Biology and Cell Biology, Northwestern University, Evanston, Illinois 60201

CONTENTS

INTRODUCTION

The pathway of expression of genetic information in animal cells, from DNA to protein via RNA intermediates, is highly complex and tightly regulated (Darnell 1982; Nevins 1983). Particularly intricate are the post-transcriptional processing events required to convert the primary gene transcripts into the functional intermediates designated messenger RNAs (mRNAs). The primary transcripts of RNA polymerase II, except for a

459

0743–4634/86/1115–0459$02.00

distinct group of small RNAs, are referred to as heterogeneous nuclear RNAs (hnRNAs). In higher eukaryotes all hnRNAs have a 5'-cap structure (m⁷Gppp-), and up to one-quarter of them also acquire a 3'-polyadenylate [poly(A)] tail. Only a subset of hnRNAs, about one-fifth to one-quarter of the hnRNAs in higher eukaryotes, are actually precursors to translatable cytoplasmic mRNA, and these are designated pre-mRNAs (Brandhorst & McConkey 1974; Herman & Penman 1977; Harpold et al 1979; Salditt-Georgieff et al 1981, 1982). Typically, pre-mRNAs contain polyadenylated tails, and the majority of them contain intervening sequences that are later spliced out (Abelson 1979; Green 1986; Padgett et al 1986). Little is known about the ensuing events except that the RNA is translocated through nuclear pores and that spliced mRNAs accumulate in the cytoplasm.

mRNAs conserve the 5'- (cap) and 3'- [typically a poly(A) tail; Brawerman 1981] ends of the respective pre-mRNAs and contain an uninterrupted reading frame for translation. At any given time, not all translatable mRNAs in the cell are actually translated. The actively translated mRNAs are engaged with ribosomes to form polyribosomes and can be readily separated from the untranslated mRNA by velocity sedimentation in sucrose gradients. The nonpolysomal-to-polysomal ratio is not the same for all mRNAs nor is it fixed. It can drastically change for specific mRNAs in response to specific signals and for some or all of the mRNAs under a variety of environmental (e.g. heat shock, virus infection) and developmental circumstances. Thus, although the relative abundance of specific mRNAs is typically the major factor that determines the relative amounts of various proteins synthesized, the translation repertoire of the eukaryotic cell can vary even for a given set of mRNAs due to differential selection of mRNAs for translation. This process is referred to as translational regulation or translational control. Unlike in prokaryotes, in eukaryotes the majority of mRNAs are quite stable, and many have a half-life of the order of the cell cycle time itself (Brandhorst & Humphries 1971; Singer & Penman 1972; Greenberg 1972; Brandhorst & McConkey 1974). Different mRNAs have different half-lives, and these also can be modulated for specific mRNAs in a given cell (e.g. globin mRNA, tubulin mRNA). Modulation of mRNA stability is an extremely important process because it can drastically affect the level of a specific mRNA; however, little is known about the elements that control this stability. Untranslated mRNAs are sometimes stabilized and stored for very long periods of time, as is the case in many oocytes. One of the ultimate goals of molecular and cell biology is to understand all of these processes in terms of both molecular detail and cellular topology.

Experimentally, hnRNAs can be distinguished from other RNAs on the basis of their size, their subcellular compartmentation, and the characteristics (e.g. antibiotic sensitivity) of the RNA polymerase that transcribes them. With this operational definition, it was found that hnRNAs, like other polynucleotides in the nucleus, form complexes with specific proteins. The unique particles thus generated are termed hnRNP particles or hnRNP complexes. The hnRNP particles are one of the most abundant structures in the nucleus and are the sites of RNA processing. Similarly, mRNAs in the cytoplasm are associated with specific proteins to form mRNP complexes. The mRNP and hnRNP proteins are different, and an exchange of proteins therefore accompanies mRNA nucleocytoplasmic transport. The mRNP proteins are likely to be important in the translation, stability, and localization of mRNAs and perhaps also in mRNA nucleocytoplasmic transport. Interest in RNP complexes stems from the fact that they are the structural entities within which hnRNA and mRNA exist in the cell (rather than as naked polynucleotides). Therefore more needs to be learned about them to understand how the posttranscriptional portion of the pathway of expression of genetic information operates in the cell.

This review outlines major recent developments and significant earlier observations that led to current knowledge of the structure and function of the ribonucleoprotein complexes of hnRNA and mRNA. Space limitations preclude comprehensive citation of the numerous publications in the field, many of which are worthy of extensive discussion. Additional discussion and references are found in several recent reviews on these subjects (Martin et al 1980; Samarina & Krichevskaya 1981; Knowler 1983; Spirin & Ajtkhozhin 1985). The organization of hnRNP and mRNP particles in the nucleus and cytoplasm, respectively, and their possible association with underlying subcellular structures are issues of tremendous interest and potential significance but are beyond the scope of this review.

NUCLEAR RIBONUCLEOPROTEIN PARTICLES

Evidence for, and Isolation of, hnRNP Particles

MORPHOLOGICAL STUDIES   The original ideas about the existence of non-ribosomal nuclear RNA as a component of RNP complexes emanated from electron microscopic observations. In the 1950s Gall (1955, 1956) and Swift (1963) described ribonuclease-sensitive granules associated with or near chromosomes in several different cell types. Microscopic observations of RNA polymerase II transcripts on the large lampbrush chromosomes of amphibian oocytes suggested that the hnRNA becomes associated with proteins to form ribonucleoprotein structures (hnRNPs) immediately upon transcription (Gall & Callan 1962; Malcolm & Sommerville 1974). Using refined chromatin spreading techniques Miller and colleagues (Miller & Bakken 1972 ; McKnight & Miller 1976) examined the morphology of nascent transcripts of specific genes. With these spreading techniques the chromatin is dispersed and the nascent RNP molecules, both extended and protein deficient, were visualized with remarkable clarity. They showed a linear array of beaded globular protein units, about 20 nm in diameter, connected by RNAase-sensitive strands (Foe et al 1976 ; Lamb & Daneholt 1979 ; Malcolm & Sommerville 1977 ; McKnight & Miller 1979 ; Sommerville 1981). More recent studies revealed an orderly arrangement of the protein particles on specific nascent hnRNAs in situ (Beyer et al 1980, 1981 ; Osheim et al 1985 ; Tsanev & Djondjurov 1982).

BIOCHEMICAL STUDIES   Biochemical studies, beginning with the pioneering work of Samarina, Georgiev and their colleagues in the 1960s (Samarina et al 1966, 1968 ; reviewed in Samarina & Krichevskaya 1981), provided additional evidence for hnRNA-protein complexes and much information about the hnRNP particle. If protein denaturants (chaotropic reagents or ionic detergents) and high salt concentrations are avoided, most of the hnRNA can be released from nuclei with considerable amounts of protein associated with it. The most common methods of releasing hnRNA-protein complexes from nuclei are by mechanical disruption (e.g. sonication) or by leaching out after limited RNAase digestion (endogenous or exogenous RNAases) (Samarina et al 1968 ; Lukanidin et al 1972 ; Pederson 1974 ; Martin et al 1975 ; Beyer et al 1977 ; Karn et al 1977 ; Stevenin et al 1977 ; Maundrell & Scherrer 1979 ; Walker et al 1980). The bulk of the chromatin and nucleoli in mechanically ruptured nuclei are usually first removed by low-speed centrifugation, and the clarified fraction is defined as the nucleoplasm. The hnRNA in the nucleoplasm sediments in sucrose gradients (at moderate salt concentration of 50–100 mM NaCl) as a heterodispersed material between 30 and 250 S. It has a buoyant density (after fixation with formaldehyde or glutaraldehyde) of 1.3–1.4 g/ml, which indicates that it is composed of about 75–90% protein (Samarina & Krichevskaya 1981). In contrast, the protein-free hnRNA sediments much more slowly under the same conditions, at about 30 S or less, and has a buoyant density of 1.8 g/ml. All of the heterodispersed fast-sedimenting hnRNA in the nucleoplasm is converted by mild RNAase digestion to slower sedimenting material. Much of the latter forms relatively discrete homodispersed particles that sediment under the same conditions at 30–40 S (Pederson 1974 ; Beyer et al 1977 ; Karn et al 1977) (referred to hereafter as 30-S particles or monoparticles). The hnRNA-containing material that is released from nuclei by mild exogenous nuclease digestion or by prolonged incubation at 37°C (which presumably allows digestion with endogenous RNAase), also sediments as 30- to 40-S particles (Samarina et al 1968 ; Martin et al 1978 ; LeStourgeon et al 1978).

Each of the different cell fractionation methods has considerable limitations. When prepared as above, the nucleoplasm contains only about 50% of the total nuclear hnRNA and the hnRNP proteins; the rest are associated with the chromatin pellet and cannot be readily analyzed. The analysis of nucleoplasm therefore necessarily excludes a sizeable portion of the hnRNPs, and the sheer force may disrupt large hnRNP complexes. In the RNAase release method, the hnRNA and the hnRNPs are fragmented, and proteolysis and protein rearrangements (cf Stevenin et al 1979) can occur in the course of the prolonged incubations at high temperature. The specific portion of the total hnRNPs analyzed in these studies is not well documented and likely also represents only about 50% of the rapidly labeled nuclear RNA. It has, in fact, been suggested that in gently lysed nuclei all of the hnRNP particles are somehow associated with chromatin (Kimmel et al 1976).

An interesting approach that circumvents the need for any nuclear fractionation was utilized by Lahiri & Thomas (1985) to examine the hnRNPs in mitotic cells. In these cells gentle lysis of the plasma membrane is sufficient to release hnRNPs. The hnRNP particles from mitotic cells were found to be similar to those prepared from interphase cells by the above methods. Although the precise sedimentation properties of the released RNP complexes depend on the monovalent ion concentration and on the tissue of origin, complexes of similar general properties have been prepared from a wide and diverse range of cells in culture and in tissues. These cells include (in addition to those from mammals, avians, and amphibians) those from *Drosophila* (Risau et al 1983), *Artemia salina* (Marvil et al 1980; Nowak et al 1980), *Dictyostelium* (Firtel & Pederson 1975), and *Physarum* (Christensen et al 1977). There is little or no information in the literature about the RNP complexes in yeast, protozoa, or plants.

The existence of hnRNA in hnRNA-protein complexes was easily accepted, but a clear and consistent picture of the composition and structure of hnRNP complexes did not readily emerge. Progress in unambiguously identifying the hnRNP proteins has been hampered, perhaps to the largest extent, by the limitations of the experimental methods (such as velocity sedimentation and isopycnic banding) commonly used to prepare hnRNP complexes. These methods have not resulted in the complete separation of intact and pure hnRNPs. They rely on copurification of proteins with RNA from fractionated cells as the criterion for the identification of RNP proteins and therefore have several serious shortcomings. First, because adventitious RNA-protein associations can occur, it is difficult to ascertain, in the absence of other data, that a protein was an authentic RNP protein in the cell. Second, authentic RNP proteins may dissociate under the conditions used for the isolation of the RNA-protein complex. Third, contaminating structures of similar sedimentation and physical properties cannot be separated from actual hnRNPs. And fourth, the identification of RNPs is limited by the ability to unambiguously identify labeled hnRNA.

In spite of these inherent limitations, reproducible patterns on sucrose gradients did provide the basis for the consensus that gradually emerged that a group of proteins in the 30–43 kDa range is associated with hnRNA in 30-S monoparticles. These proteins, first described by Samarina et al (1968) as one or two proteins, were later, with improved electrophoretic techniques, shown to be considerably more complex (Martin et al 1974, 1978; Pederson 1974; Billings & Martin 1978; Beyer et al 1977; Karn et al 1977) and to consist of six bands by one-dimensional SDS-poly-acrylamide gel electrophoresis (SDS-PAGE). These proteins, some or all of which are also occasionally referred to as "core proteins," are a distinct, nonchromatin subset of nuclear proteins. This specificity further argues for the authenticity of hnRNP particles. The A, B, C nomenclature (in HeLa cells A1 = 34 kDa; A2 = 36 kDa; B1 = 37 kDa; B2 = 38 kDa; C1 = 41 kDa; C2 = 43 kDa; Table 1) of Beyer et al (1977) is the most widely accepted and will be used here. Numerous other proteins of higher molecular weight were also reported, but their authenticity remained controversial; it was even suggested that there may be several different types of hnRNP monoparticles (Gattoni et al 1978; Stevenin et al 1977; Jacob et al 1981). The several-hour-long sedimentation could have affected the hnRNPs present via the effects of proteolysis, RNAases, centrifugal drag force, or protein rearrangements. Hence, at that time researchers understandably reasoned (e.g. Beyer et al 1977) that only proteins that tracked precisely with and only with hnRNA or hnRNA fragments could be classified as hnRNP proteins. Thus proteins that sedimented under the same conditions also outside of the hnRNA peak were not considered hnRNP proteins because it was assumed that hnRNP proteins must all be contained only in stable hnRNP particles. The criterion of cosedimentation is also subject to the limited resolution of one-dimensional SDS-PAGE, in which different proteins with similar mobilities can be erroneously regarded as the same protein.

PHOTOCHEMICAL CROSS-LINKING IN INTACT CELLS   Decisive evidence for the existence of hnRNA in the nucleus in distinct hnRNA-protein complexes was recently obtained using UV-induced RNA-protein cross-linking in intact cells (Mayrand et al 1981; Mayrand & Pederson 1981; Van Eekelen et al 1981a,b; Dreyfuss et al 1984a,b; Bag 1984; Greenberg & Carroll 1985). This method of RNP identification overcomes the problems encountered with the preparation methods described above, because it involves the identification of proteins that are in direct contact with RNA in vivo. The photochemical UV cross-linking method relies on the fact that UV light photoactivates RNA and converts it to an extremely reactive, short-lived molecule that reacts virtually indiscriminately with other molecules, including proteins, in direct contact with it. In effect this is photo-affinity labeling of the RNA binding protein in vivo. The cross-linked hnRNA-protein and mRNA-protein complexes can then be isolated from the nuclear and cytoplasmic fractions, respectively, after boiling in SDS and mercaptoethanol. The hnRNAs or mRNAs can be isolated by affinity chromatography on oligo(dT)-cellulose, to which they bind through their 3'-poly(A) tails. The method is general, simple and clean. The protein-denaturing conditions (boiling in SDS) ensure that only proteins covalently linked to the RNA are purified with it. This process eliminates proteins nonspecifically associated with the RNA and prevents the loss of genuine RNP proteins which may occur during cell fractionation. The proteins can be released from the RNA-protein cross-linked complexes by digestion with RNAases and analyzed by SDS-PAGE. This method is highly specific and very efficient; under proper conditions the yield of the reaction (in terms of both RNA recovery and protein cross-linking) is high, and RNA breakage is minimal (Adam et al 1986a). Thus one can examine the proteins that interact with essentially the entire polyadenylated RNA population, rather than a subset of it.

The major [^{35}S]methionine-labeled proteins that become cross-linked to polyadenylated hnRNA in the HeLa cells have molecular weights of 120, 68, 53, 43, 41, 38, and 36 kDa (Mayrand et al 1981; Van Eekelen et al 1981a; Economides & Pederson 1983; Dreyfuss et al 1984a,b; Choi & Dreyfuss 1984b). The cross-linked proteins at 36 and 38 kDa probably correspond to A and B proteins, and the bands at 41 and 43 kDa are the C1 and C2 of the 30-S hnRNP subparticles described by Beyer et al (1977). The cross-linking of hnRNA to a unique set of proteins in vivo indicates that the hnRNA is associated with a specific set of RNA binding proteins.

ISOLATION OF THE hnRNP COMPLEX BY SPECIFIC IMMUNOADSORPTION   By immunizing mice with UV cross-linked RNA-protein complexes obtained in vivo, monoclonal antibodies to genuine RNA-contacting hnRNP proteins were recently generated (Dreyfuss et al 1984b; Choi & Dreyfuss 1984a). Several of these antibodies have been used as immunoaffinity reagents to isolate the hnRNP complex from vertebrate cell nucleoplasm employing rapid immunoadsorption (Choi & Dreyfuss 1984b). Immuno-adsorptions with two different monoclonal antibodies (4F4 and 2B12) to the hnRNP C proteins both isolate a similar complex from Hela cells. The complex contains proteins and large hnRNA of up to ~10 kilobases (kb) in length (Choi & Dreyfuss 1984b). More than 50% of the rapidly labeled nucleoplasmic RNA can be readily immunoprecipitated. By SDS-PAGE, the major proteins of the isolated complexes labeled by [^{35}S]methionine to steady state are of 34, 36 (A1 and A2), 37, 38 (B1 and B2), 41, and 43 kDa (C1 and C2); and doublets of 68 and 120 kDa (Figure 1). Additional proteins of 45 kDa and much larger are also seen, but so far little is known about them. In SDS-PAGE the major proteins of the complex appear identical to those hnRNP proteins that become cross-linked to the hnRNA upon UV-light exposure in vivo. Immunoprecipitation with different, non-cross-reacting monoclonal antibodies to the 120-kDa protein (Choi & Dreyfuss 1984b) and to the A1 protein (Pinol-Roma et al, unpublished results) isolates the same complex of proteins in a similar stoichiometry. Similar hnRNP complexes were isolated from rodent and avian cells.

The advantages of the immunoaffinity procedure for the isolation of the hnRNP complex are that it is specific, rapid, and mild and that it is not dependent on radioactive labeling for detection of the hnRNA. It yields very large hnRNP particles that appear to be intact and pure. The co-immunoprecipitation of the hnRNA and all of these proteins with anti-

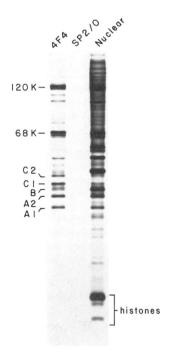

Figure 1   The proteins of human hnRNP particles. hnRNP particles were isolated from growing HeLa cell nucleoplasm by immunoabsorption with a monoclonal antibody (4F4) to the hnRNP C1 and C2 proteins. The proteins were labeled by culturing the cells with [^{35}S]methionine overnight (Choi & Dreyfuss 1984b).

bodies to different genuine hnRNP proteins strongly suggests that the hnRNP complex is a unitary structure that comprises highly conserved protein components. The isolation of the hnRNP complex has further defined the components of this structure and will make it possible to study them in detail. Based on results from SDS-PAGE of [^{35}S]methionine-labeled proteins in isolated hnRNP and nuclear fractions, it has been estimated that the hnRNP complex is one of the most abundant structures in the nucleus, accounting for at least one-third of the protein in the nucleoplasm. The evidence accumulated so far indicates that there is only one type of hnRNP complex, but in principle, minor forms comprised of other proteins may also exist.

General Structural Features of hnRNP Particles

As described above, early sedimentation data and RNA and protein composition data indicated that large hnRNP complexes are heterodispersed polyparticles that are converted to monoparticles (30-S complexes) by mild nuclease digestion. A "beads-on-a-string" structure for the hnRNA and its associated proteinaceous particles (see Samarina & Krichevskaya 1981) was suggested by the sedimentation data and by the overall shape and dimensions visualized by the electron microscopy of sedimented polyparticles, 30-S particles (Samarina et al 1968; Beyer et al 1977; Karn et al 1977; Martin et al 1978), and of particles seen on nascent transcripts in spread chromatin preparations (Miller spreads; e.g. McKnight & Miller 1976; Beyer et al 1981; Tsanev & Djondjurov 1982). Each polyparticle appears to be a unit of one hnRNA chain with proteinaceous monoparticles. Although the hnRNA alone may by itself provide the link necessary to hold the 30-S particles together as a polyparticle, additional factors and protein-protein, protein-RNA, or RNA-RNA interactions may serve ancillary roles in holding the large hnRNP complex together. Most parts of the hnRNA, with the clear exception of the poly(A) tail, are engaged with or are between 30-S particles. The heterodispersed sedimentation observed for the bulk of nucleoplasmic hnRNA results from the differences in lengths and structure of specific hnRNAs, which determines the number of monoparticles associated with each. Specific hnRNAs may sediment as more discrete forms (Sperling et al 1985). Further RNAase digestion of monoparticles causes them to dissociate, which suggests that the RNA associated with monoparticles is essential for maintaining their integrity. RNA-protein interactions are therefore important in holding the monoparticles together, and protein-protein interactions are not sufficient to do so.

The rapidly labeled RNA that cosediments with 30-S monomers can range in length from about 100–1000 nucleotides (usually 500 ± 100) depending on the extent of RNAase digestion used in the preparation. This RNA includes pre-mRNA sequences, as determined by hybridization to total cytoplasmic polyribosomal cDNA (Kinniburgh & Martin 1976) and to specific mRNAs (Maundrell & Scherrer 1979; Pederson & Davis 1980; Munroe 1982; Stevenin et al 1982). Several reports (Sekeris & Niessing 1975; Deimel et al 1977; Gallinaro & Jacob 1979; Howard 1978; Zieve & Penman 1981) describe the detection of small nuclear RNPs (snRNPs) associated with hnRNP complexes. However, in most cases the specificity of the association is not certain. Evidence for interaction between snRNAs and hnRNAs in vivo has been presented (Calvet & Pederson 1981, 1982; Setyono & Pederson 1984).

Proteins of hnRNP Particles

The major steady-state [^{35}S]methionine-labeled proteins of HeLa hnRNP particles recognized so far are the A, B, and C proteins and doublets of 68 and 120 kDa (Figure 1 and Table 1). Except for some reduction in the 68K doublet, most of these proteins are also seen in the 30- to 40-S fraction after sedimentation in sucrose gradients. The major proteins cosediment with [^{3}H]hnRNA in the 30-S fraction in sucrose gradients after limited RNAase digestion. The protein complex is held together as long as it is associated with hnRNA fragments of about 125 ± 25 nucleotides or longer (Choi & Dreyfuss 1984b). These proteins are therefore part of the monomer particles, and they comprise almost the entire protein complement of large (polyparticle) hnRNP complexes. The proteins A1, A2, C1, 68-kDa, and 120-kDa are the most abundant by Coomassie blue staining and

Table 1 Proteins of human hnRNP monoparticles[a]

Protein	Molecular weight	Isoelectric point (± 0.5)	Relative amount	Affinity for RNA[b]	Posttranslational modification[c]
A1	34,000	9.0	3–4	+	DMA
A2	36,000	8.0	3–4	+	DMA
B1	37,000	8.5	1	+	nd
B2	38,000	9.0	1	+	nd
C1	41,000	6.0	3–4	+ + +	Pi
C2	43,000	6.0	1–2	+ + +	Pi
68K$_{1,2}$	68,000	6–8	1–2	+ +	nd
120K$_{1,2}$	120,000	6.5	1–2	+ +	Pi

[a] In growing HeLa cell nucleoplasm.
[b] Determined by resistance to dissociation by salt and by binding to ssDNA (Y. D. Choi et al, unpublished).
[c] DMA = dimethylarginine; Pi = phosphorylation; nd = not determined.

[^{35}S]methionine steady-state labeling. Many, if not all, of these proteins are in contact with the hnRNA and can be cross-linked to it in intact cells by UV light (Choi & Dreyfuss 1984b). The picture derived from two-dimensional gel electrophoresis (Brunel & Lelay 1979; Suria & Liew 1979; Peters & Comings 1980; Knowler 1983; Wilk et al 1985) is considerably more complex: Many of the proteins resolve into several spots, and there is evidence for posttranslational modifications of many of them. It is quite certain (from examination of the electrophoretic patterns) that not all of the protein components of hnRNPs have been identified.

The hnRNP proteins are very abundant in the nucleus of growing cells, as abundant as histones. But unlike histones, the overall amount of hnRNP proteins can vary substantially depending on the growth or transcriptional state of the cell. With increased RNA polymerase II transcription the amount of hnRNP proteins increases. Indications of such correlations have been described in steroid-responsive tissues (Knowler 1976), and quiescent cells have been reported to contain less of some of the hnRNP proteins than do growing cells (Stunnenberg et al 1978; LeStourgeon et al 1978; Celis et al 1986). HeLa (human) hnRNP proteins are the best characterized, and most of the discussion here refers to them.

A AND B PROTEINS   The A and B proteins are members of two related families of basic proteins that share common antigenic determinants (Leser et al 1984; Leser & Martin 1986). Proteins of this group are associated with the RNP fibers of transcriptionally active chromatin, which suggests that they become associated with nascent hnRNA (Martin & Okamura 1981). The A group proteins, A1 and A2, have isoelectric points of about 9.2 and 8.4, respectively (Beyer et al 1977; Wilk et al 1985). A1 and A2 have similar (but different) amino acid compositions containing a very high percentage of glycine (25%) and the rare modified amino acid N^G,N^G-dimethyl arginine (Boffa et al 1977; Christensen et al 1977; Beyer et al 1977; Karn et al 1977). The A proteins dissociate from the RNA in vitro in 0.13–0.15 M NaCl (Beyer et al 1977; LeStourgeon et al 1981). Results of mobility tests in SDS-PAGE after UV cross-linking in vivo indicate that A1 is efficiently cross-linked to hnRNA and that A2 is either not at all, or is much less, cross-linked (Choi & Dreyfuss 1984b). The ratio of A1 to A2 in hnRNP particles is about 1 : 1 in rapidly growing tissue culture cells; in total cell material, however, considerably more A2 is detected (Celis et al 1986). Much less A1 is found in cells that are stationary than in proliferating cells or tissues of adult animals (LeStourgeon et al 1978; Celis et al 1986). Similar proteins are found in all mammals (Beyer et al 1977; Karn et al 1977; Choi & Dreyfuss 1984b), but the size of A proteins in divergent vertebrates can vary (Leser et al 1984). Immunofluorescence evidence suggests that the A proteins are confined to the nucleus (Leser et al 1984).

Recent immunological (Valentini et al 1985; Pandolfo et al 1985), peptide mapping, and cDNA sequencing data (Chase & Williams 1986; K. Williams, personal communication; S. Riva, personal communication) demonstrate that a 24-kDa single-stranded DNA binding protein from calf thymus (Herrick & Alberts 1976a), UP1, is a fragment of the A1 protein. The amino acid sequence of UP1 (Williams et al 1985) is identical to the first 195 amino acids of the amino acid sequence of the A1 predicted by the cDNA clone. Based on this DNA sequence, the A1 protein contains another 124 amino acids that constitute a very glycine-rich (about 40%) domain at the COOH terminus of the protein (K. Williams, personal communication). In addition to A1, polyclonal antibodies to UP1 recognize other hnRNP proteins, which are presumably A and B proteins (Valentini et al 1985). Thus antigenic determinants similar to A1 probably exist in these other hnRNP proteins. Previous studies on UP1 revealed that it binds single-stranded DNA (ssDNA) tightly but apparently without sequence specificity and that it is a very effective helix-destabilizing protein (Herrick & Alberts 1976b). Perhaps other proteins previously classified as ssDNA binding proteins in various cells and viruses are also hnRNP proteins.

A helix-destabilizing hnRNP protein related to the A proteins of mammals, HD40, was isolated from the brine shrimp *Artemia salina* (Marvil et al 1980; Nowak et al 1980). This 40-kDa protein binds to and disrupts residual secondary structure of single-stranded nucleic acids at a stoichiometry of about one protein per 12–15 nucleotides (Thomas et al 1981, 1983). A cDNA clone for HD40 was isolated, and genomic DNA crosshybridizing with it was detected in divergent animals and plants (Cruz-Alvarez et al 1985).

The B proteins, B1 and B2 (37 and 38 kDa), have isoelectric points of about 8.3 and 9.2, respectively (Beyer et al 1977; Wilk et al 1985). Their prevalence in hnRNP particles is about one-third that of the A1, A2, and C1 proteins and about the same as that of C2 (Beyer et al 1977; LeStourgeon et al 1981; Choi & Dreyfuss 1984b). The B proteins dissociate from the RNA at moderate salt concentrations similar to those at which the A proteins dissociate (0.13–0.15 M) (Beyer et al 1977). Proteins of molecular weights that correspond to those of the B proteins are cross-linked to RNA by exposure to UV light in intact cells (Economides & Pederson 1983), but without specific antibody tests it is not certain that these are the B proteins. The B proteins are antigenically related to each other and to the A proteins, as shown by their reactions to monoclonal antibodies (Leser et al 1984). Immunofluorescence shows that they are confined to the nucleus during interphase. Like the A proteins, the B proteins possess unusual amino acid compositions of 25% glycine and several moles each of the unusual residue dimethyl arginine (Beyer et al 1977; Wilk et al 1985). Two-dimensional gel electrophoresis has shown that the B proteins exist in several posttranslationally modified forms (Brunel & Lelay 1979; Peters & Comings 1980; Suria & Liew 1979; Wilk et al 1985), but the specific modifications have not been determined.

c PROTEINS   The C proteins (C1, 41 kDa, and C2, 43 kDa) are major constituents of 30-S hnRNP monoparticles (Beyer et al 1977) and are the two most prominent proteins cross-linked to hnRNA by UV light treatment in vivo (Dreyfuss et al 1984b; Choi & Dreyfuss 1984b). The C proteins bind RNA more tightly than do the other major hnRNP proteins, as determined by resistance to dissociation from RNA in high salt concentrations (Beyer et al 1977; LeStourgeon et al 1981; Y. D. Choi & G. Dreyfuss, unpublished results) and by binding to ssDNA (S. Pinol-Roma & G. Dreyfuss, unpublished results). The two C proteins are highly similar to each other: They are both recognized by the same monoclonal antibodies, and antibodies raised against purified C1 also react with C2 (Dreyfuss et al 1984b; Choi & Dreyfuss 1984a). In humans the two proteins have different but related partial peptide maps and the same acidic isoelectric points (6.0 ± 0.5) (Dreyfuss et al 1984b). A monoclonal antibody raised against the human C proteins reacts with the C proteins in widely divergent species ranging from humans to reptiles (Choi & Dreyfuss 1984b). In all species examined there are two C proteins in the range from 39 to 42 kDa for C1 and from 40 to 45 kDa for C2. The C proteins are phosphorylated in vivo (Dreyfuss et al 1984b; Choi & Dreyfuss 1984a; Holcomb & Friedman 1984) and in vitro by a casein kinase type II (Holcomb & Friedman 1984). Like the other components of 30-S particles, they are associated with both poly(A)-containing and non-poly(A)-containing hnRNAs. Immunofluorescence microscopy demonstrated that the C proteins are segregated to the nucleus. Within the nucleus the C proteins are not found in nucleoli and are not associated with chromatin, as seen in cells in prophase (Choi & Dreyfuss 1984a). Nakagawa et al (1986) recently isolated cDNA clones for the human C proteins. These clones hybridize to genomic DNA sequences in divergent eukaryotes, including yeast, which suggests that C proteins are ubiquitous components of hnRNPs in eukaryotes. So far, the C proteins are the only hnRNP proteins shown to have a role in pre-mRNA splicing (Choi et al 1986). Van Eekelen & Van Venrooij (1981) suggested that the C proteins are associated with the nuclear matrix, but subsequent experiments with specific antibodies showed that only a very small fraction of these proteins remain with a detergent-, nuclease-, and high salt–resistant nuclear substructure (Dreyfuss et al 1984b).

68- AND 120-kDa PROTEINS   In one-dimensional SDS gel electrophoresis, doublets of proteins of 68 and 120 kDa are coimmunoprecipitated with other hnRNP proteins and are in contact with hnRNA as shown by photochemical UV cross-linking in intact cells (Dreyfuss et al 1984b; Choi & Dreyfuss 1984b). They are, therefore, authentic hnRNP components but are so far less well characterized than the A, B, and C proteins. From results on two-dimensional gels it is apparent that there are more than two proteins of these molecular weights (Y. D. Choi & G. Dreyfuss, unpublished results). The lower band of the 120-kDa doublet is recognized by a monoclonal antibody, 3G6 (Dreyfuss et al 1984b), and this antibody can immunoprecipitate the hnRNP complex (Choi & Dreyfuss 1984b). Like the C proteins, 120-kDa doublet is seen by immunofluorescence to be confined to the nucleus, is extensively phosphorylated, is conserved across vertebrates, and is associated with both polyadenylated and non-polyadenylated hnRNA (Dreyfuss et al 1984b).

3′-POLY(A) AND 5′-CAP BINDING PROTEINS   All hnRNAs contain a $5'$-m^{7}Gppp-cap structure. The protein(s) that are bound to the hnRNA cap in hnRNPs have not been identified. Using photoaffinity labeling with a cap analog, three proteins of about 120, 89, and 80 kDa were detected in the nuclear fraction (Patzelt et al 1983). The significance of these findings is uncertain; moreover, it is not clear whether or not the hnRNA cap is associated with 30-S monoparticles.

Unlike other parts of the hnRNA, the 3′-poly(A) tail is not associated with 30-S particles. Instead, it forms a distinct particle that sediments at 15 S. This indicates that the tail is associated with considerable amounts of protein, because the protein-free form sediments at 4 S (Quinlan et al 1974). The protein to RNA ratio is actually higher in these 15-S particles than in 30-S particles, as judged by their lower buoyant density (Quinlan et al 1977). However, these proteins do not confer any protection to the poly(A) tail from nucleases (Baer & Kornberg 1980; Tomcsanyi et al 1983). The protein composition of the hnRNA poly(A)-ribonucleoprotein complex isolated by sucrose gradients was investigated by several groups (Quinlan et al 1974, 1977; Samarina & Krichevskaya 1981; Firtel & Pederson 1975; Kish & Pederson 1975; Tomcsanyi et al 1983). These studies revealed several proteins of 60 and 70–90 kDa. Using UV cross-linking in nuclear extracts from HeLa cells, Setyono & Greenberg (1981) identified a 60-kDa protein that is in direct contact with the poly(A) sequence. However, additional proteins that do not become cross-linked to RNA may be found in the complex. Sachs & Kornberg (1985) recently identified a 50–55 kDa poly(A) binding protein in nuclei of yeast. The physiological role of the hnRNA poly(A) binding proteins or the poly(A)-ribonucleoprotein complex is not known, but it is reasonable to assume that they may have a role in the formation of the poly(A) tail, such as in serving as a length-measuring system for the poly(A) polymerase.

Arrangement of Proteins in hnRNP Monomer Particles

The relative amounts of the major proteins of 30-S particles (see Table 1) of growing cells can be estimated from Coomassie blue staining and from steady-state [^{35}S]methionine labeling. A1 and A2 are present in similar amounts. The two B proteins are also present in equal amounts but at about one-third the level of the A proteins. C1 is present in the same amount as the A polypeptides; C2 (like the B proteins) is present at about one-third that amount. The 68 and 120-kDa proteins are about as abundant in mass (not necessarily stoichiometry) in isolated complexes as the A and C1 proteins are. Based on the estimated molecular mass of 30-S monoparticles isolated by sucrose gradient sedimentation, it has been suggested that these particles are composed of three or four repeating units each composed of 3A1, 3A2, 1B1, 1B2, 3C1, and 1C2 (Lothstein et al 1985). However, this suggestion does not take into account the high molecular weight proteins (e.g. 68 and 120 kDa) and other uncharacterized proteins that may be part of the structure.

Protease and RNAase digestion experiments and chemical and UV cross-linking experiments permit some conclusions about the relative positions of the RNA and proteins in hnRNP particles. In intact 30-S particles, proteins A2 and B1 occupy an internal, protease-protected position (Lothstein et al 1985), and they may exist as three or four tetramers of $(A2)_3(B1)$ or as pentamers of $(A2)_3(B1)(B2)$. After RNAase digestion of the hnRNP, A2, B1, and B2 remain associated in nuclease-resistant structures that readily aggregate. If digestion is performed in low Mg^{2+} concentration, twelve such residual tetramers assemble to form highly regular 20-nm 43-S particles.

The proteins A1, C1, and C2, on the other hand, are sensitive to mild proteolysis and dissociate from the internal A2, B1, and B2 complexes, which suggests that they are peripheral (LeStourgeon et al 1981). Consistent with this position are findings showing that A1, C1, and C2 are in direct contact with RNA, as determined by photochemical cross-linking, and that monoclonal antibodies against C1 and C2 and against A1 efficiently immunoprecipitate intact hnRNP particles and 30-S particles (Dreyfuss et al 1984b; Choi & Dreyfuss 1984b; Y. D. Choi et al, unpublished results). In these studies the 68-kDa proteins are also seen in contact with the hnRNA. They probably occupy peripheral positions in the monoparticles because they are readily lost during sucrose gradient sedimentation (Y. D. Choi & G. Dreyfuss, unpublished results). The 120-kDa proteins also occupy peripheral positions: An antibody (3G6) that can bind them precipitates intact hnRNPs (Dreyfuss et al 1984b; Choi & Dreyfuss 1984b). From chemical cross-linking experiments with isolated 30-S particles (Lothstein et al 1985) it appears that A1, A2, and C1 each exist as homotypic trimers. These findings also indicate that the A1 and A2 proteins are next neighbors and that B proteins are in contact with A and C proteins.

The ratio of C1 to A1 in the immunoprecipitates can be used as an index of the intactness of the 30-S particle. When the hnRNA that is associated with monoparticles is degraded to segments of 125 ± 25 nucleotides, the monoparticles are still intact. However, when these segments of hnRNA are degraded further to 60–75 nucleotides, most monomers are no longer intact (Y. D. Choi et al, unpublished results). The C proteins, together with 68- and 120-kDa proteins, remain as a complex after digestion with a nuclease that yields intact RNA stretches shorter than 60–75 nucleotides. However, this residual structure dissociates upon further digestion, which suggests that it is held together by short stretches of RNA. By immunoprecipitation with the cognate monoclonal antibody, it appears that A1 is released as a single protein without the other major hnRNP proteins (Y. D. Choi et al, unpublished results). Because the 30-S monoparticle dissociates upon nuclease digestion, protein-protein interactions alone are clearly not sufficient to hold it together. Thus hnRNP proteins probably do not pre-exist as complexes of proteins only.

Arrangement of hnRNA in hnRNP Particles

Studies of protein composition, nuclease digestion, and sedimentation properties suggest that native large hnRNP complexes are composed mostly of multiple 30-S particles connected by highly nuclease-sensitive stretches of hnRNA. Data from nuclease digestion experiments and cross-linking of RNA to peripheral monoparticle proteins indicate that most, if not all, of the particle-associated hnRNA occupies a peripheral, nuclease-accessible position in intact 30-S particles. However, it is also possible that only parts of the hnRNA chain are initially exposed on the surface of hnRNP particles and that when these are cleaved, a structural change results that exposes the rest of the hnRNA. The fragments of hnRNA recovered with gradient-purified monoparticles average ~ 500–800 bases in length (Martin et al 1978; LeStourgeon et al 1981; Steitz & Kamen 1981). Studies on immunopurified hnRNP particles have shown that monoparticles are associated with ~ 500 ± 100 nucleotides of hnRNA and that this stretch of RNA can be further cleaved and trimmed to two or three stretches of 125 ± 25 nucleotides before the monoparticle dissociates (Choi & Dreyfuss 1984b; Y. D. Choi & G. Dreyfuss, unpublished results). The monoparticle proteins do not dissociate until the average RNA fragment is cleaved below 125 ± 25 bases in length (Choi & Dreyfuss 1984b). This could be the length of RNA in close contact with the proteins of the monoparticles, or it could be the length of RNA associated with protein subdomains within monoparticles. When the RNA is cleaved down to about 60–75 nucleotides, a residual particle containing C1, C2, and the 68- and 120-kDa proteins can be isolated with these short nucleotide segments. After cleavage of the hnRNA within the 30-S particle the hnRNA fragments and the proteins remain associated in a single complex, which suggests that the hnRNP monomer can serve as an "operating table" for hnRNA processing. From the amount of hnRNA lost upon nuclease conversion of large hnRNPs to 30-S monoparticles (Y. D. Choi & G. Dreyfuss, unpublished results) the average length of the intermonoparticle linker RNA is about 250 ± 50 nucleotides. As discussed above, the poly(A) segment of polyadenylated hnRNAs is not associated with 30-S particles but rather is bound to a 60-kDa protein and, probably together with several other proteins, forms a distinct 15-S particle.

The position of hnRNA intron and exon sequences in hnRNP particles and the specific arrangement of monoparticles and monoparticle proteins on specific hnRNAs have been the subjects of intense investigation, but as yet no conclusions can be drawn regarding these important issues. To understand the assembly and function of hnRNP particles we must answer two important questions: Are the positions of 30-S particles and of individual 30-S proteins specific and fixed on the primary transcript? If so, is this the result of sequence-specific features or of an assembly process that also acts as a measuring device? Electron microscopic studies of specific *Drosophila* transcripts (Beyer et al 1981; Osheim et al 1985) in highly dispersed preparations using Miller's chromatin spreading techniques revealed an orderly arrangement of protein particles of about 20 nm in diameter (about the size of isolated 30-S particles) along the hnRNA. The particles are neither randomly nor uniformly distributed; rather, their location correlates with nascent transcript cleavage. A class of RNPs stable under the preparation conditions are associated specifically with splice junctions. The significance of these findings rests on the identity of the protein particles seen in these preparations. This is not yet known, but it seems doubtful that typical 30-S hnRNP particles could survive the specific detergent, pH, and ionic conditions used for the preparation of the specimens. In contrast, examination of less dispersed (less deproteinized) preparations (Osheim et al 1985) showed that the transcripts are completely covered with protein in a roughly particulate form (average particle diameter ~ 22.5 nm). This suggests little or no specificity in the packaging of hnRNA with protein in particles, presumably 30-S particles. This is consistent with the findings obtained by nuclease digestions (Steitz & Kamen 1981; Munroe 1982). Differential sensitivity to nuclease, which suggests nonrandom distribution of proteins on specific pre-mRNAs, has been reported in the case of polyoma (Steitz & Kamen 1981) and β-globin (Patton et al 1985; Patton & Chae 1985). The usefulness of nuclease digestion mapping is limited because it is not certain which proteins generate the observed pattern and because the hnRNP structures are very labile and could vary depending on the isolation conditions. Using UV cross-linking, proteins associated with specific regions of adenovirus pre-mRNAs have been detected (Van Eekelen et al 1982; Ohlsson et al 1982). The significance of these patterns is not obvious from these or other studies (Stevenin et al 1982; Huang & Chae 1983; Munroe & Pederson 1981), thus the question remains unanswered. Clearly, more mapping of specific proteins on specific hnRNAs is needed.

Also of considerable interest are the precise location of specific hnRNA sequence features [such as double-stranded regions and oligo(A) and oligo(U) stretches] within hnRNP particles and the contribution of such features to the particle structure. The presence of such sequences has been documented, but their significance is not yet apparent. Internal oligo(A) sequences of pre-mRNA, 20–40 nucleotides long, were found in 30-S particles (Kinniburgh & Martin 1976; Martin et al 1978). Kish & Pederson (1977) found oligo(U) sequences of ~ 15–50 nucleotides in HeLa hnRNP particles. These sequences may be complexed with oligo(A) or poly(A) sequences. They are resistant to nuclease and may be covered with proteins. Small amounts of double-stranded RNA (dsRNA) were detected in 30-S particles (Ryskov et al 1973) by several methods (see Samarina & Krichevskaya 1981), including nuclease digestion (Calvet & Pederson 1978). These sequences do not seem to bind 30-S particle proteins or other proteins; they may protrude from the surface of 30-S monoparticles as well as being located between monoparticles (Martin et al 1978).

Assembly and Disassembly of hnRNP Particles

The assembly of hnRNP particles apparently occurs as the hnRNA is still a nascent transcript. This assumption is based primarily on the above-cited microscopic observations showing that chromatin-associated transcripts are bound with proteins and on analysis of the rapidly labeled RNA in this fraction (Augenlicht & Lipkin 1976). Martin & Okamura (1981) used immunocytochemical procedures to show that the A and B groups of the hnRNP proteins are localized to regions containing actively transcribed

RNA, and they suggested that these proteins become associated with nascent transcripts. Economides & Pederson (1983) showed by UV cross-linking that proteins in the 30–40 kDa range, which presumably correspond to the A, B, and C proteins, become cross-linked to rapidly labeled hnRNA soon after transcription. The 30-S monoparticle hnRNP proteins interact with RNA even if it is not polyadenylated (Dreyfuss et al 1984b; Pullman & Martin 1983; Wilk et al 1983). These findings further suggest that assembly of hnRNP particles is an early posttranscriptional event that precedes polyadenylation and splicing of pre-mRNA.

hnRNP proteins must be reutilized since protein synthesis is not immediately required for incorporation of hnRNA into hnRNP particles and because the hnRNP proteins examined are very stable relative to the half-life of the hnRNAs (Martin & McCarthy 1972). hnRNA turnover and the turnover of the major hnRNP proteins are therefore not tightly coupled. There does not seem to be (at least when analyzed in nucleoplasm by sedimentation) a very large pool of free (not hnRNA-bound) hnRNP proteins. Therefore, disassembly of hnRNPs prior to transport of mRNA to the cytoplasm is likely to be a major source of hnRNP proteins for reassembly on nascent hnRNAs.

What are the signals for hnRNP assembly? Possible determinants include the RNA polymerase II complex itself, the cap structure (m^7Gppp in hnRNAs) or other posttranscriptional modifications, the size of the RNA, and possibly the subnuclear localization of the transcripts. Clearly, small nuclear RNAs (snRNAs) are not associated with hnRNP proteins but rather with distinct snRNP proteins. This, however, does not mean that it is not the polymerase complex itself (some of the snRNAs are also transcribed by RNA polymerase II) that determines the specificity because the snRNP proteins and the trimethyl cap of the snRNAs are acquired in the cytoplasm, and it is therefore possible that the snRNAs initially are bound with hnRNP proteins. The contribution, if any, of introns and factors that interact with intron- or intron/exon-junction sequences is not yet known. It has been suggested (Pederson 1983) that intronless transcripts may not be assembled into hnRNP complexes. This conclusion is based on observations in heat-shocked *Drosophila* cells (Mayrand & Pederson 1983), but these findings have recently been questioned (Kloetzel & Schuldt 1986). The order of assembly of hnRNP proteins on the hnRNA is not known.

Protein-RNA reconstitution experiments so far have demonstrated that proteins from the 30–40 S nucleoplasmic region of sucrose gradients bind RNA and form complexes similar to hnRNP particles in sedimentation properties and general appearance by electron microscopy (Kulguskin et al 1980; Wilk et al 1983; Pullman & Martin 1983). While this is a very promising approach, the criteria used so far to assess reconstitution are not sufficient because there are no specific functional assays for hnRNP monoparticles. It will be necessary to show that the composition, stoichiometry, and specific arrangement of the individual proteins in these reconstituted particles are similar to those found in hnRNP particles isolated from cells.

Little is known about the process of disassembly of hnRNP particles. It must occur in the nucleus prior to or coincident with mRNA transport and is likely to be the source of most of the hnRNP proteins which are probably recycled to form new particles with newly synthesized hnRNA. Allosteric effectors or covalent posttranslational modification of hnRNP proteins by specific enzymes in the vicinity of the nuclear-pore complexes may be involved in decreasing the normally high affinity of hnRNP proteins for the RNA. Specific conditions within the nuclear pore complexes may favor dissociation of the hnRNP particle. In vitro reconstituted systems should be very useful tools for exploring these possibilities.

Functions of hnRNP Particles and hnRNP Proteins

Two extreme functions for hnRNP complexes are likely and not mutually exclusive: hnRNPs may have a predominantly structural role, namely, they may be involved in the packaging of hnRNA. The packaging could serve several functions, such as preventing tangling of transcripts, compacting the hnRNA, facilitating RNA strand displacement and release from template DNA, and protecting the hnRNA from degradation by endogenous nucleases. Alternatively, hnRNPs may be more directly and actively involved in the posttranscriptional processing of hnRNA and in mRNA production, including splicing of pre-mRNA and transport to the cytoplasm. hnRNPs may provide correct substrate presentation and possibly process enzymatic activities. Although a complete account of the physiological significance of hnRNP particles is well beyond our present knowledge and may take a long time to obtain, it has recently become possible to experimentally address specific questions about the role of hnRNP particle proteins in the biogenesis of mRNA.

The idea that pre-mRNA processing occurs in the nucleus in hnRNP particles and that the hnRNP complex itself is a critical element in the formation of mRNA has been a major theme in this field of research since the earliest observations of RNP complexes. The hnRNP monoparticle appears to have the necessary properties to serve as an "operating table" for RNA splicing, and hnRNPs are indeed associated with pre-mRNA and spliced mRNA in the nucleus (Y. D. Choi & G. Dreyfuss, unpublished results). Until recently, however, it was difficult to test experimentally the functional significance of hnRNPs because there were no definitive probes for hnRNP proteins and no in vitro assay for pre-mRNA splicing. With the development of in vitro cell-free systems that faithfully splice mRNA precursors (Hernandez & Keller 1983; Padgett et al 1983, 1984; Krainer et al 1984; reviewed in Green 1986; Padget et al 1986) and with the availability of specific antibodies to proteins of hnRNP complexes it has become possible to examine the question of whether hnRNP proteins play a role in the splicing of pre-mRNA.

Choi et al (1986) investigated the effect of several monoclonal antibodies to hnRNP proteins on pre-mRNA splicing. It was found that splicing in vitro in HeLa nuclear extract of a mRNA precursor was inhibited by a monoclonal antibody to the hnRNP C proteins. The inhibition with the anti-C antibody, 4F4, is at an early step of the reaction—cleavage at the 3′-end of the upstream exon and the formation of the intron lariat. In contrast, preboiled 4F4, or a different anti-C monoclonal antibody (designated 2B12), or antibodies to other hnRNP proteins (the 120- and 68-kDa proteins), and nonimmune mouse antibodies have no inhibitory effect. Sedimentation experiments showed that the 4F4 antibody diminishes, but does not prevent, the ATP-dependent formation of a 60-S splicing complex (spliceosome) that contains pre-mRNA, proteins, and snRNAs. This complex is probably necessary for the progression of the splicing reaction (Brody & Abelson 1985; Grabowski et al 1985). Furthermore, the 60-S splicing complex contains C proteins, and it can be immunoprecipitated with the antibody to the C proteins. Moreover, depletion of C proteins from the splicing extract by immunoadsorption with 4F4 or 2B12 results in the loss of splicing activity, whereas mock depletion with nonimmune mouse antibodies has no effect. A 60-S splicing complex does not form in a C protein–depleted nuclear extract. These results indicate an essential role for the proteins of the hnRNP complex in the splicing of mRNA precursors. The splicing complex therefore appears to be a modified hnRNP monoparticle or monoparticle subdomain that contains, in addition to hnRNP proteins, other components specifically necessary for splicing, including the snRNPs U1 and U2 (Black et al 1985; Krainer & Maniatis 1985) and possibly also U5 (Chabot et al 1986), U4, and U6 (D. L. Black & J. A. Steitz, personal communication) (see also Green 1986 and Padgett et al 1986). The stabilized particles seen by Osheim et al (1985) may correspond to these complexes. Because very short pre-mRNAs can be spliced in vitro, it may be that a spliceosome is composed not of a complete monoparticle but only of one of its subdomains containing a C-protein. Since hnRNP formation precedes polyadenylation, hnRNP proteins may also be important in this process. Tentative schematic presentations of the structure of hnRNP particles and their involvement in mRNA biogenesis are shown in Figure 2.

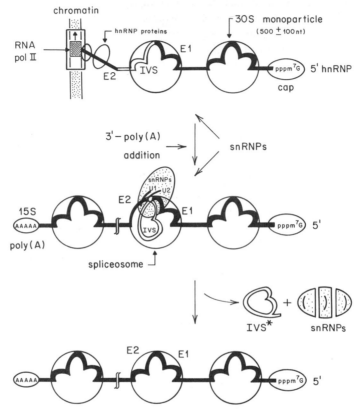

Figure 2 Schematic presentation of a tentative generalized model of hnRNP particle structure and its involvement in pre-mRNA splicing. E = exon coding sequences. IVS = intervening sequence (intron). The figure depicts splicing of a small intron of a size that can
be accommodated in a 30-S particle. For much larger introns two general possibilities can
be envisioned: the intron loops out of the surface of the same monoparticle (possibly
associated with hnRNP proteins) and splicing occurs within the context of that monoparticle.
Alternatively, the donor (5′) and acceptor (3′) exon/intron junctions could be positioned on
two different 30-S monoparticles, and splicing could occur by bringing the two together. The
latter would require considerably more complicated spatial maneuvering of monoparticles.
At present it is not known if snRNPs interact with the hnRNA (pre-mRNA) during or after
hnRNP particle assembly or how they affect the packing of the hnRNA.

CYTOPLASMIC RIBONUCLEOPROTEIN PARTICLES

Evidence for, and Isolation of, mRNP Complexes

As noted earlier, not all mRNAs are translated at the same time or with the same efficiency. Actively translated cytoplasmic mRNAs are engaged with ribosomes (polyribosomal mRNAs) and untranslated mRNAs (nonpolysomal mRNAs) are not. In addition to defining the translational state of mRNA, association with polyribosomes is the basis for the physical separation of these two functional classes of mRNAs by fractionation of cytoplasmic extracts on sucrose density gradients.

Spirin and collaborators provided the first evidence that mRNA can exist apart from polyribosomes in the form of a nucleoprotein complex. They demonstrated that in early stages of development in fish and sea urchin embryos mRNA was found in RNP particles that sedimented more slowly (20–75 S) than did ribosomes (80 S). The nonribosomal nature of the particles was demonstrated by their distinct buoyant density in CsCl gradients after fixation with formaldehyde (Spirin et al 1965): 1.42–1.45 g/ml versus 1.52 g/ml for ribosomal subunits. Evidence that the RNA in these particles, which the authors termed informosomes, contained mRNA was drawn from their ability to direct protein synthesis in vitro (Spirin & Nemer 1965; Spirin 1969). The buoyant density of fixed particles became an important criterion for defining and characterizing ribonucleoprotein complexes (Spirin et al 1965). This method was subsequently used to demonstrate that nonpolysomal mRNAs exist in diverse eukaryotic cells (Perry & Kelley 1968; Henshaw & Loebenstein 1970; Spohr et al 1972; Gander et al 1973), and its application to polysomal mRNAs helped establish the generality of the concept that mRNA in the cytoplasm is always found in the form of mRNP complexes. mRNAs released from polyribosomes by a variety of treatments (Lebleu et al 1971; Schochetman & Perry 1972; Lee et al 1971; Blobel 1972; Henshaw 1968; Perry & Kelley 1968) were also shown to be complexed with protein and to have a buoyant density similar to that of mRNPs in CsCl gradients.

Following the detection of poly(A) segments on mRNAs (reviewed in Brawerman 1983), it became possible to rapidly resolve mRNA with its associated proteins from ribosomal material by oligo(dT) or poly(U) chromatography without prior fixation and without CsCl gradients (Lindberg & Sundquist 1974). This important advance spawned numerous observations of mRNPs in a wide variety of organisms. Unlike deproteinized mRNA, which can be eluted from an oligo(dT) column at a low salt concentration, the column-bound mRNPs can only be eluted with formamide or at elevated temperatures (Lindberg & Sundquist 1974;

Jain et al 1979). Even when prepared by these methods, mRNPs are not completely free of ribosomal and other cytoplasmic proteins. Thus the assignment of these proteins to the mRNP fraction remains uncertain in the absence of other data because it cannot be ascertained that the eluted proteins were retained by the column strictly through specific association with the poly(A)$^+$ mRNA.

Early on, the physiological significance of mRNP particles was called into question (Baltimore & Huang 1970) because it was found that RNA incubated with cytoplasmic extract forms RNA-protein complexes of similar sedimentation properties and buoyant density. However, these artificial complexes, unlike mRNPs isolated from cells, do not form at NaCl concentrations of 0.15 M or higher. This finding influenced subsequent investigators to isolate mRNP complexes at 0.5 M NaCl. This procedure, though necessary, only decreases the likelihood of nonspecific associations, it does not eliminate them. In addition, it carries the risk of promoting the dissociation of genuine mRNP proteins from the complex.

The first reports of the protein composition of mRNPs came from work done on reticulocytes from duck and rabbit (Morel et al 1971; Blobel 1972). Both reports found two similar major proteins of $M_r \sim 52,000$ and 78,000. Although many different proteins have been identified in various studies, proteins of similar molecular weights (the larger being the poly(A) binding protein, see below) seem to be consistent mRNP components and have been detected in many different cell types, including HeLa cells (Kumar & Pederson 1975), KB cells (Lindberg & Sundquist 1974; Van der Marel et al 1975; Sundquist et al 1977), Ehrlich ascites cells (Jeffrey 1977; Van Venrooij et al 1977), kidney cells (Irwin et al 1975), and muscle cells (Bag & Sarkar 1975, 1976; Heywood et al 1975). Because a polyadenylate sequence, unlike other RNA stretches, is not degraded by some ribonucleases (pancreatic and T1 RNAases), poly(A) tails can be prepared and purified by chromatography on oligo(dT). Kwan & Brawerman (1972) first suggested that the poly(A) segment of mRNA is associated with protein on the basis of sedimentation experiments and binding to nitrocellulose. Blobel (1973) analyzed the composition of the poly(A)-protein RNP complex and found a single protein of 78 kDa. This protein is the most highly conserved and most extensively studied RNP protein. A protein of similar molecular weight (72,000–78,000, referred to hereafter as 72,000) is found to be a consistent and major component of mRNP complexes in almost all reports. The tight association of this protein with the poly(A) tail of mRNA has been reported in diverse cells (e.g. Barrieux et al 1975; Schwartz & Darnell 1976; Kish & Pederson 1975, 1976; Gaedigk et al 1985; Jain et al 1979; Jeffrey 1978; Vincent et al 1981).

Other methods for preparing mRNPs have also been introduced but have not become as widely used. These include centrifugation in Cs_2SO_4 without prior fixation (Greenberg 1977) and electrophoresis in non-denaturing agarose gels (Tasseron-De-Jong et al 1979). It is reassuring that they revealed a similar pattern of prominent mRNP proteins. The same proteins were more recently shown to be tightly associated with mRNA by photochemical cross-linking in fractionated cells (Greenberg 1980, 1981; Setyono & Greenberg 1981). The detection of a simple and consistent pattern of tightly associated proteins in diverse mRNAs (Bryan & Hayashi 1973) from widely divergent organisms and cell types, using different methodologies, gives considerable credence to the authenticity and biological significance of mRNPs. Blobel's (1973) observation that the poly(A) segment of mRNA is associated with the 72,000 poly(A) binding protein was of particular significance because it was the first indication of sequence specificity.

PHOTOCHEMICAL CROSS-LINKING IN INTACT CELLS   The copurification of proteins with RNA does not prove that they interact with one another in the intact cell. The proteins bound to mRNA in intact vertebrate cells were identified using UV cross-linking (Wagenmakers et al 1980; Van Eekelen et al 1981c; Van Venrooij et al 1982; Dreyfuss et al 1984a; Bag 1984; Greenberg & Carroll 1985; Adam et al 1986a). The predominant [35S]-methionine-labeled polypeptides have approximate molecular weights of 72,000, 68,000, 53,000, and 50,000. A considerable number of less abundant proteins were also detected. By label transfer from [3H]nucleotide-labeled RNA the major proteins have molecular weights of 72,000, 68,000, and 53,000 (Wagenmakers et al 1980; Van Eekelen et al 1981b; Adam et al 1986a; Dreyfuss et al 1986). The 68-kDa protein, unlike the 72- and 53-kDa proteins, may dissociate from mRNA in uncross-linked complexes at 0.5 M KCl, which would explain why it was not previously considered a major mRNP protein. By Coomassie blue or silver staining, the 72-kDa protein is the most abundant cross-linked mRNP. The poly(A) tail of mRNA is selectively cross-linked to the 72-kDa protein (Greenberg & Carroll 1985; Adam et al 1986a). The cross-linked proteins are not ribosomal, and their cross-linking is strictly dependent on UV irradiation of intact cells. The sequence selectivity and the lack of random cross-linking to abundant cytoplasmic proteins underscore the specificity of the photochemical cross-linking procedure. Under typical irradiation conditions (15 W germicidal lamp, 4.5 cm distance, 3 min) about 87% of the poly(A)[+] sequences are recovered and very little RNA chain breakage occurs (Adam et al 1986a).

Structure and Function of mRNP Complexes

The major proteins shown to interact with the mRNA both in vivo and in vitro have molecular weights of 72,000, 68,000, 53,000, and 50,000. The 72-kDa protein is bound to the poly(A) tail. The binding sites of the other proteins have not been determined. Several proteins that bind specifically, but probably transiently, to the 5'-cap structure of the mRNA have been found. These have molecular weights of 24,000, 50,000, and 80,000 (Sonnenberg 1981; Griffo et al 1982; Pelletier & Sonnenberg 1985). The 24-kDa protein, the best-characterized mRNA cap binding protein, is not detected by UV cross-linking in vivo, but it can be readily detected by UV cross-linking in vitro if the label is placed in the cap itself (S. A. Adam et al, unpublished results). The cap binding activity is part of a cap binding complex; it contains eIF4b proteins, and it functions in the initiation of protein synthesis. The cap binding proteins were recently reviewed by Shatkin (1985) and will not be discussed further here. To date there is no direct evidence that sequences other than the poly(A) tail and the cap are bound to proteins. Specific functions for other mRNP proteins have not been directly shown. In addition to the major mRNP proteins, some of which are almost consistently found in all studies, a multitude of other proteins have been described. At this point, enumeration of these proteins is fruitless because practically no two papers record the same patterns. Thus it is not possible to evaluate the authenticity or significance of any of them. In addition to the major proteins common to all mRNAs, there probably are proteins that bind specific mRNAs and also proteins that bind the same mRNA in different subcellular compartments (e.g. polyribosomal and nonpolyribosomal). This may explain the numerous minor proteins that become cross-linked with the mRNA upon exposure to UV light or other agents. What is needed to sort out this complexity is definitive evidence for the authenticity of the mRNP protein (such as proof that they interact specifically with mRNA in the cell), demonstration of a function for these proteins, or evidence for a specific binding site on the mRNA. Specific probes for the proteins, including antibodies and cDNA clones, will be essential in addressing these questions.

Little is known about the structure of the mRNP complex other than about the 3'-poly(A) segment and the 5'-cap. The poly(A) tail appears to form a particle of unique periodicity with the specific poly(A) binding protein (Baer & Kornberg 1980). There are as yet no indications that the proteins of mRNPs interact with other proteins to form higher order structures analogous to the monoparticles of hnRNPs. Altogether the structure of the mRNP complex appears to be much simpler than that of hnRNP. Indeed, the function of mRNA in translation suggests that the mRNA must be exposed, accessible, and able to thread through ribosomes. In principle it is also possible that the translational machinery may be able to move through some loosely associated binding proteins. However, mRNA packaging may be different in special circumstances such as in the storage of mRNAs in early embryogenesis, when packing, sequestering, and protecting the RNA (rather than its immediate translation) may be important. There are also reasons to believe that in other special circumstances (e.g. heat shock) additional specific mRNA segments, e.g. 5' and 3' untranslated sequences are bound by specific proteins.

POLY(A)-RIBONUCLEOPROTEIN COMPLEX   The poly(A) segment of mRNA was found to cross-link to the 72-kDa poly(A) binding protein upon exposure to UV light in vitro (Setyono & Greenberg 1981; Greenberg 1981; Greenberg & Carroll 1985) and in vivo (Greenberg & Carroll 1985; Adam et al 1986a). Recently, Sachs & Kornberg (1985) also identified cytoplasmic poly(A) binding activity in yeast in the molecular weight region between 66,000 and 79,000. Moreover, Adam et al (1986b) identified a 72-kDa protein that is cross-linked to the poly(A) tail of yeast mRNA in vivo and has poly(A)-specific binding activity in vitro. The fact that such a protein was found in diverse cells suggests that it is common to many, and perhaps all, mRNAs. In fact, the 72-kDa protein is also cross-linked to the poly(A) tail of vesicular stomatitis virus mRNAs in infected cells (Adam et al 1986a). This protein may also be associated with other parts of the mRNA. There are conflicting reports as to whether the poly(A) tail is associated with the 72-kDa protein only in polyribosomal mRNAs or also in nonpolyribosomal mRNAs (e.g. Van Venrooij et al 1977; Vincent et al 1981; Butcher & Arenstein 1983; Greenberg & Carroll 1985).

Baer & Kornberg (1980), using digestion with a non-base-specific RNAase (T2), detected a repeating structure in the cytoplasmic poly(A)-ribonucleoprotein complex. The repeating structure consists of multiples of about 25–27 adenosine residues bound to protein and can form spontaneously by incubation of poly(A) with cytoplasmic but not with nuclear extract. The protein responsible for this repeating structure was subsequently isolated from rat liver and found to be of molecular weight 75,000 (Baer & Kornberg 1983). It is probably the same protein that binds poly(A) in vivo. The absence of similar poly(A) organizing activity in the nuclear extract (Baer & Kornberg 1980) and the lack of cross-linking of the 72-kDa protein to nuclear RNA (Setyono & Greenberg 1981; Dreyfuss et al 1984a,b) suggest that this protein is not present in the nucleus, in contrast to the finding of Kumar & Pederson (1975). Kelly & Cox (1982) observed that the size distribution of the poly(A) tail of globin mRNA in vivo shows peaks at intervals of ∼25 residues, which agrees with the expected periodicity. This appears to reflect the fashion in which the poly(A) binds to the proteins, and these interactions may control mRNA degradation. The ability of the poly(A) binding protein to protect the poly(A) tail from nuclease suggests a role for the protein in mRNA stability.

Although the poly(A) binding protein is the most abundant and was the first mRNA binding protein described, little is known about its structure or function. The intimate association of the poly(A) binding protein with the poly(A) tail probably indicates that their functions are interrelated. It has been suggested that the poly(A) tail-protein complex is involved in various key aspects of mRNA metabolism, including nucleocytoplasmic transport (Schwartz & Darnell 1976), mRNA stability (Zeevi et al 1982), and translation (Van Venrooij et al 1977; Vincent et al 1981; Schmid et al 1983; Jacobson & Favreau 1983), and that it may be related to a poly(A) polymerase (Rose et al 1979). It has been difficult to directly address these questions because antibody and gene probes for the protein were not available. The mRNA poly(A) binding protein from vertebrates is poorly immunogenic in mice and rabbits, and previous attempts by a number of laboratories to produce antibodies to it were not successful. Adam et al (1986b) recently identified, and produced antibodies to, the major proteins that interact with polyadenylated RNAs in the yeast *Saccharomyces cerevisiae*. The poly(A) segment of the mRNA in yeast is also selectively cross-linked to a 72-kDa protein. Mice immunized with purified, UV cross-linked RNA-protein complexes produced antibodies to the major yeast mRNP proteins, including the poly(A) binding proteins. A yeast genomic DNA library constructed in an expression vector was screened immunologically, and a recombinant phage producing a large β-galactosidase-RNP fusion protein bearing the gene for the poly(A) binding protein was isolated. The expressed fusion protein had specific poly(A) binding activity. DNA blot analysis suggested a single gene for the poly(A) binding protein, and mRNA blot analysis detected an mRNA of 2.1 kb in length (Adam et al 1986b). A. Sachs & R. Kornberg (personal communication) also produced antibodies to the yeast protein and isolated the gene. These findings open the way for molecular and genetic characterization of the mRNA poly(A) binding protein.

mRNP PROTEINS OF SPECIFIC mRNAs   The studies discussed so far isolated and characterized the proteins associated with all or with a large number of different mRNAs. Important insights can be obtained from analysis of the proteins that interact with specific mRNAs. This approach presents the considerable difficulty of selecting specific mRNAs with bound proteins. For some mRNAs, e.g. globin and protamine, partial purification can be accomplished by velocity sedimentation alone because of their distinct size and abundance in a specific cell type. Morel et al (1971), Blobel (1972), Burns & Williamson (1975), and Vincent et al (1981) prepared partially purified globin mRNPs from chicken and mouse erythroblasts, and Gedamu et al (1977) isolated protamine mRNP particles from trout testis. Several proteins were detected, and the 72-kDa poly(A) binding protein was common. Several recent studies used immobilized cDNA for hybrid selection of specific mRNAs after UV irradiation of intact cells to purify cross-linked mRNP complexes. Van Venrooij et al (1982) selected adenovirus 2 mRNAs and found that they are bound in infected cells to typical major host mRNP proteins. Ruzdijic et al (1984) used the same method to isolate the nonpolyadenylated histone H4 mRNA from HeLa cells. They found that in the polyribosomal fraction it is cross-linked to 49- and 52.5-kDa proteins, whereas the nonpolyribosomal H4 mRNA is cross-linked to 43- and 57-kDa proteins.

Adam et al (1986a) employed a different approach to examine the proteins that are associated in vivo with a small and unique set of mRNAs, the mRNAs of vesicular stomatis virus (VSV). This method does not require isolation of these mRNAs from the total mRNA as the mRNAs of VSV can be selectively labeled in vivo because the polymerase of the virus, unlike host RNA polymerase II, is not inhibited by actinomycin D (at 5 μg/ml). The proteins that were cross-linked in vivo specifically to the five mRNAs of VSV were labeled by incorporating radioactive nucleotides into VSV mRNAs only. The same major proteins that become cross-linked to host mRNAs also became cross-linked to VSV mRNAs (Adam et al 1986a). The poly(A) segment of VSV mRNAs, like that of host mRNAs, was also associated with the 72-kDa poly(A) binding protein. The major mRNPs are therefore ubiquitous and are common to different mRNAs in the same cell. Furthermore, that the VSV mRNAs are transcribed in, and are entirely confined to, the cytoplasm argues that mRNAs can acquire the major mRNP proteins in the cytoplasm, presumably without nuclear processes.

Reconstitution experiments with specific mRNAs represent another useful and promising avenue for studying mRNP structure and for identifying sequence-specific mRNA binding proteins. Gaedigk et al (1985) and Greenberg & Carroll (1985) examined the proteins that bind purified globin mRNA in mouse erythroleukemia cells and rabbit reticulocyte lysate, respectively. The major proteins that bound the globin mRNA were similar to those normally detected when the entire mRNA population is analyzed.

One conclusion from all of these studies is that the major mRNP proteins are common to different mRNAs and therefore must recognize common mRNA features. These include the poly(A) tail, the cap, the poly-adenylation signal (AAUAAA), and non-sequence-specific ssRNA or dsRNA binding proteins. However, it is very likely that there are also specific proteins for specific mRNAs. These could be of two general classes: (*a*) sequence-specific proteins that would bind only to mRNAs containing a particular sequence (e.g. the leader sequence of heat-shock mRNAs), and (*b*) compartment-specific mRNA binding proteins that would bind all mRNAs in a particular subcellular topological or metabolic compartment or in a special physiological state (e.g. translated, untranslated, membrane-bound, or putative cytoskeleton-associated mRNAs). These sequence-specific and compartment-specific proteins superimposed on the major mRNP proteins that are common to all mRNAs could explain the multiplicity of components of mRNPs seen in different studies even for specific mRNAs.

mRNP COMPLEXES IN VIRUS-INFECTED CELLS   Virus-infected animal cells are unique systems in which to study mRNP and hnRNP proteins. During lytic infection, copious amounts of a small number of well-defined viral genes are transcribed, and drastic changes in mRNA formation and translation in the host usually occur. These changes in host mRNA metabolism are a consequence of the expression of virus proteins. Furthermore, because the abundant mRNAs of the virus are structurally similar to those of the host, they can be considered prototypes of host mRNA.

A tight association of a virus-encoded protein with mRNA was first observed by Lindberg & Sundquist (1974) in HeLa cells infected with adenovirus (Ad) 2. mRNP complexes isolated from these cells at high salt concentrations contained large amounts of the 100-kDa late nonstructural viral protein (Van der Marel et al 1975; Tasseron-De-Jong et al 1979). The 100-kDa protein could not, however, be detected by nucleotide label transfer after UV cross-linking (Van Venrooij et al 1982). The significance of these observations is not yet clear.

A complex of a viral protein with mRNA was also found in VSV-infected cells. Grubman & Shafritz (1977) showed that mRNP particles from VSV-infected cells contain the viral N protein. Adam et al (1986a), using UV cross-linking in intact VSV-infected cells, also detected the N protein in mRNP complexes and found that VSV mRNAs are associated with host mRNPs, including the poly(A) binding protein. Rosen et al (1982) isolated from VSV-infected cells a unique mRNP particle that contained the five VSV mRNAs and almost exclusively the N protein. Although the function of the N-VSV mRNA interaction is not known, it is probably functionally relevant: Rosen et al (1984) demonstrated that

N-VSV mRNA particles inhibit protein synthesis in rabbit reticulocyte lysate and wheat germ extracts. The inhibition is at an early step of initiation of protein synthesis, i.e. the formation of the ternary complex eIF-2.GTP.Met-tRNA. The N protein–mRNA complex may therefore be involved in the shutting off of total protein synthesis that occurs in VSV-infected cells. In general, viral protein-mRNA complexes are potentially of great importance because they may give rise to new RNP forms that may modify elements of normal host pathways and facilitate viral functions.

TRANSLATED AND UNTRANSLATED mRNAs   The search for differences in the proteins that bind translated and untranslated mRNAs is motivated by the interest in identifying the factors that control the state of translation of mRNA. It has received much attention, and over the years numerous studies have compared polyribosome-bound and "free," non-polyribosomal mRNPs. Although some differences were detected (e.g. Butcher & Arenstein 1983; Schmid et al 1983; Ruzdijic et al 1984; Jeffrey 1977; Blobel 1973; Liautard et al 1976; Bag 1984; Vincent et al 1981; Van Venrooij et al 1977), no consistent differences that can account for the functional state of the mRNA were found, and a coherent picture did not emerge. This task is complicated by the very fine and blurred line of distinction between mRNP proteins and translational factors that may cofractionate with mRNPs. Specific functional assays will be necessary to establish a direct role for mRNP proteins in translation.

Stored mRNAs can be considered to be a special class of untranslated mRNPs. The early work of Spirin and colleagues demonstrated that these mRNAs in sea urchin and fish embryos are found in mRNP complexes. Several proteins that are associated with stored mRNPs have been identified.

DYNAMIC STRUCTURE OF mRNP COMPLEXES   In addition to the changes found in mRNP complexes after viral infections, structural changes in these complexes have been observed in cells treated with inhibitors of mRNA synthesis. VSV infection, actinomycin D (5 μg/ml), camptothecin, and DRB (5,6-dichloro-1-β-ribofuranosyl benzimidazole), all of which inhibit transcription by RNA polymerase II, cause a prominent protein of $\sim$38 kDa that cannot be normally cross-linked to mRNAs to become cross-linkable to mRNAs in vivo (Dreyfuss et al 1984a). The onset of the effect is rapid, and it is completely and rapidly reversible. Inhibitors of protein synthesis, rRNA synthesis, and polyadenylation do not affect the cross-linking of the 38-kDa protein to mRNA. These agents that promote the cross-linkable interaction of the 38-kDa protein with mRNA do not affect proteins in contact with poly(A)$^+$ hnRNA and do not markedly affect protein synthesis. Although the significance of the interaction of the 38-kDa protein with mRNA is not known, these observations demonstrate that commonly used inhibitors of transcription bring about a structural change in mRNA-ribonucleoprotein complexes in vivo.

Greenberg's work (1980) indicates that the mRNP complex is a dynamic structure in which bound proteins can exchange with an unbound pool. In vitro the same proteins became cross-linked to mRNAs (upon UV exposure) in cytoplasmic extracts from cells treated with actinomycin D as in control cytoplasms. It was therefore suggested that mRNA-associated proteins can exchange with a free pool of proteins because the cytoplasm of the actinomycin-D-treated cells contained no newly made mRNAs and ongoing mRNA synthesis was not required for mRNP formation. However, this may not happen in the intact cell. The cross-linking of mRNA and a protein that may correspond to the 38-kDa protein discussed above was also detected in Greenberg's (1980) study after treatment with actinomycin D. The exchange of mRNP proteins, which does not occur with proteins in other RNA-containing structures such as ribosomes, may be important in converting mRNAs from one functional state to another, as for instance, in the activation of stored maternal mRNAs upon egg fertilization.

RIBONUCLEOPROTEINS AND NUCLEOCYTOPLASMIC TRANSPORT OF mRNA

The transport of mRNA from the nucleus to the cytoplasm is a critical process in eukaryotes. Little is known about this process, and it is still somewhat difficult to design useful experiments because the important mechanistic questions are not clear. However, it has now been shown by UV cross-linking and through the use of specific antibodies that the major hnRNP proteins are confined to the nucleus in vivo (Jones & Martin 1980; Dreyfuss et al 1984a,b; Choi & Dreyfuss 1984a; Martin & Okamura 1981; Leser et al 1984) and that the mRNA in the cytoplasm is associated with a different set of proteins. These findings are consistent with data previously obtained from fractionated cells (Kumar & Pederson 1975; Liautard et al 1976) and suggest that mRNAs must exchange the proteins with which they are associated in the nucleus upon transport to the cytoplasm. The translocation of the mRNA across the nuclear envelope through nuclear pore complexes is, therefore, accompanied by a protein exchange process. The nuclear proteins must dissociate prior to, or at the time of, mRNA translocation. This suggests that the dissociation of hnRNP proteins from the mRNA must be an early event in the transport pathway, and it is thus a distinct process that can now be addressed experimentally, both in vivo and in vitro.

Acknowledgments

I thank the many colleagues and the members of my laboratory for helpful discussions and for sharing unpublished results. Work in the author's laboratory was supported by grants from the NIH (GM31888), NSF (PCM 8216052), the Leukemia Foundation, Inc., and the Searle Leadership Fund.

Literature Cited

Abelson, J. 1979. RNA processing and the intervening sequence problem. *Ann. Rev. Biochem.* 48 : 1035–63

Adam, S. A., Choi, Y. D., Dreyfuss, G. 1986. The interaction of mRNA with proteins in VSV infected cells. *J. Virol.* 57 : 614–22

Adam, S. A., Nakagawa, T. Y., Swanson, M. S., Woodruff, T., Dreyfuss, G. 1986b. Isolation and expression of the gene for the mRNA polyadenylate binding protein. *Mol. Cell. Biol.* In press

Augenlicht, L. H., Lipkin, M. 1976. Appearance of rapidly labeled, high molecular weight RNA in nuclear ribonucleoprotein: Release from chromatin and association with protein. *J. Biol. Chem.* 251 : 2592–99

Baer, B. W., Kornberg, R. D. 1980. Repeating structure of cytoplasmic poly(A)-ribonucleoprotein. *Proc. Natl. Acad. Sci. USA* 77 : 1890–92

Baer, B. W., Kornberg, R. D. 1983. The protein responsible for the repeating structure of cytoplasmic poly(A) ribonucleoprotein. *J. Cell Biol.* 96 : 717–21

Bag, J. 1984. Cytoplasmic mRNA-protein complexes of chicken muscle cells and their role in protein synthesis. *Eur. J. Biochem.* 141 : 247–54

Bag, J., Sarkar, S. 1975. Cytoplasmic nonpolysomal messenger ribonucleoprotein containing actin messenger RNA in chicken embryonic muscles. *Biochemistry* 14 : 3800–7

Bag, J., Sarkar, S. 1976. Studies on a nonpolysomal ribonucleoprotein coding for myosin heavy chains from chick embryonic muscle. *J. Biol. Chem.* 251 : 7600–9

Baltimore, D., Huang, A. S. 1970. Interaction of HeLa cell proteins with RNA. *J. Mol. Biol.* 47 : 263–72

Barrieux, A., Ingraham, H. A., David, D. N., Rosenfeld, M. G. 1975. Isolation of messenger-like ribonucleoproteins. *Biochemistry* 14 : 1815–21

Beyer, A. L., Bouton, A. H., Miller, O. L. Jr. 1981. Correlation of hnRNP structure and nascent transcript cleavage. *Cell* 26 : 155–65

Beyer, A. L., Christensen, M. E., Walker, B. W., Le Stourgeon, W. M. 1977. Identification and characterization of the packaging proteins of core 40S hnRNP particles. *Cell* 11 : 127–38

Beyer, A. L., Miller, O. L. Jr., McKnight, S. L. 1980. Ribonucleoprotein structure in nascent hnRNA is nonrandom and sequence-dependent. *Cell* 20 : 75–84

Billings, P. B., Martin, T. E. 1978. Proteins of nuclear ribonucleoprotein subcomplexes. *Meth. Cell Biol.* 17 : 349–76

Black, D. L., Chabot, R., Steitz, J. A. 1985. U2 as well as U1 small nuclear ribonucleoproteins are involved in premessenger RNA splicing. *Cell* 42 : 737–50

Blobel, G. 1972. Protein tightly bound to globin mRNA. *Biochem. Biophys. Res. Commun.* 47 : 88–95

Blobel, G. 1973. A protein of molecular weight 78,000 bound to the polyadenylate region of eukaryotic messenger RNAs. *Proc. Natl. Acad. Sci. USA* 70 : 924–28

Boffa, L. C., Karn, J., Vidali, G., Allfrey, V. G. 1977. Distribution of N^G, N^G-dimethylarginine in nuclear protein fraction. *Biochem. Biophys. Res. Commun.* 74 : 969–76

Brandhorst, B. P., Humphries, T. 1971. Synthesis and decay rates of major classes of deoxyribonucleic acid like RNAs in sea urchin embryos. *Biochemistry* 10 : 877–81

Brandhorst, B. P., McConkey, E. H. 1974.

Stability of nuclear RNA in mammalian cells. *J. Mol. Biol.* 85: 451–63

Brawerman, G. 1981. The polyadenylate tail of mRNA. *CRC Crit. Rev. Biochem.* 8: 1–38

Brody, E., Abelson, J. 1985. The "spliceosome": Yeast pre-messenger RNA associates with a 40S complex in a splicing-dependent reaction. *Science* 228: 963–67

Brunel, C., Lelay, M. N. 1979. Two-dimensional analysis of proteins associated with heterogeneous nuclear RNA in various animal cell lines. *Eur. J. Biochem.* 99: 273–83

Bryan, R. N., Hayashi, M. 1973. Two proteins are bound to most species of polysomal mRNA. *Nature New Biol.* 244: 271–74

Burns, A. T. H., Williamson, R. 1975. Isolation of mouse reticulocyte globin messenger ribonucleoprotein by affinity chromatography using oligo(dT)-cellulose. *Nucl. Acids Res.* 2: 2251–55

Butcher, P. D., Arenstein, H. R. V. 1983. Efficient translation and polyribosome binding of ^{125}I-labelled rabbit globin messenger ribonucleoprotein. *FEBS Lett.* 153: 119–24

Calvet, J. P., Meyer, L. M., Pederson, T. 1982. *Science* 217: 456–58

Calvet, J. P., Pederson, T. 1978. Nucleoprotein organization of inverted repeat DNA transcripts in heterogeneous nuclear RNA-ribonucleoprotein particles from HeLa cells. *J. Mol. Biol.* 122: 361–78

Calvet, J. P., Pederson, T. 1981. Base pairing interactions between small nuclear RNAs and nuclear RNA precursors as revealed by Psoralen cross-linking in vivo. *Cell* 26: 363–70

Celis, J. E., Bravo, R., Arenstorf, H. P., LeStourgeon, W. M. 1986. Identification of proliferation-sensitive human proteins amongst components of 40S hnRNP particles. *FEBS Lett.* 194: 101–9

Chabot, B., Black, D. L., LeMaster, S. M., Steitz, J. A. 1986. The 3′ splice in pre-messenger RNA is recognized by a small nuclear ribonucleoprotein. *Science* 230: 1344–49

Chase, J. W., Williams, K. R. 1986. Single-stranded DNA binding proteins required for DNA replication. *Ann. Rev. Biochem.* 55. In press

Choi, Y. D., Dreyfuss, G. 1984a. Monoclonal antibody characterization of the C proteins of heterogeneous nuclear ribonucleoprotein complexes in vertebrate cells. *J. Cell Biol.* 99: 1997–2004

Choi, Y. D., Dreyfuss, G. 1984b. Isolation of the heterogeneous nuclear RNA-ribonucleoprotein complex (hnRNP): A unique supra molecular assembly. *Proc.*

Natl. Acad. Sci. USA 81: 7471–75

Choi, Y. D., Grabowski, P. J., Sharp, P. A., Dreyfuss, G. 1986. Heterogeneous nuclear ribonucleoproteins: Role in RNA splicing. *Science* 231: 1534–39

Christensen, M. E., Beyer, A. L., Walker, B., LeStourgeon, W. M. 1977. Identification of N^G,N^G-dimethylarginine in a nuclear protein from the lower eukaryote *Physarum polycephalum* homologous to the major proteins of mammalian 40S ribonucleoprotein particles. *Biochem. Biophys. Res. Commun.* 74: 621–29

Cruz-Alvarez, M., Szer, W., Pellicer, A. 1985. Cloning of cDNA sequences for an *Artemia Salina* hnRNP protein. *Nucl. Acids Res.* 11: 3917–30

Darnell, J. E. 1982. Variety in the level of gene control in eukaryotic cells. *Nature* 297: 365–71

Deimel, B., Louis, C., Sekeris, C. 1977. The presence of small molecular weight RNAs in nuclear ribonucleoprotein particles carrying hnRNA. *FEBS Lett.* 73: 80–84

Dreyfuss, G., Adam, S. A., Choi, Y. D. 1984a. Physical change in cytoplasmic messenger ribonucleoproteins in cells treated with inhibitors of mRNA transcription. *Mol. Cell. Biol.* 4: 415–23

Dreyfuss, G., Choi, Y. D., Adam, S. A. 1984b. Characterization of hnRNA-protein complexes in vivo with monoclonal antibodies. *Mol. Cell. Biol.* 4: 1104–14

Dreyfuss, G., Choi, Y. D., Adam, S. A. 1986. The ribonucleoprotein structures along the pathway of mRNA formation. In *Mechanism of Action of Thyroid Hormones*, ed. L. DeGroot. New York: Academic

Economides, I. V., Pederson, T. 1983. Structure of nuclear ribonucleoprotein: Heterogeneous nuclear RNA is complexed with a major sextet of proteins in vivo. *Proc. Natl. Acad. Sci. USA* 80: 1599–1602

Firtel, R. A., Pederson, T. 1975. Ribonucleoprotein particles containing heterogeneous nuclear RNA in the cellular slime mold *Dictyostelium discoideum. Proc. Natl. Acad. Sci. USA* 72: 301–5

Foe, V. E., Wilkinson, L. E., Laird, C. D. 1976. Comparative organization of active transcription units in *Oncopeltus fasciatus. Cell* 9: 131–46

Gaedigk, R., Oehler, S., Kohler, K., Setyono, B. 1985. *In vitro* reconstitution of messenger ribonucleoprotein particle from globin messenger RNA and cytosol proteins. *FEBS Lett.* 179: 201–7

Gall, J. G. 1955. On the submicroscopic structure of chromosomes. *Brookhaven Symp. Biol.* 8: 17–32

Gall, J. G. 1956. Small granules in the amphibian oocyte nucleus and their relationship to RNA. *J. Biophys. Biochem. Cytol.* 2: 393–96 (Suppl.)

Gall, J. G., Callan, H. G. 1962. ³H-Uridine incorporation in lampbrush chromosomes. *Proc. Natl. Acad. Sci. USA* 48: 562–70

Gallinaro, H., Jacob, M. 1979. An evaluation of small nuclear RNA in hnRNP. *FEBS Lett.* 104: 176–82

Gander, E. S., Stewart, A. G., Morel, C. M., Scherrer, K. 1973. Isolation and characterization of ribosome-free cytoplasmic messenger ribonucleoprotein complexes from avian erythroblasts. *Eur. J. Biochem.* 38: 443–52

Gattoni, R., Stevenin, J., Devilliers, G., Jacob, M. 1978. Size heterogeneity of monoparticles from nuclear ribonucleoproteins containing premessenger RNA. *FEBS Lett.* 90: 318–23

Gedamu, L., Dixon, G. H., Davies, P. L. 1977. Identification and isolation of protamine messenger ribonucleoprotein particles from rainbow trout testis. *Biochemistry* 16: 1383–91

Grabowski, P. J., Seiler, S. R., Sharp, P. A. 1985. A multicomponent complex is involved in the splicing of messenger RNA precursors. *Cell* 42: 345–53

Green, M. 1986. Pre-mRNA splicing. *Ann. Rev. Genet.* 20: In press

Greenberg, J. R. 1972. High stability of messenger RNA in growing cultured cells. *Nature* 240: 102–4

Greenberg, J. R. 1977. Isolation of messenger ribonucleoproteins in cesium sulfate density gradients: Evidence that polyadenylate and non-polyadenylate messenger RNAs are associated with protein. *J. Mol. Biol.* 108: 403–16

Greenberg, J. R. 1980. Proteins crosslinked to messenger RNA by irradiating polyribosomes with UV light. *Nucl. Acids Res.* 8: 5685–5701

Greenberg, J. R. 1981. The polyribosomal mRNA-protein complex is a dynamic structure. *Proc. Natl. Acad. Sci. USA* 78: 2923–26

Greenberg, J. R., Carroll, E. III 1985. Reconstitution of functional mRNA-protein complexes in a rabbit reticulocyte cell-free translation system. *Mol. Cell. Biol.* 5: 342–51

Griffo, J. A., Tahara, S. M., Leas, J. P., Morgan, M. A., Shatkin, A. J. 1982. Characterization of eukaryotic initiation factor 4A, a protein involved in ATP-dependent binding of globin mRNA. *J. Biol. Chem.* 257: 5246–53

Grubman, M., Shafritz, D. A. 1977. Identification and characterization of messenger ribonucleoprotein complexes from

vesicular stomatitis virus-infected cells. *Virology* 81: 1–16

Harpold, M. M., Evans, R. M., Salditt-Georgieff, M., Darnell, J. E. 1979. Production of mRNA in Chinese hamster cells: Relationship of the rate of synthesis to the cytoplasmic concentration of nine specific mRNA sequences. *Cell* 17: 1025–35

Henshaw, E. C. 1968. Messenger RNA in rat liver exists as ribonucleoprotein particles. *J. Mol. Biol.* 36: 401–11

Henshaw, E. C., Loebenstein, J. 1970. Rapidly labeled, polydisperse RNA in rat liver cytoplasm: Evidence that it is contained in ribonucleoprotein particles of heterogeneous size. *Biochem. Biophys. Acta* 199: 405–20

Herman, R. C., Penman, S. 1977. Multiple decay rates of heterogeneous nuclear RNA in HeLa cells. *Biochemistry* 16: 3460–65

Hernandez, N., Keller, W. 1983. Splicing of in vitro synthesized messenger RNA precursors in HeLa cell extracts. *Cell* 35: 89–99

Herrick, G., Alberts, B. 1976a. Purification and physical characterization of nucleic acid helix-unwinding proteins from calf thymus. *J. Biol. Chem.* 251: 2124–32

Herrick, G., Alberts, B. 1976b. Nucleic acid helix-coil transitions mediated by helix-unwinding proteins from calf thymus. *J. Biol. Chem.* 251: 2133–41

Heywood, S. M., Kennedy, D. S., Bester, A. J. 1975. Stored myosin messenger in embryonic chick muscle. *FEBS Lett.* 53: 69–72

Holcomb, E. R., Friedman, D. L. 1984. Phosphorylation of the C-proteins of HeLa cell hnRNP particles. *J. Biol. Chem.* 259: 31–40

Howard, E. F. 1978. Small nuclear RNA molecules in nuclear ribonucleoprotein complexes from mouse erythroleukemia cells. *Biochemistry* 17: 3228–36

Huang, H. M., Chae, C. B. 1983. Different RNA patterns of globin and nonglobin 40S heterogeneous nuclear RNA-protein complexes in chicken reticulocyte nuclei. *Nucl. Acids Res.* 11: 7057–67

Irwin, D., Kumar, A., Malt, R. A. 1975. Messenger ribonucleoprotein complexes isolated with oligo(dT)-cellulose chromatography from kidney polysomes. *Cell* 4: 157–65

Jacob, M., Devillers, G., Fuchs, J. P., Gallinaro, H., Gattoni, R., et al. 1981. In *The Cell Nucleus*, ed. H. Busch, 8: 194–259. New York: Academic

Jacobson, A., Favreau, M. 1983. Possible involvement of poly(A) in protein synthesis. *Nucl. Acids Res.* 11: 6353–68

494 DREYFUSS

Jain, S. K., Pluskal, M. G., Sarkar, S. 1979. Thermal chromatography of eukaryotic messenger ribonucleoprotein particles on oligo(dT)-cellulose. *FEBS Lett.* 97: 84–90

Jeffrey, W. R. 1977. Characterization of polypeptides associated with messenger RNA and its polyadenylate segment in Ehrlich ascites messenger ribonucleoproteins. *J. Biol. Chem.* 252: 3525–32

Jeffrey, W. R. 1978. Composition and properties of messenger ribonucleoprotein fragments containing and lacking polyadenylate. *Biochim. Biophys. Acta* 521: 217–28

Jones, R. E., Okamura, C. S., Martin, T. E. 1980. Immunofluorescent localization of the proteins of nuclear ribonucleoprotein complexes. *J. Cell Biol.* 86: 235–43

Karn, J., Vidali, G., Boffa, L. C., Allfrey, V. G. 1977. Characterization of the non-histone nuclear proteins associated with rapidly labeled heterogeneous nuclear RNA. *J. Biol. Chem.* 252: 7307–22

Kelly, J. M., Cox, R. A. 1982. Periodicity in the length of 3'-poly(A) tails from native globin mRNA of rabbit. *Nucl. Acids Res.* 10: 4173–79

Kimmel, C. B., Sessions, S. K., MacLeod, M. C. 1976. Evidence for an association of most nuclear RNA with chromatin. *J. Mol. Biol.* 102: 177–91

Kinniburgh, A. J., Martin, T. E. 1976. Detection of mRNA sequences in nuclear 30S ribonucleoprotein subcomplexes. *Proc. Natl. Acad. Sci. USA* 73: 2725–29

Kish, V. M., Pederson, T. 1975. Ribonucleoprotein organization of polyadenylate sequences in HeLa cell heterogeneous nuclear RNA. *J. Mol. Biol.* 95: 227–38

Kish, V. M., Pederson, T. 1976. Poly(A)-rich ribonucleoprotein complexes of HeLa cell messenger RNA. *J. Biol. Chem.* 251: 5888–94

Kish, V. M., Pederson, T. 1977. Heterogeneous nuclear RNA secondary structure: Oligo(U) sequences base-paired with poly(A) and their possible role as binding sites for heterogeneous nuclear RNA-specific proteins. *Proc. Natl. Acad. Sci. USA* 74: 1426–30

Kloetzel, P.-M., Schuldt, C. 1986. The packaging of nuclear ribonucleoprotein in heat-shocked *Drosophila* cells is unaltered. *Biochim. Biophys. Acta.* In press

Knowler, J. T. 1976. The incorporation of newly synthesized RNA into nuclear ribonucleoprotein particles after oestrogen administration to immature rats. *Eur. J. Biochem.* 64: 161–65

Knowler, J. T. 1983. An assessment of the evidence for the role of ribonucleoprotein particles in the maturation of eukaryote mRNA. *Int. Rev. Cytol.* 84: 103–53

Krainer, A. R., Maniatis, T. 1985. Multiple factors including the small nuclear ribonucleoproteins U1 and U2 are necessary for pre-mRNA splicing in vitro. *Cell* 42: 725–36

Krainer, A. R., Maniatis, T., Ruskin, B., Green, M. R. 1984. Normal and mutant human β-globin pre-mRNAs are faithfully and efficiently spliced in vitro. *Cell* 36: 993–1005

Kulguskin, V. V., Krichevskaya, A. A., Lukanidin, E. M., Georgiev, G. P. 1980. Studies on dissociation and reconstitution of nuclear 30S ribonucleoprotein particles containing pre-mRNA. *Biochim. Biophys. Acta* 609: 410–24

Kumar, A., Pederson, T. 1975. Comparison of proteins bound to heterogeneous nuclear RNA and messenger RNA in HeLa cells. *J. Mol. Biol.* 96: 353–65

Kwan, S. W., Brawerman, G. 1972. A particle associated with the polyadenylate segment in mammalian messenger RNA. *Proc. Natl. Acad. Sci. USA* 69: 3247–50

Lamb, M. M., Daneholt, B. 1979. Characterization of active transcription units in Balbiani rings of *Chironomus tentans*. *Cell* 17: 835–48

Lahiri, D. K., Thomas, J. O. 1985. The fate of heterogeneous nuclear ribonucleoprotein complexes during mitosis. *J. Biol. Chem.* 260: 598–603

Lebleu, B., Marbaix, G., Huez, G., Temmerman, J., Burney, A., et al. 1971. Characterization of the messenger ribonucleoprotein released form reticulocyte polyribosomes by EDTA treatment. *Eur. J. Biochem.* 19: 264–69

Lee, S. Y., Krsmanovic, V., Brawerman, G. 1971. Initiation of polysome formation in mouse sarcoma 180 ascites cells. Utilization of cytoplasmic messenger ribonucleic acid. *Biochemistry* 10: 895–900

Leser, G. P., Escara-Wilke, J., Martin, T. E. 1984. Monoclonal antibodies to heterogeneous nuclear RNA-protein complexes. *J. Biol. Chem.* 259: 1827–33

Leser, G. P., Martin, T. E. 1986. The major protein components of hnRNP complexes. In *DNA: Proteins Interactions and Gene Regulation*, ed. E. B. Thompson, J. Papaconstantinon. Univ. Texas Press. In press

LeStourgeon, W. M., Beyer, A. L., Christensen, M. E., Walker, B. W., Poupore, S. M., et al. 1978. The packaging proteins of core hnRNP particles and the maintenance of proliferative cell states. *Cold Spring Harbor Symp. Quant. Biol.* 42: 885–98

LeStourgeon, W. M., Lothstein, L., Walker, B. W., Beyer, A. L. 1981. The composition and general topology of RNA and protein in monomer 40S ribonucleoprotein particles. In *The Cell Nucleus*, ed. H. Busch, 9: 49–87. New York: Academic

Liautard, J. P., Setyono, B., Spindler, E., Kohler, K. 1976. Comparison of proteins bound to the different functional classes of messenger RNA. *Biochim. Biophys. Acta* 425: 373–83

Lindberg, U., Sundquist, B. 1974. Isolation of messenger ribonucleoproteins from mammalian cells. *J. Mol. Biol.* 86: 451–68

Lothstein, L., Arenstorf, H. P., Wooley, J. C., Chung, S. Y., Walker, B. W., et al. 1985. General organization of protein in HeLa 40S ribonucleoprotein particles. *J. Cell Biol.* 100: 1570–81

Lukanidin, E. M., Zalmanson, E. S., Komaromi, L., Samarina, O. P., Georgiev, G. P. 1972. Structure and function of informofer. *Nature New Biol.* 238: 193–96

Malcolm, D. B., Sommerville, J. 1974. The structure of chromosome derived ribonucleoprotein in oocytes of *Triturus cristatus carnifex*. *Chromosoma* 48: 137–58

Malcolm, D. B., Sommerville, J. 1977. The structure of nuclear ribonucleoprotein of amphibian oocytes. *J. Cell Sci.* 24: 143–65

Martin, T. E., Billings, P., Levey, A., Ozarsian, S., Quinlan, I., et al. 1974. Some properties of RNA: Protein complexes from the nucleus of eukaryotic cells. *Cold Spring Harbor Symp. Quant. Biol.* 38: 921–32

Martin, T. E., Billings, P. B., Pullman, J. M., Stevens, B. J., Kinniburgh, A. J. 1978. Substructures of nuclear ribonucleoprotein complexes. *Cold Spring Harbor Symp. Quant. Biol.* 42: 899–909

Martin, T. E., McCarthy, B. J. 1972. Synthesis and turnover of RNA in the 30S nuclear ribonucleoprotein complexes of mouse ascites cells. *Biochim. Biophys. Acta* 277: 354–67

Martin, T. E., Okamura, C. S. 1981. In *The Cell Nucleus*, ed. H. Busch, 9: 119 44. New York: Academic

Martin, T. E., Pullman, J. M., McMullen, M. E. 1980. Structure and function of nuclear and cytoplasmic ribonucleoprotein complexes. In *Cell Biology: A Comprehensive Treatise*, ed. D. M. Prescott, L. Goldstein, 4: 137–74. New York: Academic

Marvil, D. K., Nowak, L., Szer, W. 1980. A single-stranded nucleic acid-binding protein from *Artemia salina*. I. *J. Biol. Chem.* 255: 6466–72

Maundrell, K., Scherrer, K. 1979. Characterization of pre-mRNA-containing nuclear ribonucleoprotein particles from avian erythroblasts. *Eur. J. Biochem.* 99: 225–38

Mayrand, S., Pederson, T. 1981. Nuclear ribonucleoprotein particles probed in living cells. *Proc. Natl. Acad. Sci. USA* 78: 2208–12

Mayrand, S., Pederson, T. 1983. Heat shock alters nuclear RNP assembly in *Drosophila* cells. *Mol. Cell. Biol.* 3: 161–71

Mayrand, S., Setyono, B., Greenberg, J. R., Pederson, T. 1981. Structure of nuclear ribonucleoprotein: Identification of proteins in contact with poly(A)⁺ heterogeneous nuclear RNA in living HeLa cells. *J. Cell Biol.* 90: 380–84

McKnight, S. L., Miller, O. L. 1976. Ultrastructural patterns of RNA synthesis during early embryogenesis of *Drosophila melanogaster*. *Cell* 8: 305–19

McKnight, S. L., Miller, O. L. 1979. Postreplication non-ribosomal transcription units in *D. melanogaster* embryos. *Cell* 17: 551–63

Miller, O. L., Bakken, A. H. 1972. Morphological studies of transcription. *Karolinska Symp. Res. Meth. Reprod. Endocrinol.* 5: 155–67

Morel, C., Kayibanda, B., Scherrer, K. 1971. Proteins associated with globin messenger RNA in avian erythroblasts: Isolation and comparison with the proteins bound to nuclear messenger-like RNA. *FEBS Lett.* 18: 84–88

Munroe, S. H. 1982. Ribonucleoprotein structure of adenovirus nuclear RNA probed by nuclease digestion. *J. Mol. Biol.* 162: 585–606

Munroe, S. H., Pederson, T. 1981. Messenger RNA sequences in nuclear ribonucleoprotein particles are complexed with protein as shown by nuclease protection. *J. Mol. Biol.* 147: 437–49

Nakagawa, T. Y., Swanson, M. S., Wold, B. J., Dreyfuss, G. 1986. Molecular cloning of cDNA for the nuclear ribonucleoprotein particle C proteins: A conserved gene family. *Proc. Natl. Acad. Sci. USA* 83: 2007–11

Nevins, J. R. 1983. The pathway of eukaryotic mRNA formation. *Ann. Rev. Biochem.* 52: 441–66

Nowak, L., Marvil, D. K., Thomas, J. O., Szer, W. 1980. A single-stranded nucleic acid-binding protein from *Artemia salina*. II. *J. Biol. Chem.* 255: 6473–78

Ohlsson, R. I., van Eekelen, C., Philipson, L. 1982. Non-random localization of ribonucleoprotein (RNP) structures within an adenovirus mRNA precursor. *Nucl. Acids Res.* 10: 3053–68

Osheim, Y. N., Miller, O. L., Beyer, A. L. 1985. RNP particles at splice junction sequences on *Drosophila* transcripts. *Cell* 43: 143–51

Padgett, R. A., Grabowski, P. J., Konarska, M. M., Seiler, S., Sharp, P. A. 1986. Splicing of messenger RNA precursors. *Ann. Rev. Biochem.* 55: 1119–50

Padgett, R. A., Hardy, S. F., Sharp, P. A. 1983. Splicing of adenovirus RNA in a cell-free transcription system. *Proc. Natl. Acad. Sci. USA* 80: 5230–34

Padgett, R. A., Konarska, M., Grabowski, P. J., Hardy, S. F., Sharp, P. A. 1984. Lariat RNAs as intermediates and products in the splicing of messenger RNA precursors. *Science* 225: 898–903

Pandolfo, M., Valentini, O., Biamonti, G., Morandi, C., Riva, S. 1985. Single stranded DNA binding proteins derive from hnRNP proteins by proteolysis in mammalian cells. *Nucl. Acids Res.* 13: 6577–90

Patton, J. R., Chae, C. B. 1985. Specific regions of the intervening sequence of β-globin RNA are resistant to nuclease in 50S heterogeneous nuclear RNA-protein complexes. *Proc. Natl. Acad. Sci. USA* 82: 8414–18

Patton, J. R., Ross, D. A., Chae, C. B. 1985. Specific regions of β-globin RNA are resistant to nuclease digestion in RNA-protein complexes in chicken reticulocyte nuclei. *Mol. Cell. Biol.* 5: 1220–28

Patzelt, E., Blaas, D., Kuechler, E. 1983. Cap binding proteins associated with the nucleus. *Nucl. Acids Res.* 11: 5821–35

Pederson, T. 1974. Proteins associated with heterogeneous nuclear RNA in eukaryotic cells. *J. Mol. Biol.* 83: 163–83

Pederson, T. 1983. Nuclear RNA-protein interactions and messenger RNA processing. *J. Cell Biol.* 97: 1321–26

Pederson, T., Davis, N. G. 1980. Messenger RNA processing and nuclear structure: Isolation of nuclear ribonucleoprotein particles containing β-globin messenger RNA precursors. *J. Cell Biol.* 87: 47–54

Pelletier, J., Sonnenberg, N. 1985. Photochemical crosslinking of cap binding proteins to eukaryotic mRNAs: Effect of mRNA 5' secondary structure. *Mol. Cell. Biol.* 11: 3222–30

Perry, R. P., Kelley, D. E. 1968. Messenger RNA-protein complexes and newly synthesized ribosomal subunits: Analysis of free particles and components of polyribosomes. *J. Mol. Biol.* 35: 37–59

Peters, K. E., Comings, D. E. 1980. Two-dimensional gel electrophoresis of rat liver nuclear washes, nuclear matrix, and hnRNA proteins. *J. Cell Biol.* 86: 135–55

Pullman, J. M., Martin, T. E. 1983. Reconstitution of nucleoprotein complexes with mammalian heterogeneous nuclear ribonucleoprotein (hnRNP) core proteins. *J. Cell Biol.* 97: 99–111

Quinlan, T. J., Billings, P. B., Martin, T. E. 1974. Nuclear ribonucleoprotein complexes containing polyadenylate from mouse ascites cells. *Proc. Natl. Acad. Sci. USA* 71: 2632–36

Quinlan, T. J., Kinniburgh, A. J., Martin, T. E. 1977. Properties of a nuclear polyadenylate protein complex from mouse ascites cells. *J. Biol. Chem.* 252: 1156–61

Risau, W., Symmons, P., Saumwerber, H., Frasch, M. 1983. Nonpackaging and packaging proteins of hnRNA in *Drosophila melanogaster*. *Cell* 33: 529–41

Rose, K. M., Jacob, S. T., Kumar, A. 1979. Poly(A) polymerase and poly(A) specific mRNA binding protein are antigenically related. *Nature* 279: 260–62

Rosen, C. A., Ennis, H. L., Cohen, P. S. 1982. Translational control of vesicular stomatitis virus protein synthesis: Isolation of an mRNA sequestering particle. *J. Virol.* 44: 932–38

Rosen, C. A., Siekierka, J., Ennis, H. L., Cohen, P. S. 1984. Inhibition of protein synthesis in vesicular stomatitis virus infected chinese hamster ovary cells: Role of virus mRNA-ribonucleoprotein particle. *Biochemistry* 23: 2407–11

Ruzdijic, S., Bog, J., Sells, B. H. 1984. Cross-linked proteins associated with a specific mRNA in the cytoplasm of HeLa cells. *Eur. J. Biochem.* 142: 339–45

Ryskov, A. P., Saunders, G. F., Farashyn, V. R., Georgiev, G. P. 1973. Double helical regions in nuclear precursor of mRNA. *Biochim. Biophys. Acta* 312: 152–64

Sachs, A. B., Kornberg, R. D. 1985. Nuclear polyadenylate binding protein. *Mol. Cell. Biol.* 5: 1993–96

Salditt-Georgieff, M., Darnell, J. E. Jr. 1982. Further evidence that the majority of primary nuclear RNA transcripts in mammalian cells do not contribute to mRNA. *Mol. Cell. Biol.* 2: 701–7

Salditt-Georgieff, M., Harpold, M. M., Wilson, M. C., Darnell, J. E. Jr. 1981. Large heterogeneous nuclear ribonucleic acid has three times as many 5' caps as polyadenylic acid segments, and most caps do not enter polysomes. *Mol. Cell. Biol.* 2: 179–87

Samarina, O. P., Krichevskaya, A. A. 1981. Nuclear 30S RNP particles. In *The Cell Nucleus*, ed. H. Busch, 9: 1–48. New York: Academic

Samarina, O. P., Krichevskaya, A. A., Georgiev, G. P. 1966. Nuclear ribonucleoprotein particles containing messenger ribonucleic acid. *Nature* 210: 1319–22

Samarina, O. P., Lukanidin, E. M., Molnar, J., Georgiev, G. P. 1968. Structural organization of nuclear complexes containing DNA-like RNA. *J. Mol. Biol.* 33 : 251–63

Schmid, H. P., Schonfelder, M., Setyono, B., Kohler, K. 1983. 76-kDa poly(A) protein is involved in the formation of 48S initiation complexes. *FEBS Lett.* 157 : 105–10

Schochetman, G., Perry, R. P. 1972. Characterization of messenger RNA released from L cell polyribosomes as a result of temperature shock. *J. Mol. Biol.* 63 : 577–90

Schwartz, H., Darnell, J. E. 1976. The association of protein with polyadenylic acid of HeLa cell messenger RNA : Evidence for a "transport" role of a 75,000 molecular weight polypeptide. *J. Mol. Biol.* 104 : 833–51

Sekeris, C. E., Niessing, J. 1975. Evidence for the existence of a structural RNA component in the nuclear ribonucleoprotein particles containing heterogeneous RNA. *Biochem. Biophys. Res. Commun.* 62 : 642–50

Setyono, B., Greenberg, J. R. 1981. Proteins associated with poly(A) and other regions of mRNA and hnRNA molecules as investigated by crosslinking. *Cell* 24 : 775–83

Setyono, B., Pederson, T. 1984. Ribonucleoprotein organization of eukaryotic RNA. XXX. Evidence that UI small nuclear RNA is a ribonucleoprotein when base-paired with pre-messenger RNA *in vivo*. *J. Mol. Biol.* 174 : 285–95

Shatkin, A. J. 1985. mRNA cap binding proteins : Essential factors in initiating translation. *Cell* 40 : 223–24

Singer, R. H., Penman, S. 1972. *Nature* 240 : 99–102

Sommerville, J. 1981. In *The Cell Nucleus*, ed. H. Busch, 7 : 1–57. New York : Academic

Sonnenberg, N. 1981. ATP/Mg^{++}-dependent crosslinking of cap binding proteins to the 5' end of eukaryotic mRNA. *Nucl. Acids Res.* 9 : 1643–50

Sperling, R., Sperling, J., Levine, A. D., Spann, P., Stark, G. R., Kornberg, R. D. 1985. Abundant nuclear ribonucleoprotein form of CAD RNA. *Mol. Cell. Biol.* 5 : 569–75

Spirin, A. S. 1969. Informosomes. *Eur. J. Biochem.* 10 : 20–35

Spirin, A. S., Ajtkhozhin, M. A. 1985. Informosomes and polyribosome-associated proteins in eukaryotes. *Trends Biochem. Sci.* 10 : 162–65

Spirin, A. S., Belitsina, N. V., Lerman, M. I. 1965. Use of formaldehyde fixation for studies of ribonucleoprotein particles by caesium chloride density-gradient centrifugation. *J. Mol. Biol.* 14 : 611–15

Spirin, A. S., Nemer, M. 1965. Messenger RNA in early sea urchin embryos : Cytoplasmic particles. *Science* 150 : 214–17

Spohr, G., Kayibanda, B., Scherrer, K. 1972. Polyribosome-bound and free cytoplasmic-hemoglobin-messenger RNA in differentiating avian erythroblasts. *Eur. J. Biochem.* 31 : 194–208

Steitz, J. A., Kamen, R. 1981. Arrangement of 30S heterogeneous nuclear ribonucleoprotein on polyoma virus late nuclear transcripts. *Mol. Cell. Biol.* 1 : 21–34

Stevenin, J. H., Gallinaro-Matringe, R., Gattoni, R., Jacob, M. 1977. Complexity of the structure of particles containing heterogeneous nuclear RNA as demonstrated by ribonuclease treatment. *Eur. J. Biochem.* 74 : 589–602

Stevenin, J., Gattoni, R., Divilliers, G., Jacob, M. 1979. Rearrangements in the course of ribonuclease hydrolysis of pre-messenger ribonuclear proteins. *Eur. J. Biochem.* 95 : 593–606

Stevenin, J., Gattoni, R., Keohavong, P., Jacob, M. 1982. Mild nuclease treatment as a probe for a non-random distribution of adenovirus-specific RNA sequences and of cellular RNA in nuclear ribonucleoprotein fibrils. *J. Mol. Biol.* 155 : 185–205

Stunnenberg, H. G., Louis, C., Sekeris, C. E. 1978. Depletion in nuclei of proteins associated with hnRNA, as a result of inhibition of RNA synthesis. *Exp. Cell. Res.* 112 : 335–44

Sundquist, B., Persson, T., Lindberg, U. 1977. Characterization of mRNA-protein complexes from mammalian cells. *Nucl. Acids Res.* 4 : 899–915

Suria, D., Liew, C. C. 1979. Characterization of proteins associated with nuclear ribonucleoprotein particles by two-dimensional polyacrylamide gel electrophoresis. *Can. J. Biochem.* 57 : 32–42

Swift, H. 1963. Cytochemical studies on nuclear fine structure. *Exp. Cell Res.* 9 : 54 67 (Suppl.)

Tasseron-De-Jong, J. G., Bronwer, J., Rietveld, K., Zoetemelk, C. E. M., Bosch, L. 1979. Messenger ribonucleoprotein complexes in human KB cells infected with adenovirus 5 contain tightly bound viral-coded 100K proteins. *Eur. J. Biochem.* 100 : 271–83

Thomas, J. O., Glowacka, S. K., Szer, W. 1983. Structure of complexes between a major protein of heterogeneous nuclear ribonucleoprotein particles and polyribonucleotides. *J. Mol. Biol.* 171 : 439–55

Thomas, J. O., Razziuddin, M., Sobota, A., Boublik, M., Szer, W. 1981. *Proc. Natl. Acad. Sci. USA* 78 : 2888–92

Tsanev, R. G., Djondjurov, L. P. 1982.

Ultrastructure of free ribonucleoprotein complexes in spread mammalian nuclei. *J. Cell Biol.* 94: 662–66

Tomcsanyi, T., Molnar, J., Tigyi, A. 1983. Structural characterization of nuclear poly(A)-protein particles in rat liver. *Eur. J. Biochem.* 131: 283–88

Valentini, O., Biamonti, G., Pandolfo, M., Morandi, C., Riva, S. 1985. Mammalian single-stranded DNA binding proteins and heterogeneous nuclear RNA proteins have common antigenic determinants. *Nucl. Acids Res.* 13: 337–46

Van der Marel, P., Tasseron-De-Jong, J. G., Bosch, L. 1975. The proteins associated with mRNA from uninfected and adenovirus type 5-infected KB cells. *FEBS Lett.* 51: 330–34

Van Eekelen, C. A. G., Mariman, E. C. M., Reinders, R. J., Van Venrooij, W. 1981a. Adenoviral hnRNA is associated with host proteins. *Eur. J. Biochem.* 119: 461–67

Van Eekelen, C., Ohlsson, R., Philipson, L., Mariman, E., Van Beek, R., et al. 1982. Sequence dependent interaction of hnRNP proteins with late adenoviral transcripts. *Nucl. Acids Res.* 10: 7115–31

Van Eekelen, C. A., Riemen, T., Van Venrooij, W. J. 1981b. Specificity in the interaction of hnRNA and mRNA with proteins as revealed by *in vivo* crosslinking. *FEBS Lett.* 130: 223–26

Van Eekelen, C. A. G., Van Venrooij, W. J. 1981. hnRNA and its attachment to a nuclear protein matrix. *J. Cell Biol.* 88: 554–63

Van Venrooij, W. J., Riemen, T., van Eekelen, C. A. G. 1982. Host proteins are associated with adenovirus specific mRNA in the cytoplasm. *FEBS Lett.* 145: 62–65

Van Venrooij, W. J., van Eekelen, C. A. G., Jansen, R. T. P., Princen, J. M. G. 1977. Specific poly-A-binding protein of 76,000 molecular weight in polyribosomes is not present on poly A of free cytoplasmic mRNP. *Nature* 270: 189–91

Vincent, A., Goldenberg, S., Scherrer, K. 1981. Comparisons of the proteins associated with duck-globin mRNA and its polyadenylated segment in polyribosomal and repressed free messenger ribonucleoprotein complexes. *Eur. J. Biochem.* 114: 179–93

Wagenmakers, A. J. M., Reinders, R. J., Van Venrooij, W. J. 1980. Cross-linking of mRNA to proteins by irradiation of intact cells with ultraviolet light. *Eur. J. Biochem.* 112: 323–30

Walker, B. W., Lothstein, L., Baker, C. L., LeStourgeon, W. M. 1980. The release of 40S hnRNP particles by brief digestion of HeLa nuclei with micrococcal nuclease. *Nucl. Acids Res.* 8: 3639–57

Wilk, H. E., Angeli, G., Schaefer, K. P. 1983. *In vitro* reconstitution of 35S ribonucleoprotein complexes. *Biochemistry* 22: 4592–4600

Wilk, H. E., Werr, H., Friedrich, D., Kiltz, H. H., Schaefer, K. P. 1985. The core proteins of 35S hnRNP complexes: Characterization of nine different species. *Eur. J. Biochem.* 146: 71–81

Williams, K. R., Stone, K. L., LoPresti, M. B., Merrill, B. M., Planck, S. R. 1985. Amino acid sequence of the UP1 calf thymus helix-destabilizing protein and its homology to an analogous protein from mouse myeloma. *Proc. Natl. Acad. Sci. USA* 82: 5666–70

Zeevi, M., Nevins, J. R., Darnell, J. E. Jr. 1982. Newly formed mRNA lacking polyadenylic acid enters the cytoplasm and the polyribosomes but has a shorter half-life in the absence of polyadenylic acid. *Mol. Cell. Biol.* 2: 517–25

Zieve, G., Penman, S. 1981. Subnuclear particles containing a small nuclear RNA and heterogeneous nuclear RNA. *J. Mol. Biol.* 145: 501–23

Ann. Rev. Cell Biol. 1986. 2 : 499–516

MECHANISM OF PROTEIN TRANSLOCATION ACROSS THE ENDOPLASMIC RETICULUM MEMBRANE

Peter Walter and Vishwanath R. Lingappa

Department of Biochemistry and Biophysics and Departments of Physiology and Medicine, University of California Medical School, San Francisco, California 94143

CONTENTS

INTRODUCTION

In this review we attempt a timely survey of issues concerning protein translocation across the membrane of the endoplasmic reticulum of eukaryotic cells. We focus on recent developments, open questions and current controversies. Due to limited space, this review cannot be and is not

499

0743–4634/86/1115–0499$02.00

intended to be comprehensive. Where appropriate, reference to more detailed reviews is given in the text.

Eukaryotic cells contain a multiplicity of membrane-delimited compartments. The selective localization of particular proteins provides the basis for each of these compartments to serve various specialized functions. Thus, for example, the mitochondrion is the exclusive residence of enzymes involved in oxidative phosphorylation; similarly, oxidative detoxification takes place exclusively in the endoplasmic reticulum (ER). The proteins that compose, and are contained within, particular membrane systems are kept there by the impermeability of the lipid bilayer to diffusion of proteins across membranes. How then is compartmentalization of newly synthesized proteins achieved, in view of the fact that the cytosol is the common site of synthesis for the majority of proteins, though they are destined for distinct subcellular locations? The term intracellular protein topogenesis has been coined (Blobel 1980) to describe the specialized mechanisms by which newly synthesized proteins selectively overcome the permeability barrier of specific intracellular membranes to achieve their correct subcellular localization. This review addresses the question of how proteins that pass through or reside in the intracisternal space are specifically synthesized on membrane-bound ribosomes and translocated into the ER lumen.

As in the study of other protein translocation events (e.g. across mitochondrial membranes) there are two fundamental issues to resolve regarding transport across the ER membrane: (a) How is the target membrane recognized and distinguished from all other membrane systems? (b) Once it has been targeted, how is the polypeptide chain translocated across the lipid bilayer into the lumen of the organelle?

HISTORICAL BACKGROUND

The work of Palade and coworkers on the secretory pathway (reviewed by Palade 1975) focused attention on ribosomes bound to the rough endoplasmic reticulum as the site of synthesis of secretory proteins. The subsequent demonstration of vectorial discharge of puromycin-released polypeptides into the lumen of isolated rough microsomal vesicles (Redman & Sabatini 1966) suggested that a specialized mechanism was responsible for translocation across the ER membrane: Nascent polypeptides emerged into the lumen of the microsomal vesicles concomitant with their synthesis. These results raised the intriguing question of how the cell could distinguish the mRNAs for secretory proteins from those for cytoplasmic or mitochondrial proteins and selectively translate the former on ER-bound ribosomes.

The signal hypothesis (Blobel & Dobberstein 1975) was proposed to account for these phenomena. Over the last 15 years overwhelming evidence has accumulated from a plethora of experimental systems in favor of this model. As it specifically relates to secretory proteins, the essential tenets of an updated version of this hypothesis (for a recent review see Walter et al 1984) are that: (a) the information for localization of newly synthesized proteins into the lumen of the ER is encoded in a discrete segment of the nascent polypeptide, the signal sequence; (b) this signal sequence interacts with a series of receptors, some of them cytoplasmic, others integral to the ER membrane. Some of these receptors function in targeting the chain to the ER membrane, others function in its actual translocation across that membrane. These latter receptors, together with associated proteins in the ER membrane, constitute the "translocon," a postulated engine able to drive signal sequence–bearing chains across the ER membrane through a proteinaceous pore or channel.

More recently, the concepts of the signal hypothesis have been expanded to describe a general framework for intracellular protein topogenesis (Blobel 1980). According to this model, "topogenic sequences" within discrete segments of targeted proteins are decoded by specific receptors, either during (cotranslational) or shortly after (posttranslational) their biosynthesis. The specificity of such signal sequence–receptor interactions targets the proteins to the correct intracellular membranes where they are fed into translocons that move them across the hydrophobic core of the lipid bilayer. Similarly, it has been proposed that another class of topogenic sequences   termed stop-transfer sequences—interacts with the translocon to arrest further transport and thereby achieve an asymmetric transmembrane orientation of integral membrane proteins. Thus many of the concepts developed in this review for soluble ectoplasmic proteins are directly applicable to the problem of integration of transmembrane proteins. Recent developments reviewed below suggest that translocons in different intracellular membrane systems may function more similarly than previously thought.

MECHANISM OF TARGETING

With the availability of in vitro systems that faithfully reproduce the translocation of nascent proteins [secretory proteins (Blobel & Dobberstein 1975), lysosomal proteins (Erickson et al 1983), and certain classes of integral membrane proteins (Katz et al 1977)], it became feasible to investigate the molecular requirements for protein translocation across the ER membrane. So far, two components, the signal recognition particle

(SRP) and the SRP receptor, have been purified and shown to function in the targeting events preceding the actual translocation event.

Signal Recognition Particle

SRP is an 11S small cytoplasmic ribonucleoprotein (Walter & Blobel 1982). In our current view, SRP functions as an adapter between the protein synthetic machinery in the cytoplasm and the protein translocation machinery in the ER membrane.

STRUCTURE OF SRP   SRP was first recognized by its ability to restore the translocation activity of salt-extracted microsomes in vitro (Warren & Dobberstein 1978). It was purified to homogeneity from a salt extract of canine pancreatic microsomal vesicles using this activity as an assay (Walter & Blobel 1980). SRP consists of a small (300 nucleotide) 7SL RNA (Walter & Blobel 1982) and six nonidentical polypeptide chains organized into four SRP proteins. These proteins are two monomers, a 19-kDa polypeptide and a 54-kDa polypeptide, and two heterodimers, one composed of a 9-kDa and a 14-kDa polypeptide, and the other comprised of a 68-kDa and a 72-kDa polypeptide (Siegel & Walter 1985). When SRP is disassembled under nondenaturing conditions, the RNA and the protein fractions are inactive by themselves, but together they can readily be reconstituted into an active particle (Walter & Blobel 1983; Siegel & Walter 1985).

Recent studies revealed that different assayable functions of SRP in the targeting process can be assigned to specific structural domains of the particle. These separable functions include the recognition of signal sequences and the ability of SRP to arrest specifically the translation of nascent signal sequence–bearing proteins (Siegel & Walter 1986b). These domains are schematically indicated in Figure 1 superimposed on the secondary structure of 7SL RNA. This model is supported by recent evidence demonstrating that SRP is a rod-shaped, elongated structure (Andrews et al 1985) and that the RNAs—visualized directly by electron spectroscopic imaging—span the entire length of the particle (D. W. Andrews et al, submitted for publication).

SIGNAL RECOGNITION   Once SRP had been purified to homogeneity it became possible to study its activity in greater detail. Results of experiments testing both the effects of SRP on the translation of secretory proteins *and* its binding properties with various components in the translation-translocation system have led to the model of the SRP cycle shown in Figure 2.

In brief, SRP is thought to bind in a signal-sequence-independent manner with relatively low affinity to biosynthetically inactive ribosomes (Figure 2a, b) (Walter et al 1981). Upon emergence of a signal sequence as part of the nascent polypeptide chain, the affinity of SRP for the ribosome increases (Figure 2c); in the case of preprolactin synthesized on wheat germ ribosomes this increase amounts to three to four orders of magnitude. The SRP-ribosome-nascent chain complex is then targeted to the membrane of the ER via a direct interaction of SRP with the SRP receptor (Walter & Blobel 1981b), an integral membrane protein that is restricted in its subcellular localization to this membrane system (Hortsch et al 1985). At this point SRP and the SRP receptor detach from the ribosome and can reenter the cycle, i.e. both molecules are thought to act catalytically in the targeting process. The ribosome-nascent chain complex engages in a functional ribosome membrane junction, and the translocation of the nascent polypeptide proceeds (see below). (For a more detailed description of the SRP cycle see Walter et al 1984.)

ELONGATION ARREST   When SRP is included in in vitro translation systems in the absence of microsomal membranes, it blocks protein synthesis concomitant with the increase in its affinity for the ribosome just after the signal peptide becomes exposed outside the large ribosomal subunit (Walter & Blobel 1981b; Meyer et al 1982a). In some cases a discretely sized protein fragment that corresponds to the elongation-arrested secretory protein can be detected by gel electrophoresis; in other cases the arrested forms appear as a broader smear on gels, which indicates that SRP can recognize signal sequences and arrest elongation within a certain range of chain lengths. It is also observed that some nascent polypeptides are arrested, while others transiently pause in chain growth (P. Walter, unpublished results). Therefore, in these latter cases arrest is often difficult to detect (Meyer 1985). Interestingly, while elongation arrest has been demonstrated as a kinetic delay of elongation in translation systems reconstituted from mammalian components (K. Matlack & P. Walter, unpublished results), the same effect is more pronounced (as a strict blockage of elongation) when signal-bearing proteins are translated in a heterologous wheat germ system. Thus while the general phenomenon of arrested elongation is ubiquitous, different in vitro systems reflect it to a different degree. Therefore it remains to be established whether SRP acts in vivo as a strict "on-off" switch or functions as a more graded rate-controlling factor.

Two distinct biochemical approaches were employed to map the elongation-arrest function to a separate and separable domain of SRP. One functional domain was shown to consist of the 9/14-kDa SRP proteins and those 7SL RNA sequences that are homologous to repetitive Alu DNA (see Figure 1, left). One experimental approach employed single omission experiments in which SRPs were reconstituted from fractionated and purified protein and RNA components (Siegel & Walter 1985). A second approach involved the preparation of a subparticle obtained after nucleolytic dissection of SRP (Siegel & Walter 1986). These perturbed SRPs lacking the elongation-arrest domain are still active in signal recognition and targeting; therefore, elongation arrest *cannot* be a prerequisite for protein translocation across the membrane. In the absence of elongation arrest, however, most signal-bearing nascent proteins lose their ability to

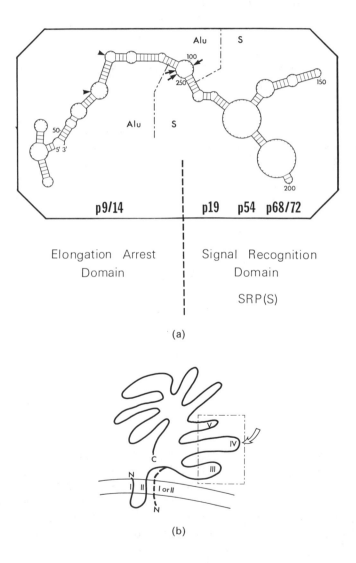

(a)

(b)

be translocated if elongation proceeds beyond a critical point in the absence of membranes. Thus elongation arrest seems to maintain the nascent chain in a translocation-competent state by preventing (or delaying) its further elongation into the cytoplasmic space and thereby adds to the fidelity of the reaction. The particular length range in which a nascent protein remains translocation competent may vary for different proteins (see below).

Since SRP contains an RNA as a structural component, it is tempting to speculate that this RNA engages in base-pairing interactions with other nucleic acids during the SRP's functional cycle. The RNA components in the translational apparatus are likely candidates for participants in such interactions (Walter & Blobel 1982; Zwieb 1985). However, there is at present no direct evidence for such interactions. A possible mechanism for elongation arrest could involve the binding of 7SL RNA to the A-site on the ribosome, thus preventing the next amino acyl tRNA from binding. Indeed, the secondary structure of 7SL RNA in the elongation-arrest

---

←

Figure 1   Domain structure of SRP (*left*) and the SRP receptor (*right*). (*a*) (From Siegel & Walter 1986a): SRP is composed of two separable domains. A possible phylogenetically conserved secondary structure for 7SL RNA is shown (Siegel & Walter 1986a). Similar secondary structures have been proposed by Gundelfinger et al (1984), E. Ullu (personal communication), and Zwieb (1985). Connecting lines between the RNA strands indicate base pairs; G-U pairs are included. (For an extensive description of SRP structure see Siegel & Walter 1986b.) Micrococcal nuclease cleaves the particle at the point indicated by arrows, removing the elongation-arresting domain. Additional cuts mapped by Gundelfinger et al (1983) are indicated by arrowheads. The elongation-arresting domain includes both ends of the RNA (labeled 5' and 3') and is comprised of sequences that are homologous to the repetitive Alu DNA sequence family. Evolutionary considerations suggest that 7SL RNA is the parent molecule for repetitive Alu DNA (Ullu & Tschudi 1985). The thin dashed lines indicate the boundaries of homology between 7SL RNA and an Alu consensus sequence. The elongation-arresting domain also contains the 9/14-kDa SRP protein. The other domain, termed SRP(S), retains signal recognition and translocation promoting function and is comprised of the middle portion of 7SL RNA (the S-segment) and the remaining three SRP proteins. As mentioned in the text, the 54-kDa SRP protein can be selectively cross-linked to signal peptides and may therefore provide the signal binding pocket (*h*) (From Lauffer et al 1985). A model of the disposition of the SRP receptor α-subunit in the membrane of the ER is shown. Putative structural and functional features as deduced from the primary sequence (Lauffer et al 1985) are indicated. Regions I and II are putative membrane-spanning regions; whether both of them or either one alone functions as the membrane anchor of the receptor or if additional hydrophobic regions are contributed by the β-subunit is presently not known. Regions III–V contain the charge clusters described in the text. The boxed domain contains regions strongly resembling RNA binding proteins; their presence suggests that the SRP–SRP receptor interaction may include binding of 7SL RNA to this domain. The arrow indicates the position of the protease-sensitive site. Cleavage of the receptor at this position results in the release of the 52-kDa cytoplasmic fragment. This fragment does not have two properties of the intact receptor: the binding affinity for SRP and the ability to release elongation arrest (Lauffer et al 1985; Gilmore et al 1982a).

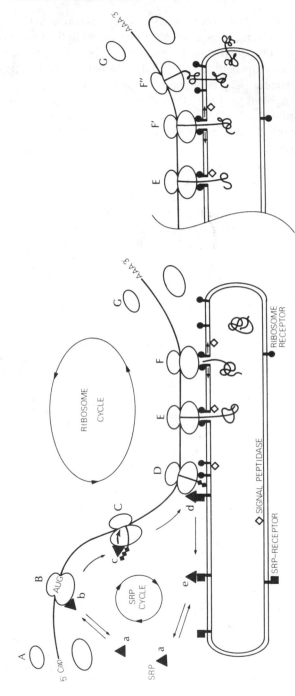

Figure 2 Model (from Walter et al 1984) for protein translocation across the ER membrane for soluble intracisternal proteins (*left*) and integral membrane proteins that possess a structural domain on the intracisternal face of the membrane (*right*). The key features of the model are outlined in the text. (For a more extensive description see Walter et al 1984.)

domain of SRP resembles that of a tRNA that is missing the anticodon stem. In addition, the physical dimensions of SRP would easily allow the particle to bridge the distance between the nascent chain exit site on the ribosome (where the signal sequence emerges) and the peptidyl transferase activity known to be located between the two ribosomal subunits (Andrews et al 1985).

Signal Sequences

What constitutes the essential features of a signal sequence and how such sequences are recognized by SRP remain unsolved problems. Signal sequences show no recognizable primary sequence homology, and a recent compilation shows that sequence variation can be rather extreme (von Heijne 1985). Yet studies on a variety of systems both in vivo and in vitro demonstrate conservation of signal sequence function over the widest evolutionary distances (Muller et al 1982). As a consequence we are still not able to predict with confidence which regions in proteins might function as internal signal sequences. Nevertheless, internal signal sequences have been demonstrated unequivocally (Bos et al 1984). Moreover, cleavage by signal peptidase is not required for translocation (Palmiter et al 1978).

One of the few characteristic features of signal sequences is a variable stretch of hydrophobic amino acids in the core of the sequence. Point mutations in the hydrophobic core in bacterial signal sequences have been shown to abolish function (Lee & Beckwith 1986, this volume). Based on the hydrophobicity of these regions and on evidence from biophysical studies with synthetic signal peptides (reviewed by Briggs & Gierasch 1986), it has been suggested that these sequences act as amphiphiles that are integrated into and possibly perturb lipid bilayers. There is, however, still no evidence that the general mechanism for translocation involves a direct interaction of signal sequences with the hydrophobic core of the lipid bilayer. Indeed, several lines of evidence suggest direct interactions of signal sequences with *proteins*.

The clearest evidence for such interactions involve SRP. Since SRP is a soluble ribonucleoprotein, its interactions with signal sequences can be studied in the absence of membranes by measuring binding or by observing the SRP-mediated modulation of protein synthesis. For example, when signal sequences that are rich in leucine are translated in the presence of the amino acid analog β-hydroxy-leucine, SRP signal recognition is abolished (Walter et al 1981; Walter & Blobel 1981b). This demonstrates that SRP directly recognizes features in the nascent chain. Moreover, the finding conclusively rules out the possibility that sequences in the mRNA alone are responsible for the observed effect. (After the discovery of an RNA component in SRP the latter notion was considered attractive because of the possibility of recognition via putative base-pairing interactions.) Direct proof of an SRP–signal sequence interaction was recently provided by cross-linking experiments. Two groups independently showed that a photoactivable cross-linking reagent was selectively incorporated into the amino-terminal region of the signal peptide for nascent preprolactin. Each group found that the signal peptide is in *direct* contact with the 54-kDa SRP protein (Kurzchalia et al 1986; Krieg et al 1986).

SRP Receptor

Using the same in vitro protein translocation assays that led to the purification of SRP, two distinct approaches were taken to identify the corresponding *membrane* components involved in targeting of signal sequence–bearing nascent chains to the ER membrane. These approaches eventually led to the discovery and purification of the SRP receptor, the first membrane protein proven to play a vital role in this process.

One of these approaches was based on the early observation that proteolysis of microsomal membranes completely abolishes their protein translocation activity but that, most importantly, the activity can be restored by addition to an extract prepared by limited proteolysis of the original microsomal membrane fraction (Walter et al 1979; Meyer & Dobberstein 1980a). This proteolytic dissection and functional reconstitution provided the assay for the purification of the protease-solubilized component. The activity was purified as a basic 52-kDa protein (apparent mobility on SDS PAGE is 60 kDa) (Meyer & Dobberstein 1980b), which was subsequently demonstrated (by immunological techniques) to be a proteolytic fragment derived from a 69-kDa integral membrane protein (apparent mobility 72 kDa) restricted in its subcellular localization to the endoplasmic reticulum (Meyer et al 1982b).

The second approach took advantage of the observations that, when assayed in the absence of microsomal membranes, SRP causes a site-specific elongation arrest in the synthesis of presecretory proteins and that microsomal membranes contain an activity that releases the elongation arrest. Based on these observations, the elongation-arrest-releasing activity was predicted to reside in a membrane protein termed the SRP receptor (Walter & Blobel 1981b) [subsequently named the docking protein (Meyer et al 1982a)]. Fractionation of a detergent extract of microsomal membranes employing affinity chromatography on SRP-Sepharose as a key step allowed purification of the SRP receptor. The purified fraction contained a predominant 69-kDa membrane protein and the arrest-releasing activity. Using both immunological and peptide-mapping techniques, the SRP receptor was shown to be identical to the membrane protein identified via the proteolytic dissection methods described above (Gilmore et al 1982a,b).

Recently, the primary structure of the 69-kDa SRP receptor protein was determined from its cognate cloned cDNA, and its relationship to the cytoplasmic SRP receptor fragment was determined (Lauffer et al 1985). This fragment was shown to begin with residue 152 of the intact protein. Thus, it is sequences within the 151 amino acids at the amino terminal that anchor the SRP receptor in the lipid bilayer. Two distinctly hydrophobic regions have been identified that constitute putative α-helical trans-membrane segments. Since either of these segments would position a positively charged amino acid in the hydrophobic core of the lipid bilayer, the receptor probably interacts with other integral membrane proteins that neutralize these charges. Recent evidence suggests the existence of proteins that can be copurified with the 69-kDa SRP receptor protein or isolated by affinity techniques. In particular, an ER membrane protein with an apparent molecular weight of 30 kDa was found by a variety of techniques to be tightly associated with the 69-kDa protein (Tajima et al 1986). Thus the SRP receptor appears to be a hetero-dimeric protein that in addition to the 69-kDa polypeptide (the SRP receptor α-subunit) contains a second 30-kDa subunit (β-subunit). Carboxy-terminal to the putative trans-membrane regions in the α-subunit is an unusually hydrophilic domain. In particular, unusually large clusters of charged amino acids are found surrounding the site of proteolytic cleavage that severs the 52-kDa cyto-plasmic domain (see Figure 1, *right*). This domain of the SRP receptor strongly resembles nucleic acid binding proteins, which suggests that the receptor may transiently interact directly with the 7SL RNA in SRP and that the SRP–SRP receptor affinity could be mediated, at least in part, by a protein–nucleic acid interaction.

The SRP receptor is unlikely to be part of the translocon itself, because the receptor is present in the ER membrane in substoichiometric amounts with respect to membrane-bound ribosomes. Thus it was suggested that the SRP receptor functions "catalytically" and is recycled once correct targeting of the ribosome has been achieved (Gilmore & Blobel 1983). There is also evidence for an additional activity that is distinct from SRP and the SRP receptor and may interact with the targeted signal sequence and act as a secondary signal receptor(s) in the ER membrane (Gilmore & Blobel 1985; Prehn et al 1980). However, a protein serving this function has not yet been identified.

MECHANISM OF TRANSLOCATION

Machinery

Cell-free systems provided a detailed molecular description of the targeting machinery, but have yet to allow insights into the molecular details of the translocation process. In part this difficulty results from the apparent obligate coupling of translocation and translation: Transport across the ER membrane takes place cotranslationally; completed precursors are not detectable in vivo in the cytoplasm. In cell-free systems translocation proceeds only during a limited time and under the fastidious conditions required for the synthesis of the very molecule whose translocation is being studied. As a result, although several specific polypeptides have been implicated as functional components of the translocon, the direct role of any of these proteins remains to be demonstrated. For example, two integral membrane proteins, termed ribophorins, have been suggested to act as ribosome receptors (Kreibich et al 1978); the recent purification of signal peptidase, a relatively abundant complex of six polypeptides, suggests that these proteins are involved in other functions besides signal cleavage (Evans et al 1986).

Translocation Substrates

Although we know little about the actual machinery involved, insight into certain aspects of the mechanism of translocation has recently been obtained by approaches involving manipulation of the translocation substrates. For example, expression of engineered cDNAs encoding fusion proteins in transcription-linked translation systems demonstrated that a signal sequence was sufficient to direct translocation of normally cytoplasmic globin, both in vitro (Lingappa et al 1984) and in vivo (K. Simon et al, submitted for publication). Thus, the specific information for translocation was contained within the signal sequence and not the "passenger" protein.

A more complex version of these experiments raised interesting questions as to the mechanism of translocation (Perara & Lingappa 1985). The DNA sequence coding for globin, normally a cytosolic protein, was fused with the 5' end of the DNA sequence for preprolactin, a secretory protein that has an amino-terminal signal sequence. This fusion protein thus contained the preprolactin signal sequence at an internal position, 117 amino acids from the initiator methionine. When expressed in a transcription-linked translation system, this internal signal sequence was not only cleaved by signal peptidase, but directed the translocation of both flanking protein domains. Surprisingly, carbonate extraction demonstrated that neither the globin domain with the signal sequence attached at its carboxy terminus nor the prolactin domain were integrated into the membrane. Instead, both resided in the vesicle lumen either free or bound to proteins. This result suggests that signal sequences are not buried in the bilayer directly but perform their function by interacting with a proteinaceous machinery in the membrane. Moreover, translocation of the globin domain by a subsequently emerging signal sequence suggests that the energy used for the globin domain's synthesis is not required for its translocation. Thus the commonly observed coupling of translocation and translation may not be an obligate requirement for transport across the ER membrane.

The notion that the translocation machinery can function independently of protein synthesis has now received direct support from different experimental systems.

Posttranslational Translocation in Yeast

Recently, in vitro translation-translocation systems from the yeast *Saccharomyces cerevisiae* have been established (Hansen et al 1986; Waters & Blobel 1986; Rothblatt & Meyer 1986). The precursor to the yeast pheromone α-factor has been used as a model secretory protein. Contrary to all expectations, this precursor, an ~18.5 kDa protein, is translocated across yeast ER membranes posttranslationally, i.e. after it has been completely synthesized and has been released from ribosomes. Prepro-α-factor has no particularly hydrophobic or amphipathic stretches in its primary sequence (other than a typical signal sequence), making it unlikely that its posttranslational translocation is due to some passive partitioning of the protein across the lipid bilayer. Furthermore, the posttranslational translocation reaction is ATP-dependent and requires protein elements both in the membrane and the soluble fraction. Whether these protein components are related in any way to the putative yeast SRP and SRP receptor analogs remains to be established by biochemical analysis. It is clear from these data, however, that translocation of prepro-α-factor does not require coupling to protein synthesis. Therefore, the translocon can, in principle, accept its substrate posttranslationally and in the absence of the ribosome.

It should be kept in mind that the posttranslational translocation of prepro-α-factor was observed in vitro in a system artificially depleted of ER membranes during synthesis. This finding does not prove that prepro-α-factor ever crosses the ER membrane posttranslationally in vivo, where ER membranes are always present during translation. Rather, the actual degree of coupling of translocation and protein synthesis will depend on the relative rates of the respective processes. If targeting and translocation are fast with respect to protein elongation, a strictly vectorial cotranslational translocation mode will result, as appears to be the rule in mammalian cells in vivo (Bergman & Kuehl 1979; Glabe et al 1980).

Posttranslational Translocation of Genetically Engineered Substrates

Similar findings also emerged from the use of engineered clones in mammalian cell-free translation systems (Perara et al 1986; Mueckler & Lodish 1986). Using a procedure that generates a truncated mRNA lacking a termination codon, secretory polypeptide chains could be synthesized and presented to membranes in the absence of further chain elongation while still held by the ribosome that effects their synthesis. It was demonstrated that such chains could be translocated and that nucleotide triphosphates were required as the energy source for this process. In contrast to the situation in the yeast system described above, in most of these cases translocation could be abolished by releasing the nascent chain from the ribosome by artificial termination with the amino acyl tRNA analog puromycin. As expected, translocation was abolished by deletion of the coding region for the signal sequence. In some cases, however, it was also found that some short chains could translocate in a ribosome-independent condition analogous to that found for prepro-α-factor in the yeast system (E. Perara & V. R. Lingappa, submitted for publication). Thus it appears that, at least for the proteins investigated, polypeptide chain growth proceeds through stages in which translocation competence is a property of the chain itself or is maintained by interaction with the ribosome (see Figure 3).

These results show cotranslational translocation in a new light: The role of the membrane-bound ribosome is not to extrude or push the chain through the bilayer as suggested by some observers (Wickner & Lodish 1985). Rather, translocation is catalyzed by an energy-consuming protein engine in the ER membrane, and the ribosome acts, in most but not all cases, as a ligand that maintains the translocation competence of the nascent chain.

CONCEPTS AND CONTROVERSIES

We have surveyed the development of ideas on the problem of translocation of newly synthesized proteins across the ER membrane. Initially, attention was focused on the coupling of translocation to translation, a feature unique to translocation across the ER membrane. This has given way to the realization that obligate coupling to translation is not a prerequisite for translocation and that transport across membranes of a variety of organelles may share common features. These include the involvement of a targeting receptor to discriminate among proteins intended for different destinations, a translocon that somehow transports the targeted protein across the bilayer, and a requirement for energy (derived from hydrolysis of nucleoside triphosphates or from an electrochemical gradient) to drive translocation. The recognition of these steps has resulted from the study of diverse proteins in a variety of organisms and from the study of "artifacts" generated in vitro, i.e. biochemically or genetically altered translocation machinery (Siegel & Walter 1986b) and substrates (Perara & Lingappa 1985), whose aberrant behavior has provided insight into fundamental details of the targeting and translocation problem. Even as new questions emerge, many old ones (e.g. the molecular nature of the signal sequence–receptor interaction) remain unanswered.

Other questions must now be reformulated. For example, in spite of the recent demonstration that the translocon in the ER membranes can, in principle, accept translocation substrates posttranslationally, translocation most likely occurs cotranslationally in vivo. The observation that most posttranslational translocation across the ER membrane appears to be ribosome dependent in vitro supports this notion. As described earlier, ribosome-independent and ribosome-dependent modes of posttranslational translocation across the ER membrane probably reflect the requirements for maintenance of the "translocation competent state" of the nascent chain (see Figure 3). Loss of translocation competence may be due to folding (aberrant or normal) or oligomerization of the protein, or entanglement of the signal sequence with the rest of the chain such that the resulting structure can no longer functionally interact with either the targeting or translocation machinery. A few proteins (such as yeast prepro-α-factor) retain translocation competence even as free, completed polypeptides. For most proteins, however, translocation competence is restricted to a generally narrow range of chain lengths. This range can be extended if the polypeptide is targeted to the membrane while still attached to the ribosome. However, eventually most proteins reach a point in chain elongation where translocation competence is no longer maintained, even when the protein is associated with the ribosome. One of the roles of the SRP-induced elongation arrest may therefore be to extend the effective range of translocation competence for the nascent polypeptide chains.

Previously, the nascent chain was thought to be vectorially translocated across the membrane as it emerged from the ribosome; the finding of posttranslational translocation raises the possibility that the translocon may be sufficiently pliable to accept (partially) folded domains rather than exclusively linear polypeptide chains. Alternatively, the translocon may effect unfolding of such domains prior to translocation. In either case the molecular environment traversed by the protein as it passes through the bilayer remains to be investigated. The finding that translocation is driven by nucleoside triphosphate hydrolysis is a direct demonstration of a protein

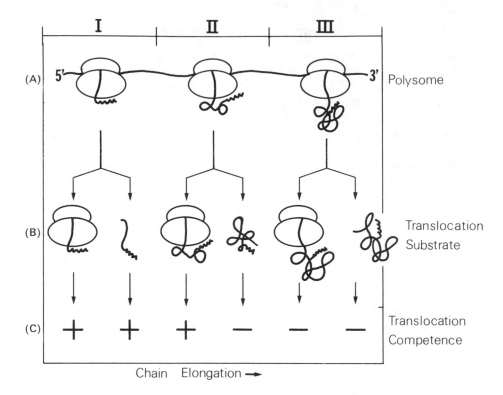

Figure 3 Ribosome dependence of translocation competence. This figure depicts the natural history of the relationship of chain growth (*A*) to translocation competence (*C*). The ribosome dependence of posttranslational translocation was assayed for various lengths of polypeptide synthesized. Progressively shorter polypeptides were synthesized by translating mRNA transcripts in vitro that were progressively truncated at their 3′ end and therefore lacked termination codons (Perara et al 1986; E. Perara & V. R. Lingappa, manuscript in preparation). Ribosomes that have reached the 3′ end of such a truncated mRNA appear unable to release the newly synthesized polypeptide. Release can be artificially achieved by treatment with puromycin. Such translocation substrates, either with or without release from the ribosomes (as indicated in *B*), can be assayed for translocation competence upon presentation to a microsomal membrane preparation in the presence of nucleoside triphosphate to supply energy. In this assay the ribosome dependence or independence of the translocation competence is reflected in the ability or inability of puromycin pretreatment to abolish translocation by releasing the chain from the ribosome (see right arms of branched arrows). (*A*) depicts three ribosomes on a polysome at various stages (I, II, and III) during the synthesis of a hypothetical secretory polypeptide chain. In (*C*) translocatin competence as assayed posttranslationally (see above) is indicated (+). At stage I, the nascent chain is translocation competent, and this competence is independent of the presence of the ribosome, as experimentally demonstrated. As chain growth proceeds, the polypeptide enters stage II where its translocation competence requires the ribosome. Finally, late in chain growth (stage III) the chain is no longer competent to interact with receptors and other proteins involved in translocation. Whether loss of translocation competence in stage III involves a loss of targeting function or loss of a productive interaction with the translocon remains to be determined. It is not known whether SRP is required for posttranslational translocation in either case.

engine in the membrane and rules out a spontaneous process previously suggested (Wickner 1979; Engelman & Steitz 1980). It remains to be established how the energy of hydrolysis is used by the translocon.

Old controversies regarding co- versus posttranslational translocation appear to be resolved. In retrospect it could be concluded that many prokaryotic proteins (targeted to the plasma membrane) do not require ribosomes to maintain their translocation competence. This also appears to be the case for all proteins (so far studied) that are translocated across the peroxisomal membrane and the mitochondrial and chloroplast envelopes. The most challenging problems for future research now include the further fractionation and purification of all the essential, as well as modulatory, components of the targeting and translocation machinery. This should ultimately allow their reconstitution in in vitro systems for the mechanistic analysis of their functions. Finally, our goal must be the understanding of how these components function in vivo. This should include elucidation of the regulatory or homeostatic mechanisms involved in harnessing such a remarkable set of protein machines as the translocons.

ACKNOWLEDGMENTS

We wish to thank David Andrews, Patricia Hoben, and Leander Lauffer for many helpful comments on the manuscript. This work was supported by NIH grants GM-32384 to PW and GM-31626 to VRL. PW is a recipient of support from the Chicago Community Trust/Searle Scholars Program.

Literature Cited

Andrews, D. W., Walter, P., Ottensmeyer, F. P. 1985. *Proc. Natl. Acad. Sci. USA* 82: 785–89

Bergman, L. W., Kuehl, L. M. 1979. *J. Biol. Chem.* 254: 8869–76

Blobel, G. 1980. *Proc. Natl. Acad. Sci. USA* 77: 1496–1500

Blobel, G., Dobberstein, B. 1975. *J. Cell Biol.* 67: 835–51

Bos, T. J., Davis, A. R., Nayak, D. P. 1984. *Proc. Natl. Acad. Sci. USA* 81: 2337–41

Briggs, M. S., Gierasch, L. M. 1986. *Adv. Protein Chem.* 38: In press

Engelman, D. M., Steitz, T. A. 1981. *Cell* 23: 411–22

Erickson, A. H., Walter, P., Blobel, G. 1983. *Biochem. Biophys. Res. Commun.* 115: 275–80

Evans, E., Gilmore, R., Blobel, G. 1986. *Proc. Natl. Acad. Sci. USA* 83: 581–85

Gilmore, R., Blobel, G., Walter, P. 1982a. *J. Cell Biol.* 95: 463–69

Gilmore, R., Blobel, G. 1983. *Cell* 35: 677–85

Gilmore, R., Blobel, G. 1985. *Cell* 42: 497–505

Gilmore, R., Walter, P., Blobel, G. 1982b. *J. Cell Biol.* 95: 470–77

Glabe, C. G., Hanover, J. A., Lennarz, W. J. 1980. *J. Biol. Chem.* 255: 9236–41

Gundelfinger, E. D., Carlo, M. D., Zopf, D., Melli, M. 1984. *EMBO J.* 3: 2325–32

Gundelfinger, E. D., Krause, E., Melli, M., Dobberstein, B. 1983. *Nucleic Acids Res.* 11: 7363–73

Hansen, W. B., Garcia, P. D., Walter, P. 1986. *Cell* 45: 397–406

Hortsch, M., Griffiths, G., Meyer, D. I. 1985. *Eur. J. Cell Biol.* 38: 271–79

Katz, F. N., Rothman, J. E., Lingappa, V. R., Blobel, G., Lodish, H. F. 1977. *Proc. Natl. Acad. Sci. USA* 74: 3278–82

Kreibich, G., Freienstein, C. M., Pereyra, B. N., Ulrich, B. L., Sabatini, D. D. 1978. *J. Cell Biol.* 77: 488–506

Krieg, U., Walter, P., Johnson, A. 1986. *Proc. Natl. Acad. Sci. USA.* In press

Kurzchalia, T. V., Wiedmann, M., Girshovich, A. S., Bochkareva, E. S., Bielka, H., Rapoport, T. A. 1986. *Nature* 320: 634–36

Lauffer, L., Garcia, P. D., Harkins, R. N., Coussens, L., Ullrich, A., Walter, P. 1985. *Nature* 318: 334–38

Lee, C., Beckwith, J. 1986. *Ann. Rev. Cell Biol.* 2: 315–36

Lingappa, V. R., Chaider, J., Yost, C. S., Hedgpeth, J. 1984. *Proc. Natl. Acad. Sci. USA* 81: 456–60

Meyer, D. I. 1985. *EMBO J.* 4: 2031–33

Meyer, D. I., Dobberstein, B. 1980a. *J. Cell Biol.* 87: 498–502.

Meyer, D. I., Dobberstein, B. 1980b. *J. Cell Biol.* 87: 503–8

Meyer, D. I., Krause, E., Dobberstein, B. 1982a. *Nature* 297: 647–50

Meyer, D. I., Louvard, D., Dobberstein, B. 1982b. *J. Cell Biol.* 92: 579–83

Mueckler, M., Lodish, H. F. 1986. *Cell* 44: 629–37

Muller, M., Ibrahimi, I., Chang, C. N., Walter, P., Blobel, G. 1982. *J. Biol. Chem.* 257: 11860–63

Palade, G. 1975. *Science* 189: 347–58

Palmiter, R. D., Gagnon, J., Walsh, K. A. 1978. *Proc. Natl. Acad. Sci. USA* 75: 94–98

Perara, E., Lingappa, V. R. 1985. *J. Cell Biol.* 101: 2292–2301

Perara, E., Rothman, R. E., Lingappa, V. R. 1986. *Science* 232: 348–52

Prehn, S., Nurnberg, P., Rapaport, T. A. 1980. *Eur. J. Biochem.* 107: 185–95

Redman, C. M., Sabatini, D. D. 1966. *Proc. Natl. Acad. Sci. USA* 56: 608–15

Rothblatt, M., Meyer, D. I. 1986. *Cell* 44: 619–28

Siegel, V., Walter, P. 1985. *J. Cell Biol.* 100: 1913–21

Siegel, V., Walter, P. 1986a. *Nature* 320: 81–84

Siegel, V., Walter, P. 1986b. In *Genetic Engineering*, Vol. 8, pp. 179–94, ed. J. K. Setlow. New York: Plenum

Tajima, S., Lauffer, L., Rath, V., Walter, P. 1986. *J. Cell Biol.* 103: In press

Ullu, E., Tschudi, C. 1985. *Nature* 312: 171–72

von Heijne, G. 1985. *J. Mol. Biol.* 184: 99–105

Walter, P., Blobel, G. 1980. *Proc. Natl. Acad. Sci. USA* 77: 7112–16

Walter, P., Blobel, G. 1981a. *J. Cell Biol.* 91: 551–56

Walter, P., Blobel, G. 1981b. *J. Cell Biol.* 91: 557–61

Walter, P., Blobel, G. 1982. *Nature* 299: 691–98

Walter, P., Blobel, G. 1983. *Cell* 34: 525–33

Walter, P., Gilmore, R., Blobel, G. 1984. *Cell* 38: 5–8

Walter, P., Ibrahimi, I., Blobel, G. 1981. *J. Cell Biol.* 91: 545–50

Walter, P., Jackson, R. C., Marcus, M. M., Lingappa, V. R., Blobel, G. 1979. *Proc. Natl. Acad. Sci. USA* 76: 1796–99

Warren, G., Dobberstein, B. 1978. *Nature* 273: 569–71

Waters, G., Blobel, G. 1986. *J. Cell Biol.* 102: 1543–50

Wickner, W. T. 1979. *Ann. Rev. Biochem.* 48: 23–45

Wickner, W. T., Lodish, H. F. 1985. *Science* 230: 400–7

Zwieb, C. 1985. *Nucleic Acids Res.* 13: 6105–24

Ann. Rev. Cell Biol. 1986. 2 : 517–46

REGULATION OF THE SYNTHESIS AND ASSEMBLY OF CILIARY AND FLAGELLAR PROTEINS DURING REGENERATION

Paul A. Lefebvre

Department of Genetics and Cell Biology, University of Minnesota, St. Paul, Minnesota 55108

Joel L. Rosenbaum

Department of Biology, Yale University, New Haven, Connecticut 06511

CONTENTS

INTRODUCTION

A large and rapidly expanding literature indicates that the regeneration of cilia and flagella is being used increasingly as a system to study the control

517

0743–4634/86/1115–0517$02.00

of synthesis and assembly of specific proteins in a eukaryotic organelle. Cilia and flagella are uniquely suited for such studies because their position at the cell surface allows them to be experimentally removed under conditions in which the cell survives and regenerates new cilia or flagella (e.g. Grebecki & Kuznicki 1959; Auclair & Siegel 1966; Rosenbaum & Child 1967). Moreover, they can be easily isolated and purified, which facilitates biochemical analysis (Child 1959; Witman et al 1972). In several systems— the most extensively studied is *Chlamydomonas*—mutants with aberrant control of flagellar length, assembly, motion, stability, and regeneration have been isolated; some of these are described in this review.

Here we consider the types of experiments that have been used to describe the regeneration of cilia and flagella in lower eukaryotes. We pay particular attention to defining the questions that can be addressed in these systems: How does flagellar amputation trigger specific expression of the genes for flagellar proteins? How is the synthesis of ciliary and flagellar proteins controlled? Where does assembly occur, and how is it controlled? How do cells regulate the disassembly of flagellar microtubules? How do these organelles regulate their size (length)? We restrict our discussion to the synthesis and assembly of ciliary and flagellar components after amputation. The synthesis and assembly of the tubulins during the cell cycle and during development have been reviewed previously (Fulton 1977; Raff 1979; Cleveland & Sullivan 1985).

Cilia and flagella are complex organelles, both structurally and biochemically. They consist of an axoneme, which is an array of microtubules and their accessory structures, enclosed in an extension of the cell membrane (Manton & Clarke 1952; Warner 1972). In cilia and flagella from most organisms, the microtubules are arranged as a ring of 9 doublet microtubules. Each doublet consists of a complete A tubule with 13 protofilaments and a B subfiber with 10 or 11 protofilaments. These doublets surround a core of 2 singlet microtubules (the central pair or central tubules). Bound to the outer doublet microtubules are the inner and outer dynein arms, which generate motion by forming cross-bridges with adjacent doublets and causing them to slide against one another (reviewed by Gibbons 1981). The outer doublets are held in a ring by "nexin" links (Stephens 1970). The central pair of microtubules are connected to the outer doublet microtubules by radial spokes, each of which consists of a stalk and an enlarged spoke head (Hopkins 1970; Warner 1972).

The protein composition of cilia and flagella is as complex as their ultrastructure. In *Chlamydomonas*, for example, more than 150 proteins have been detected in isolated flagella examined on two-dimensional gels (Piperno et al 1977). As an example of the biochemical complexity of eukaryotic flagella, consider that the radial spokes of *Chlamydomonas* flagella consist of 17 specific polypeptides, 5 of which comprise the spoke head and 12 of which form the stalk (Piperno et al 1977; Huang et al 1981). The elegant genetic and biochemical experiments that have been used to assign many of these proteins to particular axonemal structures have recently been reviewed (Luck 1984; Huang 1986). The biochemistry of the best understood flagellar proteins, the dyneins and tubulins, has also been reviewed (Gibbons 1981; Dustin 1978; Roberts & Hyams 1979; McKeithan & Rosenbaum 1984).

ASSEMBLY OF CILIA AND FLAGELLA

Remarkably, cilia and flagella, for all of their structural and biochemical complexity, are replaced rapidly by cells after deflagellation or deciliation. The regeneration of cilia or flagella after amputation has been studied in several systems. Using *Peranema*, Chen (1950) was the first to observe that flagella can regenerate after amputation. Lewin (1953) was the first to measure the kinetics of regeneration. He used *Chlamydomonas moewesii* that had lost their flagella during incubation in the dark on agar plates and were then resuspended in liquid and exposed to light. Flagellar regeneration after amputation has also been described in *Astasia* and *Euglena* (Rosenbaum & Child 1967), *Polytomella* (Brown & Rogers 1978), and *Volvox* (Coggins & Kochert 1980); cilia regeneration has been described in *Paramecium* (Grebecki & Kuznicki 1959), *Tetrahymena* (Child 1965; Rosenbaum & Carlson 1969), and sea urchin embryos (Auclair & Siegel 1966; Stephens 1972; Burns 1973).

Kinetics of Regeneration

In almost every case of ciliary or flagellar growth that has been studied, rapid, decelerating kinetics have been found. An example of flagellar regeneration kinetics (the regrowth of *Chlamydomonas* flagella after amputation) is shown in Figure 1. In this case regeneration begins without a significant lag period at an initially high rate ($\sim 0.2\ \mu m/min$), which gradually slows as the flagella approach their original length (12 μm). These regeneration kinetics are qualitatively similar to the kinetics seen in most regeneration systems. In a careful study of single cells of *Peranema*, Tamm (1967) amputated flagella with a microprobe and found that the rate of regrowth depended on the length of the flagella left after amputation. The curve generated by measuring the regeneration rates of flagella cut at many different lengths matched that of the regeneration of a cell whose flagellum had been completely removed. Child (1963) reported the same observation in *Chlamydomonas*. He deflagellated whole populations of cells and then studied individual cells that had been deflagellated to various lengths.

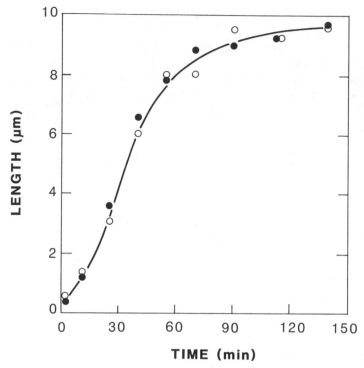

Figure 1 The flagellar regeneration of *Chlamydomonas* vegetative cells after deflagellation. Cells were deflagellated by mechanical shearing, and samples were fixed at various times during regeneration. Flagellar length (in micrometers) was measured for at least 30 samples per time point.

These results suggest that the control of flagellar growth kinetics resides in the flagella, because the cell bodies presumably contain the same amount of flagellar components no matter where the flagella are cut and the rate of regeneration depends on the length of the flagellar stump. Levy (1974) and Child (1978) have developed models to explain the kinetics of flagellar regeneration based on these results.

Two other experiments, however, argue that flagella cannot limit their own regeneration. When *Chlamydomonas* cells were experimentally induced to resorb 75% of their flagellar length by treatment with various drugs (see section on disassembly of cilia and flagella, below), the pool of flagellar proteins in the cytoplasm increased because the resorbed flagellar proteins were assembly competent. When the cells were washed out of the drug medium and allowed to regenerate, the flagella regrew at a rate faster than the fastest observed after deflagellation (0.4 μm/min after resorption versus 0.12 μm/min for control cells) (Lefebvre et al 1978). This rapid regeneration occurred on flagella that had already reached 25% of their normal length, which indicates that rapid distal growth can occur.

The second experiment utilized the temporary quadriflagellates formed during mating in *Chlamydomonas*. When a mutant with short flagella (*shf-1*) was mated to a wild-type, a temporary dikaryon cell with two short and two normal length flagella was formed. Almost immediately, the short flagella began to grow (Jarvik et al 1984). The rate of this growth was at least as rapid as the most rapid rate seen after deflagellation, even though it occurred on flagella that were already half-length. As in the previous experiment, no intrinsic flagellar regeneration rate dictated by flagellar length was seen in these experiments. The difference between these latter two results and the deflagellation experiments of Tamm and Child may be due to the effects of deflagellation. Possibly the "healing" of the membrane after deflagellation or the replacement of some soluble factor lost into the medium accounts for the differences in regeneration rate. It is clear, however, that there is no strict relation between flagella length and the possible rate of flagellar growth at that length.

Length Control

Several lines of research indicate that the final length attained by cilia and flagella is under active control. For example, in a number of organisms the amount of assembly-competent flagellar protein in the cell is greater than that used to assemble the flagella. Rosenbaum et al (1969) found that if *Chlamydomonas* vegetative cells were deflagellated in cycloheximide, an inhibitor of protein synthesis, one-half length flagella (6 μm) were regenerated, which indicates that the cells maintain a pool of essential flagellar proteins in unassembled form (see also Farrell 1976). The existence of a pool of flagellar tubulin in the cell has been shown directly by immunochemical methods. Piperno & Luck (1977), using an antibody to flagellar β tubulin, found that *Chlamydomonas* contained a pool of β tubulin that was just sufficient to form one-half length flagella, but little more. Because not all of the assembly-competent flagellar proteins are assembled into the flagella, the amount of flagellar protein in the cell must not be the limiting factor controlling flagellar length. In several other organisms, such as sea urchins (Auclair & Siegel 1966), *Ochromonas* and *Euglena* (Rosenbaum & Child 1967), *Polytomella* (Brown & Rogers 1978), and *Tetrahymena* (Rannestad 1974), unassembled pools of ciliary or flagellar proteins are also maintained, which suggests that in each of these systems the length of cilia or flagella is controlled by factors other than the availability of assembly-competent proteins.

The length of the cilia or flagella of several organisms can be manipulated experimentally. The most striking example is the effect of trypsin or concanavalin A on sea urchin blastulae. These embryos are covered with cilia of uniform length, averaging 25 μm, with a small tuft of extra-long (40–70 μm), immotile cilia at the animal pole (Lallier 1975). Riederer-Henderson & Rosenbaum (1979) found that if embryos were treated with trypsin or concanavalin A for 18 hr, all of the cilia on the embryo grew to lengths of 40–70 μm. This effect involved some irreversible change in the embryo: After the embryo was covered with long cilia and the trypsin or concanavalin A was washed out, the cilia remained extra long. If these extra-long cilia were then amputated, the embryos regenerated long cilia of 40–70 μm (Riederer-Henderson & Rosenbaum 1979; Burns 1979).

Two less dramatic examples of experimentally induced flagellar elongation beyond normal length are found in *Chlamydomonas* and *Naegleria*. Telser (1977) found that if *Chlamydomonas* cells were placed in distilled water, the flagella elongated as much as 30% in less than 2 hr. By growing *Chlamydomonas* in media of different osmolarity, Solter & Gibor (1978) were able to produce cultures of cells with different flagellar lengths. In *Naegleria*, Fulton & Walsh (1980) found that if an inhibitor of transcription, actinomycin D, was added at an appropriate time between the onset and completion of flagellar outgrowth during the transformation of cells from amoebae to flagellates, the resulting flagella were 10% longer than normal.

Genetic Analysis of Flagellar Length

The flagellar length control mechanism in *Chlamydomonas* can be genetically altered to produce mutants with either long or short flagella. A number of mutants exist that grow flagella 2 to 3 times the wild-type length. The first of these to be described were unlinked, nonconditional mutants, *lf 1* and *lf 2* (McVittie 1972). A third mutant, *cs 89* (Jarvik et al 1976), has normal length flagella as a vegetative cell at temperatures of 20°C or higher but has extra-long flagella at 13°C. Starling & Randall (1971) showed that the excess flagellar length phenotype of *lf 1* is recessive in crosses with wild-type cells using an experiment similar to the one shown in Figure 2 (for mutant *cs 89*). Wild-type length control was rapidly restored when a *cs 89* cell with long flagella (Figure 2b) fused with a wild-type cell (Figure 2a) during mating. Immediately after mating a quadriflagellate cell with 2 long (20–30 μm) and 2 normal length (12 μm) flagella was produced (Figure 2c). Within 30 min the two long flagella shortened to the wild-type length of 12 μm (Figure 2d), which indicates that the wild-type cell must take control of the mutant flagella to restore normal length (Jarvik et al 1976).

Another type of long flagella mutant has been described, which retains its long flagella in quadriflagellates after mating with wild-type cells (Jarvik

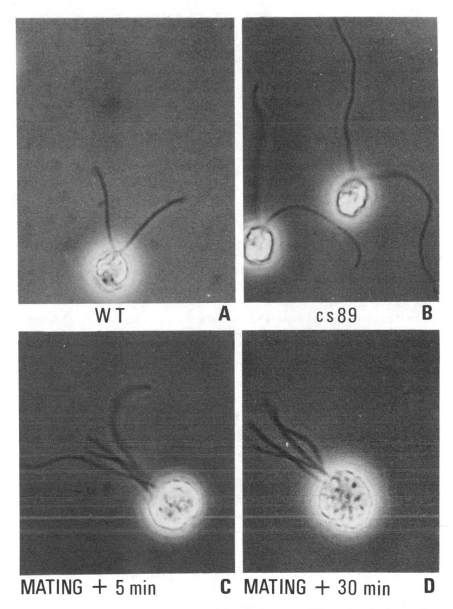

Figure 2 Resorption of the long flagella of mutant *cs 89* to wild-type length after mating to wild type. Before mating, wild-type cells have an average flagellar length of 12 μm and *cs 89* flagella an average of 20–30 μm. Within 30 min after cell fusion during mating the flagella all have an average length of 10–12 μm.

et al 1980). If the four flagella are amputated, however, the quadriflagellate cell regenerates four normal length flagella. This result suggests that the long flagella contains some component that blocks the wild-type cell from gaining control of flagellar length but that this block can be removed by amputating the flagella.

A mutant with variable numbers of flagella has been valuable in the study of flagellar length control in *Chlamydomonas* (Kuchka & Jarvik 1982). These studies of mutant *vfl-2* analyzed the relationship between the length of *Chlamydomonas* flagella and the pool of unassembled flagellar proteins. *vfl-2* cells can have 0–4 flagella, but the cell bodies are of a uniform size regardless of flagellar number. No matter how many flagella each cell has, all of the flagella are of the wild-type length. If a population of *vfl-2* cells is deflagellated in cycloheximide to block protein synthesis, variable numbers of flagella are regenerated, and these flagella are all one half the normal length. As described earlier, wild-type cells regenerate two one-half length flagella in the absence of protein synthesis. Thus, *vfl-2* cells that regenerate three one-half length flagella must have a pool of unassembled flagellar proteins 50% larger than in the wild-type, and cells that regenerate a single one-half length flagellum must have a pool 50% smaller than in the wild-type. These results demonstrate a linkage between the number of flagella in a cell and the size of the unassembled pool of flagellar proteins maintained in the cell. Clearly, flagellar length control in *Chlamydomonas* is not simply a matter of maintaining a constant size pool of flagellar proteins and partitioning these equally to the two flagella.

Sites of Assembly During Flagellar Growth

The regeneration of cilia and flagella requires the assembly of the microtubule components of the axoneme, the attachment of all of the accessory structures (e.g. the radial spokes and the dynein arms) to the outer doublets, and the expansion of the flagellar membrane around the axoneme as it grows. A number of studies have localized the site of assembly of flagellar components during regeneration.

FLAGELLAR MEMBRANE   Bouck (1971) studied the site of growth of the flagellar membrane in *Ochromonas*. He took advantage of the fact that the mastigonemes (hair-like structures that extend from the outer doublets, through the flagellar membrane, and into the medium) are assembled in the cytoplasm in morphologically identifiable stages. Mastigoneme assembly begins near the nucleus and continues in the Golgi; preassembled mastigonemes leave the Golgi in membrane vesicles, which are then added to the flagellar membrane. Bouck found that these membrane vesicles, clearly identified by their mastigoneme contents, were inserted at a specific location at the base of the flagella during flagellar regeneration. Therefore, at least in *Ochromonas*, flagellar membrane components appear to be added at the proximal end.

OUTER DOUBLET MICROTUBULES   The site of incorporation of most newly synthesized proteins in regenerating flagella has been shown to be the distal tip using pulse-labeling experiments (Rosenbaum & Child 1967; Rosenbaum et al 1969). Cells were deflagellated, allowed to regenerate to approximately one-half length, and then pulse-labeled with ^{3}H-containing amino acids. Label incorporation was shown by autoradiography and light microscopy to be predominantly in the distal half of the growing flagella. Witman (1975) repeated these experiments, but he removed the membrane and analyzed label incorporation into the axonemes using electron microscopy and autoradiography. Figure 3 is an example of a typical autoradiographic image from Witman's work; 65% of the silver grains detected by autoradiography were over the distal half of the axonemes. Although the conclusion of these studies was that most newly synthesized flagellar proteins are added at the distal end of the growing flagellum, there is significant incorporation of flagellar components along the whole length of the structure. Whether this nondistal assembly represents uniform incorporation of proteins along the axoneme or specific incorporation of certain flagellar proteins at the proximal end is not known.

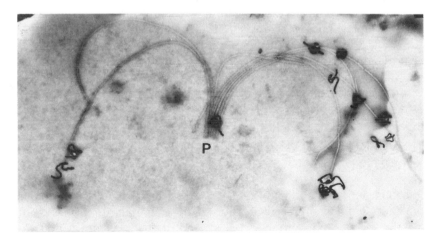

Figure 3   Visualization of label (^{3}H) incorporation into the distal half of axonemes from cells pulse-labeled during regeneration. The sample is an autoradiograph of a negatively stained axoneme. Note the bundling of the microtubules, marking the proximal end (*P*). (Reprinted from Witman 1975 with permission.)

CENTRAL PAIR MICROTUBULES   The possibility that central pair micro-tubules grow from their proximal end was raised by the observation that the central pair microtubules of *Chlamydomonas* (Dentler & Rosenbaum 1977) and *Tetrahymena* (Sale & Satir 1977) are capped at their distal tip and that the cap is attached to the inner surface of the flagellar membrane. This cap may prevent the addition of tubulin subunits onto the distal tip of the central pair, as suggested by Dentler & Rosenbaum (1977), who showed that tubulin subunits can only be added to the distal end of the central pair in vitro if the cap structure is missing. They also found that the caps are in place on the central pair at all stages of flagellar regen-eration, which indicates that if the distal tip is indeed unavailable for subunit addition during flagellar assembly, then proximal growth of the central pair may occur.

POLARITY OF FLAGELLAR MICROTUBULES   When techniques for in vitro microtubule assembly were developed, they were used to determine that the two ends of flagellar microtubules demonstrate different assembly polarity, i.e. the proximal and distal ends of axonemes nucleate micro-tubule assembly in vitro at different rates (Allen & Borisy 1974; Binder et al 1975). For both outer doublet and central pair microtubules, assembly of chick brain tubulin occurred five times faster at the distal end than at the proximal end. This polarity was confirmed by Euteneuer & McIntosh (1981) using a method in which partial microtubule sheets are assembled onto preformed microtubules to determine the intrinsic polarity. The par-tial microtubules form "hooks" whose orientation indicates the polarity of the template microtubule. Using this method it was shown that all of the A and B subfibers of the outer doublets of an axoneme, as well as the central pair microtubules, are oriented in the same direction, with the plus end (that with the highest assembly rate) distal to the basal body. Thus, if central pair microtubules grow from their proximal end, this growth occurs at the slow growth end.

IN VITRO ASSEMBLY OF FLAGELLAR MICROTUBULES

A complete understanding of the growth of cilia and flagella will require an understanding of the regulation of flagellar microtubule assembly. To that end, several groups have attempted to break down and then reassemble flagellar microtubules. This effort has met with only partial success: The reassembly of singlet, but not doublet, microtubules has been reported. A major difficulty in this effort has been the stability of ciliary and flagellar microtubules. Stephens (1968) reported that the detergent

Sarkosyl could break down outer doublet microtubules and that these could reassemble into protofilament sheets. Complete microtubules were rarely formed, however, and the sheet assembly occurred at 0°C or in colchicine, conditions that would be expected to block assembly of bona fide microtubules. Kuriyama (1976) found that sonication of outer doublets from sea urchin flagella partially solubilized the microtubules. The resulting tubulin preparation could assemble into single microtubules on added doublet microtubule seeds in the presence of glycerol. Binder & Rosenbaum (1978) and Farrell and his colleagues (Farrell & Wilson 1978; Farrell et al 1979) studied the polymerization of sonication-derived tubulin from outer doublets of both cilia and flagella and described in some detail the conditions for reassembly and the composition of the resulting microtubules. Tubulin from outer doublet microtubules solubilized in the French pressure cell has also been shown to reassemble in vitro (Pfeffer et al 1978). Linck & Langevin (1981) used thermal fractionation to solubilize the B subfiber specifically, and they described its reassembly properties.

The major conclusion from such studies is that the properties of outer doublet microtubules that distinguish them from single, cytoplasmic microtubules, such as their greater stability and characteristic structure, are not conferred by the tubulins alone. The assembly and disassembly properties of solubilized outer doublet tubulin were very similar to those of brain tubulin with regard to such parameters as the optimal ionic conditions, temperatures, and nucleoside triphosphate requirements for assembly. Podophyllotoxin was found to inhibit polymerization at stoichiometries similar to those seen for brain tubulin assembly (Farrell et al 1979). Solubilized doublet tubulin, like brain tubulin, assembled preferentially at the distal end of growing microtubules (Binder & Rosenbaum 1978): Most importantly, in all of the above studies the reassembled microtubules were singlets, not doublets.

Flagellar tubulin in these studies could be purified by several rounds of assembly and disassembly. Outer doublet–derived tubulin that was greater than 94% pure could be assembled into microtubules with a low critical tubulin concentration (less than 1 mg/ml). This finding indicates that if nontubulin microtubule associated proteins (MAPs) are necessary for the assembly of outer doublet tubulin, they must act at very low concentration.

Binder & Rosenbaum (1978) attempted to induce assembly of doublet microtubules in vitro by adding *Chlamydomonas* basal bodies or pieces of preformed outer doublet microtubules to the assembly reaction. The basal bodies and doublet pieces did act as nucleating centers for the polymerization reaction. Microtubule polymerization occurred on both A tubules and B subfibers, which indicates that both tubules in the doublet can serve as nucleating centers for the polymerization of the solubilized flagellar tubulin. Even with this addition to the experiment, however, only singlet microtubules were assembled.

The above experiments should serve as a basis for approaching the long term goals of assembling doublet microtubules in vitro and studying the regulation of their assembly. Linck & Langevin (1981), using tubulin solubilized primarily from the B subfiber, found that the helical lattice characteristic of B subfibers was formed by in vitro assembly. Therefore, at least part of the outer doublet—the B subfiber—may have been reconstituted in vitro. It is not known what factors are missing from the in vitro experiments that are necessary for the formation of complete doublets. Possibly the procedures used to solubilize the tubulin do not solubilize some accessory protein necessary for the formation of doublets. Another possibility is that not all of the different types of tubulin present in the doublet are solubilized or at least are not present after sonication in stoichiometries appropriate for doublet assembly. As discussed by Stephens (1978), there are a number of different forms of both α and β tubulin present in outer doublet microtubules. A newly described posttranslational modification of tubulin (below) may be involved in conferring the unique properties of doublet microtubules.

ACETYLATION OF FLAGELLAR α TUBULIN

Recently, pulse-labeling and in vitro translation experiments have shown that flagellar α tubulin is posttranslationally modified in a number of organisms including *Chlamydomonas* (Lefebvre et al 1980), *Polytomella* (McKeithan et al 1983), *Crithidia* (Russell & Gull 1984), and *Physarum* (Green & Dove 1984). In *Chlamydomonas*, the modification has been shown to be an acetylation of an ε-amino group of a lysine residue (L'Hernault & Rosenbaum 1985b). Although the modification in the other species has not been shown to be acetylation, in each case the change in isoelectric point of the protein after modification is consistent with acetylation. Monoclonal antibodies specific for the acetylated form of α tubulin have been used to show that the species distribution of acetylated tubulin is very wide and includes sea urchins, *Drosophila*, humans, and *Tetrahymena* (Piperno & Fuller 1985). The acetylating enzyme, tubulin acetyl-transferase, has been isolated from *Chlamydomonas* flagella and partially purified (Greer et al 1985). Several observations suggest that the acetylation is correlated to the assembled state of the flagella or cilia: (*a*) If flagellar regeneration is prevented by the addition of colchicine after deflagellation of *Chlamydomonas*, acetylation does not occur (Brunke et al 1982b; L'Hernault & Rosenbaum 1983). (*b*) If *Chlamydomonas* cells are induced to resorb their flagella by the addition of any of several chemical agents (see below), the α tubulin is deacetylated during the flagellar disassembly (L'Hernault & Rosenbaum 1985a). (c) Tubulin isolated from the membrane-matrix fraction of flagella (which should contain only unassembled tubulin) is not acetylated, which indicates that acetylation is coupled to flagellar tubulin assembly. (d) In *Polytomella*, a single-celled flagellate from which both cytoplasmic and flagellar microtubules can be isolated in large quantities, the cytoplasmic microtubules contain the unacetylated form of α tubulin and the flagellar microtubules contain predominantly the acetylated form (McKeithan et al 1983). (e) Monoclonal antibodies specific for acetylated α tubulin react with axonemal tubulin from all sources tested but do not react with soluble, cytoplasmic tubulin (Piperno & Fuller 1985). Clearly, acetylation is not strictly associated with doublet microtubules since central pair microtubules of axonemes and some nonaxonemal microtubules are acetylated. The acetylation is found largely in very stable microtubules. While the acetylation of α tubulin does not directly result in the assembly of doublet microtubules, it may potentiate their formation when other flagellar proteins and unidentified assembly conditions are present.

DISASSEMBLY OF CILIA AND FLAGELLA

The stability of flagellar doublet microtubules in vitro is reflected by their stability in vivo. In many organisms, cytoplasmic microtubules are easily depolymerized by cold, colchicine, or pressure, whereas flagellar or ciliary microtubules are comparatively stable to any of these treatments (see, for example, Behnke & Forer 1967). Both ciliary and flagellar microtubules can, however, be readily broken down in vivo by a number of organisms, and the flagellar proteins thus released can be reused to regrow new cilia or flagella. Flagellar resorption occurs in *Chlamydomonas* and many other organisms during the cell cycle as the flagella are resorbed before cell division (for a review of predivision flagellar resorption see Bloodgood 1974). Rosenbaum et al (1969) showed that if one of the two *Chlamydomonas* flagella is amputated, the other is rapidly resorbed with linear kinetics and the new flagellum is elongated. Flagellar proteins from the resorbed flagellum contributed to the regeneration of the new flagellum (Coyne & Rosenbaum 1970). When the resorbed flagellum and the regenerated flagellum reach the same length, both elongate to the original length. Such resorption of the flagella must, therefore, be a carefully regulated process that is coupled in some way to regeneration. In *Polytomella*, axonemes can be forced intact into the cytoplasm using hydrostatic pressure. These axonemes are slowly broken down during the growth of new flagella during recovery from the pressure treatment; measurements of regeneration in the absence of protein synthesis showed that the material from the disassembling axonemes is used in the regrowth of the new flagella. Surprisingly, if the pool size of flagellar proteins in *Polytomella* is depleted by a previous cycle of flagellar amputation and regrowth in cycloheximide, the axonemes forced in by pressure treatment break down much faster, as if the cell disassembles axonemes in response to a need to replenish pools of flagellar proteins or grow new flagella (Brown & Rogers 1978). In *Tetrahymena*, Rannestad (1974) found that if only some of the cilia were amputated, the remaining cilia were resorbed into the cell, and the protein subunits from the resorbed cilia were used to regenerate new cilia. Thus in *Tetrahymena*, *Chlamydomonas*, and *Polytomella*, the cell is able to disassemble one set of cilia or flagella at the same time that other flagella or cilia are elongating. If we assume that the conditions (e.g. Ca^{2+} concentrations) at the sites of assembly and disassembly of microtubules are different, cells clearly must be able to locally alter conditions in the cytoplasm to allow both processes to occur simultaneously.

A number of chemical treatments have been found to induce flagellar shortening in *Chlamydomonas*. If Ca^{2+} is removed from the medium and the monovalent cation concentration is increased (to 12.5 mM Na^+ or K^+ from the original 3 mM), flagella undergo shortening with linear kinetics and are completely resorbed into the cell after 120–150 min (Lefebvre et al 1978; Quader et al 1978). This resorption is immediately reversed by re-adding Ca^{2+} to the medium, even in the presence of elevated levels of Na^+ or K^+. Flagellar resorption is also induced by elevated Na^+ or K^+ concentrations of 50 mM or higher, even in the presence of Ca^{2+}, and this resorption is reversed when the excess monovalent cation is removed (Bloodgood et al 1979). Solter & Gibor (1978) reported that increasing the tonicity of the medium causes flagellar resorption in *Chlamydomonas*. If the osmolarity of the medium is increased threefold (from 81 to 240 mOsmol/kg) by addition of the sugars mannitol or sucrose, the flagella shorten. In fact, cells grown in medium supplemented with 20 mM NaCl (a 12% increase in medium tonicity) have flagella only one-half normal length (6 μm versus 12 μm). If the osmolarity of the medium is *lowered* by suspending cells in distilled water, flagella elongate as much as 30% within 3 hr (Telser 1977).

Flagellar resorption can also be experimentally induced with low concentrations of various drugs. The herbicide amiprophosmethyl (APM) (Collis & Weeks 1978), the methylxanthines caffeine (Hartfiel & Amrhein 1976) and isobutyl methylxanthine (IBMX) (Lefebvre et al 1980), and the anesthetic halothane (Telser 1977), all induce flagellar resorption. When IBMX is added to the medium, the flagella are resorbed to one-half length within two hr. None of these drugs induces complete resorption, as one-third to one-half length flagella persist on the cells even after several hours.

No common feature of these treatments has been suggested to explain their common ability to induce flagellar resorption.

SYNTHESIS OF FLAGELLAR AND CILIARY PROTEINS

A major focus of work on flagellar regeneration has been the synthesis of tubulin and other ciliary and flagellar proteins. One major impetus for the study of regeneration comes from the realization that this process provides an opportunity to experimentally stimulate the synthesis of ciliary or flagellar proteins and thus to study the induced expression of a particular set of genes.

Requirements for New Protein and RNA Synthesis in Regeneration

Inhibitors of protein synthesis and RNA synthesis have been used to establish whether or not new synthesis is needed to regenerate flagella or cilia in several systems. Inhibitor studies showed that new protein synthesis is needed to regenerate full length flagella in *Euglena* (Rosenbaum & Child 1967), *Chlamydomonas* (Rosenbaum et al 1969), and *Polytomella* (Brown & Rogers 1978), as well as the hundreds of cilia of *Tetrahymena* (Child 1965). Cilia regeneration in *Tetrahymena* was shown to be sensitive to inhibition of RNA synthesis by actinomycin D (Child 1965; Guttman & Gorovsky 1979).

In sea urchin embryos, experiments using two different protein synthesis inhibitors gave conflicting results on whether or not new synthesis of ciliary proteins is necessary to regenerate a new set of cilia after deciliation. In the presence of puromycin, which inhibits protein synthesis by 90%, sea urchin embryos regenerated cilia up to five times (Auclair & Siegel 1966; Burns 1973). Pactamycin, however, which inhibits 95% of protein synthesis, completely inhibited regeneration (Child & Aptner 1969). A possible resolution to this discrepancy comes from the work of Stephens (1977), who found by in vivo labeling that large pools of most ciliary proteins, in amounts sufficient for several regenerations, exist in sea urchin embryos, but after deciliation de novo synthesis of five or six minor proteins is needed to allow regeneration to occur (see below). Perhaps the small residual synthesis of proteins in puromycin allows production of sufficient amounts of the hypothetical "limiting proteins" to enable regeneration to occur. New RNA synthesis is apparently not necessary to regenerate a full complement of cilia in sea urchin embryos after deciliation. Actinomycin D has no effect on the rate or final extent of cilia regeneration in late gastrula of the sea urchin *Paracentrotus* (Auclair & Siegel 1966).

The dependence of flagellar regeneration on new mRNA production was shown most convincingly by Tamm (1969), who enucleated cells at different stages of the cell cycle and then measured the flagellar regeneration kinetics after deflagellation. Even without a nucleus, *Peranema* deflagellated early in the cell cycle could regenerate one-half length flagella. If the cells were enucleated and deflagellated late in the cell cycle, full-length flagella were regenerated with normal kinetics. Tamm concluded from these experiments that new transcription is not needed for regeneration late in the cell cycle, but that some nuclear function is required early in the cell cycle to allow formation of full length flagella.

Direct Measurement of Ciliary and Flagellar Protein Synthesis

The above experiments with inhibitors of transcription and translation indicate that in a number of systems, synthesis of RNA and proteins is needed to assemble new cilia or flagella. In some of these same systems the synthesis of individual ciliary or flagellar proteins has been directly assayed.

SEA URCHINS   Stephens (1977) used [^{14}C]leucine labeling to study the synthesis of many ciliary proteins after deciliation of sea urchin (*Strongylocentrotus*) blastulae. Embryos were deciliated and were labeled during regeneration of the cilia. After one regeneration in label, the embryos were deciliated again and chased with unlabeled leucine. Changes in the specific activity of individual ciliary proteins during the chase were monitored to estimate the pool size and rate of synthesis of the ciliary proteins. These experiments indicate that the tubulins, dyneins, and many other ciliary proteins are present in the sea urchin embryo in amounts sufficient for multiple regenerations in the absence of new protein synthesis. These results confirm the observation of Raff et al (1971) that a large pool of tubulin exists in sea urchin embryos.

Stephens (1977) found that the tubulins in the A tubule and the B subfiber of the outer doublets behave as if they are derived from different compartments. In the in vivo labeling experiment described above, the specific activity of the tubulins in the B subfiber decreased 15% during the regeneration under chase conditions, but the specific activity of tubulins in the A tubule actually increased 30–35% during the chase. The difference in specific activity between the tubulins from the A tubule and the B subfiber is consistent with the possibility that these tubulins are separate gene products. This possibility received strong biochemical support from the discovery of Stephens (1978) that these different tubulins have distinctly different tryptic peptide maps. There are enough α and β tubulin genes in sea urchin (as many as 20 each; Cleveland et al 1980; Alexandraki & Ruderman 1981) to produce many different tubulin gene products.

Certain proteins are made in small amounts in sea urchin embryos after deciliation and are not maintained in a pool. By the experimental protocol described above, at least five minor ciliary proteins showed greatly reduced specific activity after the chase. The specific activity of these proteins was at least 85% less after the chase than after the regeneration in the presence of label. Thus, for each of these proteins, almost the total amount synthesized during the regeneration in the presence of label was assembled into the cilia, such that during the second regeneration these proteins were newly made and unlabeled. One of these proteins comigrated with "nexin," which connects the outer doublets in the axoneme (Stephens 1970). Another, called component 20, is a major nontubulin protein found in the stable protofilament structure that remains after extensive solubilization of outer doublet microtubules (Linck 1973).

A synthesis control model is suggested by these results in which large pools of assembly-competent ciliary proteins, such as the tubulins and dyneins, exist in the embryo, but ciliary assembly, either after deciliation or during development, is controlled by the synthesis of a limiting amount of certain proteins (Stephens 1977). Only enough of these limiting proteins would be synthesized to assemble one set of normal length cilia. The pools of the other, nonlimiting ciliary proteins would slowly be replenished and maintained, but deciliation would not stimulate the synthesis of these proteins. If the availability of a certain protein in the embryo limits the amount of tubulin that can be assembled into doublet microtubules, the size of the pool of this protein would determine the total amount (and therefore the length) of cilia that could be assembled.

Although deciliation did not induce the synthesis of tubulin in sea urchin embryos, Merlino et al (1978) were able to induce a detectable increase in tubulin synthesis by three successive deciliations. The increase, two- to threefold for both α and β tubulin, was detected by isolation and in vitro translation of poly(A) RNA. The observation that this increase was inhibited by actinomycin D indicates that it may have involved increased transcription of the tubulin genes.

TETRAHYMENA   Another useful model system for studying the stimulation of ciliary protein synthesis after deciliation is *Tetrahymena*, because it can synchronously and rapidly regenerate 600–700 cilia per cell after deciliation (Child 1965; Rosenbaum & Carlson 1969). Guttman & Gorovsky (1979) found, using pulse-labeling experiments, that starved, nondeciliated *Tetrahymena pyriformis* synthesized essentially no tubulin but that between 80 and 140 min after deciliation tubulin synthesis increased dramatically, accounting for 7–8% of the cell's total protein synthesis. In contrast, Nelsen (1975) found little or no stimulation of tubulin synthesis following deciliation of *Tetrahymena*. The latter experiments, however, used cells subjected to less stringent starvation conditions than those used by Guttman & Gorovsky (1979). The different findings of the two groups could therefore reflect different metabolic states of the cells. At least some of the increased tubulin synthesis in starved *Tetrahymena* may be due to new mRNA production; Marcaud & Hayes (1979) showed that deciliation led to an increased level of translatable mRNA for a protein comigrating with tubulin on one-dimensional gels.

Studies of the timing of synthesis of tubulin after deciliation show that *Tetrahymena* must contain a pool of tubulin available for assembly in the absence of protein synthesis (Guttman & Gorovsky 1979). Within 60 min after deciliation, 60–80% of cells regain motility, which indicates at least partial cilia regeneration. By 60 min, however, no tubulin synthesis can be detected in *Tetrahymena*. Thus some tubulin must be present in unassembled form before deciliation to allow regeneration of cilia before the onset of tubulin synthesis. This conclusion supports those derived from experiments involving deciliation in the presence of protein synthesis inhibitors (Rannestad 1974).

The partitioning of newly synthesized tubulin between the cytoplasmic pool and the regenerating cilia was shown by labeling cells during cilia regeneration after deciliation and then examining the distribution of labeled tubulin between the cilia and the cytoplasm. Of the newly synthesized tubulin, 45% was found in the cilia and the remaining 55% was located in the cell bodies, presumably refilling the pool of unassembled ciliary tubulin. In *Tetrahymena thermophila*, however, Calzone et al (1979) found that eventually all of the newly synthesized tubulin made after deciliation was incorporated into the growing cilia. In starved *T. thermophila*, the rate of tubulin synthesis was shown to be 0.03 pg/hr. After deciliation, this rate of synthesis increased 160-fold, to 4.9 pg/hr, which was nearly identical to the rate of accumulation of tubulin during cilia regeneration.

CHLAMYDOMONAS   Amputating *Chlamydomonas* flagella stimulates a rapid and coordinate synthesis of flagellar proteins (Weeks & Collis 1976; Lefebvre et al 1978). This induction of synthesis is particularly striking in gametic cells, a differentiated cell type produced after nitrogen starvation. Gametes have greatly reduced numbers of ribosomes and a low basal level of protein synthesis (Kates & Jones 1964). Because the level of protein synthesis in gametes is so low, it was possible to label and distinguish in vivo the synthesis of many flagellar proteins. Only low levels of synthesis of flagellar proteins were detectable in nondeflagellated cells, but within the first 30 min after deflagellation the synthesis of the α and β tubulins, the dyneins, the major flagellar membrane protein, and at least 20 other flagellar proteins was greatly stimulated (Lefebvre et al 1978). Flagellar proteins were the major proteins synthesized in gametes after deflagellation; they accounted for 80% of total protein synthesis between 30 and 60 min after deflagellation. By comparison, flagellar proteins accounted for less than 10% of the total protein synthesis in nondeflagellated cells. Synthesis rates for the major flagellar proteins remained maximal for 60 to 90 min, then returned to predeflagellation levels by 120–240 min, depending on the protein examined.

Remillard & Witman (1982) extended these studies by measuring both the rate of synthesis of individual flagellar proteins after deflagellation and the rate of their transport into the flagella. Using pulse-labeling and two-dimensional gel analysis, they concentrated on proteins identified as residing in particular flagellar structures, such as the central tubule complex and the radial spokes. An important conclusion from these studies was that the rates of synthesis of proteins contained in the same flagellar structure were similar and could differ from the kinetics of synthesis of proteins in different structures. An example of this difference is shown in Figure 4. In this figure, the rate of incorporation of ^{35}S into proteins of the radial spoke complex (proteins R1, R2, R3, and R4) is compared to the rate of incorporation into α and β tubulin and into a flagellar membrane protein (m). The synthesis of radial spoke proteins increased rapidly after deflagellation, peaked by 15 min, and returned to undetectable levels after 30 min. Synthesis of α and β tubulin was stimulated within 15 min postdeflagellation, but did not peak until 50 min, and still occurred at elevated levels 2 hr after deflagellation. A second conclusion from these studies was that the rate of transport of individual proteins does not necessarily parallel the rate of their synthesis. For example, transport of α and β tubulin into regenerating flagella peaked 15–30 min post-deflagellation, although as discussed above, synthesis peaked at 50 min. The rate of transport of the tubulins closely paralleled the rate of elongation of the regenerating flagella. At the opposite extreme, one of the central tubule proteins and the actinlike flagellar protein (Piperno & Luck 1979) were not detectably labeled in regenerating flagella, even though deflagellation stimulated synthesis of these proteins. This result suggests that a very large pool of these proteins exists in the cell, which was confirmed by isotope dilution experiments (Remillard & Witman 1982).

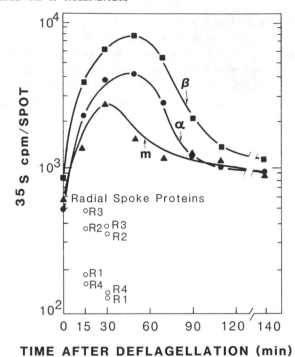

Figure 4   Label incorporation into various flagellar proteins at different times after deflagellation of *Chlamydomonas. R1–R4* are radial spoke proteins, *m* is a flagellar membrane protein, α and β are α and β tubulins. (Reprinted from Remillard & Witman 1982 with permission.)

The increase in synthesis of the major flagellar proteins after deflagellation is accompanied by increases in the amounts of translatable mRNAs for these proteins. In vitro translation of polysomes isolated from control and deflagellated cells demonstrated that deflagellation stimulated the loading of mRNAs onto polysomes (Weeks & Collis 1976). Translation of total RNA yielded the same results, which indicates that the tubulin mRNA was not present in nondeflagellated cells in a stored, nontranslatable form (Lefebvre et al 1980). Increased mRNA levels for at least 25 proteins were seen after deflagellation using poly(A) RNA, and most of these comigrated on two-dimensional gels with proteins from purified flagella. Thus deflagellation stimulated the accumulation of mRNAs not only for the tubulins, but for many of the flagellar proteins.

Cloning of tubulin and other flagellar sequences   Cloned probes for the tubulins, and more recently a number of other flagellar proteins, have been isolated and used to directly measure the level of transcripts from the individual genes for flagellar proteins. Silflow & Rosenbaum (1981) and

Minami et al (1981) isolated cDNA clones for both α and β tubulin from *Chlamydomonas*, using as starting material poly(A) RNA isolated from cells after deflagellation. Hybridization of the α and β tubulin cDNA probes with total genomic DNA digested with restriction enzymes showed that there are two different genes for α tubulin and two for β tubulin in the *Chlamydomonas* genome (Silflow & Rosenbaum 1981; Brunke et al 1982a).

Analysis of cloned genomic DNA fragments confirmed that the tubulin gene family in *Chlamydomonas* contains four members and showed that the expression of each of the genes is induced coordinately following deflagellation (Brunke et al 1982a; Youngblom et al 1984; Silflow et al 1985). The DNA sequence was determined for each of the four genes, providing a complete picture of tubulin gene structure and heterogeneity in *Chlamydomonas*. The two β tubulin genes code for exactly the same protein (Youngblom et al 1984), while the two predicted α tubulin proteins differ by only two amino acids (Silflow et al 1985). These findings indicate that all of the microtubules in *Chlamydomonas* are composed of only three different primary gene products, two α tubulins and one β tubulin. A similar parsimony in tubulin gene number has been observed in *Tetrahymena*, in which only one α tubulin gene has been found (Callahan et al 1984), even though multiple isoforms of the α tubulin protein are present in the cell (Suprenant et al 1985).

Small regions of homology between the various tubulin genes in *Chlamydomonas* could be candidates for sequences involved in the coordinate regulation of expression of these genes. For example, a GC-rich region to the 3' side of the TATA box and a 16-base-pair consensus sequence located to the 5' side of the TATA box could play a role in the coordinated transcription of the four genes (Brunke et al 1984). There is also a striking conservation of sequence between the third introns of the two β tubulin genes ($>90\%$) and conservation of small portions of the second intron of the α tubulin genes (Youngblom et al 1984; Silflow et al 1985).

Recently, cloned probes for other flagellar proteins were isolated. Schloss et al (1984) prepared a cDNA library from RNA isolated from cells after deflagellation. By differential hybridization with RNA from deflagellated and nondeflagellated cells, clones were isolated containing sequences whose expression increased after deflagellation. Fourteen different non-tubulin cDNA clones were shown by restriction enzyme analysis, RNA blot analysis, and Southern blot hybridization to genomic DNA to be the products of different genes. These clones represent genes whose expression increases after deflagellation; these genes should include those coding for flagellar proteins, as well as nonflagellar "stress-induced" sequences whose expression increases after deflagellation. Such sequences were described by Guttman et al (1980) in *Tetrahymena* and by May & Rosenbaum (1980) in *Chlamydomonas*.

Cloned probes for specific flagellar proteins were isolated from *Chamydomonas* by preparing antibodies against individual, purified flagellar proteins and then screening an expression library to isolate genomic fragments corresponding to the original antigen (Williams et al 1986). Using this method, genes for two of the radial spoke proteins and two of the dynein proteins were isolated. One of the radial spoke proteins corresponds to the *pf 26* locus (Huang et al 1981), as shown by in vitro translation of RNA isolated by hybrid selection from *pf 26* cell RNA.

RNA transcript levels for different flagellar protein genes   Using the cloned probes described above, the expression of flagellar protein genes before and after deflagellation was compared. The major conclusion from these studies is that although there are some slight differences between genes, the kinetics of accumulation of the transcripts of different flagellar protein genes is coordinate. Within the first 15 min after deflagellation, transcript levels for each of the genes increased at least twofold, and as much as eightfold, above predeflagellation levels. In fact, transcript accumulation for the radial spoke and dynein genes above predeflagellation levels was seen within 5 min after flagellar amputation. Although the patterns of RNA accumulation were very similar for various flagellar proteins, slight but reproducible differences were observed in the rate and the extent of accumulation of transcripts for different genes for flagellar proteins after deflagellation (Schloss 1984). As discussed below, both increased transcription and altered mRNA stability affect the level of transcripts for the flagellar protein genes. It is not yet known which mechanisms account for the subtle differences seen in the postdeflagellation accumulation of the different transcripts.

At what level is the synthesis of flagellar proteins controlled?   Much of the increase in tubulin transcript levels is due to increased transcription of the genes. Keller et al (1984) measured the rate of synthesis of tubulin RNA from isolated nuclei in vitro (run-off transcription experiments). These experiments, which measure the completion in vitro of nascent RNA chains, suggested that the rate of transcription of the α and β tubulin genes increased fivefold within 20 min after deflagellation. Experiments in which tubulin RNA synthesis was measured in vivo by pulse labeling (Baker et al 1984) showed that the labeling of α and β tubulin transcripts increased sixfold within 10 min after deflagellation, peaked by 15 to 20 min, and returned to predeflagellation levels within 90 to 100 min. Both experiments indicate that deflagellation induces a rapid and transient increase in the transcription of the α and β tubulin genes.

Comparison of the rate of tubulin RNA accumulation with the rate of tubulin gene transcription suggests that a specific stabilization of tubulin transcripts occurs after deflagellation. By comparing the rate of tubulin RNA synthesis with the rate of accumulation of the RNA in the same experiment, Baker et al (1984) concluded that there was approximately a twofold increase in the stability of tubulin RNA transcripts at the time of peak accumulation. For technical reasons, it was not possible to measure directly mRNA stability during this period, so this is at best an approximation. However, specific stabilization of tubulin mRNA would explain its accumulation at levels above those predicted by its rate of synthesis.

Destabilization of mRNAs for tubulin and other flagellar proteins   By 60 to 90 min after deflagellation, tubulin mRNA levels and flagellar protein synthesis drop rapidly back toward predeflagellation levels. Baker et al (1984) showed that during this period tubulin mRNA is destabilized: The half life for β tubulin is <23 min and that of α tubulin is <27 min. (The predeflagellation half lives are 39–47 min for β tubulin and 48–65 min for α tubulin.) The possibility that this destabilization is an active process that requires synthesis of new proteins is raised by experiments using cycloheximide to block protein synthesis. When cells were deflagellated in cycloheximide and the levels of tubulin mRNA were measured at various times after deflagellation, it was found that tubulin mRNA levels increased normally after deflagellation in the absence of protein synthesis; however, cycloheximide prevented the mRNA levels from decreasing at later times. For example, by 3 hr postdeflagellation, when control cell tubulin mRNA levels had returned to predeflagellation levels, cycloheximide-treated cells had levels equal to the maximum accumulation observed after deflagellation (Baker et al 1986).

One candidate for the protein(s) whose synthesis is needed for the rapid decay of flagellar mRNAs is tubulin itself. In other systems, high levels of unpolymerized tubulin have been shown to specifically inhibit the synthesis of additional tubulin (Ben-Ze'ev et al 1979; Cleveland et al 1981), and this autofeedback regulation appears to operate at a posttranscriptional level (e.g. Cleveland & Havercroft 1983). Possibly the rapid breakdown of flagellar mRNAs seen after deflagellation is caused by the accumulation of tubulin during flagellar regeneration, and cycloheximide prevents this breakdown by preventing the synthesis of tubulin.

Specific inhibition of flagellar protein synthesis   A number of agents that cause *Chlamydomonas* flagella to be resorbed into the cell have been shown to specifically inhibit the synthesis of flagellar proteins. These include amiprophosmethyl (APM) (Collis & Weeks 1978), isobutyl methyl-xanthine (IBMX), and sodium pyrophosphate (PPi) (Lefebvre et al 1980). Although the mechanism by which these inhibitors block synthesis is not known, it does involve rapid decreases in the mRNA levels for tubulin and many other flagellar proteins (Minami et al 1981; Silflow et al 1981). Possibly the specific decrease in flagellar protein mRNA levels during resorption occurs by the same mechanism as the destabilization of flagellar mRNAs seen after deflagellation. The destabilization in this model would be caused by an accumulation of tubulin or other flagellar protein(s) in the cytoplasm.

It should be noted that APM, IBMX, and PPi, each of which completely inhibits flagellar regeneration, all inhibit flagellar protein synthesis after deflagellation, as well as during resorption. Cells treated with these agents respond to deflagellation by inducing flagellar protein synthesis, but the induction is short lived. The feedback regulation model predicts that these agents inhibit flagellar protein synthesis after deflagellation by preventing the regrowth of flagella, causing newly synthesized tubulin and other flagellar proteins to accumulate rapidly in the cytoplasm and destabilizing mRNAs for flagellar proteins. As discussed below, *Chlamydomonas* does turn on the synthesis of tubulin and other flagellar proteins after deflagellation in the presence of colchicine. This agent prevents flagellar regeneration by binding directly to tubulin. In colchicine, however, RNA transcript levels for the flagellar protein genes are reduced to as little as 30% of the levels seen after deflagellation of untreated cells (J. Schloss, unpublished observations).

HOW DOES AMPUTATION OF FLAGELLA STIMULATE SYNTHESIS OF FLAGELLAR PROTEINS?

As described above, the response of several organisms to amputation of their cilia or flagella is induction of synthesis of those proteins needed to replace the organelles. Although the signal that triggers this synthesis is unknown, a number of experiments provide some clues as to possible signal mechanisms.

One attractive model for the signal is that cells recognize the removal of flagellar proteins from an unassembled pool during regeneration and respond to decreased levels of some component in the pool by initiating new synthesis of flagellar proteins. Thus the trigger for stimulating new synthesis would be the regrowth of cilia or flagella. However, if regrowth of *Chlamydomonas* flagella is prevented by treatment with colchicine, stimulation of flagellar protein synthesis still occurs after deflagellation (Weeks et al 1977; Lefebvre et al 1978), although the synthesis is reduced relative to untreated cells, as discussed above. Furthermore, some mutants of *Chlamydomonas* that cannot regenerate their flagella after amputation are still able to induce flagellar protein synthesis, which indicates again that flagellar regeneration and the accompanying pool depletion are not involved in the induction signal (Lefebvre et al 1984). In sea urchin blastulae, however, repeated deciliation and regeneration cycles are needed to induce the synthesis of the tubulins (Merlino et al 1978). This finding indicates that in this system the depletion of the ciliary protein pool during regeneration may be coupled to the induction of ciliary protein synthesis.

The possibility exists that some specific damage associated with amputation triggers the synthesis of ciliary or flagellar proteins. In *Chlamydomonas*, however, it has been shown that new synthesis of flagellar proteins can be experimentally induced by simply shortening the flagella using IBMX or other resorption-inducing agents. As discussed earlier, flagellar protein synthesis is specifically decreased during resorption, but after the resorption-inducing agent is removed from the medium, synthesis of flagellar proteins is stimulated to greater than preresorption levels. This effect is seen even if flagella are shortened by only 25% (Lefebvre et al 1980). The "turn on" signal in this experiment could reflect the ability of the cells to monitor the length of their flagella and respond to either flagellar shortening (by resorption) or removal (by amputation) by inducing new synthesis of flagellar proteins. Alternatively, the induction of flagellar protein synthesis after resorption could reflect some undefined "rebound" response to the specific inhibition of flagellar protein synthesis during resorption.

Stimulation of flagellar protein synthesis after deflagellation may be mediated by some signal protein whose synthesis is induced early after deflagellation. This hypothetical protein would then stimulate the synthesis of the other flagellar proteins. In *Tetrahymena*, for example, deciliation induces the synthesis of a specific nonciliary protein rapidly, followed somewhat later by increased synthesis of the tubulins (Guttman & Gorovsky 1979). However, this protein has also been shown to be synthesized in response to heat shock, even though heat shock does not stimulate synthesis of ciliary proteins (Guttman et al 1980). Similarly, in *Chlamydomonas* deflagellation has been shown to induce the synthesis of a subset of the heat shock proteins (May & Rosenbaum 1980). Experiments with cycloheximide in *Chlamydomonas* have shown that protein synthesis is not needed to trigger the rapid accumulation of mRNAs for flagellar proteins induced by deflagellation, as discussed above (Baker et al 1986). Thus whatever trigger stimulates flagellar mRNA accumulation after deflagellation does not require new protein synthesis.

FUTURE PROSPECTS

It has been possible to isolate mutants of *Chlamydomonas* that cannot stimulate the synthesis of flagellar proteins after deflagellation (Gealt & Weeks 1980; Lefebvre et al 1984). These mutants have flagella but cannot regenerate them in response to amputation. The genetic analysis of this useful class of mutants should help define the nature and complexity of the mechanism that regulates flagellar protein synthesis.

The tools needed for a detailed molecular description of flagellar gene expression are becoming available. Sequence analysis of the cloned genes for the tubulins and other flagellar proteins should suggest common features of the genes that are the target sites for the mechanisms that regulate expression. Testing the role of such sequences directly, however, will require the ability to transform one or more of the systems discussed in this article with cloned genes, a technique not yet available in any of these systems. Low efficiency transformation of *Chlamydomonas* has been reported (Rochaix & van Dillewijn 1982), and efforts to improve the efficiency of the procedure are underway.

ACKNOWLEDGMENTS

We would like to thank Dr. Carolyn Silflow for helpful discussions and editorial comments, and Gail Hall for skillful preparation of the manuscript.

Literature Cited

Alexandraki, D., Ruderman, J. 1981. Sequence heterogeneity, multiplicity, and genome organization of α and β tubulin genes in sea urchins. *Mol. Cell. Biol.* 1: 1125–37

Allen, C., Borisy, G. 1974. Structural polarity and directional growth of microtubules of *Chlamydomonas* flagella. *J. Mol. Biol.* 90: 381–402

Auclair, W., Siegel, B. W. 1966. Cilia regeneration in the sea urchin embryo: Evidence for a pool of ciliary proteins. *Science* 154: 913–15

Baker, E. J., Keller, L. R., Schloss, J. A., Rosenbaum, J. L. 1986. Protein synthesis is required for rapid degradation of tubulin mRNA and other deflagellation-induced RNAs in *Chlamydomonas rein-*

hardi. Mol. Cell. Biol. 6: 54–61

Baker, E. J., Schloss, J. A., Rosenbaum, J. L. 1984. Rapid changes in tubulin RNA synthesis and stability induced by deflagellation in *Chlamydomonas. J. Cell Biol.* 99: 2074–81

Behnke, O., Forer, A. 1967. Evidence for four classes of microtubules in individual cells. *J. Cell Sci.* 2: 169–92

Ben-Ze'ev, A., Farmer, S., Penman, S. 1979. Mechanisms of regulating tubulin synthesis in cultured mammalian cells. *Cell* 17: 319–25

Binder, L., Dentler, W., Rosenbaum, J. 1975. Assembly of chick brain tubulin onto flagellar axonemes of *Chlamydomonas* and sea urchin sperm. *Proc. Natl. Acad. Sci. USA* 72: 1122–26

Binder, L., Rosenbaum, J. 1978. The in vitro assembly of flagellar outer doublet tubulin. *J. Cell Biol.* 79: 500–15

Bloodgood, R. A. 1974. Resorption of organelles containing microtubules. *Cytobios* 9: 143–61

Bloodgood, R. A., Leffler, E. M., Bojczuk, A. T. 1979. Reversible inhibition of *Chlamydomonas* flagellar surface motility. *J. Cell Biol.* 82: 664–74

Bouck, G. B. 1971. The structure, origin, isolation, and composition of the tubular mastigonemes of the *Ochromonas* flagellum. *J. Cell Biol.* 50: 362–84

Brown, D. L., Rogers, K. A. 1978. Hydrostatic pressure-induced internalization of flagellar axonemes, disassembly, and reutilization during flagellar regeneration in *Polytomella. Exp. Cell Res.* 117: 313–24

Brunke, K. J., Anthony, J. G., Sternberg, E. J., Weeks, D. P. 1984. Repeated consensus sequence and pseudopromoters in the four coordinately regulated tubulin genes of *Chlamydomonas reinhardi. Mol. Cell. Biol.* 4: 1115–24

Brunke, K. J., Collis, P. S., Weeks, D. P. 1982b. Post-translational modification of tubulin dependent on organelle assembly. *Nature* 297: 516–19

Brunke, K. J., Young, E. E., Buchbinder, B. U., Weeks, D. P. 1982a. Coordinate regulation of the four tubulin genes of *Chlamydomonas reinhardi. Nucleic Acids Res.* 10: 1295–1310

Burns, R. G. 1973. Kinetics of the regeneration of sea urchin cilia. *J. Cell Sci.* 13: 55–67

Burns, R. G. 1979. Kinetics of the regeneration of sea urchin cilia. II. Regeneration of animalized cilia. *J. Cell Sci.* 37: 205–15

Callahan, R. C., Shalke, G., Gorovsky, M. A. 1984. Developmental rearrangements associated with a single type of expressed α tubulin gene in *Tetrahymena. Cell* 36: 441–45

Calzone, F., Allis, C., Angerer, R., Gorovsky, M. 1979. Regulation of protein synthesis during cilia regeneration in starved *Tetrahymena. J. Cell Biol.* 83: 405a

Chen, Y. T. 1950. Investigations of the biology of *Peranema trichophorum. Q. J. Microbiol. Sci.* 91: 279–308

Child, F. M. 1959. The characterization of the cilia of *Tetrahymena pyriformis. Exp. Cell Res.* 18: 258–67

Child, F. M. 1963. Flagellar regeneration: A mechanism accounting for its initiation and regulation. *Biol. Bull.* 125: 361

Child, F. M. 1965. Mechanisms controlling the regeneration of cilia in *Tetrahymena. J. Cell Biol.* 27: 18a

Child, F. M. 1978. The elongation of cilia and flagella: A model involving antagonistic growth zones. In *Cell Reproduction: Daniel Mazia Dedicatory Volume (ICN-UCLA Symp. Mol. Cell. Biol.)*, ed. E. R. Dirkson, D. Prescott, C. Fox, 12: 337–50. New York: Academic

Child, F. M., Aptner, M. N. 1969. Experimental inhibition of ciliogenesis and ciliary regeneration in *Arbacia* embryos. *Biol. Bull.* 137: 394–95

Cleveland, D. W., Havercroft, J. C. 1983. Is apparent autoregulatory control of tubulin synthesis non-transcriptionally regulated? *J. Cell Biol.* 97: 919–24

Cleveland, D. W., Lopata, M. A., MacDonald, R. J., Cowan, N. J., Rutter, W. J., Kirschner, M. W. 1980. Number and evolutionary conservation of α and β tubulin and cytoplasmic β and γ actin genes using specific cloned cDNA probes. *Cell* 20: 95–105

Cleveland, D. W., Lopata, M. A., Sherline, P., Kirschner, M. W. 1981. Unpolymerized tubulin modulates the level of tubulin mRNAs. *Cell* 25: 537–46

Cleveland, D. W., Sullivan, K. F. 1985. Molecular biology and genetics of tubulin. *Ann. Rev. Biochem.* 54: 331–66

Coggins, S., Kochert, G. 1980. Flagellar growth and regeneration in the life cycle of *Volvox carteri. J. Cell Biol.* 80: 39a

Collis, P. S., Weeks, D. P. 1978. Selective inhibition of tubulin synthesis by amiprophosmethyl during flagellar regeneration in *Chlamydomonas reinhardi. Science* 202: 440–42

Coyne, B., Rosenbaum, J. L. 1970. Flagellar elongation and shortening in *Chlamydomonas*: II. Re-utilization of flagellar proteins. *J. Cell Biol.* 47: 777–81

Dentler, W., Rosenbaum, J. L. 1977. Flagellar elongation and shortening in *Chlamydomonas*: III. Structures attached to the tips of flagellar microtubules and their relationship to the directionality of flagellar microtubule assembly. *J. Cell Biol.* 74: 747–59

Dustin, P. 1978. *Microtubules*. Berlin: Springer-Verlag

Euteneuer, U., McIntosh, J. R. 1981. Polarity of some motility related microtubules. *Proc. Natl. Acad. Sci. USA* 78: 372–76

Farrell, K. W. 1976. Flagellar regeneration in *Chlamydomonas reinhardtii*: Evidence that cycloheximide pulses induce a delay in morphogenesis. *J. Cell Sci.* 20: 639–54

Farrell, K. W., Morse, A., Wilson, L. 1979. Characterization of the in vitro reassembly of tubulin derived from stable *Strongylocentrotus purpuratus* outer doublet microtubules. *Biochemistry* 18: 905–11

Farrell, K. W., Wilson, L. 1978. Microtubule reassembly in vitro of *Strongylocentrotus purpuratus* sperm tail outer doublet tubulin. *J. Mol. Biol.* 121: 393–410

Fulton, C. 1977. Cell differentiation in *Naegleria gruberi*. *Ann. Rev. Microbiol.* 31: 597–629

Fulton, C., Walsh, C. 1980. Cell differentiation and flagellar elongation in *Naegleria gruberi*: Dependence on transcription and translation. *J. Cell Biol.* 85: 346–60

Gealt, M. A., Weeks, D. P. 1980. Tubulin synthesis in a temperature-sensitive mutant of *Chlamydomonas reinhardtii*. *Exp. Cell Res.* 127: 329–39

Gibbons, I. 1981. Cilia and flagella of eukaryotes. *J. Cell Biol.* 91: 107s–24s

Grebecki, A., Kuznicki, L. 1959. Autoprotection in *Paramecium caudatum* by influencing the biochemical properties of its medium. *Acta Biol. Exp.* 17: 71–107

Green, L. L., Dove, W. F. 1984. Tubulin proteins and RNA during the myxamoeba-flagellate transformation of *Physarum polycephalum*. *Mol. Cell. Biol.* 4: 1706–11

Greer, K., Moruta, H., L'Hernault, S. W., Rosenbaum, J. L. 1985. α-tubulin acetylase activity in isolated *Chlamydomonas* flagella. *J. Cell Biol.* 101: 2081–84

Guttman, S. D., Glover, C. V., Allis, C. D., Gorovsky, M. A. 1980. Heat shock, deciliation, and release from anoxia induce the synthesis of the same set of polypeptides in starved *Tetrahymena pyriformis*. *Cell* 22: 299–307

Guttman, S. D., Gorovsky, M. A. 1979. Cilia regeneration in starved *Tetrahymena*: An inducible system for studying gene expression and organelle biogenesis. *Cell* 17: 307–17

Hartfiel, G., Amrhein, N. 1976. The action of methylxanthines on motility and growth of *Chlamydomonas reinhardtii* and other flagellated algae. Is cyclic AMP involved? *Biochem. Physiol. Pflanzen* 169: 531–56

Hopkins, J. M. 1970. Subsidiary components of the flagella of *Chlamydomonas reinhardtii*. *J. Cell Sci.* 7: 823–39

Huang, B. 1986. *Chlamydomonas reinhardtii*, a model system for the genetic analysis of flagellar structure and motility. *Int. Rev. Cytol.* 99: 181–215

Huang, B., Piperno, G., Ramanis, Z., Luck, D. J. L. 1981. Radial spokes of *Chlamydomonas* flagella: Genetic analysis of assembly and function. *J. Cell Biol.* 88: 80–88

Jarvik, J., Lefebvre, P. A., Rosenbaum, J. L.

1976. A cold-sensitive mutant of *Chlamydomonas reinhardi* with aberrant control of flagellar length. *J. Cell Biol.* 70: 149a

Jarvik, J., Reinhart, F., Adler, S. 1980. Length control in the *Chlamydomonas* flagellum. *J. Cell Biol.* 87: 38a

Jarvik, J., Reinhart, F., Kuchka, M., Adler, S. 1984. Altered flagellar size control in *shf-1* short-flagella mutants of *Chlamydomonas reinhardtii*. *J. Protozool.* 31: 199–204

Kates, J. R., Jones, R. J. 1964. The control of gametic differentiation in liquid cultures of *Chlamydomonas*. *J. Cell. Comp. Physiol.* 63: 157–64

Keller, L. R., Schloss, J. A., Silflow, C. D., Rosenbaum, J. R. 1984. Transcription of α- and β-tubulin genes in vitro in isolated *Chlamydomonas reinhardi* nuclei. *J. Cell Biol.* 98: 1138–43

Kuchka, M. R., Jarvik, J. W. 1982. Analysis of flagellar size control using a mutant of *Chlamydomonas reinhardtii* with a variable number of flagella. *J. Cell Biol.* 92: 170–75

Kuriyama, R. 1976. *In vitro* polymerization of flagellar and ciliary outer fiber tubulin into microtubules. *J. Biochem.* 80: 153–65

Lallier, R. 1975. Animalization and vegetalization. In *The Sea Urchin Embryo—Biochemistry and Morphogenesis*, ed. G. Czihak, pp. 473–509. New York: Springer-Verlag

Lefebvre, P. A., Barsel, S.-E., Stuckey, M., Swartz, L. 1984. Isolation and characterization of mutants of *Chlamydomonas* defective in the regulation of flagellar gene expression. *J. Cell Biol.* 99: 185a

Lefebvre, P. A., Nordstrom, S. A., Moulder, J. E., Rosenbaum, J. L. 1978. Flagellar elongation and shortening in *Chlamydomonas*. IV. Effects of flagellar detachment, regeneration, and resorption on the induction of flagellar protein synthesis. *J. Cell Biol.* 78: 8–27

Lefebvre, P. A., Silflow, C. D., Wieben, E. D., Rosenbaum, J. L. 1980. Increased levels of mRNAs for tubulin and other flagellar proteins after amputation or shortening of *Chlamydomonas* flagella. *Cell* 20: 469–77

Levy, E. M. 1974. Flagellar elongation: An example of controlled growth. *J. Theor. Biol.* 43: 133–49

Lewin, R. 1953. Studies on the flagella of algae. II. Formation of flagella by *Chlamydomonas* in light and darkness. *Ann. NY Acad. Sci.* 56: 1091–93

L'Hernault, S. W., Rosenbaum, J. L. 1983. *Chlamydomonas* α-tubulin is post-translationally modified in the flagella during flagellar assembly. *J. Cell Biol.* 97: 258–63

L'Hernault, S. W., Rosenbaum, J. L. 1985a. Reversal of the post-translational modification on *Chlamydomonas* flagellar α-tubulin occurs during flagellar resorption. *J. Cell Biol.* 100: 457–62

L'Hernault, S. W., Rosenbaum, J. L. 1985b. *Chlamydomonas* α-tubulin is post-translationally modified by acetylation on the ε-amino group of a lysine. *Biochemistry* 24: 473–78

Linck, R. W. 1973. Comparative isolation of cilia and flagella from lamellibranch mollusc *Aequipecten irradiens*. *J. Cell Sci.* 12: 345–67

Linck, R. W., Langevin, G. 1981. Reassembly of flagellar B (αβ) tubulin into singlet microtubules. Consequences for cytoplasmic structure and assembly. *J. Cell Biol.* 89: 323–37

Luck, D. J. L. 1984. Genetic and biochemical dissection of the eucaryotic flagellum. *J. Cell Biol.* 98: 789–94

Manton, I., Clarke, B. 1952. An electron microscopic study of the spermatozoid of *Sphagnum*. *J. Exp. Bot.* 3: 204–15

Marcaud, L., Hayes, D. 1979. RNA synthesis in starved deciliated *Tetrahymena pyriformis*. *Eur. J. Biochem.* 98: 267–73

May, G., Rosenbaum, J. L. 1980. Induction and synthesis of heat shock proteins in *Chlamydomonas reinhardtii*. *J. Cell Biol.* 87: 272a

McKeithan, T. W., Lefebvre, P. A., Silflow, C. D., Rosenbaum, J. L. 1983. Multiple forms of tubulin in *Polytomella* and *Chlamydomonas*: Evidence for a precursor of flagellar α-tubulin. *J. Cell Biol.* 96: 1056–63

McKeithan, T. W., Rosenbaum, J. L. 1984. The biochemistry of microtubules: A review. *Cell Muscle Motil.* 5: 255–88

McVittie, A. 1972. Flagellar mutants of *Chlamydomonas reinhardtii*. *J. Gen. Microbiol.* 71: 525–40

Merlino, G. T., Chamberlain, J. P., Kleinsmith, L. J. 1978. Effects of deciliation on tubulin messenger RNA activity in sea urchin embryos. *J. Biol. Chem.* 253: 7078–85

Minami, S. A., Collis, P. S., Young, E. E., Weeks, D. P. 1981. Tubulin induction in *C. reinhardii*: Requirement for tubulin mRNA synthesis. *Cell* 24: 89–95

Nelsen, E. M. 1975. Regulation of tubulin during ciliary regeneration in non-growing *Tetrahymena*. *Exp. Cell Res.* 94: 152–58

Pfeffer, T. A., Asnes, C. F., Wilson, L. 1978. Polymerization and colchicine binding properties of outer doublet microtubules solubilized by the French pressure cell. *Cytobiologie* 16: 367–72

Piperno, G., Fuller, M. 1985. Monoclonal antibodies specific for an acetylated form of α tubulin recognize the antigen in cilia and flagella from a variety of organisms. *J. Cell Biol.* 101: 2085–94

Piperno, G., Huang, B., Luck, D. J. L. 1977. Two-dimensional analysis of flagellar proteins from wild-type and paralyzed mutants of *Chlamydomonas reinhardtii*. *Proc. Natl. Acad. Sci. USA* 74: 1600–4

Piperno, G., Luck, D. J. L. 1977. Microtubular proteins of *Chlamydomonas reinhardtii*. An immunochemical study based on the use of an antibody specific for the β-tubulin subunit. *J. Biol. Chem.* 252: 383–91

Piperno, G., Luck, D. J. L. 1979. An actin-like protein is a component of axonemes from *Chlamydomonas* flagella. *J. Biol. Chem.* 254: 2187–90

Quader, H., Cherniack, J., Filner, P. 1978. Participation of calcium in flagellar shortening and regeneration in *Chlamydomonas reinhardii*. *Exp. Cell Res.* 113: 295–301

Raff, E. C. 1979. The control of microtubule assembly in vivo. *Int. Rev. Cytol.* 59: 1–96

Raff, R. A., Greenhouse, G., Gross, K. W., Gross, P. R. 1971. Synthesis and storage of microtubule proteins by sea urchin embryos. *J. Cell Biol.* 50: 516–27

Rannestad, J. 1974. The regeneration of cilia in partially deciliated *Tetrahymena*. *J. Cell Biol.* 63: 1009–17

Remillard, S. P., Witman, G. B. 1982. Synthesis, transport, and utilization of specific flagellar proteins during flagellar regeneration of *Chlamydomonas*. *J. Cell Biol.* 93: 615–31

Riederer-Henderson, M. A., Rosenbaum, J. L. 1979. Ciliary elongation in blastulae of *Arbacia punctulata* induced by trypsin. *Dev. Biol.* 70: 500–9

Roberts, K., Hyams, J. S., eds. 1979. *Microtubules*. London: Academic. 595 pp.

Rochaix, J. D., van Dillewijn, J. 1982. Transformation of the green alga *Chlamydomonas reinhardii* with yeast DNA. *Nature* 296: 70–72

Rosenbaum, J., Carlson, K. 1969. Cilia regeneration in *Tetrahymena* and its inhibition by colchicine. *J. Cell Biol.* 40: 415–25

Rosenbaum, J., Child, F. 1967. Flagellar regeneration in protozoan flagellates. *J. Cell Biol.* 34: 345–64

Rosenbaum, J., Moulder, J., Ringo, D. 1969. Flagellar elongation and shortening in *Chlamydomonas*. I. The use of cycloheximide and colchicine to study the synthesis and assembly of flagellar proteins. *J. Cell Biol.* 41: 600–19

Russell, D. G., Gull, K. 1984. Flagellar regeneration of the trypanosome *Crithidia fasciculata* involves post-translational modification of cytoplasmic α tubulin. *Mol. Cell. Biol.* 4: 1182–85

Sale, W., Satir, P. 1977. The termination of the central microtubules from the cilia of *Tetrahymena pyriformis. Cell Biol. Int. Rep.* 1: 45–49

Schloss, J. A. 1984. Subroutines in the programme of *Chlamydomonas* gene expression induced by flagellar regeneration. *J. Embryol. Exp. Morphol.* 83: 89–101 (Suppl.)

Schloss, J. A., Silflow, C. D., Rosenbaum, J. L. 1984. mRNA abundance changes during flagellar regeneration in *Chlamydomonas reinhardtii. Mol. Cell Biol.* 4: 424–34

Silflow, C. D., Chisholm, R. L., Conner, T. W., Ranum, L. P. W. 1985. The two α-tubulin genes of *Chlamydomonas reinhardi* code for slightly different proteins. *Mol. Cell. Biol.* 5: 2389–98

Silflow, C. D., Lefebvre, P. A., McKeithan, T. W., Schloss, J. A., Keller, L. R., Rosenbaum, J. L. 1981. Expression of flagellar protein genes during flagellar regeneration in *Chlamydomonas. Cold Spring Harbor Symp. Quant. Biol.* 46: 157–69

Silflow, C. D., Rosenbaum, J. L. 1981. Multiple α- and β-tubulin genes in *Chlamydomonas* and regulation of tubulin mRNA levels after deflagellation. *Cell* 24: 81–88

Solter, K. M., Gibor, A. 1978. The relationship between tonicity and flagellar length. *Nature* 275: 651–52

Starling, D., Randall, J. 1971. The flagella of temporary dikaryons of *Chlamydomonas reinhardii. Genet. Res.* 18: 107–13

Stephens, R. E. 1968. Reassociation of microtubule protein. *J. Mol. Biol.* 33: 517–19

Stephens, R. E. 1970. Isolation of nexin—the linkage protein responsible for maintenance of the nine-fold configuration of flagellar axonemes. *Biol. Bull.* 139: 438

Stephens, R. E. 1972. Studies on the development of the sea urchin *Strongylocentrotus draebachiensis.* III. Embryonic synthesis of flagellar proteins. *Biol. Bull.* 142: 489–504

Stephens, R. E. 1977. Differential protein synthesis and utilization during cilia formation in sea urchin embryos. *Dev. Biol.* 61: 311–29

Stephens, R. E. 1978. Primary structural differences among tubulin subunits from flagella, cilia, and the cytoplasm. *Biochemistry* 17: 2882–91

Suprenant, K. A., Hays, E., LeCluyse, E., Dentler, W. L. 1985. Multiple forms of tubulin in the cilia and cytoplasm of *Tetrahymena thermophila. Proc. Natl. Acad. Sci. USA* 82: 6908–12

Tamm, S. 1967. Flagellar development in the protozoan *Peranema trichophorum. J. Exp. Zool.* 164: 163–86

Tamm, S. 1969. The effect of enucleation on flagellar regeneration in the protozoon *Peranema trichophorum. J. Cell Sci.* 4: 171–78

Telser, A. 1977. The inhibition of flagellar regeneration in *Chlamydomonas reinhardii* by the inhalational anesthetic halothane. *Exp. Cell Res.* 107: 247–54

Warner, F. D. 1972. Macromolecular organization of eukaryotic cilia and flagella. *Adv. Cell Mol. Biol.* 2: 193–235

Weeks, D. P., Collis, P. S. 1976. Induction of microtubule protein synthesis in *Chlamydomonas reinhardi* during flagellar regeneration. *Cell* 9: 15–27

Weeks, D., Collis, P., Gealt, M. 1977. Control of induction of tubulin synthesis in *Chlamydomonas reinhardi. Nature* 268: 667–68

Williams, B. D., Mitchell, D. R., Rosenbaum, J. L. 1986. Molecular cloning and expression of flagellar radial spoke and dynein genes of *Chlamydomonas. J. Cell Biol.* 103: 1–12

Witman, G. B. 1975. The site of in vivo assembly of flagellar microtubules. *Ann. NY Acad. Sci.* 253: 178–91

Witman, G., Carlson, K., Berliner, J., Rosenbaum, J. 1972. *Chlamydomonas* flagella. I. Isolation and electrophoretic analysis of microtubules, matrix, membranes, and mastigonemes. *J. Cell Biol.* 54: 507–39

Youngblom, J., Schloss, J. A., Silflow, C. D. 1984. The two β-tubulin genes of *Chlamydomonas reinhardtii* code for identical proteins. *Mol. Cell. Biol.* 4: 2686–96

SUBJECT INDEX

A

Accessory cells
  T-cell activation and, 238-40
Acetylcholine
  inositol phospholipids and,
    149
Acetylcholine receptor
  muscarinic
    G proteins and, 392
1-oleoyl-2-Acetylglycerol
  protein kinase C activation
    and, 153
Acrosomal granule
  bindin and, 14
  sea urchin spermatozoa and,
    2, 5
Acrosomal reaction
  bindin and, 14-15
  fucose sulfate glycoconjugate
    and, 8, 10
  induction of, 9
  sea urchin spermatozoa and,
    2-3, 5
Actin
  cell adhesion and, 356-57
  cell migration and, 339-42
  microtubules and, 440
  sea urchin spermatozoa and, 5
Actinomycin D
  ciliary regeneration and, 531
Actomyosin system
  cell migration and, 342
Adenosine triphosphate
  See ATP
Adenovirus E1A
  nuclear proteins of, 381-82
Adenylate cyclase
  cell surface receptors and,
    407
  G proteins and, 392
  GTP and, 395
  hormonal control of, 397-99
  olfaction and, 407
  protein kinase C and, 162
Adrenergic receptors
  adenylate cyclase and, 392
  protein kinase C and, 159
Aequorin
  sea urchin eggs and, 18
Alkaline phosphatase
  E. coli, 322
Alzheimer's disease
  microtubule-associated pro-
    teins and, 440

Amino acids
  uptake of
    membrane potentials and,
      192
Aminoisobutyric acid
  nuclear envelope and, 368
Aminopeptidase A
  brush border membrane and,
    286-89
  homing route of, 298
  peptide bonds and, 285
Aminopeptidase N
  brush border membrane and,
    286-89
  homing route of, 298
  intestinal distribution of, 258
  papain solubilization and, 259
  substrate specificity of, 285
  subunit interactions of, 263-
    64
Aminopeptidase W, 286
Amiprophosmethyl
  flagellar protein synthesis
    and, 540
Amoebae
  directed migration of, 337,
    341
  nuclear envelope of, 369
Ankyrin
  microtubules and, 428
Antibodies
  ATPases and, 181
  B cells and, 232
  cell adhesion molecules and,
    32
  microtubule-associated pro-
    teins and, 446
  sea urchin spermatozoa and,
    12-13
  T-cell activation and, 235-36
  See also Monoclonal antibod-
    ies, Polyclonal antibodies
Antigens
  activation
    T cells and, 240
  polyoma virus large T, 381
  SV40 large T
    cell nucleus and, 377-79
Aphidicolin
  blastula cell nucleus and, 384
A proteins, 469-71
Arachidonate
  protein kinase C and, 161
Arachidonic acid
  sea urchin eggs and, 16-17

Arbacia punctulata
  egg jelly of, 6-7
  spermatozoa of
    guanylate cyclase and, 11
Arrestin
  transducin cycle and, 402
Artemia salina
  A protein of, 470
  ribonucleoprotein complexes
    of, 463
Arthropods
  phototransduction in, 403-4
Asialoglycoprotein receptor
  biosynthesis of, 294
Astasia
  flagellar regeneration in, 519
Astrocytes
  L1 antigen and, 33
  microtubule-associated pro-
    teins and, 433-34
ATP
  membrane potentials and, 180
  nucleoplasmin transport and,
    386
  proton ATPases and, 180
  sea urchin spermatozoa and,
    4-5
  transducin cycle and, 402
ATPases
  See specific type
Atractyloside
  mitochondrial ADP-ATP ex-
    changer and, 193

B

Bacterial toxins
  G proteins and, 394
  See also specific type
Basement membrane
  components of, 28-29
  deposition of, 28
  histogenesis and, 28
  laminin in, 37
B-cell differentiation factors
  immune response and, 232
B cells
  functions of, 232
  growth promotion for, 158
Bile acids
  transport systems for, 258
Bindin
  sea urchin spermatozoa and,
    14-15

CUMULATIVE INDEXES

CONTRIBUTING AUTHORS, VOLUMES 1–2

A

Al-Awqati, Q., 2:179–99
Anderson, R. G. W., 1:1–39

B

Beckwith, J., 2:315–36
Bourne, H. R., 2:391–419
Brinkley, B. R., 1:145–72
Brown, M. S., 1:1–39

C

Chaponnier, C., 1:353–402

D

Dingwall, C., 2:367–90
Dreyfuss, G., 2:459–98
Duband, J. L., 1:91–113

E

Edelman, G. M., 2:81 116
Ekblom, P., 2:27–47
Ezzell, R., 1:353–402

F

Farquhar, M. G., 1:447–88
Finer-Moore, J., 1:317–51
Fujiki, Y., 1:489–530
Fuller, S., 1:243–88

G

Garoff, H., 1:403–45
Gerhart, J., 2:201–29
Goldstein, J. L., 1:1–39

H

Hartwig, J. H., 1:353–402
Hynes, R. O., 1:67–90

J

Janmey, P., 1:353–402

K

Keller, R., 2:201–29
Kemler, R., 2:27–47
Kikkawa, U., 2:149–78
Kupfer, A., 2:337–65
Kwiatkowski, D., 1:353–402

L

Laskey, R. A., 2:367–90
Lazarow, P. B., 1:489–530
Lee, C., 2:315–36
Lefebvre, P. A., 2:517–46
Lind, S., 1:353–402
Lingappa, V. R., 2:499–516

M

MacDonald, H. R., 2:231–53
Marchesi, V. T., 1:531–61
Mooseker, M. S., 1:209–41
Murray, A., 1:289–315

N

Nabholz, M., 2:231–53
Nishizuka, Y., 2:149–78

O

O'Farrell, P. H., 2:49–80
Olmsted, J. B., 2:421–57

P

Parry, D. A. D., 1:41–65
Pederson, D. S., 2:117–47

R

Rosenbaum, J., 2:517–46
Russell, D. W., 1:1–39

S

Schekman, R., 1:115–43
Schneider, W. J., 1:1–39
Scott, M. P., 2:49–80
Semenza, G., 2:255–313
Shapiro, L., 1:173–207
Simons, K., 1:243–88
Simpson, R. T., 2:117–47
Singer, S. J., 2:337–65
Smith, D., 1:353–402
Southwick, F. S., 1:353–402
Steinert, P. M., 1:41–65
Stossel, T. P., 1:353–402
Stroud, R. M., 1:317–51
Stryer, L., 2:391–419
Szostak, J. W., 1:289–315

T

Thiery, J., 1:91–113
Thoma, F., 2:117–47
Trimmer, J. S., 2:1–26
Tucker, G. C., 1:91–113

V

Vacquier, V. D., 2:1–26
Vestweber, D., 2:27–47

W

Walter, P., 2:499–516

Y

Yin, H. L., 1:353–402

Z

Zaner, K. S., 1:353–402

CHAPTER TITLES, VOLUMES 1–2

Annual Reviews Inc.

A NONPROFIT SCIENTIFIC PUBLISHER

4139 El Camino Way
P.O. Box 10139
Palo Alto, CA 94303-0897 • USA

Annual Reviews Inc. publications may be ordered directly from our office by mail or use our Toll Free Telephone line (for orders paid by credit card or purchase order, and customer service calls only); through booksellers and subscription agents, worldwide; and through participating professional societies. Prices subject to change without notice. ARI Federal I.D. #94-1156476

- **Individuals:** Prepayment required on new accounts by check or money order (in U.S. dollars, check drawn on U.S. bank) or charge to credit card — American Express, VISA, MasterCard.
- **Institutional buyers:** Please include purchase order number.
- **Students:** $10.00 discount from retail price, per volume. Prepayment required. Proof of student status must be provided (photocopy of student I.D. or signature of department secretary is acceptable). Students must send orders direct to Annual Reviews. Orders received through bookstores and institutions requesting student rates will be returned.
- **Professional Society Members:** Members of professional societies that have a contractual arrangement with Annual Reviews may order books through their society at a reduced rate. Check with your society for information.
- **Toll Free Telephone orders:** Call 1-800-523-8635 (except from California) for orders paid by credit card or purchase order and customer service calls only. California customers and all other business calls use 415-493-4400 (not toll free). Hours: 8:00 AM to 4:00 PM, Monday-Friday, Pacific Time.

Regular orders: Please list the volumes you wish to order by volume number.
Standing orders: New volume in the series will be sent to you automatically each year upon publication. Cancellation may be made at any time. Please indicate volume number to begin standing order.
Prepublication orders: Volumes not yet published will be shipped in month and year indicated.
California orders: Add applicable sales tax.
Postage paid (4th class bookrate/surface mail) **by Annual Reviews Inc.** Airmail postage or UPS, extra.

ANNUAL REVIEWS SERIES	Prices Postpaid per volume USA/elsewhere	Regular Order Please send:	Standing Order Begin with:
		Vol. number	Vol. number
Annual Review of ANTHROPOLOGY			
Vols. 1-14 (1972-1985)	$27.00/$30.00		
Vol. 15 (1986)	$31.00/$34.00		
Vol. 16 (avail. Oct. 1987)	$31.00/$34.00	Vol(s). _____	Vol. _____
Annual Review of ASTRONOMY AND ASTROPHYSICS			
Vols. 1-2, 4-20 (1963-1964, 1966-1982)	$27.00/$30.00		
Vols. 21-24 (1983-1986)	$44.00/$47.00		
Vol. 25 (avail. Sept. 1987)	$44.00/$47.00	Vol(s). _____	Vol. _____
Annual Review of BIOCHEMISTRY			
Vols. 30-34, 36-54 (1961-1965; 1967-1985)	$29.00/$32.00		
Vol. 55 (1986)	$33.00/$36.00		
Vol. 56 (avail. July 1987)	$33.00/$36.00	Vol(s). _____	Vol. _____
Annual Review of BIOPHYSICS AND BIOPHYSICAL CHEMISTRY			
Vols. 1-11 (1972-1982)	$27.00/$30.00		
Vols. 12-15 (1983-1986)	$47.00/$50.00		
Vol. 16 (avail. June 1987)	$47.00/$50.00	Vol(s). _____	Vol. _____
Annual Review of CELL BIOLOGY			
Vol. 1 (1985)	$27.00/$30.00		
Vol. 2 (1986)	$31.00/$34.00		
Vol. 3 (avail. Nov. 1987)	$31.00/$34.00	Vol(s). _____	Vol. _____